D0754730

PALAEOECOLOGICAL EVENTS DURING THE LAST 15 000 YEARS

INTERNATIONAL GEOLOGICAL CORRELATION PROGRAMME

PROJECT 158 **B**

612(500)
P343
1996

PALAEOECOLOGICAL EVENTS DURING THE LAST 15 000 YEARS

Regional Syntheses of Palaeoecological Studies of Lakes and Mires in Europe

Edited by

B.E. BERGLUND
Department of Quaternary Geology, Lund University, Sweden

H.J.B. BIRKS
Botanical Institute, University of Bergen, Norway and Environmental Change Research Centre, University College London, UK

M. RALSKA-JASIEWICZOWA
Institute of Botany, Polish Academy of Sciences, Crakow, Poland

H.E. WRIGHT
Limnological Research Center, University of Minneapolis, USA

JOHN WILEY & SONS
Chichester • New York • Brisbane • Toronto • Singapore

Copyright © 1996 by John Wiley & Sons Ltd,
Baffins Lane, Chichester,
West Sussex PO19 IUD, England
National   01243 779777
International   (+44)1243 779777

All rights reserved.

No part of this book may be reproduced by any means,
or transmitted, or translated into a machine language
without the written permission of the publisher.

Other Wiley Editorial Offices

John Wiley & Sons, Inc., 605 Third Avenue,
New York, NY 10158–0012, USA

Jacaranda Wiley Ltd, 33 Park Road, Milton,
Queensland 4064, Australia

John Wiley & Sons (Canada) Ltd, 22 Worcester Road,
Rexdale, Ontario M9W IL1, Canada

John Wiley & Sons (SEA) Pte Ltd, 37 Jalan Pemimpin #05-04,
Block B, Union Industrial Building, Singapore 2057

British Library Cataloguing in Publication Data

A catalogue record for this book is available from the British Library

ISBN   0-471-95840-9

Typeset in 10/12pt Times from editor's disks by Thomson Press (India) Ltd, New Delhi
Printed and bound in Great Britain by Bookcraft (Bath) Ltd
This book is printed on acid-free paper responsibly manufactured from sustainable forestation,
for which at least two trees are planted for each one used for paper production.

Contents

_____

† = deceased

List of Contributors

B. AABY   The National Museum, Ny Vestergade 11, DK-1471 Copenhagen K, Denmark

T. ALM   Institute of Museal Sciences, University of Tromsø, N-9037 Tromsø, Norway

B. AMMANN   Systematisch-Geobotanisches Institut, University of Bern, Altenbergrain 21, CH-3013 Bern, Switzerland

S.Th. ANDERSEN   Geological Survey of Denmark, Thoravej 8, DK-2400 Copenhagen NV, Denmark

S. AUBERT   Laboratoire d'Ecologie terrestre de Toulouse, UMR 9964, Université Paul Sabatier, 39, allées Jules Guesde, F-31062 Toulouse Cedex, France

J.L. de BEAULIEU   Laboratoire de Botanique historique et Palynologie, URA CNRS D 1152, Faculté des Sciences et Techniques St-Jérome, F-13397 Marseille Cedex 20, France

K.-E. BEHRE   Niedersächsisches Institut für historische Küstenforschung, Postfach 2062, D-26360 Wilhelmshaven, Germany

J. M. BELET   Centre des Faibles Radioactivités, Laboratoire mixte CNRS- CEA, Domaine du CNRS, Avenue de la Terrasse, F-91198 Gif sur Yvette, France

B.E. BERGLUND   Department of Quaternary Geology, Lund University, Tornavägen 13, S-223 63 Lund, Sweden

H.J.B. BIRKS   Botanical Institute, University of Bergen, Allégaten 41, N-5007 Bergen, Norway and Environmental Change Research Centre, University College London, 26 Bedford Way, London WC1H 0AP UK

S. BORTENSCHLAGER   Institut Für Botanik der Leopold-Franzens Universität Innsbruck, Sternwartestraße 15, A-6020 Innsbruck, Austria

E. BOZILOVA   Department of Botany, Faculty of Biology, University of Sofia "St Kliment Ohridski", Sofia, Bulgaria

R.H.W. BRADSHAW   Southern Swedish Forest Research Centre, Swedish University of Agricultural Sciences, Box 49, S-230 53 Alnarp, Sweden

A. BRANDE   Institute of Ecology, Technical University Berlin, Schmidt-Ott-Strasse 1, D-12165 Berlin, Germany

F.M. CHAMBERS   Centre for Environmental Change and Quaternary Research, Department of Geography and Geology, Cheltenham and Gloucester College, Francis Close Hall, Swindon Road, Cheltenham GL50 4 AZ, UK

M. CULIBERG   Biološki inštitut, ZRC SAZU, Goposka 13, SL-61000 Ljubljana, Slovenia

F. DAVID   Laboratoire de Botanique historique et Palynologie, URA CNRS D 1152, Faculté des Sciences et Techniques St-Jérome, F-13397 Marseille Cedex 20, France

L. DENYS — Department of Biology, University of Antwerp (UCA), Groenenborgerlaan 171, B-2020 Antwerpen, Belgium

G. DIGERFELDT — Department of Quaternary Geology, Lund University, Tornavägen 13, S-223 63 Lund, Sweden

G.A. ELINA — Institute of Biology, Karelian Research Centre, Russian Academy of Sciences, Petrozavodsk, Russia

R. ENGELMARK — Institute of Archaeology, Umeå University, S-901 87 Umeå, Sweden

L.V. FILIMONOVA — Institute of Biology, Karelian Research Centre, Russian Academy of Sciences, Petrozavodsk, Russia

M. FILIPOVA — Museum of Natural History, Varna-9000, Bulgaria

L. FILIPOVICH — Institute of Botany, Bulgarian Academy of Sciences, Sofia-1113, Bulgaria

S. FIMREITE — N-3800 Bø, Telemark, Norway

M. FONTUGNE — Centre des Faibles Radioactivités, Laboratoire mixte CNRS-CEA, Domaine du CNRS, Avenue de la Terrasse, F-91198 Gif sur Yvette, France

M.-J. GAILLARD — Department of Quaternary Geology, Lund University, Tornavägen 13, S-223 63 Lund, Sweden

D. GALOP — Laboratoire d'Ecologie terrestre de Toulouse, UMR 9964, Université Paul Sabatier, 39, allées Jules Guesde, F-31062 Toulouse Cedex, France

G. GLÜCKERT — Department of Quaternary Geology, University of Turku, FIN-20500 Turku, Finland

J. GREIG — Department of Ancient History & Archaeology, Birmingham University, Edgbaston, Birmingham B15 2TT, UK

G.E. HANNON — Department of Quaternary Geology, University of Lund, Tornavägen 13, S-223 63 Lund, Sweden

S. HICKS — Department of Geology, University of Oulu, Linnanmaa, FIN-90570 Oulu, Finland

M.F. HUAULT — Laboratoire de Géologie, Université de Rouen, Faculté des Sciences, F-76821 Mont-Saint-Aignan Cedex, France

H. HYVÄRINEN — Department of Geology, University of Helsinki, P.O. Box 11, FIN-00014 Helsinki, Finland

E. ILVES — Institute of Zoology and Botany, Estonian Academy of Sciences, 181 Riia Road, EE 2400 Tartu, Estonia

G. JALUT — Laboratoire d'Ecologie terrestre de Toulouse, UMR 9964, Université Paul Sabatier, 39, allées Jules Guesde, F-31062 Toulouse Cedex, France

C.R. JANSSEN — Laboratory of Palaeobotany and Palynology, University of Utrecht, Heidelberglaan 2, NL-3584 CS Utrecht, The Netherlands

J. JÓHANSEN[†] — Natural History Museum, FR-100 Tórshavn, Faroe Islands

M. KABAILIENÉ — Department of Geology and Mineralogy, Vilnius University, M.K. Ciurlionis Str., 21/27, 2009 Vilnius, Lithuania

S. KARLSSON — Department of Quaternary Geology, Stockholm University, Odengatan 63, S-113 22 Stockholm, Sweden

P. KIDEN — Rijks Geologische dienst, Distrikt Zuid, Vincent van Goghstraat 78, PO Box 35, NL-5670 AA Nuenen, The Netherlands

H. KÜSTER — Institute für Vor- und Frühgeschichte der Universität München, Feldmochinger Strasse 7, D-80992 München, Germany

M. LATAŁOWA — Department of Plant Ecology and Nature Protection, Gdansk University Al. Legionow 9, 80-441 Gdansk, Poland

A.F. LOTTER — Systematisch-Geobotanisches Institut, University of Bern, Altenbergrain 21, CH-3013 Bern, Switzerland

H. MÄEMETS — Institute of Zoology and Botany, Estonia Academy of Sciences, 181 Riia Road, EE 2400 Tartu, Estonia

U. MILLER — Department of Quaternary Geology, Stockholm University, Odengatan 63, S-113 22 Stockholm, Sweden

F.J.G. MITCHELL — School of Botany, Trinity College, Dublin 2, Ireland

D. MOE — Botanical Institute, University of Bergen, Allégaten 41, N-5007 Bergen, Norway

B. MØRKVED — Institute of Museal Sciences, University of Tromsø, N-9037 Tromsø, Norway

E. NILSSEN — Institute of Clinical Medicine, University of Tromsø, N-9037 Tromsø, Norway

M. O'CONNELL — Department of Botany, University College, Galway, Ireland

B. ODGAARD — Geological Survey of Denmark, Thoravej 8, DK-2400 Copenhagen NV, Denmark

K. OEGGL — Institut für Botanik der Leopold-Franzens Universität Innsbruck, Sternwartestraße 15, A-6020 Innsbruck, Austria

AA. PAUS — Botanical Institute, University of Bergen, Allégaten 41, N-5007 Bergen, Norway

J.R. PILCHER — Palaeoecology Laboratory, Queen's University, Belfast, UK

R. PIRRUS — Institute of Geology, Estonian Academy of Sciences, 7 Estonia Ave., EE 0100 Tallinn, Estonia

M. RALSKA-JASIEWICZOWA — W. Szafer Institute of Botany, Polish Academy of Sciences, Lubicz 46, 31-512 Cracow, Poland

H. RAMFJORD — University of Trondheim, AVH, N-7055 Dragvoll, Norway

J. RISBERG — Department of Quaternary Geology, Stockholm University, Odengatan 63, S-113 22 Stockholm, Sweden

M. RÖSCH — Landesdenkmalamt Baden-Würtemberg, Fischensteig 9, D-78343 Gaienhofen-Hemmenhofen, Germany

A.-M. RÔUK — Institute of History, Estonian Academy of Sciences, 6 Rüütli Street, EE 0001 Tallinn, Estonia

K. RYBNÍČEK — Institute of Botany, Academy of Sciences, Bělidla 4a, CZ-603 00 Brno, Czech Republic

E. RYBNÍČKOVA — Institute of Botany, Academy of Sciences, Bělidla 4a, CZ-603 00 Brno, Czech Republic

L. SAARSE — Institute of Geology, Estonian Academy of Sciences, 7 Estonia Ave., EE 0100 Tallinn, Estonia

A. SARV — Institute of Geology, Estonian Academy of Sciences, 7 Estonia Ave., EE 0100 Tallinn, Estonia

S.F. SELVIK — University of Trondheim, AVH, N-7055 Dragvoll, Norway

A. ŠERCELJ — Slovenska akademiji znanosti in umetnosti, Novi trg 3, 61000 Ljubljana, Slovenia

H. SIMOLA — Karelian Institute, University of Joensuu, P.O. Box 111, FIN-80101 Joensuu, Finland

R. SØRENSEN — Geological Institute, Agricultural University of Norway, Postbox 28, N-1432 Ås, Norway

S. TONKOV — Department of Botany, Faculty of Biology, University of Sofia "St Kliment Ohridski", Sofia, Bulgaria

Y. Vasari — Department of Ecology and Systematics, University of Helsinki, P.O. Box 7, FIN-00014 Helsinki, Finland

C. Verbruggen — Geological Institute, University of Ghent, Krijgslaan 281, B-9000 Ghent, Belgium

L. Visset — Faculté des Sciences et des Techniques, Laboratoire d'Ecologie, 2 rue de la Houssinière, F-44072 Nantes Cedex 03, France

D. Voeltzel — Faculté des sciences de Nantes et UPR 403 CNRS, Laboratoire de Paléoenvironements atlantiques, F-44072 Nantes Cedex 03, France

K.-D. Vorren — Institute of Biology and Geology, University of Tromsø, N-9037 Tromsø, Norway

I. Vuorela — Geological Survey of Finland, FIN-02150 Espoo, Finland

N. Wahlmüller — Institut Für Botanik der Leopold-Franzens Universität Innsbruck, Sternwartestrabe 15, A-6020 Innsbruck, Austria

W.A. Watts — School of Botany, Trinity College, Dublin 2, Ireland

H.E. Wright Jr. — Limnological Research Center, University of Minnesota, 310 Pillsbury Drive S.E., Minneapolis, MN 55455-0219, USA

Preface

The International Geological Correlation Project IGCP 158, named "Palaeohydrological Changes in the Temperate Zone in the Last 15,000 Years", included two subprojects: A, dealing with fluvial environments and B, dealing with lake and mire environments. The project was initiated in 1977 and planned to last for a decade. Meetings with field conferences were organized every year in Europe until 1988 (Berglund 1990 with list of meetings and conference volumes). The main leader of IGCP 158 and subproject A was Professor Leszek Starkel, Krakow, while I was the leader of subproject B, assisted by the project secretary Magdalena Ralska-Jasiewiczowa, co-editor of this book.

Project aims

The aim of this project has been to detect biotic as well as physical environmental changes on the basis of multidisciplinary, stratigraphical studies of lake and mire deposits. It attempted to provide well-documented data on palaeoecological changes since deglaciation, related to vegetation, lake development, mire and lake hydrology, soil erosion, human impact, etc. The aim was much wider than only the documentation of palaeohydrological changes. After regional and continental correlations were established, palaeoecological patterns could be traced and hopefully correlated with results from the river basins. The causes were searched for in terms of climate, human impact, and biotic conditions.

Research strategy

It was proposed to subdivide each country into physical-geographically uniform type regions and to select representative reference sites within them. These field study sites were of two kinds: (1) secondary reference sites for palaeovegetation and chronology, (2) primary reference sites, within palaeoecological reference areas, for a more detailed interpretation of environmental changes. We define palaeoecological references sites as lakes (often ancient lakes) or mires (bogs) carefully chosen to obtain long cores with continuous sedimentation, preferably covering the entire late-glacial time 15/13000 to 10000 years BP, the entire Holocene time 10000 BP to present, or even the entire time since the last glaciation. This is quite easy in former glaciated terrain where lake and mire basins are frequent, while in flat landscapes of Central Europe several short-time cores have to be combined to get a complete record. Altogether, a network of reference sites was obtained for palaeoecological correlations. We emphasized the application of uniform stratigraphical, chronological, and palaeobiological methods in order to obtain comparable results. As a result, project handbooks were published (Berglund 1979–82, 1986).

Geographical area

Geographically, the project was intended to cover the temperate and subarctic regions of the North American and the Eurosiberian continents. However, the project collaboration, following the guidelines mentioned, has been concentrated to Europe. Some countries in the former USSR joined the project in the final phase

Data compilation and publication

The results from the project were to be compiled and published at three different geographical levels:

- Reference sites, i.e. local studies published by individual scientists, in journals or in some cases in national monographic compilations (e.g. Poland, Switzerland, USSR). The first list of such reference sites was published in 1986 (Ralska-Jasiewiczowa). It comprised about 500 sites from 18 countries. In the present book we present the same number of sites from 21 countries, with references to publications (Appendix).
- Type regions, i.e. compilations of so-called "Regional syntheses of palaeoecological events". The main results related to vegetation events, other biotic events, soil events, hydrological events, climatic events anthropogenic events, etc., are presented in a uniform way. A standardized scheme has been used for the event stratigraphy. Preliminary syntheses were presented at the project symposium in Sweden in 1987. They have now been revised and completed. About 95 regional syntheses from 21 countries are presented in this volume. The geographical coverage is shown in Fig. 1. For different reasons there are many gaps which do not imply that white areas on the map have no palaeoecological information, simply they are not represented in our selection of data. We believe that on the continental scale the representation of investigated sites is good enough for palaeoecological correlations and for illustrating the present status of research.
- Continental syntheses. These have to be done in the future, on the basis of uniform palaeoecological data available through databases. This step has been beyond the aims of this project.

Future applications

The time-stratigraphical information presented here is the fundamental core of the European Pollen Database (EPD) founded in 1989 with Arles in Frances as the institutional centre. The relationship between IGCP 158 B and EPD is described by Dr Jacques-Louis de Beaulieu in the Foreword of this book. When this database is fully developed, correlation and mapping of biotic data, on regional as well as continental scales, will make it possible to understand patterns of the past, such as tree immigration, timberline changes, palaeofloristic changes, diversity changes, species competition, deforestation, secondary successions, soil development, soil erosion, eutrophication, acidification, etc. The causes behind such vegetational and hydrological changes can then be found in climatic, biotic, and anthropogenic processes. The interaction between the physical and the biotic systems is in focus for the current research on the geosphere–biosphere, particularly within the International Geosphere Biosphere Program (IGBP) and its core project on Past Global Changes (PAGES).

Compilation and editorial procedure

The idea of a presentation of the individual and national research efforts in a simple and uniform way was born during the project meeting in Switzerland 1985—during a warm summer day in an alpine meadow the co-editor John Birks brought up the phrase "Palaeoecological events". He also formulated the first "Suggestions for Regional Syntheses" in 1985, later revised by us in 1986 using the Danish synthesis as a case study (formulated and illustrated by Bent Aaby). However, during the time of compilation we found problems in obtaining uniform reports and getting an overall response. Thanks to the activities of the fourth co-editor Herbert Wright we distributed a new "Memorandum" in 1990. For different reasons there were further delays, mainly caused by filling in gaps. The dynamic political situation in Europe after 1989 was another cause for this delay. This means that drafts have been delivered to the main office in Lund over a period of five years. During the same period we have witnessed an explosion of computer technology in Europe! This delay, together with the different research facilities in the European countries, have caused a certain heterogeneity in the syntheses published here.

Thanks to the willingness and generosity of Herbert Wright's collaborators Linda C.K. Shane and Erica Haas it has been possible to produce a series of simplified pollen diagrams based on raw pollen data. Some diagrams were improved in a similar way by my colleague Thomas Persson here in Lund and by Sylvia Peglar and John Birks in Bergen. Other drawings were improved by my drawing assistants Christin Andréasson and Britt Nyberg. After scientific revision, linguistic checking, and overall editing by John Birks, Herbert Wright, and myself, the final corrections were transferred onto discs by my son Mårten Berglund, and to some extent also by Jonas Ekström and Thomas Persson. Some word processing and the substantial letter communication throughout the years of this project have been in the hands of Karin Price, secretary in Lund. I would like to express my thanks to all these persons for their

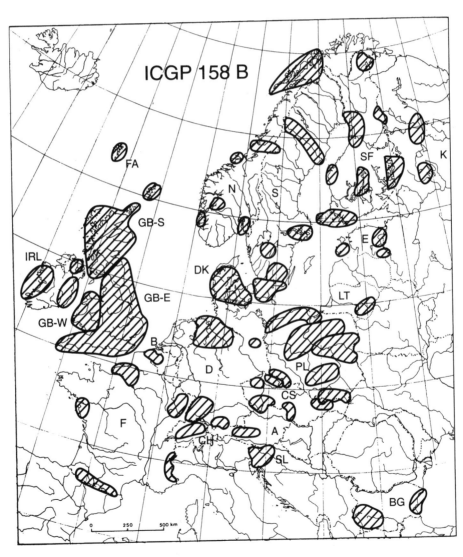

Fig. 1 Map showing geographical areas covered by type-region syntheses described in this book

Preface

willingness to help with these tasks which never seemed to come to an end.

Finally, my thanks go to all enthusiastic colleagues in the whole of Europe who have contributed to our joint research during the period of the project and during the production of this book. We all belong to the Family of European Palaeoecologists. Thank you for your patience and your friendship during many years. I am also deeply grateful to my close friends, the co-editors who gradually joined me and helped carrying the burden of bringing the project to an end!

BJÖRN E. BERGLUND

References

Berglund, B.E. (ed.) 1979–82: *Palaeohydrological changes in the temperate zone in the last 15,000 years. Subproject B. Lake and mire environments. Project guide. Vols. I–III.* Dept. Quaternary Geology, Lund. 123, 340, and 163 pp.

Berglund, B.E. (ed.) 1986: *Handbook of Holocene palaeo-ecology and palaeohydrology.* Wiley, Chichester. 869 pp.

Berglund, B.E. 1990: IGCP 158 B 1977–1988—retrospect and prospect. International collaboration on environmental changes during the last 15,000 years. *Quaternary Studies in Poland* 10, 33–40.

Ralska-Jasiewiczowa, M. 1986: *Palaeohydrological changes in the temperate zone in the last 15,000 years. Subproject B. Lake and mire environments. Project catalogue for Europe.* Dept. Quaternary Geology, Lund. 161 pp.

Foreword

Neoecology and palaeoecology are two complementary facets of the same edifice. They support each other, since modern ecosystems are a key for interpreting past environments while an "objective" knowledge of the latter may help the ecologists to avoid making serious misinterpretations. Ecologists often neglect the perspective of time when discussing ecological processes of today.

The paradigm of ecology is that ecosystems are governed by climatic, biotic, and edaphic factors. But palaeoecology tempers the notion of climax by revealing the instability which constantly affects the biosphere. Moreover, it shows the fundamental importance of historical parameters (e.g. climate cycles during the Pleistocene) in the establishment of ecosystems. Today the conformation between models of vegetation dynamics and empirical forest successions inferred from pollen analysis leads to a better understanding of plant distribution. In recent years pollen analysis has also been a powerful tool to demonstrate that vegetation communities have changed through time in their species combinations. In other words, it is hard to find modern analogues for past communities.

These new approaches require both a great precision in historical reconstructions at the regional scale and syntheses at the continental scale (assessment of phenomena such as refuges, migration, and human impact). This justified the coordination proposed by the IGCP 158 B programme, which implied a continuation of traditional regional studies but with a uniform sampling strategy and high quality laboratory methods. A combination of different palaeoecological techniques should make it possible to detect climatic/hydrological as well as biotic changes.

IGCP 158 B leads to a European database

The IGCP project 158 B played a prominent role in launching a pollen database for Europe. For participants in the closing session of the IGCP 158 B project in June 1988 in Cracow it was obvious that the European Pollen Database (EPD) was initiated during this meeting. It was agreed that in order to develop joint research programmes it was necessary to set up a homogeneous pollen database which would collect available information from regional working groups. The practical basis of the database was established during a meeting organized, at Björn Berglund's initiative, in Frostavallen near Lund (August 1989) as a prolongation of the Cracow meeting.

From the outset, the objective of the IGCP 158 project was to harmonize the terminologies and methods of its participants in order to make the results of each regional group compatible and thus enable the construction of a coherent synthesis at the continental scale. This synthesis was intended to lead to a mapping of palaeovegetation and palaeoenvironment, reproducing with more precision previous initiatives such as the atlas by Huntley & Birks (1983). The orange "Handbook" (1986), which answered the first objective of methodological definitions, was a first outcome of the project. But once data have been collected, the best way to harmonize information is to enter them into a common database and thus submit them to a homogeneous processing.

Inasmuch as the IGCP project was based on a multidisciplinary approach, a structure assembling all types of data (diatoms, molluscs, insects, etc.) could have logically been envisaged, but this would have been an enormous task. So, as pollen analysis was a common denominator to all researches involved in the project, a pollen database (also including macroscopic plant remains) was chosen as a priority. An important characteristic of the EPD is that its structure is completely compatible with the North American Pollen Database (NAPD); this explains the important role played by George L. Jacobson and Eric C. Grimm during the building phase of the EPD. This perfect coordination between the two databases is a good starting point for a future development of pollen databases in other continents.

It soon become obvious that such an enterprise does not only concern the participants in the IGCP project but also a wider community of palynologists and scientists interested in global changes, as expressed in the introduction of the first Newsletter of the EPD: "As we all know, pollen analysis is a time-consuming activity. Pollen-analytical data are thus a valuable scientific resource that sould be permanently archived for future generations of researchers. We are living in the "information revolution" with an ever-increasing availability of powerful personal computers, with the development of computer-based databases and data-retrieval systems, and with the enormous expansion of the primary scientific literature. It is against this background that a European pollen database (EPD) is now being developed to provide for all palynologists a permanent archive of the basic data generated by pollen analysis in Europe, a tool for further research on palaeoecological and biogeographical problems at a variety of temporal and spatial scales, and a primary data-source for furthering our understanding of past environmental history at a time when research on Global Change is becoming important". Therefore, the EPD was no longer exclusively the outcome of the IGCP 158 B project; many participants in its development originated from other groups, e.g. the palaeoclimatology EPOCH programme and the Environment programme of the Commission of European Communities which provided financial support to the project from 1991. The question of a central or federal structure for the EPD has been much debated.

The Frostavallen choice of a central structure located in Arles, South France, ensured a homogeneous compilation and a coherent management of data exchanges, but the EPD also involves a number of national compilation centres, and this inevitably leads to some compromises in the EPD management.

Future collaboration in Europe

After three years of construction EPD was opened to users in the spring of 1994. It then contained 450 compiled Holocene and Late-glacial sites, and 70 new sites have been compiled since. It seems that three additional years of effort (... and a little more good-will from palynologists!) are still necessary for the EPD to become a really efficient tool. Despite the above-mentioned affiliation and the call for contributions made by Björn Berglund, it appears that only 20% of the reference sites of the IGCP 158 B project are included in the EPD. The entering of all the IGCP sites in the EPD would improve the database enormously!

This volume presents the regional syntheses made within the framework of the IGCP 158 B programme: its publication took as long a time as was necessary for the construction and opening of the EPD. The reason for that is the great difficulty in arriving at a homogeneous representation of results. A database is the ideal solution to solve the types of problems that considerably delayed the publication of such a book.

Although the present volume gives an idea of the excellent data obtained within the IGCP 158 B network, its final goal, i.e. a synthesis at a continental scale, is not fully achieved. It seems that the EPD tool could be used by the community of European pollen analysts to develop and implement a common project: "Mapping of palaeovegetation in Europe during the last 15000 years".

JACQUES LOUIS DE BEAULIEU

References

Berglund, B.E. (ed.) 1986: *Handbook of Holocene palaeoecology and Palaeohydrology*. Wiley, Chichester. 869 pp.
Huntley, B. & Birks, H.J.B. 1983: *An atlas of past and present pollen maps*. Cambridge University Press, Cambridge. 650 pp.

1

Ireland

F.J.G. Mitchell, R.H.W. Bradshaw, G.E. Hannon,
M. O'Connell, J.R. Pilcher and W.A. Watts

INTRODUCTION

Ireland is shaped like a saucer, with its flat central plain surrounded by chiefly coastal mountains rising to over 1000 m in the southwest (Fig. 1.1). Quaternary deposits dominate the surficial geology, and only small parts of the country were ice-free at the glacial maximum (Ehlers *et al.* 1991). The bedrock geology is rather varied (Holland 1981). Carboniferous limestone underlies the central plain and other low-lying areas. The coastal uplands are dominated by Devonian sandstone in the south, granite in the east, Dalradian schist and gneiss in the northwest, and basalt in the northeast (Fig.1.2).

Ireland has a restricted flora compared to the rest of Europe (Webb 1983). This is primarily due to the limited range of climate and habitat, but the almost complete glaciation of the island, followed by early isolation during the Holocene, are further factors. The earliest date for human presence in Ireland is *ca.* 8900 BP, and human impact on the landscape has been significant since *ca.* 5500 BP (Mitchell 1986).

Our three type regions are based not only on geologic and edaphic features, but also on contemporary phytogeographical considerations. These three regions reflect the physical attributes of the environment, and they appear to be meaningful divisions for at least some of the Holocene.

IRL-a is the Atlantic fringe of the country, with a varied geology but relatively uniform climate (Fig. 1.3). Extensive blanket peat is the main feature of the region. It also harbours many of the rarer plant species, including Lusitanian and arctic–alpine elements. IRL-b includes the northeastern part and is the region with the shortest growing season. IRL-c covers southeast and central Ireland; it contains the most favourable agricultural soils, with extensive influence of limestone and limestone drift.

IRL-b contains the best-dated sites in the country thanks to the radiocarbon-dating facility in Belfast. Recent research effort has produced well-dated sites in regions IRL-a and IRL-b. The main reference sites are from lowland areas, but we refer to characteristic upland diagrams as well (Fig. 1.3). Variations in former tree-pollen distribution across the country are best shown in the form of contour maps (Mitchell 1986, p. 88; Birks 1989).

TYPE REGION IRL-a, WESTERN IRELAND

Altitude: 0–1041 m.

Climate: Mean January temperature 5°C, July 15°C. Precipitation 1400 mm yr^{-1}, exceeding 2400 mm in mountains. Approx. 200 wet days yr^{-1}. Oceanic. Strong westerly winds (>14 m sec^{-1}) frequent in winter.

Palaeoecological Events During the Last 15 000 Years: Regional Syntheses of Palaeoecological Studies of Lakes and Mires in Europe.
Edited by B.E. Berglund, H.J.B. Birks, M. Ralska-Jasiewiczowa and H.E. Wright. © 1996 John Wiley & Sons Ltd.

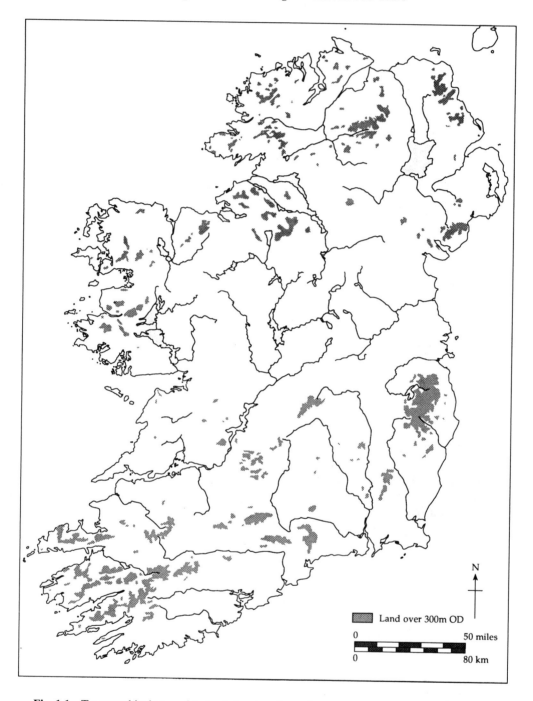

Fig. 1.1   Topographical map of Ireland showing the major watercourses and land over 300 m

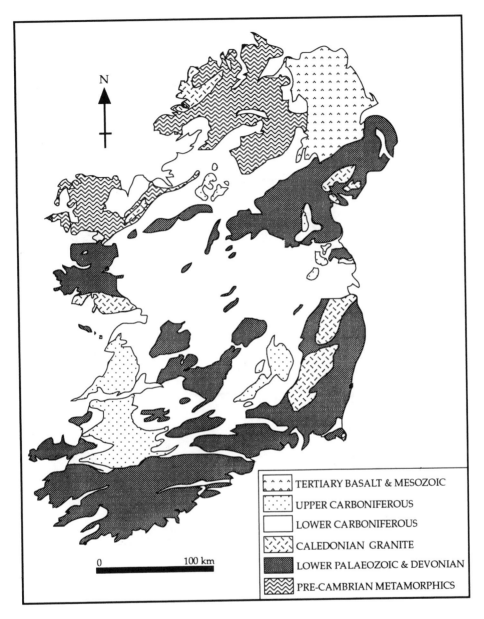

Fig. 1.2 Simplified geological map of Ireland

Geology: Glacial drifts derived mainly from granite and lower Carboniferous/Devonian red sandstone. Bedrock in driftless areas consists mainly of acid rocks except for the karstic Burren area of northern Co. Clare.

Topography: Mainly upland. Low relief about Donegal and Galway Bays and Shannon Estuary.

Population: Distinct coastal settlement pattern includes the most thinly settled regions of Ireland.

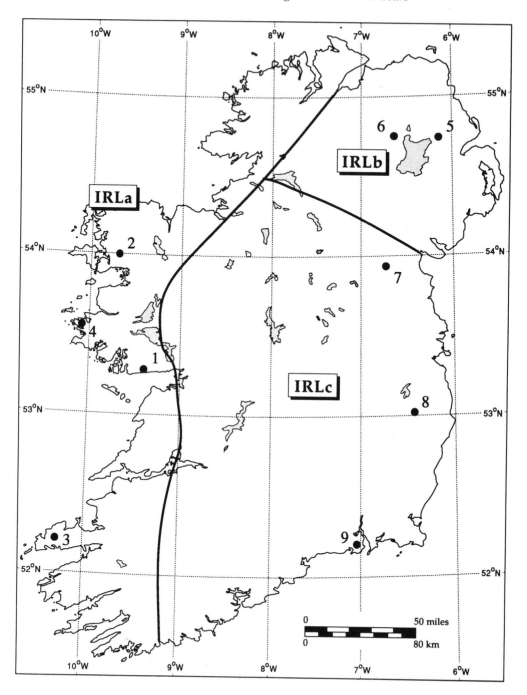

Fig. 1.3   Map of Ireland showing the three type regions (IRL-a, IRL-b, IRL-c) and the reference sites

Vegetation: Predominantly blanket bog and poor pasture. Young/immature coniferous plantations frequent, with pockets of *Quercus*-dominated natural woodland in isolated valleys or on former estates. *Corylus* scrub frequent in the Burren.

Soils: Peats and peaty podzols and gleys. Some brown podzols in valleys in the southwest.

Land use: Sheep farming especially in the uplands. Cattle rearing and dairying important in the southwest. Blanket bog under increasing threat from mechanical turf cutting and afforestation. National Parks in Counties Kerry, Clare, Galway, and Donegal.

Reference site 1. Lough Namackanbeg, Spiddal, Co. Galway (O'Connell *et al.* 1988)

Latitude 53°16′N, Longitude 9°17′W. Elevation 90 m. Age range 9200–0 BP. Lake, now a *schwingmoor*. Nine local pollen-assemblage zones (paz). Uppermost two zones complicated by inwash of older peat (Fig. 1.4). Additionally the upland reference site 2, Lough Corslieve, Co. Mayo, at 320 m (Bradshaw & Browne 1987).

NMKI-1	Pre 9200 BP	*Betula–Salix* peak zone
NMKI-2	9200–8800 BP	*Corylus* peak zone
NMKI-3	8800–7100 BP	*Pinus–Corylus–Quercus–Ulmus*
NMKI-4	7100–5100 BP	*Quercus–Corylus–Alnus–Ulmus–Pinus*
NMKI-5	5100–4400 BP	*Quercus–Betula–Pinus*
NMKI-6	4400–3300 BP	*Quercus–Taxus–Pinus*
NMKI-7	3300–1950 BP	*Quercus–Corylus–Fraxinus–Calluna*
NMKI-8	1950–*ca*.1300 BP	*Calluna–Quercus–Betula–*Cerealia
NMKI-9	*ca*.1300–0 BP	*Calluna–Plantago lanceolata*

Common patterns

(1) Early Holocene succession of *Salix, Betula*, and *Corylus*.

(2) *Quercus* and *Ulmus* arrive more or less simultaneously at *ca.* 8800 BP.

(3) Expansion of *Alnus* at or shortly after 7000 BP, although local expansion may be delayed by up to 1000 yr. At lake sites *Alnus* expansion is preceded by expansion of *Pinus* and followed by its decline; lake-level changes appear to be involved.

(4) Elm decline is a distinctive feature even though *Ulmus* representation is low.

(5) Expansion of *Fraxinus* begins at *ca.*4500 BP.

(6) Increase in anthropogenic impact occurs in Bronze Age (*ca.* 4000 BP onwards) and especially in early Christian time (AD 400 and later). The latter results in large-scale inwash of organics, with consequent complication of the palaeoecological record from lake sites.

(7) Secondary rise of *Pinus* dates from 16th century.

Unique patterns

(1) At beginning of the Holocene, *Juniperus* scrub and/or *Empetrum* heaths are important.

(2) Early rise to dominance of *Pinus* (with *Corylus* in the Burren).

(3) Dominance of *Pinus* throughout earlier Holocene except in upland areas, where it began to succumb to blanket-bog growth at *ca.* 7000 BP. In the later Holocene it invades lowland bog surfaces. In West Connemara and Clare Island this dates to *ca.* 4000 BP and may result from a climatically-induced drying out of bog surfaces. *Ulmus* always less than 10% of arboreal pollen.

(4) *Taxus* is important in the early post-glacial of Killarney (Co. Kerry), Clare, West Galway, and Mayo at various times in the post-elm decline period. In the latter areas it ceases to be of regional importance after AD 300.

(5) *Pinus* at lowland sites is infrequent from 3500 BP onwards, and it probably became extinct in early Christian times at the few sites where it persisted.

(6) Expansion of blanket bog at lowland sites is a feature of the later Holocene. Its spread was probably accelerated by human impact, especially from the mid-Bronze Age (*ca.* 3500 BP) onwards.

6

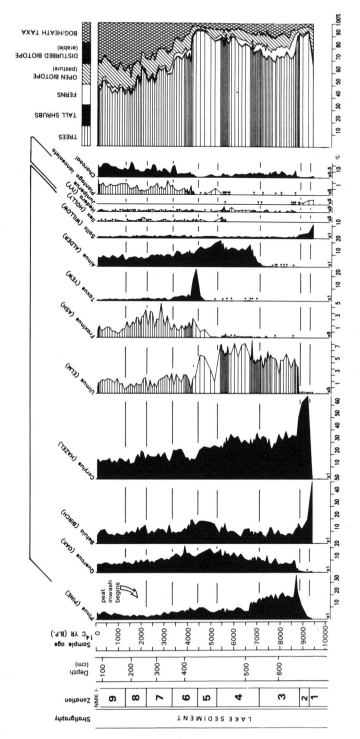

Fig. 1.4 Holocene pollen diagram from Lough Namackanbeg, Spiddal (profile NMK I). The main percentage pollen curves are shown, based on a total terrestrial pollen sum (TTP). The charcoal curve refers to charcoal particles ≤37μm in size; the values are expressed as percentages of TTP + charcoal particles. The horizontal scale magnification is given at the base of each curve (e.g. ×10). A closed circle is used to indicate low representation which may not be readily legible because of the horizontal scale employed; "+" is used to indicate presence outside the count

TYPE REGION IRL-b, NORTHERN IRE-LAND

Altitude: 0–850 m.

Climate: Mean January temperature 4°C, July 15°C. Precipitation 1000 mm yr^{-1}. Approx. 170 wet days yr^{-1}. Oceanic.

Geology: Glacial drifts derived from Lower Palaeozoic slate and shale, Tertiary basalt, and granite. Some sands and gravels.

Topography: Chiefly upland, except around Lough Neagh and in the large valleys of the Bann and Blackwater. Low-relief coastline in Co. Down.

Population: Dense settlement around Belfast, with more scattered patterns in the lowland valleys. Sparse population in mountain areas.

Vegetation: Managed grasslands in lowland areas. Blanket peat in the mountains. Some coniferous plantations and a few isolated *Quercus*-dominated natural woodlands.

Soils: Gleys and acid brown earths in lowland areas. Peats and peaty podzols in the mountains.

Land use: Dairy farming in the north and southwest of the region. Pig farming and poultry in the centre, and sheep grazing in the mountains.

Reference site 5. Sluggan Bog, Co. Antrim
(Smith & Goddard 1991)

Latitude 54°46′N, Longitude 6°18′W. Elevation 65 m. Age range 12 350–0 BP. Raised bog (fig. 1.5). Ten pollen-assemblage zones (paz). Additionally the upland reference site 6, Slieve Gallion, Co. Tyrone at 320 m (Pilcher 1973).

SB-1	Pre 12350 BP	Cyperaceae
SB-2	12350–11000BP	Cyperaceae–Rosaceae–*Betula*
SB-3	11000–9700 BP	Cyperaceae-Gramineae–*Rumex*
SB-4	9700–9200 BP	*Betula–Juniperus*
SB-5	9200–8570 BP	*Corylus–Betula*
SB-6	8570–7020 BP	*Ulmus–Quercus–Corylus–Pinus*
SB-7	7020–4900 BP	*Ulmus–Quercus–Alnus*
SB-8	4900–3840 BP	*Quercus–Alnus–Corylus*
SB-9	3840–1560 BP	*Quercus-Alnus–Plantago lanceolata*
SB-10	1560–0 BP	Gramineae–*Quercus*

Common patterns

The general pattern of vegetational development is similar to that described for the west of Ireland (IRL-a). The beginning of the Holocene is marked by the expansion of *Juniperus* pollen coinciding with a fall in Gramineae and Cyperaceae pollen. *Salix* and *Betula*, succeeded by *Corylus*, form the first closed forest. The dominance of *Corylus* ends with the establishment of tall canopy woodland with *Pinus*, *Ulmus*, and *Quercus*. *Alnus* is present from *ca*. 7000 BP and *Fraxinus* from *ca*. 4000 BP. The decline of *Ulmus* is associated with a decline in *Pinus*. The effects of man are first seen *ca*. 5000 BP and become progressively more important.

Unique patterns

(1) *Juniperus* is more persistent at the beginning of the Holocene in high-altitude areas. *Betula* expansion may be delayed in upland areas, contributing to the persistence of *Juniperus*.
(2) The arrival of *Quercus* in upland areas is late compared with lowland sites. *Ulmus* expands 820 years earlier than *Quercus* at Meenadoan (Pilcher & Larmour 1982) and 880 years earlier at Slieve Gallion.
(3) The final decline of *Pinus* appears to be *ca*. 4000 BP.

TYPE REGION IRL-c, SOUTHEAST AND CENTRAL IRELAND

Altitude: 0–926 m.

Climate: Mean January temperature 5°C, July 15.5°C. Precipitation 900 mm yr^{-1}. Approx.160 wet days yr^{-1}. Oceanic.

Geology: Sands, gravels, and limestone-derived

8

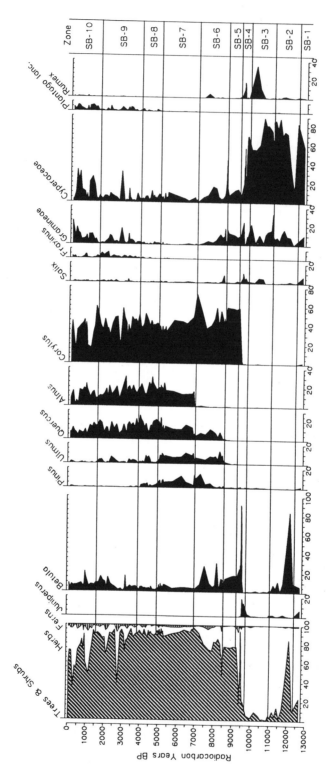

Fig. 1.5 Late-glacial and Holocene pollen diagram from Sluggan Bog, Co. Antrim, showing percentage representation of selected taxa based on an AP pollen sum

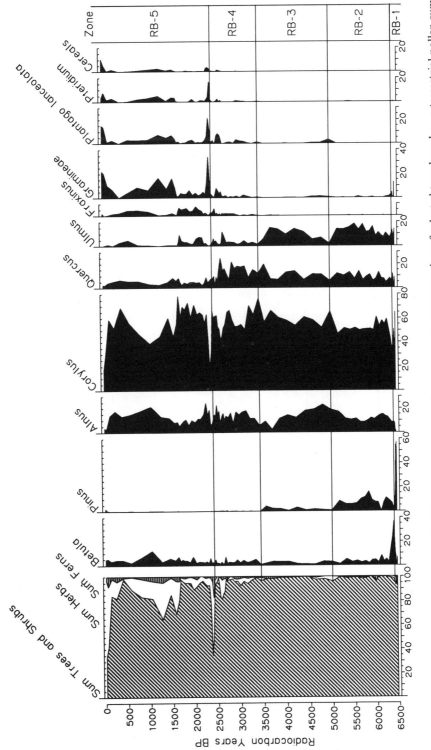

Fig. 1.6 Holocene pollen diagram from Red Bog, Co. Louth, showing percentage representation of selected taxa based on a terrestrial pollen sum which excludes peatland taxa

(a) Holocene environmental change in Ireland. I. Vegetation

Date (^{14}C yr BP)	Date (cal.-AD/BC)[1]	Blytt-Sernander	Jessen (1949)	Mitchell (1956)	Regional Pollen-Assemblage Zones	Other diagnostic species and events
0	1950			X (modern)	AP ↓ : 2° ↗ *Pinus*[2]	Planting of exotics[2]
1000	1020	Sub Atlantic	VIII	IX (Christian)	*Q-Alnus*-Gramineae	final ↘ in *Ulmus* and *Fraxinus*
2000	10			VIIIb (pagan)	*Q-Fraxinus-Betula*	Late Iron Age Woodland regeneration
	AD ↑ BC ↓					
3000						Woodland ↘ sharply *Q* and *Alnus* ↘ sharply
	1230					*Pinus* ± ↓ in W. Ireland
4000	2480/2550	Sub-Boreal	VIIb	VIIIa (pagan)	*Q-Corylus-Fraxinus* (*Ulmus* still important in Midlands)	*Pinus* ± ↓ in N. Ireland *Fraxinus* ↑ *Taxus* (W. Ireland) **Neolithic Landnam ELM DECLINE**
5000	3780					
6000	4865/4910	Atlantic	VIIa	VII	*Ulmus-Q-Corylus-Alnus* *Q-Alnus-Pinus-Corylus* in W. Ireland	*Ilex* expands
7000	5825					*Alnus* curve ↑[3]
						Pinus peaks prior to establishment of *Alnus* (seen mainly in lake profiles)
	6810/6910	Boreal	VI	VI	*Corylus-Q-Ulmus* *Corylus-Q-Pinus-Ulmus* in W. Ireland	
9000	8030				*Corylus* peak zone	*Ulmus and Q* ↑ *Corylus* post-glacial maximum (*ca.* 80%)
			V	V	*Betula* peak zone	*Corylus* invading
10 000	9160/9240/9370	Pre-Boreal	IV	IV	*Salix* *Juniperus*	Tall shrub replaces herb-dominated vegetation
				Late-glacial		

AP↓: Arboreal pollen decline; *Q*: *Quercus*; ↑: curve expands for first time ; ↗: increase ±↓: almost extinct

1: two or more dates given where there is a bad wiggle in the ^{14}C calibration curve the greater part of which is based on 7272-year Irish-German oak chronology (Pilcher *et al.* (1984) Nature *312*, 150–152)

2: AD 1600: 1/8 of Ireland under woodland; 1800: only 1/50 (McCracken, E (1971) *Irish woods since Tudor times.* David and Charles, Newton Abbot)

3: *Alnus* expansion not synchronous; 6700 BP in Connemara; upland areas, N. Ireland, as late as 5200 BP

Fig. 1.7 Event stratigraphical scheme for entire Ireland (M. O'Connell)

(b) Holocene environmental change in Ireland. II. Geology, archaeology, and climate

Date (^{14}C yr BP)	Date (cal.-AD/BC)[1]	Geology	Archaeology/History	Climate
0	1950	Drainage and widescale peat cutting	The Great Hunger (1845–46) 1600s: widescale felling of oak and planting of exotics begins	Little Ice Age
1000	1020		Normans in Ireland (AD 1169) first Viking settlements (AD 790s)	Lesser climatic optimum
		Relative rise in sea level in SW and W coasts	Irelands "Golden Age" Early christian period and renewed framing activity	Cool (little building; few oak timbers AD 524–648)
2000	10		Late Iron Age lull (mainly in pollen record)	Warm
	AD↑ BC↓	Bad wiggle in ^{14}C cal. curve, the so-called "Halstatt disaster" Hekla III (3100 cal.-BP)[3]	LBA/EIA[2] arable farming at Carrowanglogh, NE Mayo	
3000	1230		High population pressures (farming on marginal land)	Narrow rings in Irish bog oak[3]
		Recurrence surfaces in rised bogs Widespread initiation of blanket bog Hekla IV (4200 cal.-BP); ^{14}C years substantially longer than calendar years	LN/EBA[2]: wedge tombs (cf. Burren) LN[2]: Passage tombs (Newgrange)	Drier periods facilitating establishment of pine on bog surfaces in W. Ireland and Midlands
4000	2480/ 2550		Céide fields, N. Mayo-1000+ ha of Neolithic fields	
5000	3780	High sea levels on E and NE coasts[4]	Oldest human bones (Neolithic) from Poulnabone portal tomb; court tomb construction; first evidence of farming (Landnam) at L. Sheeauns	
6000	4865 4910		Larnian culture (L. Mesolithic)	Post-glacial climatic optimum Summer temperatures 2–3°C >today?
			Mesolithic settlement at Newferry	
7000	5825	Diatomite deposition begins at Newferry, L. Bann		↗ ppt facilitating ↑ of *Alnus*? ↘ ppt or ↗ evapo-transpiration resulting in lower lake tevels
		Low lake levels		
8000	6810 6910	End of landbridge of GB Reedswamp and fen is replaced by *Sphagnum*-dominated raised bog vegetation in Midlands	L. Boora, Offaly (Early Mesolithic)	
9000	8030	Large areas of open water in mid-Shannon basin and Midlands where raised bog later developed	Mount sandel, coleraine (Early Mesolithic)	
10000	9160/ 9240/ 9370	Solifluction ceases	No evidence of Palaeolithic peoples	Max. summer insolation in N. Hemisphere Rapid rise in temperature
			Late-glacial	

↑: curve expands for first time; ↗: increase; ↘: decrease ppt: precipitation

1: two or more dates given where there is a bad wiggle in the ^{14}C calibration curve, the greater part of which is based on 7272-year Irish-German oak chronology (Pilcher *et al.* (1984) *Nature 312*, 150–152)

2: LN: Late Neolithic; EBA: Early Bronze Age; LBA: Late Bronze Age; EIA: Early Iron Age (Halstatt)

3: Very narrow rings in Irish Bog oak 1159–1140 cal.-BC (Baillie and Munro (1988) *Nature 332*, 334–346)

4: Carter, Devoy and Shaw (1989) *Journal of Quaternary Science 4*, 7–24

origin elsewhere. Granite and Silurian sediments exposed in mountain areas.

Topography: Large, flat central plain. More mountainous in the southeast.

Population: Dense settlement around Dublin and Cork. Thin, even distribution in north and east of region. More scattered small populations in the south and west.

Vegetation: Raised bogs in Midlands, blanket peat in mountains. Lowland grasslands in southeast. Marginal grasslands in the west. Conifers planted on wetter marginal land.

Soils: Acid brown earths in south and east. Gleys and podzols to the west.

Land use: Some arable land in south and east. Cattle grazing on grasslands and sheep in upland areas.

Reference site 7. Red Bog, Co. Louth
(Mitchell 1956; Watts 1985)

Latitude 53°57′N, Longitude 6°38′W. Elevation 50 m. Age range ca. 6500–0 BP. Raised bog (Fig. 1.6). Five pollen-assemblage zones (paz). Additionally the upland reference site 8, Arts Lough, Co. Wicklow at 490 m (Bradshaw & McGee 1988).

RB-1	Pre 6400BP	*Pinus–Ulmus–Corylus*
RB-2	6400–5000BP	*Ulmus–Quercus–Corylus–*
		Alnus
RB-3	5000–3450 BP	*Quercus–Corylus–Ulmus*
RB-4	3450–2430 BP	*Quercus–Fraxinus–*
		Plantago lanceolata
RB-5	2430–0 BP	*Corylus–Fraxinus–*
		Plantago lanceolata

Common patterns

The general pattern of vegetational development is similar to that of the other two regions described above, although there are considerable differences in the level of representation of a number of taxa.

(1) The Holocene opens with a major expansion of *Juniperus* (Craig 1978).
(2) This is succeeded by *Betula* and then *Corylus*.

(3) *Quercus* and *Ulmus* both arrive in the south at ca. 9000 BP.
(4) High values of *Ulmus* and *Corylus* pollen characterize the earlier Holocene.
(5) *Ulmus* declines at ca. 5200 BP.

Unique patterns

(1) Development of raised bogs on central plain from the onset of the Holocene.
(2) Relatively high percentage values for *Ulmus* pollen at high altitude in the Wicklow Mountains in the early Holocene.
(3) In the mid-Holocene, *Ulmus* falls from exceptionally high to low values. This is normally followed by a series of recoveries and subsequent declines (e.g. O'Connell 1980).
(4) *Fraxinus* attains a higher representation following the *Ulmus* decline than in IRL-a and IRL-b.

CONCLUSIONS

Large variations in topography and soil type over short distances, especially in coastal areas, make it difficult for single Irish reference sites to be representative of large regions. Radiocarbon dating has also illustrated the highly diachronous nature of major vegetation changes during the Holocene, with the exception of the *Ulmus* decline (Smith & Pilcher 1973; Edwards 1985). In the light of these factors, the establishment of a national chronosequence would be misleading. A clearer image of the Holocene vegetation changes in Ireland can be obtained from isopoll and isochrone maps (Mitchell 1986, p. 88; Birks 1989) and analysis of rates of palynological change (Mitchell 1995). These data may be summarized thus: *Pinus* and *Corylus* were the most abundant trees at 7000 BP and were fairly evenly distributed across the country. *Ulmus* had a centre of distribution on limestone drift in the southern part of IRL-b, where *Pinus* was rather uncommon. *Quercus* was most frequent in western and mountain areas and also in the northwest and southeast, where *Ulmus* had a relatively minor role. The expansion of *Alnus* was particularly diachronous, being delayed to as late as 5500 BP at upland sites in IRL-b (Smith & Pilcher 1973).

BP at upland sites in IRL-b (Smith & Pilcher 1973).

Rather different patterns had developed in the vegetation by 5200 BP. *Ulmus* and *Corylus* still dominated large parts of IRL-c, whereas *Quercus*, *Pinus,* and *Betula* characterized IRL-a. Following the *Ulmus* decline, human activity was the dominant factor (cf. Molloy & O'Connell 1993), but there is considerable evidence that climatic change was also important. Such evidence includes frequent *Pinus* and *Quercus* stumps in blanket and raised bogs (cf. McNally & Doyle 1984) and stratigraphical changes that are particularly pronounced in the upper layers of raised bogs. By about 2000 BP or shortly afterwards, *Pinus* ceased to form woodlands and probably became extinct (Bradshaw & Browne 1987). The final large-scale demise of woodland came with the spread of Christianity at about AD 400. From this time onwards, *Ulmus, Fraxinus*, and *Taxus* were of only minor importance in the remaining woodlands. The complete dominance of *Quercus* in present-day woodland fragments is a recent development (Mitchell 1988).

REFERENCES

Birks, H.J.B. 1989: Holocene isochrone maps and patterns of tree-spreading in the British Isles. *Journal of Biogeography 16*, 505–540.

Bradshaw, R.H.W. & Browne, P. 1987: Changing patterns in the post-glacial distribution of *Pinus sylvestris* in Ireland. *Journal of Biogeography 14*, 237–248.

Bradshaw, R.H.W. & McGee, E. 1988: The extent and time-course of mountain blanket peat erosion in Ireland. *New Phytologist 108*, 219–224.

Craig, A.J. 1978: Pollen percentage and influx analyses in southeast Ireland: a contribution to the ecological history of the Late-glacial period. *Journal of Ecology 66*, 297–324.

Dodson, J.R. 1990: The Holocene vegetation of a prehistorically inhabited valley, Dingle peninsula, Co. Kerry. *Proceedings of the Royal Irish Academy 90B*, 151–174.

Edwards, K.E. 1985: Chronology. *In* Edwards, K.J. & Warren, W.P. (eds) *The Quaternary history of Ireland*, 279–293. Academic Press, London.

Ehlers, J., Gibbard, P.L., Rose, J. 1991: *Glacial deposits in Great Britain and Ireland*. Balkema, Rotterdam.

Holland, C.H. 1981: *A geology of Ireland*, Scottish Academic Press, Edinburgh.

McNally, A. & Doyle, G.J. 1984: A study of subfossil pine layers in a raised bog complex in the Irish midlands. I. Palaeowoodland extent and dynamics. II. Seral relationships and floristics. *Proceedings of the Royal Irish Academy 84B*, 57–80.

Mitchell, F.J.G. 1988: The vegetational history of the Killarney Oakwoods, SW Ireland: evidence from fine spatial resolution pollen analysis. *Journal of Ecology 76*, 415–436.

Mitchell, F.J.G.1995: The dynamics of Irish Post-glacial forests. *In* Pilcher, J.R. & Mac an T-Saoir, S.S. (eds) *Wood, Trees and Forests in Ireland*, 13–23, Royal Irish Academy, Dublin.

Mitchell, G.F. 1956: Post-Boreal pollen-diagrams from Irish raised-bogs. *Proceedings of the Royal Irish Academy 54B*, 185–251.

Mitchell, G.F. 1986: *Shell guide to reading the Irish landscape*. Country House, Dublin.

Molloy, K. & O'Connell, M. 1991: Palaeoecological investigations towards the reconstruction of woodland and land-use history at Lough Sheeauns, Connemara, western Ireland. *Review of Palaeoecology and Palynology 67*, 75–113.

Molloy, K. & O'Connell, M. 1993: Early land use and vegetation history at Derryinver Hill, Renvyle Peninsula, Co. Galway, Ireland. *In* Chambers, F.M. (ed.) *Climatic change and human impact on the landscape*, 185–199. Chapman & Hall, London.

O'Connell, M. 1980: The developmental history of Scragh Bog, Co. Westmeath and the vegetational history of its hinterland. *New Phytologist 85*, 301–319.

O'Connell, M., Molloy, K. & Bowler, M. 1988: Post-glacial landscape evolution in Connemara, western Ireland with particular reference to woodland history. *In* Birks, H.H., Birks, H.J.B., Kaland, P.E. & Moe, D. (eds) *The cultural landscape — past, present and future*, 487–514. Cambridge University Press, Cambridge.

Pilcher, J.R. 1973: Pollen analysis and radiocarbon dating of a peat on Slieve Gallion, Co. Tyrone, N. Ireland. *New Phytologist 72*, 681–689.

Pilcher, J.R. & Larmour, R. 1982: Late-glacial and post-glacial vegetational history of the Meenadoan nature reserve, County Tyrone. *Proceedings of the Royal Irish Academy 82B*, 277–295.

Smith, A.G. & Goddard, I.C. 1991: A 12500 year record of vegetational history at Sluggan Bog, Co. Antrim, N. Ireland (incorporating a pollen zone scheme for the non-specialist). *New Phytologist 118*, 167–187.

Smith, A.G. & Pilcher, J.R. 1973: Radiocarbon dates and vegetational history of the British Isles. *New Phytologist 72*, 903–914.

Watts, W.A. 1985: Quaternary vegetation cycles. *In* Edwards, K.J. & Warren, W.P. (eds) *The Quaternary history of Ireland*, 155–185. Academic Press, London.

Webb, D.A. 1983: The flora of Ireland in its European context. *Journal of Life Sciences of the Royal Dublin Society 4*, 143–160.

2

Great Britain—England

J. GREIG

With contributions by David Bartley, Keith Bennett, Petra Day, Elizabeth Huckerby, Jim Innes, Sylvia Peglar, Winifred Pennington and Martyn Waller

INTRODUCTION

Syntheses of pollen results from the British Isles have a long history, starting with Godwin's *History of the British flora* (1956) and followed by Pennington's *History of British vegetation* (1974). Regional syntheses have also been prepared, for example for the northwest of England (Pennington 1970), as well as treatments of particular regional problems such as the chalklands (Waton 1982a, 1982b) and the immigration of trees in the Holocene (Birks 1989).

It is not easy to summarize vegetational developments of any region, particularly the varied landscapes of England, from a few "typical" pollen diagrams. Nor is it within the scope of this regional synthesis to make a concise summary of all the available information. The object of the present effort is rather to compare the major features of representative pollen diagrams and to provide reasonably concise and simple information on the subject.

It is quite clear that, when more data are examined, more regional variation will be found, as can be seen in detailed local syntheses such as that for northwest England (Pennington 1970). No such thing as a "typical" pollen diagram may exist, for the variation among sites is so great in sediment type, catchment area and characteristics, elevation, aspect, underlying geology, and other factors, that each pollen diagram is unique. Also, in many regions pollen diagrams are too few to justify comparisons of sites of similar type. However, even if the pollen diagrams selected here are not fully "typical", they should show enough of the main vegetational changes of the past to be useful.

In the following descriptions of the English landscape, several references have been much used, and to avoid repetition they are mentioned here. The landscape descriptions draw heavily on Trueman (1971), and Carter *et al.* (1974). For the remnants of natural vegetation, Hywel-Davies & Thom (1984) has been consulted.

Geology and topography

The geology of England (Fig. 2.1) includes examples of most of the geological periods. The earliest rocks are the Precambrian slate of Charnwood Forest in the east Midlands (GB-j) (for type regions, see Fig. 2.2). Lower Palaeozoic (Cambrian, Ordovician, Silurian) is represented by the slate of the Lake District (GB-q). Devonian rocks are represented by the Old Red Sandstone of Devon (GB-a). Carboniferous rocks, including limestone, sandstone, and coal, make up an important part of the Pennines (GB-o). Permian and Triassic rocks, including the New Red Sandstone and the Mercian Mudstone (formerly

Palaeoecological Events During the Last 15 000 Years: Regional Syntheses of Palaeoecological Studies of Lakes and Mires in Europe.
Edited by B.E. Berglund, H.J.B. Birks, M. Ralska-Jasiewiczowa and H.E. Wright. © 1996 John Wiley & Sons Ltd.

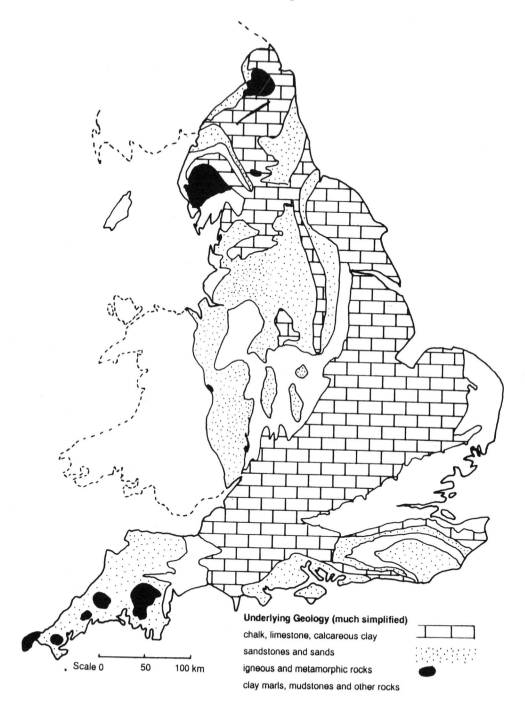

Underlying Geology (much simplified)

chalk, limestone, calcareous clay

sandstones and sands

igneous and metamorphic rocks

clay marls, mudstones and other rocks

Scale 0　　　50　　　100 km

Fig. 2.1　Simplified geological map of England

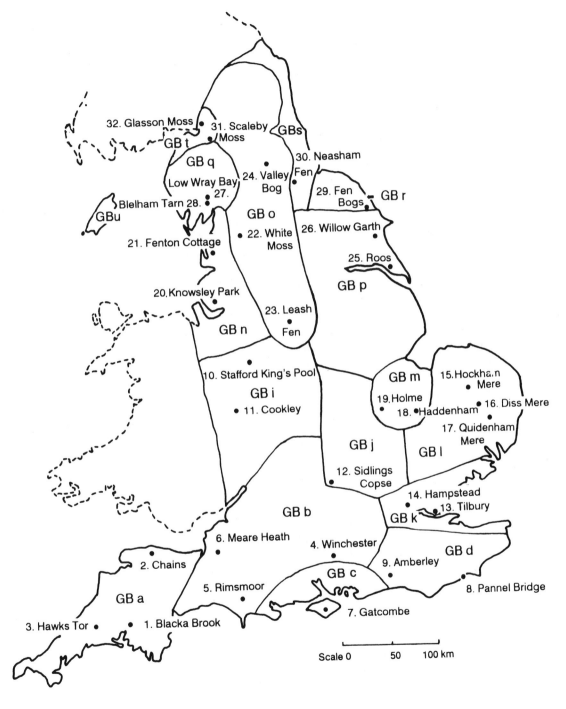

Fig. 2.2 Type regions of England and pollen reference sites (also listed in the Appendix)

Keuper Marl), are easily eroded and have given rise to deep soils in the vales of Eden (GB-t), York (GB-p), and the Midlands (GB-i, GB-j). The Magnesian Limestone, which outcrops east of the Pennines, also belongs here (GB-o). Mesozoic rocks extend over much of a band from northeast to southwest England, including Jurassic rocks, the Oxford Clays, and Cretaceous limestone or chalk (GB-b, GB-j, GB-p). Tertiary sediments are found in southern and eastern England in the Hampshire and London basins (GB-c, GB-k).

The Weichselian (=Devensian) glaciation covered most of northern England with ice during its maximum and deposited glacial drift over much of central and eastern England.

The topography of the "rolling English landscape" (Fig. 2.3) typically consists mainly of small hills divided by a few plains, such as Salisbury Plain (GB-b), the Cheshire and east Lancashire plains (GB-n), and the Fens (GB-m). Some of the hills are on chalk or limestone, such as the South Downs (GB-d), North Downs (GB-k), Chilterns (GB-k), Cotswolds (GB-b), Yorkshire Wolds (GB-p), Mendips (GB-b), and Wenlock Edge (GB-i). Harder rocks give rise to higher hills, such as the granite of Dartmoor and Exmoor (GB-a), Carboniferous sandstone and limestone of the Pennines (GB-o), slate in the Lake District (GB-q), and granite that intrudes sandstone in the Cheviots (GB-s). The highest mountains in England, about 1000 m, are in the Lake District (GB-q).

River systems divide this landscape into a number of large drainage basins, such as the valleys of the Severn (GB-i), Trent (GB-j and p), Thames (GB-k), Yare (GB-l), Dee, Mersey (GB-n), Ribble, Wharfe, Nidd, (GB-o), Tees (GB-r), Tyne, Wear (GB-s), and Eden (GB-t).

Climate

The climate of England is strongly influenced by the Gulf Stream, which brings warm oceanic currents to the western coasts, with an average temperature of 12° C in summer and 8° C in winter, thus making the climate warmer on land too. The climate is also modified by the pattern of atmospheric depressions that bring generally changeable weather with prevailing mild southwest winds and high rainfall, particularly to the north and west. The resulting climate is mostly oceanic and mild in the west but more continental in the east. The growing season is around 7–8 months in the lowlands, 5–6 months in the uplands more than about 200 m asl, and 9–12 months in the extreme southwest and along the south coast (Gregory 1974). The land can be divided into a "highland zone" above about 200 m and a "lowland zone" below.

Extensive records of recent climatic events show that the major differences from the present were during the Little Ice Age in the 16th–18th centuries when the climate was more continental with colder winters in particular, and during a climatic optimum of the early medieval period, when the climate was more favourable for farming than today (Lamb 1982, 1984). During the Bronze Age a period of warmer and drier climate is shown by the reduced growth of peat and by thermophilous beetle faunas (Osborne 1984). Drastic climatic change took place in the late Weichselian and early Holocene.

Soils

The soils are of several main types. Podzols, found on suitable substrates in most regions, particularly in GB-a, c, l, o, q, r, and u, were created by degradation of brown earths following woodland clearance and other human activities, mainly since the Bronze Age (Dimbleby 1962, 1985). Blanket peat has also developed in many of these regions, also largely as a response to human activities (Moore 1975).

Brown earths predominate in the south and east, varying in content according to the parent material. Clay-rich soils occur around London and over parts of the Midlands, but the overlying loess has largely disappeared.

Brown forest soils and rendzina occur where the parent material is calcareous, as in GB-b, GB-d, GB-k, and parts of GB-o.

In some low-lying areas extensive fens have formed, such as the Somerset Levels (GB-b), the Fens (GB-m), and the Humberside wetlands (GB-p). Raised bogs are present in areas with the highest rainfall, such as the Lancashire plain (GB-n, q), the Lake District (GB-q), the Pennines (GB-o), and the Cheviots (GB-s).

Fig. 2.3 Topographical map of England showing the main rivers, named landscapes and land over 100 m

Potential natural vegetation

English vegetation (and indeed that of the British Isles as a whole) has a general similarity to that of the Atlantic and North Sea coast from northern Spain north through France and the Low Countries to northern Germany, Denmark, and western Norway, which Polunin & Walters (1985) call "Atlantic vegetation". An overall account of British vegetation is given by Tansley (1949), and a phytosociological classification is in preparation (e.g. Rodwell 1991). The terminology follows Rackham (1980), who uses "wood" and "woodland" to describe vegetation with trees, "wildwood" for its original undisturbed form, and "forest" for heavily managed woodland, particularly plantations.

The original vegetation of much of England from *ca.* 7500 BP was a wildwood extending from the lowlands to above the present tree-line. The last of this wildwood has long gone, but relics of old woodland occur particularly in areas long protected as wood pasture, such as in the royal parks, Windsor Old Park and the New Forest.

A good account of English woodlands is given by Rackham (1980), from which the following is mainly derived. At present *Quercus* is the main tree in the secondary woodland and managed forests, which supplied the favourite building timber and also the raw material for charcoal. Some *Tilia* woods still survive in scattered places. *Ulmus*, however, has not recovered from the disastrous outbreak of disease in 1976. *Fraxinus excelsior* woods occur especially on limestone as the typical secondary woodland, and this tree is also found frequently in hedges as a result of its colonising ability. *Fagus sylvatica* spread into southeast and south-central England more than 2000 years ago, where it is characteristic of the modern downs and managed woodlands such as Epping Forest. *Castanea sativa* is another tree of managed woodland, being traditionally coppiced on a 7-year cycle to supply fence posts. It was probably introduced into England long ago. *Ilex aquifolium* woods are possible survivors from wood pasture. Recent arrivals include *Aesculus hippocastanum*, introduced about 400 years ago and widespread although usually planted, and *Acer pseudoplatanus*, which spreads rapidly and lets little else grow in its dense shade. Plantation forest established in the 20th century has concentrated on exotic conifers such as *Picea abies* and *Larix* spp.

Wet woodland with *Alnus glutinosa* carr and usually *Salix* spp. and sometimes also *Quercus robur* occurs in such suitable wetlands. *Betula* woodland occurs on heaths and mires, as well as in colonized waste land.

Woody vegetation, including trees, is otherwise widely present in the form of hedges, which are traditionally used for dividing fields and which provide habitat for many plants and animals. Another common habitat of woody plants is scrub around wood margins and in waste places, which favour thorny trees that colonize readily, such as *Crataegus monogyna*, *Prunus spinosa*, *Sambucus nigra*, and on calcareous soils *Viburnum lantana*, *Frangula alnus*, and *Euonymus europaeus*.

Extensive heathlands occur on the lighter acid soils as well as on upland moors. The most characteristic plant here is *Calluna vulgaris*. Oceanic heaths also commonly have *Ulex* spp., *Vaccinium myrtillus*, *Deschampsia flexuosa*, *Molinia caerulea*, and *Pteridium aquilinum*.

On better soils, human settlement causes woodland to be replaced by grassland as the secondary vegetation, thanks to the oceanic climate. The main kinds are the neutral grasslands (Arrhenatherion), those on wetter and more acid soils (Molinion), and those on drier calcicolous soils (Mesobromion). Some traces of these semi-natural grasslands still survive beside the uniform greenery of the agricultural grasslands.

Even more transient in time, but nonetheless very widespread, are the weed communities. Almost all the weeds have apparently been introduced since the Mesolithic period, and the range of taxa is similar to that found in continental Europe. Many annual weeds have a distribution largely controlled by habitat needs and are common in gardens. The traditional cornfield weeds are now mostly extremely rare or extinct in England (*Bupleurum rotundifolium*, for example). Longer-lived weed communities grow in waste places and road edges. Weeds from outside Europe have become established, such as *Solidago canadensis* and *Matricaria matricarioides*.

The arctic–alpine element in the flora of upper Teesdale, including *Minuartia stricta, Gentiana,* and *Galium boreale,* seems to be a relict from the early Holocene.

Neutral and acid mires develop as blanket bogs on poorly drained upland slopes where rain is abundant. Raised bogs, wholly dependent on high precipitation, occur mainly in the wetter northwest England. A typical bog flora includes *Calluna vulgaris, Menyanthes trifoliata, Eriophorum* spp., *Molinia caerulea,* and several *Carex* and *Sphagnum* species.

Fens have a richer nutrient status (neutral to base-rich). They occur along rivers and around lakes, where the groundwater contains enough minerals. They have a richer flora than do acid mires; *Myrica gale* is typical of the transition from acid bog to poor fen. Fen floras typically include many *Carex* and *Juncus* species, as well as *Cladium mariscus.*

Aquatic and bankside plant communities occur in (usually shallow) lakes and rivers. Typical taxa include *Potamogeton* spp., *Myriophyllum* spp., *Utricularia* spp., *Typha* spp., *Sparganium* spp., and *Alisma plantago-aquatica.*

Human settlement

Human settlement has greatly affected the landscape, starting perhaps during the Mesolithic period *ca.* 7000 BP (Smith *et al.* 1989). In the early Neolithic, *ca.* 5500 BP, life based mainly on hunting and gathering seems to have continued, with occasional cereal pollen grains showing that crops were also grown (Edwards & Hirons 1984). With the elm decline (Later Neolithic) *ca.* 5000 BP the signs of woodland clearance increase (for a detailed discussion see Peglar 1993a), as do the records of cereal pollen. However, macrofossil remains show that wild plants were still an important part of the economy (Moffett *et al.* 1989). Wildwood seems to have been extensively cleared in most regions in the Bronze Age, *ca.* 3000 BP, and the lime wood virtually disappeared from central and southern England. Much alluvium was deposited in the river valleys as a result of erosion. Woodland clearance continued in phases in the Iron Age *ca.* 2500 BP, the Roman period *ca.* 2000 BP, and the medieval period.

In the last 200 years the cultural landscape has been changed first into an industrial one in regions with suitable raw materials (water power, coal, iron ore, limestone) such as the Midlands, and then in recent years into a post-industrial one. Recent landscape changes include the maintenance of farming by a system of grants to maintain the cultural landscape, some afforestation, often with exotic conifers, and serious threats to remaining wetlands from continued peat extraction.

Population and land use

England is densely populated, with more than 200 people km^{-2} over large parts of the northwest, Midlands, and southeast, and few areas have less than 100 people km^{-2} (Carter *et al.* 1974). This overall view does not show the great concentrations in conurbations such as Tyneside, West Yorkshire, Greater Manchester, Merseyside, the West Midlands, and a large area around London. Before industrialization the population was more evenly spread because work was both labour-intensive and scattered; the drift from countryside to town has been happening for centuries. The uplands must always have had a smaller population, because the land is less productive.

The landscape as a whole is mainly farmed. Building, industry, and transport systems take up 7% of the land, while 6% remains wooded. Some of this woodland consists of semi-natural, managed relicts of former forest, and some has been planted with conifers by the Forestry Commission. The remainder of the land (87%) is mostly farmed, although only 5% of the population works on the land. The main farming over much of lowland England is mixed, with grass pastures for dairy and beef cattl e and some sheep and meadow for various forms of grassy fodder. Some pigs and hens are kept (usually indoors). The main field crops are cereals (wheat, barley, and oats), potatoes, sugar beet, oilseed rape, and recently flax. On higher ground and in the north the balance changes towards more livestock farming. In eastern England arable farming is much more common, especially of cereals. Fruit, vegetables, and hops are grown in favoured areas such as around the Vale of Evesham in the Midlands (GB-i), the east (GB-i, m, and p), and the southeast (GB- d and k) (data from Carter *et al.* 1974). About half the food is home-grown, or 80% by value; imported products include fruit and other food from hotter countries, and wheat suitable for bread from more continental ones such as Canada. In order to counteract the problems of agricultural over-production, some agricultural land is being planted with trees under a set-aside programme.

Type regions

Seventeen type regions have been identified on the basis of geology, climate, and vegetation (Fig. 2. 2). These type regions were defined before the writer was assigned to this project. The definition of type regions is difficult because England contains so much diversity in quite a small area. Thus the geology and hence the soils vary from chalk to granite. The varied topography affects the climate which ranges from the oceanic west to the more continental east and is cooler farther north and in uplands. These factors in turn affect the potential vegetation.

The type regions are as follows:

GB-a   Southwest peninsula
GB-b   South-central England
GB-c   Hampshire–Dorset Tertiary basin
GB-d   Southeast England
GB-i   West Midlands and Severn basin
GB-j   East Midlands
GB-k   London Basin
GB-l   East Anglia
GB-m   Fenlands
GB-n   Lancashire–Cheshire plain
GB-o   Pennine uplands
GB-p   The Peak District and Lincolnshire–Yorkshire lowlands
GB-q   Lake District
GB-r   North York Moors
GB-s   Northumbrian plain
GB-t   Solway lowlands
GB-u   Isle of Man

Four type regions are hilly or mountainous and are thus topographically defined: GB-a, which includes the Dartmoor, Exmoor, and Bodmin Moor uplands; GB-o, the Pennines, although geologically rather varied; GB-q, the Lake District; and GB-r, the

North York Moors. Another two regions are defined by the surface geology: the peats of the Fens (GB-m) and the Tertiary sands of the New Forest and its surroundings (GB-c). The other lowland regions were then divided on a geographical basis, for example the Thames basin (GB-k) and the east Lancashire and Cheshire plains (GB-n). Some of these have common features, such as the underlying chalk or limestone in the case of much of GB-b (south-central England), GB-d (southeast England), and GB-p (Lincolnshire–Yorkshire lowlands).

Available palaeoecological data

Valley mires (including fens) are perhaps the most widespread potential source of palaeoecological data, although often buried under alluvium. On calcareous substrates a few solution hollows (dolines) exist, such as at Rimsmoor (Waton 1982a, 1982b). In addition, coastal and lowland peat deposits occur in the Fenlands and the Somerset Levels. From the Midlands north, periglacial features are sometimes associated with solution hollows, resulting in the formation of many small lakes and wetlands, such as the Shropshire and Cheshire meres. A few remains of raised bogs survive. In the uplands are blanket bogs, especially in the oceanic west, and lakes and tarns (ponds) occur in the Lake District.

The availability and quality of data from this range of sites is rather varied; some of the early pollen work was not done in as much detail or dated as well as now, and some sites have been re-analysed, such as Hockham Mere (Bennett 1983) and Hatfield Moors (Smith 1985). Some work is also of variable quality, with results lacking sufficient detail or radiocarbon dates. The availability of modern work varies as well, and many results are incompletely published (summary diagrams only, for example) or not at all, as in the case of doctoral dissertations. As much as one third of the information is thus difficult to obtain.

Additional studies are often made on other material, particularly plant macrofossils and beetles, notably the work of P. Osborne.

It has been possible to present at least some data from all but two regions, namely GB-t and GB-u.

TYPE REGION GB-a, SOUTHWEST PENINSULA

This region covers the peninsula west of Exeter and the counties of Cornwall, most of Devon, and part of Somerset. Almost all results are from the uplands. Recent work has been done on Bodmin Moor (Brown 1977), Exmoor (Merryfield & Moore 1974; Merryfield 1977; Moore *et al.* 1984; Francis & Slater 1991, 1993) and on Dartmoor (Simmons 1964; Beckett 1981a). Other relevant work includes that on the late Weichselian (=Devensian) and early Holocene (=Flandrian) (Caseldine & Maguire 1986). Far fewer data exist for the lowland part of this region, such as Hatton & Caseldine (1992) and the multidisciplinary results from the coastal Mesolithic site at Westward Ho! (Balaam *et al.* 1987).

Climate: Mean January 6–7°C, July 16–17°C at sea level, uplands much cooler. Precipitation on Dartmoor and other uplands 1200–2200 mm yr^{-1}, elsewhere 700–1200 mm yr^{-1}. Snow is infrequent in the lowlands, but heavy snowfalls may occur on the uplands.

Geology: Carboniferous and Devonian sandstones, with granite forming the hills such as the various moors (see below). Some Permo-Trias sandstone in the east.

Topography: Hilly, with extensive uplands over 200 m asl, such as Dartmoor (highest point 621 m), Exmoor (520 m), and Bodmin Moor (419 m). Much of the coast is lined with cliffs.

Population: 50–100 km^{-2} except on uplands where it is 6–12 km^{-2}.

Vegetation: Heather moors on the uplands, or planted forest of exotic conifers as on Dartmoor. Several taxa are particular to this region, probably because of the favourable climate and sheltered valleys.

Soils: Sandy fertile soils on the Devonian and Permo-Trias. Poorer soils elsewhere. Serpentine soils very poor.

Land use: Dairy farming and horticulture in lowlands. Grazing on uplands. Tin mining, china clay extraction. Tourism. 25% of the land area is National Park.

Reference site 1. Blacka Brook (Beckett 1981a)

Latitude 50°30′N, Longitude 4°01′W. Elevation *ca.* 270 m. Age range 9000–0 BP. Mire. Seven radiocarbon dates. 130 cm sediment. Eight local and regional pollen-assemblage zones (Fig. 2.4). The site is a mire close to the head of a tributary of the River Plym, on the south edge of Dartmoor near Plymouth.

Local paz	Age	Pollen-assemblage zone
BB1	*ca.* 9000–8500 BP	*Betula–Gramineae*
BB2	*ca.* 8500–8300 BP	*Betula–Quercus Coryloid–Gramineae*
BB3	8300–6000 BP	*Coryloid–Betula–Quercus–Ulmus*
BB4	6000–4500 BP	*Gramineae–Coryloid–Quercus*
BB5	4500–2900 BP	*Coryloid–Gramineae–Alnus–Calluna–Quercus*
BB6	2900–2000 BP	*Coryloid–Gramineae–Alnus–Calluna–Quercus–Cerealia*
BB7	2000–*ca.* 800 BP	*Gramineae–Calluna–Plantago lanceolata*
BB8	*ca.* 800–0 BP	*Coryloid–Gramineae–Calluna–Alnus–Quercus*

Radiocarbon dates from Blacka Brook

Depth (cm)	^{14}C age	Laboratory code
13–15	2530±70	HAR-3377
24–26	740±80	HAR-3359
26–28	1130±70	HAR-3593
49–51	4070±70	HAR-3379
51–53	5690±80	HAR-3594
89–91	8250±80	HAR-3380
117–119	8520±120	HAR-3360

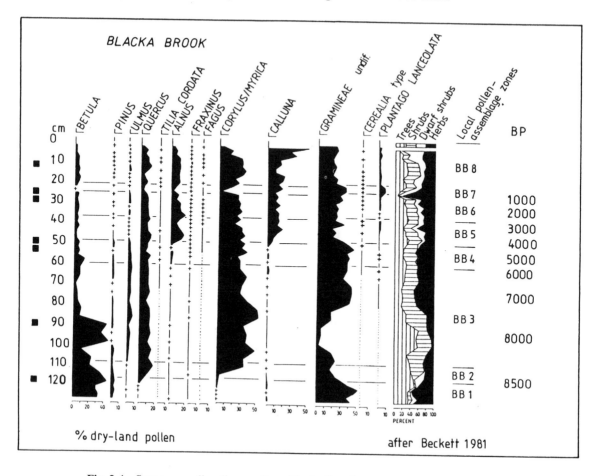

Fig. 2.4   Summary pollen diagram from Blacka Brook, Dartmoor, after Beckett (1981a)

Discussion

(1) The early Holocene woodland *ca.* 9000–7000 BP was dominated by *Betula*, with only a small amount of *Pinus*.

(2) The fully developed woodland *ca.* 7000–5000 BP (later part of BB 3) was mainly of *Quercus* and *Ulmus*, together with *Betula*. The woodland appears to have grown as high as 240 m asl in some places, such as at this site, and on deep soil, since eroded away. There are only traces of *Tilia*, which would probably have grown lower down in the valleys.

(3) *Ulmus* declines *ca.* 4500 BP, also possibly *Pinus* (BB5). Distinct woodland clearances are apparent *ca.* 1000 BP (BB7) and at the top of the diagram.

(4) Some woodland grew back again *ca.* 4500–3500 BP in the diagrams from Lee Moor and Wotter Common (Beckett 1981a) and in the reference site *ca.* 2000 BP (BB 6).

(5) Secondary woodland with *Alnus* expanded first from *ca.* 4500 BP (BB 4) and then greatly *ca.* 4500 BP (BB5). *Fagus* appears in traces from *ca.* 3500 BP onwards.

(6) Ericales present throughout the sequence, increase slightly *ca.* 6000 BP (BB3) and then greatly *ca.* 4500 BP (BB5) and again in the subrecent part of the diagram (top part of BB8).

(7) Cultural indicators are present in traces from 6500 BP (BB3) and increase (BB5) *ca.* 4000 BP. Peaks occur at *ca.* 3500 BP and *ca.* 900 BP (BB7). The Cerealia record is very slight but

more or less continuous from *ca.* 4500 BP. The main land use was apparently pasture. Archaeological evidence from prehistoric field boundaries (reaves) dates mainly from the Bronze Age and shows that farming was possible at greater elevations than today.

(8) Very little information exists about the vegetation history of the low-lying parts of this region. There is evidence for clearance of secondary woodland of *Fraxinus* at *ca.* 1800 BP and arable farming at Aller Farm, 80 m asl (Hatton & Caseldine 1992).

(9) Events are similar to these in the other mainly upland regions, GB-o, GB-q, and GB-r.

TYPE REGION GB-b, SOUTH-CENTRAL ENGLAND

This includes the counties of Avon, Gloucestershire, and parts of Somerset, Dorset, Oxfordshire, Wiltshire, Hampshire, and Berkshire. Pollen analysis in this region started with Godwin's work on the Somerset Levels fenland associated with trackways in the 1930s. More recent results include reference site 6 (Beckett & Hibbert 1976, 1979; Beckett 1978). Otherwise, results have been rather scattered, e.g. a diagram from the Gordano Valley north of the Somerset Levels (Jefferies *et al.* 1968). More pollen evidence comes from the southern part of the type region, Kingswood and Okers (Waton 1982a, 1982b) and unpublished work by Haskins (1978) as well as reference site 5 (Rimsmoor). Other notable work from this region includes reference site 4 (Winchester), Snelsmore, and Woodhay (Waton 1982a, 1982b).

Climate: Mean January 5–6°C, July 16°C. Precipitation 750–1000 mm yr^{-1}. Mean daily sunshine 4–5 hr day^{-1}

Geology: Chalk and Jurassic clay and limestone. Peat in Somerset Levels.

Topography: Some plains (Somerset Levels, Taunton Dene). Some hills, with small areas above 200 m asl (Blackdowns, Mendip Hills, Cotswolds, Dorset Downs).

Population: Mostly 50–200 km^{-2}, less in some areas (on the Downs).

Vegetation: Rich wet fenland in Somerset Levels, woodlands on limestone, and extensive areas of downland grasslands.

Soils: In the lowlands partly clays derived from Jurassic clay and chalk, rendzina on limestone as in the Cotswolds and Mendips.

Land use: Dairy farming in lowlands (Cheddar cheese), cider orchards, mixed farming. The downlands were traditionally used for sheep grazing.

Reference site 4. Winchester (Winnal Moor)
(Waton 1982a, 1982b)

Latitude 51°04′N, Longitude 1°19′W. Elevation 40 m. Age range 8700–0 BP. Mire. 420 cm sediment. One radiocarbon date, chronology otherwise by correlation with other sites. five regional pollen-assemblage zones are represented by nine local subzones (Fig. 2.5). The site is an area of fen in the valley of the River Itchen north of Winchester at Winnal Moors, roughly 2 km long and 300–500 m wide.

Local paz		Age	Pollen-assemblage zone
WT1	C1	8700–8100 BP	*Corylus–Pinus–Quercus–Ulmus*
WT2	C1	8100–7000 BP	*Pinus–Corylus–Quercus–Ulmus*
WT3	C2	7000–5600 BP	*Corylus–Alnus–Quercus–Tilia–Ulmus*
WT4	C3	5600–4500 BP	*Gramineae–Corylus–Quercus–Alnus–Cerealia*
WT5	C3	4500–3450 BP	*Gramineae–Quercus–Alnus*
WT6	C3	3450–2600 BP	*Gramineae–Quercus–Cerealia*
WT7	C4	2600–900 BP	*Gramineae–Quercus–Cerealia–Fagus*
WT8	C4	900–300 BP	*Gramineae–Quercus*
WT9	C5	300–0 BP	*Gramineae–Quercus–Pinus–Aesculus*

Radiocarbon date from Winchester

Depth (cm)	^{14}C age	Laboratory code
ca. 285	5630±90	HAR-4342

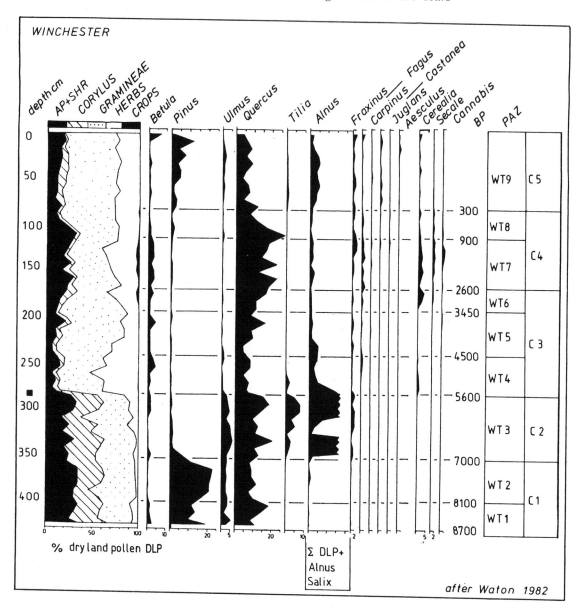

Fig. 2.5 Summary pollen diagram from Winchester, after Waton (1982b)

Discussion

(1) Early Holocene woodland developed with *Pinus, Quercus*, and *Ulmus*. The development of woodland here takes place in a sequence similar to that in other sites on non-calcareous substrata.

(2) *Alnus* and *Tilia* spread *ca.* 7000 BP. The wood-

land at its peak consisted of about 40% *Tilia*, with *Ulmus, Quercus*, and a little *Fraxinus*. *Alnus* would have dominated the vegetation of the valley mires.

(3) The main woodland clearance phase was *ca.* 5600 BP (WT4), with sharp reductions in per-

centages of *Quercus*, *Ulmus*, *Tilia*, and *Alnus*. Another clearance phase occurred at *ca.* 900 BP (WT8). Other pollen diagrams show varying patterns of woodland reduction, and woodland persisted longer around some sites such as Snelsmore (Waton 1982a, 1982b).

(4) Secondary woodland development is shown by the arrival of *Fagus ca.* 5000 BP in WT4 and its increase *ca.* 3500 BP in WT6. *Fagus* pollen percentages are usually small except at Snelsmore, where a steady 5% dryland pollen record shows that *Fagus* woods developed. The beech woods probably grew on the Tertiary sands and gravels rather than on chalk. *Pinus* and *Alnus* spread again at Winchester *ca.* 300–0 BP in WT9.

(5) A little lowland heath development is indicated by a small Ericales record from *ca.* 5600 BP.

(6) Cultural indicators including Cerealia-type appear in the first clearance phase at *ca.* 5600 BP (WT4). A reduction *ca.* 4500–3500 is followed by a peak *ca.* 3500–900 BP in WT6–7. *Cannabis* is present *ca.* 2000–1000 BP in WT7. *Castanea* and *Juglans* are present in the subrecent parts of the diagram, WT9.

(7) The soils over much of this region may originally have been loessic and different in character from the remaining calcareous and clay soils of the present.

(8) The sequence of vegetation events in this region is rather varied and not wholly represented by any one pollen diagram.

TYPE REGION GB-c, HAMPSHIRE–DORSET TERTIARY BASIN

This region covers parts of the counties of Hampshire and Dorset and is defined by Tertiary sands and gravels, although chalk also outcrops in this region, for example on the Isle of Wight. Early work in this area was done by Seagrief (1959, 1960). One of these sites, Cranes Moor, is being re-analysed (Barber, in litt.). Other work in this area has been done by Scaife (1980, 1982, 1988, Scaife & Burrin 1992), including that on the reference site, Gatcombe Withy Bed.

Climate: Mean January 5–6°C, July 16°C. Precipitation 750–1000 mm yr^{-1}. Mean daily sunshine 4–5 hr day^{-1}.

Geology: Tertiary sand in this region, as well as some chalk on the Isle of Wight.

Topography: Rather flat below 100 m.

Population: Mostly 50–200 km^{-2}.

Vegetation: Heathland in New Forest, built-up round Southampton conurbation.

Soils: Sandy.

Land use: Varied, with farming and residential; industry and residential concentrated in the conurbation around Southampton. The New Forest is an area of former wood pasture, now used for recreation.

Reference site 7. Gatcombe Withy Bed (River Medina) (Scaife 1980, 1982, 1988; Scaife & Burrin 1992)

Latitude 50°40′N, Longitude 1°17′W. Elevation 30 m. Age range *ca.* 10000–0 BP. Mire. Four radiocarbon dates. 240 cm sediment. Eight regional pollen-assemblage zones (Fig. 2.6). The site is a fen in the valley of the River Medina on the Isle of Wight, overgrown with an *Alnus/Salix* carr.

Local paz	Age	Pollen-assemblage zone
GTW1	*ca.* 10000 BP	Herbs–*Betula*–*Pinus* (not shown)
GTW2	*ca.* 10000 BP	Herbs–*Betula*–*Pinus* (not shown)
GTW3	*ca.* 10000 BP	Herbs–*Pinus*–*Betula* (not shown)
GTW4	10000–9000 BP	*Juniperus*–*Pinus*–*Betula*
GTW5	9000–7900 BP	*Betula*–*Pinus*–*Corylus*
GTW6	7900–7500 BP	*Corylus*–*Pinus*
GTW7	7500–6000 BP	*Corylus*–*Quercus*–*Pinus*–*Ulmus*
GTW8	6000–4500 BP	*Alnus*–*Quercus*–*Corylus*–*Ulmus*
GTW9	4500–1500 BP	*Alnus*–*Quercus*–*Corylus*–*Betula*–Cerealia

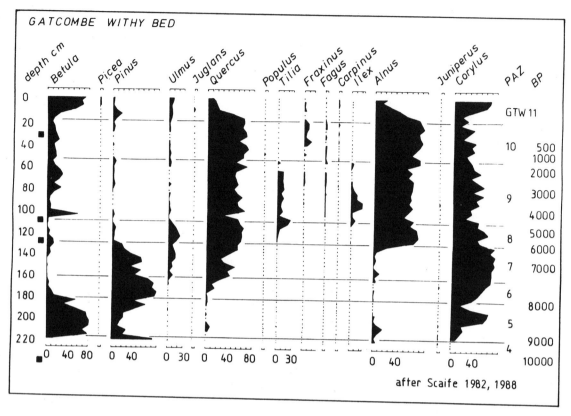

Fig. 2.6 Summary pollen diagram from Gatcombe Withy Bed, after Scaife (1982)

Local paz	Age	Pollen-assemblage zone
GTW10	1500–250 BP	*Alnus–Quercus–Corylus*
GTW11	250–0 BP	*Betula–Corylus–*
		Alnus–Quercus

Radiocarbon dates from Gatcombe Withy Bed

Depth cm	^{14}C age	Laboratory code
215	9970±50	SRR-1433
130	6386±50	SRR-1339
100	4850	?
30	490	?

Discussion

(1) Late Weichselian evidence of herbs with scattered *Betula* and *Juniperus* scrub in GTW1–3.

(2) Woodland development with *Betula* and *Corylus* in GTW5 and *Pinus* in GTW6 *ca.* 9000–7500 BP.

(3) Further woodland development, with spread of *Quercus ca.* 7800 BP, *Ulmus ca.* 7500 BP in GTW7.

(4) Spread of *Alnus*, maybe local, *ca.* 6100 BP, and of *Tilia ca.* 5500 BP in GTW8. Woodland maximum.

(5) Woodland clearance in GTW9, declines in *Ulmus* and *Tilia* and minor reductions in *Quercus* and *Alnus* at *ca.* 4500 BP. Further woodland reduction with *Tilia* decline, *ca.* 1900 BP (GTW10). Main reduction of *Quercus* and *Alnus ca.* 200–300 BP (GTW11).

(6) Secondary woodlands with *Ilex* (possibly wood pasture), *Betula,* and a little *Fagus* developed from *ca.* 4500 BP (GTW9). A trace of *Carpinus* is present from *ca.* 3000 BP (GTW9). *Fraxinus* woodland is present *ca.* 1000–500 BP (GTW 10).

(7) Cultural indicators include an occasional Cerealia record from *ca.* 4500 BP onwards (GTW9) and continuous from *ca.* 3000 BP. *Juglans* is present in the uppermost levels (GTW11).

(8) *Picea* (exotic to Britain) arrives in submodern period, introduced and planted (GTW11). *Betula* and *Pinus* also increase.

TYPE REGION GB-d, SOUTHEAST ENGLAND

This area along the southeast coast includes the counties Surrey, Sussex, and part of Kent. Most of the pollen sites are valley mires. Some early work was done by Godwin at Frogholt and Wingham (Godwin 1962). Investigations into the chalklands were done by Thorley (1981) and by Waton on Amberley (Waton 1982a, 1982b). The valleys of the Ouse and Cuckmere have also been investigated (Scaife & Burrin 1992), as has the East Sussex coast (Moffat 1978), as well as some small sections (Moore & Evans 1991). Finally, very detailed work was done by Waller (1987, 1993) on the Pannel Bridge site in the Brede valley, which is the reference site for this type region.

Climate: Mean January 5 °C, July 16–17 °C. Precipitation 625–1000 mm yr^{-1}. Sunshine 4–4.5 hr day^{-1}.

Geology: Upper Cretaceous chalk and Lower Cretaceous sandstone and clay (the Wealden strata). In addition Quaternary marine deposits and Holocene alluvium are present.

Topography: Some hills (North Downs, Weald, South Downs) with some land above 200 m. Broad valleys in between, slightly hilly. Chalk cliffs at Dover.

Population: 50–200 km^{-2} overall.

Vegetation: Calcicolous grassland on chalk hills (Downs), but much ploughed now. Extensive *Quercus* woodlands on clays and sands (Ashdown Forest in the Weald), heaths on the sands, and some mires (Thursley Common). Former marshlands, such as Romney Marsh and those in the floodplains of many rivers, have largely been drained.

Soils: Thin rendzinas and clay with flints on chalk, brown earths and stagnogleys on clay and compact sandstone, and sandy soils and locally podzols on the looser sands of the Wealden strata.

Land use: Farming (fruit and hops in Kent), sheep grazing on the Downs, and woodland management for charcoal used in iron smelting in the Weald. Building land, roads (London overspill), tourism and retirement on south coast.

Reference site 8. Pannel Bridge (Waller 1987, 1993)

Latitude 50°56′N, Longitude 0°45′E. Elevation 4 m. Age range 10000–0 BP. Mire. Four radiocarbon dates. 11.33 m peat. Four local pollen-assemblage zones with 12 local subzones (Fig. 2.7). The site of Pannel Bridge is in the small valley of the River Pannel in East Sussex, which flows into the western edge of the alluvial complex of Romney Marsh.

Local paz	Age	Pollen-assemblage zone
PB1	10000–9400 BP	Cyperaceae–*Pinus*–Gramineae–*Betula*
PB2	9400–7000 BP	*Corylus–Pinus–Quercus*– Filicales
PB2a	9400–8500 BP	*Pinus–Corylus*
PB2b	8500–7000 BP	*Corylus–Quercus–Alnus*–Cyperaceae–Gramineae–Filicales
PB3	7000–3700 BP	*Alnus–Quercus–Corylus–Tilia*–Cyperaceae
PB3a	7000–6600 BP	*Alnus–Corylus–Tilia–Quercus*
PB3b	6600–6000 BP	Cyperaceae–*Alnus*
PB3c	6000–5100 BP	*Alnus–Corylus–Quercus–Tilia*
PB3d	5100–4000 BP	*Alnus*–Cyperaceae–*Quercus–Corylus–Tilia*
PB3e	4000–3700 BP	Cyperaceae–Filicales
PB4	3700–0 BP	Cyperaceae–*Alnus–Quercus–Betula–Corylus–Myrica*–filicales
PB4a	3700–3200 BP	*Betula–Quercus–Alnus–Corylus–Typha latifolia*-type–Filicales

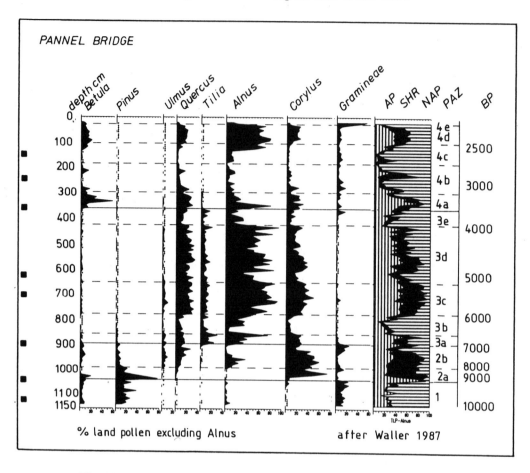

Fig. 2.7 Summary pollen diagram from Pannel Bridge, after Waller (1987)

Local paz	Age	Pollen-assemblage zone
PB4b	3200–2800 BP	Cyperaceae–*Pteridium*– *Salix*
PB4c	2800–1900 BP	*Myrica–Alnus*
PB4d	1900–400 BP	*Alnus–Corylus–Quercus– Betula*–Cyperaceae
PB4e	400–0 BP	Gramineae–Cyperaceae– *Alnus*–Filicales

Radiocarbon dates from Pannel Bridge

Depth (cm)	[14]C age	Laboratory code
142–150	2670±80	SRR-2885
240–248	2980±80	SRR-2886
356–359	3700±90	SRR-2887
620–628	5040±80	SRR-2888

Depth (cm)	[14]C age	Laboratory code
700–708	5540±80	SRR-2889
892–900	7000±90	SRR-2890
1036–1050	9380±100	SRR-2891
1116–1132	9960±110	SRR-2892

Discussion

(1) The sequence starts at the beginning of the Holocene, with *Betula, Pinus,* a little *Corylus,* and early *Alnus* pollen and macrofossils showing the existence of woodland in an open landscape from *ca.* 10 000 BP onwards (PB1). *Alnus* is present especially early here, as the rational limit is not reached at other sites in southeast England until *ca.* 8000 BP (Birks 1989).

(2) *Corylus* increases early at *ca.* 9500 BP (PB2a).

(3) *Ulmus* expands *ca.* 8400 BP, *Quercus* expands *ca.* 9100 BP (PB2b).

(4) *Alnus* expands in several phases together with *Corylus ca.* 8000 BP (PB2b), *ca.* 7000 BP (PB3a), and *ca.* 6200 BP (PB3c) to form local *Alnus* carr over site.

(5) Tilia expands from *ca.* 7000 BP (PB3a).

(6) Woodland maximum *ca.* 6000–5000 BP (Pb3c), when the mire was covered with *Alnus* carr. A mixed woodland with *Tilia, Quercus,* and *Ulmus* grew on the surrounding dry land.

(7) Woodland clearance episodes start with the first *Ulmus* decline, *ca.* 5000 BP (PB3d). Further woodland clearance is shown *ca.* 4000 BP (second *Ulmus* decline, *Quercus* decline and first *Tilia* decline) (PB3e), *ca.* 3700 BP (second *Tilia* decline, AP reduction) (PB4a), and *ca.* 3300 BP (third *Tilia* decline, AP) (PB4b), and *ca.* 2600 BP (reduction in *Quercus*) (PB4c). These clearance episodes are defined by rather small changes in pollen percentages and seem to have been either small clearances or distant ones. That these clearances are more local than regional is shown by clearance episodes at different dates at other sites, for example at *ca.* 2800 BP, 2400 BP, and 2000 BP at Amberley Wild Brooks, with major signs of human impact there at *ca.* 1400 BP (Waton 1982a, 1982b).

(8) Secondary woodland is indicated by scattered *Fraxinus* from *ca.* 6600 BP (PB3b) and a more or less consistent curve from *ca.* 5000 BP (PB3d). The *Fagus* record is scattered from *ca.* 3700 BP (PB4a), and *Fagus* records from other sites in this region are similarly small and scattered, apart from Amberley Wild Brooks, which has a consistent record of about 5% *Fagus* in the dryland pollen sum (excluding *Alnus*) from *ca.* 1400 BP (Waton 1982a, 1982b), probably representing beechwoods. *Fagus* woodland indicated in the pollen diagram from Newbridge, Sussex, probably represents medieval time (Moore & Evans 1991). Otherwise this characteristic woodland of parts of southern England is surprisingly hardly evident (Thorley 1981). Secondary woodland with *Castanea*, probably planted and subrecent in

date, is indicated in the pollen diagram from Keston, Kent, as well as evidence of *Ilex* woodland there and at Hothfield, Kent (Moore & Evans 1991). Other taxa include a scattered record of *Frangula alnus* from *ca.* 2500 BP and of *Viburnum* sp. throughout the diagram, becoming a distinct curve from *ca.* 2500 BP. Only occasional grains of *Carpinus* and *Ilex* are present.

(9) Anthropogenic indicators are rare. Only occasional grains of Cerealia pollen are present from *ca.* 5000 BP (PB3d), with a nearly continuous record from *ca.* 3500 BP (PB4a). *Fagopyrum* occurs at <2000 BP (PB4e). Some other sites, such as Amberley Wild Brooks (Waton 1982a, 1982b), show far greater signs of human impact, such as Cerealia and Cannabaceae from *ca.* 1400 BP and especially from *ca.* 900 BP.

(10) Alluviation in many river valleys in this type region took place after the phase of human activity shown by the first *Ulmus* decline at around 5000 BP, resulting in the deposition of thick inorganic fills, probably of redeposited loess (Scaife & Burrin 1992).

TYPE REGION GB-i, WEST MIDLANDS AND SEVERN BASIN

This includes the counties of Herefordshire, West Midlands, Worcestershire, Shropshire, Warwickshire, and Staffordshire. Much of this area is the drainage basin of the River Severn and its tributaries, such as the Avon and Wye. To the north, part of the Trent basin is also included. The amount of early work is small, for example Hardy (1939). Considering the number of sites available, surprisingly little work on the Shropshire meres has been done, except for Marton Pool (Rowlands 1966) and Crose Mere (Beales 1980). Recent work on the Severn valley (Brown 1988, Brown & Barber 1985) and its tributaries, such as the Stour (Brown 1984, Greig, in prep.), is summarized by Barber & Twigger (1987). Work has also been done on the Cannock Chase area of Staffordshire (Moss 1987). Detailed results have recently been published from the King's Pool,

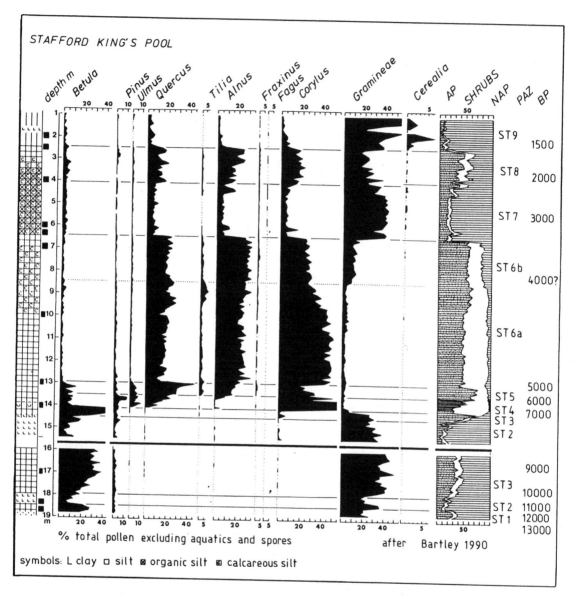

Fig. 2.8 Summary pollen diagram from Stafford King's Pool, after Bartley & Morgan (1990)

Stafford (Bartley & Morgan 1990), reference site 10. Other noteworthy palaeoecological work in this region has been on beetles, for example Osborne (1984), and on sedimentology (Shotton 1978).

Climate: Mean January 5°C, July 16°C, precipitation 500–1000 mm yr⁻¹, sunshine 3.5–4 hr day⁻¹.

Geology: Permo-Triassic rocks, either marl such as the Mercian mudstone or sandstone such as the Kenilworth conglomerate (formerly Bunter sandstone). Coal Measures. Some limestone (Wenlock Edge), some volcanics and metamorphics (Malvern Hills, Wrekin, Long Mynd). Alluvium in river valleys. Some periglacial features.

Topography: River basin, with isolated hills reaching 300 m, and some land over 200 m.

Population: 50–200 km^{-2} in countryside, >200 km^{-2} in conurbations (West Midlands).

Vegetation: Woodland (some *Tilia* woods: Shrawley wood), managed woodland (Forest of Arden), and plantation forest (Cannock Chase). Farming land, the fields divided by hedges. A little upland moor (Stiperstones) and limestone grassland (Wenlock Edge).

Soils: Sandy soils, a little podzolization. Clay soils. Alluvium forms the best soil.

Land use: Market gardening, fruit and hop growing on alluvium (Vale of Evesham, Wye valley). Otherwise mixed farming, especially stock raising.

Reference site 10. Stafford King's Pool.
(Bartley & Morgan 1990)

Latitude 52° 48′ N, Longitude 2° 06′ W. Elevation 75 m. Age range 13000–1000 BP. Former lake. Twelve radiocarbon dates. 30 m sediment. Nine regional paz (Fig. 2.8).

Local paz	Age	Pollen-assemblage zone
ST1	13000–12000 BP	Cyperaceae–Gramineae–*Betula*
ST2	12000–11000 BP	Cyperaceae–Gramineae–*Artemisia*
ST3	11000–7000 BP	*Betula*–Gramineae
ST4	7000–6500 BP	*Corylus–Quercus–Ulmus*
ST5	6500–6000 BP	*Quercus–Alnus–Corylus–Ulmus*
ST6	<6000–3500 BP	*Corylus–Alnus–Quercus*
ST7	3500–2200 BP	Gramineae–Cyperaceae–*Plantago–Rumex*
ST8	2200–1700 BP	*Corylus–Alnus–Quercus*–Gramineae–*Plantago*
ST9	1700–1000 BP	Gramineae–Cyperaceae–Cerealia

Radiocarbon dates from Stafford King's Pool

Depth (cm)	^{14}C age	Laboratory code
195–210	1370±70	WAT-275
255–265	3860±120	WAT-850
395–410	3040±120	WAT-269
595–610	2790±110	WAT-268
625–635	2500±70	WAT-849
685–700	3590±60	WAT-398
995–1010	4760±120	WAT-274
1300–1315	6200±200	WAT-397
1395–1400	9680±140	WAT-266
1700–1710	9650±400	WAT-399
1795–1805	11480±260	WAT-267
1850–1860	10500±300	WAT-396
1885–1890	12070±220	WAT-262
1890–1895	13250±300	WAT-255

Reference site 11. Cookley (Worcestershire)
(Greig in prep.)

Latitude 52° 25′ N, Longitude 2°16′ W. Elevation 60m. Age range 10000–400 BP. River valley mire. Two radiocarbon dates. 250 cm peat sediment. Four paz, seven subzones (Fig. 2.9).

Local paz	Age	Pollen-assemblage zone
CKL1	10000–9000 BP	Gramineae–*Juniperus*–Cyperaceae
CKL2	*ca.* 9000–7000BP	*Betula*–Gramineae
CKL3a	*ca.* 7000–5000 BP	*Alnus–Quercus–*Coryloid*–Tilia–Ulmus*
CKL3b	*ca.* 5000–3700 BP	*Alnus–Quercus–*Coryloid*–Tilia–Ulmus*–Cerealia
CKL4a	*ca.* 3700–? BP	Gramineae–*Quercus-Alnus–*Coryloid–Cerealia
CKL4b	?–? BP	Gramineae–*Alnus–Quercus*
CKL4c	?–1300 BP	Gramineae–*Quercus*–Cerealia–*Alnus–*Coryloid–*Centaurea cyanus*
CKL4d	1300–? BP	Gramineae–*Quercus*–Cerealia–*Filipendula*–herbs
CKL4e	?–? BP	Gramineae–Cerealia–herbs

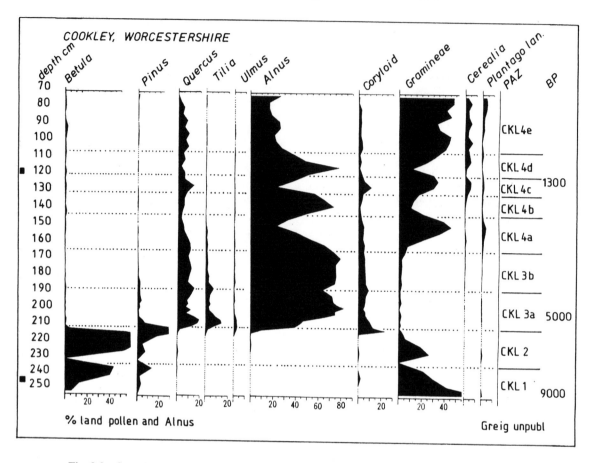

Fig. 2.9 Summary pollen diagram from Cookley, Worcestershire, after Greig (unpublished)

Radiocarbon dates (from Cookley)

Depth (cm)	^{14}C age	Laboratory code
120 cm	1260±80	HAR-3109
250 cm	9160±80	Birm-974

Discussion

This is based on results from the two reference sites, Stafford King's Pool (Bartley & Morgan 1990) and Cookley (Greig in prep.), as well as the results from Crose Mere (Beales 1980) and the synthesis in Barber & Twigger (1987).

(1) Late Weichselian stages are represented at Crose Mere CMCP1, with herbs and *Salix* present. A pioneer phase with *Juniperus* occurs at the end of CMCP1.

(2) Possible Allerød Interstadial deposits (Stafford ST1) indicate a tundra vegetation of herbs with some *Betula* from *ca.* 13000–12000 BP there (Fig. 2.8), while at Crose Mere the 40% *Betula* pollen values (CMCP2) show that woodland spread widely there.

(3) Evidence of the Younger Dryas Stadial (=Loch Lomond Stadial) is shown by *Betula nana* macrofossils and a sharp reduction in *Betula* to *ca.* 10% total pollen in Stafford ST2, and events in Crose Mere CMCP3 are similar. The Cookley diagram (Fig. 2.9) indicates an open arctic heath with *Juniperus*, Gramineae, Cyperaceae, and Ericales at an equivalent stage, CKL1.

(4) The early Holocene rise in *Betula* and pioneer phase with *Juniperus* is dated *ca.* 10300 BP at

Crose Mere (CMCP4), with similar events at Stafford ST3.

(5) Early woodland development of *Pinus* woods seems to have been very short-lived according to the evidence in ST4 but was more extensive at other sites such as Cookley CKL2 and Hartlebury (Brown 1984), areas of sandy soils.

(6) Coryloid reaches very high values at some sites at various dates, the earliest being *ca.* 9100 BP at Crose Mere. Elsewhere its rational limit is not well dated but seems to be somewhat later, for example at Stafford *ca.* 7000 BP (Fig. 2.8). This variation is probably the result of local factors.

(7) Further woodland development is shown by the expansion of *Ulmus* and *Quercus ca.* 7100 BP in Stafford ST4, then *Alnus ca.* 6200 BP (ST4–5), and *Tilia* and *Fraxinus ca.* 6000 BP (ST5). The development of woodland at Wilden marsh seems to have generally taken place earlier (Brown 1988), while *Ulmus* and *Quercus* expand at *ca.* 8800 BP at Crose Mere (Beales 1980).

(8) The woodland maximum is characterized by the highest *Alnus* pollen values in many pollen diagrams, representing local *Alnus* carr. The super-abundance of *Alnus*, together with high *Corylus* in many pollen diagrams, tends to obscure the representation of the more regional woodland in pollen diagrams, unless they are excluded from the pollen sum. This woodland on dry land, using corrected figures (Andersen 1970) seems to have been mixed woodland about half *Tilia*, together with *Quercus, Ulmus, Fraxinus* and other taxa. However, at Rush Pool in Hartlebury Common (Brown 1984), there is evidence of persisting *Betula* in the woodland; the local sandy soil is probably a factor. At two sites, Stafford (Fig. 2.8) and Cookley (Fig. 2.9), the phase of maximum tree pollen (ST5, CKL2) is immediately succeeded by lower values. This may be the result of an "Atlantic hiatus", a period when little sediment was deposited (Rybníček & Rybníčková 1987; Bartley & Morgan 1990).

(9) The earliest possible signs of human activity are microscopic pieces of charcoal *ca.* 6000–7000 BP at Stafford (ST4). More certain evidence of woodland clearance starts at Stafford (Fig. 2.8) with the first *Ulmus* decline *ca.* 5000 BP (ST6a), then with a *Tilia* decline *ca.* 4000 BP (ST6b). The main clearance is *ca.* 3400 BP (ST7), with sharp reductions in *Quercus, Alnus,* and *Corylus*, possibly from local settlement. Further clearance is indicated at *ca.* 1600 BP (ST9). The clearance episodes indicated at other sites correspond to some extent; the first *Ulmus* decline is dated around 5400 BP at Wilden Marsh (Brown 1988). The main clearance episode at Rush Pool, Hartlebury, is dated to *ca.* 2600 BP (Brown 1984), *ca.* 3200–2600 BP at Wilden Marsh (Brown 1988), and a local landscape almost totally cleared of woodland by *ca.* 2600 BP at Ripple Brook in the Severn valley (Brown & Barber 1985). Further clearance is evident at *ca.* 1300 BP at Cookley (CKL 4c) (Greig in prep.) and *ca.* 1200 BP at Wilden Marsh (Brown 1988).

(10) Cultural indicators and crops are first indicated by slight Cerealia-type pollen records at Stafford (Fig. 2.8) from *ca.* 5000 BP (ST6a). From *ca.* 3400 BP is a succession of samples with >1% Cerealia-type pollen, representing possible phases of intense or local cereal growing or processing (ST7–8). From *ca.* 1400–900 BP large amounts of Cerealia-type, *Cannabis,* and *Centaurea cyanus* pollen indicate agriculture, crop processing, or waste from the town of Stafford beside the former lake, while *Centaurea nigra* and *Trifolium* records may indicate hay meadows (ST9). At Crose Mere substantial Cerealia-type pollen records are later, starting *ca.* 4000 BP and becoming continuous from *ca.* 2200 BP. After this a large Cannabaceae record starts *ca.* 1600 BP, and *Centaurea cyanus* appears, followed by several *Fagopyrum* records, although the radiocarbon dates may be in error (Beales 1980). At Rush Pool and Wilden Marsh are fairly strong Cerealia-type records from *ca.* 2600 BP. There are few records of *Fagopyrum* and *Castanea* in the subrecent parts of the diagrams (Brown 1984, 1988).

(11) Secondary woodland is indicated by scattered records of *Fagus* and even more occasional

ones of *Carpinus* from *ca.* 3000 BP at Stafford (ST7). Pollen diagrams in the type region show no evidence that either *Fagus* or *Carpinus* developed into extensive woodlands anywhere there. *Carpinus* is near its present northern limit, although not considered to be native in Staffordshire (Edees 1972). At Cookley (Fig. 2.9), *Ilex* expands with woodland clearance in CKL3b (perhaps around 3500 BP), suggesting the development of wood pasture. At the top of the Crose Mere diagram, a record of *Picea* probably represents subrecent plantation.

(12) The development of heathland is demonstrated at Stafford (Fig. 2.8) by *Calluna* records, which increase sharply at the main clearance horizon at *ca.* 3400 BP (ST7), at Wilden after the clearance episode *ca.* 2600/3200 BP (Brown 1988), and at Rush Pool *ca.* 2600 BP (Brown 1984), a site where heathland still grows today. Some sites, such as Crose Mere, show little signs of heathland.

(13) Alluviation took place especially from *ca.* 3000–2800 BP in the Avon valley as shown by radiocarbon-dated wood in alluvium. Mollusc assemblages show the river-bed change from shallow and stony to deeper and muddy (Shotton 1978). Alluviation in the Severn valley is discussed by Brown & Barber (1985).

TYPE REGION GB-j, EAST MIDLANDS

East Midlands: Northamptonshire, Leicestershire, Rutland, Bedfordshire, Hertfordshire, parts of Buckinghamshire, Nottinghamshire, and a little of Oxfordshire. This is the one region of England from which practically no pollen diagrams exist at present, although this position is likely to change (A.G. Brown *in litt.*)

This region, like the preceding West Midlands, is one of few extremes, a mainly agricultural landscape with scattered industry in the towns.

Climate: Mean January 4–5°C, July 16–17°C. Precipitation 500–750 mm yr^{-1}, mean daily average 3.5–4 hrs sunshine day^{-1}.

Geology: Permo-Triassic rocks include Kenilworth Conglomerate (Bunter sandstone) pebble beds (used as an aquifer), Keuper Marl, and gypsum. Underlying Coal Measures outcrop in places (coal, mudstone, sandstone seams). Lias clay. Precambrian slate outcrops in Charnwood Forest. Weichselian deposits include extensive areas of gravel along the River Trent.

Topography: Dissected river valleys (of the Avon, Nene, Welland, Ouse, Soar, Erewash, and Trent) radiating from the watershed. Rolling hills < 200 m asl.

Population: Mostly 50–200 km^{-2}, concentrated in county towns (Leicester, Nottingham, Derby, Northampton, Bedford).

Vegetation: Woodland pasture in the remains of the royal forests (really hunting parks) (Sherwood, Whittlebury, Rockingham, and Salcey). Some moorland (Charnwood Forest). Vegetation otherwise not distinctive.

Soils: Wide range according to parent materials: brown earths through sands and clays to humus–iron podzols on sandstone.

Land use: Mixed farming and industry in the towns.

Reference site 12. Sidlings Copse (Oxfordshire)
(Day 1991, 1993)

Latitude 51°47′N, Longitude 1°12′W. Elevation 80 m. Age range *ca.* 9500–0 BP. Valley fen. Five radiocarbon dates. 270 cm sediment. Eight local pollen assemblage zones, three subzones (Fig. 2.10).

Local paz	Age	Pollen–assemblage zone
SC1	*ca.* 9500–9200 BP	Gramineae
SC2	*ca.* 9200–9100 BP	*Pinus–Corylus*
SC3	*ca.* 9100–6800 BP	*Corylus*
SC4	*ca.* 6800–3800 BP	*Quercus–Alnus*
SC4a		*Ulmus–Tilia* subzone
SC4b		*Corylus* subzone
SC4c		*Ulmus–Tilia–Filicales* subzone
SC5	*ca.* 3800–3500 BP	*Alnus–*Gramineae–Filicales
SC6	*ca.* 3500–1800 BP	*Pteridium–*Filicales

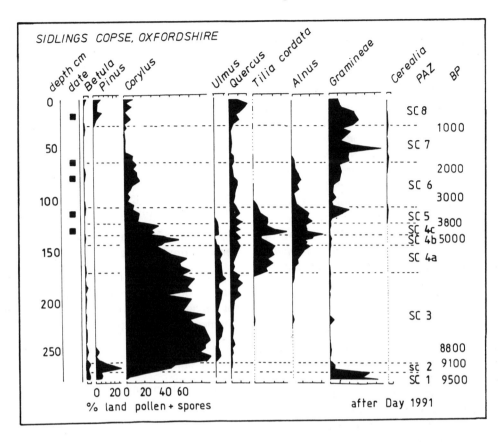

Fig. 2.10 Summary pollen diagram from Sidlings Copse, Oxfordshire, after Day (1991)

Local paz	Age	Pollen-assemblage zone
SC7	*ca.* 1800–1000 BP	Gramineae–*Plantago lanceolata*
SC8	*ca.* 1000–0 BP	*Quercus*–Gramineae

Radiocarbon dates from Sidlings Copse

Depth (cm)	[14]C age	Laboratory code
16–17	855±70	OxA-2046
61–62	1820±80	OxA-2047
77–79	2350±70	OxA-2048
113–114	3820±100	OxA-2049
128–131	3820±100	OxA-2050

Discussion

This compares reference site 12 with the nearby site at Cothill Fen (Day 1991).

(1) The earliest phase (SC1), dated to *ca.* 9500–9200 BP at Cothill Fen, suggests an open landscape with some *Pinus, Betula,* and herbs in the early Holocene.

(2) The *Corylus* spread in SC1 from *ca.* 9300 BP according to the dated sequence at Cothill Fen (Day 1991). A small peak of *Pinus* at *ca.* 9200 BP.

(3) *Ulmus* and *Quercus* spread at *ca.* 9100 BP (SC3) and *Quercus* at *ca.* 8800 BP. *Pinus* then disappears from the record and *Betula* nearly so.

(4) *Alnus*, present since *ca.* 7400 BP, increases at *ca.* 6800 BP together with *Tilia* (SC4). *Tilia* is the most abundant tree pollen, even before correction for productivity and dispersal (Andersen 1970). *Tilia*-dominated wildwood would have grown on drier land, with extensive *Corylus* and *Alnus* in the wetter parts of the valley.

(5) Woodland clearance phases start with the drop

in *Ulmus* and *Tilia* and an increase in *Corylus* percentages, interpreted as the *Ulmus* decline of *ca.* 5000 BP in SC4b, followed by recovery. The main phase of woodland clearance, with the second *Ulmus* and *Tilia* declines, is dated to *ca.* 3800 BP (SC5). Further clearances resulted in a partly open landscape by *ca.* 3100 BP (SC6) and an almost completely open one by *ca.* 1800 BP (SC7). Palaeochannels in the Nene valley reveal a landscape already mostly cleared of woodland by *ca.* 2000 BP (Brown & Keough 1992).

(6) Some woodland regeneration took place (SC4c) after the first *Ulmus* and *Tilia* declines. Another regeneration phase at *ca.* 1000 BP is shown by increased *Quercus* and *Pinus* in SC8, and *Populus* at the top of the pollen diagram represents *P. gileadensis* planted at the fen-edge in 1949.

(7) The earliest signs of human impact on the vegetation are seen in high charcoal values *ca.* 8800–7700 BP (SC3). A single *Triticum*-type pollen grain dated to *ca.* 6700 BP occurred at Cothill. Cerealia-type pollen records start *ca.* 3800 BP. The upper part of the sequence has records of a range of grassland plants such as *Centaurea nigra* type, suggesting that the land was mainly used for grassland rather than cereal crops. The Nene palaeochannels contained a nearly continuous Cerealia-type pollen record and occasional grains of *Cannabis*-type, *Juglans*, and *Fagopyrum* (Brown & Keough 1992).

(8) Secondary woodland developed by regeneration to a similar woodland flora, with little sign of other taxa such as *Fagus* (Day 1993a).

(9) Alluviation is shown by the mineral content increasing *ca.* 4000 BP and staying high for the rest of the sequence.

TYPE REGION GB-k, LONDON BASIN

The region includes Greater London and Middlesex and parts of surrounding counties: Surrey, Sussex, Kent, Berkshire, Essex, Hertfordshire, and Buckinghamshire. Most of the palynological work seems to be recent, with work on the Thames estuarine deposits Devoy (1979) and other drier areas (Baker *et al.*

1978) and on some archaeological deposits (Greig 1992).

Climate: Mean January 3–4°C, July >17°C. Precipitation 500–625 mm yr^{-1}. Mean daily average sunshine 4–4.5 hr day^{-1}.

Geology: Tertiary clay (London Clay) overlying chalk, and some Eocene sand and gravel such as the Bagshot Beds (Hampstead Heath). Periglacial river gravels.

Topography: A large river basin with a rim of chalk hills, extending out into tidal marshlands.

Population: >200 km^{-2}. This is probably the most densely populated region in Europe.

Vegetation: Despite being a very built-up region some woodland still remains (Epping Forest, Windsor Great Park, etc.). Extensive wetlands of the Thames estuary. Heathland (Hampstead).

Soils: Varied, with clays and sands (Reading Beds, Thanet Sands).

Land use: The main land use is for building and for roads—this is an extremely densely populated region.

Reference site 13. Tilbury (World's End)
(Devoy 1979)

Latitude 51°27′N, Longitude 0°22′E. Elevation 0.1 m. Age range *ca.* 8000–2000 BP. River estuary mire. Eight radiocarbon dates. 1400 cm sediment. Six local paz (Fig. 2.11).

Local paz	Age	Pollen-assemblage zone
TA	8200–7600 BP	*Quercus*–*Corylus*-type–*Alnus*–*Ulmus*–*Pinus*
TB	7600–6600 BP	*Quercus*–*Corylus*-type–*Alnus*–Cyperaceae–Gramineae–*Tilia*
TC	6200–? BP	*Quercus*–*Corylus*-type–Gramineae
TD	?–? BP	Gramineae–*Quercus*–*Corylus*-type
TE	?–3800 BP	Gramineae–*Corylus*-type–*Quercus*

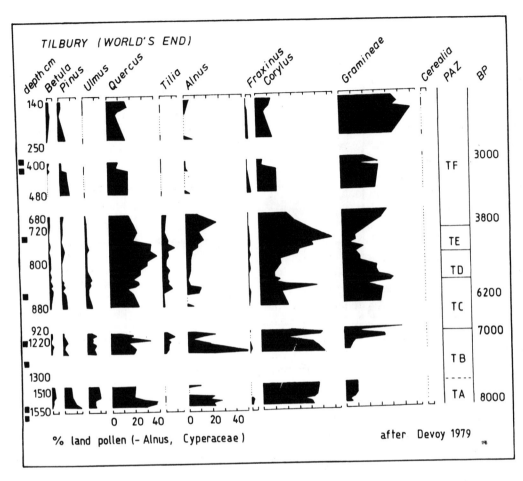

Fig. 2.11 Summary pollen diagram from Tilbury, Essex, after Devoy (1979). The gaps are for levels without countable pollen

Local paz	Age	Pollen-assemblage zone
TF	3800–?2000 BP	Gramineae–Cyperaceae–Chenopodiaceae–*Corylus*-type–*Quercus*

Radiocarbon dates from Tilbury

Depth (cm)	[14]C age	Laboratory code
392–397	3020±65	Q-1433
410–415	3240±75	Q-1432
731–735	3850±80	Q-1431
852–854	6200±90	Q-1430
1220–1224	6595±95	Q-1429

Depth (cm)	[14]C age	Laboratory code
1248–1252	7050±100	Q-1428
1533–1536	7830±110	Q-1427
1547–1550	8170±110	Q-1426

Reference site 14. West Heath Spa (Hampstead Heath, London) (Greig 1991)

Latitude 51°35′N, Longitude 0°07′W. Elevation 120 m. Spring mire. No radiocarbon dates. 130 cm sediment. Four local paz, six subzones (Fig. 2.12).

Local paz	Age zone	Pollen-assemblage paz
WHS 1	*ca.* 6000–5000 BP	*Quercus–Tilia–Ulmus–Corylus*

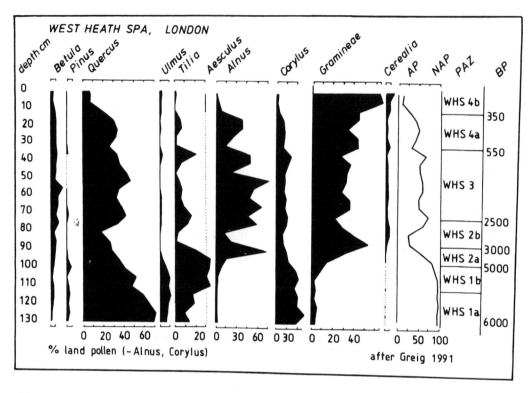

Fig. 2.12   Summary pollen diagram from West Heath Spa, Hampstead, London, after Greig (1991)

Local	Age zone	Pollen-assemblage paz
WHS1a	*ca.* 6000–5500 BP	*Quercus–Tilia–Ulmus*
WHS1b	*ca.* 5500–5000 BP	*Tilia–Quercus–Corylus–Polypodium*
WHS2	*ca.* 5000–?BP	*Gramineae–Alnus–Quercus*
WHS2a	*ca.* 5000–?	*Gramineae–Alnus–Quercus*
WHS2b	*ca.* ?–?	*Gramineae–Alnus–Quercus–Fagus*
WHS3	*ca.* ?–?	*Quercus–Alnus–Gramineae*
WHS4	*ca.* 600–0 BP	*Gramineae–Quercus–Centaurea cyanus*
WHS4a	*ca.* 600–300 BP	*Gramineae–Quercus–Centaurea cyanus*
WHS4b	*ca.* 300–0 BP	*Gramineae–Quercus–Aesculus–Centaurea cyanus*

There are no radiocarbon dates from this site, for the sediment was mainly inorganic. The chronology is derived from similar events at other sites and from historical information.

Discussion

The Tilbury sequence, coming from estuarine deposits, is heavily influenced by wetland and saltmarsh vegetation. The Hampstead Heath sequence, coming from a small site, reflects the dry-land vegetation of the gravel.

(1) Early Holocene events are not covered.
(2) By *ca.* 8000 BP woodland had developed with a large local *Alnus* component (TA), and by *ca.* 7000 BP woodland had developed on drier land with *Quercus, Ulmus, Tilia,* and traces of *Ilex aquifolium,* together with the shrubs *Cornus sanguinea, Euonymus europaeus, Rhamnus catharticus,* and *Viburnum* (TB). These last reflect the calcareous soils.

(3) Woodland clearance is first shown by an *Ulmus* decline in the Tilbury diagram (Fig. 2.11) at *ca.* 5000 BP (TD). A second clearance phase is indicated by the virtual disappearance of *Tilia, Alnus,* and *Ulmus ca.* 3800–3000 BP (TF). In the West Heath Spa diagram (Fig. 2.12) an *Ulmus* and *Tilia* decline probably took place around 5000 BP (WHS2a); at about this point the beetle vector of elm disease, *Ceratocystis ulmi,* first appeared together with beetles that feed on herbivore dung (Girling 1991). In WHS4 probable medieval activity (dated by the appearance of *Centaurea cyanus*) cleared more of the remaining woodland. A sequence at Runnymede shows that the occupied parts of the landscape were substantially cleared of *Quercus* and *Alnus* woodland by *ca.* 2800 BP (Greig 1992).

(4) Secondary woodland development is shown in WHS2 by the appearance of *Ilex* in open woodland, perhaps representing wood pasture. *Alnus* expands greatly after initial woodland clearance, perhaps in response to drainage changes. A final clearance episode (WHS4b) may be dated to *ca.* 400 BP by the appearance of *Aesculus* pollen, the tree having been introduced to England about then. None of these sites show signs of significant *Fagus* woodland; however a diagram from Lodge Road, Epping, has an unusual sequence in which *Fagus* increases to 10–20%, together with *Betula* and a little *Carpinus* in a phase of woodland regeneration following a *Tilia* decline dated to *ca.* 1100 BP (Baker *et al.* 1978). It remains to be seen whether this sequence is unique.

(5) Human impact is shown by Cerealia-type pollen before the *Ulmus* decline in the West Heath Spa diagram (Fig. 2.12), and a continuous Cerealia-type pollen record after it. There is little sign of human impact in the Tilbury sequence, perhaps because this was an area of inhospitable marshland until largely built over.

(6) Heathland developed shortly after the *Ulmus* decline in the West Heath Spa pollen diagram (WHS2) and represents podzolization of the sandy soils and probably maintenance in this state by grazing.

These sites show the difficulty of finding something "typical" of a region—the different sites simply reflect different extremes of the same landscape. The Tilbury results are typical of the wet, rather base-rich conditions in the river valley clay, while the West Heath Spa results refer to the rather dry and acid conditions of the sand and gravel land farther from the River Thames. The Lodge Road (Epping Forest) results show yet another aspect of the vegetation, rare evidence of *Fagus* woodland (Baker *et al.* 1978).

TYPE REGION GB-1, EAST ANGLIA

This type region includes the counties of Norfolk, Suffolk, and parts of Essex and Cambridgeshire. Early work in this region was done by H. and M.E. Godwin, summarized in Godwin (1956). Large amounts of work have since been done in general vegetation history (Bennett 1983; Peglar 1992, 1993a, 1993b, 1993c) and in the Devensian and Early Flandrian (=Holocene) (Hunt & Birks 1982).

Climate: Mean January 4–5°C, July 16–17°C, precipitation 500–750 mm yr^{-1}, mean daily average sunshine 4–4.5 hr day^{-1}. The most continental climate in England (hot summers, cold winters).

Geology: Chalk covered by glacial till, sand, and gravel. Tertiary sand and clay deposits. Extensive glacial features (Hoxne, Cromer, Ipswich).

Topography: Rolling countryside mainly under100m (some below sea level), with few large river systems.

Population: 100–200 km^{-2}. Few large towns (Norwich, Ipswich).

Vegetation: Some ancient woodland (Rackham 1980), a steppe element in the Breckland flora, some traditional hay meadows in Suffolk, and wetland in the Norfolk Broads.

Soils: Sands (Norfolk Breckland) and clays. Some podzols.

Land use: Good farming land, cereals, sugar beet, specialized crops include *Sinapis alba* and *Lavandula,* while *Crocus sativus* was grown in the past.

Reference site 15. Hockham Mere (Bennett 1983)

Latitude 52°30'N, Longitude 0°50'E. Elevation 33 m. Age range 12600–1600 BP. 1185 cm sediment.

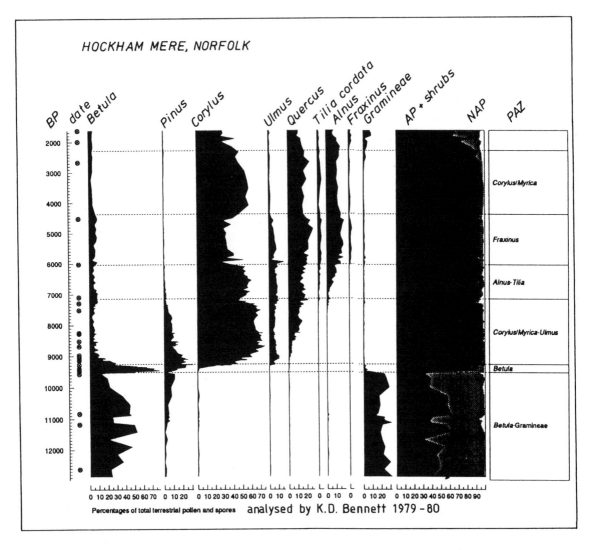

Fig. 2.13   Summary pollen diagram from Hockham Mere, Norfolk, after Bennett (1983)

Lake. 23 radiocarbon dates. Five local paz, three subzones (Fig. 2.13).

Local paz	Age	Pollen-assemblage zone
HM1	12600–9500 BP	*Betula*–Cyperaceae–Gramineae–*Juniperus*
HM2	9500–9250 BP	*Betula*–*Pinus*
HM3	9250–7150 BP	Coryloid–*Pinus*–*Quercus*–*Ulmus*
HM4	7150–6000 BP	Coryloid–*Quercus*–*Ulmus*–*Alnus*–*Tilia*
HM5	6000–1600 BP	Coryloid–*Quercus*–*Alnus*–*Fraxinus*
HM5a	6000–4400 BP	Coryloid–*Quercus*–*Alnus*–*Ulmus*
HM5b	4400–2300 BP	Coryloid–*Quercus*–*Alnus*
HM5c	2300–1600 BP	Coryloid–*Quercus*–*Alnus*–*Calluna*–herbs

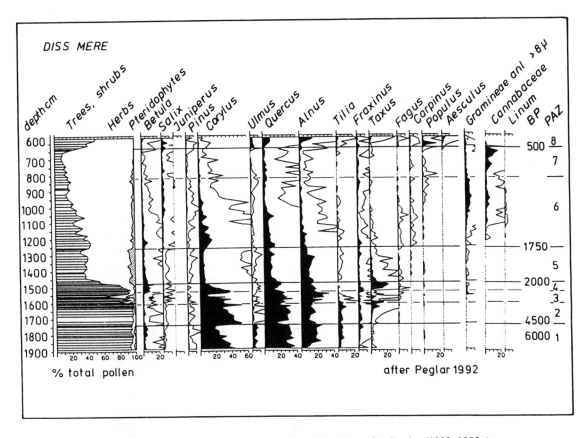

Fig. 2.14   Summary pollen diagram from Diss Mere after Peglar (1992, 1993a)

Radiocarbon dates from Hockham Mere

Depth (cm)	[14]C age	Laboratory code
129.5–134.5	1625±45	Q-2225
257.5–262.5	1980±80	Q-2224
289.5–294.5	2660±50	Q-2223
341.5–346.5	4500±100	Q-2222
429.5–434.5	6010±100	Q-2221
501.5–506.5	7080±60	Q-2220
529.5–534.5	7280±75	Q-2219
609.5–614.5	7505±90 ·	Q-2218
673.5–678.5	8230±150	Q-2217
769.5–774.5	8250±80	Q-2216
817.5–822.5	8500±80	Q-2215
937.5–942.5	8675±60	Q-2214
977.5–982.5	8960±95	Q-2213
1025.5–1030.5	9040±110	Q-2212
1063.5–1070.5	9110±115	Q-2211
1081.5–1087.5	9130±600	Q-2210

Depth (cm)	[14]C age	Laboratory code
1091.5–1096.5	9270±150	Q-2209
1100.5–1104.5	9390±140	Q-2208
1104.5–1108.5	9460±100	Q-2207
1111.5–1116.5	9560±95	Q-2206
1129.5–1134.5	10820±900	Q-2205
1153.5–1158.5	11160±190	Q-2204
1177.5–1182.5	12620±85	Q-2203

Reference site 16. Diss Mere
(Peglar 1992, 1993a, 1993b)

Latitude 52°22′N, Longitude 1°6′E. Elevation 29 m.
Age range *ca.* 7000-0 BP, no radiocarbon dates but
age estimates based on pollen-stratigraphic correlation with other ^{14}C dated sites (Peglar 1992, 1993a,
1993b). Lake. Depth of sediment 1300 cm. Eight paz
(Fig. 2.14).

Local paz	Age		Regional pollen-assemblage zone
DM1	*ca.*7000–5000 BP	5	*Quercus–Ulmus– Tilia*
DM2	*ca.*5000–3500 BP	6	*Quercus–Alnus– Betula–Corylus– Tilia–Taxus*
DM3	*ca.*3500–3000BP	6	*Quercus–Alnus– Betula–Corylus–* herbs
DM4	*ca.*3000–2500BP	6	*Quercus–Corylus– Alnus–Taxus*
DM5	*ca.* 2500–1800 BP	7	*Alnus–Corylus– Quercus–*herbs
DM6	*ca.* 1800–500 BP	8	*Alnus–Corylus– Quercus–*herbs– cereals
DM7	*ca.* 500–100 BP	8	herbs–Cannabaceae
DM8	*ca.* 100 BP–present	9	*Ulmus–Quercus– Alnus–*herbs

Discussion

(1) Late Weichselian stages are represented in the early part of HM1 (Fig. 2.13), covering *ca.* 12600–9500 BP (Bennett 1983). This covers the Allerød Interstadial and Younger Dryas Stadial up to the early Holocene. Open *Betula* and *Juniperus* scrub and much herbaceous vegetation is evident with *Artemisia*. A decrease in *Betula* and increase in Gramineae just before 11000 BP may show the vegetational response to colder conditions during the Younger Dryas Stadial period.

(2) In HM2 *Betula* expands greatly, probably in response to warming climate in the early Holocene, and *Juniperus* disappears between *ca.* 9300 and 9700 BP.

(3) Woodland development is shown in HM3 as a succession of taxa immigrate: *Pinus* from *ca.* 9400 BP, *Corylus/Myrica* by *ca.* 9300 BP, expanding rapidly to *ca.* 8800 BP. *Ulmus* also expands at *ca.* 9300 – 9000 BP. *Quercus* increases just before *ca.* 9000–8900 BP. By *ca.* 9000 BP (HM3), herb pollen decreases to around 5%, showing the almost complete woodland

cover around these lakes. *Quercus* spreads further *ca.* 8600–8200 BP. The *Tilia* curve becomes continuous (rational limit–Birks, 1989) from *ca.* 7600 BP, and expansion is complete by just after 6600 BP, with a corresponding drop in *Corylus/Myrica* percentages. *Alnus* pollen becomes continuous, from *ca.* 8200 BP in HM3, with a sharp increase from *ca.* 6800 BP, reaching a maximum by *ca.* 6300 BP in HM4.

(4) At the time of maximum woodland *ca.* 7000–6000 BP, the Holocene sequence at Diss Mere (Peglar 1993a, 1993b) starts with DM1 (Fig. 2.14), which represents the original deciduous wildwood surrounding the site, with 40% *Tilia*, 20% *Quercus*, and 10% each *Ulmus*, *Fraxinus*, *Corylus*, and *Alnus*, when corrected for pollen productivity (Andersen 1970). This corresponds to HM4.

(5) Woodland clearance episodes begin *ca.* 6000 BP with the first *Ulmus* decline (not definitely associated with human activity) at Hockham Mere and a further *Ulmus* decline *ca.* 4500 BP at Hockham Mere (Fig. 2.13), and the first one at Diss Mere (Fig. 2.14), together with other indications of woodland clearance. The latter is discussed in detail in Peglar (1993b), who suggests that it was the result of severe pathogenic attack on elms already weakened by human impact. Further episodes of woodland clearance probably took place *ca.* 2300, 2000, and 1500 BP at Diss Mere, and *ca.* 2300 BP and after at Hockham Mere.

(6) Secondary woodland with *Fraxinus* developed from *ca.* 6000 BP in HM5a, and from the beginning of DM1. *Taxus* and a little *Fagus* are present after *ca.* 4500 BP (DM2), then *Ilex aquifolium* at *ca.* 3500 BP (DM3). *Fagus* arrives *ca.* 2400 BP, and *Carpinus ca.* 1600 BP at Hockham.

(7) Heathland expansion is indicated by greater *Calluna* pollen values from *ca.* 3500 BP, and herb pollen values increase.

(8) Human impact and cultivated plants arrive late; Cerealia-type pollen does not become important until *ca.* 2000 BP (HM5c) at Hockham Mere (Fig. 2.13). Cerealia-type appears somewhat earlier, *ca.* 2300 BP (DM4), at Diss Mere

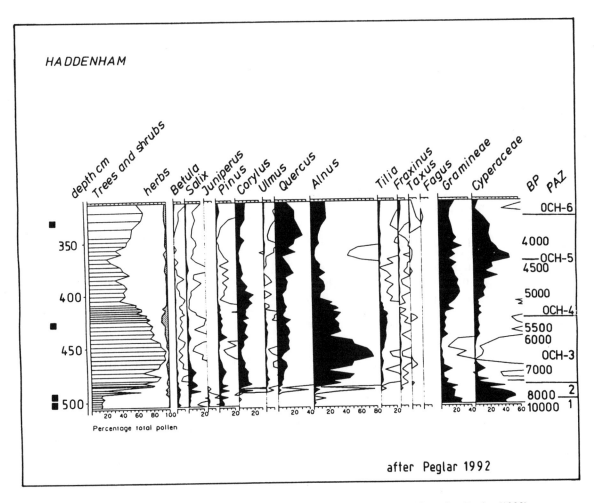

Fig. 2.15 Summary pollen diagram from Haddenham, Cambridgeshire, after Peglar (1992)

(Fig. 2.14). It increases significantly *ca.* 1600 BP together with Cannabaceae and *Linum* (DM6), and again *ca.* 700 BP (DM7).

TYPE REGION GB-m, FENLANDS

Parts of Lincolnshire, Cambridgeshire, Norfolk, and Suffolk. The Fenlands comprise a low-lying region bordering the Wash. Marine and fresh-water clay has been deposited, and mires have developed. Pollen analyses have long been done on material from this ideal environment, starting at the nearby laboratory in Cambridge. More recent examples of

this work include Godwin & Vishnu Mittre (1975) on Holme Fen and Whittlesey Mere and Peglar (1992) and Waller (1994). A Late Weichselian site has been studied by West (1993), and archaeological deposits have been investigated by Smith *et al.* (1989).

Climate: Mean January 4–5°C, July 16°C, precipitation 500–625 mm yr^{-1}, mean daily average sunshine 4–4.5 hr day^{-1}. The climate is relatively continental.

Geology: Little bedrock: the Fens are a basin filled with up to 30 m of recent alluvium together with some marine sediments on glacial tills overlying Cretaceous and Jurassic rocks.

Topography: This is a rather flat region at about sea level, drained by the rivers Witham, Welland, Nene, and Ouse into the Wash—a very "Dutch" landscape.

Vegetation: Large expanses of wetlands, of which traces still remain (Wicken and Woodwalton Fens). Coastal saltmarshes.

Soils: Peats and clays, silts, and sands. Around the Fenland are soils where peat has become incorporated into underlying pre-Holocene sediments.

Land use: Farming, especially for vegetables and sugar beet. Some grasslands exist in the washes.

Reference site 18. Haddenham (Peglar 1992)

Latitude 52°20′N, Longitude 0°3′E. Elevation 0 m. Age range 10650–3850 BP. Fen. Four radiocarbon dates. 510 cm sediment. Six paz (Fig. 2.15).

Local paz	Age	Pollen-assemblage zone
OCH1	10650–9000 BP	Cyperaceae–Gramineae–*Betula*–*Salix*–*Pinus*–*Juniperus*
OCH2	9000–8000 BP	*Salix*–*Pinus*–*Corylus*
OCH3	8000–4800 BP	*Alnus*–*Quercus*–*Corylus*–*Ulmus*–*Tilia*
OCH4	4800–4400 BP	Gramineae–*Quercus*–*Corylus*–*Alnus*
OCH5	4400–3850 BP	*Quercus*–Gramineae–*Alnus*–*Corylus*
OCH6	3850–? BP	*Quercus*–*Alnus*–Gramineae–*Corylus*

Discussion

(1) The early-Holocene phase *ca.* 10000–8000 (OCH1) is dominated by *Pinus*, with *Betula*, *Salix*, and *Juniperus*, but different phases cannot be distinguished within this.

(2) Woodland then develops with *Alnus* from *ca.* 8000 BP. *Corylus*, *Ulmus*, *Quercus*, and *Tilia* expand together abruptly, and *Fraxinus* arrives *ca.* 7000 BP. *Tilia* expands *ca.* 6300 BP. Particularly high values of *Alnus* and *Salix* represent local vegetation *ca.* 6700 BP (OCH 2–3).

(3) Woodland at its maximum *ca.* 6000 BP is indi-

cated by *Alnus*, *Quercus*, *Corylus*, *Tilia*, and *Ulmus*. This represents *Alnus* and *Salix* carr growing on the mire, and mixed woodland with *Quercus*, *Tilia*, *Pinus*, *Ulmus*, and *Fraxinus* on drier land.

(4) Tree-pollen values (mainly *Alnus*) start to fall from *ca.* 6000 BP as soon as the maximum (*ca.* 95% trees and shrubs) is reached. Because Gramineae and Cyperaceae expand at this time, it could represent a change to a fen surface. The first *Ulmus* decline is *ca.* 5400 BP (OCH3). *Tilia* and *Pinus* decline *ca.* 5200 BP (OCH3).

(5) Secondary woodland is shown by an increase in *Quercus* from *ca.* 4000 BP (OCH5). *Fagus* and *Taxus* arrive *ca.* 3700 BP (OCH6).

(6) The cultural indicators *Plantago lanceolata* and probable Cerealia-type are present from *ca.* 5400–4000 BP.

TYPE REGION GB-n, LANCASHIRE–CHESHIRE PLAIN

This region includes Cheshire, Lancashire, Merseyside, and Greater Manchester. Although much of the original landscape is now covered by urbanization, many Cheshire meres still exist, and farther north are the mosses (raised bogs), of which fragments now remain, as surveyed in the Northwest Wetlands Project (Huckerby *et al.* 1992). Some well-dated pollen diagrams from this region include Red Moss (Hibbert *et al.* 1971) and Knowsley Park (Innes 1994).

Climate: Mean January 4–5°C, July 15–16°C, precipitation 625–750 mm yr^{-1}. Mean daily average sunshine 3.5–4 hr day^{-1}. Climate fairly oceanic.

Geology: Little bedrock, except for some Triassic sandstone. Economically important salt deposits lie beneath north Cheshire plain. Much of the area is covered by glacial deposits up to 60m thick. Periglacial features together with subsidence of salt deposits have given rise to many lakes and mires.

Topography: Flat coastal plains (Cheshire plain) and the north of the River Mersey (Fig. 2.3), with a few hills.

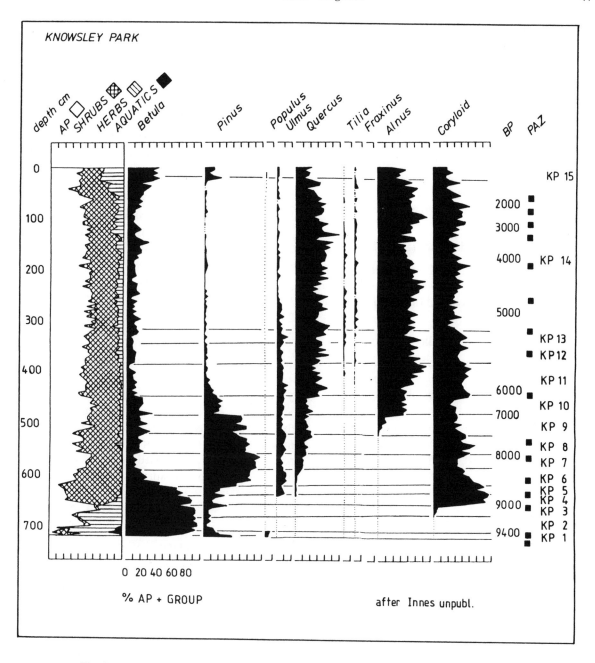

Fig. 2.16 Summary pollen diagram from Knowsley Park, Merseyside, after Innes (1994)

Population: 100–200 km^{-2}, with concentrations in conurbations such as Merseyside, Greater Manchester, Preston.

Vegetation: Coastal mudflats, sand dunes, and saltmarshes. Wetlands around meres (small lakes) on the Cheshire plain and on mosses (mires) throughout the region.

Soils: Mainly sandy, some clay.

Land use: Mixed farming, cereal and vegetable growing in rural areas. Heavy industry (chemicals, petrochemicals, glass) in conurbations.

Reference site 20. Knowsley Park (Innes 1994)

Latitude 53°28′ N, Longitude 2°55′ W. Elevation 70 m. Age range *ca*.10000–0 BP. Former lake. Eighteen radiocarbon dates. 720 cm sediment. 15 paz (Fig. 2.16).

Local paz	Age	Pollen-assemblage zone
KP1	?–9300 BP	*Pinus–Betula– Populus–Juniperus*
KP2	9400–9300 BP	*Pinus–Betula*
KP3	9300–9150 BP	*Pinus–Betula*–Coryloid
KP4	9150–*ca*. 9000 BP	Coryloid–*Betula–Pinus*
KP5	*ca*. 9000–8900 BP	*Betula*–Coryloid– *Pinus–Ulmus–Quercus*
KP6	8900–8650 BP	*Pinus*–Coryloid– *Ulmus–Quercus– Betula*
KP7	8650–8050 BP	*Pinus–Ulmus– Quercus*–Coryloid
KP8	8050–7700 BP	*Pinus*–Coryloid– *Quercus–Ulmus– Betula*
KP9	7700–*ca*. 6900 BP	*Pinus*–Coryloid– *Betula–Quercus– Ulmus–Alnus*
KP10	*ca*. 6900–6190 BP	*Alnus*–Coryloid– *Betula–Pinus– Quercus–(Ulmus)*
KP11	6190–5700 BP	Coryloid–*Alnus– Quercus–Pinus– (Tilia–Fraxinus– Ulmus)*

Local paz	Age	Pollen-assemblage zone
KP12	5700–5450 BP	Coryloid–*Alnus– Quercus–(Tilia– Fraxinus–Ulmus)*
KP13	5450–5250 BP	Coryloid–*Alnus–Quercus*
KP14	5250–300 BP	Coryloid–*Alnus– Quercus*–Ericales
KP15	300–0 BP	*Betula–Alnus–* Coryloid– *Quercus*–herbs

Radiocarbon dates from Knowsley Park

Depth (cm)	^{14}C age	Laboratory code
42–48	1680±50	Birm-1177
67–73	2670±60	Birm-1176
93–97	2860±60	Birm-1175
120–125	3490±70	Birm-1174
180–185	4370±70	Birm-1173
232–238	4600±60	Birm-1172
320–327	5290±80	Birm-1191
336–344	5440±80	Gu-5239
407–415	5750±160	Birm-1223
441–449	6190±80	Gu-5237
516–524	7700±80	Gu-5240
551–559	8060±80	Gu-5241
581–589	8650±80	Gu-5242
616–624	8880±90	Gu-5243
636–644	9030±100	Gu-5238
651–659	9160±80	Gu-5245
676–684	9280±80	Gu-5246
706–712	9305±65	SRR-4515

Reference site 21. Fenton Cottage, Lancashire (Huckerby *et al.* 1992)

Latitude 53°54′ N, Longitude 2°55′ W. Elevation 9 m. Age range 4900-0 BP. Mire. Nineteen radiocarbon dates. 480 cm sediment. Six paz (Fig. 2.17).

Local paz	Age	Pollen–assemblage zone
FCPa	4860–4370 BP	*Quercus*–Gramineae
FCPb	4370–3790 BP	*Betula–Alnus*
FCPc	3790–2080 BP	*Alnus–Quercus*– Coryloid–*Calluna*
FCPd	2080–1590 BP	Gramineae–*Pteridium– Calluna*

Local	Age zone	Pollen-assemblage paz
FCPe	1590–820 BP	*Quercus*–Coryloid–*Calluna*
FCPf	820–0 BP	Gramineae–*Calluna*

Radiocarbon dates from Fenton Cottage

Depth (cm)	¹⁴C age	Laboratory code
25–30	390±50	Gu-5141
45–50	820±50	Gu-5142
70–75	1200±70	Gu-5156
90–95	1380±60	Gu-5143
135–140	1590±50	Gu-5144
160–165	1810±90	Gu-5157
173–178	1940±110	Gu-5158
195–200	2080±90	Gu-5159
220–230	2570±100	Gu-5161
245–250	2730±110	Gu-5163
260–265	2940±60	Gu-5164

Depth (cm)	¹⁴C age	Laboratory code
280–285	3180±60	Gu-5165
305–310	3370±70	Gu-5166
345–350	3790±100	Gu-5167
380–385	4170±50	Gu-5168
400–405	4220±60	Gu-5169
420–431	4370±50	Gu-5145
438–450	4590±90	Gu-5147
450–480	4860±110	Gu-5148

Discussion

(1) The Late Weichselian and early Holocene sequence begins with a peak of *Juniperus* and *Populus*, together with *Salix* and *Empetrum*, in KP1 up to *ca.* 9400 BP.

(2) Early Holocene woodland development starts with the expansion of *Betula ca.* 9400–9300 BP (KP2). Coryloid expands *ca.* 9200 BP (KP3),

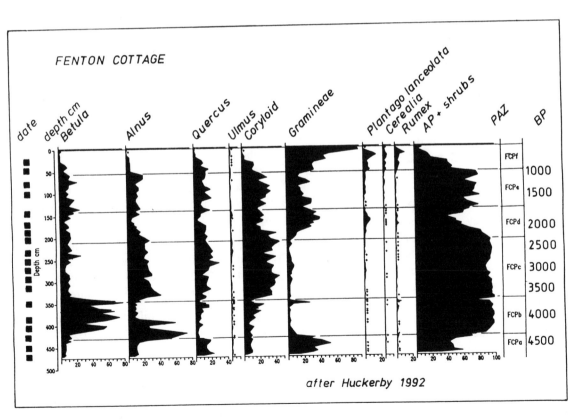

Fig. 2.17 Summary pollen diagram from Fenton Cottage, Lancashire, after Huckerby *et al.* (1992)

Pinus, Ulmus and *Quercus* from *ca.* 8700 BP (KP5).

(3) Further woodland development takes place with the replacement of *Pinus* by *Alnus ca.* 7000 BP (KP10) and expansion of *Tilia* and *Fraxinus ca.* 5800 BP (KP11).

(4) The woodland at its maximum *ca.* 6000–5000 BP (KP11–13) around Knowsley Park consisted of dense *Quercus, Ulmus,* and *Tilia,* with some *Betula* and *Corylus* as an understorey on dry land. *Alnus* carr grew in wet places around the mire, and herbs were not abundant. At Fenton Cottage Zones FCPa and FCPb represent the succession following the marine conditions of the Lytham VI transgression (Tooley 1978) to an *Alnus*-dominated community followed by a *Betula* dominance, with *Quercus* and some remaining traces of *Ulmus* in the woodland.

(5) The first evidence of woodland clearance in the Knowsley Park diagram (Fig. 2.16) is an *Ulmus* decline *ca.* 5000 BP. *Tilia* declines *ca.* 3000 BP, and finally some woodland reduction occurs *ca.* 2000 BP. The small amount of human impact in the Knowsley Park pollen diagram shows that Merseyside has a long history of being an undesirable area in which to live. There is more evidence of human activity in the Fenton Cottage diagram (Fig. 2.17), which starts *ca.* 4900 BP, probably after an *Ulmus* decline. Possible human interference with woodland is mainly seen in reduced *Alnus* and *Quercus* percentages *ca.* 4300–3700 (FCPb) and again *ca.* 2200–1700 BP (FCPd) and *ca.* 1000–500 BP (FCPf).

(6) In the Knowsley Park diagram, secondary woodland with *Fagus* only appears *ca.* 1000 BP, apart from scattered grains. Scattered grains of *Ilex* were present. *Picea* appears at the very top of the sequence, representing plantations.

(7) Heathland, as shown by the *Calluna* curve, developed early at Knowsley Park *ca.* 9000 BP and continues throughout the sequence without much change in percentage.

(8) Cultural indicators are rare in the Knowsley Park diagram. Cerealia-type pollen is present as scattered grains in the Fenton Cottage pollen diagram from the beginning of the sequence *ca.* 4500 BP. A group of Cerealia-type grains and

also a peak of *Plantago lanceolata* correspond to the clearance episode *ca.* 2200–1700 BP (FCPd). Finally, a continuous curve of Cerealia-type appears after *ca.* 1000 BP (FCPf), evidence of intensified medieval settlement in the vicinity.

TYPE REGION GB-o, THE PEAK DISTRICT AND PENNINE UPLANDS

This includes Derbyshire, parts of Staffordshire, South Yorkshire, West Yorkshire, North Yorkshire, Durham, Cumbria, and Northumberland. This region is defined by its elevation, but great variation exists within this apparent uniformity. The climate varies from the oceanic western side to the drier and more continental eastern one. The bedrock varies too, the important difference being between limestone and sandstone, and in turn giving rise either to fertile, base-rich soils or to base-poor soils with a tendency to podzolize. Examples of the results from this area include the pollen diagrams from the southwest Pennines (Tallis & Switsur 1973; Tallis 1985). The south Pennines in the Derbyshire Peak District have been studied by Hicks (1971), and the lowlands by Wiltshire & Edwards (1993). Farther north, much good work has been published by Bartley (1975), Tinsley (1975), Bartley *et al.* (1990), and Bartley & Chambers (1992). The north Pennines have been extensively studied by Chambers (1978) and Turner & Hodgson (1979, 1981, 1983). Some work has been done in the valleys and lowlands by Wiltshire & Edwards (1993).

Climate: Mean January 0–4°C, mean July 11–13°C. Precipitation 1000–2500 mm yr^{-1}. Mean daily average sunshine <3–3.5 hr day^{-1}. Climate dependent on elevation and aspect. Snow fall in winter.

Geology: Mainly Carboniferous sandstone (Millstone Grit) and limestone, with a few igneous intrusions.

Topography: Mountainous (by English standards). Much of this land is over 400 m asl and most over 200 m asl, with higher peaks (Cross Fell: 893 m, Mickle Fell: 790 m, Whernside: 704 m). Countryside rather bleak. Divided by river valleys.

Population: Mostly <12 km^{-2}. Settlement mainly in the valleys (dales) such as Wharfedale, Wensleydale, Ribblesdale, etc.

Vegetation: Varies according to the parent material: some *Fraxinus* woodland or mountain calcicolous grassland on limestone, arctic–alpine relict flora on sugar limestone in upper Teesdale. *Calluna* moorland, blanket peat, some pasture on Millstone Grit and other substrates producing acid soils. Some of the area has been planted with exotic conifers.

Soils: Various; relatively fertile on calcareous substrata, thinner soils on Millstone Grit. Blanket peat on moors.

Land use: Hill farming, partly maintained by grants. Formerly large estates managed the moorland for shooting. After the 1940s much of this land has been opened to the public, and recreation is increasingly important, as shown by the popularity of long-distance paths such as the Pennine Way. The Pennines include several National Parks, such as the Peaks and the Dales parks. Stone is quarried, especially in the limestone regions.

Reference site 22. White Moss 1 (Bartley *et al.* 1990)

Latitude 54°17′N, Longitude 3°12′W. Elevation 190 m. Age range 9900–500 BP. Mire. Four radiocarbon dates. 680 cm sediment. Seven local pollen-assemblage zones, two subzones (Fig. 2.18).

Local paz	Age	Pollen-assemblage zone
WM1	9900–9500 BP	*Betula*–Gramineae
WM2	9500–?8500 BP	*Betula–Corylus–*Gramineae–(*Quercus–Ulmus*)
WM3	?8500–7500 BP	*Corylus–Betula–Pinus–Quercus–Ulmus–(Alnus)*
WM4	7500–5000 BP	*Corylus–Alnus–Quercus*
WM4a	7500–6200 BP	*Alnus–Corylus–Quercus–Pinus*
WM4b	6200–5000 BP	*Alnus–Corylus–Quercus*

Local paz	Age	Pollen-assemblage zone
WM5	5000–2000 BP	*Alnus–Corylus–Quercus*–Gramineae–(Cerealia)
WM6	2000–1500 BP	*Corylus–Alnus–*Gramineae–*Quercus*
WM7	1500–?BP	*Corylus*–Gramineae–*Alnus–Quercus*–Cerealia

Radiocarbon dates from White Moss 1

Depth (cm)	^{14}C age	Laboratory code
167	1470±100	Birm-666
391	5080±100	Birm-665
523–527	6750±70	SRR-2486
583–587	7590±70	SRR-2487

Discussion

(1) Late-Weichselian open vegetation before *ca.* 9400 BP with *Betula* woodland (WM1) becomes more herbaceous in the beginning of WM2. Then *Corylus* seems to have covered the landscape during the early Holocene. *Pinus* is present from *ca.* 9000 BP, an early record (cf. Birks 1989).

(2) Further woodland development takes place from *ca.* 8800 BP with the spread of *Pinus, Ulmus,* and *Quercus*. Pollen sites on limestone have a different succession.

(3) Records of *Alnus* start about 8500 BP, with major expansion *ca.* 7600 BP when the species of *Sphagnum* in the peat indicates wetter conditions, providing *Alnus* with the opportunity for spreading at the expense of *Pinus*. This is *ca.* 1000 years before the rational limit for *Alnus* at other sites in this region (Birks 1989).

(4) The woodland at its maximum *ca.* 6400–5000 BP develops with *Alnus, Quercus, Betula, Ulmus,* and *Fraxinus,* but only a trace of *Tilia*.

(5) A different pattern of vegetational development, similar to that for Scottish sites, is indicated by a pollen diagram from the northern Pennines (Turner & Hodgson 1981). A long pioneer phase dominated by *Pinus* starts *ca.* 7200 BP, with a much delayed and limited expansion of *Quercus* and *Alnus*, perhaps reflecting a particularly cold climate.

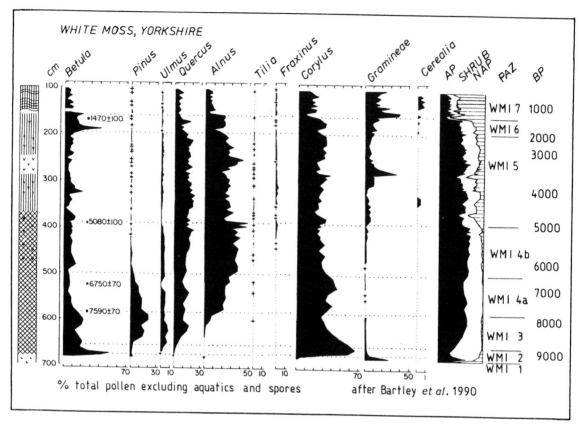

WHITE MOSS, YORKSHIRE

% total pollen excluding aquatics and spores   after Bartley *et al.* 1990

Fig. 2.18   Summary pollen diagram from White Moss, near Settle, Yorkshire, after Bartley *et al.* (1990)

(6) Woodland clearance starts with the *Ulmus* decline, dated to 5080±110 BP (WM5). After this are small phases of woodland disturbance at various times. A short clearance phase occurs at *ca.* 3500 BP at White Moss, and the main clearance phase *ca.* 2000 BP (WM7). Several other Pennine sites, however, show signs of increased woodland clearance *ca.* 4000–3600 BP, for example at Eshton Tarn (Bartley *et al.* 1990), Nidderdale (Tinsley 1975), and Rishworth (Bartley 1975). In the southern Pennines, clearance phases of increasing intensity are dated to *ca.* 2700, 2500, and 2300 BP (Hicks 1971). At White Moss the main clearance episode may be later than at other sites, because the boulder clay rendered the area unattractive to settlers.

(7) Human impact starts with some disturbance before the *Ulmus* decline shown by an increase

of herb pollen at Eshton Tarn. A cereal peak there is dated *ca.* 3600–3400 BP, showing that the fertile limestone soil was suitable for cereal cultivation as high as 144 m asl. At White Moss a Cerealia-type record occurs *ca.* 4000 BP, and once again from *ca.* 1500 BP and even more from *ca.* 1100 BP (Bartley *et al.* 1990).

(8) Possible secondary woodland development is shown by continuous records of *Fraxinus* from *ca.* 5500 BP. Occasional grains of *Fagus* and *Carpinus* in the diagram seem likely to have been blown from lower down.

(9) The pollen diagrams from this region consist of two main types, according to the local soil. The diagrams from regions of rather acid sands and clays show woods mainly of *Quercus* and *Ulmus*, some of which were replaced by moorland rather late, a pattern similar to that of upland GB-a.

The diagrams from regions of base-rich soils show usually earlier and greater woodland clearance and farming with signs of secondary *Fraxinus* woodland and calcicolous grassland.

TYPE REGION GB-p, LINCOLNSHIRE AND YORKSHIRE LOWLANDS

The type region includes parts of Lincolnshire, Humberside, parts of Nottinghamshire, South Yorkshire, and parts of North Yorkshire. Few pollen diagrams exist for this region. Early work includes that on Late Weichselian–early Holocene deposits at Tadcaster by Bartley (1962) and on Hatfield Moors (A.G. Smith 1958), recently continued (B.M. Smith 1985). More recently, work on the Bog, Roos, and Hornsea by Beckett (1975, 1981b), on Willow Garth by Bush (1986, 1988), and archaeologically-oriented work on Seamer Carr by Cloutman & Smith (1988) and on Star Carr by Day (1993b), have greatly increased the amount of data available.

Climate: Mean January 5°C, July 15–16°C. Precipitation 500–750 mm yr^{-1}. Mean daily average sunshine 3.5–4 hr day^{-1}.

Geology: Mainly Cretaceous (chalk) in Humberside, Jurassic sandstone and limestone elsewhere (Fig. 2.1). The valleys and plains are covered by glacial deposits mainly of clay in the vales of the Rivers Ancholme and Witham and in Holderness. Cover sands and gravel over chalk occur near Scunthorpe. The river-valley systems include extensive wetlands along the Ouse, Hatfield, and Thorne Moors, and on Seamer Carr, Star Carr, and other sites along the southern edge of the North York Moors.

Topography: Largely gently rolling land with flat clay plains and river valleys <100 m asl with hills to about 200 m (Lincolnshire Wolds, Yorkshire Wolds).

Population: Ranges around 50–100 km^{-2}.

Vegetation: Chalk grassland, some wetlands (Hatfield, Thorne Moors), coastal vegetation. Some heathland (Skipwith Common).

Soils: Very varied: sands, rendzinas, or gleys,

according to underlying material.

Land use: Mixed farming, especially cereals, beets, potatoes.

Reference site 25. The Roos Bog,
(Beckett 1975, 1981b)

Latitude 53°42′N, Longitude 0°7′W. Elevation 5m. Age range *ca.* 13000–2000 BP. Mire, former lake. five radiocarbon dates. Depth of sediment 1150 cm. Ten paz.

Local paz	Age	Pollen-assemblage zone
RB1	*ca.* 13100–11500 BP	Cyperaceae–*Betula*–Gramineae–*Hippophaë*
RB2	*ca.* 11500–11000 BP	*Betula–Juniperus–Salix*
RB3	*ca.* 11000–10000 BP	Herbs–*Betula-Pinus*
RB4	*ca.* 10000–9000 BP	*Betula–Pinus*-herbs
RB5	*ca.* 9000–7000 BP	*Corylus–Betula–Pinus–Ulmus*
RB6	*ca.* 7000–5000 BP	*Alnus–Ulmus–Tilia-Quercus*
RB7	*ca.* 5000–4000 BP	*Alnus–Quercus*–herbs
RB8	*ca.* 4000–? BP	*Alnus*–Gramineae–*Calluna*
RB9	?–? BP	Gramineae–herbs
RB10	?–2000 BP	Gramineae–*Betula*–*Alnus*–herbs

Radiocarbon dates from Roos Bog

Depth (cm)	^{14}C age	Laboratory code
920–925	10120±180	Birm-405
1091–1095	11220±220	Birm-406
1102–1105	11450±170	Birm-407
1110–1115	11500±170	Birm-318
1133–1140	13045±270	Birm-317

Reference site 26. Willow Garth (Bush 1986, 1988)

Latitude 54°05′N, Longitude 0°17′W. Elevation 18 m. Age range 9500–0 BP. Mire. Ten radiocarbon dates. 136 cm sediment. Five paz (Fig. 2.19).

Local paz	Age	Pollen–assemblage zone	Local paz	Age	Pollen–assemblage zone
WG1	9500–? BP	*Pinus*–Gramineae	WG4	?7000–2900 BP	*Corylus*–Gramineae–*Quercus*
WG2	?–8800 BP	*Betula*–*Pinus*–Gramineae			
WG3	8800–?7000 BP	*Betula*–*Pinus*–*Corylus*–Gramineae	WG5	2900–0 BP	Gramineae–*Pinus*

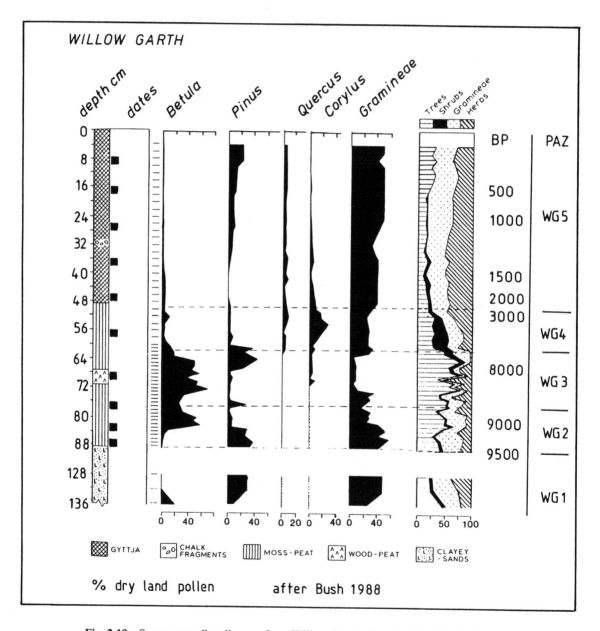

Fig. 2.19 Summary pollen diagram from Willow Garth, Humberside, after Bush (1988)

Radiocarbon dates from Willow Garth

Depth (cm)	^{14}C age	Laboratory code
8–10	110±60	SRR-2665
16–18	700±50	SRR-2666
26–28	1170±50	SRR-2667
36–38	1300±50	SRR-2668
46–48	2120±50	SRR-2669
56–58	3970±50	SRR-2670
68–70	8290±80	SRR-2671
76–78	8910±80	SRR-2672
82–84	9380±80	SRR-2673
86–88	9460±80	SRR-2674

Discussion

(1) The Late Weichselian sequence at Roos Bog starts before 13000 BP (RB1, RB2) with tree *Betula* present as well as *B.* cf. *nana, Salix, Juniperus, Hippophaë,* and a range of arctic herbs. This woodland spread *ca.* 12000 BP in the Allerød Interstadial. Another Late Weichselian–early Holocene sequence has been studied at Tadcaster (Bartley 1962).

(2) Tree and shrub pollen are reduced in RB3 from after *ca.* 11200 BP until before *ca.* 10100 BP, corresponding to the Younger Dryas Stadial period.

(3) Further expansion of *Betula* and *Pinus* after *ca.* 10100 BP represents the start of the Holocene. No *Juniperus* peak exists here (and the scattered records of *Ulmus, Quercus, Alnus,* and *Corylus* are not thought to represent local vegetation). The *Betula–Pinus* phase seems to end *ca.* 8000 BP according to the dates on the Willow Garth sequence.

(4) Early woodland development is shown by an abrupt increase in *Corylus* at Roos (RB5) and Hornsea (HO2), *ca.* 8300 BP according to the date at Willow Garth (WG3). *Quercus* and *Ulmus* spread at Roos, Hornsea, and Willow Garth.

(5) Later woodland development is shown by the increase in *Alnus* first at Hornsea, much later at Roos, apparently the result of differences in local habitat. *Tilia* increases just before the phase of maximum tree pollen; a corresponding reduction in *Corylus* is undated but is probably *ca.* 7000 BP.

(6) The phase of maximum woodland (RB6, HO3) shows evidence of probably local *Alnus* carr and mixed woodland with *Quercus, Tilia, Ulmus,* and a little *Fraxinus* on drier land. Little evidence exists for *Pinus* and *Betula* woodland and herb vegetation. This phase is not clear at Willow Garth, because a discontinuity is apparent in that profile.

(7) Woodland clearance is shown by the first *Ulmus* decline together with the first *Tilia* decline. The *Ulmus* decline has been dated to 5099±50 BP at Gransmoor Quarry in this region (Beckett 1975). A second *Tilia* decline takes place together with almost all other AP at Roos (RB8), and a *Tilia* and *Quercus* decline at Hornsea (HO5) may represent the same woodland clearance episode. The Willow Garth results indicate a landscape largely clear of woodland from *ca.* 4000–0 BP (WG4–5).

(8) Secondary woodland development is shown by scattered pollen records of taxa such as *Fagus, Carpinus, Ilex, Hippophaë, Viburnum,* and Rhamnaceae in the diagrams. Woodland regeneration is apparent between the first and second declines of *Ulmus* and *Tilia.*

(9) Heathland is shown by a peak in RB8, which represents the vegetation on the bog itself; *Calluna* is hardly shown in the Hornsea diagram (HO5). Few habitats are suitable for heathland in this region.

(10) Human impact was slight, with little evidence from *ca.* 5000 BP. Cultural indicators and crops are indicated by Cerealia-type records in RB 8–10 and HO5, corresponding with the main phase of woodland clearance, which probably took place at some time before *ca.* 2000 BP as these sequences appear to end.

(11) Development of grassland communities is shown in the later phases of the Roos and Hornsea diagrams by records of taxa such as *Centaurea nigra*-type and *Plantago lanceolata.* A few taxa such as *Polygala, Poterium*-type (*Sanguisorba minor*), and *Thalictrum* may indicate grassland on base-rich soils. In the Willow Garth results pollen (and macrofossils) of grassland taxa are always abundant.

TYPE REGION GB-q, LAKE DISTRICT

The type region covers the county of Cumbria. A 50-year tradition of pollen analysis in this region is summarized in Pennington (1970, 1991), together with many limnological studies carried out at the Freshwater Biological Association at Windermere. Reference sites typical of the region are hard to select, for the topography is very varied, going from coastal habitats at sea level (Wimble 1986) to mountain tops around 1000 m inland.

Climate: Mean January 5°C at sea level (similar to that on the south coast of England), 0°C in the hills. Precipitation >2500 mm yr^{-1} (England's rainiest place). Mean daily average sunshine 3–3.5 hr day^{-1}.

Geology: Slate, sandstone, and especially igneous rocks in mountains.

Topography: Mountainous, much land >400 m asl, most over 200 m asl, with notable peaks such as England's highest mountains Scafell (978 m), Helvellyn (951 m), and Skiddaw (938 m). In between are the large lakes and numerous smaller lakes (tarns). Fringe of lowland around coast.

Population: 12–25 km^{-2} overall, >200 km^{-2} along part of coast.

Vegetation: Partly montane and somewhat poor on the mainly acid soils. Relicts of the northernmost *Tilia* woodland are found in the southern Lake District (Pigott 1991).

Land use: Mainly sheep farming, with grazing in the extensive fells. Today herds of tourists mainly roam the fells.

Reference site 27. Low Wray Bay (Pennington 1977)

Latitude 54°4'N, Longitude 2°58'W. Elevation 33m. Age range 14500–10000 BP. Lake. Fifteen radiocarbon dates. 180 cm. Eight pollen-assemblage zones. Event stratigraphy, Fig. 2.20. The site is in shallow water at the northern end of Windermere.

Local paz	Age	Pollen-assemblage zone
1.	*ca.* 14500–13000 BP	*Rumex–Gramineae– Salix herbacea*

Local paz	Age	Pollen-assemblage zone
2.	*ca.* 13000–12000 BP	*Betula–Juniperus*
3.	*ca.* 12000–11500 BP	*Betula–Rumex– Gramineae*
4.	*ca.* 11500–11000BP	*Betula–Juniperus*
5.	*ca.* 11000–10500 BP	*Rumex–Artemisia*
6.	*ca.* 10500–10000 BP	Herbs
7.	*ca.* 10000–9800 BP	*Juniperus*
8.	*ca.* 9800–9000 BP	*Betula*

Radiocarbon dates

Depth (cm)	^{14}C age	Laboratory code
1–2	12272±280	SRR-668
2–7.5	11344±90	SRR-669
7.5–11	12213±150	SRR-670
11–15	12132±175	SRR-671
15–19	12112±125	SRR-672
19–23	12517±150	SRR-673
23–26	12499±120	SRR-674
26–29	12441±82	SRR-675
29–32	12913±120	SRR-676
32–35	13185±170	SRR-677
35–37	12567±240	SRR-678
37–41	13938±210	SRR-679
41–44	13863±270	SRR-680
44–47	14557±280	SRR-681
47–54	14623±360	SRR-682

(0 cm is a stratigraphic boundary representing the base of the Loch Lomond Stadial.)

Reference site 28. Blelham Tarn
(Pennington 1965, 1979; Pennington *et al.* 1976)

Latitude 54°24'N, Longitude 2°59'W. Elevation 50 m. Age range *ca.* 14000–0 BP. Lake. Ten radiocarbon dates. Eleven local paz. Event stratigraphy, Fig. 2.20.

Local paz	Age	Pollen-assemblage zone
1.	14000–13000 BP	*Rumex*
2.	13000–11000 BP	*Betula–Juniperus*
3.	11000–10500 BP	*Artemisia*
4.	10500–9800 BP	*Juniperus*
5.	9800–8400 BP	*Corylus–Betula–Quercus (+Ulmus)*

REGIONAL EVENT STRATIGRAPHY, Lake District, after Pennington (1977, 1979, 1991)

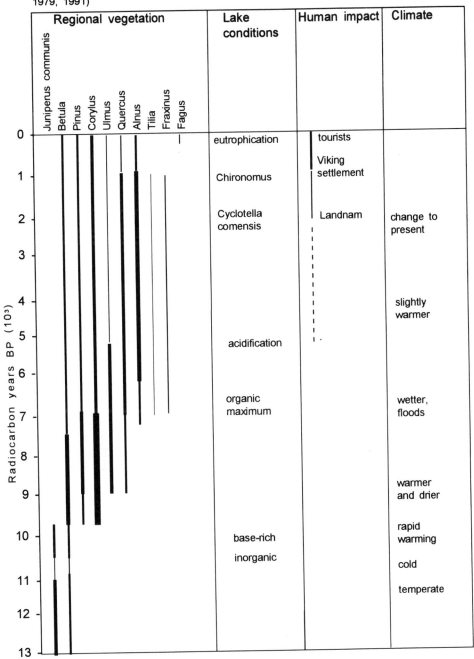

Fig. 2.20　Event stratigraphy, for Lake District, after Pennigton (1977, 1979, 1991)

Local paz	Age	Pollen-assemblage zone
6.	8400–7500 BP	*Corylus–Betula–Quercus– Pinus (+ Ulmus)*
7.	7500–5100 BP	*Quercus–Alnus–Corylus– Ulmus*
8.	5100–2300 BP	*Quercus–Alnus–Corylus (+ Fraxinus)*
9.	2300–911 BP	*Betula–Quercus–Alnus– Corylus (+ Gramineae, P. lanceolata)*
10.	911–200 BP	*Betula–Quercus–Alnus– Corylus (+Ilex, Gramineae, P. lanceolata)*
11.	200–0 BP	*Betula–Quercus–Alnus– Corylus(+Pinus, Quercus, Fagus)*

Discussion

(1) The earliest evidence is from Older Dryas Stadial sediments at Low Wray Bay, Windermere (Pennington 1977, 1991), dated to *ca.* 14500 BP and indicating open arctic vegetation with *Salix herbacea* and herbs. At Blelham there is evidence of grassland and communities with *Rumex* from *ca.* 14000 BP.

(2) Windermere Interstadial deposits are represented by *Betula* and *Juniperus* between 13000 and 11000 BP at both Low Wray Bay and Blelham Tarn, representing open woodland. This is the type site for the Windermere Interstadial.

(3) Younger Dryas Stadial deposits are shown by a return to herb vegetation with *Rumex* and *Artemisia* growing in tundra, dated from 11000–10500 BP at Low Wray Bay and at Blelham Tarn. There is evidence of solifluction.

(4) The beginning of the Holocene is marked by the return of vegetation with *Betula, Juniperus,* and herbs at *ca.* 10500 BP, with a *Juniperus* peak 10300–9800 BP. At Blelham Bog *Juniperus* is present first, followed by *Betula* from *ca.* 9800 BP. *Corylus* and *Quercus* arrive shortly after.

(5) At Blelham Tarn *Corylus* expands at *ca.* 8400

BP, *Ulmus* arrives. Woodland grew up to 760 m (Pennington 1970).

(6) *Quercus* and *Alnus* expand by *ca.* 7500 BP. *Pinus* maximum *ca.* 8400–7600 BP.

(7) The woodland at its maximum is represented by pollen of *Quercus, Alnus, Corylus, Betula,* and a little *Fraxinus. Tilia* pollen forms continuous records only at Windermere and at Whinfell Tarn, and the pollen values when corrected show that *Tilia* was up to about 10% of the tree cover there (Pennington 1979). *Alnus* grew in valley mires up to 37 m (Pennington 1970).

(8) High *Corylus* values in many diagrams seem to be from wood-margin vegetation, especially at the edges of the tarns and therefore very well represented in the lake-edge deposits.

(9) Woodland clearance phases are not evident *ca.* 5000 BP at Blelham Tarn or at Whinfell Tarn, as they are in some other diagrams such as those from the upland Blea and Angle Tarns, where *Pinus* was also cleared (Pennington 1991, p.18). The first real clearance phase in most other lowland sites and also in some upland sites is a Landnam episode *ca.* 2300 BP. The second or main clearance episode is dated to *ca.* 900 BP, with decreases in *Quercus, Betula, Alnus,* and *Fraxinus* as well as in *Corylus.* Clearance phases at various dates are also shown in other pollen diagrams.

(10) Secondary woodland with *Ilex* appears *ca.* 1700 BP, increasing at the clearance phase *ca.* 900 BP at Blelham Tarn, possibly the result of soil inwash. The scattered *Tilia* record *ca.* 1000 BP may not be from secondary woodland; an increase in *Tilia* representation *ca.* 5100–2500 BP at Whinfell Tarn could be the result of increased pollen dispersal in more open woodland after clearance or it may come from a real increase in *Tilia* woodland; soil inwash does not appear to be the cause of the pollen increase (Pennington 1979). The last part of the diagram indicates the planting of trees such as *Pinus, Quercus* and *Fagus* 200–0 BP.

(11) Human impact and crop plants are shown by a record of *Plantago lanceolata* from before *ca.* 3000 BP, which increases first at the Landnam episode at *ca.* 2300 BP and then from the main

clearance episode at *ca.* 900 BP and onwards, together with *Cannabis*. At other sites, Barfield Tarn near the coast is one of the few to show Cerealia-type and other signs of human impact as early as *ca.* 5000 BP (Pennington 1991). Perhaps in this case the settlement was very close to the pollen site. The mires by the coast record increased human impact *ca.* 3200–2700 BP, *ca.* 2000–1600 BP, and *ca.* 900–400 BP as at White Moss, Cumbria (Wimble 1986). A farming phase with *Cannabis* and *Linum* pollen occurs *ca.* 1200 BP at Ehenside Tarn (Pennington 1970).

(12) The clearance episode at *ca.* 900 BP is a feature common to many pollen diagrams in this region, representing Scandinavian settlement in the area, which is also evident in the place names such as -thwaite (=clearing), tarn (= lake), -shaw (= wood), fell (= hill).

(13) The transition from woodland to acid moorland is shown at some upland sites such as Devoke Water and Burnmoor Tarn at *ca.* 250 m asl, which took place at *ca.* 4000–3000 BP and *ca.* 1800–1600 BP. *Calluna* was present from the base of the sequence at >7600 BP, increasing together with carbon in phases of tree-pollen reduction (Pennington 1991). Blanket peat formed above 520 m from *ca.* 3900 BP (Pennin-gton 1970, p.71).

(14) GB-q pollen diagrams indicate events similar to those from other mainly acid uplands such as GB-r, GB-o, and parts of GB-a.

TYPE REGION GB-r, NORTH YORK MOORS

This covers parts of North Yorkshire and Cleveland. The North York Moors are an upland area. A considerable amount of recent work has been done in this area, for example by Atherden (1972, 1976, 1979), Jones (1976, 1977, 1978), and Innes & Simmons particularly on possible Mesolithic activity (1988). Dimbleby (1962, 1985) studied pollen in buried soils. The vegetation history of the region is summarized in Simmons *et al.* (1993).

Climate: Mean January 0–4°C, mean July 11–13°C. Precipitation 1500–2000 mm yr^{-1}. Mean daily aver-

age sunshine 3.5–4 hr day^{-1}. Climate dependent on elevation and aspect. Snow fall in winter. Can be foggy as a result of the cooling effect of the North Sea, the "Roker".

Geology: Mainly Jurassic sandstone.

Topography: Hilly with much land over 200 m, and the hill tops up to 500 m. Dissected by valleys draining into the River Tees to the north, and into the River Derwent to the south. Cliffs along much of the coast.

Population: 6–50 km^{-2}.

Vegetation: Woodlands remaining on steep valley sides. Moorland on hills, forestry plantation of exotic conifers.

Soils: Sandy soils, podzols under moors, mires where drainage is restricted.

Land use: Sheep grazing on moors, mixed farming in valleys. Potash mining. Recreation in National Park.

Reference site 29. Fen Bogs (Atherden 1972, 1976)

Latitude 54°22′N, Longitude 0°42′W. Elevation 164 m. Age range *ca.* 9000 BP. Mire. Six radiocarbon dates. 950 cm sediment. Ten paz (Fig 2.21).

Local paz	Age	Pollen-assemblage zone
FB1	*ca.* 8750 BP	*Betula–Pinus*
FB2	*ca.* 8750–6950 BP	*Pinus–Corylus/Myrica*
FB3	*ca.* 6950–4700 BP	*Quercus–Alnus– Ulmus–Tilia*
FB4	4700–3400 BP	*Quercus–Alnus*
FB5	3400–2300 BP	*Quercus–Alnus–* ruderals
FB6	2300–1500 BP	Gramineae–ruderals– aquatics
FB7	1500–1100 BP	Lower *Betula– Corylus/Myrica– Calluna*
FB8	1100–400 BP	Gramineae–ruderals
FB9	400–*ca.* 150 BP	Upper*Betula– Corylus/Myrica– Calluna*
FB10	*ca.* 150–0 BP	*Calluna*

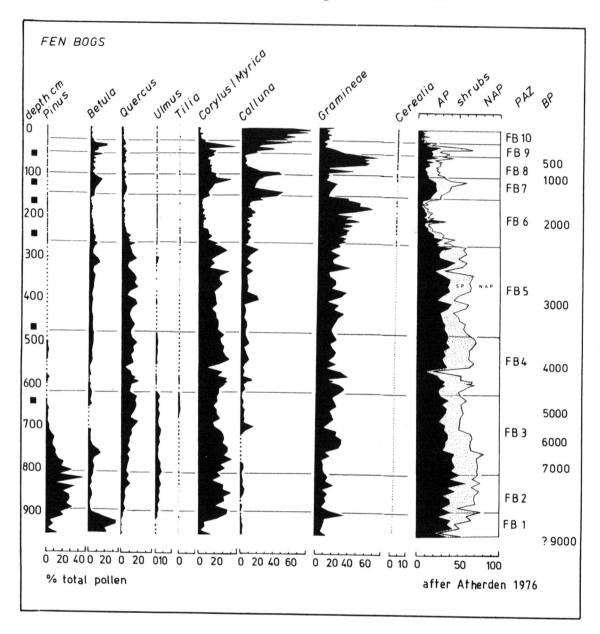

Fig. 2.21   Summary pollen diagram from Fen Bogs, after Atherden (1976)

Radiocarbon dates from Fen Bogs

Depth (cm)	¹⁴C age	Laboratory code
59–62	390±100	T-1151
113–116	1060±60	T-1087
160–163	1530±130	T-1086
255–257	2280±120	T-1085
475–481	3400±90	T-1150
632–635	4720±90	T-1084

Discussion

(1) The Late Weichselian sequence has been described by Jones (1977).

(2) Early Holocene events in the lower part of the Fen Bogs diagram *ca.* 9000–7500 BP (FB1) show woodland development first with *Betula* (FB1). *Pinus* and *Corylus/Myrica* then replaces *Betula* (FB2). *Quercus* and *Ulmus* are present as traces in FB1, and a significant amount in FB2. Gramineae, Cyperaceae, and *Calluna* present during these stages probably represent the local mire vegetation. The early Holocene phases of woodland succession are generally similar to those seen in other regions.

(3) Mixed woodland with *Quercus, Alnus,* and *Ulmus* develops from FB1 and with a small amount of *Tilia* in FB2. Occasional records of ruderal plants are interpreted as signs of possible slight human impact. At time of maximum woodland (FB3), open woodland of *Quercus* and *Corylus* exists together with herb and heath communities on the uplands. Lower down in the valleys was mixed deciduous woodland with *Quercus, Ulmus,* and only a little *Tilia* (Innes & Simmons 1988). These events are not radiocarbon-dated.

(4) Herbaceous pollen remains at *ca.* 40% throughout the woodland maximum. Although some of this would have originated in mire vegetation, the woodland was apparently open enough for herb vegetation.

(5) Minor woodland clearance episodes that may be connected with Mesolithic occupation have been studied by Innes & Simmons (1988). The first *Ulmus* decline horizon is *ca.* 4700 BP at the start of FB4. A second *Ulmus* decline and the

Tilia decline are at *ca.* 3400 BP at the start of FB5. After this are small dips in tree pollen, possible evidence of small temporary clearings for animal grazing. *Fraxinus* and *Betula* increase slightly. Major woodland clearance affecting *Betula, Corylus,* and *Quercus* defines FB6 at *ca.* 2300 BP. Further signs of woodland clearance occur at *ca.* 100–500 BP (FB9). The final clearance phase represents subrecent times. Pollen analyses of soils buried beneath barrows (burial mounds) reveal much greater evidence for woodland clearance and consequent podzolization at occupied sites progressively from *ca.* 4000 BP onwards (Dimbleby 1962, 1985).

(6) Secondary woodland development is shown in FB5, *ca.* 3700 BP by traces of *Fagus.* Some woodland regeneration is indicated in FB7 *ca.* 1400 BP and again in FB9 *ca.* 400 BP. The appearance of *Picea* in FB10 shows the recent planting of exotic conifers.

(7) Possible signs of human impact before 5000 BP have been found at a number of sites in this type region (Innes & Simmons 1988). Later disturbance to the woodland is discussed above, and the growing of crops is shown by Cerealia-type pollen records starting at *ca.* 2400 BP (FB6). The records are greater during phases of increased human impact. The cereals may have been grown more in the valleys than on what are now the moors.

(8) Moorland is shown by *Calluna* throughout the diagram, with a number of peaks that may be associated with human activities at *ca.* 1500 BP, *ca.* 1000 BP, and from *ca.* 400 BP (FB10). The final increase in *Calluna* shows expansion of moorland for sheep grazing and management of *Lagopus scoticus* for the "sport" of shooting.

TYPE REGION GB-s, NORTHUMBRIAN PLAIN

This region includes parts of the counties of Cleveland, Durham, Tyne and Wear, and Northumberland. Recent work includes Bartley (1966) on Late Weichselian deposits. Mainly Holocene profiles were studied by Bartley *et al.* (1976), Davies & Turner

(1979), Donaldson & Turner (1977). Dumayne (1992) and Dumayne & Barber (1994) concentrated on evidence of human impact and the Roman frontier. Macklin *et al.* (1991) studied alluviation.

Climate: Mean January <4°C, July 14–15°C. Precipitation 625–1250 mm yr⁻¹. Mean daily average sunshine 3.5–4 hr day⁻¹.

Geology: Coal Measures and Magnesian limestone around Durham and Newcastle, rising into Mill-

stone Grits and limestones further west. Volcanic intrusions (Whin Sill, Cheviot Hills).

Topography: A rather flat coastal plain divided by the Rivers Tees, Wear, Tyne, and Tweed, becoming hilly in the west as the land rises up to the Pennines (GB-o). To the north the Cheviot Hills rise over 600 m.

Vegetation: Moorland on the Millstone Grit hills, mountain grasslands on limestone, and agricultural land in the lowlands.

Land use: Mixed farming, hill farming in uplands.

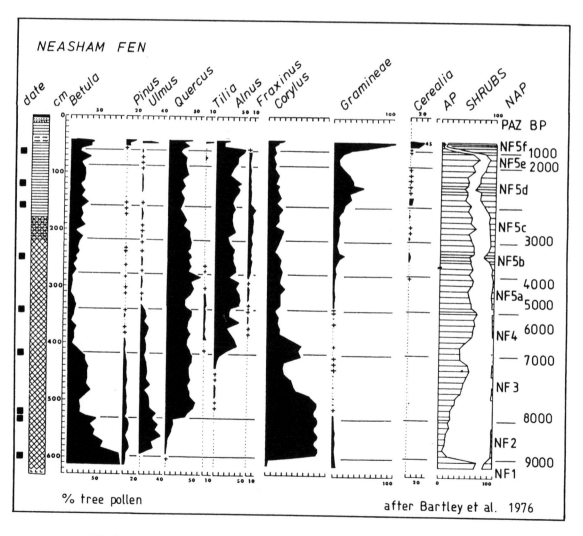

Fig. 2.22   Summary pollen diagram from Neasham Fen, after Bartley *et al.* (1976)

Conurbations along lower Rivers Tees, Wear, and Tyne.

Reference site 30. Neasham Fen (Bartley *et al.* 1976)

Latitude 54°30′N, Longitude 1°24′W. Elevation 46 m. Age range 10000-0 BP. Mire. Eleven radiocarbon dates. 620 cm sediment. Five local paz, six subzones (Fig.2. 22).

Local paz	Age	Pollen-assemblage zone
NF1	*ca.* 10000–9200 BP	*Betula–Salix– Gramineae– Cyperaceae*
NF2	*ca.* 9200–8200 BP	*Corylus–Betula–Pinus– Ulmus*
NF3	*ca.* 8200–7000 BP	*Quercus–Betula– Corylus–Alnus*
NF4	*ca.* 7000–5500 BP	*Quercus–Alnus– Corylus–Fraxinus– Tilia*
NF5	*ca.* 5500–0 BP	*Quercus–Alnus– Corylus–Gramineae*
NF5a	*ca.* 5500–3500 BP	*Quercus–Alnus– Corylus–Gramineae– Plantago lanceolata*
NF5b	*ca.* 3500–3000 BP	*Quercus–Alnus– Corylus–Gramineae– Plantago–Cerealia*
NF5c	*ca.* 3000–2800 BP	*Quercus–Alnus– Corylus– Gramineae–AP*
NF5d	*ca.* 2800–2300 BP	*Quercus–Alnus– Corylus– Gramineae–NAP*
NF5e	*ca.* 2300–1200 BP	*Quercus–Alnus– Corylus– Gramineae–AP*
NF5f	*ca.* 1200–0 BP	*Quercus–Alnus– Corylus– Gramineae–NAP*

Radiocarbon dates from Neasham Fen

Depth (cm)	¹⁴C age	Laboratory code
55–60	1213±60	SRR-96
100–105	2804±60	SRR-97
105–110	2850±60	SRR-98
135–140	2538±50	SRR-99
140–145	2488±75	SRR-100
245–250	3242±70	SRR-101
335–340	5468±80	SRR-102
410–415	6962±90	SRR-103
530–535	8202±95	SRR-104
580–585	8829±120	SRR-105
590–595	9082±90	SRR-106

Discussion

The Neasham Fen results can best be discussed in relation to a number of pollen diagrams from this immediate area such as Thorpe Bulmer, published together by Bartley *et al.* (1976).

(1) Late Weichselian evidence up to *ca.* 10300 BP from the base of the Thorpe Bulmer sequence shows that the vegetation was mainly Gramineae and Cyperaceae. The following phase with *Salix, Juniperus,* and *Empetrum* and herbs such as *Helianthemum* and *Artemisia* is correlated with the Allerød Interstadial. Then all trees and shrubs almost disappear with the exception of *Salix*, and a herb flora with *Artemisia* prevails. This probably represents the Younger Dryas Stadial.

(2) Early Holocene woodland development (NF1) from *ca.* 9400–9000 BP is indicated mainly by *Betula* pollen, together with *Corylus*, as at other sites in this region, and the arctic tundra herb flora disappears.

(3) Further woodland development is shown in NF2; thermophilous trees spread, *Corylus* from *ca.* 9000 BP, *Ulmus* from *ca.* 9100 BP and, in NF3, *Quercus* from *ca.* 8000 BP. There are corresponding reductions in *Betula* and in *Corylus* from *ca.* 7500 BP; *Pinus* values are small. The other pollen diagrams from this area show a similar sequence except that some (Bishop Middleham, Nunstainton) have much higher *Pinus* values, perhaps since they are close to Magnesian limestone on which pine would have readily grown.

(4) Further woodland development is shown in

NF4. *Alnus* rises *ca.* 7000 BP, but earlier at Mordon Carr, *Tilia* appears *ca.* 6800 BP and *Fraxinus* at *ca.* 6500 BP.

(5) The woodland at its maximum is represented by *Quercus*, *Alnus*, *Corylus*, *Ulmus*, and *Betula*, with a little *Tilia ca.* 7000–5500 BP in NF4. The small amount of herb pollen shows that woodland cover was fairly complete, and that herbs were few even at the surface of this small bog.

(6) Woodland reduction starts with the *Ulmus* decline, dated 5468±80 BP at Neasham (NF5). Peat growth was slow at this time. In a second phase of woodland clearance *Tilia* disappears completely, together with an increase in herb pollen in NF5b, *ca.* 3800–3200 BP. Tree-pollen reductions occur at this point in some other diagrams from the region, less in others. A further clearance episode defines NF5d *ca.* 2600–2000 BP. The final almost complete clearance of woodland from this area in NF5f appears to be from *ca.* 1000 BP onwards. The main woodland clearance is dated *ca.* 2000–1300 BP at Hallowell Moss near Durham (Donaldson & Turner 1977). Farther north near Hadrian's Wall, large-scale woodland clearance occurred from *ca.* 2000 BP (Barber *et al.* 1993).

(7) No significant increases in cultural indicators are apparent in the early clearance phases such as NF5a. A little Cerealia-type pollen and *Plantago lanceolata* occur together with a scattering of other herb records not present earlier, *ca.* 3800–3200 BP in NF5B. Cultural indicators are more common at some other sites (Hutton Henry and Bishop Middleham) at this time, showing the variability in amount of human activity in different areas. This may be connected with better soils on the Magnesian Limestone. Even more such records occur in NF5d *ca.* 2600–2000 BP. The greatest records are in NF5f from *ca.* 1200 BP and later. *Cannabis* in the upper part of the Thorpe Bulmer diagram dates from *ca.* 2000–850 BP. Farther north near Hadrian's Wall, a phase of increased human activity is apparent during this period (Dumayne 1992; Dumayne & Barber 1994).

(8) Secondary woodland containing *Fraxinus* seems to have spread following woodland disturbance at *ca.* 5000 BP in NF5a, and more so *ca.* 4000 BP in NF5b. Occasional grains of *Fagus* and *Carpinus* occur from NF5b onwards, and a single grain of *Ilex*.

(9) Alluviation is shown by results from Calally Moor, Northumberland, from *ca.* 3000 BP and *ca.* 400 BP (Macklin *et al.* 1991).

PALAEOECOLOGICAL PATTERNS AND EVENTS (Figs 2.23 and 2.24)

(1) Older Dryas Stadial *ca.* 14500–13000 BP and Younger Dryas Stadial *ca.* 11000–10500 BP. Arctic tundra vegetation is represented in a few pollen diagrams covering the Late Weichselian period. Herbaceous taxa are abundant at most sites, such as Crose Mere (GB-i) (Beales 1980), Low Wray Bay and Blelham with grassland and *Rumex* (GB-q) (Pennington 1977), and Thorpe Bulmer with Gramineae and Cyperaceae (GB-s) (Bartley *et al.* 1976). Tree taxa are scarcely represented: *Salix* is the only woody plant at Crose Mere (Beales 1980), identified as *S. herbacea* at Low Wray Bay, where this stage is dated to *ca.* 14500–13000 BP (Pennington 1977). A second, similar, phase of arctic tundra vegetation is indicated at a number of sites with deposits dating from the Younger Dryas Stadial period; the pollen assemblage is herb dominated by *Artemisia* as at Hockham Mere (GB-l) (Bennett 1983), *Salix* and *Artemisia* as at Thorpe Bulmer, or *Hippophaë* as at Crose Mere (Beales 1980). At Low Wray Bay the pollen assemblage is dominated by *Rumex* and *Artemisia*, and this episode is dated to *ca.* 11000–10500 BP (Pennington 1977). There is little sign of consistent similarity or difference among the pollen assemblages at the various sites.

(2) Allerød Interstadial 13000–11000 BP and early Holocene period, 10500–9000 BP. Arctic heath is characterized by the arrival of *Juniperus* at many sites such as Gatcombe (GB-c) (Scaife 1980), Stafford (GB-i)(Bartley & Morgan 1990), Hockham Mere (GB-l) (Bennett 1983), Knowsley, together with *Populus* (GB-n) (Innes, in prep.), Roos (GB-p) (Beckett 1981a), Blelham (GB-q) (Pennington 1965), Thorpe Bulmer (GB-s) (Bartley *et al.* 1976)

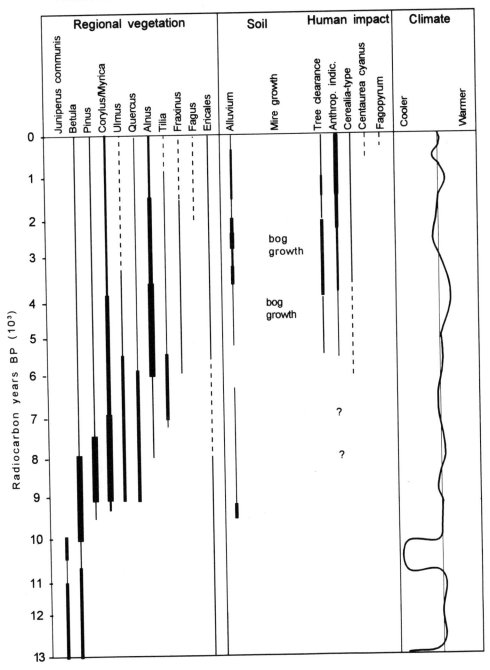

Fig. 2.23 Event stratigraphy for lowland England

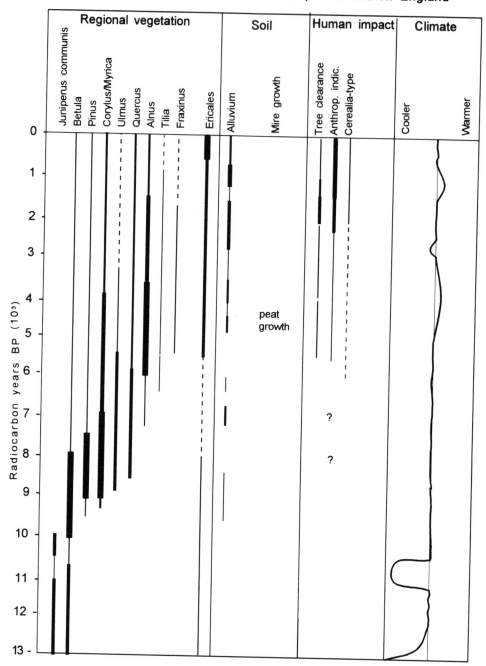

Fig. 2.24　Event Stratigraphy for upland England

and Low Wray Bay where it is dated *ca.* 13000–11000 BP (GB-q) (Pennington 1977), corresponding to the Allerød Interstadial. At this, the type site, the stage has the local name Windermere Interstadial. A second arctic heath phase occurs in early Holocene deposits, characterised by increasing values of *Betula* and *Juniperus.* Herbs are present at first, often a rich flora on calcareous substrates as at Scaleby Moss (GB-t) (Walker 1966). The only possible pattern is that *Betula* pollen reaches much greater values at some sites in the south and in the west, possibly the result of milder climate. Woodland development then squeezes out this herb flora and light-demanding shrubs such as *Juniperus* and *Hippophaë.*

(3) *Betula* (probably *B. pendula* and *B. pubescens*) spread in east-central England during the early Holocene before 10000 BP (Birks 1989), a pattern reflected by the reference sites. It is present at *ca.* 13000 BP at Low Wray Bay (GB-q) (Pennington 1977), and at *ca.* 9960 BP at Pannel Bridge, showing that *Betula* was also present in southeastern as well as eastern England then (GB-d) (Waller 1993). *Betula* values reach a maximum before 9000 BP, thereafter decreasing as the pollen values of *Corylus* and *Ulmus* rise, as at Hockham Mere (GB-l) (Bennett 1983), or values of *Pinus* and *Corylus* rise as in the Knowsley Park diagram (GB-n) (Innes in prep.). After this, *Betula* woodland does not seem to have been significant, except at a few sites such as Hartlebury Common, where the sandy soil provides a suitable habitat (GB-i) (Brown 1984) and at some upland sites such as White Moss (GB-o) (Bartley *et al.* 1990), and Fen Bogs (GB-r) (Atherden 1976). *Betula* persists in small amounts *ca.* 10% of total pollen especially at upland sites, while at others it virtually disappears from *ca.* 6000 BP. Small peaks of *Betula* pollen in later parts of some pollen diagrams appear to represent pioneer woodland spreading during phases of decreased human activity, as at Pannel Bridge PB4a, *ca.* 360–3200 BP (GB-d)(Waller 1993).

(4) *Pinus sylvestris* seems to have spread from the southeast in the early Holocene from *ca.* 9500 BP, reaching the north of England before *ca.* 9000 BP. *Pinus* is fairly abundant in the early Holocene parts of several pollen diagrams in different type regions, but only small amounts are present in others such as Stafford (GB-i) (Bartley & Morgan 1990). In GB-o, Bartley *et al.* (1990) note the pattern where *Pinus* became established by *ca.* 9500 BP mainly in areas with limestone soils, either because they offered a favourable habitat or because *Pinus* could tolerate dry conditions there. By *ca.* 7700–7000 BP *Pinus* fades into unimportance at most sites as other tree taxa increase their pollen records. Finally, subrecent peaks of *Pinus* in a number of diagrams, such as Gatcombe, Sidlings, Knowsley and Neasham, are probably the result of afforestation. *Pinus* is also discussed by Bennett (1984).

(5) *Corylus avellana* initially spread from the western coast of England (and Wales) *ca.* 9500 BP (Birks 1989), expanding rapidly eastwards at many sites by *ca.* 9000 BP with a large although often fluctuating record. *Corylus* values decrease somewhat as woodland develops its full range of trees. *Corylus* then decreases in stages corresponding to the woodland reduction phases. The term Coryloid is used by some authors to indicate the possibility that *Myrica gale* pollen was also present, but the two taxa are separated in the results from Pannel Bridge (GB-d) (Waller 1993). At a few sites Coryloid values remain rather low throughout, as at Cookley (GB-i) (Greig in prep.), Haddenham (GB-m) (Peglar 1992), and Willow Garth (GB-p) (Bush 1988), which may show the variable abundance of *Corylus* then, as today (Rackham 1980). The large and often sharply fluctuating Coryloid pollen values in many pollen diagrams make it hard to understand what was happening. They can also obscure other smaller changes in other pollen taxa unless excluded from the pollen sum. It is also hard to assess the importance of *Corylus* in the woodland, or at its edge.

(6) *Ulmus,* probably *U. glabra* with some *U. procera* and *U. minor,* arrived in southeastern England around 9500 BP and spread to the northwest, reaching the north of England about 8700 BP (Birks 1989), a pattern shown by the reference sites. *Ulmus* reached its rational limit later at some sites, such as Blacka Brook, *ca.* 8300 BP (GB-a) (Beckett 1981b) and Stafford *ca.* 7000 BP (GB-i) (Bartley & Morgan 1990). *Ulmus* values initially reach high levels,

for example *ca.* 30% tree pollen *ca.* 8500 BP at Neasham Fen (GB-s) (Bartley *et al.* 1976). Thereafter, the relative proportion of *Ulmus* pollen decreases as other tree taxa spread. *Ulmus* seems to have been a significant part of the wildwood at its peak, reaching 25% tree cover on corrected figures (Peglar 1992).

The initial *Ulmus* decline(s) are mostly dated to *ca.* 5500–5000 BP. The *Ulmus* decline and the importance of human activity, *Ulmus* disease, and other factors have been, are, and probably will be much discussed (Rackham 1980; Scaife 1988; Peglar 1993b, etc.). At some sites only one apparent *Ulmus* decline occurs, at others two or more, locally together with reductions in other tree taxa such as *Tilia*. Thereafter *Ulmus* pollen records become slight. Evidence of human impact around the *Ulmus* decline (or first *Ulmus* decline) is often rather slight, consisting of changes in the relative percentages of various pollen types and with few cultural indicators, as in the Craven (GB-o) (Bartley *et al.* 1990). Cerealia-type pollen only appears at this point in a few pollen diagrams, such as Winchester, in which a marked reduction in tree pollen also occurs (Waton 1982a, 1982b).

(7) The earliest records of *Quercus* (probably *Q. petraea* and *Q. robur*) are from the extreme southwest of England around 9500 BP, and it reached the north of England by *ca.* 8500 BP (Birks 1989). *Quercus* is the most abundant tree pollen in the woodland maximum parts of many pollen diagrams, and even when corrected for its high productivity and dispersal (Andersen 1970) *Quercus* was probably the commonest tree in the wildwood mosaic covering large parts of England, especially in type regions GB-a, n, o, p, q, r, s, and t. *Quercus* pollen decreases in stages during phases of woodland clearance, sometimes showing a partial recovery in between, but there is no clear "*Quercus* decline". These *Quercus* decreases are interpreted as phases of greater woodland clearance. They can be clearly seen and dated in diagrams with especially good resolution, such as Fenton Cottage, Lancashire (GB-n) (Huckerby *et al.* 1992). *Quercus* woods often grow on poor soils where they compete best (Rackham 1980), and this land may have been unattractive to

prehistoric settlers, who cleared this woodland last of all. Under later management practice, *Quercus* is the main tree in wood pasture, such as Windsor Forest (Rackham 1980).

(8) *Alnus glutinosa* spread erratically from both Wales and southeast England with a rational limit around 8000 BP (Birks 1989), although this is reached *ca.* 10000 BP at Pannel Bridge (GB-d) (Waller 1993) and *ca.* 9000 BP at Gatcombe (GB-c) (Scaife 1980). *Alnus* becomes abundant at rather varying dates at different sites. The possible causes for this have been discussed by Brown (1988), Bennett & Birks (1990), and Tallantire (1992). Edaphic factors may be significant, for the *Alnus* expansion is connected with the change to unhumified *Sphagnum imbricatum* and *S.* cf. *rubellum* peat, indicating increased wetness, at White Moss (GB-o) (Bartley *et al.* 1990). *Alnus* is by far the most abundant pollen type in many diagrams after its expansion *ca.* 6000 BP, probably because alder carr surrounded many lakes and wetlands where sediments were being deposited, and alder growing along streams may have further increased pollen deposition by water transport. *Alnus* pollen values decline in woodland clearance phases, usually together with those of other trees and shrubs such as *Quercus* and *Corylus*. At some sites most of the *Alnus* goes at the first clearance episode, as at Winchester *ca.* 5500 BP (GB-b) (Waton 1982a, 1982b). But at sites with several episodes the main clearance may be quite late, as at White Moss *ca.* 1500 BP. *Alnus* grows in valley communities and also as plateau woods and is not particularly associated with *Quercus* and *Corylus* (Rackham 1980), so the decline in these three pollen types together may represent clearance of several woodland communities at once.

(9) *Tilia* records, mainly *T. cordata* with some *T. platyphyllos*, reach the rational limit at Pannel Bridge *ca.* 8000 BP, expanding from *ca.* 7000 BP (GB-d) (Waller 1993). *Tilia* spread northwest, reaching northern England by *ca.* 5500 BP (Birks 1989), which is its present native limit (Pigott 1991). The percentages of *Tilia* pollen are low in most pollen diagrams, although in GB-b, c, d, i, k, l, m, part of n, q, and p there is a modest amount of *Tilia* pollen at woodland maximum in some diagrams. When

calculated as the proportion of tree cover in the woodland (Andersen 1970), *Tilia* is shown to have been important in parts of the wildwood (Rackham 1980; Greig 1982; Peglar 1992). This is particularly clear when *Alnus* and Coryloid are excluded from the reckoning. After the woodland maximum, *Tilia* pollen declines at various times after *ca.* 5000 BP, connected with woodland clearance episodes. At some sites *Tilia* declines at the same horizon as *Ulmus*, at others the main *Tilia* decline is later, for example at the clearance episode *ca.* 3700 BP in a number of diagrams. The dramatic decline in *Tilia* may reflect its having grown on good (often loessic) soils, which were also suitable for farming, and also its inability to recol-onize after clearance and erosion, for it does not reproduce very successfully from seed in England now (Pigott 1991). A number of old woods with *Tilia* still exist (Rackham 1980).

(10) *Fraxinus excelsior* records start about 7000 BP at a number of sites in the south and middle of England, *ca.* 5500 BP in the north. The amounts of pollen are usually 1–2% total pollen, and 5% is only exceeded at a few sites such as Pannel Bridge (GB-d) (Waller 1993) and Sea Mere (GB-l) (Sims, in Peglar 1992). Such values, after correction for production and dispersal (Andersen 1970), show that *Fraxinus* was important in the woodlands of some areas. At some sites there are reductions in primary *Fraxinus* woodland during woodland clearance episodes, as at Pannel Bridge *ca.* 3300 BP (Waller 1993). At some other sites, *Fraxinus* values increase after *ca.* 5000 BP as the result of the growth of secondary woodlands following clearance of the wildwood, for example *ca.* 1000–300 BP at Gatcombe and Winchester (GB-b, c) (Scaife 1980; Waton 1982a, 1982b), *ca.* 1000 BP at Aller Farm (GB-a) (Hatton & Caseldine 1992), and at Leash Fen *ca.* 4000–3700, 3500–2200, and from 1600 BP onwards (GB-o) (Hicks 1971).

(11) *Fagus sylvatica*, arrived in the southeast around 3000 BP and spread northwest as far as GB-i, GB-j, and GB-p, colonizing cleared land (Birks 1989). The amount of *Fagus* in pollen diagrams is usually less than 1% total pollen, except at a few sites such as Amberley Wild Brooks where it reaches around 5% from 1400 BP (GB-d), Snelsmore (GB-b) where the record is from *ca.* 250 BP and therefore

represents planted *Fagus* (Waton 1982a, 1982b), and 10–20% *Fagus* at Lodge Road Epping *ca.* 1100 BP (GB-k) (Baker *et al.* 1978). *Fagus* pollen is found as far north as the Lake District (GB-q) although significant amounts of pollen date from the time when it was planted there *ca.* 250 BP and later (Pennington 1979). This lack of evidence of *Fagus* woodland has surprised researchers in England (Thorley 1981), for it is well represented in Denmark (Andersen 1976). The main area of *Fagus* woods in England seems to have been on acidic sands and gravels in GB-b, c, d, and k (for example the wood pasture of Epping Forest), and the southern parts of GB-i, j, and l (Rackham 1980).

(12) Minor woodland variants are indicated by pollen records of a number of other taxa. *Ilex aquifolium* appears as a rational curve at Gatcombe *ca.* 4500–3000 BP (GB-c) (Scaife 1980) and at West Heath Spa (GB-k), where its presences may demonstrate the establishment of wood pasture after initial woodland clearance (Greig 1991). *Taxus baccata* likewise features occasionally, as at Diss Mere *ca.* 4500–2000 BP (GB-l) (Peglar 1992). *Carpinus betulus* is found from *ca.* 2000 BP, with a slight increase from *ca.* 1750 BP at Diss Mere (GB-l) (Peglar 1992, 1993b), but only occasional grains have been found at most other sites in southern England. However, these scattered records go far north (GB-s) of the present-day natural distribution up to the Midlands (Bartley *et al.* 1976), and also into the uplands (GB-o) (Eshton Tarn, 144 m asl, Bartley *et al.* 1990). The main distribution of *Carpinus* today is in GB-k and southern GB-l, with less as far north as GB-i, j, and m (Rackham 1980). *Castanea sativa* pollen is also found occasionally, with more consistent records in subrecent parts of some pollen diagrams as at Winchester (GB-b) (Waton 1982a, 1982b) and at Keston (GB-d) (Moore & Evans 1991), which appear to be from plantations of *Castanea*, since the tree is not native to Britain (Rackham 1980). *Populus* pollen is occasionally found, in early Holocene Knowsley Park (GB-n) (Innes 1994) and also in some subrecent deposits as at Sidlings Copse (GB-j) (Day 1991). *Sambucus nigra* pollen sometimes occurs as a distinct curve, probably the result of shrubbery growing in places where soils were enriched through human influence, as at Stafford

(GB-i) (Bartley & Morgan 1990). The rosaceous trees *Sorbus, Crataegus,* and *Prunus* are occasionally recorded, as is *Acer,* but they are probably too greatly underrepre-sented for a true idea of their presence to be obtained. *Hedera helix* shows a nearly continuous record in some pollen diagrams (Beckett 1981b). Occasional *Lonicera* pollen grains are found. Rhamnaceae (*Rhamnus* and *Frangula*), *Euonymus europaeus, Viburnum opulus,* and *V. lantana* occur mainly in calcareous regions (Devoy 1979).

(13) Ericales pollen, a group that includes, among other taxa, the arctic–alpine *Empetrum nigrum,* various species of *Erica,* and *Calluna vulgaris,* which is probably the most widespread taxon, represent heathlands and heather moor. Continuous Ericales records occur throughout the Holocene in many pollen diagrams, probably representing at first arctic-type heath in the early Holocene, then moorland development, as at Knowsley Park (GB-n) (Innes 1994). Ericales pollen records then usually remain small until about the *Ulmus* decline, when they become abundant. In the uplands blanket peat started growing (GB-a, o, q, and r) (Moore *et al.* 1984, Pennington 1991, Atherden 1976), reflecting human activity and also possible climatic change, as at Blacka Brook, (GB-a) where *Calluna* pollen percentages increase sharply around 4000 BP together with other "cultural indicators" (Beckett 1981a). In the lowlands, Ericales grew on mires, and also on sandy heaths that developed in response to grazing pressure, as at West Heath Spa (GB-k) (Greig 1991). In contrast, some lowland diagrams, such as the Old Mere Hornsea, contain only a few scattered Ericales pollen grains, probably the result of transport from far away, for this is a chalky region with no local heathland (Beckett 1981b).

(14) Early effects of human impact on the landscape have been much sought, both from natural sites that have signs of human habitation in the surroundings (Simmons & Innes 1987), or from occupation sites themselves, such as on the North York Moors (GB-r) (Innes & Simmons 1988), Star Carr (9600–9300 BP) (GB-p)(Day 1993b), Seamer Carr (8000 BP), and Flixton Carrs (GB-p) (Cloutman & Smith 1988), and Peacock's Farm (8500–6800 BP) (GB-m) (Smith *et al.* 1989). Changes in some pollen values

and also charred layers at some sites appear to be the result of human activities, although proof is difficult (for example, Bartley *et al.* 1990, pp. 626–627). Occasional Cerealia-type pollen grains are found from before the *Ulmus* decline (Edwards & Hirons 1984), for example at Cothill Fen (Day 1991), which may reflect early Neolithic activity. The initial *Ulmus* decline itself has been discussed in (6) above.

Evidence of human impact after the first *Ulmus* decline takes the form of declines in tree pollen (particularly *Tilia* and *Quercus*) and increases in cultural indicators such as *Plantago lanceolata* and *Artemisia,* sometimes also Cerealia-type. These can be interpreted as signs of clearance episodes and sometimes of cereal growing. Although the dates of these clearances vary somewhat, around 3800 BP extensive human activity corresponded to the late Neolithic and early Bronze Age. The palynological evidence varies from site to site, but it is generally stronger and earlier at lowland sites such as Winchester (GB-b) (Waton 1982a, 1982b). At some sites, such as Sidlings Copse, the cleared land seems mainly to have been used as pasture (GB-j) (Day 1991). In the uplands, clearance was often less extensive and later than in the lowlands, and it can sometimes be connected with soils; the fertile soils on limestone were cleared and farmed before those on boulder clay, where the site White Moss (GB-o) shows that the main clearance of woodland did not occur until *ca.* 2000 BP. The settlers must have used the uplands more for grazing than for growing cereals, as one might expect (Bartley *et al.* 1990).

The main palynological evidence of field crops is from the cereals *Triticum, Hordeum, Avena,* and *Secale,* which comprise Cerealia-type pollen, the level of identification in most cases. Cerealia are more closely identified further in only a few pollen diagrams (Beckett 1981b, Waton 1982a, 1982b). The first Cerealia-type pollen occurs as isolated finds before the *Ulmus* decline, which can be dated to the early Neolithic period. Corresponding macrofossil evidence shows that Cerealia amounted to only a small proportion of the total food available to them (Moffett *et al.* 1989). More significant finds are usually considerably later than this; thus there is a continuous Cerealia-type pollen record at Winchester from *ca.* 3500 BP (GB-b) (Waton 1982a, 1982b). Cerealia-type records are usually concentrated in phases of woodland clearance. They often seem to reflect the closeness of the pollen site to the occupied landscape, so the few scattered grains at Blacka Brook (GB-a) (Beckett 1981b) and other upland sites, and at some lowland ones, too, such as Knowsley Park (GB-n) (Innes 1994) may represent distant grainfields or that they were down in the valleys, while the greater records at other sites may represent nearer grainfields or cereal processing sites. In some pollen diagrams the first substantial finds of Cerealia-type pollen are as late as *ca.* 2000 or even 1500 BP and later, as at Fenton Cottage (Huckerby *et al.* 1992). Cerealia pollen is generally poorly dispersed.

Other crops include Cannabaceae and *Cannabis*-type records, which may represent *Cannabis sativa*, and the pollen occurs as a distinct curve in a number of pollen diagrams (GB-l) (Peglar 1992, 1993b, 1993c). Although *Cannabis* disperses pollen aerially, hemp was retted (soaked for processing to liberate the useful fibres) in pools and bogs, thus releasing large amounts of pollen directly into the sediment, and retting horizons are the likely source of large *Cannabis*-type peaks. Of the other field crops, *Linum usitatissimum* has a very sparse pollen record that gives no true idea of its past importance (*Linum* was also retted). *Vicia faba* and *Pisum sativum* produce and disperse small amounts of pollen so that only occasional grains are found. *Fagopyrum esculentum* has highly distinctive pollen, which is occasionally recorded in pollen diagrams. It seems to have been introduced about 500 BP but was never apparently very important as a crop. *Juglans* and *Castanea* were both introduced, and occasional pollen grains are found in deposits from the last 500 years or so. *Vitis* is known to have been grown in England during the climatic optimum *ca.* 800–500 BP, and a pollen record may yet emerge.

Some weeds also leave significant pollen records. *Artemisia*, found in Late Weichselian and early Holocene stages, re-appears as a cultural indicator and weed at the *Ulmus* decline and later. Another distinctive weed is *Centaurea cyanus*, the pollen of which appears *ca.* 2000 BP at a few sites, but which greatly increases *ca.* 800 BP.

(15) The first grasslands are the arctic herb communities in the Late Weichselian and early Holocene. Indicator herbs such as *Rumex, Helianthemum,* and *Thalictrum* occur, together with Gramineae and Cyperaceae. Then, following woodland development and clearance, signs of grasslands re-appear at some sites. In the later Holocene, grassland is an important semi-natural vegetation, often owing its existence and development to human activities. Gramineae pollen values do not always give an accurate picture of grasslands, because members of the Gramineae occur in most habitats, as well as in grasslands themselves. Better evidence is provided by some characteristic grassland herbs that give a recognisable pollen record, such as *Centaurea nigra, Plantago lanceolata, Trifolium repens,* and *T. pratense*, besides the ever-present Gramineae, to provide some kind of information about the development of grasslands. *Centaurea nigra*-type pollen is found as scattered records at a number of sites, including Thorpe Bulmer (GB-s), where other grassland plants such as *Trifolium* sp. and *Linum catharticum* were found from *ca.* 2000 BP onwards (Bartley *et al.* 1976).

Chalk or limestone grassland is indicated by records of calcicolous taxa such as *Sanguisorba minor*, of which there is a distinct curve soon after 3100 BP at Eshton Tarn (GB-o) (Bartley *et al.* 1990), and from about 3300 BP at Willow Garth (GB-p), together with *Centaurea nigra, Helianthemum* sp., and *Linum catharticum* (Bush 1988).

(16) Alluviation. The main phases of alluviation within the archaeological period are 9600–9200 BP, 7200–6800 BP, 4800–4200 BP, 3800–3300 BP, 2700–2400 BP, 1900–1600 BP, 800–600 BP, and 400–250 BP (Macklin & Needham (1992, p.17).

ACKNOWLEDGEMENTS

Thanks are due to Keith Barber, Keith Bennett, Frank Chambers, Petra Day, Elizabeth Huckerby, Jim Innes, Peter Moore, Sylvia Peglar, Winifred Pennington, Martyn Waller, and Colin Wells for providing data and pollen diagrams and for their comments on draught texts. Pat Wiltshire also provided useful comments. Acknowledgement is due to English Heritage, funding agency for the North West Wetlands Survey and also for the writer, for their kind permission to quote from the Fenton Cottage and Cookley results.

REFERENCES

Andersen, S. Th. 1970: The relative pollen productivity and representation of north European taxa, and correction for tree pollen spectra. *Danmarks geologiske Undersøgelse, series 2, 96,* 99 pp.

Andersen, S. Th. 1976: Local and regional vegetational development in eastern Denmark in the Holocene. *Danmarks geologiske Undersøgelse, Årbog 1976,* 5–27.

Atherden, M.A. 1972: *A contribution to the vegetation and land use history of the eastern-central North York Moors.* PhD thesis, Durham University.

Atherden, M.A. 1976: Late Quaternary vegetational history of the North York Moors. 3. Fen Bogs. *Journal of Biogeography 3,* 115–124.

Atherden, M.A. 1979: Late Quaternary vegetational history of the North York Moors. 7. Pollen diagrams from the eastern-central area. *Journal of Biogeography 6,* 63–83.

Baker, C.A., Moxey, P.A. & Oxford, P.M. 1978: Woodland continuity and change in Epping Forest. *Field Studies 4,* 645-669.

Balaam, N.D., Bell, M.G., David, A.E.U., Levitan, B., Macphail, R.I., Robinson, M. & Scaife, R.G. 1987: Prehistoric and Romano-British at Westward Ho!, Devon; archaeological and palaeoenvironmental surveys 1983 and 1984. *In* Balaam, N.D., Levitan, B. & Straker, V. (eds) Studies in palaeoeconomy and palaeoenvironment in South West England. *British Archaeological Reports (British Series) 181,* 163–264.

Barber, K.E. & Twigger, S.N. 1987: Late Quaternary palaeoecology of the Severn basin. *In* Gregory, K.J., Lewin, J., & Thornes J.B. (eds) *Palaeohydrology in practice,* 217–247. Wiley, Chichester.

Barber, K.E., Dumayne, L. & Stoneman, R. 1993: Climatic change and human impact during the late Holocene in northern Britain. *In* Chambers, F.M. (ed.) *Climate change and human impact,* 225–236. Chapman & Hall, London.

Bartley, D.D. 1962: The stratigraphy and pollen analysis of lake deposits near Tadcaster, Yorkshire. *New Phytologist 61,* 277–287.

Bartley, D.D. 1966: Pollen analysis of some lake deposits near Bamburgh in Northumberland. *New Phytologist 65,* 141–166.

Bartley, D.D. 1975: Pollen analytical evidence for prehistoric forest clearance in the upland area west of Rishworth, W. Yorkshire. *New Phytologist 74,* 375–381.

Bartley, D.D. & Chambers, C. 1992: A pollen diagram, radiocarbon ages and evidence of agriculture on Extwistle Moor, Lancashire. *New Phytologist 121,* 311–320.

Bartley, D.D. & Morgan, A.V. 1990: The palynological record of the King's Pool, Stafford, England. *New Phytologist 116,* 177–194.

Bartley, D.D., Chambers, C. & Hart-Jones, B. 1976. The vegetational history of parts of south and east Durham. *New Phytologist 77,* 437–468.

Bartley, D.D., Jones, I.P. & Smith, R.T. 1990: Studies in the Flandrian vegetational history of the Craven district of Yorkshire: the lowlands. *Journal of Ecology 78,* 611–632.

Beales, P.W. 1980: The Late Devensian and Flandrian vegetational history of Crose Mere, Shropshire. *New Phytologist 85,* 133–161.

Beckett, S.C. 1975: *The Late Quaternary vegetational history of Holderness, Yorkshire.* PhD thesis, Hull University.

Beckett, S.C. 1978: The environmental setting of the Meare Heath track. *Somerset Levels Papers 4,* 42–46.

Beckett, S.C. 1981a: Pollen analysis of the peat deposits. In Smith, K., Coppen, J., Wainwright, G.J. & Beckett, S.C., part 4 (pp. 245–273) The Shaugh Moor Project, third report: settlement and environmental investigations. *Proceedings of the Prehistoric Society 47,* 205–273.

Beckett, S.C. 1981b: Pollen diagrams from Holderness, north Humberside. *Journal of Biogeography 8,* 177–198.

Beckett, S.C. & Hibbert, F.A. 1976: An absolute pollen diagram from the Abbot's Way. *Somerset Levels Papers 2,* 24–27.

Beckett, S.C. & Hibbert, F.A. 1979: Vegetational change and the influence of prehistoric man in the Somerset levels. *New Phytologist 83,* 577–560.

Bennett, K.D. 1983: Devensian, Late-glacial and Flandrian vegetational history at Hockham Mere, Norfolk, England. *New Phytologist 95,* 457–487.

Bennett, K.D. 1984: The Post-glacial history of *Pinus sylvestris* in the British Isles. *Quaternary Science Reviews 3,* 133–155.

Bennett, K.D. & Birks, H.J.B. 1990: Postglacial history of alder (*Alnus glutinosa* [L.] Gaertn.) in the British Isles. *Journal of Quaternary Science 5(2),* 123–133.

Birks, H.J.B. 1989: Holocene isochrone maps and patterns of tree-spreading in the British Isles. *Journal of Biogeography 16,* 503–540.

Brown, A.G. 1984: The Flandrian vegetation history of Hartlebury Common, Worcestershire. *Proceedings of the Birmingham Natural History Society 25(2),* 89–98.

Brown, A.G. 1988: The palaeoecology of *Alnus* (alder) and the postglacial history of floodplain vegetation. Pollen percentage and influx data from the West Midlands, United Kingdom. *New Phytologist 110,* 425–436.

Brown A.G. & Barber, K.E. (1985) Late Holocene palaeoecology and sedimentary history of a small lowland catchment in central England. *Quaternary Research 24,* 87–102.

Brown, A.G. & Keough, M.K. 1992: Palaeochannels and palaeolandsurfaces: the geoarchaeological potential of some Midland floodplains. *In* Needham, S. & Macklin, M.G. (eds) Alluvial archaeology in Britain. *Oxbow Monographs 27*, 185–196.

Brown, A.P. 1977: Late Devensian and Flandrian vegetational history of Bodmin Moor, Cornwall. *Philosophical Transactions of the Royal Society, series B 276*, 251–320.

Bush, M.B. 1986: *The Late Quaternary palaeoecological history of the Great World Valley.* Unpublished PhD thesis, Hull University.

Bush M.B. 1988: Early Mesolithic disturbance: a force on the landscape. *Journal of Archaeological Science 15*, 453–462.

Carter, H. *et al.* 1974: *An advanced geography of the British Isles.* Hulton, Amersham.

Caseldine, C. & Maguire, D.J. 1986: Lateglacial – early Flandrian vegetation change on north Dartmoor, south west England. *Journal of Biogeography 13*, 255–264.

Chambers, C. 1978: A radiocarbon-dated pollen diagram from Valley Bog, on the Moor House national nature reserve. *New Phytologist 80*, 273–280.

Cloutman, E.W. & Smith, A.G. 1988: Palaeoenvironments in the Vale of Pickering, parts 1–3. *Proceedings of the Prehistoric Society 54*, 1–58.

Davies, G. & Turner, J. 1979: Pollen diagrams from Northumberland. *New Phytologist 82*, 783–804.

Day, S.P. 1991: Post-glacial vegetation history of the Oxford region. *New Phytologist 119*, 445–470.

Day, S.P. 1993a: Woodland origin and ancient woodland indicators: a case-study from Sidlings Copse, Oxfordshire UK. *The Holocene 3(1)*, 45–53.

Day, S.P. 1993b: Preliminary results of high-resolution palaeoecological research at Star Carr, Yorkshire. *Cambridge Archaeological Journal 3(1)*, 129–140.

Devoy, R.J.N. 1979: Flandrian sea-level changes and vegetation history of the lower Thames estuary. *Philosophical Transactions of the Royal Society of London, series B 285*, 355–407.

Dimbleby, G.W. 1962: The development of British heathlands and their soils. *Oxford Forestry Memoirs 23*.

Dimbleby, G.W. 1985: *The palynology of archaeological sites.* Academic Press, London.

Donaldson, A.M. & Turner, J. 1977: A pollen diagram from Hallowell Moss, near Durham City, UK. *Journal of Biogeography 4*, 25–33.

Dumayne, L. 1992: *Late Holocene palaeoecology and human impact on the environment of northern Britain.* PhD thesis, Southampton University.

Dumayne, L. & Barber, K.E. 1994: The impact of the Romans on the environment of northern England: pollen data from three sites close to Hadrian's Wall. *The Holocene 4(2)*, 165–173.

Edees, E.S. 1972: *Flora of Staffordshire.* David & Charles, Newton Abbot.

Edwards, K.J. & Hirons, K.R. 1984: Cereal pollen grains in pre-elm decline deposits: implications for the earliest agriculture in Britain and Ireland. *Journal of Archaeological Science 11*, 71–80.

Francis, P.D. & Slater, D.S. 1991: A record of vegetational and land use change from upland peat deposits on Exmoor. Part 2: Hoar Moor. *Proceedings of the Somerset Archaeology and Natural History Society 134*, 1–25 (for 1990).

Francis, P.D. & Slater, D.S. 1993: A record of vegetational and land use change from upland peat deposits on Exmoor. Part 3: Codsend Moors. *Proceedings of the Somerset Archaeology and Natural History Society 136*, 9–28 (for 1992).

Girling, M.A. 1991: Mesolithic and later landscapes interpreted from the insect assemblages of West Heath Spa, Hampstead. *In* Collins, D. & Lorimer, D. (eds) Excavations at the Mesolithic site on West Heath, Hampstead 1976–1981. *British Archaeological Reports (British Series) 217*, 72–88.

Godwin, H. 1956, 1975: *History of the British flora* (1st and 2nd edns). Cambridge University Press, Cambridge.

Godwin, H. 1962: Vegetation history of the Kentish chalk downs as seen at Wingham and Frogholt. *Veröffentlichungen des geobotanischen Instituts der Eidgenossischen technischen Hochschule, Stiftung Rübel, in Zürich 37*, 83–99.

Godwin, H. & Vishnu Mittre 1975: Studies of the post-glacial history of British vegetation 16. Flandrian deposits of the Fenland margin at Holme Fen and Whittlesey Mere, Hunts. *Philosophical Transactions of the Royal Society of London, series B 270*, 561–604.

Gregory, K.J. 1974: Chapter 1, The physical basis: the land, and Chapter 2, The physical basis, the air above. *In*: Carter, H., Dawson, J.A., Diamond, D.R., Gregory, K.J., Johnson, J.H., Strachan, A.J. & Thomas, D. (Eds) *An advanced geography of the British Isles*, 30–68. Hulton, Amersham.

Greig, J. 1982: Past and present lime woods of Europe. *In* Bell, M. & Limbrey, S. (eds) Archaeological aspects of woodland ecology. *British Archaeological Reports (International Series) 146*, 23–55.

Greig, J. 1991: From lime forest to heathland—five thousand years of change at West Heath Spa, Hampstead, as shown by the plant remains. *In* Collins, D. & Lorimer, D. (eds) Excavations at the Mesolithic site on West Heath, Hampstead 1976–1981. *British Archaeological Reports (British Series) 217*, 89–99 (for 1989).

Greig, J. 1992: The deforestation of London. *Review of Palaeobotany and Palynology 73*, 71–86.

Greig, J. (in prep): Cookley pollen diagram.

Hardy, E.M. 1939: Studies in the post-glacial history of British vegetation: V. The Shropshire and Flint Maelor mosses. *New Phytologist 38*, 364–396.

Haskins, L.E. 1978: *The vegetational history of SE Dorset.* Unpublished PhD thesis, Southampton University.

Hatton, M. & Caseldine, C.J. 1992: Vegetation change and land use history during the first millennium AD at Aller Farm, east Devon, as indicated by pollen analysis. *Devon Archaeological Society Proceedings 49*, 107–114 (for 1991).

Hibbert, F.A., Switsur, V.R. & West, R.G. 1971: Radiocarbon dating of Flandrian pollen zones at Red Moss, Lancashire. *Proceedings of the Royal Society of London, series B 177*, 161–176.

Hicks, S. 1971: Pollen analytical evidence for the effect of prehistoric agriculture on the vegetation of north Derbyshire. *New Phytologist 70*, 647–667.

Huckerby, E., Wells, C. & Middleton, R.H. 1992: Recent palaeoecological and archaeological fieldwork in Over Wyre, Lancashire. *In* Middleton, R.H. (ed.) *North West Wetlands Survey Annual Report 1992.* Lancaster University Archaeological Unit, Lancaster.

Hunt, T.C. & Birks, H.J.B. 1982: Devensian late-glacial vegetational history at Sea Mere, Norfolk. *Journal of Biogeography 9*, 517–538.

Hywel Davies, J. & Thom, V. 1984: *The Macmillan guide to Britain's nature reserves.* Macmillan, London.

Innes, J. B. 1994: Palaeoecological survey, *In* Cowell, R.W. & Innes, J. B. The wetlands of Merseyside. *Lancaster Imprints 2* (North West Wetlands Survey 1) 139–151.

Innes, J.B. & Simmons, I.G. 1988: Disturbance and diversity; floristic changes associated in pre-elm decline woodland recession in north-east Yorkshire. *In* Jones, M. (ed.) *Archaeology and the flora of the British Isles. Oxford University Committee for Archaeology Monograph 14*, 7–20.

Jefferies, R.L., Willis, A.J. & Yemm, E.W. 1968: The Late- and Postglacial history of the Gordano valley, north Somerset. *New Phytologist 67*, 335–348.

Jones, R.L. 1976: Late Quaternary vegetation history of the North York Moors. IV. Seamer Carrs. *Journal of Biogeography 3*, 397–406.

Jones, R.L. 1977: Late Quaternary vegetation history of the North York Moors. V. The Cleveland Dales. *Journal of Biogeography 4*, 352–362.

Jones, R.L. 1978: Late Quaternary vegetation history of the North York Moors. VI. The Cleveland Moors. *Journal of Biogeography 5*, 81–92.

Lamb, H.H. 1982: Reconstruction of postglacial climate over the world. *In* Harding, A.F. (ed.) *Climatic change in later prehistory,* 11–32. Edinburgh University Press, Edinburgh.

Lamb, H.H. 1984: Climate in the last thousand years: natural climatic fluctuations and change. *In* Flohn, H.F. & Fantechi, R. (eds) *The climate of Europe: past, present and future.* 25–64. Kluwer, Dordrecht.

Macklin, M.G. and Needham, S. 1992: Studies in British alluvial archaeology: potential and prospect. *In*

Needham, S. and Macklin, M.G. (eds) Alluvial archaeology in Britain. *Oxbow Monographs 27*, 9–23.

Macklin, M.G., Passmore, D.G., Stevenson, A.C., Cowley, D.C., Edwards, D.N., O'Brien, C.F. 1991: Holocene alluviation and land-use change on Callaly Moor, Northumberland, England. *Journal of Quaternary Science 6(3)*, 225–232.

Merryfield, D.L. 1977: *Palynological and stratigraphical studies on Exmoor.* PhD Thesis, King's College, London.

Merryfield, D.L. & Moore P.D. 1974: Prehistoric activity and blanket peat initiation on Exmoor. *Nature 250*, 439–441.

Moffat, B. 1978: The environment of Battle Abbey estates (E. Sussex). A re-evaluation using pollen analysis and sediments. *Landscape History 8*, 77–93.

Moffett, L., Robinson, M. & Straker, V. 1989: Cereals, fruit and nuts: charred plant remains from Neolithic sites in England and Wales and the Neolithic economy. *In* Milles, A., Williams, D. & Gardner, N. (eds) The beginnings of agriculture. *British Archaeological Reports (International Series) 496*, 243–261.

Moore, P.D. 1975: Origin of blanket mires. *Nature 256*, 267–269.

Moore, P.D. & Evans, A. 1991: The development of valley mires in south east England. *Aquilo, Series Botanica 30*, 25–34.

Moore, P.D., Merryfield, D.L. & Price, M.D.R. 1984: The vegetation and development of blanket mires. *In* Moore, P.D. (ed.) *European mires*, 203–235. Academic Press, London.

Moss, P.A. 1987: *Late-Quaternary pollen studies in Staffordshire.* Unpublished thesis, King's College, London.

Osborne, P.J. 1984: Some British later prehistoric insect faunas and their climatic implications. *In* Harding, A.F. (ed.) *Climatic change in later prehistory,* 68–74. Edinburgh University Press, Edinburgh.

Peglar, S.M. 1992: *The development of the cultural landscape of East Anglia,* U.K. DSc thesis, University of Bergen.

Peglar, S.M. 1993a: The mid-Holocene *Ulmus* decline at Diss Mere, Norfolk, U.K. a year-by-year pollen stratigraphy from annual laminations. *The Holocene 3(1)*, 1–13.

Peglar, S.M. 1993b: The development of the cultural landscape around Diss Mere, Norfolk, U.K., during the last 7000 years. *Review of Palaeobotany and Palynology 76*, 1–47.

Peglar, S.M. 1993c: Mid- and late-Holocene vegetation history of Quidenham Mere, Norfolk, UK interpreted using recurrent groups of taxa. *Vegetation History and Archaeobotany 2*, 15–28.

Pennington, W. 1965: The interpretation of some postglacial vegetation diversities at different Lake District sites. *Proceedings of the Royal Society series B 161*, 310–323.

Pennington, W. 1970: Vegetation history of the north-west of England: a regional synthesis. *In* Walker, D. & West R.J. (eds) *Studies in the vegetational history of the British Isles*, 47–79. Cambridge University Press, Cambridge.

Pennington, W. 1974: *The history of British vegetation*, 2nd edn., English Universities Press, London.

Pennington, W. 1975: A chronostratigraphic comparison of Late-Weichselian and Late-Devensian subdivisions, illustrated by two radiocarbon dated profiles from western Britain. *Boreas 4*, 157–171

Pennington, W. 1977: The late Devensian flora and vegetation of Britain. *Philosophical Transactions of the Royal Society of London, series B 280*, 247–271.

Pennington, W. 1979: The origin of pollen in lake sediments: an enclosed lake compared to one receiving inflow streams. *New Phytologist 83*, 189–213.

Pennington W. 1991: Palaeolimnology in the English lakes—some questions and answers over fifty years. *Hydrobiologia 214*, 9–24.

Pennington, W., Cambray, R.S., Eakins, J.D. & Harkness, D.D. 1976 Radionucleotide dating of the recent sediments at Blelham Tarn. *Freshwater Biology 6*, 317–331.

Pigott, C.D. 1991: Biological flora of the British Isles 174: *Tilia cordata* Miller. *Journal of Ecology 79*, 1147–1207.

Polunin, O. & Walters, M. 1985: *A guide to the vegetation of Britain and Europe*. Oxford University Press, Oxford.

Rackham, O. 1980. *Ancient woodland; its history, vegetation and uses in England*. Arnold, London.

Rodwell, J.S. (ed.) 1991: *British plant communities*. Cambridge University Press, Cambridge.

Rowlands, P.H. 1966: *Pleistocene stratigraphy and palynology in West Shropshire*. Unpublished PhD thesis, Birmingham University.

Rybníček, K. & Rybníčková, E. 1987: Palaeogeobotanical evidence of Middle Holocene stratigraphic hiatuses in Czechoslovakia and their explanation. *Folia Geobotanica et Phytotaxonomica 22*, 313–327.

Scaife, R.G. 1980: *Late Devensian and Flandrian palaeoecological studies in the Isle of Wight*. PhD thesis, King's College, London.

Scaife, R.G. 1982: Late Devensian and early Flandrian vegetation changes in southern England. *In* Bell, M.G. & Limbrey, S. (eds) Archaeological aspects of woodland ecology. *British Archaeological Reports (International Series) 146*, 57–74.

Scaife, R.G. 1988: The elm decline in the pollen record of South East England and its relationship to early agriculture. In Jones, M. (eds) Archaeology and the flora of the British Isles, Oxford University Committee for Archaeology Monograph *14*, 21–33.

Scaife, R.G. & Burrin, P. 1992: Archaeological inferences from alluvial sediments: some findings from southern England. *In* Needham, S. & Macklin, M.G. (eds) Alluvial archaeology in Britain. *Oxbow Monographs 27*, 75–91.

Seagrief, S.C. 1959: Pollen diagrams from southern England: Wareham, Dorset and Nursling, Hampshire. *New Phytologist 58*, 316–325.

Seagrief, S.C. 1960: Pollen diagrams from southern England: Cranes Moor, Hampshire. *New Phytologist 59*, 73–83.

Shotton, F.W. 1978: Archaeological inferences from the study of alluvium in the lower Severn–Avon valleys. *In* Limbrey, S. & Evans, J.G. (eds) *The effect of man on the landscape: the lowland zone. Council for British Archaeology Research Reports 21*, 27–32.

Simmons, I.G. 1964: Pollen diagrams from Dartmoor. *New Phytologist 63*, 165–180.

Simmons, I. & Innes, J. 1987: Mid Holocene adaptations and later Neolithic forest disturbance. *Journal of Archaeological Science 14*, 385–403.

Simmons, I.G., Atherden, M.A., Cloutman, E.W., Cundill, P.R., Innes, J.B. & Jones, R.L. 1993: Prehistoric environments. In Spratt, D.A. (ed.) *Prehistoric and Roman archaeology of NE Yorkshire*, 2nd edn., Council for British Archaeology, *Research Reports 87*, 15–49, York.

Smith, A.G. 1958: Post-glacial deposits in south Yorkshire and north Lincolnshire. *New Phytologist 57*, 19–49.

Smith, A.G., Whittle, A., Cloutman, E.G. & Morgan, L.A. 1989: Mesolithic and Neolithic activity and environmental impact on the south-east fen-edge in Cambridgeshire. *Proceedings of the Prehistoric Society 55*, 207–249.

Smith, B.M. 1985: *A palaeoecological study of raised mires in the Humberhead levels*. PhD thesis, University of Wales (Cardiff).

Tallantire, P.A. 1992: The alder [*Alnus glutinosa* (L.) Gaertn.] problem in the British Isles, a third approach to its palaeohistory. *New Phytologist 122*, 717–731.

Tallis, J.H. 1985: Mass movement and erosion of a south Pennine blanket peat. *Journal of Ecology 73*, 283–315.

Tallis, J.H. & Switsur, V.R. 1973: Studies of southern Pennine peats VI. A radiocarbon dated pollen diagram from Featherbed Moss, Derbyshire. *Journal of Ecology 61*, 743–751.

Tansley, A.G. 1949: *Britain's green mantle*, 2nd ed. Allen & Unwin, London.

Thorley, A. 1981: Pollen analytical evidence relating to the vegetational history of the Chalk. *Journal of Biogeography 8*, 93–106.

Tinsley, H. 1975: The former woodland of the Nidderdale Moors and the role of early man in its decline. *Journal of Ecology 63*, 1–26.

Tooley, M.J. 1978: *Sea-level changes in North-west England during the Flandrian Stage*. Oxford University Press, Oxford.

Trueman, A.E. 1971: *Geology and scenery in England and Wales* (revised by J.B. Whittow & J.R. Hardy). Penguin, Harmondsworth.

Turner, J. & Hodgson, J. 1979: Studies in the vegetational history of the northern Pennines. 1. Variations in the composition of the early Flandrian forests. *Journal of Ecology 67*, 629–646.

Turner, J. & Hodgson, J. 1981: Studies in the vegetational history of the northern Pennines 2. An atypical pollen diagram from Pow Hill, Co. Durham. *Journal of Ecology 69*, 171–188.

Turner, J. & Hodgson, J. 1983: Studies in the vegetational history of the northern Pennines. 3. Variations in the composition of the mid-Flandrian forests. *Journal of Ecology 71*, 95–118.

Walker, D. 1966: The late-Quaternary history of the Cumberland lowland. *Philosophical Transactions of the Royal Society of London, series B 251*, 1–210.

Waller, M. 1987: *The Flandrian vegetation history and environmental development of the Brede and Pannel valleys, East Sussex.* Unpublished PhD thesis, North London Polytechnic.

Waller, M. 1993: Flandrian vegetational history of south-eastern England. Pollen data from Pannel Bridge, East Sussex. *New Phytologist 124*, 345–369.

Waller, M. 1994: Flandrian environmental change in Fenland. The Fenland Project 9, *East Anglian Archaeology 70*.

Waton, P. 1982a: Man's impact on the Chalklands: some new pollen evidence. *In* Bell, M. and Limbrey, S. (eds.) Archaeological aspects of woodland ecology. *British Archaeological Reports (International Series) 146*, 75–91.

Waton, P.V. 1982b: *A palynological study of the impact of man on the landscape of central southern England with special reference to the chalklands.* Unpublished PhD thesis, Southampton University.

West, R.G. 1993: On the history of the late Devensian Lake Sparks in southern Fenland, Cambridgeshire, England. *Journal of Quaternary Science 8(3)*, 217–234.

Wiltshire, P.E.J. & Edwards, K.J. 1993: Mesolithic, early Neolithic and later prehistoric impacts on vegetation at a riverine site in Derbyshire, England. *In* Chambers, F.E.M. (ed.) *Climatic change and human impact on the landscape*, 157–168. Chapman & Hall, London.

Wimble, G.T. 1986: *The palaeoecology of the lowland coastal raised mires of south Cumbria.* Unpublished PhD thesis, University of Wales (Cardiff).

3

Great Britain—Wales

F.M. Chambers

INTRODUCTION

Wales is a principality of considerable landscape diversity; it has greater relief than England, yet it is only a quarter the size. Six relief regions can be defined, based on denudation chronology (Brown 1960, Bowen 1977, Caseldine 1990), but all are discontinuous (Fig. 3.1); this makes the designation of coherent regions difficult. The type regions employed here were those previously defined for Great Britain, but they are not internally consistent (see Fig. 3.1). Wales is treated as a whole for this introductory section; subsequent type-region accounts are therefore kept short. An improved regional subdivision is suggested later.

Wales has diverse bedrock geology, from Precambrian to Jurassic sedimentary formations, with some igneous and metamorphic rocks (Fig. 3.2). Post-Jurassic sediments are principally of Quaternary age, mapped as Drift (see *National Atlas of Wales*), consisting of widespread tills, head, and localized glaciofluvial sediments as well as extensive upland peat of variable depth, localized lowland raised mires, and coastal peats of the Glamorgan and Gwent levels (akin to the better-known Somerset Levels on the other side of the Bristol Channel (Smith & Morgan 1989)). Many small glacial lakes are present in the mountains, particularly in region GB-h, with some longer valley-trough lakes. Natural lowland lakes are all smaller than 200 ha. The main mountain area—Snowdonia, in region GB-h—

contains the highest peak (1085 m) but feeds no rivers of note. The two major rivers, Wye and Severn, originate within a few kilometres of each other in region GB-g in west-central Wales. They diverge and flow east and then south to the Severn Estuary in region GB-i, to be joined by the River Usk.

Uplands have been defined as areas above 244 m (800 ft) and are estimated to cover 37% of Wales (Countryside Commission 1983, Caseldine 1990). The major mountains generated their own ice-caps in the pleniglacial. Considerable dispute still exists as to the relative importance of local "Welsh" ice and more distantly generated "Irish Sea" ice in the glaciations of North Wales. Following the pleniglacial, widespread ice largely disappeared, but the higher-altitude valleys in North Wales (region GB-h) and glacial cwms (cirques) in the Brecon Beacons (in region GB-i) experienced localized readvances during the Late Weichselian stadial.

Brown earth soils probably developed in much of Wales during the Holocene, but in upland areas present-day soils are now dominated by a range of brown podzolic soils, stagnohumic gleys, and stagnopodzols, with deeper peats on some plateaux. Soils vary with topography, with peaty gleys on plateaux and more freely drained acid brown earths on slopes. Almost all the vegetation of Wales has been significantly influenced by human activity, although the scale and intensity of impact varies immensely in time and space (Taylor 1980, Caseldine 1990). Human presence dates back to the Pleistocene with early

Palaeoecological Events During the Last 15000 Years: Regional Syntheses of Palaeoecological Studies of Lakes and Mires in Europe.
Edited by B.E. Berglund, H.J.B. Birks, M. Ralska-Jasiewiczowa and H.E. Wright. © 1996 John Wiley & Sons Ltd.

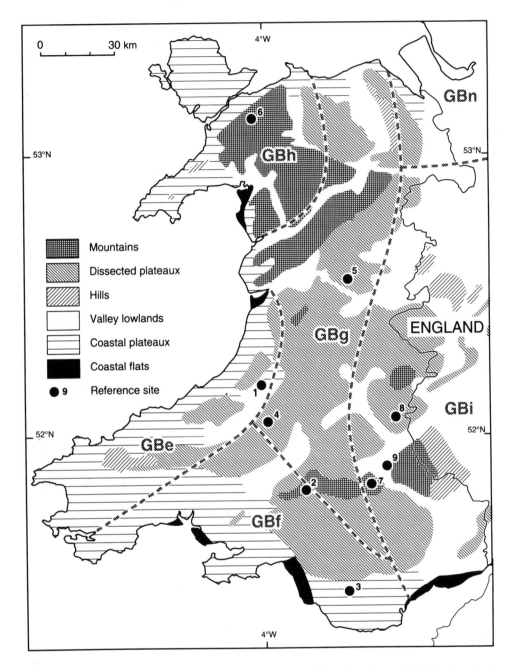

Fig. 3.1 Map of relief regions of Wales (based on map in Bowen (1977), after Brown (1960)), showing locations of reference sites and the Great Britain type Regions (compare with Fig. 3.5)

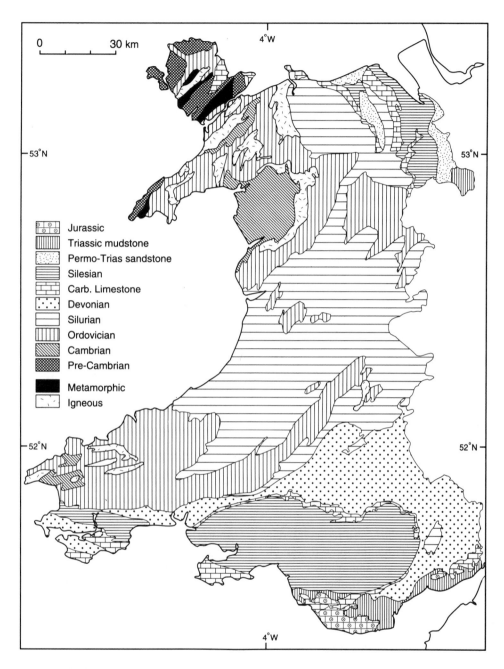

Fig. 3.2 Bedrock geology of Wales (simplified and adapted from the *National Atlas of Wales*, University of Wales Press)

cave assemblages; human impact in the Holocene can be traced from the Mesolithic onwards. Distribution maps of Mesolithic sites and artefacts display major gaps (Wymer 1977, Jacobi 1980). Sites are mainly revealed on coastal cliffs and by upland peat erosion or ploughing: the skewed and certainly incomplete distribution of Mesolithic sites in Wales has been described as "the archaeology of exposure" (R.B. White, pers. com.). Evidence for Mesolithic impact upon vegetation is circumstantial, although more persuasive in some areas (Smith & Cloutman 1988) than others (cf. Chambers *et al.* 1988, Birks 1988). Seasonal movement from coast to uplands has been suggested (Jacobi 1980).

Though dominantly a land of pastoralism, arable agriculture in Wales is attested from the Neolithic. Neolithic settlement distribution in southern Wales is largely inferred from burial monuments, whose typology and distribution up major river valleys in regions GB-i and GB-g imply a strong cultural affinity with England. Seasonal use of upland grazing has been postulated for the Neolithic in region GB-g (Moore 1981).

The Bronze Age is well represented in Wales, particularly in the uplands, with many stone-circles and burial sites. Evidence for widespread woodland clearance in this period is persuasive. Early industry also probably had localized impact (Mighall & Chambers 1993): copper mining, employing fire setting, is known from the Bronze Age at several sites in west-central and northern Wales (in regions GB-e and GB-h).

Late prehistory is particularly well represented in Gwynedd (region GB-h) with numerous stone-built hut circles and hillforts—each apparently controlling a block of territory. Former timber dwellings, including stake-walled roundhouses, may also have been abundant here and probably elsewhere in Wales, but few traces survive. Late-prehistoric ironworking is known at several North Wales sites and involved charcoal smelting in specialized furnaces.

Roman conquest in the first century AD led apparently to increased demands on the environment in some areas, notably in parts of region GB-i, with major woodland clearance (Jones *et al.* 1985). With the decline of hillforts, agricultural settlement in much of western and northern Wales during Dark

Age and Medieval times is claimed to have been based on a system of transhumance between year-round *hendref* and summer *hafod*, leading to seasonal use of upland pastures and hence to human manipulation of marginal environments. However, traces of former upland arable cultivation also exist, and the *hendref–hafod* model may be an oversimplification. In parts of Wales, Cistercian monasteries such as Strata Florida controlled extensive upland areas in Medieval times, and their wealth was largely based on pastoralism. Drovers' roads, along which cattle and other beasts were driven to markets in the east, are well known across Wales and emphasize a pastoral bias to agriculture in recent centuries. The upland landscape was by then largely deforested.

Industrialization of the last 200 years affected some areas of Wales markedly, notably the South Wales coalfield (region GB-f). The Lower Swansea Valley was the world's centre for copper smelting in the 18th century. Ironworks and coal mining dominated the economy and environment of Region GB-f from the start of the Industrial Revolution. Inland, iron-workings have ceased, but coastal steel production continues. Coal mining, which formerly employed 250000 in South Wales, largely in deep mining, now employs just 1000, mainly in opencast workings that have recently destroyed areas of semi-natural moorland.

The legacy of industrialization is very evident in the Rhondda, Merthyr, and neighbouring valleys, with numerous coal tips and slag heaps, but long-distance air pollution is also believed to have affected the vegetation of the uplands of South and Mid-Wales (Chambers *et al.* 1979). Acid precipitation has significantly affected some lakes and their catchments in west-central and North Wales (Battarbee *et al.* 1988).

The major landscape changes of recent decades have been the damming of rivers for reservoirs, the planting of large areas of conifers in the uplands and Vale of Glamorgan, principally of *Picea sitchensis,* and the ploughing and seeding of upland pasture under EU grants, often in areas that had not been tilled for some 2000 years. Sheep have come to dominate upland grazing, with high subsidized stocking rates on thin pastures.

The following type regions, as defined for Great Britain, are represented wholly in Wales (Fig. 3.1):

GB-e South–West Wales
GB-f South Wales
GB-g Mid-Wales uplands
GB-h Snowdonia

Region GB-i, designated (English) West Midlands and the Severn basin, includes significant areas of eastern Wales. Note, however, that significant areas of the upper Severn and Wye catchments are represented in region GB-g. Part of northeastern Wales comes within English type region GB-n (Lancashire–Cheshire plain).

In a recent review, Caseldine (1990) lists 228 sites with pollen data in Wales.

TYPE REGION GB-e, SOUTH-WEST WALES

This region encompasses the coast and hinterland of Dyfed, including the Teifi valley plus the Preseli Hills. It contains all of the Pembrokeshire Coast National Park.

Altitude: 0–467 m.

Climate: Mean January temperature 5°C, July 15°C. Precipitation 1000 to >1600 mm yr^{-1}, with 190 to 200 wet days per year. Oceanic.

Geology: Diverse Lower Palaeozoic sedimentary rocks, with considerable intrusive and extrusive igneous rocks, largely Ordovician.

Topography: Coastal plateaux, rising to skyline tors of Mynydd Preseli; plus the Teifi valley.

Population: Several small towns, including ports and some coastal holiday resorts; many hamlets and farmsteads.

Vegetation: Much land is down to grass for dairying. Some woodland, including planted Sitka spruce. Large raised mires of Cors-goch Glan Teifi (Tregaron Bog) and Borth Bog.

Soils: Chiefly brown earths and brown podzolic soils, with associated stagnogleys.

Land use: Chiefly dairying, with minor arable and fodder crops; some forestry.

Reference site 1. Tregaron (Southeast) Bog
(Hibbert & Switsur 1976; also, Godwin & Mitchell, 1938, Turner 1964)

Latitude 52°15′N, Longitude 3°55′W. Elevation 165 m. Age range 10200–0 BP. Ombrotrophic (raised) mire. Six local pollen-assemblage zones (paz). (Fig. 3.3, Table 3.1)

10200–9750	BP	*Betula–Pinus–Salix–Juniperus*
9750–9300	BP	*Betula–Pinus–Corylus*
9300–8150	BP	*Corylus–Pinus*
8150–6990	BP	*Pinus–Corylus–Quercus*
6990–4990	BP	*Quercus–Ulmus–Alnus*
4990–recent		*Quercus–Alnus*

TYPE REGION GB-f, SOUTH WALES

This region includes the Vale of Glamorgan, the valleys of the South Wales coalfield, the western part of the Glamorgan uplands, and Mynydd Du (Black Mountain (*sic*)) north of Swansea.

Altitude: 0–600 m.

Climate: Mean January temperature 5°C, mean July 16.5°C. Precipitation ranges from 900 mm (south coast) to over 2400 mm at high altitude. Oceanic.

Geology: Dominated by the South Wales coalfield; rocks are Carboniferous Westphalian sandstone and coal-bearing strata, fringed by Namurian Millstone Grit and Carboniferous Tournaisian and Visean Limestone. Triassic mudstone and Jurassic Lower Lias occur in the Vale of Glamorgan.

Topography: North-facing escarpments define the northern boundary of the type region, but the dominant features are the sandstone uplands, deeply dissected on their southern flanks by the South Wales valleys and bounded on the south by the undulating Vale of Glamorgan and on the southwest by the low hills of the Gower peninsula.

Population: The highest population densities of Wales are found here, locally exceeding 50 persons ha^{-1} in Cardiff, Swansea, and the Rhondda and Merthyr valleys, compared with an average population density for all of Wales of 1.2 ha^{-1}.

82

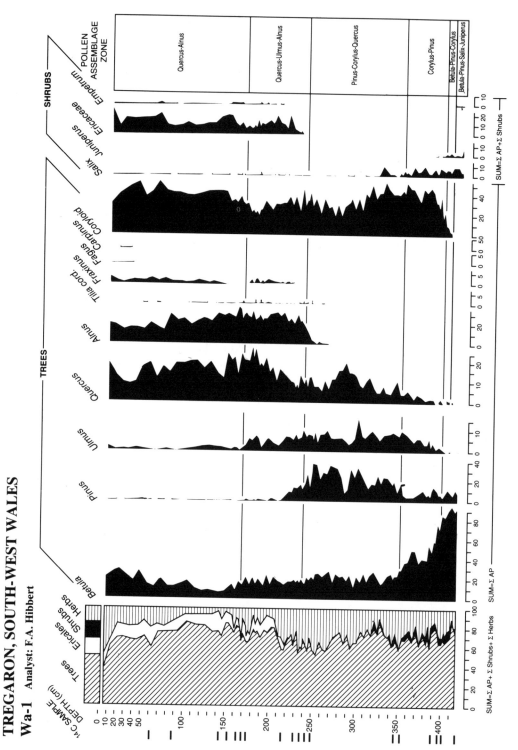

TREGARON, SOUTH-WEST WALES

Wa-1 Analyst: F.A. Hibbert

Fig. 3.3 Pollen diagram from reference site 1 in region GB-e: Tregaron (Southeast) Bog, Dyfed (after Hibbert & Switsur 1976)

Table 3.1 Tregaron. Radiocarbon dates for pollen-assemblage zone boundaries and for changes within the zones
(from Hibbert & Switsur 1976)

	Pollen-assemblage zone	Age BP	Sample depth (cm)	Laboratory reference
	Anthropogenic effects	2920±50	62– 64	Q-947
	Plantago pollen increase	3350±50	86– 88	Q-946
	Ulmus and *Fraxinus* rise	4695±55	144–146	Q-945
	Plantago pollen increase	4715±55	152–154	Q-944
	Beginning of fall in *Ulmus*	4890±70	163–165	Q-943
(f)	*Quercus–Alnus* zone begins	4990±70	167–169	Q-942
	End of fall in *Ulmus*	5110±70	172–174	Q-941
	Tilia curve becomes continuous	5980±100	214–216	Q-940
	Alnus rise completed	6530±110	228–230	Q-939
	Alnus values increase	6980±140	236–238	Q-938
(e)	*Quercus–Ulmus–Alnus* zone opens	6990±180	242–244	Q-937
	Alnus curve begins	7130±180	248–250	Q-936
(d)	*Pinus–Corylus–Quercus* zone begins	8150±150	348–350	Q-935
	Pinus values above *Betula*	8285±150	354–356	Q-934
(c)	*Corylus–Pinus* zone opens	9300±190	390–392	Q-933
	Ulmus values rise	9550±200	398–400	Q-932
(b)	*Betula–Pinus–Corylus* zone begins	9750±220	404–406	Q-931
(a)	*Betula–Pinus–Salix–Juniperus* zone begins	10200±220	413–416	Q-930

Vegetation: Upland moors dominated by *Molinia caerulea* have been extensively planted with *Picea sitchensis*; lower lands contain varied vegetation ranging from productive grasslands to limestone heath.

Soils: Uplands have stagnopodzols, stagnohumic gleys, and brown podzolic soils, whereas soils in the lowlands range from argillic brown earths to stagnogleys, with thin rendzinas on limestone.

Land use: Predominantly sheep grazing and forestry in the uplands, with dairying and beef cattle in the lowlands.

Reference site 2. Waun-Fignen-Felen
(Smith & Cloutman 1988)

Latitude 51°51′N, Longitude 3°42′W. Elevation 488 m. Age range 10000–1000 BP. Ombrotrophic (blan-

ket) mire, over former lake and reedswamp deposits. Six local pollen-assemblage zones (paz). Age ranges are approximate only, as they vary among dated profiles.

ca. 10000–*ca.* 8000 BP	*Betula*–Gramineae–*Salix*
ca. 8000–*ca.* 7000 BP	*Corylus–Calluna–Ulmus–Pinus*
ca. 7000–*ca.* 5000 BP	*Quercus–Ulmus–Alnus*
ca. 5000–*ca.* 3900 BP	*Quercus–Alnus*
ca. 3900–*ca.* 2800 BP	*Quercus–Alnus–Plantago*
ca. 2800–*ca.* 1000 BP	*Betula–Fraxinus–Narthecium*

Reference site 3. Llanilid
(Walker & Harkness 1990)

Latitude 51°31′N, Longitude 3°27′W. Elevation 60 m. Age range 13200–9300 BP. Organic muds underlying silts and clays, overlain by peat. 11 local pollen-assemblage zones (paz), eight of which cover the Late-glacial.

TYPE REGION GB-g, MID-WALES UP-LANDS

This is a composite region that contains parts of Clwyd and the Clwydian range, the Berwyn Mountains (827 m), and parts of upland west-central Wales, including Cadair Idris (893 m), the Arans (907 m), Pumlumon (752 m), and the Cambrian Mountains.

Altitude: 0–907 m.

Climate: Mean January temperature 4.5°C, mean July 15.5°C. Precipitation very variable from 700 mm in the northeast to over 2000 mm on mountains. Oceanic.

Geology: Largely Silurian with graptolitic shale; some Ordovician rocks. Localized igneous rocks.

Topography: Largely dissected plateaux, with some mountainous peaks.

Population: Generally at low density for Wales, with most areas at less than 1 person ha^{-1}. Higher densities are concentrated in north-coast holiday resorts.

Vegetation: Variable, with much of the uplands either as moorland or recently afforested.

Soils: Brown earths, brown podzolics, stagnohumic gleys, stagnopodzpols, and raw peat soils. Localized rendzinas on limestone along parts of the north coast.

Land use: Predominantly sheep grazing and forestry at altitude; beef and dairy cattle in the lowlands.

Reference site 4. Cefn Gwernffrwd
(Chambers 1982b)

Note that this site is only 15 km from Tregaron (Southeast) Bog—the reference site for type region GB-e.)

Latitude 52°07′N, Longitude 3°50′W. Elevation 395 m. Age range 10000–0 BP. Basin mire. Conventional

(Godwin–Jessen) pollen-assemblage zones: Zone IV to Zone VIII. Also, eight phases (a–h) in more detailed pollen diagram covering the period 6690 to 2220 BP.

Reference site 5. Carneddau (Walker 1993)

Latitude 52°35′N, Longitude 3°29′W. Elevation 400 m. Age range 9000–0 BP. Basin mire (Carneddau 1, 2, 4) and blanket mire profiles (Carneddau 3, 5); the blanket mire age range is 4750–0 BP. Six regional pollen-assemblage zones not well dated.

TYPE REGION GB-h, SNOWDONIA

This region encompasses most of present-day Gwynedd (apart from southern Meirionnydd) and includes Anglesey (Ynys Mon), the Lleyn Peninsula, the Arfon lowlands, and Snowdon, the Glyders, Carnedd Llywelyn, the Rhinogs, Arenig Fawr, and the vale of Conwy. Snowdonia National Park is in this region.

Altitude: 0–1085 m.

Climate: Strongly oceanic but variable, depending on altitude. Mean January temperature 5°C, mean July 15.5°C. Precipitation strongly orographic, ranging from 800 mm to over 4000 mm. Wet days range from 190 to over 230 days yr^{-1}.

Geology: Extremely complex. Precambrian rocks form part of the Lleyn Peninsula and parts of Anglesey; Lower Cambrian rocks dominate the Harlech Dome, and Ordovician rocks dominate Snowdon. Significant igneous intrusions and extrusions also occur. Much Pleistocene till plasters the lowlands, together with outwash sands and gravels.

Topography: Mountainous in Snowdon and the Glyder range, with considerable relief in most areas except coastal plateaux of Lleyn and Anglesey.

Population: Concentrated in coastal towns, but inland settlements in small towns, hamlets, and farmsteads.

Vegetation: Very variable, and strongly affected by altitude and by degree of human influence. Unvegetated rocks and scree at high altitudes.

Table 3.2 Nant Francon. Radiocarbon dates for pollen assemblage zone boundaries and for changes within the zones (from Hibbert & Switsur 1976)

	Pollen-assemblage zone	Age BP	Sample Depth (cm)	Laboratory refefence
	Major deforestation	4255 ± 50	164–166	Q-907
	Ulmus values recover	4420 ± 60	170–172	Q-906
	End of fall of *Ulmus*	4870 ± 60	190–193	Q-905
(f)	*Quercus–Alnus* zone begins	5050 ± 70	207–210	Q-904
	Last high *Ulmus* values	5160 ± 70	215–218	Q-903
	Alnus values maximum	6725 ± 100	400–403	Q-902
(e)	*Quercus–Ulmus–Alnus* zone opens	6790 ± 100	410–413	Q-901
	Alnus values rise	6880 ± 100	420–423	Q-900
(d)	*Pinus–Corylus–Quercus* zone opens	8120 ± 120	509–512	Q-899
	Check on Q-899	8160 ± 120	502–508	Q-899 bis
	Alnus curve begins	8450 ± 150	528–531	Q-898
	Ulmus curve begins	8640 ± 150	546–549	Q-897
	Corylus values rise	8810 ± 170	551–554	Q-896
(c)	*Betula–Corylus–Isoetes* zone opens	8930 ± 170	557–560	Q-895
	First record of *Alnus*	9100 ± 180	562–565	Q-894
	Near end of *Juniperus* curve	9630 ± 200	570–573	Q-893
	Decline of *Juniperus* values	9745 ± 200	584–587	Q-892
(b)	*Betula–Pinus–Corylus* zone opens	9870 ± 200	587–591	Q-891
	Peak of *Juniperus* curve	9920 ± 220	597–608	Q-890 bis
(a)	*Betula–Pinus–Juniperus* zone opens	10080 ± 220	608–620	Q-890

Soils: Stagnogleys and brown earths predominate in Anglesey and Lleyn, and rankers and Stagnopodzols dominate in mountainous areas and brown podzolic soils on the western fringes of the uplands.

Land use: Land use varies considerably, with extensive upland sheep grazing and some forestry in the mountains and uplands, contrasting with cattle-dominated areas of parts of Arfon and Anglesey.

Reference site 6. Nant Ffrancon
(Hibbert & Switsur 1976)

Latitude 53°08′N, Longitude 4°02′W. Elevation 198 m. Age range >10000–0 BP. Infilled valley, with former lake. Six local pollen-assemblage zones (paz). (Fig. 3.4, Table 3.2)

10080–9870 BP *Betula–Pinus–Juniperus(–Quercus)*
9870–8930 BP *Betula–Pinus–Corylus*
8930–8120 BP *Betula–Corylus(–Isoetes)*
8120–6790 BP *Pinus–Corylus–Quercus*
6790–5050 BP *Quercus–Ulmus–Alnus*
5050–?0 BP *Quercus–Alnus*

TYPE REGION GB-i, WEST MIDLANDS (OF ENGLAND) AND THE SEVERN BASIN

This large region includes the upland Brecon Beacons and the Black Mountains (*sic*), the catchments of the Afon Llynfi, the Usk, the lower Wye and the Severn, as well as the Gwent lowlands.

Altitude: 0–873 m.

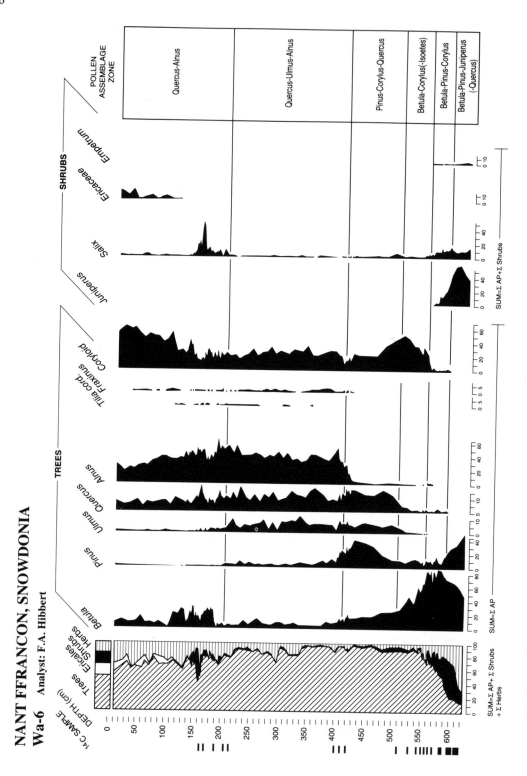

Fig. 3.4 Pollen diagram from reference site 6 in Region GB-h: Nant Ffrancon, Gwynedd (after Hibbert & Switsur 1976)

Climate: Mean January temperatures 4.5°C (west) to 4°C (east), mean July 16°C (west) to 16.5°C (east). Precipitation ranges from 2400 mm on Brecon Beacons to less than 800 mm to the east in rain shadow. Oceanic.

Geology: Extremely variable in the middle and north of the region, consisting of Silurian, Carboniferous, Permian, and Triassic rocks, including saliniferous strata, variably plastered with Pleistocene till; largely Devonian Old Red Sandstone in the south, but with substantial Pleistocene alluvium in river valleys.

Topography: Very variable. North-facing scarps of the Brecon Beacons and Black Mountains contrast with long dip slopes to the south. Superimposed antecedent drainage of the Wye is discordant, cutting deep meandering gorges, contrasting with the wide floodplain of the lower Severn. In the northern part of the region the hummocky moraines of the Shropshire lowlands contrast with the stark hills of the Long Mynd, the Stiperstones, and the escarpment of Wenlock Edge.

Population: Well-populated in the southeast of the region, in England; less populated to the north and west in Wales.

Vegetation: Very variable.

Soils: Predominantly brown earths, characteristically red in Old Red Sandstone area; stagnohumic gley soils at high altitudes in the southwest; brown alluvial soils of the Severn.

Land use: Dairy and beef cattle in the lowlands, with some arable land. Some forestry; sheep grazing at high altitude.

Reference site 7. Brecon Beacons
(Chambers 1982b)

Latitude 51°52′N, Longitude 3°23′W. Elevation 715 m. Age range *ca.* 4295–0 BP. Ombrotrophic (blanket) mire. 13 local phases (a–m), with features as follows (comparisons are made with earlier phases).

a	undated Neolithic	High *Corylus, Rumex*
b	*ca.* 4295–*ca.* 4150 BP	Reduced *Corylus, Rumex;* higher *Calluna*
c	*ca.* 4150–*ca.* 3475 BP	Increased *Quercus, Fraxinus, Corylus,* Cyperaceae; lower *Calluna*
d	*ca.* 3475–*ca.* 3245 BP	Reduced *Corylus*; higher *Calluna, Potentilla*-type
e	*ca.* 3245–*ca.* 2875 BP	High *Calluna, Pteridium*
f	*ca.* 2875–*ca.* 2320 BP	Increased *Fraxinus*; major fluctuations in NAP types
g	*ca.* 2320–*ca.* 2125 BP	High *Corylus*
h	*ca.* 2125–*ca.* 2010 BP	Decline in AP; rise in Cyperaceae
i	*ca.* 2010–*ca.* 1855 BP	High *Narthecium*
j	*ca.* 1855–*ca.* 1200 BP	High *Calluna*, low Cyperaceae
k	*ca.* 1200–*ca.* 545 BP	High Cyperaceae; low *Calluna*
l	*ca.* 545–*ca.* 325 BP	Increased *Calluna, Empetrum*
m	*ca.* 325–*ca.* present	Reduced *Empetrum*; increase in AP

Reference site 8. Rhosgoch Common
(Bartley 1960)

Latitude 52°07′N, Longitude 3°10′W. Elevation 230 m. Age range 12000–2500 BP. Infilled lake basin. Conventional (Godwin–Jessen) pollen-assemblage zones: Zone I to Zone VIIb. No radiocarbon dates.

Reference site 9. Llangorse Lake
(Chambers 1985, Jones *et al.* 1985)

Latitude 51°55′N, Longitude 3°15′W. Elevation 230 m. Age range >10,000–0 BP. Lake. Conventional (Godwin–Jessen) pollen-assemblage zones: Zone IV to Zone VIII. In addition, there are pollen, mollusc, and ostracod records from early Holocene lake marl (Walker *et al.* 1993).

[Also reference sites Crose Mere (Beales 1980) and Ashmoor Common (Brown 1982), both in England.]

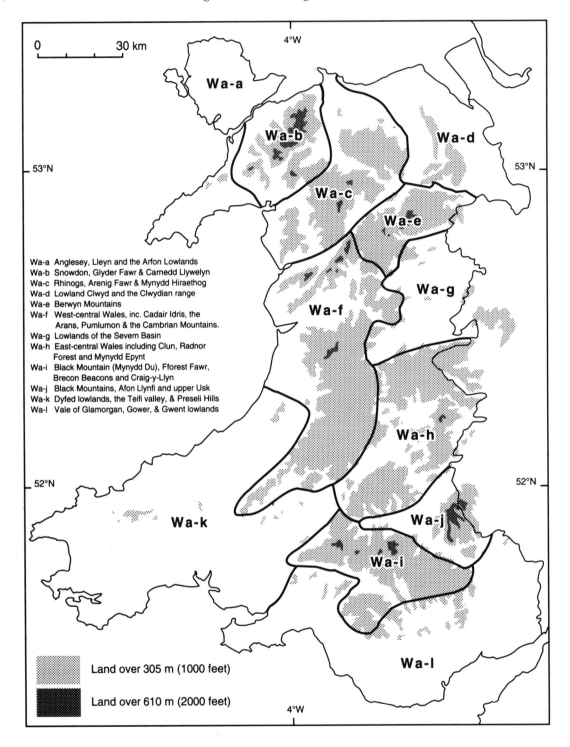

0   30 km

4°W

Wa-a

Wa-b

Wa-d

53°N   53°N

Wa-c

Wa-e

Wa-g

Wa-f

Wa-a Anglesey, Lleyn and the Arfon Lowlands
Wa-b Snowdon, Glyder Fawr & Carnedd Llywelyn
Wa-c Rhinogs, Arenig Fawr & Mynydd Hiraethog
Wa-d Lowland Clwyd and the Clwydian range
Wa-e Berwyn Mountains
Wa-f West-central Wales, inc. Cadair Idris, the
        Arans, Pumlumon & the Cambrian Mountains.
Wa-g Lowlands of the Severn Basin
Wa-h East-central Wales including Clun, Radnor
        Forest and Mynydd Epynt
Wa-i Black Mountain (Mynydd Du), Fforest Fawr,
        Brecon Beacons and Craig-y-Llyn
Wa-j Black Mountains, Afon Llynfi and upper Usk
Wa-k Dyfed lowlands, the Teifi valley, & Preseli Hills
Wa-l Vale of Glamorgan, Gower, & Gwent lowlands

Wa-h

52°N   52°N

Wa-k

Wa-j

Wa-i

Land over 305 m (1000 feet)

Wa-l

Land over 610 m (2000 feet)

4°W

Fig. 3.5 Proposed larger-scale and more coherent regions of Wales for future studies. Eastern regions are bounded
by the border with England

DISCUSSION AND CONCLUSIONS

(1) For a relatively small country, Wales has exhibited at any one time a considerable variation in Holocene vegetation. In the early and middle Holocene this vegetation was largely of woodland but also included seral communities of coastal halosere and psammosere and subalpine mountain vegetation, with some mire communities of raised, basin, and blanket type. In the middle to later Holocene, much woodland was replaced by meadow, pasture, and *ffridd* (upland heath), with some arable land, and by an expansion of bog on plateau uplands.

(2) Of native British tree genera, only one is believed to have a range limit within Wales: the eastern part of region GB-f may mark the northern and westward limit of possibly native stands of *Fagus* (cf. Godwin 1975; but see Birks 1989).

(3) Within Wales, and no less within type regions, there was considerable variation in the relative composition of early and middle Holocene forests, most notably in the patchy distribution and variable abundance of *Pinus* (Walker 1982).

(4) The relative abundance of *Tilia* varied across Wales, with high representation (presumably *T. cordata*) reported by Hyde (1936) at a site on the South Wales coast (region GB-f). Locally high representation occurred and is still attested in the lower Wye and Usk valleys (including some *T. platyphyllos*) (region GB-i). To the north and west, *T. cordata* may have grown but at much lower densities.

(5) Considerable variation exists among sites in the timing of the expansion of *Alnus* (Chambers & Elliott 1989), with some suggestion that its expansion at sites in region GB-h in North Wales preceded expansion elsewhere; one site produced evidence for one of the earliest Flandrian alder carrs in Britain, dating from *ca.* 8465 BP (Chambers & Price 1985), although the reliability of this evidence has recently been queried by Tallantire (1992). Not all sites in region GB-h show early *Alnus*, illustrating the individualistic behaviour of this taxon (cf. Bennett & Birks 1990).

(6) Records of submerged forests occur at various points on the west and south coast of Wales, submerged by (Boreal) Holocene sea-level rise.

(7) The predominant woodland vegetation in the mid-Holocene was of *Quercus petraea*, notably in the north and west of Wales. Other major tree taxa include *Alnus* and *Corylus*, with some *Ulmus*. *Betula* persisted into the mid-Holocene at some sites (cf. Chambers 1982a). Although seldom recorded, *Taxus* may have been more abundant in the southeast (Chambers 1985).

(8) The elm decline is consistent at several sites *ca.* 4950 to 5050 BP, but at a few others the elm decline is apparently younger (Taylor 1987, Smith 1991). At many sites in the north and west of Wales, the elm decline may in reality mean a decline in elm representation from a mere 3% of total land pollen to less than 1%.

(9) Almost all Welsh vegetation has been significantly affected by human activity, and much has been transformed (cf. Walker 1993). This influence commenced in prehistory and has continued at varying rates and intensities over the whole country.

(10) The initiation of blanket mires (*sensu lato*) varied in time and space. On Black Mountain (Mynydd Du in region GB-f) peat development started in the Mesolithic, but in most upland areas "blanket peats" date from after the elm decline (post-5000 BP), with some peats initiated in the Neolithic, many in the Bronze Age *ca.* 3750 BP, but others not until the Dark Ages, *ca.* 1400 BP. Peat depth is not a reliable guide to age: deep peats (up to 2 m) on the Migneint (region GB-h) are younger than peats 1 m deep on the Brecon Beacons (Region GB-i) but are the same age as peats 0.25 m deep on the Glamorgan uplands. Similar differences in age–depth relationships can be found within type regions.

(11) Although some deep high-altitude peats have been considered "climatic peats" in Wales, most so-called blanket peats are not strictly climatic in origin, nor do they carry "blanket bog"

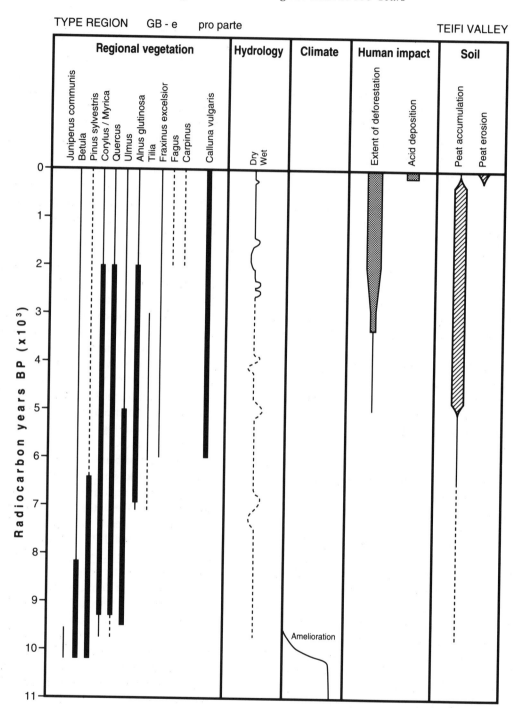

Fig. 3.6 Event stratigraphy for type region GB-e, Teifi valley

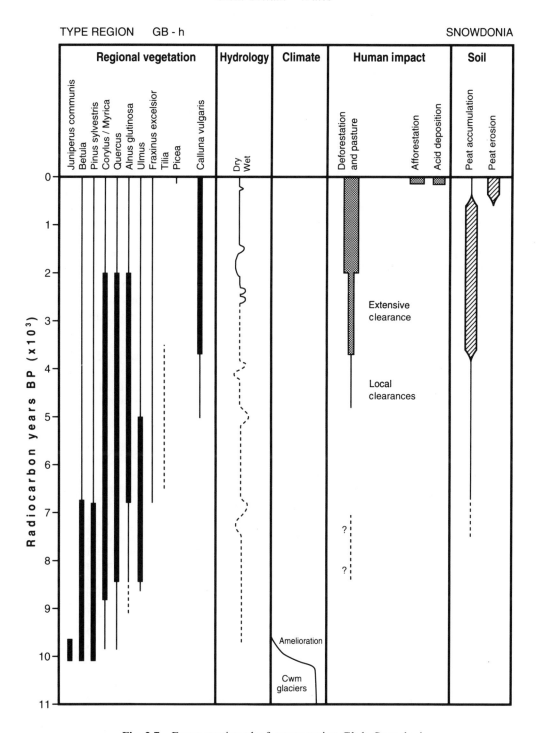

Fig. 3.7 Event stratigraphy for type region Gb-h, Snowdonia

vegetation *sensu* Tansley (1939); rather, they owe their existence more to human influence (Moore 1973, 1975, 1988) or conceivably to pedogenic factors (Taylor & Smith 1980, Smith & Taylor 1989).

(12) Evidence is accumulating from studies of subfossil pine stumps in nearby England to suggest a drier climate in the period *ca.* 4500–4000 BP, which allowed pine growth on White Moss in Cheshire (Lageard *et al.* 1992); however, in eastern Wales a major period of active pine growth on the extensive raised mires of Fenn's/Whixall Moss appears to be *ca.* 3200–3000 BP, perhaps implying a short-lived dry episode in late prehistory.

(13) Significant climatic shifts to wetter conditions seem to have taken place *ca.* 3900 BP, perhaps also at *ca.* 2800 BP (Haslam 1987) and again at *ca.* 1400 BP (Blackford & Chambers 1991).

(14) Large areas of upland Wales now carry so-called semi-natural moorland. This is dominated by *Molinia caerulea* and *Trichophorum cespitosum* in the south and central areas, with *Calluna vulgaris* more prevalent in parts of the north and in the drier east. Evidence suggests that the dominance of *Molinia* is recent (Chambers 1983) and may in many areas postdate the start of the Industrial Revolution. Burning regimes, grazing pressure, and atmospheric pollution may all have influenced the relative dominance of *Molinia* over *Calluna* (Chambers *et al.* 1979).

(15) Large areas of the deeper upland peats are now severely affected by hagging (peat erosion). On the Brecon Beacons, peat erosion is thought to have started *ca.* 425 BP (Chambers 1982b). Elsewhere evidence also suggests it is relatively recent, within the past 250 years.

(16) Although the "natural" flora of Wales would be largely dominated by trees (except at high altitudes in region GB-h and on coastal flats and inland raised mires), very little native woodland remains. The woodland type best represented is *Quercetum petraeae*, but much of this has become invaded by introduced *Rhododendron ponticum*, now naturalized in *Quercus* woods in north and west Wales. Isolated trees

of *Sorbus aucuparia* can be found at high altitudes.

(17) The type regions as constructed in the 1985 IGCP map for Great Britain are inconsistent and incoherent when applied at the scale of Wales. In future research, regional accounts might be better produced by dividing Wales into more numerous and more coherent larger-scale regions (cf. the 12 suggested in Figure 3.5). Even these regions have internal inconsistencies, largely owing to the amount of dissection, which leads to significant variation in drift geology, topography, local climate, soil type, and land use.

(18) It has to be concluded that the concept of *representative* reference sites may be somewhat inappropriate (cf. Chambers 1985) for a country as heterogeneous as Wales in geology, topography, climate, and human influence.

ACKNOWLEDGEMENTS

Thanks to A.G. Smith and to M.J.C. Walker for access to publications in press. Andrew Lawrence drew Figs 3.1–3.5.

REFERENCES

Bartley, D.D. 1960: Rhosgoch Common, Radnorshire: stratigraphy and pollen analysis. *New Phytologist 59*, 238–262.

Battarbee, R.W. (and 15 others) 1988: *Lake acidification in the United Kingdom 1800–1986.* 86 pp. ENSIS, London.

Beales, P.W. 1980: The Late Devensian and Flandrian vegetational history of Crosemere, Shropshire. *New Phytologist 85*, 133–161.

Bennett, K.E. & Birks, H.J.B. 1990: Postglacial history of alder (*Alnus glutinosa* (L.) Gaertn.) in the British Isles. *Journal of Quaternary Science 5*, 123–133.

Birks, H.J.B. 1988: Introduction. *In* Birks, H.H., Birks, H.J.B., Kaland, P.E. & Moe, D. (eds) *The cultural landscape—past, present, future*, 179–188. Cambridge University Press, Cambridge.

Birks, H.J.B. 1989: Holocene isochrone maps and patterns of tree-spreading in the British Isles. *Journal of Biogeography 16*, 503–540.

Blackford, J.J. & Chambers, F.M. 1991: Proxy records of climate from blanket mires: evidence for a Dark Age (1400 BP) climatic deterioration in the British Isles. *The Holocene 1*, 63–67.

Bowen, D.Q. 1977: The land of Wales. *In* Thomas, D. (ed.) *Wales: a new study*, 11–35. David & Charles, Newton Abbott.

Brown, A.G. 1982: Human impact on the former floodplain woodlands of the Severn. *In* Bell, M. & Limbrey, S. (eds) *Archaeological aspects of woodland ecology*, 93–104. British Archaeological Reports, International Series S146, Oxford.

Brown, E.H. 1960: *The relief and drainage of Wales*. 186 pp. University of Wales Press, Cardiff.

Caseldine, A. 1990: *Environmental archaeology in Wales*. 197 pp. St David's University College, Lampeter.

Chambers, F.M. 1982a: Environmental history of Cefn Gwernffrwd, near Rhandirmwyn, Mid-Wales. *New Phytologist 92*, 607–615.

Chambers, F.M. 1982b: Two radiocarbon-dated pollen diagrams from high-altitude blanket peats in South Wales. *Journal of Ecology 70*, 445–459.

Chambers, F.M. 1983: Three radiocarbon-dated pollen diagrams from upland peats north-west of Merthyr Tydfil, South Wales. *Journal of Ecology 71*, 475–487.

Chambers, F.M. 1985: Flandrian environmental history of the Llynfi catchment, South Wales. *Ecologia Mediterranea 11*, 73–80.

Chambers, F.M. & Elliott, L. 1989: Spread and expansion of *Alnus* Mill. in the British Isles: timing, agencies and possible vectors. *Journal of Biogeography 16*, 541–550.

Chambers, F.M. & Price, S.-M. 1985: Palaeocology of *Alnus* (alder): early post-glacial rise in a valley mire, north-west Wales. *New Phytologist 101*, 334–344.

Chambers, F.M., Dresser, P.Q. & Smith, A.G. 1979: Radiocarbon dating evidence of the impact of atmospheric pollution on upland peats. *Nature 282*, 829–831.

Chambers, F.M., Kelly, R.S. & Price, S.-M. 1988: Development of the late-prehistoric cultural landscape in upland Ardudwy, north-west Wales. *In* Birks, H.H., Birks, H.J.B., Kaland, P.E. & Moe, D. (eds) *The cultural landscape—past, present, future*, 333–348. Cambridge University Press, Cambridge.

Countryside Commission 1983: *What future for the uplands*? Countryside Commission, Cheltenham.

Godwin, H. 1975: *History of the British flora*, 2nd edn. 541 pp. Cambridge University Press, Cambridge.

Godwin, H. & Mitchell, G.F. 1938: Stratigraphy and development of two raised bogs near Tregaron, Cardiganshire. *New Phytologist 37*, 425–454.

Haslam, C.J. 1987: *Late-Holocene peat stratigraphy and climatic change—a macrofossil investigation from the raised mires of north-western Europe*. Unpublished PhD thesis, University of Southampton.

Hibbert, F.A. & Switsur, V.R. 1976: Radiocarbon dating of Flandrian pollen zones in Wales and Northern England. *New Phytologist 77*, 793–807.

Hyde, H.A. 1936: On a peat bed at the East Moors, Cardiff. *Transactions of the Cardiff Naturalists' Association 69*, 39–42.

Jacobi, R.M. 1980: The early Holocene settlement of Wales. *In* Taylor, J.A. (ed.) *Culture and environment in prehistoric Wales*, 131–206. British Archaeological Reports, British Series 76, Oxford.

Jones, R., Benson-Evans, K. & Chambers, F.M. 1985: Human influence upon sedimentation in Llangorse Lake, Wales. *Earth Surface Processes and Landforms 10*, 227–235.

Lageard, J.G.A., Chambers, F.M. & Thomas, P.A. 1992: Palaeoforest reconstruction from peat exhumations at White Moss, South Cheshire, UK. *In* Bartholin, T.S., Berglund, B.E., Eckstein, D. & Schweingruber, F.H. (eds) *Tree rings and environment, Proceedings of the International Dendrochronological Symposium, Ystad, South Sweden (1990)*, 172–176. Lundqua Report Vol. 34, Dept Quaternary Geology, Lund University, Lund.

Mighall, T. & Chambers, F.M. 1993: The environmental impact of prehistoric mining at Copa Hill, Cwmystwyth, Wales. *The Holocene 3*, 260–264.

Moore, P.D. 1973: The influence of prehistoric cultures upon the initiation and spread of blanket bog in upland Wales. *Nature 241*, 350–353.

Moore, P.D. 1975: Origin of blanket mires. *Nature 256*, 267–269.

Moore, P.D. 1981: Neolithic land use in mid-Wales. *Proceedings of the 4th International Palynological Conference (Lucknow) 3*, 279–290.

Moore, P.D. 1988 The development of moorlands and upland moors. *In* Jones, M. (ed.) *Archaeology and the flora of the British Isles*, 116–122. Oxbow Books, Oxford.

Smith, A.G. (with contributions from Girling, M.A., Green, C.A., Hillman, G.C. & Limbrey, S.) 1991: Buckbean Pond: environmental studies. *In* Musson, C.J. (ed., assisted by Britnell, W.J. & Smith, A.G.) *The Breiddin Hillfort: a later prehistoric settlement in the Welsh Marches*, 91–107. Council for British Archaeology, *Research Report 76*, London.

Smith, A.G. & Cloutman, E.W. 1988: Reconstruction of Holocene vegetation history in three dimensions at Waun-Fignen-Felen, an upland site in South Wales. *Philosophical Transactions of the Royal Society of London B 332*, 159–219.

Smith, A.G. & Morgan, L. 1989: A succession to ombrotrophic bog in the Gwent Levels, and its demise: a Welsh parallel to the peats of the Somerset Levels. *New Phytologist 112*, 145–167.

Smith, R.T. & Taylor, J.A. 1989: Biopedological processes in the inception of peat formation. *International Peat Journal 3*, 1–24.

Tallantire, P.A. 1992: The alder (*Alnus glutinosa* (L.) Gaertn.) problem in the British Isles: a third approach to its palaeohistory. *New Phytologist 122*, 717–731.

Tansley, A.G. 1939 (reprinted 1949): *The British Islands and their vegetation*. Vol. 2. 930 pp. Cambridge University Press, Cambridge.

Taylor, J.A. 1980: Environmental changes in Wales during the Holocene period. *In* Taylor, J.A. (ed.) *Culture and environment in prehistoric Wales,* 101–130. British Archaeological Reports, British Series 76, Oxford.

Taylor, J.A. 1987: *Timescales of environmental change.* University College of Wales, Aberystwyth.

Taylor, J.A. & Smith, R.T. 1980: The role of pedogenic factors in the initiation of peat formation and in the classification of mires. *Proceedings of the 6th International Peat Congress (Duluth),* 109–118.

Turner, J. 1964: The anthropogenic factor in vegetational history. I. Tregaron and Whixall Mosses. *New Phytologist 63,* 73–90.

Walker, M.J.C. 1982: Early- and mid-Flandrian environmental history of the Brecon Beacons, South Wales. *New Phytologist 91,* 147–165.

Walker, M.J.C. 1993: Flandrian vegetation change and human activity in the Carneddau area of upland mid-Wales. *In* Chambers, F.M. (ed.) *Climate change and human impact on the landscape,* 169–183. Chapman & Hall, London.

Walker, M.J.C. & Harkness, D.D. 1990: Radiocarbon dating the Devensian Lateglacial in Britain: new evidence from Llanilid, South Wales. *Journal of Quaternary Science 5,* 135–144.

Walker, M.J.C., Griffiths, H.I., Ringwood, V. & Evans, J.G. 1993: An early-Holocene pollen, mollusc and ostracod sequence from lake marl at Llangorse Lake, South Wales, UK. *The Holocene 3,* 138–149.

Wymer, J.J. 1977: *A gazetteer of Mesolithic sites in England and Wales.* 511 pp. Council for British Archaeology Research Report No. 22, Geo Abstracts & CBA, London.

Additional bibliography

Type-region accounts are based on data and information in the *National Atlas of Wales*, published by University of Wales Press, Cardiff. For full bibliography of palynological sites in Wales, see Caseldine (1990).

4

Great Britain—Scotland

H.J.B. BIRKS

With contributions by H.H. Birks, P.D. Kerslake, S.M. Peglar and W. Willams

INTRODUCTION

Scotland contains the largest tract of mountainous terrain, the highest peak (Ben Nevis 1344 m), the oldest rocks, and the most extensive areas of natural and semi-natural vegetation in the British Isles. It is an area of very considerable geologic and topographic and hence edaphic, floristic, and vegetational diversity. Many low-lying areas, particularly in the extreme west, rise steeply to high mountain peaks (1000 m or more) over very short distances.

Scotland consists, geologically and topographically, of three major areas resulting from crustal movements. These are the Highlands and Islands, the Midland Valley, and the Southern Uplands. The Highlands and Islands are by far the largest area and include the whole of the country north of the Highland Boundary Fault (Fig. 4.1). The Midland Valley lies to the south of the Highland Boundary Fault and north of the Southern Uplands Boundary Fault (Fig. 4.1). The Southern Uplands comprise the area between the Southern Uplands Boundary Fault and the border between England and Scotland.

Geology and topography

The Highlands and Islands consist primarily of a block of pre-Palaeozoic schist and gneiss overlain in places by younger rocks and with intrusions of several ages (Fig. 4.2). The major pre-Palaeozoic rock types are: (1) Lewisian gneiss forming the lowest member in northwest Scotland as the oldest rocks in the country; (2) Moine gneiss and schist occupying a large part of Scotland from the northeast almost to the southern edge of the Highlands, and from the large Moine Thrust Plane in the northwest eastwards to the Eastern Highlands; (3) Dalradian schist, quartzite, gneiss, and metamorphosed limestone in the southern and central parts of the area; (4) Torridonian sandstone, conglomerate, and shale confined to the northwest as a strip about 30 km wide and 185 km long.

Small areas of Cambrian and early Ordovician rocks are restricted to the northwest and consist of limestone, quartzite, and grit. Many igneous intrusions occur within the Highlands and Islands, probably dating to a time earlier than the lowest Devonian Old Red Sandstone. These intrusions are mainly biotite and hornblende granite and quartz diorite, as well as small areas of more acid and ultrabasic rocks. Extensive areas of Old Red Sandstone occur in the northeast and the southeast as well as in the Orkneys and Shetland. The only other rocks in the Highlands and Islands that occupy large areas are the Tertiary olivine basalt, gabbro, and granite in the west, particularly the Inner Hebrides, and some small areas of Permo-Triassic New Red Sandstone and Jurassic sedimentary rocks.

Topographically the Highlands and Islands consist of mountains up to 1344 m and narrow steep-

Palaeoecological Events During the Last 15 000 Years: Regional Syntheses of Palaeoecological Studies of Lakes and Mires in Europe.
Edited by B.E. Berglund, H.J.B. Birks, M. Ralska-Jasiewiczowa and H.E. Wright. © 1996 John Wiley & Sons Ltd.

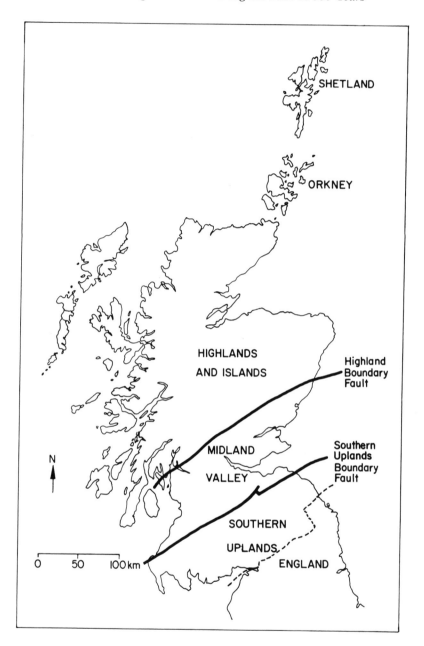

Fig. 4.1   Scotland showing the three major topographic areas (Highlands and Islands, Midland Valley, and Southern Uplands), the Highland Boundary Fault, the Southern Uplands Boundary Fault, and the border with England

sided valleys. Striking contrasts exist between the west and east coasts due to geologic structure and erosion. The west coast has innumerable islands and is indented by narrow lochs, many of which are fjords. In contrast, the east coast has few off-shore islands and is almost entirely unbroken; the gentle eastwards slope grades to a low-lying coastal plain in the east and northeast.

The Midland Valley is a generally low-lying area of subdued topography with some hills and gentle

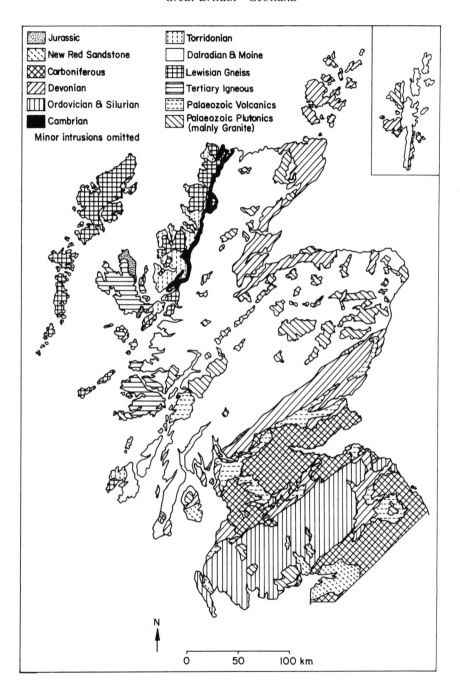

Fig. 4.2 Geologic map of Scotland showing the principal rock types

upland areas. It lacks any pre-Palaeozoic rocks and consists primarily of Devonian Old Red Sandstone, Carboniferous sandstone, limestone, and volcanic basalt and trachyte, and Permian New Red Sandstone. The Old Red Sandstone consists of sandstone, conglomerate, and shale, and within the Lower Old Red Sandstone there are several volcanic rocks and intrusions, including lava, hornblende granite, and quartz diorite.

The Southern Uplands has a distinctive structure, being composed of Ordovician and Silurian shale, greywacke, mudstone, and grit forming a plateau between 300 and 600 m elevation. The rocks are intensively folded but dip generally to the southeast. At the northern edge of the area, where the oldest rocks occur, there are locally areas of Old and New Red Sandstone and Carboniferous rocks. In the southwest several granite intrusions form tops over 800 m elevation in the Galloway Hills. Topographically the area is generally similar to the Midland Valley, and these two areas are often grouped together as the "Lowlands" of Scotland in contrast to the "Highlands" to the north and west.

Further details of the geography and topography of Scotland are given by Fitzpatrick (1964), Sissons (1967), Price (1976), Whittow (1977), and Murray (1987).

Climate

The climate of Scotland is determined by latitude and by its position between a great ocean and a large continental mass. Its notoriously changeable weather results from the meeting of continental and oceanic and of polar and tropical air masses. This meeting leads to a rapid succession of fronts and depressions driven by the prevailing westerly winds from across the warm waters of the North Atlantic Drift. Western Scotland and the Western Islands therefore experience an extremely humid but mild climate. On the west coast, where moist winds are forced upwards by the high mountain ranges, the annual rainfall on the coast is about 150 cm increasing to about 300 cm in the coastal mountains. The summit of Ben Nevis receives about 400 cm rain per year whereas only 6 km away to the west at sea-level Fort William

has only 195 cm rain per year. In contrast, the east coast of Scotland generally receives about 75 cm rain per year. More important ecologically than total precipitation (Ratcliffe 1968) is the number of wet days per year (measurable rain >1 mm): 220 or more a year in many areas in the west (Fig. 4.3), resulting in a constantly moist atmosphere. The prevalance of hard acid rocks in such an oceanic climate results in the landscape of western Scotland being dominated by acid peaty soils and extensive blanket bogs.

The Atlantic currents have a warming influence along the whole western seaboard (Fig. 4.4). The winters do not show great extremes of cold in the west and are much less severe than in eastern Scotland. Summers, however, are distinctly cooler, with a difference of 0.5–1°C at the same latitude between the west and east coasts (Fig. 4.5). Extreme windiness is another feature of the oceanic climate, and the west coast has many severe gales, especially during winter. The decrease in average summer temperature from south to north (Fig. 4.5) is in addition to the major west–east gradient.

These climatic factors and gradients interact so that the overall climate becomes increasingly unfavourable for plant growth in a northwesterly direction. The combined effects of increasing windiness, moisture, and lack of summer warmth are shown most clearly by the downward shift of the upper and lower limits of the natural vegetation zones. Although only represented today by fragments, the upper limit of woodland falls from over 600 m in the Cairngorms in the Eastern Highlands to about 300 m in the northwest. The lower limit of dwarf-shrub alpine heath drops even more, from about 760 m to only 300 m, whereas on the most storm-swept coasts of the northwest these heaths occur almost at sea-level (McVean and Ratcliffe 1962).

Further details of the climate of Scotland are in McVean and Ratcliffe (1962), Green (1964), Ratcliffe (1968), Birse and Dry (1970), Birse and Robertson (1970), and Birse (1971).

Soils

In an area as geologically, topographically, and climatically diverse as Scotland, a considerable range

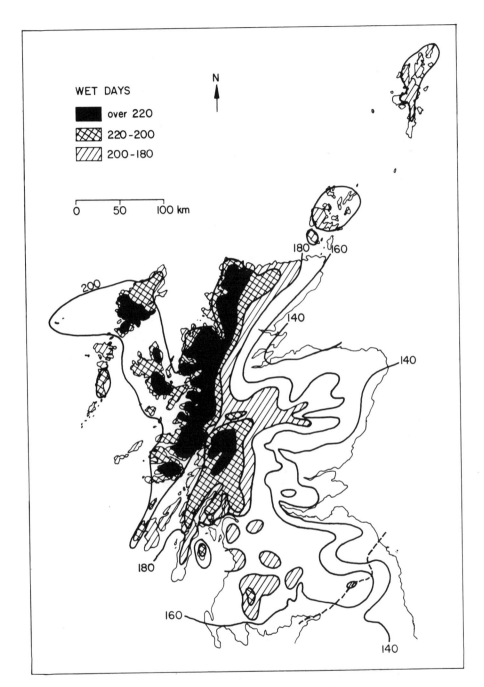

Fig. 4.3 Number of wet days (a day with at least 1 mm rain) per year in Scotland (modified from Ratcliffe 1968)

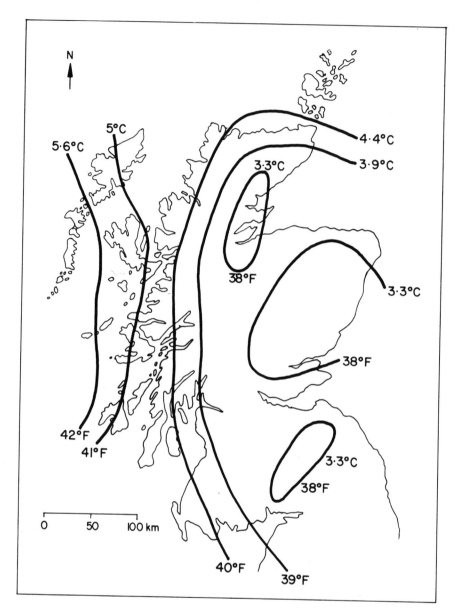

Fig. 4.4 Isotherms of mean January temperatures at sea-level for Scotland (excluding Shetland) (modified from Murray 1987)

of soil types is to be expected (Fitzpatrick 1964). Some general tendencies in soil development, however, can be discerned (McVean and Ratcliffe 1962).

Over almost all of Scotland the preponderance of acidic rocks gives rise to soils with low availability of plant nutrients. Fertile, base-saturated soils are locally associated with limestone, dolomite, calcareous schist, calcareous basalt, and coastal shell-sand. Hard siliceous rocks such as granite and quartzite tend to produce shallow coarse-textured and porous soils, whereas impermeable clayey soils are associated with soft argillaceous and calcareous rocks

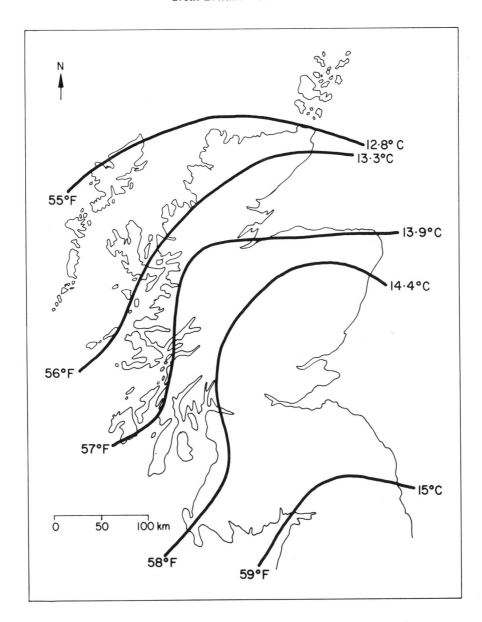

Fig. 4.5 Isotherms of mean July temperatures at sea-level for Scotland (excluding Shetland) (modified from Murray 1987)

in the Midland Valley and with low-lying areas of the Southern Uplands and the Eastern Highlands.

Under the cool wet climate of the west, podsolization is the dominant pedogenic process. In combination with the extensive occurrence of acidic rocks it is responsible for the prevalence of base-poor soils. The associated tendency for raw acid humus and peat development is widespread, especially towards the west coast and along the gradient of increasing climatic wetness.

The prevalent soil types are acidic peat, podsol, brown earth, and gley. Calcareous loams of rendzina affinity, alpine rankers, and skeletal and solifluction soils are much rarer. A general pedogenic uniformity of climate is apparent, and differentiation within the groups is largely due to topography and bedrock (McVean and Ratcliffe 1962). Podsols and brown earths vary altitudinally, whereas peats and gleys tend to be similar at different elevations.

Potential natural forest vegetation

Scottish forests have undergone considerable destruction over the last 5000 years. The extent of this is clear from McVean and Ratcliffe's (1962) maps of the actual and potential distributions of forest in Scotland. The potential distribution is summarized in Fig. 4.6, redrawn from Map 3 in McVean and Ratcliffe (1962). It is based on the occurrence of existing natural and semi-natural woodlands, on surviving woodland fragments on ungrazed islands in lochs, in ravines and other steep places, and on historical records and place-name evidence. Bennett (1989) has prepared a map of possible forest types at 5000 **BP** for the whole of the British Isles. It corresponds, in general terms, with the potential forest regions of McVean and Ratcliffe (1962; Fig. 4.6).

South and west Scotland are potentially areas of *Quercus petraea* forest with *Betula pubescens*. The scarcity of relict pine and derived *Calluna vulgaris* moorland suggests that *Pinus sylvestris* was unimportant in the natural forests. Existing woodlands consist mainly of *Q. petraea* and *B. pubescens*. On basic soils *Ulmus glabra* and *Fraxinus excelsior* also occur, with *Corylus avellana* in more open areas. *Corylus* is not restricted to rich soils, and birch and hazel frequently form extensive stands where oak has been removed. Other native trees and shrubs in southern and western Scottish woods today include *Alnus glutinosa*, *Hedera helix*, *Ilex aquifolium*, *Lonicera periclymenum*, *Populus tremula*, *Prunus padus*, *P.spinosa*, *Salix cinerea*, *S.caprea*, *Sorbus aucuparia,* and *Viburnum opulus*. Many of these woods occur on steep boulder-strewn slopes, and in the moist, oceanic climate of the extreme west bryophytes, lichens, and filmy ferns luxuriate on boulders and tree trunks. Several species with Macaronesian–Tropical distributions reach their

northernmost world localities in western Scotland (Ratcliffe 1968).

The Central and Eastern Highlands and the east coast of the Highland area are potentially areas of *Pinus sylvestris*-dominated forest, often with *Betula pendula* and with *Quercus petraea* and *Q.robur* in favourable low-lying south-facing sites. Pinewoods extend locally into the North-West Highlands, where they have abundant tall *Juniperus communis* and *Calluna vulgaris*, thick carpets of bryophytes, and some *Ilex aquifolium*, *Sorbus aucuparia*, *Prunus padus,* and *Betula* spp. in the canopy.

Birchwoods (mainly *B.pendula*) are widespread in central Scotland, locally on fertile soils but also in openings in the acid pinewoods. Clearance of pine or birch forest followed by grazing and/or burning has lead to *Calluna vulgaris*-dominated moorland or species-poor *Agrostis–Festuca* grassland.

The elevation and composition of the natural tree-line in Scotland are unknown, due to the virtual elimination of subalpine scrub communities by clearance, burning, and grazing. The potential tree-line in the Central Highlands may be about 750 m (Spence 1960), 400 m in the North-West Highlands, and descending to about sea-level in the Shetlands, associated with the progressively more severe climate at equivalent altitudes.

Northern Scotland and the Western Isles are potentially areas of birch–hazel forest. Exposure prevents the growth of nearly all trees except birch and hazel, with alder and willows in wet habitats. Birch woods (mainly *Betula pubescens*) occur locally in steep block-strewn areas, with *Ilex aquifolium*, *Sorbus aucuparia*, *Prunus padus*, *Salix cinerea,* and *Lonicera periclymenum*. *Quercus petraea* may occur very locally. *Corylus avellana*-dominated stands tend to occur in wind-exposed areas protected to some degree from grazing animals, such as on talus below steep sea-cliffs. *Fraxinus excelsior* occurs locally on shallow limestone soils and *Alnus glutinosa* is locally widespread on basic waterlogged or flushed soils.

On the exposed coasts of the far north, parts of Orkney and Shetland, and many of the smaller Hebridean islands, closed forest may not have developed extensively. *Betula pubescens*, *Corylus avellana*, *Salix* spp., *Populus tremula,* and *Sorbus aucuparia*

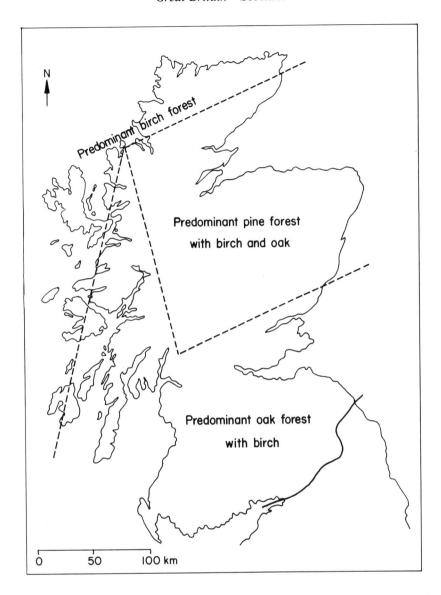

Fig. 4.6 Potential natural forest regions in Scotland (modified from McVean and Ratcliffe 1962)

may have formed stands in sheltered areas, possibly with some oak and alder. Coastal grasslands and heaths may have predominated in the most exposed coastal areas.

The phytosociology of the present-day woodlands and other vegetation types in Scotland is described by McVean and Ratcliffe (1962), Birks (1973), Birse (1980, 1984), and Robertson (1984).

Land use

Climate and soils have a strong influence on the type and extent of land use. The most favourable areas for farming are in the lowlands of the south and east, where the summer warmth and sunshine give a long growing season (mean annual accumulated temperatures over 6°C are in excess of 833; Green 1964) and the low rainfall limits soil leaching and peat formation. In these areas arable farming ascends the lower foothills to at least 300 m. In some areas *Calluna* moors are adjacent to fields of oats and barley. Root crops grow well, and rich grasslands support dense cattle stocks (Ratcliffe 1977a).

In the north and west little suitable ground may be available for arable farming, as the mountains descend steeply to the sea lochs. Even where the coastal areas are less steep they may be rugged, with much bare rock, open water, or blanket peat. Only in favourable situations, for example on river alluvium or near coastal shell sand, has farming with the cultivation of oats and potatoes been possible.

The uplands are heavily exploited as sheep walk, and sheep densities are invariably highest on the more basic mountains. In the drier eastern Scotland, large areas of heather moorland are managed by rotational burning for red grouse. Deer forests occur all over the Highlands, particularly on the higher mountain ranges with hard acidic bedrock. Cattle have recently gained some popularity in certain upland areas (Ratcliffe 1977a).

The last 50–60 years have witnessed massive reforestation on the lower hill slopes throughout Scotland, nearly all of alien conifers grown in such dense stands that the intense shade and deep accumulation of acid litter and raw humus exclude almost all plant life except for a few common ferns, bryophytes, and fungi.

Type regions

It is extremely difficult to apply the type-region concept in any mountainous country because "uniform areas as regards geology, morphology, climate and biota" (Ralska-Jasiewiczowa 1986) do not really exist. In western Scotland within an area of 10 km × 10 km, the standard unit for plant recording, there can be an elevational range from sea-level to over 1200 m, a variety of contrasting landforms and geologic types, strong altitudinal climatic gradients, and large differences in flora and vegetation. Recognizing the arbitrariness of imposing any such simple classification of type regions onto a landscape that

varies continually latitudinally, longitudinally, and altitudinally, 15 type regions have been delimited in Scotland (Fig. 4.7) on the basis of geology, climate, and present-day semi-natural and potential vegetational patterns.

The type regions are as follows:

GB-v   South-West Scotland
GB-w   South-East Scotland
GB-x   Western Scotland
GB-y   Grampian Mountains
GB-z   Cairngorms
GB-za   Buchan Lowlands
GB-zb   Southern Outer Hebrides
        (Uists and Barra)
GB-zc   Isle of Skye
GB-zd   Ross and Cromarty
GB-ze   Great Glen
GB-zf   Northern Outer Hebrides
        (Lewis and Harris)
GB-zg   North-West Scotland
GB-zh   Central Sutherland
GB-zi   Caithness Lowlands and Orkneys
GB-zj   Shetland

Regions GB-v and GB-w represent the western and eastern areas south of the Highland Boundary Fault, which separates the Highlands and Islands on the north and the Midland Valley and Southern Uplands (the "Lowlands") on the south. Within the Highlands and Islands the major climatic and vegetational patterns run from west to east (mainly winter climatic parameters). Within the extreme oceanic west, the major gradient of decreasing summer warmth is reflected in the south-to-north series of type regions GB-x, GB-zd, and GB-zg. Distinct geologic and topographic areas are distinguished in the east, where regional-scale climatic gradients are less pronounced, with type regions GB-y, GB-z, GB-za, GB-ze, and GB-zh corresponding to areas of relatively distinct geology and topography. The major island groups comprise type regions Gb-zc, GB-zb, GB-zf, GB-zi, and GB-zj. Type region GB-zi includes part of the Scottish mainland, whereas GB-x includes the southern Inner Hebridean islands and southern and eastern Skye.

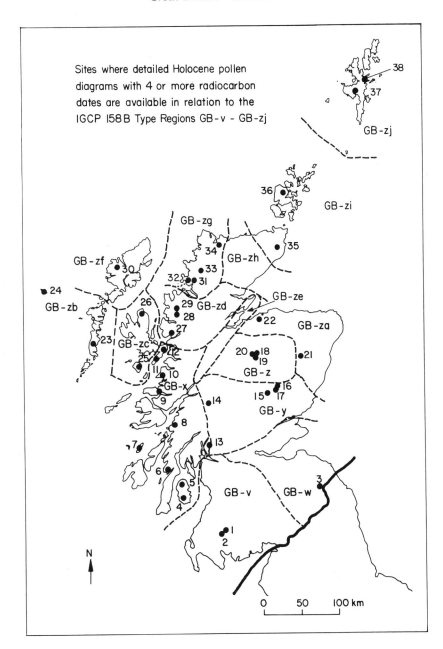

Sites where detailed Holocene pollen
diagrams with 4 or more radiocarbon
dates are available in relation to the
IGCP 158B Type Regions GB-v - GB-zj

Fig. 4.7 IGCP Project 158B type regions in Scotland (GB-v–GB-zj). The site numbers refer to biostratigraphic
Holocene reference sites. These are listed in the Appendix

Available palaeoecological data

Despite the abundance of lakes and bogs in Scotland, comparatively little is known about the detailed vegetational and environmental history of any of the type regions in Scotland. This contrasts with the enormous amount of information available from parts of England which results in large part from the historical development of Quaternary palynology and palaeoecology in Britain, with its original geographical distribution centred on East Anglia and Somerset and subsequently on the English Lake District. An advantage of this early neglect of Scotland by pollen analysts is that many studies in Scotland are comparatively recent and hence have involved modern techniques such as ^{14}C-dating and identification of herb pollen types.

There are now over 50 published Holocene and Weichselian (= Devensian) late-glacial pollen diagrams with associated radiocarbon dates from all the type regions in Scotland except for GB-zh and GB-ze (see Birks 1989), although for some of the type regions (e.g. GB-zj, GB-zf, GB-zb, GB-w, GB-zi) the number of detailed well-dated pollen diagrams is only one or two. Very few type regions exist where both lakes and bogs have been studied, where palaeoecological investigations other than pollen-analytical studies have been performed, and hence where any useful attempt at synthesizing the environmental history of a type region can be made. The known Weichselian late-glacial and Holocene vegetational history, based on pollen analysis, has been synthesized several times in different ways by, for example, Birks (1977, 1988, 1989), Bennett (1984, 1989), and Walker (1984a). Rather than repeating these syntheses the attempt here is to provide consistent syntheses of the environmental history of these type regions that have (a) several ^{14}C-dated pollen diagrams from both lakes and bogs, (b) palaeoecological studies on other groups of fossils such as diatoms and plant macrofossils, and (c) geomorphic and stratigraphic studies on the onset of important pedogenic processes.

Syntheses are presented for four palaeoecological reference areas within type regions Gb-v (Galloway Hills within South-West Scotland), GB-z (Cairngorm Mountains), GB-x and GB-zc (Isle of Skye and part of Western Scotland), and GB-zd (the Loch Maree area within Ross and Cromarty).

TYPE REGION GB-v *pro parte*, GALLOWAY HILLS (H. J. B. Birks with contributions from Hilary H. Birks and Sylvia M. Peglar)

Galloway is that part of southwest Scotland lying west of Dumfries and south of Ayr. Although predominantly a low-lying region, the central part of Galloway attains an elevation of 842 m in the so-called Galloway Hills. The Hills are of considerable palaeoecological interest because (1) fossil pine stumps commonly occur in the blanket peats (Lewis 1905) although pine is absent from the modern native flora and vegetation of Galloway and (2) several lochs (lakes) in these Hills provided the first unambiguous palaeolimnological evidence for recent surface-water acidification in the United Kingdom (Flower & Battarbee 1983; Battarbee *et al.* 1985a). Several palaeoecological studies have been completed in the Galloway Hills involving pollen analysis, macrofossils, peat and charcoal stratigraphy (Moar 1969; Birks 1972a, 1975; Edwards 1989, 1990; Edwards *et al.* 1991), and palaeolimnology (Flower & Battarbee 1983; Battarbee *et al.* 1985a, 1985b, 1988, 1989; Jones *et al.* 1986, 1989, 1990, 1993a; Flower *et al.* 1987; Birks *et al.* 1990; Stevenson *et al.* 1990, Allott *et al.* 1992; Brodin & Gransberg 1993). It is one of the most intensively studied areas for Holocene environmental history in Scotland. This account attempts to synthesize these investigations.

Altitude: 230–842 m. Much of the area lies between 350 and 650 m.

Climate: Mean January temperature 6.7°C; mean July temperature 19°C; corresponding minima are 2°C and 11°C; precipitation (mainly as rain) varies from 178 cm yr^{-1} at 305 m elevation to more than 254 cm yr^{-1} on the summits; 200–220 rain days yr^{-1} (a day with at least 0.2 mm rain) and 165–180 wet days yr^{-1} (a day with at least 1 mm rain); high average relative humidity and a high incidence of cloud, often occurring as fog on the Hills; snowfall is relatively unimportant, with only about 20 days yr^{-1} with snow falling between December and mid-April; winds are predominantly from the west, and gales are frequent on the summits; average daily sunshine rarely exceeds five hours even in mid-summer; mild oceanic climate.

Due to topography and the prevailing wind direction, the eastern Hills are generally drier and sunnier than the western Hills.

Geology: Ordovician and Silurian shale, greywacke,

and grit metamorphosed at their contact with the central granite intrusions.

Topography: Rugged, rocky, and heavily influenced by glaciation. The highest mountain (Merrick 842 m) and adjacent hills are all craggy, with impressive corries, extensive cliffs, numerous boulder fields and unfissured granite pavements with numerous perched blocks, and some screes, separated by glaciated valleys lined with uneven mounds of lateral moraine. The Ordovician and Silurian hills are relatively smooth in outline, but corries have also been carved from them, resulting in some precipitous cliffs and screes.

Population: Generally at a low density, even for Scotland, with most areas at much less than 1 person ha^{-1}. Concentrated in two or three towns around the Hills, with a few scattered farmsteads in the Hills. The population has decreased nearly 20% in the last 30 years.

Vegetation: Almost totally devoid of native trees. Blanket-bog communities widespread, except where replaced by extensive plantations of exotic conifers, mainly *Picea sitchensis*. Where water movement exists, *Molinia caerulea* communities are dominant, sometimes with *Myrica gale*. Burning has favoured the almost total dominance of *Molinia* at the expense of *Calluna vulgaris*. On steeper, well-drained slopes, especially on the Silurian and Ordovician hills, *Agrostis–Festuca* grasslands predominate with abundant *Pteridium aquilinum* and *Thelypteris limbosperma*. *Calluna* heaths are rare, probably due to centuries of grazing and burning (Stevenson & Thompson 1993). Summit plateaux are covered by *Nardus stricta–Juncus squarrosus* grassland, fragmentary *Racomitrium lanuginosum* heath, or open boulder communities with little or no soil. Patterned bogs with well-developed pools and hummocks survive in The Silver Flowe National Nature Reserve (Ratcliffe & Walker 1958).

The once remote and wild appearance of the Galloway Hills has been drastically altered by the extremely extensive planting of exotic conifers in the last 30–40 years, even in the remote central parts of the Hills.

Soils: Mildly acidic brown earths in locally favourable localities; leached podsols widespread; extensive peat accumulation in poorly drained hollows and on most gentle slopes and plateaux. Most deep (>2 m) blanket peats are currently undergoing erosion.

Land use: Commercial forestry predominant, with some sheep-walk. Red deer and feral goats present. Moor-burning and grouse-moor formerly practised locally.

Reference site 1. Loch Dungeon (Birks 1972a)

Latitude 55°8′N, Longitude 4°19′W. Elevation 305 m. Age range *ca.* 10200–150 BP. Lake. No ^{14}C-dates but age estimates are based on pollen-stratigraphic correlations with ^{14}C-dated pollen sequences (Birks 1975; Jones *et al.* 1989). 383 cm sediment. Six local pollen-assemblage zones (paz) and five regional pollen-assemblage zones represented (Fig. 4.8).

Local paz	Approximate age	Regional paz
LD-1	10200–*ca.* 9500 BP	Gramineae–*Salix*
LD-2	9500–8700 BP	*Betula–Corylus/ Myrica*
LD-3	8700–7500 BP	*Ulmus–Quercus*
LD-4	7500–5000 BP	*Alnus–Ulmus*
LD-5 +	5000–150 BP	*Alnus–Quercus–*
LD-6		*Plantago lanceolata*

An additonal uppermost regional paz (*Pinus–Fagus*, 0–150 BP) is defined by Birks (1972a) from nearby Snibe Bog. The top 20 cm at Loch Dungeon was not sampled.

The chronology used here is based entirely on conventional bulk-sample ^{14}C assays. AMS dates from chironomid remains from the Round Loch of Glenhead suggest that conventional ^{14}C dates from that site (Jones *et al.* 1993a) may be 400–1300 ^{14}C years older than the AMS dates from the equivalent horizons.

Event stratigraphy   This is presented in Figure 4.9.

Discussion

(1) No Weichselian (= Devensian) late-glacial pollen sequences are known from the Galloway

108

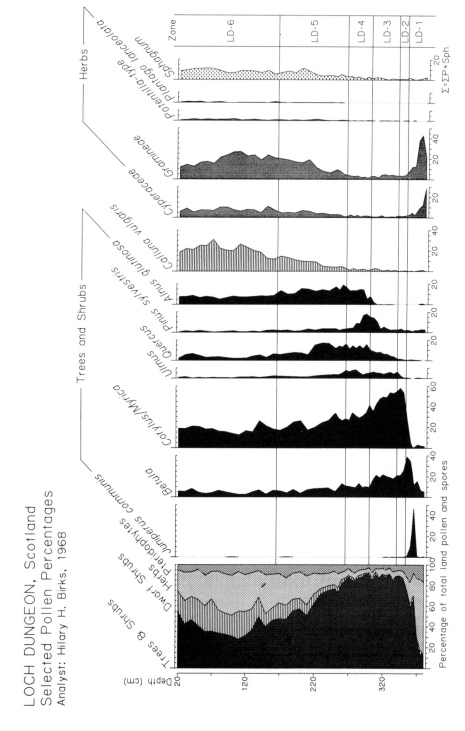

Fig. 4.8 Summary pollen diagram from Loch Dungeon, Galloway. Selected major pollen and spore taxa only are shown. The complete diagram is in Birks (1972a). Sph = *Sphagnum*

TYPE REGION   GB - v   pro parte   GALLOWAY HILLS

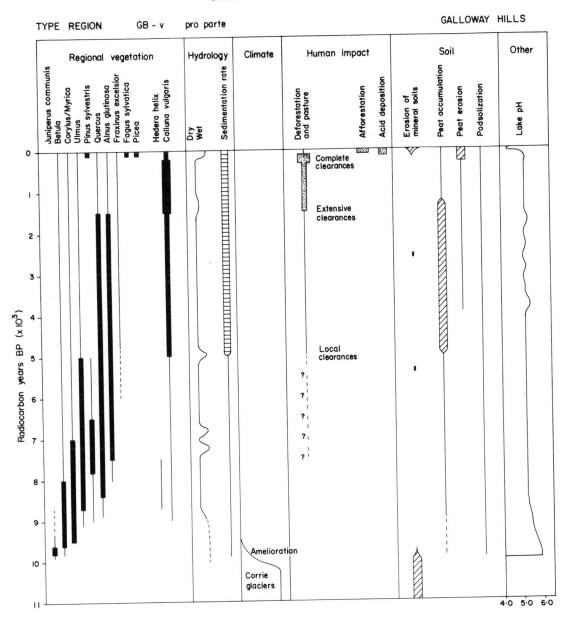

Fig. 4.9 Event stratigraphy for the Galloway Hills

Hills, presumably because of local corrie glaciation during the Loch Lomond (=Younger Dryas) Stadial (Cornish 1981) and because of extensive periglaciation. Moar (1969) provides palynological evidence to indicate that the last corrie glaciation occurred during the Younger Dryas. Cornish (1981) maps the extent of the 11

Younger Dryas corrie glaciers. Lowest ice elevation was 250 m (range = 250–550 m), equilibrium firn line ranged from 328 to 670 m, and glacier area from 0.12 to 3.18 km².

(2) The Galloway Hills are mapped within the potential "oak forest with birch" forest region by McVean & Ratcliffe (1962). The Holocene

pollen data (Fig. 4.8) confirm this, with abundant pollen of *Betula*, *Ulmus*, *Quercus*, *Alnus*, and *Corylus/Myrica* (largely *Corylus avellana*) prior to 5000 BP (LD-4). Mixed deciduous forests must have dominated in the mid-Holocene. The forest limit extended to at least 460 m (Birks 1988) and the tree-line probably consisted of *Betula*, *Pinus sylvestris*, *Populus tremula*, and *Sorbus aucuparia*.

(3) *Pinus sylvestris* occurred locally from about 7800 to 6500 BP (LD-4), probably growing on shallow acidic soils and cliffs free from competition from other trees and on dried peat surfaces (Birks 1972a). It declined regionally at about 6500 BP but persisted locally until 5000 BP (Birks 1975). Birks (1989) suggests, on the basis of the occurrence of pine elsewhere in the British Isles at about 7500 BP, that pine may have spread to Galloway not from northern England or western Scotland but from Northern Ireland.

(4) Deforestation and subsequent development and/or expansion of grassland, moorland, and blanket bog began about 5500–5000 BP (LD-5) as a result of Neolithic and Bronze Age activities. The first extensive forest clearances (LD-6) may have begun *ca.* 1500 BP followed by local Norse Viking settlements *ca.* 1100 BP. Woodlands are known, however, to have survived in the Hills until 200–300 BP. Complete deforestation may not have occurred until the 18th century with its "agricultural revolution". Large flocks of sheep were pastured in the Hills in the 19th and early 20th centuries (Birks 1972a), leading to decreases in the extent of *Calluna* moorland (Stevenson & Thompson 1993). Edwards (1989, 1990) discusses palynological and charcoal evidence for possible fine-scale Mesolithic impacts in the Hills, matching the archaeological evidence for Mesolithic penetration (Edwards *et al.* 1984).

(5) Lake-water pH changes during the Holocene have been reconstructed from diatom assemblages at the Round Loch of Glenhead (Jones *et al.* 1986, 1989; Birks *et al.* 1990). The loch was acid (pH 5.4–5.8) even in the closing stages of the Weichselian late-glacial and early Holocene. By 9200 BP pH was 5.2–5.4 and it changed little

(5.3–5.7) in the subsequent 9000 years, except for short-lived fluctuations after 3500 BP due to inwashed material from the catchment. Lake pH never fell below 5.0 until *ca.* 1900 AD, when it dropped to 4.3–4.5. Dissolved organic carbon (DOC) was low throughout the Holocene except when pine grew in the area. DOC values rose with the spread of blanket bog in the catchment from *ca.* 5000 to *ca.* 1350 BP. DOC dropped to its lowest values with recent lake acidification. Total Al, in contrast, greatly increased at *ca.* 1900 AD (Birks *et al.* 1990).

(6) All lakes situated on granite bedrock in the Galloway Hills that have been investigated palaeolimnologically show unambiguous evidence for recent lake acidification, with pH changes from 5.3–6.2 to < 5.0 in the last 100–150 years paralleled by increased concentrations of Pb, Zn, Cu, and carbonaceous "soot" particles (Battarbee *et al.* 1985a, 1989; Flower *et al.* 1987, Jones *et al.* 1990). Investigations designed to test alternative hypotheses for the cause of this recent acidification all indicate the overriding importance of deposition of strong acids ("acid rain") in the last 100–150 years in altering the chemistry of the lakes. Evidence for recent reversibility of acidification following decreased sulphate deposition (Battarbee *et al.* 1988; Allott *et al.* 1992) further supports the acid-deposition hypothesis.

(7) Minerogenic soil erosion occurred during the closing phases of the Weichselian late-glacial (Moar 1969; Birks 1972a; Jones *et al.* 1989). Soil erosion was reactivated following deforestation, leading to deposition of minerogenic layers in peats at various times since *ca.* 5000 BP (Edwards *et al.* 1991). Inwashing of peat and humus into lakes has been occurring since *ca.* 4000 BP (Jones *et al.* 1989; Birks *et al.* 1990). Extensive sheet erosion of peat, common in much of the area today, appears to have been initiated about 200–300 BP, probably due to land-use changes or, more tentatively, to climatic shifts associated with the "Little Ice Age" (Stevenson *et al.* 1990). Minerogenic soil erosion resulting from recent afforestation has increased lake-sediment accumulation rates from

0.1 cm yr^{-1} to over 2 cm yr^{-1} (Battarbee *et al.* 1985b).

(8) The widespread, almost blanket afforestation of the Galloway Hills with exotic conifers in the last 30–40 years is likely to have important permanent ecological impacts, with greater podsolization on the already acid, impoverished soils, greater accumulation of acid raw humus, increased water run-off, increased water sediment load and acidity, thereby exacerbating acid-deposition effects by the enhanced capture of dry deposition by the canopy, and further decreases in the extent of heather moorland, upland grassland, and blanket bog.

TYPE REGION GB-z, CAIRNGORMS (H.J.B. Birks with contributions from Hilary H. Birks and Sylvia M. Peglar)

The Cairngorms in the Eastern Highlands contains the largest area of really high alpine ground in Scotland and the most extensive area of native woodland, of Scots pine (*Pinus sylvestris* var. *scotica*), remaining in Britain. It experiences the most continental climate of any part of Scotland. It is an area of considerable palaeoecological interest because of (1) the native pine forests, (2) the climatic severity at high elevations, suggesting the area may have been particularly sensitive to climatic shifts associated with, for example, the "Little Ice Age", (3) the wide range of natural and semi-natural vegetation types occurring from the lowlands to the high summits; the range is larger here than anywhere else in Scotland, and (4) the extremely varied present-day flora. Several palaeoecological studies have been made in the area involving pollen analysis (Pears 1968a, 1972; Vasari & Vasari 1968; Birks 1970, 1975; O'Sullivan 1973, 1974, 1975a, 1975b, 1976, 1977; Walker 1975; Birks & Mathewes 1978; Macpherson 1980), macrofossils (Vasari & Vasari 1968; Birks 1970, 1975; Birks & Mathewes 1978), peat stratigraphy (Pears 1968a, 1969, 1972, 1975a, 1975b; Birks 1975; Dubois & Ferguson 1985), and palaeolimnology (Jones *et al.* 1993b). This account attempts to synthesize these investigations in conjunction with several geomorphological studies.

Altitude: 170–1309 m. Lower wooded areas 170–640 m, extensive mountain plateaux above 1100 m. Four summits above 1220 m, including the second highest peak in Britain (Ben Macdhui 1309 m).

Climate: Mean January temperature at Braemar (339 m elevation) 0.6°C, mean July temperature 13.1°C; mean annual temperatures are 6.9°C at 305 m, 4.7°C at 762 m, and 2.3°C at 1090 m, with annual ranges of 7.9°C, 5.5°C, and 4.9°C, respectively; days with air frost range from 103 yr^{-1} at 305 m elevation to 196 yr^{-1} at 1090 m and the number of days continuously below freezing point varies from 2 yr^{-1} at 305 m elevation to 83 yr^{-1} at 1090 m; precipitation ranges from 93 cm yr^{-1} at Braemar to more than 150 cm yr^{-1} on the high summits; 225 rain days yr^{-1} (a day with at least 0.2 mm rain) and 160–180 wet days yr^{-1} (a day with at least 1 mm rain) at Braemar; generally rather low (60–70%) average relative humidity and low incidence of cloud; potential water deficit is insignificant on the hills due to low evaporation, but it can become important on low ground; snowfall is relatively important, with about 35 days yr^{-1} with snow falling on low ground, increasing to about 90–100 days yr^{-1} on the summits; snow lies on low ground for about 90 days yr^{-1}, increasing to over 200 on the summits; snow-line (the lowest general level of snow, not necessarily covering half the ground but excluding drifts) at 1296 m usually lies continuously from early January to mid-April for about 100 days; the permanent snow-line is estimated to be about 1615 m; winds are predominantly from the southwest and winds of gale force (average for one hour) are very common on high ground (Green 1974; Pears 1967a, 1967b); the windiness leads to heavy drifting of snow, with drifts up to 20 m in north-facing corries; some depressions remain under snow semi-permanently; growing season is short (0–556 day degrees C) decreasing with altitude; soil temperatures frequently fall below 0°C above 610 m elevation; air temperature decreases 0.72°C 100 m^{-1} (lapse rate); average daily sunshine is 3.25 hours day^{-1}; experiences the most northern continental climate of any part of Scotland.

The broad low-lying valleys within the Cairngorms have less than 90 cm yr^{-1} precipitation due to rain-shadow effects.

Further details of the climate are given by Green (1974).

Geology: Two extensive (360 km², 170 km²) granite plutons intruded into schist, gneiss, quartzite, and limestone. The lower Monadhliath range to the north consists of Precambrian Moine psammitic granulite with small granite intrusions. Small diorite intrusions occur locally. The granite masses are cut by veins and small masses of aplite and pegmatite. See Sugden (1974) for further details.

Topography: Extensive undulating mountain summit plateaux with scattered tors and broad watersheds and gently rolling drift-covered smooth slopes rising from the lowland broad river valleys containing extensive fluvioglacial deposits. The high plateaux are truncated by several magnificent corries with cliffs and boulder-strewn slopes and dissected by a few deeply cut glaciated valleys and troughs with major glacial breaches, steep headwalls, precipitous massive cliffs, and valley moraines. See Sugden (1970, 1974) for further details.

Population: Generally at low density, even for Scotland, with most areas at much less than 1 person ha⁻¹. Concentrated in a few towns around the mountains (Aviemore, Carrbridge, Kingussie, Braemar), with scattered farms and lodges in the mountains.

Vegetation: The potential vegetation below about 640 m is forest, extensive areas of which remain in the Spey and Dee valleys. The most important type is Scots pine forest, and the area supports the largest native forests remaining in Britain. The pine forests show a complete range of variation in age class, growth form, structure, and density. Dense pole stands have uniform age and sparse field vegetation. Younger, fairly even-aged pines surround scattered, older, and more spreading parent trees with a well-developed field layer. Open growths of old, spreading trees or clumps of old trees, form a pine-heath on moorland. *Betula* (mainly *B. pendula*) and *Juniperus communis* are widespread and locally abundant. *Sorbus aucuparia* and *Populus tremula* are frequent.

Birchwoods (mainly *Betula pendula* but with some *B. pubescens*) are widespread, locally on more fertile soils but also recolonizing openings in the pinewoods. *Sorbus aucuparia*, *Populus tremula*, and *Prunus padus* are locally common. The virtual absence of oakwoods and mixed deciduous forests with *Quercus* spp., *Ulmus glabra*, and *Fraxinus excelsior* in the area is very striking. *Alnus glutinosa* occurs locally on damp, richer soils (especially alluvium) and in carr woodland on fen peat.

Pine normally forms the tree-limit in the Cairngorms (Pears 1967b). The upper limits are mostly artificially depressed by grazing and burning, but a presumed natural altitudinal limit still occurs at 640 m (Creag Fhiaclach), where bushy stunted pine mixed with juniper of similar stature grades into heather moor. The present tree-limit is probably maintained there by a balance between strong winds (Pears 1967a, 1967b, 1968b) and browsing by deer (Miller & Cummins 1982). The status of the tree-limit at Creag Fhiaclach is currently under investigation by Jennifer McConnell. Ecological and palaeoecological studies there seriously question the long-held view that this is a natural altitudinal limit. High-altitude birchwoods also occur, for example on calcareous drift at Morrone (Huntley & Birks 1979). Montane willows (e.g. *Salix lapponum*, *S. lanata*) are very rare and are restricted to damp basic soils on ungrazed cliff ledges.

The lower drift-covered slopes support large areas of dry *Calluna vulgaris* moorland, passing on wetter ground to shallow blanket bog. With increasing elevation, *Calluna vulgaris* occurs as a prostrate montane *Calluna* heath rich in lichens. This fades out at about 1000 m. With above-average snow cover, *Vaccinium myrtillus* and *Empetrum nigrum* ssp. *hermaphroditum* heaths occur, along with *Nardus stricta* grasslands. On the highest exposed ground *Juncus trifidus* heath, often with *Carex bigelowii*, *Racomitrium lanuginosum*, and lichens, is widespread. It forms a mosaic with long-lasting or semi-permanent snow patches in sheltered hollows dominated by *Salix herbacea*, mosses, liverworts, and lichens.

The Cairngorms is the richest area for calcifuge and soil-indifferent mountain plants in Scotland and the second richest area for all mountain plants in Britain. See Ratcliffe (1974, 1977b) for further details.

Soils: Soils are acid, with mor humus accumulation in the lowlands. Mainly formed on coarse, sandy, gravelly drift derived from granite, with a local admixture of schistose material. Markedly base-deficient. Podsol profiles occur widely in the forest zone. At higher elevations, profiles are commonly truncated and skeletal soils on exposed ground are often affected by periglacial activity. Deep accumulations of blanket peat are rare, probably because of the porosity of granitic soils. Moine Mhor, at over 900 m, is probably the highest blanket mire in Britain. Soils in the Monadhliaths tend to be less well drained than those on the Cairngorm granites, and on several plateaux at about 610 m there are deep accumulations of peat, much of which is currently undergoing erosion.

Land use: Nature conservation, deer forest and stalking, grouse moor and shooting, tourism (walking, skiing), some commercial forestry, and sheep farming.

Reference site 18. Abernethy Forest (Birks 1970; Birks & Mathewes 1978)

Latitude 56°14′N, Longitude 3°43′W. Elevation 220 m. Age range 12200–0 BP. Bog overlying infilled channel between Loch Garten and Loch Mallachie. Seven ^{14}C dates. 550 cm sediment. Seven local pollen-assemblage zones (paz) and seven regional pollen-assemblage zones represented. The pollen diagrams presented cover the Weichselian (=Devensian) late-glacial and early to mid-Holocene (12200–5600 BP; zones AF-1– AF-6 *p.p.*; Birks & Mathewes 1978) (Fig. 4.10) and the mid-Holocene to present day (*ca.* 7700 –0 BP; zones AF-5 *p.p.*, AF-6, AF-7; Birks 1970) (Fig. 4.11). The diagrams are from two separate but nearby cores.

Local paz	Approximate age	Regional paz
AF-1	>12200–11600 BP	*Rumex–Salix*
AF-2	11600–11200 BP	*Empetrum*
AF-3	11200–9700 BP	*Artemisia*
AF-4	9700–8700 BP	*Betula–Juniperus*
AF-5	8700–7200 BP	*Betula–Corylus/ Myrica*
AF-6	7200–*ca.*1000 BP	*Pinus*
AF-7	*ca.*1000–0 BP	*Calluna–Plantago lanceolata*

Event stratigraphy. This is presented in Figure 4.12.

Discussion

(1) The Weichselian (= Devensian) late-glacial vegetational history is well established for the area as a result of pollen-analytical (Vasari & Vasari 1968; O'Sullivan 1974; Walker 1975; Birks & Mathewes 1978; Macpherson 1980) and plant macrofossil (Vasari & Vasari 1968; Birks & Mathewes 1978) studies. The ^{14}C datings, however, are open to question (e.g. Sissons & Walker 1974; Clapperton *et al.* 1975; Vasari 1977; Birks & Mathewes 1978) as all the available ^{14}C dates are based on conventional ^{14}C dating of bulk samples of lake sediment. Such samples are liable to a variety of ^{14}C dating errors.

The main vegetational development (Fig. 4.10) was initiated by a pioneer *Salix*–grass–sedge vegetation with abundant *Rumex acetosella* and a range of other open-ground herbs such as *Draba norvegica*-type, *Veronica*, and *Thalictrum*. This developed into a dwarf-shrub-tundra dominated by *Empetrum* and *Betula nana* with some *Salix* and *Juniperus communis* as a result of climatic amelioration during Allerød time. Climatic deterioration during Younger Dryas time led to the formation of species-rich open vegetation in which *Artemisia* may have been very abundant, along with a range of open-ground taxa such as *Silene acaulis*, *Cerastium alpinum*, *Minuartia rubella*, *Armeria maritima*, *Saxifraga cespitosa*, *S. oppositifolia*, and *Luzula spicata*.

The extent of glacial ice in the Cairngorms during the Loch Lomond (= Younger Dryas) Stadial has been a matter of controversy. An early hypothesis (Sugden 1970, 1973) proposed ice-sheet downwastage *in situ* during the Weichselian late-glacial and with minor fluctuations in the overall ice wastage during the Younger Dryas. An alternative hypothesis (Sissons & Grant 1972; Sissons 1979a) proposes that small glaciers developed within the Cairngorm corries during the Stadial. These glaciers ranged in area from 0.12 to 8.89 km² and had equilibrium firn lines of 755–1089 m. They were fed by southerly winds, and snowfall diminished towards the north and northwest (Sissons 1979b). Although

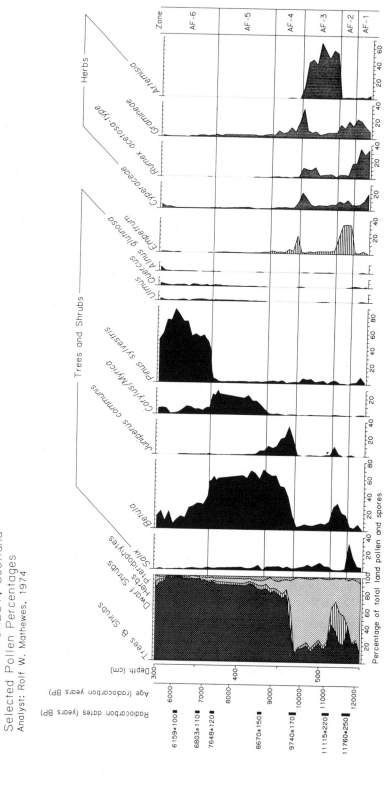

ABERNETHY FOREST, Scotland
Selected Pollen Percentages
Analyst: Rolf W. Mathewes, 1974

Fig. 4.10 Summary pollen diagram from Abernethy Forest (300–550 cm). Selected pollen and spore taxa only are shown. The complete diagram is in Birks and Mathewes (1978)

ABERNETHY FOREST, Scotland
Selected Pollen Percentages
Analyst: Hilary H. Birks, 1966-67

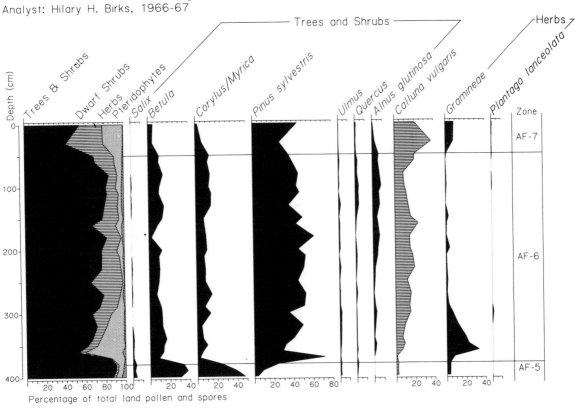

Fig. 4.11 Summary pollen diagram from Abernethy Forest (0–400 cm only). Selected pollen and spore taxa only are shown. The complete diagram is in Birks (1970)

these two competing hypotheses have not been totally resolved (Sissons 1973; Sugden 1973, 1980), the bulk of the available evidence favours Sissons' (1979a) reconstruction of small corrie glaciers.

(2) After a series of rapid and relatively short-lived changes in pollen dominance between *ca*. 9800 and 8000 BP from *Artemisia* to Cyperaceae followed by phases of Gramineae, *Empetrum*, *Juniperus communis*, *Populus tremula*, *Betula*, and *Corylus/ Myrica* (Fig. 4.10), *Pinus* pollen values rose between 8000 and 7500 BP (Birks 1989) and maintained high values up to the present day, particularly in the lowlands (Fig. 4.11). O'Sullivan (1974, 1975a, 1976) discusses apparent variations in the timings and

rates of pine expansion from site to site in the Cairngorm area. The small number of available ^{14}C dates, coupled with the inevitable problems of conventional ^{14}C datings of limnic sediments, limit detailed analysis of these apparent variations.

Stratigraphic changes in the *Alnus glutinosa* pollen curve at sites in the area are consistently very small and often rather slight compared with other regions in Scotland. There is thus considerable scope for differences in opinion in what constitutes "arrival of alder", "main expansion", "the *Alnus rise*", and "main rise" (cf. O'Sullivan 1974, 1975a, 1976, 1977). Problems of definition coupled with a shortage of ^{14}C dates from relevant narrowly defined sedimentary intervals prevent any resolution of the "true

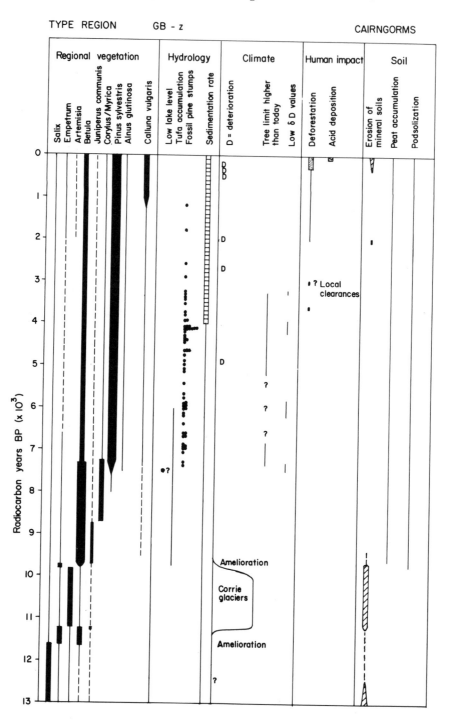

Fig. 4.12   Event stratigraphy for the Cairngorms

Middle Flandrian sequence" (O'Sullivan 1974). Available data suggest that alder began to grow locally between *ca.* 7500 and 5500 BP but that it never attained high values in the area during the Holocene (Birks 1989; Bennett & Birks 1990). As elsewhere in Britain, there is considerable variation from site to site, emphasizing the strongly individualistic site-imprints on the observed pollen stratigraphy at different sites.

(3) Available pollen data all indicate that the mid-Holocene forests were dominated by *Pinus sylvestris* with some *Betula*, *Populus tremula*, *Sorbus aucuparia,* and *Juniperus communis* (Figs 4.10 and 4.11). *Corylus avellana* may have occurred locally on more basic soils, whereas *Ulmus*, *Quercus*, and *Fraxinus excelsior*, if present at all, were extremely rare. *Salix* and *Alnus glutinosa* were also rare and confined to streamsides, waterlogged hollows, and lake margins. The natural tree-limit was probably formed by pine or birch, depending on local conditions, and may have been fringed by a belt of juniper scrub before passing into montane heaths and summit vegetation. Wood remains have been found in peats up to 793 m elevation, indicating that tree-limit must have reached this elevation at least in the mid-Holocene (Pears 1968a; Dubois & Ferguson 1985; Birks 1988).

(4) The unambiguous detection of prehistoric forest clearance in the area from pollen-stratigraphic data is extremely difficult because the local, extra-local, and regional components of pollen deposition are almost all totally dominated by pine pollen, particularly in the lowlands. Criteria adopted here for forest clearance are decreases in total tree pollen to less than 50% of total pollen associated with large rises in non-arboreal pollen, particularly *Calluna vulgaris* and Gramineae. Using these criteria, forest clearances may have occurred in the lowlands *ca.* 3600 BP (O'Sullivan 1974, 1975b), *ca.* 3000 BP, and *ca.* 1000 BP (O'Sullivan 1976). Cereal cultivation may have been present from *ca.* 1850 BP (O'Sullivan 1976). Extensive deforestation in the lowlands may have been comparatively recent, possibly *ca.* 300 BP (Birks 1988). Local-scale pollen data from mor-humus profiles within presumed native pine forests today indicate, however, that *Calluna*-dominated heathland had existed locally since *ca.* 1500 BP,

perhaps associated with the influx of Iron Age Picts (O'Sullivan 1973). Extensive forest clearance occurred in the hills earlier, leading to the widespread development of heather moor since *ca.* 2000–2500 BP (Birks 1975). The lower boundary for the *Calluna–Plantago lanceolata* paz is clearly non-synchronous within the area.

(5) Five lochs in the area have been studied palaeolimnologically (Jones *et al.* 1993b). Four of them show clear evidence for recent lake acidification, with pH changes from 5.3–6.4 to 4.7–5.0 in the last 90–150 years, paralleled by increased concentrations of Pb, Zn, and carbonaceous "soot" particles derived from industrial processes (Jones *et al.* 1993b). The most likely cause of this recent acidification in these remote high-elevation lakes is acid deposition.

(6) Evidence for possible climatic changes during the Holocene comes from a variety of sources — fossil pine stumps preserved in blanket peats, lake-level changes, datings of renewed amorphous solifluction at high elevations, lichenometry in the high corries, occurrence of tufa, and analysis of D/H isotope ratios in fossil pine stumps.

O'Sullivan (1976) has tentatively inferred a lake-level lowering *ca.* 7500 BP on the basis of lithological changes and an apparently anomalous ^{14}C date. Tufa accumulation occurred in Glen Avon between *ca.* 9700 and *ca.* 6000 BP and ceased for no obvious reason (Preece *et al.* 1984; cf. Goudie *et al.* 1993). Over 50 ^{14}C dates exist for fossil pine stumps in the area (Pears 1969, 1972, 1975a, 1975b; Birks 1975; Dubois & Ferguson 1985), with a range 7380–1190 ^{14}C yr BP. Dates cluster around 7200–6500 BP, 6000 BP, and 5500–3500 BP (Bridge *et al.* 1990). Although many problems exist in the interpretation of fossil pine stumps in terms of climatic change because of local site factors, preservation, etc. (Birks 1975; Dubois & Ferguson 1985, 1988; Pears 1988), the striking absence of any fossil stumps younger than 3250 BP above 530 m elevation strongly suggests some climatic depression of tree-line (Dubois & Ferguson 1985). The absence of stumps, however, may reflect an absence of suitable material or environment for preservation of wood remains at that time. D/H ratios preserved in the pine stumps suggest phases of increased precipitation *ca.* 7300–

7500, 6200–5800, 4200–3900, and 3300 BP (Dubois & Ferguson 1985, 1988). Renewed amorphous solifluction occurred after 4800 and 2700 BP and during the "Little Ice Age" (*ca.* 1500–1750 AD) (Sugden 1971), suggesting some climatic changes at high elevation. Tentative evidence from lichen measurements inside and outside boulder moraines suggests that permanent snow-beds formed in some of the high corries during the "Little Ice Age" (Sugden 1977; Rapson 1985).

(7) Debris-cone formation occurred *ca.* 2000 BP and 350 BP–present (Brazier & Ballantyne 1989), possibly in response to periods of exceptional rainstorms and floods that acted as a "trigger" for initiating flow. Innes (1983a) suggests, however, that recent debris flow activity is a result primarily of increased sheep grazing and burning in the last 250 years.

(8) The present presumed natural tree-limit of 640 m at Creag Fhiaclach consists of stunted pine trees showing "krummholz" growth *ca.* 2 m tall and ages of < 100–250 years (Miller & Cummins 1982; Grace & Norton 1990). Tree-ring widths at sites about 150 m below the present tree-line are significantly and positively correlated with late-winter and summer temperatures (Grace & Norton 1990). Establishment and growth of saplings at Creag Fhiaclach may be limited by repeated browsing of red deer (Miller & Cummins 1982) or by extended snowlie. The tree-limit appears to be set by a combination of browsing and climate. Although generally regarded as a natural tree-limit, there are several lines of evidence to suggest that the tree-limit here is artificially depressed (Jennifer McConnell, personal communication).

ACKNOWLEDGEMENT

We are indebted to Jennifer McConnell for reading and commenting on this account and for sharing her unpublished findings on the current tree-limit in the Cairngorms.

TYPE REGIONS GB-zc and GB-x *pro parte*, ISLE OF SKYE (H.J.B. Birks with contributions from Sylvia M. Peglar and W. Williams)

The Isle of Skye lies off the northwest coast of Scotland between latitudes 57°3′N and 57°44′N and longitudes 6°46′W and 5°38′W. It is the largest island in the Inner Hebrides at 78 km long and 84 km wide with a total area of 1720 km². Its coast is deeply indented, so that no part of the island is more than 8 km from the sea. Despite its comparatively small size, it is an island of very considerable geologic, topographic, and hence floristic and vegetational diversity. It is logically divisible (Fig. 4.13) into six major regions on the basis of topography and underlying geology. These regions are Sleat (low-lying, generally acid), Kyleakin (low- and high-altitude acid sandstone), Suardal (low-altitude limestone), Red Hills (acid granite), Cuillin Hills (gabbro), and northern Skye (low- and high-altitude basic basalt overlying limestone and shale).

At present Holocene pollen stratigraphies are available for the Sleat, Kyleakin, and northern Skye regions (Vasari & Vasari 1968; Williams 1977; Birks & Williams 1983) and Weichselian (= Devensian) late-glacial stratigraphies are available for the Cuillin, Kyleakin, Suardal, and Red Hills regions and possibly for the northern Skye and Sleat regions (Birks 1973; Walker *et al.* 1988; Walker & Lowe 1990). Sediment geochemistry has been studied for Weichselian late-glacial sediments by Walker and Lowe (1990). In addition, some palynological analyses of basal sediments relate to studies on deglaciation (Walther 1984; Walker *et al.* 1988; Benn *et al.* 1992). The geology and Quaternary history of Skye are summarized by Birks (1973) and Ballantyne *et al.* (1991), whereas the modern flora, vegetation, and ecology are discussed by Birks (1973).

The island falls ecologically and palaeoecologically into two IGCP type regions—GB-zc (Isle of Skye) and GB-x (Western Scotland). However, it is not possible to present a regional synthesis for GB-x because comparatively few sites have been investigated in detail within this vast type region. A regional synthesis is presented here for the entire Isle of Skye even though it lies within two type regions. Northern Skye belongs to GB-zc, whereas Sleat, Kyleakin, Suardal and possibly the Red Hills and Cuillin Hills belong to GB-x.

Altitude: 0–1014 m. The island consists of extensive low-lying areas between sea-level and 400 m in all regions and several upland areas ranging from

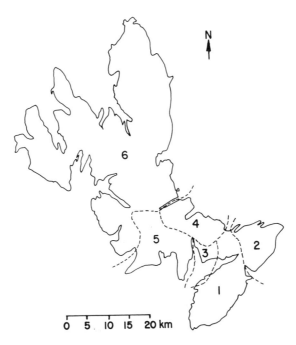

Fig. 4.13 Map of the Isle of Skye showing the six major geologic, topographic, and ecological regions. 1 = Sleat, 2 = Kyleakin, 3 = Suardal, 4 = Red Hills, 5 = Cuillin Hills, 6 = Northern Skye

400 m up to 1000+ m in the Northern Skye, Kyleakin, Red Hills, and Cuillin Hills regions. The palaeoecological sites studied are all low-lying at about 50–150 m elevation.

Climate: Mean January temperature in lowland Skye 4.2°C; mean July temperature 13.6°C; corresponding minima are 2.2°C and 10.6°C; temperature decreases with elevation and mean January and July temperature at the highest point (1014 m) are estimated to be −2.7°C and 6.2°C, respectively; precipitation (mainly as rain) ranges from 140 to 203 cm yr^{-1} in the lowlands increasing with elevation to at least 228 cm yr^{-1} on The Storr (723 m) and 340 cm yr^{-1} in the Cuillin (*ca.* 900 m); rainfall does not show a simple relation with elevation, because although tending to increase with altitude, the zone of maximum precipitation is usually some distance lee of the highest ground. Rain-shadow effects also occur; about 250 rain days yr^{-1} (a day with at least 0.2 mm rain) and 200–220 wet days yr^{-1} (a day with at least 1 mm rain) throughout the island, increasing to 220+ yr^{-1} in the uplands; high average relative humidity and a high incidence of cloud often occur as fog in the uplands; snowfall is relatively unimportant with only about 20–25 days yr^{-1} with snow falling on low ground, increasing to 75 days yr^{-1} at 920 m elevation; winds are predominantly from the south and northwest, and average wind speed is *ca.* 6.7 m s^{-1} at 33 m; gale-force winds (wind speed greater than 16.4 m s^{-1}) occur 20–30 days yr^{-1} at sea-level, increasing to 150–200 days at 1000 m elevation; average daily sunshine rarely exceeds four hours, even in mid-summer; short growing season, decreasing with elevation; striking west–east gradient in average daily duration of bright sunshine, with southern lowland Skye, especially Sleat, receiving more total sunshine than elsewhere on the island; mild northern oceanic climate.

Geology: Geologically complex and extremely varied. Palaeozoic acid Lewisian gneiss, Torridonian sandstone, mildly basic Moine schistose grit and phyllite, and small areas of Cambrian quartzite and grit overlain by Durness (? Early Ordovician) limestone and dolomite characterize the Sleat region. Cambrian quartzite and grit and Durness limestone form the small but distinct Suardal region. The Kyleakin region nearest to the Scottish mainland consists almost exclusively of Torridonian sandstone. Almost all of northern Skye is formed of considerable thicknesses of Tertiary tuff and basaltic lava overlying Jurassic limestone, sandstone, shale, and ironstone. The mountains of central Skye are formed mainly of basic and ultrabasic hard gabbro (Cuillin Hills) rich in olivine intruded into the surrounding low-lying lavas, and of acid granite and granophyre (Red Hills). In addition there are numerous horizontal dolerite sills and basic dyke swarms throughout the island formed as minor intrusions in the last phase of igneous activity. See Birks (1973) and Slesser (1975) for further details.

Topography: Very varied; everywhere rocky, generally rugged, and locally spectacular. The Sleat region is rocky and rugged despite no areas above 305 m and has steep bare rock-outcrops, peat-filled hollows and small lochans, and several deep ravines with waterfalls. The Suardal region contains limestone outcrops as low scarps, with areas of bare limestone pavement containing karstic landforms such as clint and grike structures. The Kyleakin region is rather subdued topographically, with gentle slopes reaching over 610 m and with few extensive cliffs or rock outcrops. Many deep ravines contain waterfalls. Northern Skye has spectacular inland and coastal basalt cliffs, massive landslips, and extensive screes and includes the most extensive inland cliff system in Scotland. The highest point is 723 m. To the west of these massive cliffs are flat and gently sloping basalt slopes following the low westward dip of the lavas. The coastline is extremely precipitous with spectacular vertical sea-cliffs up to 310 m high. The Cuillin region is extremely mountainous with very steep slopes, 23 serrated peaks over 920 m elevation (highest point 1014 m), narrow summit ridges, and deeply cut corries often with high cliffs and extensive scree- and boulder-strewn slopes. In contrast, the adjacent granitic Red Hills are topographically more subdued with gently contoured slopes, broad watersheds, flat summit plateaux, few corries or extensive cliffs, and large areas

of boulder-strewn and scree slopes. Much of the lowland in all regions is relatively flat and extensively covered by blanket peat.

Population: Generally at a low density (less than 1 person ha^{-1}), even for Scotland, with about 8000 people in an area of 1720 km^2. Nearly 40% of the population live in two towns (Portree, Broadford), and the rest live in crofting townships mainly around the coast and in scattered farmsteads inland. At the turn of the 19th century the population was over 23000 people prior to the so-called "Highland Clearances", marked by many evictions of crofters. A small (*ca.* 5–10%) increase in population has occurred in the last 20 years.

Vegetation: Extremely varied. No other area of comparable size in western Europe has such a wide range of vegetation types. Predominantly blanket bog communities in the lowlands and a range of montane vegetation types in the uplands. *Scirpus cespitosus–Calluna vulgaris* blanket bog on shallow (2 m or less) peat is widespread. *Scirpus–Eriophorum angustifolium* bog occurs on deeper peat. *Molinia caerulea* dominates wherever some soligenous influence exists and commonly grows with *Myrica gale*. On better drained sites *Calluna* heather-moor and *Agrostis–Festuca* grassland dominate. This grassland is the principal sheep pasture on Skye. It is derived from heather moor by extensive grazing and burning. The Sleat and Kyleakin regions are more sheltered from Atlantic storms than elsewhere on Skye. Woodland occurs frequently on slopes that are too rocky or too steep for bog development. *Betula pubescens* and *Corylus avellana* dominate, with some *Sorbus aucuparia*, *Ilex aquifolium*, and *Quercus petraea*. *Alnus glutinosa* is local in damp sites, and *Fraxinus excelsior* is confined to limestone areas. The woodlands are extremely rich in Atlantic ferns, bryophytes, and lichens, including several species with Macaronesian–Tropical distributions that reach their northernmost known world localities in the Sleat and Kyleakin regions. The Kyleakin Hills support widespread bog, heather moor, and grassland. Above 500 m dwarf, wind-pruned *Calluna* heath occurs. Summit vegetation is *Carex bigelowii–Racomitrium lanuginosum* heath.

The vegetation of the granitic Red Hills is generally similar to the vegetation of the Kyleakin Hills except for the tall *Calluna–Vaccinium myrtillus* heaths on the extensive north-facing block-screes of the Red Hills and the local abundance of dwarf *Juniperus communis* ssp. *nana* heath on the Red Hill summits. The lower slopes of the Cuillin Hills are covered by *Molinia caerulea–Calluna vulgaris* soligenous bog, with a range of flush and spring communities. Above about 400 m the steeper slopes are extensive screes with little or no vegetation. The summits are largely devoid of vegetation except for small areas of *Juncus trifidus–Festuca ovina* "fell-field". *Nardus stricta*-dominated snow patches occur in some north-facing gullies.

The Suardal limestone supports *Corylus avellana* and *Fraxinus excelsior* woodland, *Dryas octopetala* heaths (to sea-level), species-rich *Agrostis–Festuca* grasslands, and open limestone pavements with a diverse flora growing in crevices and deeper grikes and on patches of soil developed on the clints.

The low-lying areas of northern Skye support blanket bog, *Calluna* heather-moor, or *Agrostis–Festuca* grassland. Enclosed hay-meadows characterize the crofting areas. Higher elevations support an abundance of *Agrostis–Festuca* grassland rich in *Alchemilla alpina* and other low-growing montane herbs. Summit vegetation is varied, with *Carex bigelowii–Racomitrium lanuginosum* heath, *Nardus stricta* snow-patches, open "fell-fields", and gravel flushes with *Koenigia islandica* growing in one of its few British localities.

The spectacular coastline of northern Skye supports a range of coastal vegetation types, mainly cliff-ledge vegetation. Salt-marsh vegetation occurs at the heads of some of the more sheltered sea-lochs. Weed and ruderal communities occur locally but frequently throughout lowland Skye. Aquatic, swamp, and marginal fen vegetation-types associated with numerous freshwater lochs are common. See Birks (1973) for further details.

Soils: Extremely varied due in part to the diverse geology and the range of topography. Mildly acidic brown earths frequent in lowland Skye; leached podsols widespread; extensive peat accumulation in poorly drained hollows and on most gentle slopes and plateaux; ranker soils with raw humus frequent in the uplands; shallow rendzinas very rare and confined to limestone areas; bare skeletal soils, often disturbed by cryoturbation, frequent at high elevations. See Birks (1973) for further details.

Land use: Upland sheep pasture; some cattle farming; crofting and cultivation of potatoes, oats, and hay for livestock; small areas of deer forest and grouse moor; tourism; fishing; increasing commercial forestry with the planting of exotic conifers.

Late-glacial reference sites

Despite a considerable number of pollen-stratigraphical studies on presumed Weichselian (= Devensian) late-glacial deposits on Skye (Birks 1973; Walther 1984; Walker *et al.* 1988; Walker & Lowe 1990), there is still no satisfactorily ^{14}C-dated late-glacial sequence on Skye, with several dates seemingly too young and some dates seemingly too old. All ^{14}C dates so far have been based on conventional dating of bulk sediment samples. The widespread dating problems encountered by several investigators may result from the complex geology of the island, the low organic content of the sediments, contamination by redeposition of older sediments, or infiltration of modern carbon through soligenous seepage.

Reference site 12. Loch Ashik, Kyleakin region
(Walker *et al.* 1988; Walker & Lowe 1990)

Latitude 57°15′N, Longitude 5°50′W. Elevation 40 m. Age range 13000 BP–*ca.* 9000 BP. Fen over infilled lake. Six ^{14}C dates. 92 cm sediment. Eleven local and eight regional pollen-assemblage zones (paz). (Fig. 4.14)

Local paz	Approximate age	Regional paz
LA-2	13000–12500 BP	*Rumex–Salix–Gramineae*
LA-3–5	12500–11800 BP	*Juniperus–Empetrum–Gramineae–Rumex*

Local paz	Approximate age	Regional paz
LA-6	11800–10800 BP	*Empetrum–Ericaceae–Gramineae–Cyperaceae*
LA-7	10800–10000 BP	*Rumex–Artemisia–Gramineae–Cyperaceae*
LA-8	10000–9900 BP	*Cyperaceae–Rumex*
LA-9	9900–9700 BP	*Empetrum–Gramineae*
LA-10	9700–9500 BP	*Betula–Juniperus–Gramineae*
LA-11	9500–9200 BP	*Betula*

For further details of the pollen stratigraphy and local and regional pollen zonation, see Walker *et al.* (1988) and Walker and Lowe (1990).

Holocene reference sites

[14]C-dated Holocene pollen sequences are only available for three regions on Skye—northern Skye, Sleat, and Kyleakin (Williams 1977; Birks & Williams 1983).

Reference site 26. Loch Cleat (Northern Skye)
(Williams 1977; Birks & Williams 1983)

Latitude 57°41′N. Longitude 6°20′W. Elevation 38 m. Age range 10400–0 BP. Fen over infilled lake. Ten ^{14}C dates. 940 cm sediment. Six local and five regional pollen-assemblage zones (paz). (Fig. 4.15)

Local paz (subzones in parentheses)	Approximate age	Regional paz
LCT-1	10400–10100 BP	*Gramineae–Rumex*
LCT-2	10100–9500 BP	*Betula–Juniperus–Gramineae*
LCT-3	9500–8900 BP	*Betula–Corylus–Polypodiaceae*
LCT-4 (i)	8900–6550 BP	*Betula–Corylus–Polypodiaceae*
LCT-4 (ii)	6550–4900 BP	*Betula–Corylus–Polypodiaceae*
LCT-5 (i)	4900–3100 BP	*Gramineae–Betula–Plantago lanceolata*

Local paz (subzones in parentheses)	Approximate age	Regional paz
LCT-5 (ii)	3100–2600 BP	*Gramineae–Betula–Plantago lanceolata*
LCT-5 (iii)	2600–700 BP	*Gramineae–Betula–Plantago lanceolata*
LCT-6	700–0 BP	*Gramineae–Cyperaceae*

Reference site 11. Loch Meodal (Sleat region)
(Birks 1973; Williams 1977; Birks & Williams 1983)

Latitude 57°8′N. Longitude 5°5′W. Elevation 110 m. Age range 10300–0 BP. Fen over infilled lake. Ten ^{14}C dates. 970 cm sediment. Four local and six regional pollen-assemblage zones (paz). (Fig. 4.16)

Local paz (subzone in parentheses)	Approximate age	Regional paz
LML-4	10300–9600 BP	*Gramineae–Rumex*
LML-5	9600–6600 BP	*Betula–Corylus–Polypodiaceae*
LML-6 (i)	6600–5200 BP	*Betula–Alnus–Ulmus*
LML-6 (ii)	5200–4300 BP	*Betula–Alnus*
LML-6 (iii)	4300–2700 BP	*Betula–Alnus–Plantago lanceolata*
LML-6 (iv)	2700–1600 BP	*Betula–Alnus–Plantago lanceolata*
LML-6 (v)	1600–320 BP	*Betula–Alnus–Plantago lanceolata*
LML-7	320–0 BP	*Gramineae–Cyperaceae*

Reference site 12. Loch Ashik (Kyleakin region)
(Williams 1977; Birks & Williams, 1983)

Latitude 57°15′N. Longitude 5°50′W. Elevation 40 m. Age range 10325–0 BP. Fen over infilled lake. Ten ^{14}C dates. 490 cm sediment. Seven local and eight regional pollen-assemblage zones (paz). (Fig. 4.17)

Local paz (subzones in parentheses)	Approximate age	Regional paz
LAK-1	10700*–10300* BP	Cyperaceae–*Rumex*
LAK-2	10300*–9700 BP	*Betula–Juniperus–*Gramineae
LAK-3 (i)	9700–8900 BP	*Betula–Corylus–*Polypodiaceae
LAK-3 (ii)	8900–6350 BP	*Betula–Corylus–*Polypodiaceae
LAK-4 (i)	6350–4500 BP	*Betula–Alnus–Ulmus*
LAK-4 (ii)	4500–4200 BP	*Betula–Alnus–*Gramineae
LAK-5	4200–3900 BP	*Pinus–Alnus*
LAK-6 (i)	3900–2700 BP	*Betula–Alnus–Calluna*
LAK-6 (ii)	2700–200 BP	*Betula–Alnus–Calluna*
LAK-7	200–0 BP	Gramineae–Cyperaceae

(* the ^{14}C dates that these estimates are based on are almost certainly too old due to "hard-water" effects, as are the ^{14}C dates obtained by Walker & Lowe (1990) for the Weichselian late-glacial sediments at this site).

Event stratigraphy This is presented for the northern Skye, Sleat, and Kyleakin regions in Figs. 4.18–4.20.

Discussion

(1) Although differences of opinion exist about the likely extent of ice during the Loch Lomond (= Younger Dryas) Stadial on Skye (Sissons 1977; Ballantyne 1989), there is no doubt that the Stadial at the close of the Weichselian (= Devensian) late-glacial witnessed widespread corrie glaciation in the Cuillin, Red, and Kyleakin Hills and small corrie glaciers in northern Skye (Sissons 1977; Ballantyne 1988, 1989, 1990). Birks (1973), Walther (1984), Walker *et al.* (1988), and Benn *et al.* (1992) present palynological evidence to indicate that this last corrie glaciation occurred just before the earliest

Holocene, namely during the Younger Dryas. Sissons (1977) and Ballantyne (1989) map the likely extent of the seven Younger Dryas corrie glaciers on the west side of the Cuillin. Lowest ice elevation was 60 m (range = 60–460 m), equilibrium firn line ranged from 392 to 643 m, and glacier area ranged from 0.67 to 4.59 km^2. Three corrie glaciers formed in the Red Hills (lowest ice elevation 60–230 m, equilibrium firn line 286–352 m, glacier area 0.42–1.51 km^2) and three in the Kyleakin Hills (lowest ice elevation –5–220 m, equilibrium firn line 293–372 m, glacier area 2.06–5.64 km^2). Ballantyne (1989) and Walker *et al.* (1988) propose that a Cuillin ice-field developed during the Younger Dryas in the central and northern Cuillin Hills and adjacent Red Hills, with a total area of 155 km^2, lowest ice elevation of –20 m, and an equilibrium firn line of 308 m. Glaciation was very restricted in northern Skye with only two small corrie glaciers forming in the Younger Dryas (area 0.68–1.67 km^2, lowest ice elevation 230–280 m, equilibrium firn line 291–331 m).

The mean equilibrium firn line of 319 m fits into a regional eastward rise in firn-line elevations, indicating predominantly westerly airstreams during the Stadial. The firn-line elevations of the corrie glaciers also suggest that the dominant snow-bearing winds were southerlies (Sissons 1977, 1979b; Ballantyne 1989).

Extensive areas of so-called "hummocky moraine" characterize the valleys leading from the Cuillin and Red Hills. Benn (1992) has shown that these areas consist of recessional moraines, chaotic ice-stagnation moraines, drumlins, and fluted moraines. The formation of this distinctive drift landform is largely unclear, and the relationships of the various types of features to glacial history require elucidation (Benn *et al.* 1992).

The main vegetational development (Fig. 4.14) appears to have been from a pioneer *Rumex* (including *Oxyria digyna*)–Gramineae assemblage with abundant *Salix* to an interstadial (Allerød + ? Bølling times) characterized by widespread *Empetrum* heaths, with *Racomitrium languinosum* and a variety of open-ground herbs including *Rumex*, *Thalictrum*, and Caryophyllaceae. The Younger Dryas Stadial was a period of intensive soil erosion and solifluction, often leading to the reworking and

124

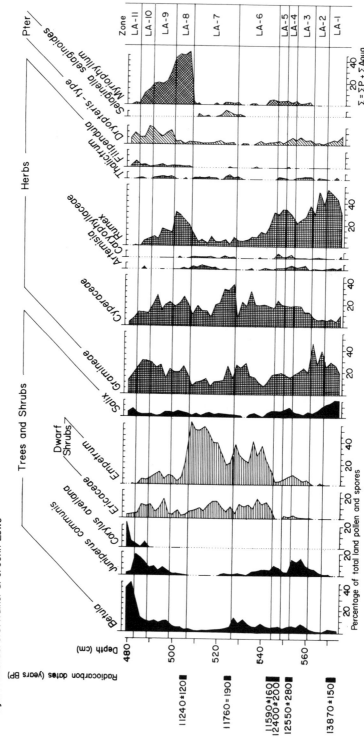

LOCH ASHIK, Scotland

Selected Pollen Percentages

Analysts: Michael J.C. Walker & J. John Lowe

Fig. 4.14 Summary Weichselian late-glacial pollen diagram from Loch Ashik, Isle of Skye. Selected major pollen and spore taxa are shown. The complete diagram is in Walker *et al.* (1988) and Walker & Lowe (1990). Aqua = Aquatic pollen and spores, Pter = pteridophytes

125

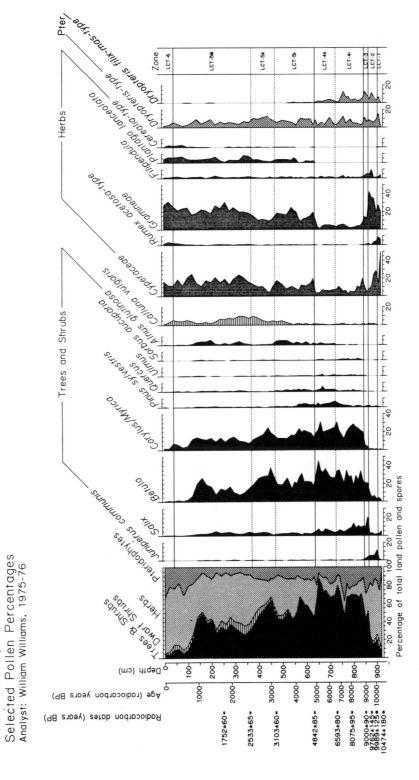

Fig. 4.15 Summary Holocene pollen diagram from Loch Cleat, Isle of Skye. Selected major pollen and spore taxa are shown. The complete diagram is in Williams (1977). Pter = pteridophytes

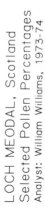

LOCH MEODAL, Scotland
Selected Pollen Percentages
Analyst: William Williams, 1973-74

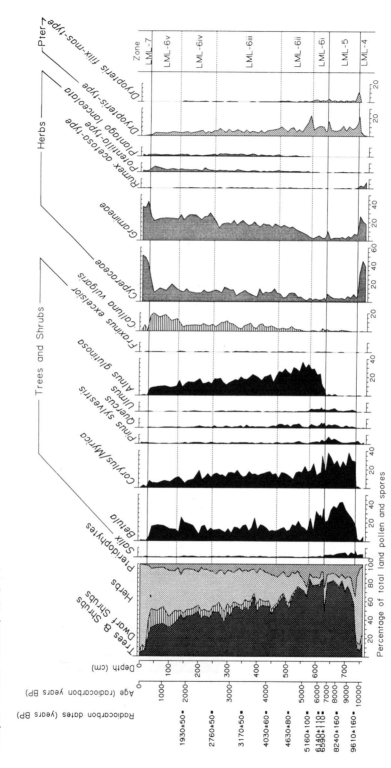

Fig. 4.16 Summary Holocene pollen diagram from Loch Meodal, Isle of Skye. Selected major pollen and spore taxa are shown. The complete diagram is in Williams (1977). Pter = pteridophytes

127

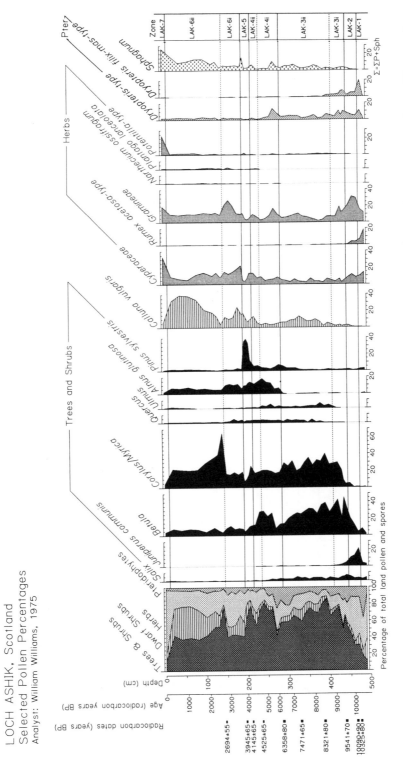

Fig. 4.17 Summary Holocene pollen diagram from Loch Ashik, Isle of Skye. Selected major pollen and spore taxa are shown. The complete diagram is in Williams (1977). Pter = pteridophytes, Sph = *Sphagnum*

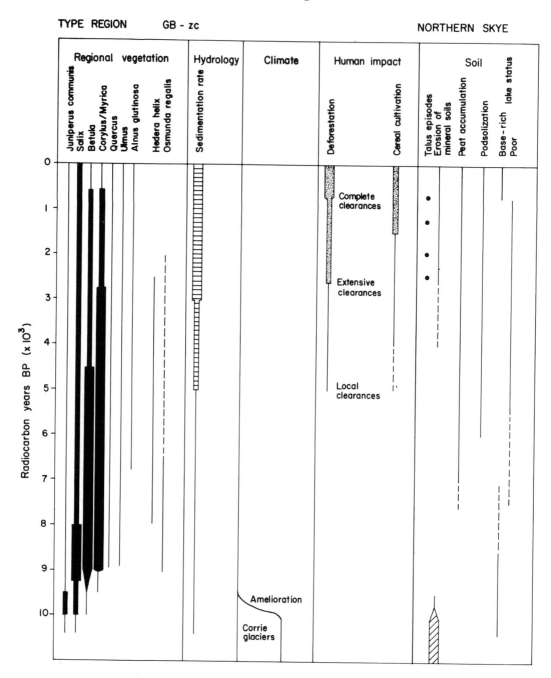

Fig. 4.18 Event stratigraphy for northern Skye

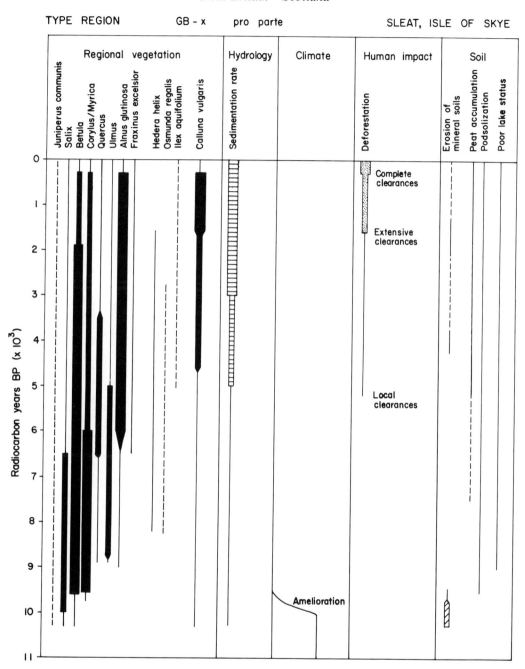

Fig. 4.19 Event stratigraphy for the Sleat region, Isle of Skye

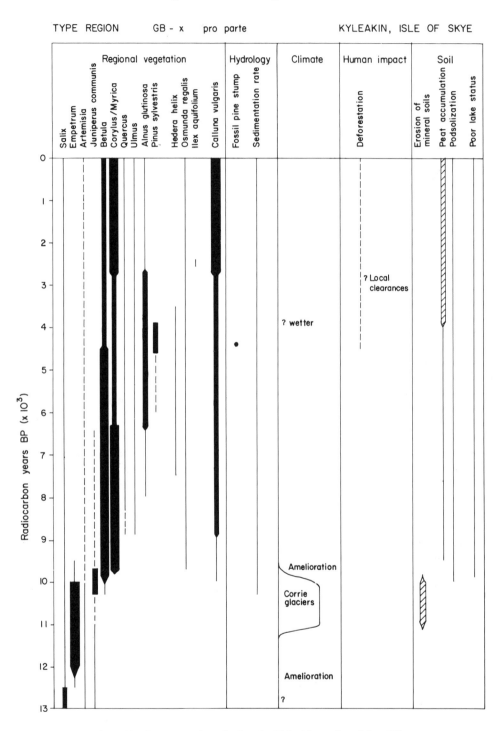

Fig. 4.20 Event stratigraphy for the Kyleakin region, Isle of Skye

redeposition of Allerød-age soils and their contained pollen and hence to the increased input of deteriorated pollen (especially *Empetrum*) and terrestrial fungal hyphae (Birks 1973). The nature of the Younger Dryas vegetation is poorly known because of this redeposition of older material, but it was probably very open and grass- and sedge-dominated with a range of open-ground taxa such as *Rumex*, *Artemisia*, *Thalictrum*, Caryophyllaceae, and *Koenigia islandica*. See Birks (1973), Walker *et al.* (1988) and Walker & Lowe (1990) for further details.

Benn *et al.* (1992) propose that deglaciation at the close of the Younger Dryas occurred in two distinct steps. The first involved numerous glacier stillstands and readvances. The second involved uninterrupted retreat and local glacier stagnation. As the first stage appears to have been occurring prior to any marked climatic amelioration at the onset of the Holocene (as inferred palynologically), Benn *et al.* (1992) suggest that early ice retreat was a response to decreased precipitation in the later part of the Younger Dryas. The second stage appears to have occurred in response to rapid and sustained temperature increases early in the Holocene.

(2) The early Holocene revegetation was characterized by a series of relatively short-lived phases of dominance of *Rumex*, Gramineae, Cyperaceae, *Empetrum*, *Juniperus*, *Betula*, *Salix*, and *Corylus* pollen between about 10000 and 9000 BP (Figs. 4.14–4.17). However, considerable variation from site to site exists in the expression of these phases and their relative abundance. In general, sites in more sheltered localities in the lee of the Cuillin and Red Hills show a development from open vegetation through juniper scrub to birch–hazel woods, whereas in exposed areas along the southern, southwestern, and northern coasts the early Holocene landscape was predominantly open grassland, with some dwarf-shrub heaths and with stands of juniper scrub and birch–hazel–willow woodland confined to sheltered sites protected from the onshore winds (Birks 1973; Williams 1977; Walker & Lowe 1990).

(3) The mid-Holocene had similar geographical variation in forest extent and composition (Fig. 4.21). The Sleat region appears to have supported the most extensive woodlands dominated by *Betula* and *Corylus avellana*, with some *Quercus*, *Ulmus*, *Sorbus aucuparia*, *Populus tremula*, *Prunus padus*, and *Ilex aquifolium*. (Fig. 4.16). *Alnus glutinosa* and *Salix* were locally frequent in damp areas. The Kyleakin region supported more open birch and hazel woods with some *Ulmus*, *Quercus*, *Sorbus aucuparia*, *Populus tremula*, and *Prunus padus*. *Salix* and *Alnus glutinosa* were locally frequent in wet areas (Fig. 4.17). In contrast, northern Skye appears to have supported scattered areas of birch and hazel scrub with *Populus tremula*, *Sorbus aucuparia*, and *Prunus padus*. *Salix* was frequent in damp areas (Fig. 4.15). It is not known if oak, elm, or *Alnus glutinosa* grew in northern Skye during the mid-Holocene.

(4) *Pinus sylvestris* was never a component of the forest vegetation on Skye except for a short period (*ca.* 4600–3900 BP) in the Kyleakin region (Fig. 4.17). The expansion of pine pollen at Loch Ashik may reflect the local growth of pine on dried peat surfaces, a widespread phenomenon in northwest Scotland at this time (see Loch Maree in type region GB-zd). Pine stumps near Loch Ashik are dated to 4420±75 BP (Q-1309). The sharp decline in pine pollen at Loch Ashik correlates with the widespread demise of pine throughout northwest Scotland at about 4000 BP (Birks 1972b, 1975; Bennett 1984) and the widespread development and expansion of blanket bog with *Calluna vulgaris*, *Sphagnum*, and *Narthecium ossifragum* and of heathland in eastern Skye. The most likely explanation for this widespread and spectacular decline of pine *ca.* 4000 BP is a shift to a more oceanic climate with increased precipitation and strong winds at this time. Such a climatic change would have increased waterlogging, encouraged the expansion of bog, and inhibited the regeneration of pine by reducing the number of good seed years.

(5) Forest clearance and associated spread of grassland and heath occurred *ca.* 5200 BP in Sleat (Fig. 4.16). By 4200 BP the landscape there was still mainly wooded but with frequent acid grasslands, heaths, and bogs. *Calluna* heath expanded considerably *ca.* 1600 BP as a result of extensive deforestation. Widespread woodland destruction and spread

132

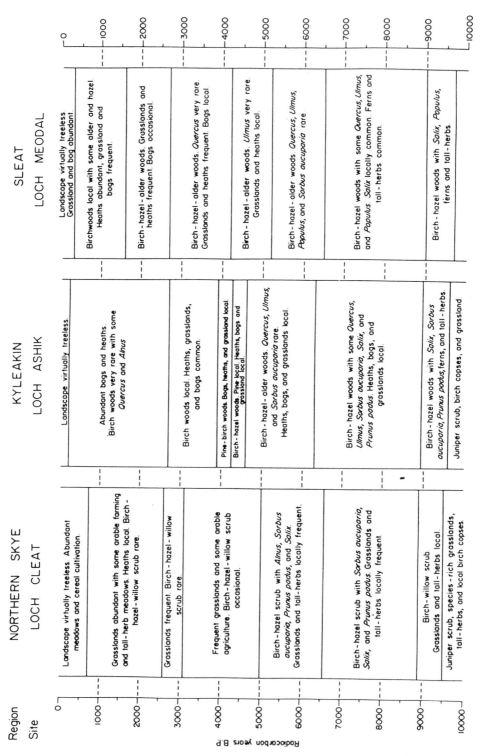

Fig. 4.21 Comparison of the inferred Holocene vegetational history for the northern Skye, Kyleakin, and Sleat regions on the Isle of Skye (from Birks & Williams 1983)

of grassland at the expense of *Calluna* heath occurred in the last 300 years, possibly as a result of the onset of cattle breeding in 1650 AD.

Clearance of scrub and the local development of arable and pastoral agriculture began in northern Skye *ca.* 5000 BP (Fig. 4.15). By 2600 BP the landscape was probably mainly treeless, with small patches of scrub restricted to rocky areas. In the last 700 years widespread destruction of the remnants of birch and hazel scrub produced the entirely treeless landscape of northern Skye today.

In eastern Skye (Fig. 4.17) *Calluna* heath, grassland, and bog appear to have been present from *ca.* 9000 BP. However, no unambiguous evidence exists for human interference *ca.* 5000 BP (cf. northern Skye and the Sleat region). By 2700 BP bog and heath were widespread and woods, mainly birch, were rare and presumably confined to slopes too steep for blanket-bog development. This has continued to the present day (prior to recent commercial forestry and planting of exotic conifers), suggesting that the pre-forestry bog-dominated landscape of the Kyleakin area was of considerable antiquity.

(6) Talus accumulation on the basalt slopes of northern Skye began *ca.* 2450 BP (Innes 1983b) at a rate of 0.2 –1.7 mm yr^{-1}, a result either of human interference, particularly burning and intensive grazing, or of climatic shifts, in particular, climatic extremes.

TYPE REGION GB-zd *pro parte*, LOCH MAREE AREA (H.J.B. Birks with contributions from Hilary H. Birks, P.D. Kerslake, and Sylvia M. Peglar)

The Loch Maree area within type region GB-zd (Ross and Cromarty) is an area of very considerable ecological interest today (Ratcliffe 1977b). It supports the northwesternmost native pinewoods in the British Isles, the northernmost extensive oakwood, a range of treeless vegetation types occurring from sea-level to over 1000 m elevation, and a rich flora of contrasting plant-geographical affinities (arctic, arctic–alpine, southern, Atlantic). It contains the Beinn Eighe National Nature Reserve, including the Loch Maree woods of Coille na Glas-Leitire, the Loch Maree islands, and the Letterewe oakwoods.

Several palaeoecological studies have been completed in the area involving pollen analysis (Durno & McVean 1959; Birks 1972b; Pennington *et al.* 1972; Kerslake 1982; Ballantyne & Whittington 1987), macrofossils (Birks 1972b; Kerslake 1982), peat stratigraphy (Durno & McVean 1959; Birks 1975, Kerslake 1982), and sediment geochemistry (Pennington *et al.* 1972).

Altitude: 0–1008 m. The palaeoecological sites studied are all low-lying at about 30–50 m elevation.

Climate: Mean January temperature 3.3°C; mean July temperature 13.4°C; corresponding minima are 0.5°C and 9.7°C; precipitation (mainly as rain) ranges from 201 cm yr^{-1} at Anacaun (28 m elevation at southeast end of Loch Maree) to 152 cm yr^{-1} towards the west coast and to over 254 cm yr^{-1} on higher ground; over 200 wet days yr^{-1} (a day with at least 1 mm rain); high average relative humidity and a high incidence of cloud, often occurring as fog on the mountains; snowfall is relatively unimportant with only about 28 days yr^{-1} with snow lying at low elevation; snow lie increases with altitude; winds are predominantly from the west although Loch Maree is relatively sheltered from winds, with only 2.7 gales yr^{-1}; gales are frequent near the coast and on the high tops; average daily sunshine is only 2.4 hours day^{-1}; mild northern oceanic climate.

Geology: The Moine Thrust Plane passes near the eastern end of Loch Maree. Acid Moine schist and granulite occur east of the Plane, whereas a complex of acidic Precambrian and Cambrian quartzite and Torridonian sandstone form the rugged landscape to the west of the Plane. Cambrian quartzite and pipe-rock and dolomitic shale and mudstone outcrop locally. Basic Lewisian gneiss and hornblende, graphitic, and mica schist characterize the northern side of Loch Maree.

Topography: Rugged and mountainous, heavily influenced by glaciation. Loch Maree lies in a long, narrow, and deep glacially eroded valley. The wider and shallower section towards the western end contains several islands. The catchment of the loch is very large. Slopes around the loch are steep and rocky, rising up to the high and spectacular hills of Beinn Eighe (1008 m), Meall Ghiubhais (878 m), and Slioch (980 m), with impressive cliffs, deeply cut corries, and extensive screes. Mountains decrease in height and individuality towards the sea. All the low-lying valleys and slopes are plastered with glacial drift, particularly the so-called "hummocky moraine" thought to be deposited during the Loch Lomond (= Younger Dryas) Stadial.

Population: Generally at a very low density, even for the Scottish Highlands, with much less than 1 person ha⁻¹. There has been nearly a 20% decrease in population in the last 30 years. Small villages only.

Vegetation: The area is famous for its native pinewoods on the southern shore of Loch Maree and on the islands in the loch and around Loch Clair. These are some of the most northerly and westerly native pinewoods in Scotland. The most northerly extensive oakwood in Scotland occurs on the north shore of Loch Maree (Ratcliffe 1977b).

The pinewoods extend to about 305 m elevation and are dominated by *Pinus sylvestris* var. *scotica*, with *Betula pubescens* and *B. pendula*, especially on flushed soils, and some *Ilex aquifolium*, *Sorbus aucuparia*, *Prunus padus*, and *Juniperus communis*, particularly on the islands. *Alnus glutinosa* occurs locally along streams, on river alluvium, and by loch shores. The oakwoods are dominated by trees intermediate between *Quercus robur* and *Q. petraea*, along with *Ulmus glabra*, *Fraxinus excelsior*, *Populus tremula*, *Alnus glutinosa*, and *Corylus avellana*.

Low-lying treeless morainic ground is covered by blanket-bog communities, soligenous mires, and species-poor *Agrostis–Festuca* grassland. Tree-line is about 305 m and above this *Juniperus communis* ssp. *nana* scrub occurs locally. On extremely wind-blasted cols and plateaux, prostrate *Arctostaphylos alpina–Calluna vulgaris* heaths commonly occur. On sheltered north- or east-facing boulder-strewn slopes tall *Vaccinium myrtillus–Calluna vulgaris* stands occur with a wide variety of Atlantic bryophytes. Species-rich grasslands occur locally associated with outcrops of dolomitic mudstone, and ungrazed rock ledges support tall-herb and fern-dominated communities.

The higher ground has some moderately late snow-beds but these are generally rare and local. Summit vegetation consists of *Racomitrium lanug-*

inosum heath, with some *Carex bigelowii* and *Nardus stricta*. On basic rocks this moss heath is rich in cushion herbs such as *Armeria maritima* and *Silene acaulis*. A sparse summit vegetation of wind-blasted erosion surfaces is a *Luzula spicata–Juncus trifidus* "fell-field".

The area as a whole has a very rich and phytogeographically diverse flora of vascular plants, bryophytes, and lichens. The cryptogams particularly display the close juxtaposition of arctic, arctic–alpine, and northern species on the hills and Atlantic and southern species in sheltered localities in the woods, reflecting the wide range of climate, geology, and habitats within this comparatively small area (Birks 1972b).

Soils: Soils of well-drained sites on sandstone and quartzite may be sandy and strongly leached with well-developed podsol profiles. Subalpine and alpine soils tend to be skeletal with a layer of raw humus overlying stones or bedrock. On poorly drained sites gleys and peats are widespread, especially on the lower gentle slopes, where blanket bog has developed extensively in treeless areas. Soils on the more basic rocks are acidic or mildly basic brown earths. Extensive alluvial deposits have accumulated by some of the rivers as they enter Loch Maree.

Land use: Nature conservation with some sheep walk and deer forest. Small areas of cattle farming in the low-lying valleys. Hotels by the loch.

Reference site 29. Loch Maree (Birks 1972b)

Latitude 57°40′N, Longitude 5°30′W. Elevation 20 m. Age range 9500–0 BP. Lake. Six ¹⁴C dates. 467 cm sediment. Five local and five regional pollen-assemblage zones (paz). (Fig. 4.22)

Local paz	Approximate age	Regional paz
LME-1	*ca.* 9500–8950 BP	*Juniperus*
LME-2	8950–8250 BP	*Betula–Corylus/ Myrica*
LME-3	8250–6500 BP	*Pinus –Pteridium*
LME-4	6500–4200 BP	*Pinus–Alnus*
LME-5	4200–0 BP	*Calluna –Pinus*

In addition an *Empetrum*-dominated paz occurs below the *Juniperus* paz at Loch Clair (Pennington *et al.* 1972) and at Subhainn Lochan (Kerslake 1982), dating from *ca.* 9500 to 9800 BP or older. Although not distinguished by Birks (1972b), *Myrica gale* pollen appears to be particularly characteristic of the *Calluna–Pinus* paz (Kerslake 1982).

Event stratigraphy   This is presented in Figure 4.23.

Discussion

(1) Weichselian (= Devensian) late-glacial sediments are either absent from the larger lochs investigated (Loch Maree; Birks 1972b; Loch Clair; Pennington *et al.* 1972) or, where present, have not been studied in any detail palynologically (Kerslake 1982), for example Subhainn Lochan, where there is a basal ^{14}C date of 11140 BP (Q-2304). Loch Lomond (= Younger Dryas) Stadial glaciers were confined to the high corries and some of the valleys where glaciers descended from the centres of ice accumulation. Glaciers ranged in size from less than 0.25 km^2 to over 25 km^2. Equilibrium firn-lines elevation ranged from 409 to 474 m and rose from west to east (Ballantyne & Sutherland 1987). Rock glaciers and protalus ramparts (Sissons 1975, 1976) probably formed during the Stadial.

The basal minerogenic sediments at Lochs Maree and Clair may represent outwash from the local Loch Lomond Stadial glaciers. Small basins such as Subhainn Lochan on the Loch Maree islands may have not received this glacial outwash, as they were protected from outwash deposition by being on rock islands.

(2) This area appears to have witnessed the earliest expansion of *Pinus sylvestris* in Scotland between 8900 and 8250 BP (Birks 1989). However, considerable variation exists in the timing of pine-pollen expansion from site to site within this small area (8900 BP Eilean Subh na Sroine, 8250 BP Loch Maree, 8010 BP Subhain Bog, 7795 BP Subhain Lochan, 7500 BP Loch Clair) (Birks 1972b; Pennington *et al.* 1972; Kerslake 1982). The reasons for this diachronism are unknown. The Loch Maree pine populations today are very distinct genetically from other Scottish and European mainland populations, as assessed by monoterpene and isozyme analyses (Forrest 1980, 1982; Kinloch *et al.*

1986), suggesting that the history of pine in the Loch Maree area was different from that in other areas in the Scottish Highlands.

(3) *Pinus* was the dominant pollen type in the mid-Holocene (Fig. 4.22), along with some *Betula* and *Alnus glutinosa*. *Quercus, Ulmus, Populus tremula, Sorbus aucuparia, Fraxinus excelsior, Ilex aquifolium*, and *Salix* were present in low amounts throughout.

(4) A gradual decline in *Pinus* pollen values began at Loch Maree *ca.* 7000 BP but occurred rapidly at *ca.* 4200 BP (Fig. 4.22; Birks 1972b). Considerable variation from site to site in the timing of the first expansion of pine is matched in the occurrence, extent, and rapidity of declining pine-pollen values in the late Holocene. Loch Clair shows no pronounced rapid decline but only gradually falling values from *ca.* 6500 to 5000 BP, followed by consistent values to the present day (Pennington *et al.* 1972). On the Loch Maree islands pine pollen does not decline at all on Eilean Subh na Sroine, a steep, rocky island with extensive pinewoods and very little mire vegetation today (Kerslake 1982). In contrast, on Eilean Subhain, an island with abundant mires and scattered pinewood today, a well-marked pine decline occurs at Subhain Bog at 3800 BP. This is accompanied by an expansion of *Rhynchospora alba*, Cyperaceae undiff., *Myrica gale*, and *Menyanthes trifoliata* pollen. A distinct pine decline also occurs at Subhain Lochan between 4200 and 4000 BP, accompanied by an expansion of *Myrica gale* and *Calluna vulgaris* pollen (Kerslake 1982).

The most likely hypothesis to explain these patterns proposes a shift towards increased waterlogging and associated paludification beginning at *ca.* 7000 BP. This shift accelerated at *ca.* 4000 BP. At sites with well-drained soils on steep slopes (Clair, Eilean Subh na Sroine), there was little or no pine decline *ca.* 4000 BP. At sites where topography was suitable for blanket-bog development (Maree, Eilean Subhain), pine declined *ca.* 3800–4000 BP. Increased waterlogging, presumably due to an increase in precipitation and associated shifts to greater oceanicity, diminished the number of good seed years and successful pine regeneration. The increase in precipitation:evaporation ratio caused

136

LOCH MAREE, Scotland
Selected Pollen Percentages
Analyst: Hilary H. Birks, 1969

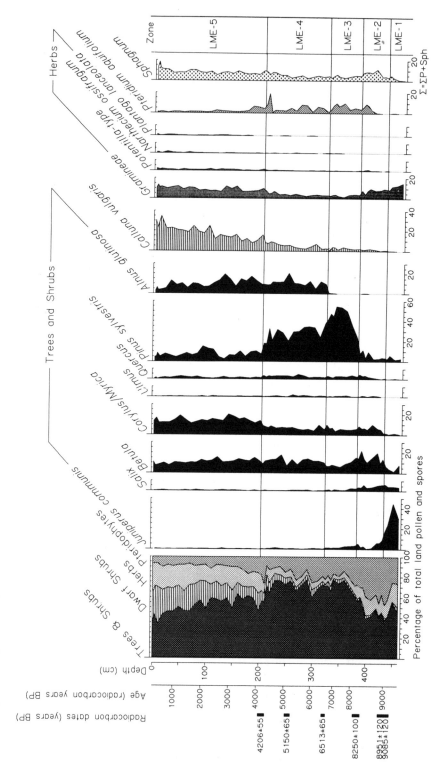

Fig. 4.22   Summary pollen diagram from Loch Maree, West Ross. Selected major pollen and spore taxa only are shown. The complete diagram is in Birks (1972 b). Sph = *Sphagnum*

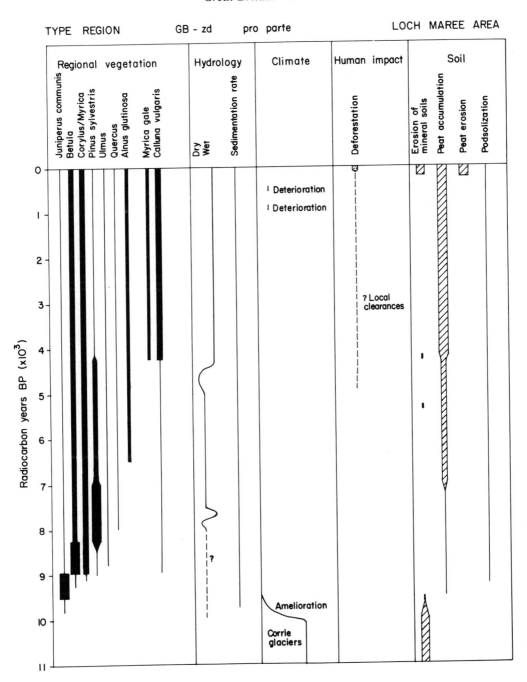

Fig. 4.23   Event stratigraphy for the Loch Maree area

waterlogging and paludification, thereby reducing the extent of suitable habitat for pine (Birks1972b).

Frequent fossil pine stumps in the area range in ^{14}C age from 7705±55 BP (Q-2236) to 4450±100 BP (Q-1150) (Birks 1975; Kerslake 1982).

(5) Archaeological evidence for human habitation is sparse until the time of Norse settlement on the west and north coasts of Scotland *ca.* 1100 BP. Palynological evidence for human activity is slight. Intense human activity centred on the local oakwoods for charcoal burning from medieval times to 300 BP (Durno & McVean 1959).

(6) Structured and amorphous solifluction features occur at high elevations in the area. Evidence exists for renewed amorphous solifluction between 900 and 500 BP (Ballentyne 1986). Modern rates of down-slope movement of solifluction lobes are *ca.* 6–35 mm yr^{-1}, whereas mid-Holocene rates probably did not exceed 6 mm yr^{-1} (Ballantyne 1986). Deflation surfaces, fell-field vegetation, and deposits of windblown sands formed by niveo-aeolian activity occur locally on the Torridonian sandstone mountains. Sand accumulation began *ca.* 8000 yr BP but was greatly reduced by mid-Holocene vegetational development. Massive erosion and redeposition have occurred in the last 200 years as a result of excessive sheep grazing (Ballantyne & Whittington 1987). Many hillslope debris flows similarly originated from the last 200 years (Innes 1983a). Their formation is often initiated by intense rainstorms (Ballantyne 1991), but formation is much more frequent and more destructive today than in the past, probably as a result of recent land-use changes, such as increased burning and grazing (Innes 1983a).

(7) Sediment accumulation rates in Lochs Maree and Clair are remarkably constant throughout the Holocene, implying little or no increase in inwashing of catchment material during the mid or late Holocene (Birks 1972b; Pennington *et al.* 1972). Support for this hypothesis comes from geochemical analyses at Loch Clair where the C, Na, K, and Mg stratigraphies all indicate no increase in erosional transport of soil material. It is thus possible that much if not all of the deforestation of the area occurred naturally by the death of trees by wind, fire,

and waterlogging, by lack of regeneration due to unfavourable edaphic and climatic conditions and lack of seed parents, and by the growth of blanket bog (Birks 1972b).

CONCLUSIONS

The emphasis of the synthesis for Scotland has been on the environmental history (biotic, edaphic, climatic, etc.) over the last 10000–13000 years in four relatively small but well-studied areas in Scotland, rather than on the vegetational history of Scotland as a whole. The major patterns of Holocene tree-spreading in Scotland based on all the available ^{14}C-dated pollen sequences in Scotland have been discussed by Birks (1989) and are not repeated here.

A major problem in evaluating the event stratigraphies for the four regions presented is to try to identify the likely causes for particular events. Clearly all the regions have experienced considerable environmental changes during the last 10000–11000 years, with major climatic shifts at about 10000 BP at the Weichselian–Holocene boundary. Many biotic and edaphic changes occurred during the Holocene, and presumably some of these were a result of climatic changes, whereas others were a result of human impact or natural biotic processes. The available data provide few clear and unambiguous indications of climatic change during the Holocene. Presumably climatic changes have been occurring throughout the Holocene, but these changes would only have been important in influencing biotic change in particular ecologically critical situations. The history of *Pinus sylvestris* in Scotland highlights these problems. Independent reconstructions from D/H isotope ratios of pine stumps in the Cairngorms suggest increases in precipitation at *ca.* 7300–7500, 6200–5800, 4200–3900, and 3300 BP. However, no major regional vegetational changes occurred in the Cairngorms at these times, presumably because no critical ecological threshold was crossed there. In contrast, the presumed increase in oceanicity at *ca.* 4200–3900 BP had major ecological effects in the Loch Maree area, in the Kyleakin region of Skye, and elsewhere in northwest Scotland, and may have resulted in the extinction of pine in much of northwest Scotland. Bradshaw (1993) suggests, however, that the decline of pine, at least in western Ireland, may have resulted from a decline in fire frequency and hence in the natural disturbance regime rather than as a direct response to climatic change. Fire is itself strongly influenced by climate, so disentangling the two will be a formidable task. Microscopic charcoal data associated with the decline of pine in northwest Scotland (Gear and Huntley 1991) do not, however, suggest any change in fire frequency after the pine decline.

Another problem that the regional syntheses for Scotland have highlighted concerns the reliance on ^{14}C dates from a variety of sedimentary materials for establishing an event stratigraphy for a region. If the available ^{14}C dates are reliable, there is a striking and surprising variation in the timing of expansion of the same pollen type within a small region (e.g. *Pinus* in the Loch Maree area).

The third problem that has arisen continually is the question whether a palaeoecological reconstruction for a site reflects the environmental history for that site only or for a wider geographical region. Clearly in mountainous areas such as Scotland, with large variations in soil, geology, climate, and topography over short distances, it is potentially dangerous to extrapolate from a single site to an entire type region. As more sites are studied in increasing detail and as more sites within a type region are studied, it appears that only a small part of the variance in the observed palaeoecological stratigraphy may be common to several sites and that considerable variance is associated with each individual site. An important and numerically large local site-imprint is apparently superimposed on the more regional-scale changes. The event stratigraphies almost certainly compound these local- and regional-scale changes.

ACKNOWLEDGEMENTS

I am indebted to Hilary Birks, Paul Kerslake, Sylvia Peglar, and Willie Williams for their co-operation in the production of this synthesis, to Hilary Birks, Sylvia Peglar, and Anne Birgit Ruud Hage for their valuable assistance, and to Herb Wright for his meticulous editing.

REFERENCES

Allott, T.E.H., Harriman, R. & Battarbee, R.W. 1992: Reversibility of lake acidification at the Round Loch of Glenhead, Galloway, Scotland. *Environmental Pollution 77*, 219–225.

Andrews, M. 1987: The past and present vegetation of Oronsay and Colonsay. *In* Mellars, P.A. (ed.) *Excavations on Oronsay: prehistoric human ecology on a small island*, 63–71. Edinburgh University Press, Edinburgh.

Ballantyne, C.K. 1986: Late Flandrian solifluction on the Fannich mountains, Ross-shire. *Scottish Journal of Geology 22*, 395–406.

Ballantyne, C.K. 1988: Ice-sheet moraines in southern Skye. *Scottish Journal of Geology 24*, 301–304.

Ballantyne, C.K. 1989: The Loch Lomond Readvance on the Isle of Skye, Scotland: glacier reconstruction and palaeoclimatic implications. *Journal of Quaternary Science 4*, 95–108.

Ballantyne, C.K. 1990: The Late Quaternary glacial history of the Trotternish Escarpment, Isle of Skye, Scotland, and its implications for ice-sheet reconstruction. *Proceedings of the Geological Association 101*, 171–186.

Ballantyne, C.K. 1991: Late Holocene erosion in upland Britain: climatic deterioration or human influence? *The Holocene 1*, 81–85.

Ballantyne, C.K. & Sutherland, D.G. 1987: *Wester Ross Field Guide*. 184 pp. Quaternary Research Association, Cambridge.

Ballantyne, C.K. & Whittington, G. 1987: Niveo-aeolian sand deposits on An Teallach, Wester Ross, Scotland. *Transactions of the Royal Society of Edinburgh: Earth Sciences 78*, 51–63.

Ballantyne, C.K., Benn, D.I., Lowe, J.J. & Walker, M.J.C. 1991: *The Quaternary of the Isle of Skye Field Guide*. 172 pp. Quaternary Research Association, Cambridge.

Battarbee, R.W., Flower, R.J., Stevenson, A.C. & Rippey, B. 1985a: Lake acidification in Galloway: a palaeoecological test of competing hypotheses. *Nature 314*, 350–352.

Battarbee, R.W., Appleby, P.G., Odell, K. & Flower, R.J. 1985b: ^{210}Pb dating of Scottish lake sediments, afforestation and accelerated soil erosion. *Earth Surface Processes and Landforms 10*, 137–142.

Battarbee, R.W., Flower, R.J., Stevenson, A.C., Jones, V.J., Harriman, R. & Appleby, P.G. 1988: Diatom and chemical evidence for reversibility of acidification of Scottish lochs. *Nature 332*, 530–532.

Battarbee, R.W., Stevenson, A.C., Rippey, B., Fletcher, C., Natanski, J., Wik, M. & Flower, R.J. 1989: Causes of lake acidification in Galloway, south-west Scotland: a palaeo-ecological evaluation of the relative roles of atmospheric contamination and catchment change for two acidified sites with non-afforested catchments. *Journal of Ecology 77*, 651–672.

Benn, D.I. 1992: The genesis and significance of "hummocky moraine": evidence from the Isle of Skye, Scotland. *Quaternary Science Reviews 11*, 781–799.

Benn, D.I., Lowe, J.J. & Walker, M.J.C. 1992: Glacier response to climatic change during the Loch Lomond Stadial and early Flandrian: geomorphological and palynological evidence from the Isle of Skye, Scotland. *Journal of Quaternary Science 7*, 125–144.

Bennett, K.D. 1984: The post-glacial history of *Pinus sylvestris* in the British Isles. *Quaternary Science Reviews 3*, 133–155.

Bennett, K.D. 1989: A provisional map of forest types for the British Isles 5000 years ago. *Journal of Quaternary Science 4*, 141–144.

Bennett, K.D. & Birks, H.J.B. 1990: Postglacial history of alder (*Alnus glutinosa* (L.) Gaertn.) in the British Isles. *Journal of Quaternary Science 5*, 123–133.

Bennett, K.D., Fossitt, J.A., Sharp, M.J. & Switsur, V.R. 1990: Holocene vegetational and environmental history at Loch Lang, South Uist, Western Isles, Scotland. *New Phytologist 114*, 281–298.

Bennett, K.D., Boreham, S., Sharp, M.J. & Switsur, V.R. 1992: Holocene history of environment, vegetation and human settlement on Catta Ness, Lunnasting, Shetland. *Journal of Ecology 80*, 241–273.

Birks, H.H. 1970: Studies in the vegetational history of Scotland. I. A pollen diagram from Abernethy Forest, Inverness-shire. *Journal of Ecology 58*, 827–846.

Birks, H.H. 1972a: Studies in the vegetational history of Scotland. II. Two pollen diagrams from the Galloway Hills, Kirkcudbrightshire. *Journal of Ecology 60*, 183–217.

Birks, H.H. 1972b: Studies in the vegetational history of Scotland. III. A radiocarbon-dated pollen diagram from Loch Maree, Ross and Cromarty. *New Phytologist 71*, 731–754.

Birks, H.H. 1975: Studies in the vegetational history of Scotland. IV. Pine stumps in Scottish blanket peats. *Philosophical Transactions of the Royal Society of London B, 270*, 181–226.

Birks, H.H. 1984: Late-Quaternary pollen and plant macrofossil stratigraphy at Lochan an Druim, northwest Scotland. *In* Haworth, E.Y. & Lund, J.W.G. (eds) *Lake sediments and environmental history*, 377–405. University of Leicester Press, Leicester.

Birks, H.H. & Mathewes R.W. 1978: Studies in the vegetational history of Scotland. V. Late Devensian and early Flandrian pollen and macrofossil stratigraphy at Abernethy Forest, Inverness-shire. *New Phytologist 80*, 455–484.

Birks, H.J.B. 1973: *Past and present vegetation of the Isle of Skye—a palaeoecological study*. 415 pp. Cambridge University Press, London.

Birks, H.J.B. 1977: The Flandrian forest history of Scotland. A preliminary synthesis. *In* Shotton, F.W. (ed.) *British Quaternary Studies Recent Advances*, 119–135.

Clarendon Press, Oxford.

Birks, H.J.B. 1988: Long-term ecological change in the British uplands. *In* Usher, M.B. & Thompson D.B.A. (eds) *Ecological Change in the Uplands,* 37–56. Blackwell Scientific Publications, Oxford.

Birks, H.J.B. 1989: Holocene isochrone maps and patterns of tree-spreading in the British Isles. *Journal of Biogeography 16*, 503–540.

Birks, H.J.B. & Madsen, B.J. 1979: Flandrian vegetational history of Little Loch Roag, Isle of Lewis, Scotland. *Journal of Ecology 67*, 825–842.

Birks, H.J.B. & Williams, W. 1983: Late-Quaternary vegetational history of the Inner Hebrides. *Proceedings of the Royal Society of Edinburgh, B 83*, 269–292.

Birks, H.J.B., Juggins, S. & Line, J.M. 1990: Lake surface-water chemistry reconstructions from palaeolimnological data. In Mason, B.J. (ed.) *The Surface Waters Acidification Programme*, 301–313. Cambridge University Press, Cambridge.

Birse, E.L. 1971: *Assessment of climatic conditions in Scotland 3. The bioclimatic sub-regions.* 12 pp. Soil Survey of Scotland, Aberdeen.

Birse, E.L. 1980: Plant communities of Scotland: A preliminary phytocoenonia. *Soil Survey of Scotland Bulletin 4*, 235 pp.

Birse, E.L. 1984: The phytocoenonia of Scotland. additions and revision. *Soil Survey of Scotland Bulletin 5*, 120 pp.

Birse, E.L. & Dry, F.T. 1970: *Assessment of climatic conditions in Scotland 1. Based on accumulated temperature and potential water deficit.* 25 pp. Soil Survey of Scotland, Aberdeen.

Birse, E.L. & Robertson, L. 1970: *Assessment of climatic conditions in Scotland 2. Based on exposure and accumulated frost.* 41 pp. Soil Survey of Scotland, Aberdeen.

Boyd, W.E. & Dickson, J.H. 1987: A post-glacial pollen sequence from Loch a' Mhuilinn, north Arran: a record of vegetation history with special reference to the history of endemic *Sorbus* species. *New Phytologist 107*, 221–244.

Bradshaw, R. 1993: Forest response to Holocene climatic change: equilibrium or non-equilibrium. *In* Chambers, F.M. (ed.) *Climate change and human impact on the landscape*, 57–65. Chapman & Hall, London.

Brazier, V. & Ballantyne, C.K. 1989: Late Holocene debris cone evolution in Glen Feshie, western Cairngorm Mountains, Scotland. *Transactions of the Royal Society of Edinburgh: Earth Sciences 80*, 17–24.

Bridge, M.C., Haggart, B.A. & Lowe, J.J. 1990: The history and palaeoclimatic significance of subfossil remains of *Pinus sylvestris* in blanket peats from Scotland. *Journal of Ecology 78*, 77–99.

Brodin, Y.-W. & Gransberg, M. 1993: Responses of insects, especially Chironomidae (Diptera), and mites to 130 years of acidification in a Scottish lake. *Hydrobiologia 250*, 201–212.

Clapperton, C.M., Gunson, A.R. & Sugden, D.E. 1975: Loch Lomond Readvance in the eastern Cairngorms. *Nature 253*, 710–712.

Cornish, R. 1981: Glaciers of the Loch Lomond Stadial in the western Southern Uplands of Scotland. *Proceedings of the Geological Association 92*, 105–114.

Donner, J.J. 1957: The geology and vegetation of late-glacial retreat stages in Scotland. *Transactions of the Royal Society of Edinburgh 63*, 221–264.

Dubois, A.D. & Ferguson, D.K. 1985: The climatic history of pine in the Cairngorms based on radiocarbon dates and stable isotope analysis, with an account of the events leading up to its colonization. *Review of Palaeobotany and Palynology 46*, 55–80.

Dubois, A.D. & Ferguson, D.K. 1988: Additional evidence for the climatic history of pine in the Cairngorms, Scotland, based on radiocarbon dates and tree ring D/H ratios. *Review of Palaeobotany and Palynology 54*, 181–185.

Durno, S.E. & McVean, D.N. 1959: Forest history of the Beinn Eighe Nature Reserve. *New Phytologist 58*, 228–236.

Edwards, K.J. 1989: Meso-Neolithic vegetational impacts in Scotland and beyond: palynological considerations. *In* Bonsall, C. (ed.) *The Mesolithic in Europe*, 143–155. John Donald, Edinburgh.

Edwards, K.J. 1990: Fire and the Scottish Mesolithic: evidence from microscopic charcoal. *In* Vermeersch, P.M. & van Peer, P. (eds) *Contributions to the Mesolithic in Europe*, 71–79. Leuven University Press, Leuven.

Edwards, K.J. & Rowntree, K.M. 1980: Radiocarbon and palaeoenvironmental evidence for changing rates of erosion at a Flandrian stage site in Scotland. *In* Cullingford, R.A., Davidson, D.A. & Lewin, J (eds) *Timescales in Geomorphology*, 207–223. Wiley, Chichester.

Edwards, K.J., Ansell, M. & Carter, B.A. 1984: New Mesolithic sites in south-west Scotland and their importance as indicators of inland penetration. *Transactions of the Dumfriesshire and Galloway Natural History and Antiquarian Society 58 (1983)*, 9–15.

Edwards, K.J., Hirons, K.R. & Newell, P.J. 1991: The palaeoecological and prehistoric context of minerogenic layers in blanket peat: a study from Loch Dee, southwest Scotland. *The Holocene 1*, 29–39.

Fitzpatrick, E.A. 1964: The soils of Scotland. *In* Burnett, J.H. (ed.) *The vegetation of Scotland*, 36–63. Oliver & Boyd, Edinburgh.

Flower, R.J. & Battarbee, R.W. 1983: Diatom evidence for recent acidification in two Scottish lochs. *Nature 305*, 130–133.

Flower, R.J., Battarbee, R.W. & Appleby, P.G. 1987: The recent palaeolimnology of acid lakes in Galloway, southwest Scotland: diatom analysis, pH trends, and the rôle of afforestation. *Journal of Ecology 75*, 797–824.

Forrest, G.I. 1980: Genotypic variation among native Scots Pine populations in Scotland based on mono-terpene analysis. *Forestry 53*, 101–128.

Forrest, G.I. 1982: Relationship of some European Scots Pine populations to native Scottish woodlands based on mono-terpene analysis. *Forestry 55*, 19–37.

Gear, A.J. & Huntley, B. 1991: Rapid changes in the range limits of Scots Pine 4000 years ago. *Science 251*, 544–547.

Goudie, A.S., Viles, H.A. & Pentecost, A. 1993: The late-Holocene tufa decline in Europe. *The Holocene 3*, 181–186.

Grace, J. & Norton, D.A. 1990: Climate and growth of *Pinus sylvestris* at its upper altitudinal limit in Scotland: evidence from tree growth-rings. *Journal of Ecology 78*, 601–610.

Green, F.H.W. 1964: The climate of Scotland. *In* Burnett, J.H. (ed.) *The vegetation of Scotland*, 15–35. Oliver & Boyd, Edinburgh.

Green, F.H.W. 1974: Climate and weather. *In* Nethersole-Thompson, D. & Watson, A. (eds) *The Cairngorms—their natural history and scenery*, 228–236. Collins, London.

Hibbert, F.A. & Switsur, V.R. 1976: Radiocarbon dating of Flandrian pollen zones in Wales and northern England. *New Phytologist 77*, 793–807.

Hirons, K.R. & Edwards, K.J. 1990: Pollen and related studies at Kinloch, Isle of Rhum, Scotland, with particular reference to possible early human impacts on vegetation. *New Phytologist 116*, 715–727.

Hulme, P.D. & Shirriffs, J. 1985: Pollen analysis of a radio-carbon-dated core from North Mains, Strathallan, Perthshire. *Proceedings of the Society of Antiquaries of Scotland 115*, 105–113.

Huntley, B. 1981: The past and present vegetation of the Caenlochan National Nature Reserve, Scotland II. Palaeoecological investigations. *New Phytologist 87*, 189–222.

Huntley, B. & Birks, H.J.B. 1979: The past and present vegetation of the Morrone Birkwoods National Nature Reserve, Scotland. I. A primary phytosociological survey. *Journal of Ecology 67*, 417–446.

Innes, J.L. 1983a: Lichenometric dating of debris flow deposits in the Scottish Highlands. *Earth Surface Processes and Landforms 8*, 579–588.

Innes, J.L. 1983b: Stratigraphic evidence of episodic talus accumulation on the Isle of Skye, Scotland. *Earth Surface Processes and Landforms 8*, 399–403.

Johansen, J. 1975: Pollen diagrams from the Shetland and Faroe Islands. *New Phytologist 75*, 369–387.

Jones, V.J., Stevenson, A.C. & Battarbee, R.W. 1986: Lake acidification and the land-use hypothesis: a mid-post-glacial analogue. *Nature 322*, 157–158.

Jones, V.J., Stevenson, A.C. & Battarbee, R.W. 1989: Acidification of lakes in Galloway, south west Scotland: a diatom and pollen study of the post-glacial history of the Round Loch of Glenhead. *Journal of Ecology 77*, 1–23.

Jones, V.J., Kreiser, A.M., Appleby, P.G., Brodin, Y.-W.,

Dayton, J., Natanski, J.A., Richardson, N., Rippey, B., Sandøy, S. & Battarbee, R.W. 1990: The recent palaeolimnology of two sites with contrasting acid-deposition histories. *Philosophical Transactions of the Royal Society of London B 327*, 397–402.

Jones, V.J., Battarbee, R.W. & Hedges, R.E.M. 1993a: The use of chironomid remains for AMS ^{14}C dating of lake sediments. *The Holocene 3*, 161–163.

Jones, V.J., Flower, R.J., Appleby, P.G., Natakanski, J., Richardson, N., Rippey, B., Stevenson, A.C. & Battarbee, R.W. 1993b: Palaeolimnological evidence for the acidification and atmospheric contamination of lochs in the Cairngorm and Lochnagar areas of Scotland. *Journal of Ecology 81*, 3–24.

Keatinge, T.H. & Dickson, J.H. 1979: Mid-Flandrian changes in vegetation on Mainland, Orkney. *New Phytologist 82*, 585–612.

Kerslake, P.D. 1982: *Vegetational history of wooded islands in Scottish lochs*. 202 pp. PhD thesis, University of Cambridge.

Kinloch, B.B., Westfall, R.D. & Forrest, G.I. 1986: Caledonian Scots Pine: origins and genetic structure. *New Phytologist 104*, 703–729.

Lewis, F.J. 1905: The plant remains in the Scottish peat mosses. Part I. The Scottish Southern Uplands. *Transactions of the Royal Society of Edinburgh 41*, 699–723.

Macpherson, J.B. 1980: Environmental change during the Loch Lomond Stadial: evidence from a site in the Upper Spey Valley, Scotland. *In* Lowe, J.J., Gray, J.M. & Robinson, J.E. (eds) *Studies in the Lateglacial of North-West Europe*, 89–101. Pergamon Press, Oxford.

McVean, D.N. & Ratcliffe, D.A. 1962: *Plant communities of the Scottish Highlands*. 445 pp. Her Majesty's Stationery Office, London.

Miller, G.R. & Cummins, R.P. 1982: Regeneration of Scots pine *Pinus sylvestris* at a natural tree-line in the Cairngorm Mountains, Scotland. *Holarctic Ecology 5*, 27–34.

Moar, N.J. 1969: Late Weichselian and Flandrian pollen diagrams from south-west Scotland. *New Phytologist 68*, 433–467.

Murray, W.H. 1987: *Scotland's mountains*. 305 pp. Scottish Mountaineering Trust, Edinburgh.

O'Sullivan, P.E. 1973: Pollen analysis of Mor humus layers from a native Scots pine ecosystem, interpreted with surface samples. *Oikos 24*, 259–272.

O'Sullivan, P.E. 1974: Two Flandrian pollen diagrams from the east-central Highlands of Scotland. *Pollen et Spores 26*, 33–57.

O'Sullivan, P.E. 1975a: Early and Middle-Flandrian pollen zonation in the Eastern Highlands of Scotland. *Boreas 4*, 197–207.

O'Sullivan, P.E. 1975b: Radiocarbon-dating and prehistoric forest clearance on Speyside (East-Central Highlands of Scotland). *Proceedings of the Prehistoric Society 40*, 206–208.

O'Sullivan, P.E. 1976: Pollen analysis and radiocarbon dating of a core from Loch Pityoulish, Eastern Highlands of Scotland. *Journal of Biogeography 3*, 293–302.

O'Sullivan, P.E. 1977: Vegetation history and the native pine-woods. *In* Bunce, R.G.H. & Jeffers. J.N.R. (eds) *Native pinewoods of Scotland*, 60–69. Institute of Terrestrial Ecology, Cambridge.

Pears, N.V. 1967a: Wind as a factor in mountain ecology: some data from the Cairngorm Mountains. *Scottish Geographical Magazine 83*, 118–124.

Pears, N.V. 1967b: Present tree-lines of the Cairngorm Mountains, Scotland. *Journal of Ecology 55*, 815–829.

Pears, N.V. 1968a: Post-glacial tree-lines of the Cairngorm Mountains, Scotland. *Transactions of the Botanical Society of Edinburgh 40*, 361–394.

Pears, N.V. 1968b: The natural altitudinal limit of forest in the Scottish Grampians. *Oikos 19*, 71–80.

Pears, N.V. 1969: Post-glacial tree-lines of the Cairngorm Mountains, Scotland: Some modifications based on radio-carbon dating. *Transactions of the Botanical Society of Edinburgh 40*, 536–544.

Pears, N.V. 1972: Interpretation problems in the study of tree-line fluctuations. *In* Taylor, J.A. (ed.) *Forest meteorology*, 31–45. University College of Wales, Aberystwyth.

Pears, N.V. 1975a: Radiocarbon dating of plant macrofossils in the Cairngorm Mountains, Scotland. *Transactions of the Botanical Society of Edinburgh 42*, 255–260.

Pears, N.V. 1975b: Tree stumps in the Scottish hill peats. *Scottish Forestry 29*, 255–259.

Pears, N.V. 1988: Pine stumps, radiocarbon dates and stable isotope analysis in the Cairngorm Mountains: some observations. *Review of Palaeobotany and Palynology 54*, 175–185.

Peglar, S.M. 1979: A radiocarbon-dated pollen diagram from Loch of Winless, Caithness, north-east Scotland. *New Phytologist, 82*, 245–263.

Pennington, W. (Mrs T.G. Tutin), Haworth, E.Y., Bonny, A.P. & Lishman, J.P. 1972: Lake sediments in northern Scotland. *Philosophical Transactions of the Royal Society of London B 264*, 191–294.

Preece, R.C., Bennett, K.D. & Robinson, J.E. 1984: The biostratigraphy of an early Flandrian tufa at Inchrory, Glen Avon, Banffshire. *Scottish Journal of Geology 20*, 143–159.

Price, R.J. 1976: *Highland landforms*. 109 pp. Highlands and Islands Development Board, Inverness.

Ralska-Jasiewiczowa, M. 1986. *Palaeohydrological changes in the temperate zone in the last 15 000 years. Subproject B. Lake and mire environments. Project Catalogue for Europe*. 161 pp. Lund University.

Rapson, S.C. 1985: Minimum age of corrie moraine ridges in the Cairngorm Mountains, Scotland. *Boreas 14*, 155–159.

Ratcliffe, D.A. 1968: An ecological account of Atlantic

bryophytes in the British Isles. *New Phytologist 67*, 365–439.

Ratcliffe, D.A. 1974: The vegetation. *In* Nethersole-Thompson, D. & Watson, A. (eds) *The Cairngorms—their natural history and scenery*, 42–76. Collins, London.

Ratcliffe, D.A. 1977a: *Highland flora.* 111 pp. Highlands and Islands Development Board, Inverness.

Ratcliffe, D.A. (ed.) 1977b: *A nature conservation review. Volume 2: Site accounts.* 320 pp. Cambridge University Press, Cambridge.

Ratcliffe, D.A. & Walker, D., 1958: The Silver Flowe, Galloway, Scotland. *Journal of Ecology 46*, 407–445.

Robertson, J.S. 1984: *A key to the common plant communities of Scotland.* 95 pp. Soil Survey of Scotland Monograph, Aberdeen.

Robinson, D.E. & Dickson, J.H. 1988: Vegetational history and land use: a radiocarbon-dated pollen diagram from Machrie Moor, Arran, Scotland. *New Phytologist 109*, 223–251.

Sissons, J.B. 1967: *The evolution of Scotland's scenery.* 259 pp. Oliver & Boyd, Edinburgh.

Sissons, J.B. 1973: Delimiting the Loch Lomond Readvance in the eastern Grampians. *Scottish Geographical Magazine 89*, 138–139.

Sissons, J.B. 1975: A fossil rock glacier in Wester Ross. *Scottish Journal of Geology 11*, 83–86.

Sissons, J.B. 1976: A remarkable protalus rampart complex in Wester Ross. *Scottish Geographical Magazine 92*, 182–190.

Sissons, J.B. 1977: The Loch Lomond Readvance in southern Skye and some palaeoclimatic inferences. *Scottish Journal of Geology 13*, 23–36.

Sissons, J.B. 1979a: The Loch Lomond Advance in the Cairngorm Mountains. *Scottish Geographical Magazine 95*, 66–82.

Sissons, J.B. 1979b: Palaeoclimatic inferences from former glaciers in Scotland and the Lake District. *Nature 278*, 518–521.

Sissons, J.B. & Grant, A.J.H. 1972: The last glaciers in the Lochnagar area, Aberdeenshire. *Scottish Journal of Geology 8*, 85–93.

Sissons, J.B. & Walker, M.J.C. 1974: Late-glacial site in the central Grampian Highlands. *Nature 249*, 822–824.

Slesser, M. 1975: *The Island of Skye.* 192 pp. Scottish Mountaineering Club District Guide, Edinburgh.

Spence, D.H.N. 1960: Studies in the vegetation of Shetland. III. Scrub in Shetland and in South Uist, Outer Hebrides. *Journal of Ecology 48*, 73–95.

Stevenson, A.C. & Thompson, D.B.A. 1993: Long-term changes in the extent of heather moorland in upland Britain and Ireland: palaeoecological evidence for the importance of grazing. *The Holocene 3*, 70–76.

Stevenson, A.C., Jones, V.J. & Battarbee, R.W. 1990: The cause of peat erosion: a palaeolimnological approach. *New Phytologist 114*, 727–735.

Stewart, D.A., Walker, A. & Dickson, J.H. 1984: Pollen diagrams from Dubh Lochan, near Loch Lomond. *New Phytologist 98*, 531–549.

Sugden, D.E. 1970: Landforms of deglaciation in the Cairngorm mountains, Scotland. *Transactions of the Institute of British Geographers 51*, 201–219.

Sugden, D.E. 1971: The significance of periglacial activity on some Scottish mountains. *The Geographical Journal 137*, 388–392.

Sugden, D.E. 1973: Delimiting Zone III glaciers in the eastern Grampians. *Scottish Geographical Magazine 89*, 63–64.

Sugden, D.E. 1974: Landforms. *In* Nethersole-Thompson, D. & Watson, A. (eds) *The Cairngorms—their natural history and scenery*, 210–221. Collins, London.

Sugden, D.E. 1977: Did glaciers form in the Cairngorms in the 17th–19th centuries? *Cairngorm Club Journal 18*, 189–201.

Sugden, D.E. 1980: The Loch Lomond Advance in the Cairngorms (a reply to J.B. Sissons). *Scottish Geographical Magazine 96*, 18–19.

Vasari, Y. 1977: Radiocarbon dating of the late glacial and early Flandrian vegetational succession in the Scottish Highlands and the Isle of Skye. *In* Gray, J.M. & Lowe, J.J. (eds) *Studies in the Scottish Lateglacial environment*, 143–162. Pergamon, Oxford.

Vasari, Y. & Vasari, A. 1968: Late- and post-glacial macrophytic vegetation in the lochs of northern Scotland. *Acta Botanica Fennica 80*, 120 pp.

Walker, M.J.C. 1975: Late Glacial and Early Postglacial environmental history of the central Grampian Highlands, Scotland. *Journal of Biogeography 2*, 265–284.

Walker, M.J.C. 1984a: Pollen analysis and Quaternary research in Scotland. *Quaternary Science Reviews 3*, 369–404.

Walker, M.J.C. 1984b : A pollen diagram from St. Kilda, Outer Hebrides, Scotland. *New Phytologist 97*, 99–113.

Walker, M.J.C. & Lowe, J.J. 1990: Reconstructing the environmental history of the last glacial–interglacial transition: evidence from the Isle of Skye, Inner Hebrides, Scotland. *Quaternary Science Reviews 9*, 15–49.

Walker, M.J.C., Ballantyne, C.K., Lowe, J.J. & Sutherland, D.G. 1988: A reinterpretation of the lateglacial environmental history of the Isle of Skye, Inner Hebrides, Scotland. *Journal of Quaternary Science 3*, 135–146.

Walther, M. 1984: *Geomorphologische Unters-uchungen zum Spätglazial und Frühholozän in den Cuillin Hills (Insel Skye, Schottland).* PhD thesis, Free University of Berlin.

Whittow, J.B. 1977: *Geology and scenery in Scotland.* 362 pp. Penguin, London.

Williams, W. 1977: *The Flandrian vegetational history of the Isle of Skye and the Morar Peninsula.* PhD thesis, University of Cambridge.

5

Faroe Islands

J. Jóhansen[†]

INTRODUCTION

The Faroe Islands are situated in the North Atlantic between Norway, Scotland and Iceland, between 61°20′ and 62°24′ N and 6°15′ and 7°41′ W (Fig. 5.1). The distance from the southernmost to the northernmost point is about 113 km, and the west–east distance is approx 75 km. All but one of the 18 islands are inhabited. Total area is about 1400 km².

Out of many essentially similar pollen diagrams two well-dated ones have been selected for this synthesis.

TYPE REGION FAROE ISLANDS

Altitude: 0–882 m.

Climate: Extreme oceanic with mild winters and cool summers. Mean February temperature 4°C, July 11°C. Precipitation 1500 mm yr⁻¹. Lying on the main route of the North Atlantic cyclones, the weather is very changeable.

Geology: The Faroes were formed during the Tertiary (Upper Paleocene and partly lower Eocene) by volcanic activity. Geologically the islands belong to the plateaubasalts of the North Atlantic. During a pause in the volcanism coal deposits were formed in some places (Fig. 5.2). The Faroes had their own ice cap during last glaciation. It is supposed that some

of the higher mountain peaks were ice-free during the last glaciation due to the small area of the shield and the comparatively high mountains (Jørgensen & Rasmussen 1986). It can therefore be assumed that some arctic elements survived on the highest tops (*Papaver radicatum*, *Omalotheca supina* a.o.). Interglacial deposits have been found at least in two places. Holocene deposits are well represented: extensive peat bogs cover most of the lowlands and are also found at considerable heights (several hundred metres).

Topography: Small islands with very steep mountains and many vertical cliffs facing the open ocean as well as inland. At the coasts, there are flat or sloping areas with habitation and cultivation. A special feature are the *ca.* 800 ravines found all over the islands. They are especially interesting botanically because sheep have not had access, and the vegetation is therefore protected from grazing.

Population: 45000 inhabitants in 1993.

Vegetation: The islands are completely devoid of trees except for those in gardens and plantations. Following Nordic Vegetation Types, published by the Nordic Ministers Council, the following are the main types: Alpine vegetation, Bogs, Sea-shore vegetation, Culture conditioned vegetation, Fresh water vegetation, and Substrate-dependent vegetation.

Soils: Peat, podzols, and pure mineral soils.

Palaeoecological Events During the Last 15 000 Years: Regional Syntheses of Palaeoecological Studies of Lakes and Mires in Europe.
Edited by B.E. Berglund, H.J.B. Birks, M. Ralska-Jasiewiczowa and H.E. Wright.   © 1996 John Wiley & Sons Ltd.

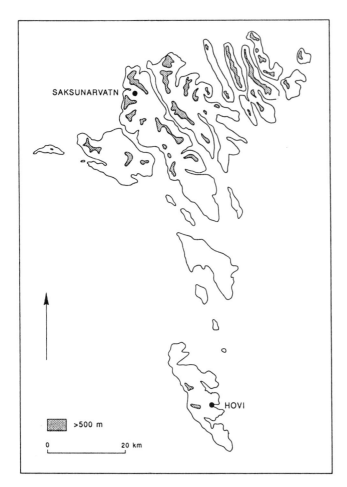

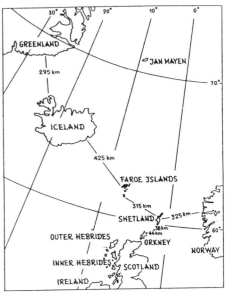

Fig. 5.1 Maps showing the location of the Faroe Islands in the North Atlantic, and the location of the two pollen sites within the islands

Land use: The only crops are grass (winter fodder for cattle and sheep) and potatoes for private use. Approximately 6% of the total area is cultivated.

Reference site 1. Saksunarvatn (Jóhansen 1982)

Latitude 62°14′N, Longitude 7°9′W. Elevation 25 m. Age range about 10000/950–2900 BP. Lake, five local pollen-assemblage zones (paz). (Fig. 5.3)

1. 10000–9100 BP   *Betula nana–Sedum –*
                            Salix–Oxyria–Huperzia–
                            Gramineae
2. 9100–8900 BP   Gramineae–Cyperaceae

3. 8900–7000 BP   *Juniperus*–Gramineae–
                            Cyperaceae
4. 7000–6000 BP   *Juniperus–Calluna–*
                            Gramineae–Cyperaceae
5. 6000–2900 BP   *Calluna–Juniperus –*
                            Gramineae–Cyperaceae

Reference site 3. Hovi B (Jóhansen 1982)

Latitude 61°30′N, longitude 6°46′W. Elevation 12 m. Age range about 3000–0 BP. Mire with open peat profile, two local pollen-assemblage zones (paz). (Fig. 5.4)

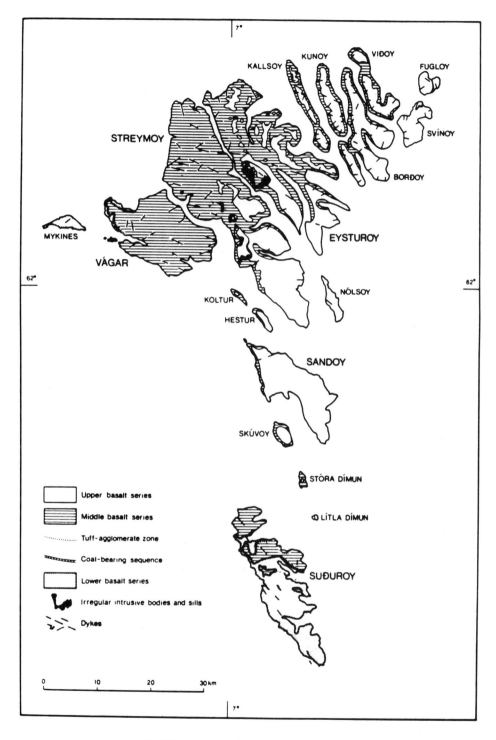

Fig. 5.2   Geological map of the Faroe Islands

Saksunarvatn lake, FAROE ISLANDS
Pollen percentage diagram (boring 2)
Analyst: Johs. Johansen

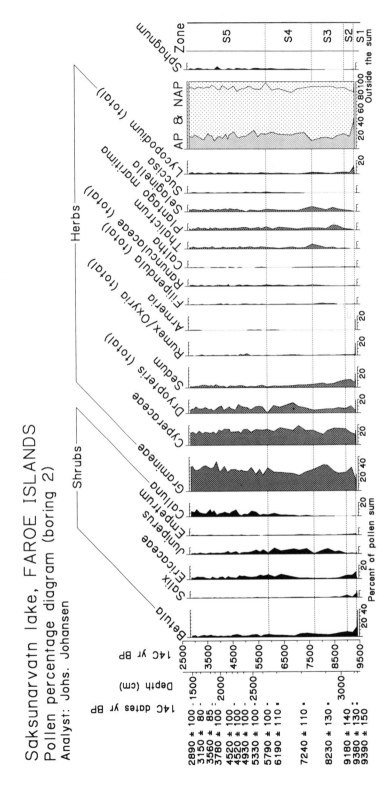

Fig. 5.3 Pollen diagram from lake Saksunarvatn. Only selected pollen and spore types are shown

1. 3000–1100 BP   *Juniperus*–Ericales–*Caltha*– *Dryopteris*–*Filipendula*– Gramineae –Cyperaceae
2. 1100–0 BP   Gramineae–Cyperaceae– Ericaceae–*Potentilla*– *Plantago lanceolata*

The pollen diagrams

Two pollen diagrams are presented. They cover the time period from *ca.* 10000 to the present. Type-region pollen-assemblage zones for the Faroe Islands (FR):

1. 10000–9000 BP   *Sedum/Rhodiola*– *Huperzia*– *Betula nana*–*Salix*– *Oxyria* assemblage. This is a period of unstable soil with a pioneer vegetation. Besides the above-mentioned taxa, pollen of *Koenigia,Cerastium, Saxifraga, Angelica, Archangelica*, and *Armeria* are found.

2. 9000–7000 BP   *Juniperus*–*Salix* (probably *S. arctica* and *S.phylicifolia*) assemblage. These shrubs cover the lowlands up to about 300 m. Gramineae and Cyperaceae are also very important.Together with *Juniperus* and *Salix* the tall-herb flora includes *Filipendula, Ranunculus, Caltha, Angelica, Archangelica, Dryopteris, Athyrium*, and different species of *Luzula*, Cyperaceae and Gramineae.

3. 7000–1100 BP   *Juniperus* –*Calluna*– Gramineae–Cyperaceae assemblage. This period is characterized by increased leaching and acidification of the soil,

with a slow but consistent increase of *Calluna, Juniperus, Salix*, and tall-herb vegetation which persist until the land occupation.

4. 1100–0 BP   Gramineae–Cyperaceae–Ericaeae–*Calluna* assemblage. Heathland vegetation characterized the landscape.

PALAEOECOLOGICAL EVENTS (Fig. 5.5)

Unique patterns

Especially to be noted is the total absence of trees during the entire Holocene. Only in one place has there been found wood remains of *Betula pubescens*, dated between 3891 and *ca.* 1000 BP (unpublished).

Climatic events

There are no precise climate indications in the pollen diagrams. The taxa all have a broad amplitude of temperature tolerance. They are distributed today towards the north as well as south.

Betula nana in the Preboreal, however, indicates a continental climate, and its disappearance consequently suggests a change towards oceanic conditions.

We do not know the temperatures of the Faroe Islands in the past. The recent vegetation is a result of sheep grazing and human interference and to a very limited extent of climate.

The absence of forest, however, is a consequence of a climate unfavourable for trees but probably also because the islands' position in the North Atlantic and their small size are unfavourable for immigration.

Human impact

Three landnam phases probably occurred in the Faroe Islands. The first, about 2300 BC, is only based on the fact that *Plantago lanceolata* pollen first occurs at this time (Jóhansen 1971, 1979). The

150

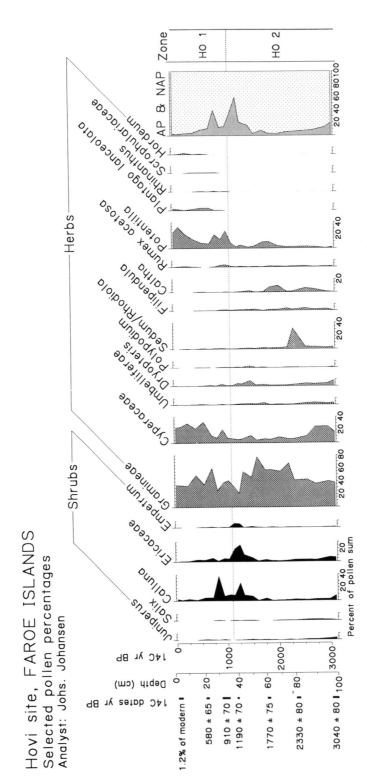

Fig. 5.4 Pollen diagram from Hovi B. Only selected pollen and spore types are shown

TYPE REGION FAROE ISLANDS   EVENT STRATIGRAPHY

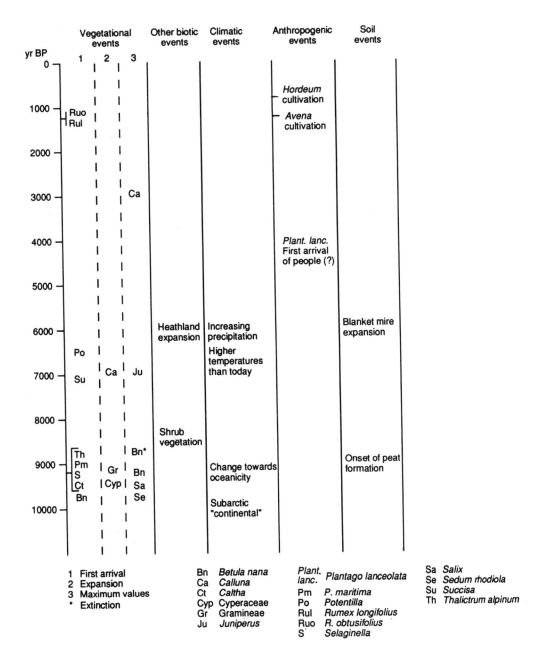

Fig. 5.5 Palaeoecological events on the Faroe Islands

second about AD 600–650 was connected with the cultivation of *Avena* and the introduction of *Rumex longifolius* and *R. obtusifolius* (Jóhansen 1971, 1979). The third, made by Norwegian Vikings about AD 800–850, is characterized by large-scale clearances (but not with fire). Tall herbs were eradicated where sheep had access. Shrubs of *Salix* and *Juniperus* also decrease, and only a few plants are found today in the islands. *Hordeum* was cultivated in historical times (Jóhansen 1982).

REFERENCES

Jóhansen, J. 1971: A palaeobotanical study indicating a previking settlement in Tjørnuvik, Faroe Islands. *Fróðskaparrit 19*, 147–157.

Jóhansen, J. 1979: Cereal cultivation in Mykines, Faroe Islands AD 600. *Danmarks Geologiske Undersøgelse Årbog 1978*, 93–103.

Jóhansen, J. 1982: Vegetational development in the Faroes from 10000 BP to the present. *Danmarks Geologiske Undersøgelse Årbog 1981*, 111–136.

Jóhansen, J. 1985: Studies in the vegetational history of the Faroe and Shetland Islands. *Annales Societatis Scientiarum Faeroensis Supplementum 11*, 117 pp.

Jóhansen, J. 1989: Jóansøkugras (*Plantago lanceolata*) og forsøgulig búseting í Føroyum. (*Plantago lanceolata* and its significance as indication of prehistoric settlement in the Faroe Islands). *Fróðskaparrit 34–35* (1986–87), 68–75.

Jørgensen, G. & Rasmussen, J. 1986: Glacial striae, roches moutonnées and ice movements in the Faroe Islands. *Geological Survey of Denmark, DGU series C no. 7*, 114 pp.

6

Norway

D. Moe, K.-D. Vorren, T. Alm, S. Fimreite, B. Mørkved,
E. Nilssen, Aa. Paus, H. Ramfjord, S.F. Selvik and
R. Sørensen

INTRODUCTION
(K.-D. Vorren and D. Moe)

Norway covers an area of 324000 km² of the north-western part of the Eurasian continent and borders the Atlantic Ocean. Over long distances it is separated from the rest of the Scandinavian peninsula by the Scandic mountain range, the Scandes. In the north and south of the country areas east of the Scandes are included in the Norwegian territory.

Norway has a great geological and topographical diversity. Its proximity to the oceans and its latitudinal extent from 58 to 71°N (1780 km) and altitudinal range from sea level up to 2468 m asl within rather short distances contributes to its climatic and vegetational complexity.

Geology and topography

In a geological sense the country is very old. A major part of the geology of the South Norwegian mountains, and also the interior of Finnmark and the Lofoten archipelago in the north, are dominated by Precambrian rocks (Fig. 6.1) (cf. Holtedahl 1960). The Caledonian mountain chain consists of metamorphosed rocks, originally sediments which filled a geosyncline mainly during the Cambro-Silurian period. The pressure of the mountain range formation was directed towards the Precambrian foreland to the east and southeast. As a result of this pressure the mountains were formed in longitudinal zones with chronologically defined rock types and structures.

In the zone of the eastern foreland (the Oslofjord region and eastern Finnmark), sediments were only slightly folded and metamorphosed. The neighbouring zone to the west is characterized by a highly complicated system of overthrusts. It includes parts of eastern Norway, the greater part of the Swedish Scandes, and parts of Finnmark. Mica schist, phyllite, and basic basalt are frequent in this area. Therefore many rare mountain plants have their isolated occurrences here. The third zone, between Trøndelag and Troms, consists mainly of the same rock types as in zone two; however, they are more intensively folded and have no overthrust massifs. The fourth zone extends along the west coast of Norway from Hordaland to Lofoten. The rocks are partly of Archaean origin and have been strongly metamorphosed.

Since the Caledonian foldings (except for the Tertiary uplift) erosional processes have dominated. Sediments cover only minor areas like the Devonian Old Red Sandstone in Møre and Romsdal, Permian sediments in the Oslo region, and Mesozoic sediments on Andøy in Nordland.

Palaeoecological Events During the Last 15 000 Years: Regional Syntheses of Palaeoecological Studies of Lakes and Mires in Europe.
Edited by B.E. Berglund, H.J.B. Birks, M. Ralska-Jasiewiczowa and H.E. Wright. © 1996 John Wiley & Sons Ltd.

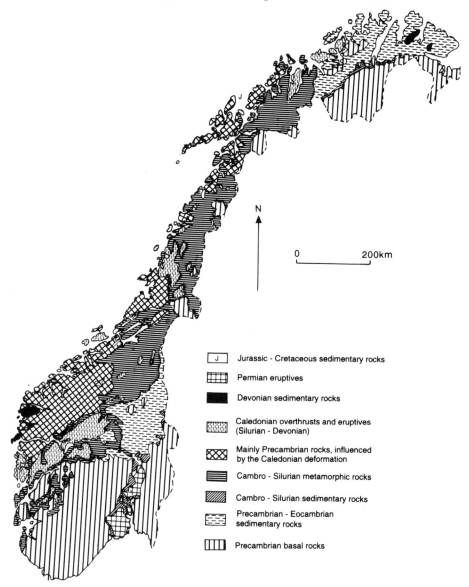

N

0   200km

	Jurassic - Cretaceous sedimentary rocks
	Permian eruptives
	Devonian sedimentary rocks
	Caledonian overthrusts and eruptives (Silurian - Devonian)
	Mainly Precambrian rocks, influenced by the Caledonian deformation
	Cambro - Silurian metamorphic rocks
	Cambro - Silurian sedimentary rocks
	Precambrian - Eocambrian sedimentary rocks
	Precambrian basal rocks

Fig. 6.1   Geological map of Norway, based on Holtedahl (1960)

During the Pleistocene ice ages Scandinavia was almost entirely covered by the inland ice. The maximal extension of the ice margin towards the west cannot be determined with certainty, and it may not have been synchronous along all the margin. Some botanists have suggested the presence of refugial areas for arctic–alpine species along the Norwegian coast during the last glaciation (Dahl 1990).

The present landscape of Norway has been clearly influenced by glacial erosional processes. Accumulation processes are of minor importance. Morainic material covers minor areas in valleys all over the country. More extensive Quaternary sediment accumulations occur in the southeastern lowland region of Norway and in Jæren (SW Norway), and the southeastern part of Finnmark. In the area of

Oslofjord and Trondheimsfjord, extensive sand, silt, and clay deposits of glacial origin occur.

Typical for the larger part of Norway are the fjords, cutting deeply into the mountain massifs of the western coast. More than 50% of the terrestrial area is situated above 500 m asl, and about 25% above 1000 m asl (Fig. 6.2). One may distinguish between glacially formed, rounded mountains and slopes of the middle levels, plateau mountains with a slightly hilly surface (e.g. Hardangervidda), and finally high mountains with alpine forms. Only in Trøndelag, central Norway, are the mountain massifs 400–500 m lower than elsewhere in the country. The relief is especially strong in western Norway, Nordland, and Troms. In southeastern Norway, southwestern Norway, and the Trondheimsfjord area, lowlands below 500 m asl dominate. Here the most extensive farming areas of Norway occur. Also Finnmarksvidda has an extensive area below 500 m asl, but with subalpine to alpine landscape characteristics.

Climate

Special for the general climatic situation in Norway are two factors:

(1) Its location at the western margin of the Eurasian continent, which results in the oceanic character of the climate.
(2) Because of latitude, the annual radiation balance decreases from south to north, but it is compensated by a longer insolation period in summer. The equable and mild climate on the western coast has an extraordinarily high positive temperature anomaly related to the latitude.

The climate of Norway exhibits great regional variations, caused by strong and locally rapidly shifting differences in relief and by the pronounced effect of the Scandes. The annual precipitation sums (Fig. 6.3) are highest near the western coast, with maximal values in southwestern Norway, in Nordland near the Svartisen glacier, and in the Lofoten area. Obvious precipitation minima are localized on the leeward side of the Scandes and in southeastern Norway, as well as in central and southern Finnmark.

The annual mean temperatures are highest along the southeastern coast and decline with increasing height above sea level and towards the east in general. The gradient from south to north is considerably smaller.

The declining degree of oceanicity in the different regions of Norway is correlated with the isolines of the annual amplitude of the mean monthly temperatures (cf. the temperature isotherms for July and January, Figs 6.4 and 6.5). The tendency revealed by a comparison of precipitation and temperature sums is as follows: The temperate climate in the coastal areas has the smallest variations; east of the Scandes, on the contrary, the differences between summer and winter temperatures are larger (more than 25°C difference between the mean January and July temperatures). These areas are also distinguished by high daily temperature variations. Similar to the alpine and subalpine region, the eastern and northern parts of Norway frequently have frost shifts during the vegetational season.

Soils

Depending on the varying relief over short distances, several soil types occur within small areas. A regional division of soil types is therefore hardly practicable. In general, the cool and moist climate especially in the coastal areas causes podzolization. On level to gently sloping plains and in depressions on waterlogged subsoils acid raw humus and peat accumulate. In the alpine vegetation belts, especially in the more continental and northern regions, lessivation and peat formation are overruled or prevented by frost action and solifluction processes, and soils with a weak upper layer of raw humus are formed. In the eastern parts of the Scandes, in some areas of southeastern Norway, and in southern Trøndelag, easily weathered rocks with strongly base-saturated soils occur. Extra-zonal grey or brown forest soils occur on the south-facing slopes on basic rocks under a vegetation of broad-leaved deciduous trees in the south and *Alnus incana* in the north. In southeastern Norway and along the coast, ombrotrophic bogs occur, whereas in the mountainous regions sloping fens dominate. In the continental areas large string-fens (aapa-mires)

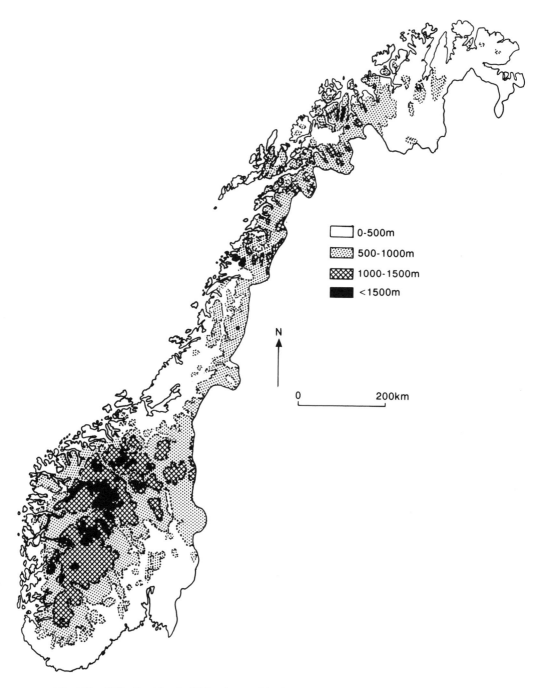

Fig. 6.2   Altitudinal levels (500 m intervals) of Norway according to Holtedahl (1960)

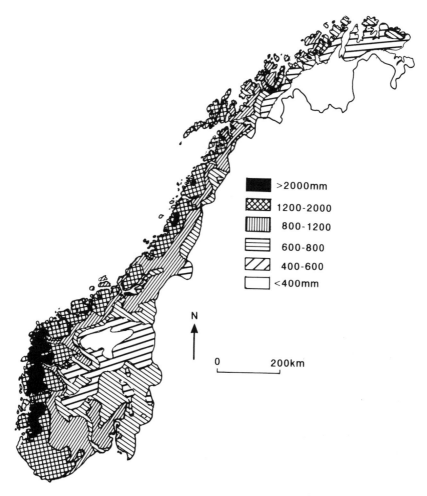

Fig. 6.3 Annual precipitation in Norway, 1931–1960 standard normal

form a climatic climax. Mires with sporadic permafrost occur primarily in the eastern and southeastern parts of Finnmark and Troms in northern Norway.

Vegetation

The potential natural vegetation of Norway, with the exception of the alpine regions, consists of boreal forests. In the oceanic areas, and increasingly towards the north, common birch (*Betula pubescens*) is the prominent woodland or forest tree. On dry and shallow soil it is replaced by Scots pine (*Pinus sylvestris*). In central and southeastern Norway forests of spruce (*Picea abies*) prevail on deep soil in moderately moist climatic regions. In the southern

part of Norway and northwards to Sogn (61°N) in western Norway a zone is characterized by oak (*Quercus robur*). Vegetational zones are based mainly on the distribution of the forest formations (biota) (Fig. 6.6). The floristic and vegetational gradients depend largely on the orographic impact on the one hand (Gjærevoll 1990, 1992) and the degree of "oceanicity" on the other (Fægri 1960).

Habitation and anthropogenic impact on the vegetation

Norway became inhabited by Mesolithic hunters and gatherers mainly during Preboreal time. Evident

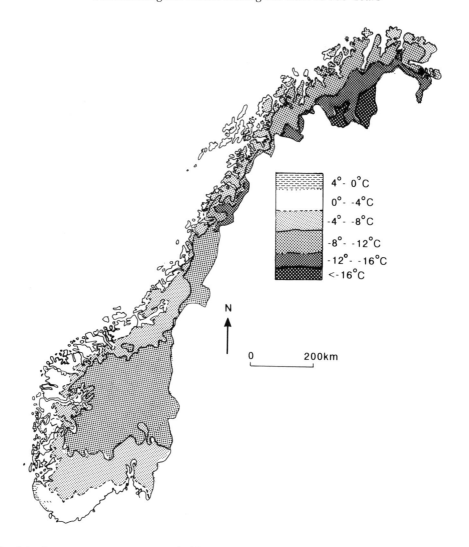

Fig. 6.4   January mean temperatures in Norway, 1961–1990 standard normals, after Aune (1993b)

anthropogenic changes caused by farming and animal husbandry occurred in early Subboreal in southern Norway, a little later in western and central Norway, and in the later part of Subboreal in northern Norway. In the northernmost, subarctic part of the country (Finnmark), agriculture developed mainly after AD 1200. At present, the regions around Oslo, Stavanger, and Trondheim are most intensively exploited for cultivation of cereals, vegetables, fruits, and grass. Pasturing was until recent times a landscape-forming feature, especially in coastal areas and in the valley and mountain areas in southern Norway. Since the 1930s pasturing has decreased, and former grazing areas have been reforested, either spontaneously by birch, alder, pine, and aspen, or by introduced trees, mainly spruce.

In the alpine regions several water courses with parts of their catchment areas have been dammed and inundated for energy (electrical power) demands.

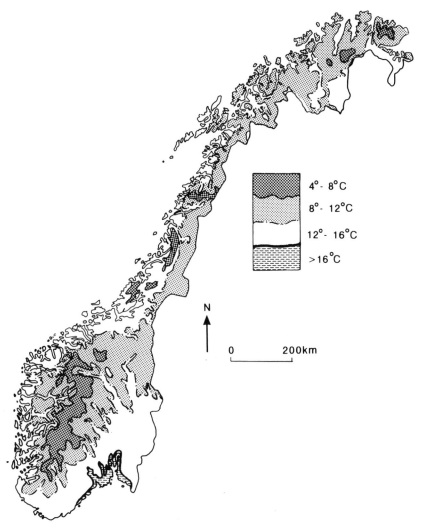

Fig. 6.5 July mean temperatures in Norway, 1961–1990 standard normals, after Aune (1993b)

Type regions

The concept of natural-geographic areas, or type regions (cf. Påhlsson 1984) is in reality not adaptable for mountainous areas such as most of Norway. However, bearing in mind the great amplitudes in geology, topography, climate, and vegetation within all regions, the following division is considered to be useful when compared with maps of biological and environmental variables, see especially Fig. 6.7:

N-f   The Oslofjord region, belonging to the boreo–nemoral (hemiboreal) vegetation zone (Sjörs 1967, Moen 1987). *The central part of this region is represented by a regional palaeoecological synthesis.*

N-g   The interior of the Telemark–Buskerud–Oppland region, belonging mainly to the middle boreal zone. Palaeoecological evidence is provided by Helge

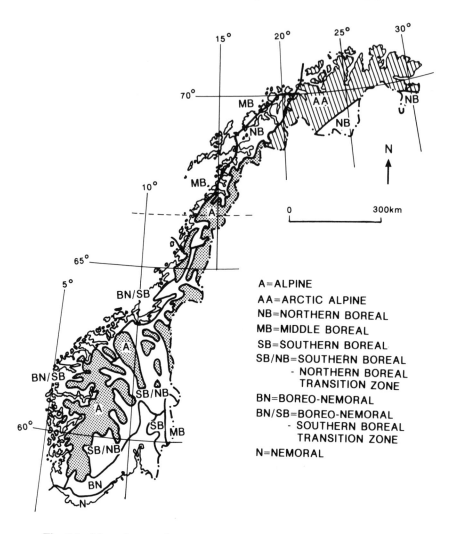

Fig. 6.6 Map of vegetation zones in Norway according to Påhlsson (1984)

A=ALPINE
AA=ARCTIC ALPINE
NB=NORTHERN BOREAL
MB=MIDDLE BOREAL
SB=SOUTHERN BOREAL
SB/NB=SOUTHERN BOREAL
 - NORTHERN BOREAL
 TRANSITION ZONE
BN=BOREO-NEMORAL
BN/SB=BOREO-NEMORAL
 - SOUTHERN BOREAL
 TRANSITION ZONE
N=NEMORAL

Høeg (pers. comm.).

N-h The Agder region, southernmost Norway. The coastal ring belongs to the nemoral zone and the rest of the lowlands mainly to the boreo–nemoral zone. Palaeoecological evidence from the region is supplied by Helge Høeg (pers.comm.).

N-j The coastal region of Vestlandet (western Norway), mainly belonging to the boreo–nemoral zone, but characterized by atlantic heathlands. *The south-*

ern part of this region is represented by a palaeoecological synthesis.

N-k The South Norwegian mountain plateau, generally belonging to the arctoalpine zone. *The western part of this mountain plateau is presented in a regional synthesis.*

N-mo The coastal area of central Norway (the Trøndelag counties). The region belongs to the boreo–nemoral zone according to Moen (1987) (north of the oak zone), or to the south boreal zone

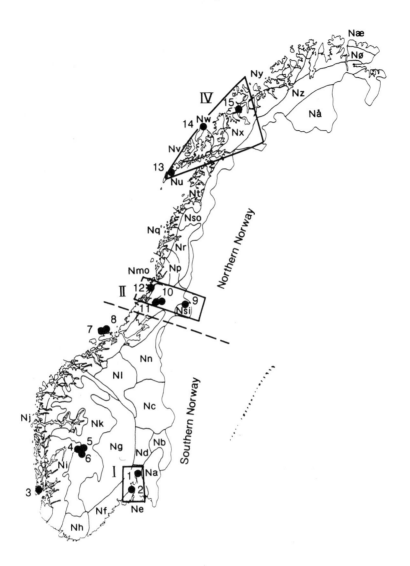

Fig. 6.7 Reference sites 1–14 of the IGCP 158B project in Norway. Na–Nå: Type regions or natural geographic regions. I–III: Palaeoecological reference areas. 1–17: Reference sites: 1: Korsegården, 2: Kjeldemyr, 3: Sandvikvatn, 4: Hadlemyrane, 5: Normannslågen, 6: Våtenga, 7: Grasvatn, 8: Krokvatn, 9: Kalvikmyr, 10: Formofoss, 11: Koltjønn, 12: Blåvasstjønn and Gorrtjønn, 13: Dønvoldmyra, 14: Endletvatn and Aeråsvatn, 15: Prestvann and Tjernet, 16: Bruvatnet, 17: Domsvatnet

according to Sjörs (1967). *The northern part of the region is presented in a type-region synthesis together with the neighbouring middle boreal zones of N-n and N-si to the east. The southwestern part of this region is presented in an independent synthesis.*

N-v, N-w, N-x
The coastal area of Lofoten, Vesterålen and southern Troms in northern Norway. *A regional synthesis from this area includes also the neighbouring regions N-u, N-y, and N-z, and covers the northern part of the middle boreal and the north boreal zone*

(cf. Eurola & Vorren 1980).

N-p-N-ø   Individual AP-diagrams are presented from the N-p, N-r, N-z, N-ø, and N-æ regions in northern Norway.

In addition to these regions, two Norwegian regions are described in the present contribution from Finland:

N-ae   Coastal area of Finnmark (the northeasternmost part of northern Norway). The region belongs to the hemi-arctic zone (cf. Hyvärinen 1975, 1976; Prentice 1981, 1982).

N-z   Inland area of Finnmark. Mainly hemi-arctic to north boreal (Hyvärinen 1975, 1976).

In this survey we present syntheses for southern Norway and northern Norway (Fig. 6.7) separately, due to the fundamental differences between these two regions.

Available palaeoecological data

Soligenous mires are distributed practically all over the country up to the low-alpine vegetation belt. Ombrogenous bogs prevail along the coast, and all over the country lakes and tarns constitute widespread potential sources of palaeoecological data. The oldest continuous sediments occur along parts of the coastal fringe and date back to *ca.* 22000 BP, and in the mountainous areas that were deglaciated latest, sedimentation started between 9500 and 8500 BP.

Most data from this range of sites have been produced during the last two decades, and a considerable part of this again after the work on the IGCP syntheses had started 15 years ago. Only part of the total information concerns cores representing the whole sequence from start of sedimentation to present. The quality and availability of this palaeoecological material also varies. Information is difficult to obtain from nearly 50% of the diagrams.

For Norway it must be pointed out that important geographical areas like central southeastern Norway, southern Norway, central western Norway and southern central Norway are not represented in this survey. This fact makes it very difficult to draw any broad-scale, nation-wide conclusions about immigration and extinction of forest trees and the long-term dynamics of vegetation formations.

SOUTHERN NORWAY
(D. Moe, Aa. Paus, and R. Sørensen)

Southern Norway is geologically mainly of Precambrian to Silurian age with generally metamorphosed rocks (Sigmond *et al.* 1984, Sigmond 1985), and with strong climatic (Aune 1993a,b,c,d) and vegetational (Dahl *et al.* 1986) gradients, latitudinally and altitudinally as well as west–east. Land uplift during the Holocene has played an important role in coastal areas, especially in the southeastern and central parts of Norway, with a maximum uplift of about 220 m in the Oslofjord region. Most of the country has a bedrock with little or no cover of surficial deposits (Thoresen 1990). The deposits are of Late-Glacial and Holocene age, although a few sites with interglacial (Eemian) deposits occur (Thoresen 1990). The existence of a North Sea land area during Late-glacial time up to 8000 BP may have been important for climate and plant immigration from the south and southwest to Norway (e.g. Oele *et al.* 1979).

Except for the inland regions, most defined "type" regions in Norway are rather diverse, with strong gradients in nearly all physical and biological respects, and not comparable with most regions used within the IGCP project.

TYPE REGION N-f, OSLOFJORD AREA
(R. Sørensen)

The Oslofjord region in this context covers an area bounded by *ca.* 59° to 60°15′N and 9°30′ to 10°30′E (Fig. 6.7, sites 1 and 2). The area was deglaciated 12000–9500 BP. A considerable part of the area lies below the past marine limit, which in the inner part was 220 m asl, and the palaeoecological development has been strongly influenced by Holocene sea-level changes.

Climate: Mean January temperature 0–4°C, mean July temperature 14–18°C. Precipitation 600–1000 mm yr^{-1}.

Geology: A large part of the area lies within the Permian Oslo Rift, with intrusive and extrusive volcanic rocks and some Cambro-Silurian sedimentary rocks. East of the eastern boundary of the Oslo Rift (east of the Oslofjord) the bedrock is Precambrian gneiss. The main Quaternary deposits are marine clays below the marine limit at 150 to 200 m asl. Ice-marginal ridges cross the region from east to west. The most prominent is the Ra ridge. The hilly areas, particularly those above the marine limit, have very thin surficial deposits.

Topography: The Permian faulting with predominantly north–south alignment has strongly affected the whole region. Hills, valleys, drainage patterns, and coastal features are controlled by fault and fracture lines, bedrock boundaries, and folding in the Cambro-Silurian sediments, thus giving the Oslo Rift area a broken topography. The gneiss province has a more subdued topography. Below the marine limit most of the depressions are filled with marine clay. From *ca.* 150 m asl to the coast, clay plains are common.

Population: Approximately 50 persons km^{-2} (Oslo not included). This is the most densely populated region in Norway.

Vegetation: Mixed forest (*Picea abies–Pinus sylvestris–Betula pubescens/B. verrucosa*) dominates, with the coniferous element increasing with distance from the coast. In the coastal areas, deciduous forests are more common, with *Quercus, Populus tremula, Alnus* (both *A. incana* and *A. glutinosa*), *Acer, Ulmus glabra*, and some *Tilia cordata*. On the west side of the Oslofjord, *Fagus sylvatica* is common.

Soils: Brown earths predominate, particularly in the coastal region and on rich parent material. True podzols are rare, even above the marine limit. Transitional "semi-podzols" (grey forest soils), however, are common. Inside the Ra moraine, bogs and mires are common.

Land use: In the lowlands near the coast, farmland covers approximately 50% of the area. Urban areas cover *ca.* 15%, and forests the rest. Further inland the coverage is *ca.* 15% agricultural land, *ca.* 5% urban areas, and *ca.* 80% forest.

Physical geographical references: Nordgård (1972), Throndsen (1977), Øy (1978), Møller (1980), and Mamen (1981).

Sub-region, Follo in southern Akershus

Altitude: 0–300 m.

Climate: Slightly continental. Annual precipitation *ca.* 800 mm. July mean temperature *ca.* 17°C. January mean temperature *ca.* −5°C, yearly mean temperature: 5.5°C.

Geology: Precambrian gneiss, with amphibolite and calc-silicate schist. The area was deglaciated between 10500 and 10000 BP. The main Quaternary deposits are marine clays below the marine limit at approximately 200 m asl. Ice-marginal ridges cross the district from east to west. The Ås and the Ski moraines are the most prominent. The hilly area, particularly north of the Ski moraines, has very little Quaternary cover.

Topography: Clay plains between 60 and 120 m fall off to the present coast. The northeastern part of the area is more hilly .

Population: Ca. 150 persons km^{-2}.

Vegetation: The coastal zone: Partly forested, with mixed *Pinus sylvestris, Quercus,* and other deciduous trees. Otherwise mixed *Picea abies, Pinus sylvestris,* and *Betula verrucosa/B. pubescens* forest.

Soils: Brown earths and transitional "semi-podzols". Bogs are common north of the Ski moraine.

Land use: Ca. 65% forest, *ca.* 25% farmland (mainly grain production), and *ca.* 10% urban areas.

A climatic and vegetational gradient exists from the outer coast inland. Each pollen diagram is divided into chronozones (Mangerud *et al.* 1974) either based on ^{14}C-datings in the diagram presented or on available diagrams close by. Local pollen-assemblage zones are established. Metachronism occurs.

164

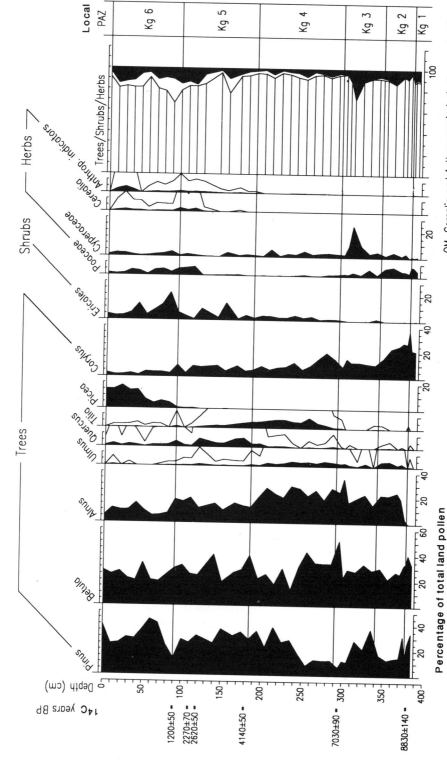

Korsegarden bog, Akershus, Norway
Selected pollen percentages
Analyst: R. Sørensen 1985

Fig. 6.8/6.9 Pollen diagram from the Korsegården raised bog. Ås, Akershus. Charcoal layer at 4000 BP. (Redrawn after Sørensen 1990)

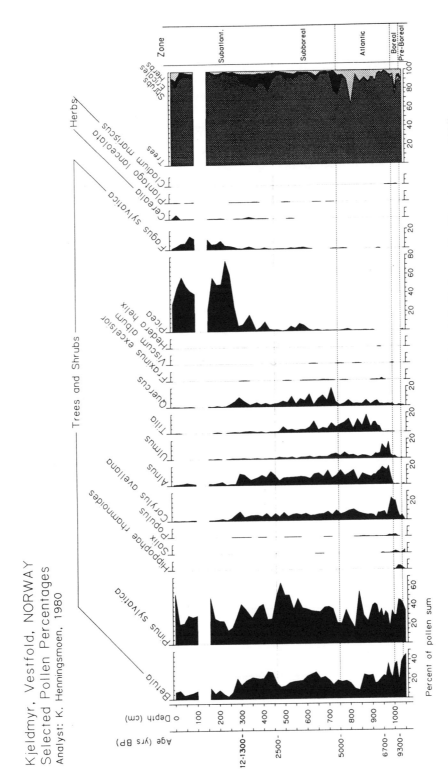

Fig. 6.10 Pollen diagram from Kjeldemyr, Sandefjord, Vestfold. (Redrawn after Henningsmoen 1980)

Reference site 1. Korsegården, Ås Akershus

Latitude 59°42′N, Longitude 10°43′ E. Elevation 85 m asl. Age range *ca.* 8500 BP to present day (*ca.* 9400 BP calibrated (see Figs 6.8 and 6.9)). Ombrotrophic bog, core 4.20 m. The reference site Korsegården was located in a narrow fjord system at 9000 BP. The vegetational development therefore in addition to climatic/successional changes may also express changes caused by changes of the local coastline up to 7000 BP. Six radiocarbon dates are included. Six local pollen-assemblage zones (paz).

Age	Local paz
Kg 1 8700–8300 BP	*Betula–Corylus*
Kg 2 8300–7500 BP	*Alnus–Corylus–Betula*
Kg 3 7500–7000 BP	*Alnus–Betula*
Kg 4 7000–4200 BP	*Tilia–Alnus–Betula*
Kg 5 4200–1200 BP	*Quercus–Pinus*
Kg 6 1200–0 BP	*Picea–Pinus–Betula*

Reference site 2. Kjeldemyr, Sandefjord
(Sandar), Vestfold

Latitude 59°10′N, Longitude 10°15′E. Elevation 87 m asl. Age range *ca.* 10000/9500 BP to present day.

The pollen diagram (Henningsmoen 1980), Fig.6. 10, shows the same geographical and vegetational development as the reference site Korsegården, southern Vestfold. However, this site has a slightly more humid and warmer climate. The soil is richer. Five radiocarbon dates are included. Seven local pollen-assemblage zones (paz).

Age	Local paz
Deglaciation–9000 BP	*Betula–Hippophaë*
9000–8000 BP	*Corylus–Betula*
8000–6700 BP	*Alnus –Ulmus*
6700–5000 BP	*Tilia–Ulmus–Pinus*
5000–2500 BP	*Quercus–Pinus*
2500–1300 BP	*Quercus–Pinus–Betula*
1300–0 BP	*Picea–Pinus–Fagus*

Regional stratigraphic events (Fig. 6. 11)

(1) Deglaciation–9000 BP: *Salix* and *Hippophaë rhamnoides* predominant. During Preboreal chronozone, the whole region was characterized by scattered islands and a coastal environment. Early immigrants were *Hippophaë, Salix, Juniperus,* and *Populus. Betula* (tree-type) and

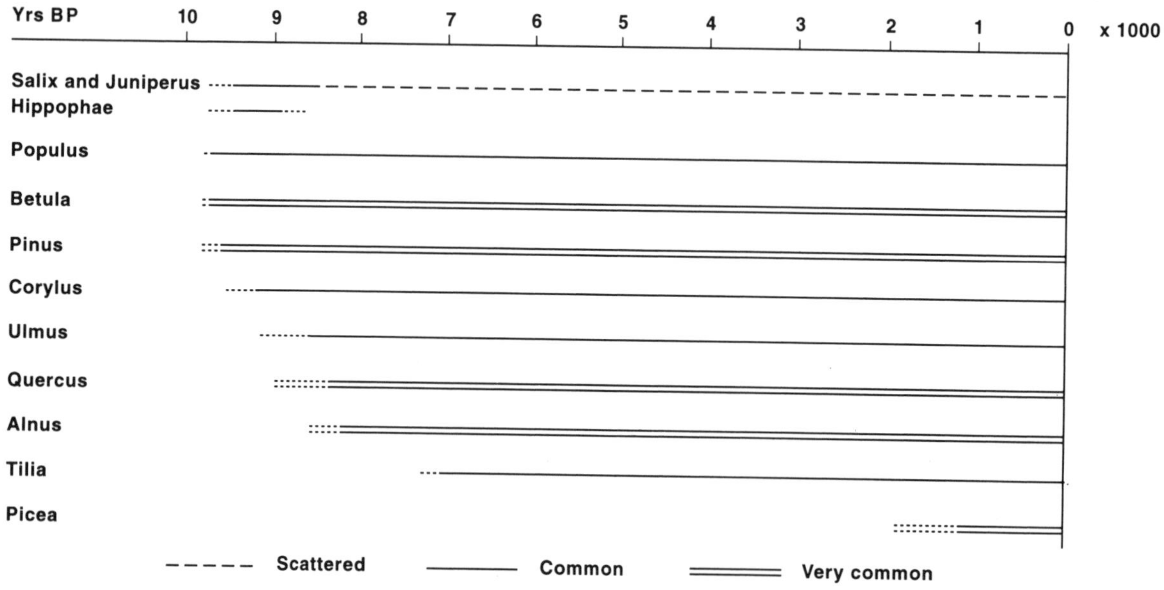

Fig. 6.11   General event of tree and shrub arrival and presence in southern Akershus county, southeastern Norway (surroundings of reference site: Korsegårdsmyra). (Data from Høeg 1982)

Pinus are present in the oldest diagrams dated back to 9800–9700 BP.

(2) Early Boreal: *Betula* and *Hippophaë* predominant, marine overrepresentation of *Pinus*. Local *Cladium mariscus*.

(3) Boreal: local *Corylus*. *Corylus* expanded between 9500 and 9300 BP; *Ulmus* and shortly after also *Quercus* appeared in amounts in the area between 8500 and 8300 BP.

(4) About 8000 BP: increase of *Alnus* and *Ulmus*. *Viscum album* pollen (site 2). *Alnus* expanded between 8400 and 8200 BP and *Tilia* between 7100 and 1100 BP.

(5) 5200–4800 BP: earliest anthropogenic interaction.

(6) 3700–2800 BP: records of early cereals.

(7) About 1300 BP: local *Picea*. Local stands of *Picea* may have occurred as early as 2500 BP in the northern part of the region (cf. Hafsten *et al.* 1979).

Local events

(1) 6700 BP increase of *Tilia*, local *Quercus* (site 2).

(2) About 5200 BP elm fall, *Tilia* reduction about 5000 BP (site 2).

(3) About 4800 BP earliest *Plantago lanceolata* occurs, slight opening of the forest (site 2).

(4) About 4500–4000 BP latest record of *Viscum album* (site 2).

(5) First cereal pollen appears at 4200–4100 BP (site 2)

(6) A small charcoal layer, and a small roundish stone were found approximately at the same level as event (3) (180 cm depth) (site 1).

(7) From 2700 BP main agricultural phase starts (site 1).

(8) About 2500 BP decrease of *Pinus*, occasional pollen of cereals (site 2).

(9) Marked "recurrence surface" is found at 2600–2300 BP (115 cm depth), indicating very little or no growth of the bog/mire (site 1).

(10) 1200/1300 BP expansion of *Picea* (site 1 and 2).

(11) About 1000 BP increase of *Fagus* (site 2).

(12) About 800–400 BP temporarily reduced agricultural activity (site 1).

Comments on local/unique events

The Preboreal vegetational patterns are basically controlled by the palaeogeography of the region, with many small, scattered, and exposed islands. At the transition to the Boreal zone, at 9000 BP, large islands and land areas had appeared in the northwestern and northeastern parts of the region.

The earliest dated traditional "weed" pollen is dated at about 6000 BP, and may be interpreted either as dispersed from nearby coastal natural stands or by human activity. The earliest continuous record of agriculture may be dated at between 4500 and 4200 BP. The "elm decline" is not well defined in the Korsegården diagram (site 1).

A major agricultural expansion occurred from about 2700 to *ca.* 2200 BP.

Palaeoecological references: Hafsten (1956), Danielsen (1970), Henningsmoen (1980), Dahl *et al.* (1986), Sørensen (1990a, b).

TYPE REGION N-j, COASTAL AREA, SOUTHWESTERN NORWAY (NORTHERN ROGALAND–SOUTHERN HORDALAND)
(Aa. Paus and D. Moe)

The coastal area within this type region N-j in southwestern Norway covers an area bounded by *ca.* 59°00′ to 60°15′N and 5°30′E.(Fig. 6.7, site 3). The coastal relief is rugged with numerous islands and skerries in the west followed by an undulating strandflat and with coastal mountains in the east with altitudes reaching about 600–700 m. The main altitudinal range is 0–200 m asl.

Climate: Climatic gradients exist from the outer coast inland. Winter temperature decreases and summer temperature and annual precipitation increase towards the east. The reference area has an oceanic climate with July and January mean temperatures of 15°C and 0°C, respectively (Bruun 1967). Annual precipitation is 1300 to 1500 mm yr^{-1}, coming mainly from the southwest/west (Bruun & Håland 1970).

Geology: Prevailing within the area is Cambro-Silurian phyllite, gabbro/amphibolite mainly of Caledonian age, phyllite and quartz-mica schist of assumed Ordovician age, and Precambrian basement of gneiss and granite (Sigmond *et al*. 1984).

The reference site, Sandvikvatn, at an altitude of 128 m asl was deglaciated close to 14000 BP (Eide & Paus 1982, Paus 1988), while moraine systems including glacier advances of younger ages are dated at lower levels and in areas farther east (Younger Dryas moraine; Anundsen 1985). The marine limit within the area is from *ca*. 20 m in the western part to *ca*. 40 to 70m in the eastern part. The palaeoecological development below this level has been strongly influenced by the Holocene sea-level changes.

The main Quaternary deposits are glaciolacustrine and glaciomarine. Different marine terraces occur. The soil cover is thin. Bogs and mires are common.

Vegetation: Traditionally the landscape was dominated by a treeless open *Calluna* vegetation (e.g. Kaland 1986). During the last 100 years, birch and pine have started to recolonize because of the decreasing land use of the heath. Locally dwarf-shrub heath still dominates with *Calluna vulgaris, Empetrum nigrum, Erica tetralix*, and *Juniperus communis*. In addition grass heaths, bogs, and bare rock form a vegetational mosaic (e.g. Røsberg 1982). On better and deeper soil downslope, tree species like *Quercus, Populus tremula, Alnus* (both *A. incana* and *A. glutinosa*), *Fraxinus excelsior, Ilex aquifolium*, and some *Taxus baccata* occur.

Population and land use: The area is sparsely populated. The nearest town, Haugesund is, together with farmland, situated on marine/glaciomarine deposits. Other areas are mostly used for pastures: in the outer region for all-year pasturing, in the inner and upper part only seasonal.

Reference site 3. Sandvikvatn, Tysvaer, Rogaland

Latitude 59°17′N, Longitude 5°30′E. Elevation 127.5 m asl. Age range: *ca*. 14000–0 BP.

The reference site, Sandvikvatn, is located in the southernmost part of the Tysvaer peninsula, extending southwards into Boknafjord. The local bedrock is Cambro-Silurian phyllite, changing to a Precambrian basement of gneiss and granite *ca*. 1 km north of the lake (Sigmond *et al*. 1984). The marine limit close to the studied site is 32–34 m asl (Rønnevig 1971). The lake lies within a hilly range rising from the surrounding strandflat. The basin and its catchment are about 50000 m² and 320000 m², respectively (Paus 1988). The core studied is from 16.7 m to 13.8 m below water surface. The site was deglaciated before 14000 BP.

The local vegetation is dominated by *Calluna vulgaris, Empetrum nigrum, Erica tetralix,* and *Juniperus communis*. In addition there are grass heaths, bogs, and bare rock forming a vegetational mosaic (e.g. Røsberg 1982).

The division of the pollen diagram (Fig.6.12) into chronozones (Mangerud *et al*. 1974) is mainly based on ^{14}C-datings in the diagram presented (Eide & Paus 1982, Paus 1988). The Late-glacial chronology, however, is adjusted according to Paus (1989a (Liastemmen), 1989b (Eigebakken), 1990 (Utsira)). Local and regional pollen-assemblage zones are established. 13 radiocarbon dates are included in the sequence. 14 local-pollen assemblage zones (paz).

	Age	Local paz
S-1	*ca*. 14000–13000 BP	*Artemisia*
S-2	13000–12700 BP	*Salix–Rumex*
S-3	12700–12000 BP	*Betula–Salix–Hippophaë*
S-4	12000–11000 BP	*Betula–Empetrum*
S-5	11000–10100 BP	*Artemisia–Rumex–Sedum*
S-6	10100–9700 BP	*Betula–Empetrum–Urtica*
S-7	9700–9500 BP	*Betula–Juniperus*
S-8	9500–8400 BP	*Corylus*
S-9	8400–7600 BP	*Alnus*
S-10	7600–5300 BP	*Quercus*
S-11	5300–3000 BP	*Pinus*
S-12	3000–1700 BP	*Calluna–Betula*
S-13	1799–200 BP	*Calluna–Potentilla–Sphagnum*
S-14	200–0 BP	*Calluna–Pinus–Sphagnum*

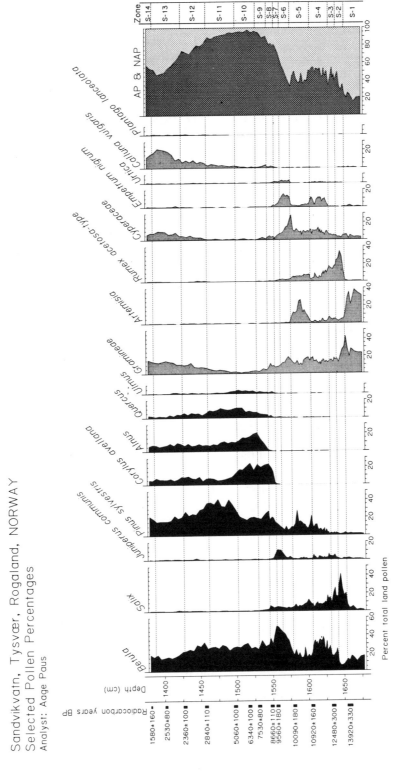

Fig. 6.12 Pollen diagram from Sandvikvatn, Tysvær (Eide & Paus 1982, Paus 1988)

Regional palaeoecological events

No distinction has been made between common and unique events as there are too few well-dated pollen diagrams from this region.

(1) Shortly after deglaciation, a tundra-like vegetation became established on well-drained mineral soils, including *Artemisia* spp., Gramineae, *Saxifraga* spp., and *Papaver* sect. *Scapiflora*.

(2) The Late-glacial Interstadial climatic warming at 13000 BP initiated a three-step successional development: (i) a *Salix* shrubs/grass vegetation including *Rumex* spp., *Filipendula*, and a few other tall herbs, (ii) from *ca.* 12700 BP, a park-tundra stage with tree birch, *Salix* shrubs, *Juniperus*, grasses and *Rumex*, and (iii) from 12000 BP, an open birch-forest stage with *Salix*, *Juniperus*, *Filipendula*, *Urtica*, and other tall herbs.

(3) The Arctic treeline crossed North Rogaland in the period 12700–12000 BP.

(4) In North Rogaland between 12300 and 12000 BP, the park tundra disappeared, and there was an increase in mineral-soil vegetation. This reflects the "Older Dryas" climatic cooling. No distinct traces of the cooling have been found elsewhere within the type region N-j. This pattern probably relates to the existence of ecotonal boundaries in North Rogaland at the time of the onset of the cooling.

(5) The Younger Dryas climatic cooling involved soil erosion, the disappearance of tree birch, and the re-occurrence of tundra with, for example, *Artemisia*, *Rumex*, *Papaver* sect. *Scapiflora*, and *Koenigia*.

(6) At 10500 BP, a slight warming is indicated by small rises in tree birch, *Filipendula*, and *Empetrum*.

(7) The Holocene warming about 10000 BP, caused the rapid establishment of dense birch forests with an understorey of dominant *Empetrum*, tall herbs, and ferns, throughout the type region.

(8) For unknown reasons, the Preboreal *Juniperus* rise is recorded after the *Betula* rise, reversing the order documented in South Sweden and on the European continent.

(9) *Corylus* and *Pinus* arrived in North Rogaland about 9500 BP or shortly after. In the outer coastal areas of South Hordaland, the local arrival of *Pinus* was delayed for up to 1000 years.

(10) *Ulmus* and *Quercus* became established locally about 9000 BP.

(11) *Alnus* (presumably *A. glutinosa*) expanded about 8400 BP.

(12) The regional broad-leaved forests reached their maximum density between 8000 and 5000 BP.

(13) A gradual deforestation started at about 5000 BP, including a decrease in *Corylus* and *Ulmus*, an increase in *Pinus*, *Calluna*, and Gramineae, and the appearance of *Plantago lanceolata* and *Narthecium ossifragum*.

(14) At about 3000 BP, human impact (burning, grazing) accelerated heath formation, involving increases in *Calluna*, *Myrica*, *Potentilla*, *Rumex*, Cyperaceae, *Sphagnum*, and charcoal, and decreases in *Pinus*, *Quercus*, and *Fraxinus*.

(15) At about 2500 BP, the earliest traces of cereal-growing *(Hordeum)* are recorded.

(16) The management of heathland, involving burning, mowing, grazing and peat cutting, caused the total deforestation of the area shortly after 2000 BP. The extensive utilization of heathland caused soil erosion and a lake sediment change from gyttja to dy.

(17) Rises in *Betula* and *Pinus*, and a *Calluna* decrease, during the last 100/200 years reflect the decreasing utilization of the healthland.

REGION N-k, CENTRAL INLAND, SOUTH NORWEGIAN ARCTIC/ALPINE AREA
(D. Moe)

A major part of the area lies between 1050 and 1300 m, but the total altitudinal range is 950–2469 m asl. Both north–south and east–west mountain ridges and isolated peaks exist. Especially in the west and north the mountain relief is extraordinary.

Climate: Climatic conditions vary within the region as a result of topography, with suboceanic conditions in the west and more continental conditions in the east. Relatively dry areas occur in the west (e.g. Veigdalen). Mean unreduced January temperature (1931–1960) in the west at 780 m / 748 m (Maurseth/ Liseth) is –7.6°C, at 1300 m (Slirå) –9.8°C, and in the east at 988 m (Haugastøl III) –9.6°C. Mean July temperature in the west is 11.7°C, at 1300 m 7.3°C, and in the east 10.7°C (Bruun 1962). The annual mean precipitation (1931–1960) is 915 mm in the west and 669 mm in the east (DNMI 1991b). Snow or hail may occur in July–August every year. At middle altitudes, the growing season starts in the second week of July and ends at the end of August.

Geology: The bedrock in the southern area is mainly of Precambrian and Eocambrian–Ordovician age. Mica schist and metamorphosed sandstone are predominant. In the northern part, the basal gneiss complex, mainly of granitic and quartz dioritic composition, is of great importance, along with overthrust eruptive rocks of basic and intermediate types.

Deglaciation took place after 10000 BP and was nearly complete close to 9000 BP. Within the region three major and several minor glaciers exist today. It is assumed that some glaciers or parts of them have existed during the entire Holocene. On flat areas morainic sediments exist, whereas in the valley bottoms fluvial or glacio-fluvial deposits occur. Elsewhere screes or bare bedrock abound.

Population: The area is permanently populated by around 200 people (hotels and Norwegian Rail).

Vegetation: Betula pubescens forms the treeline in the west at 950 m, in the east at 1050–1130 m. Within the region low-, middle- and high-alpine vegetation exists. Flat areas are peaty (mainly rather mesotrophic caused by mica schist in the bedrock and soil), and at low altitudes with willow and tall-herb vegetation. The altitudinal limits for several species in South Norway occur on south and south-west slopes and on screes.

During hundreds of years the treeline has been depressed 100–150 m by summer-farming. Reduced pastoral activities during recent decades have allowed the treeline to expand, and a succession is proceeding today. The climatic limit of the treeline is generally around 950–1000 m in the west and 1100–1150 m in the east.

Soils: Most of the area is covered by morainic sediments. Different glacially formed sedimentary features exist on the slopes and screes. Many areas are peaty.

Land use: Area for sheep and cattle grazing, no arable land. In former days extended summer-farming, some peat cutting. A wild reindeer population amounts to around 15000 animals, with some hunting. Recreation area for people in spring, summer, and autumn.

Strong altitudinal gradients exist within the region, and four reference altitudinal levels are selected: 1000 m, 1250 m, 1300 m, and 1480 m. The lowest level is just above the present treeline in the west, about 100 to 150 m below in the east; the level 1250 m is within the low alpine zone; 1300 m is in the upper part of the low alpine zone; 1480 m is in the middle alpine vegetation zone.

Each pollen diagram is divided into chronozones (Mangerud *et al.* 1974). In addition, local pollen-assemblage zones are established. Metachronism in the local zones may result from local and/or human influences.

Reference site 4. Hadlemyrane

Total age range: 9500 BP to present day. Sites: Svartavatnet (1000 m), Ulvik, Hordaland (Simonsen 1980); and Hadlemyrane (1005 m), Eidfjord, Hordaland (Moe 1978). Two radiocarbon dates are included in the Hadlemyrane profile (Fig. 6.13). Six pollen-assemblage zones (paz), the upper four of which are present at the reference site (Fig.6.13):

	Age	Local paz
1.	Ice/9500–9000 BP	Gramineae–*Rumex*– *Hippophaë*–*Salix*
2.	9000–8700 BP	*Filipendula*– *Hippophaë*–*Salix*– *Betula*
3.	8700–8000 BP	*Pinus*
4.	8000–5500 BP	Cyperaceae
5.	5500–5000/4800 BP	*Pinus*–*Ulmus*
6.	5000/4800BP	Cyperaceae– Gramineae–herbs

172

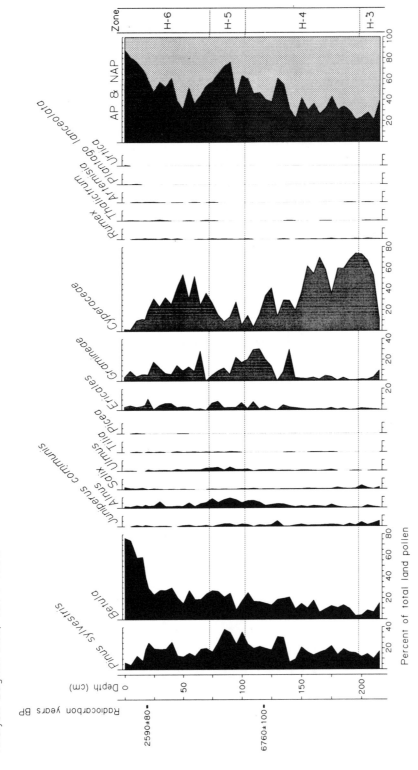

Fig. 6.13 Simplified pollen diagram from Hadlemyrane (loc. 18), Eidfjord, Hordaland. Reference level 1000 m asl (Redrawn after Moe 1978)

Event stratigraphy

(1) Introduction of a light-demanding flora, e.g. *Hippophaë* from about 9300 BP, followed by tree birch.
(2) *Pinus* predominated locally from 8700–8500 BP.
(3) Cyperaceae dominated from about 8000 BP. Wetter conditions.
(4) *Alnus* expanded locally between 7000 and 6000 BP.
(5) *Ulmus* reached its Holocene altitudinal limit 5500–4900 BP. Drier conditions.
(6) Treeline in the west at about 1050 m (*Pinus*).
(7) Traces of human disturbance from about 5000 BP.

Special references: Moe (1978), Simonsen (1980), Bang-Andersen *et al.* (1986), and Moe & Odland (1992).

Reference site 5. Nordmannslågen, Hordaland

Total age range: 9000 BP to present day. Sites: Bog 13, Hol, Buskerud, 1190 m (Moe 1978); and Nordmannslågen, Hordaland, 1245 m (Moe 1978). Three radiocarbon dates are included in the reference profile (Fig. 6.14). Four pollen-assemblage zones (paz), the upper two of which are present at the reference site (Fig. 6.14):

Age	Local paz
1. Ice/9000–8600 BP	*Salix–Hippophaë–fern–herb*
2. *ca.* 8700–8600 BP	*Betula–Hippophaë*
3. *ca.* 8600–8000 BP	*Pinus*
4. 8000–0 BP	*Salix*–herbs–Cyperaceae

Event stratigraphy

(1) The area was free of ice from about 9000 BP.
(2) A local light-demanding herb and shrub flora followed by *Hippophaë rhamnoides* and birch from 9000 to 8600 BP.
(3) Local *Pinus* from 8500 to about 8000 BP.
(4) Open treeless areas from about 8000 to the present day.
(5) Some mires date back to 7900–8000 BP.
(6) Local *Alnus* up to 1205 m asl. 7500–7000 BP.

(7) Traces of human disturbance from about 5200/5000 BP.

Special references: Moe (1978), Moe & Odland (1992). Diagram from Nordmannslågen also presented in Indrelid & Moe (1982).

Reference site 6. Våtenga, Stigstuv, Eidfjord, Hordaland

Total age range: 8200 BP to present day. Sites: Bog 11, Hol, Buskerud, 1310 m (Moe 1973); and Våtenga, Stigstuv, Eidfjord, Hordaland, 1310 m (Moe 1978). Two radiocarbon dates are included in the reference profile (Fig. 6.15). Two pollen-assemblage zones (paz).

Age	Local paz
1. *ca.* 8200–8000 BP	Compositae–fern–Gramineae
2. 8000–0 BP	Cyperaceae–*Salix*–herbs

Event stratigraphy

(1) Continuous bog development from about 8000 BP.
(2) Mire vegetation from 8000 BP to present time.
(3) No positive evidence for local tree vegetation.
(4) Traces of human interference (grazing by domestic animals) on vegetation during the last 5000 years.

Special references: Moe (1973, 1978), and Lye & Lauritzen (1975).

From the altitudinal level of 1480 m asl Caseldine (1984) has presented a pollen diagram from the site Vestre Memurubreen, Lom, Oppland, with three pollen-assemblage zones (paz):

Age	Local paz
1. 4000–*ca.* 2800 BP	Ericales–Gramineae–*Lycopodium*
2. *ca.* 2800–300 BP	Gramineae–*Salix*
3. 300–*ca.* 200 BP	Ice

174

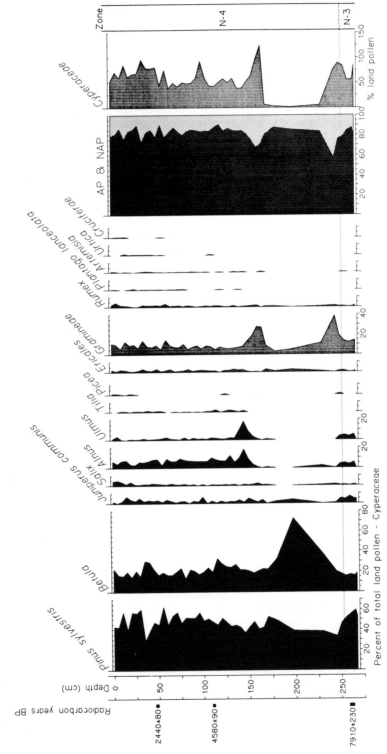

Normannslågen, Hordaland, NORWAY
Selected Pollen Percentages
Analyst: Dagfinn Moe, 1971-1973

Fig. 6.14 Simplified pollen diagram from bog at Normannslågen (loc. 35), Ullensvang Hordaland. Reference level 1250 m asl (Redrawn after Moe 1978)

Våtenga, Eidfjord, Hordaland, NORWAY
Selected Pollen Percentages
Analysts: Dagfinn Moe & K. Johannesson, 1970–1973

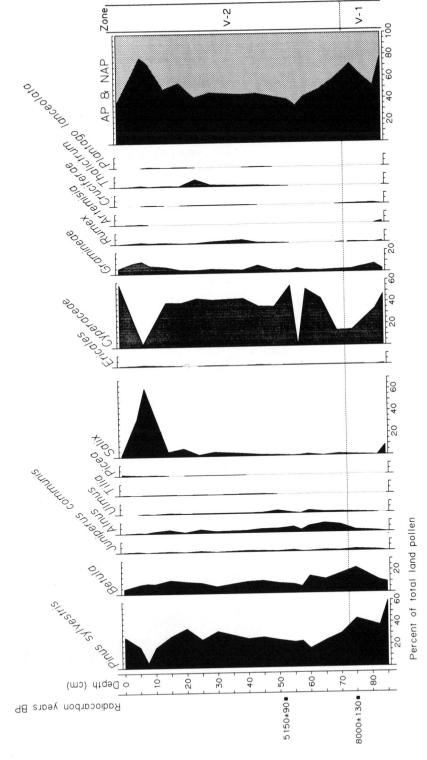

Fig. 6.15 Simplified pollen diagram from **Våtenga** (loc. 24), Stigstuv, Eidfjord, Hordaland. Reference level 1300 m asl (Redrawn after Moe 1978)

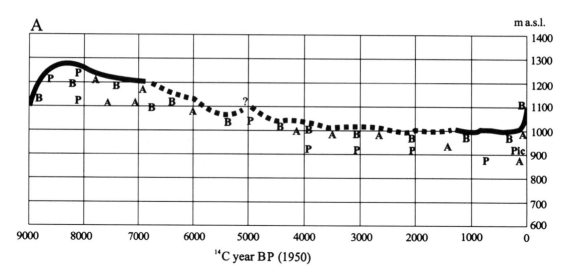

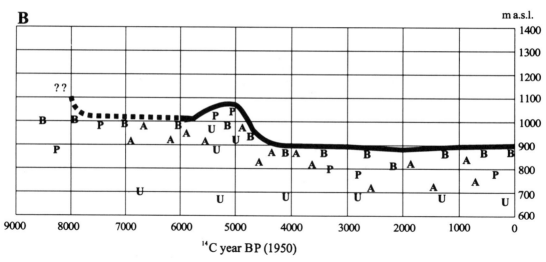

Fig. 6.16 Suggested changes of the tree line in the region studied (Arctic–Alpine area) with information on the tree species dominant in the uppermost forest: a. The central and eastern parts; b. The western parts. Abbr.: A = *Alnus*, B = *Betula*, P = *Pinus*, Pic = *Picea*, and U = *Ulmus*. (From Moe & Odland 1992. See also Kvamme 1993)

Event stratigraphy

(1) Undisturbed stratigraphical sediments are limited.
(2) Between 4000 and 2800 BP a higher altitude for mid-alpine vegetation is suggested.
(3) Grassland with some herbs and willow occurred 2800–300 BP.
(4) Ice covered the site during the Little Ice Age.

Special references: Selsing (1972), and Caseldine & Matthews (1985).

Regional palaeoecological events

General vegetational events (Figs 6.16 and 6.17)

(1) Deglaciation to 8500 BP: light-demanding flora up to 1250 m with, for example, *Hippophaë rhamnoides*.

(2) 8500–8000 BP: maximum treeline within Holocene, about 1250 m asl, pine predominated.

(3) 8000–5500 BP: reduced treeline, *Betula* and *Alnus* (*A. incana*) predominated up to 1135 m.

(4) 5500–5000/4900 BP: temporarily increased treeline, 1050 m in the west with, for example, *Pinus* and *Ulmus*.

(5) 5000/4900–3400 BP: lowering of treeline to near present-day level, ice readvance. Human impact on vegetation from about 5200 BP.

(6) 3400–2800 BP: possibly a temporary higher treeline in the west.

(7) 2800–2300 BP: low treeline, possibly to same altitude as event (5).

(8) During recent decades, from about AD 1950, increase of treeline.

Human events

(1) General human impact on vegetation since Early Neolithic time (5200 BP) at all levels.

(2) Stronger impact locally at (a) traditional hunting sites, (b) sites for bog-iron activities (from about 2200 BP), and (c) in connection with intensive grazing/summer-farming from 2800/1900 BP.

(3) Peat-cutting dated back to about 1500 BP.

(4) Reduced and/or absence of grazing. Summer-farms mostly abandoned from about AD 1950.

Soil events

(1) Peat development started at about 8000 BP and at 5000–4900 BP.

(2) Peat/soil erosion may have occurred locally (a) in all dry periods: deglaciation–8000 BP, 5500–5000/4900 BP, and perhaps 3500–2800/2700 BP; (b) in periods with more local intensive human impact from about 5200 BP, and (c) in periods with glacial expansions, from about 5000 BP and 200–300 BP (Little Ice Age).

(3) The age of the onset of present-day peat erosion from an altitude around 600 m up to at least 1300 m is unknown. Agencies like water (surplus of water), wind (exposed localities), and human disturbance are likely to be important locally, either alone or in combination.

Climatic events (Fig. 6.17)

(1) The last glacier advance is dated to about 9300 BP.

(2) Optimum climate before 8000 BP. Dry and warm (mean temperature minimum 1.7°C above the present day).

(3) Reduced summer temperature and/or increased precipitation from about 8000/7900 BP to 5500 BP. A total reduction about 1.6–1.7°C, close to present-day values.

(4) Temporary drier conditions 5500–5000/4900 BP.

(5) Reduced summer temperature and/or increased precipitation from about 5000/4900 to present day. Reduction of about 0.3°C, suggested to be near present-day values (see event (3)).

(6) Possibly temporary amelioration 3200–2800 BP.

(7) Temporary deterioration during the Little Ice Age.

TYPE REGION N-mo, SOUTHERN COASTAL AREA IN MØRE AND TRØNDELAG (CENTRAL NORWAY)
(Aa. Paus)

The altitudinal range at the Møre coast is 0–*ca.* 1000 m and at the Trøndelag coast 0–*ca.* 600 m. In outer coastal areas, the strandflat includes numerous islands, islets, and skerries. In inner coastal areas, deep fjords and steep mountains dominate.

Climate: From the outer coast to the inner fjord area, the mean January temperature changes from 1.5 to −1.5°C, and the July mean temperature changes from 13 to 14°C. The yearly mean precipitation changes from 850 to 1600 mm, with 150–190 wet days yearly. Cool oceanic.

Geology: Precambrian gneiss, mainly migmatitic. Some areas of granite and diorite. Marine deposits occur below the Late-glacial marine limit, from *ca.* 75 m to 175 m asl from outer to inner parts.

Population: About 100000, concentrated in the outer coastal areas and along fjords.

Vegetation: Mainly treeless heathland, mires, and bogs with some extensive areas of blanket bog. Small

TYPE REGION N-k   SOUTH NORWEGIAN MTS

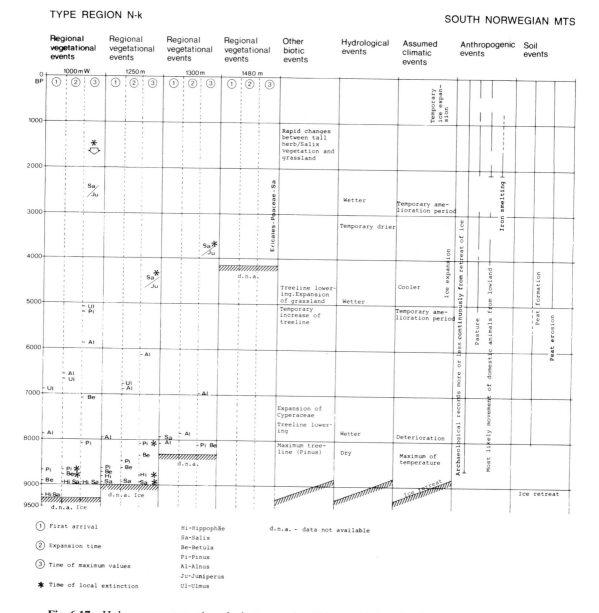

Fig. 6.17 Holocene event stratigraphy in type region N-k, the Alpine–Arctic zone, South Norway

areas of hay meadows and arable cultivation (potatoes, vegetables, and, more rarely, barley occur locally). Scattered areas of planted forests of *Pinus* spp., *Picea* spp., and *Abies* spp. In the inner coastal and more sheltered areas forests of *Pinus sylvestris* and deciduous trees (*Betula, Populus, Alnus incana*). Natural *Picea abies* in the northern part of the

region. Some *Ulmus glabra* and *Corylus avellana* in the most sheltered and fertile sites.

Soils: Mostly infertile acid peats and podzols. Smaller areas of soils fertilized by sediments/shell banks, base-rich basement, or cultivation. Large areas of bare rock.

Land use: Sheep and cattle grazing, small arable areas, remnants of peat-cutting, forestry in conifer forests, recreation and sport.

Climatic gradients exist within the region. Four sites/lakes are considered, and two site diagrams are presented (Figs. 6.18 and 6.19).

Reference sites 7 and 8. Grasvatn and Krokvatn

Reference area: The island Frøya, Sør-Trøndelag. Age range (*ca.* 11000–0 BP) (Paus 1982), with the lakes: Grasvatn (*ca.* 11000–0 BP) 63°42′N, 8°42′E, 44.5 m asl (Fig. 6.18), Røssvatn (9600–2500 BP) 63°42′N, 8°37′N, 39.0 m asl, Krokvatn (9200–0 BP) 63°42′N, 8°45′E, 27.5 m asl (Fig. 6.19), Litjklingert-jern (6000–2000 BP) 63°46′N, 8°48′E, 14.5 m asl.

Each diagram is divided into pollen-assemblage zones (paz). Altogether seven regional pollen-as-semblage zones are distinguished. Metachronous zone boundaries between the four lake diagrams suggest local influences. Radiocarbon dates exist for all diagrams, in Grasvatn four dates and in Krokvatn two dates.

Age	Local paz
11000–10300 BP	*Artemisia*–Gramineae
10300–9100 BP	*Betula*
9100–8000 BP	*Pinus*
8000–6200 BP	*Alnus*
6200–4700/4100 BP	*Ulmus–Corylus*
4700/4100–3300/2400 BP	*Pinus–Calluna*
3300/2400–0 BP	*Calluna*

Regional palaeoecological events

Common and metachronous patterns of events in time within the type region N-mo (based on a few dates only and rather fragmentary material).

Common events (Fig. 6.20)

(1) Deglaciation of outer coastal areas about 12500 BP.
Inner coastal areas deglaciated during the Allerød chronozone (12000–11000 BP).

Preboreal/Boreal *Hippophaë rhamnoides* repre-sentation is negligible in the southernmost part because of a slower Holocene land-uplift and hence smaller areas newly exposed and suitable for *Hippophaë* invasion in the south.

(2) Except for the late Allerød establishment of trees and shrubs in the southern area, no Late-glacial tree-vegetation development in the region.

(3) About 11000 BP to 10300 BP a light-demand-ing herb-dominated vegetation with, for example, Gramineae, *Artemisia*, *Rumex*, and *Helianthemum* established within the region.

(4) 10300–9100 BP small birch trees, *Salix*, and *Juniperus* dominated.

(5) About 9000 BP *Pinus* expansion, a possible delay of a few hundred years as one moves from the centre of the region towards the north and south.

(6) Expansion of *Alnus* (presumably *A. incana*) about 8000 BP.

(7) From about 5000/4700 BP opening of vegeta-tion.

(8) About 5000/4700 BP increase of *Calluna* and Gramineae (heath-formation).
Increase of charcoal and expansion of *Plantago lanceolata*. Heath formation occurs earlier in the southern coastal areas of the region (4700–3300 BP) than in the northern part (2600 BP).

(9) From about 3000/2600 BP continuous records of *Picea* pollen, possibly resulting from long-dis-tance dispersal. Local arrival of *Picea abies* in the northern coastal areas, but no *Picea* arrival in the south.

(10) From about 3500–1000 BP gradual deforesta-tion in the outer, coastal areas. No total deforesta-tion in the inner coastal areas.

Metachronous events

Indications by ^{14}C-dates and comparisons with synchronous pollen curves of wind-pollinated/long-distance taxa.

(1) Preboreal *Juniperus* maximum oldest in basins isolated from the sea in the early Preboreal.

180

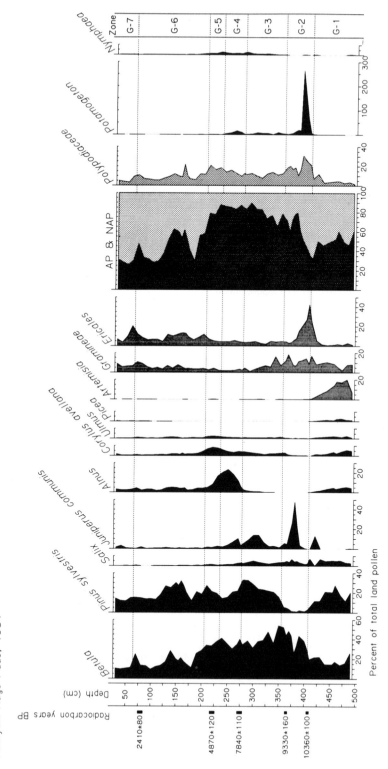

Fig. 6.18 Simplified pollen diagram from Grasvatn, Frøya. (Paus 1982)

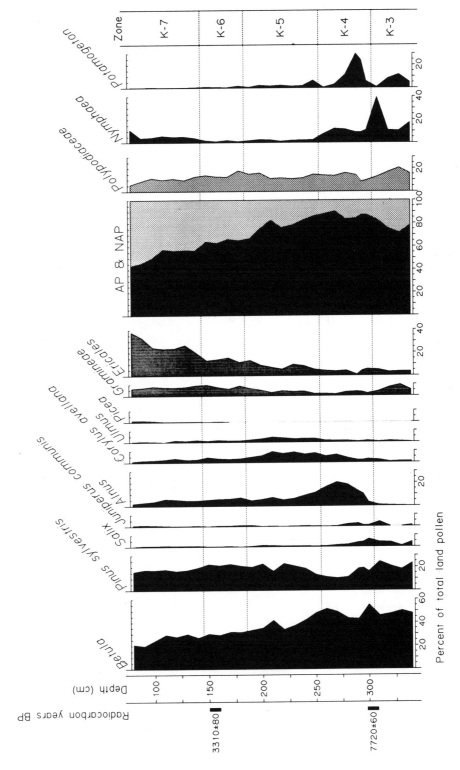

Fig. 6.19 Simplified pollen diagram from Krokvatn, Frøya. (Paus 1982)

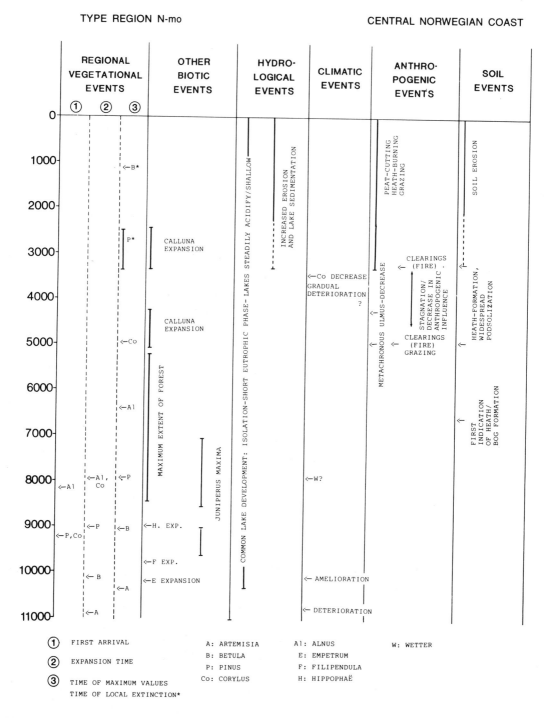

Fig. 6.20 Holocene event stratigraphy in type region N-mo, coastal areas in Møre and Trøndelag, Central Norway

(2) Boreal/Atlantic *Juniperus* maxima metachronous (cause unknown).

(3) Metachronous *Ulmus* decline (about 5000 to 4200 BP).

(4) Metachronous starts of local anthropogenic phases. First phase (clearing, burning, grazing) about 5000 or 4200 BP. Second phase (clearing, heath-burning/cutting, grazing, peat-cutting) about 3300 or 2400 BP.

SOUTHERN NORWAY—A SYNTHESIS
(D. Moe)

Deglaciation and topographical events

The Late-Pleistocene to early-Holocene topographical coastal patterns are basically controlled by the palaeogeography of the region, eustatic as well as isostatic sea-level changes. Present-day submerged seabanks both in the south and the north may have played an important role in the plant colonization of the coastal area. A large number of small, scattered, exposed islands existed. At the transition to the Boreal zone, at 9000 BP, large permanent islands and land-areas had appeared along the coast. Submerged terrestrial peat is found (e.g. Holmboe 1909, Sollesnes & Fægri 1951).

Soil events

Phases of enhanced mire formation are found at:

About 8000–7700 BP in south Norwegian mountains (Moe 1975).

About 5000–4700 BP in southern Norwegian mountain areas (Moe 1974, 1978, 1979a).

About 3000–2000 BP in the western Norwegian coastal area, soil-erosion landscape (Kaland 1986).

Glacial events

About 9700–9100 BP: glacial retreat with different relatively smaller advances. The last advance dated to 9100 BP (Nesje & Dahl 1992). Oldest organic sediments on Hardangervidda, 1250 m asl dated to about 8700 and 7600 BP. A possible small glacial advance (Nesje & Dahl 1992), comparable with a similar event in North Sweden between 7700 and 7200 BP.

About 5000–4700 BP: glacial advance (?).

About 3000–2800 BP: evidence for gelifluction processes in Jostedal glacier area, western Norway (Nesje *et al.* 1989).

300–200 BP: temporary glacial advance (Little Ice Age).

Selected floral events

For present-day distribution see Fægri (1960), Lid (1985), and Gjærevoll (1990).

Osmunda regalis (Birks & Paus 1991)
Earliest record 9500 BP. Between 9500 and 8300 BP several localities through a large part of the southwestern Norwegian coastal area.

Hippophaë rhamnoides
Deglaciation to 10400/9000 BP: The taxon occurs rather early after deglaciation in the southwestern part of the country, at the coast, and at higher altitudes: in mountain valleys and above 1250 m from about 9300 BP. In the lowland together with *Artemisia* and *Helianthemum*.

9000 to early Boreal (*ca.* 8500): *Hippophaë* predominates but is succeeded by tree-forming birch.

8500–8300 BP: Distribution of the taxon reduced drastically. The reason is a combination of reduced sea-level fluctuation and strong competition from shading.

Cladium mariscus
9500–9000 BP: Earliest record from southeast Norway.

9000–8000 BP: Occurrence in coastal areas to central Norway.

After mid-Holocene, reduction of *Cladium*-pollen (Hafsten 1965).

Pinus sylvestris
9300–9000 BP: Introduction in southern Norway, delay towards the north and south at the outer coast.

About 8500 BP: Expansion up to an altitude about 1250 m asl in central southern Norway. The main pine introduction in the north was from the east, in the south most likely from the south to east.

8000 BP: Lowering of treeline in the south-Norwegian mountains.

7500–6000 BP: A slower expansion towards the west is seen in some outer coastal areas in the north.

5500–5000 BP: A temporary expansion at lower levels in south-Norwegian mountains.

4600–4200 BP: A final expansion in the coastal areas in the north.

About 3200–2700 BP: A temporary minor pine expansion may have occurred in the western south-Norwegian mountains (Moe 1979a).

About 3000–2500 BP: A major pine retreat in the north.

Alnus spp.
Evidence for local growth in southern Norway most likely at the end of Boreal (Tallantire 1974, Moe & Odland 1992).

Frangula alnus (Moe 1984)
Earliest records found in southeastern Norway in Preboreal.

9000–8000 BP: Expansion to most of south Norway, also inner parts of the fjords.

8000–5000 BP: In southern Norway at maximum altitudes at 670 m asl.

About 5000 BP: Disappeared from high altitudes.

About 2500–1500 BP: Disappeared at Beiarn (the second most northern occurrence in northern Norway) (Moe 1991).

Ulmus glabra
Earliest records of presence at the border Boreal/Atlantic.

7500–6000 BP: Expanded to most of southern and somewhat later to central Norway.

6000–5000 BP: Reached the area near the Arctic Circle. In the same period reached maximum altitude in south-Norwegian mountain areas (Moe & Odland 1992).

5000 BP: Retreat in altitudinal distribution.

Picea abies.
2500–1500 BP: Earliest record of local growth, earliest in the southeast, latest in the central and

western Trøndelag region (Moe 1970, Hafsten *et al.* 1979, Hafsten 1985).

The species is still spreading in most of the marginal areas of its distribution, latitudinal as well as altitudinal. The greatest expansion potential seems to be in the middle parts of the country.

Vegetational events

8500–8000 BP: Maximum treeline within Holocene, about 1250 m asl, pine predominated. Evidence for birch forest above the pine is not available for southern Norway.

8000–5500 BP: Lowered treeline, *Betula* and *Alnus* (*A. incana*) predominated up to 1135 m.

5500–5000/4900 BP: Temporarily raised treeline, 1050 m in the west with, for example, *Pinus* and *Ulmus*.

5000/4900–3400 BP: Lowering of treeline to near present-day level, ice readvance.

5000–4700 BP: Anthropogenic opening of coastal areas, increase of *Calluna* and Gramineae.

3400–2800 BP: Possibly a temporarily higher treeline in the west.

2800–2300 BP: Low treeline, possibly to same altitude as in the period 5000/4900–3400 BP. Start of heather landscape development.

2500–1500 BP: A new, marked opening of the coastal forest, the heather landscape developed (Kaland 1986).

During recent decades, from about AD 1950, increased treeline and a reforestation at the coast, especially in south and central Norway.

Coastal changes

For sea-level changes, including transgressions, see Hafsten (1983), and Kaland (1984).

Climatic events

9500–9000 BP: Holocene climatic optimum, evidence for early spread of the thermophilous *Osmunda regalis*, partly also *Cladium mariscus* in south Norway.

8500–8000 BP: Maximum altitudinal distribution

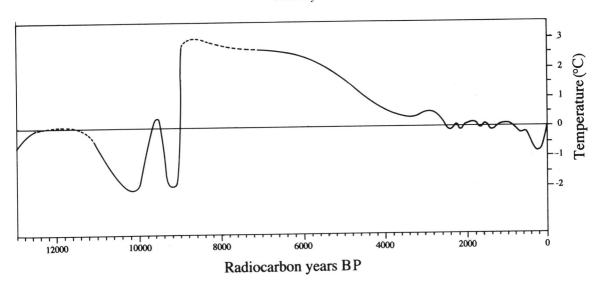

Fig. 6.21 Temperature variation through the last 13000 years based on Quaternary geological and flora-historical studies relative to the present-day situation. (Moe 1994, Kvamme 1993, and Nesje & Kvamme 1991)

of pine in south Norway coincides with maximum latitudinal distribution in the north.

8000 BP: Climatic deterioration, reduced temperature, wetter conditions, bog formation. Humid climate-demanding taxa expanded.

5500–5000/4800 BP: In some areas a temporarily drier climate.

About 5000/4800 BP: Major climatic deterioration. Glacier advance. Bog development in most of the country.

3400–2800 BP: Temporarily drier, increased summer temperature, reforestation (Fig. 6.21).

Little Ice Age is not properly documented by palynological data.

Anthropogenic events

(1) About 5200 to 4200 BP: In southern Norway metachronous *Ulmus* decline combined with earliest record of anthropogenic indicator taxa such as *Plantago lanceolata*, *Artemisia*, and *Urtica* in some coastal as well as alpine areas.

(2) About 5000 BP: Metachronous *Tilia* decline in some south-Norwegian lowland areas (inside the distribution area).

(3) About 5000 or 4200 BP: Increase of charcoal combined with first phase of clearing, burning (charcoal), grazing.

(4) 4200 BP: Earliest records of pollen of cereals.

(5) About 3300–2400 BP: Second phase of anthropogenic interaction (clearing, heath-burning/cutting, grazing, peat-cutting). Coastal heathland development. Start of summer-farming.

(6) 2200–1900 BP: Strong disturbance of treeline by man.

(7) About 800 to 500/400 BP: Temporarily reduced human interaction.

(8) 500/400 BP: Strong disturbance by man.

(9) Last 50 years reforestation both in coastal and mountainous areas caused by change in land use with, for example, less grazing and summer-farming.

ACKNOWLEDGEMENT

The authors express their sincere thanks to John Birks, University of Bergen, for his suggested improvements of the text and its contents.

NORTHERN NORWAY
(K.-D. Vorren and B. Mørkved)

Northern Norway normally includes the three northernmost counties only. However, we here also include the neighbouring district to the south, Namdalen, which is part of the county of Nord-Trøndelag.

According to Fig. 6.6 the northernmost county of northern Norway is mainly arctic–alpine, whereas the main part of the coastal area north of 65°N. belongs to the middle boreal to north boreal vegetational zone, with dominant *Betula pubescens* woodlands, and with frequent occurrences of *Pinus sylvestris* forests and *Alnus incana* stands. The southernmost part of northern Norway (*s.str.*) and Namdalen belong mainly to the southern boreal vegetation zone, with frequent occurrences of nemoral vegetation units and *Ulmus glabra* and a dominance of *Picea abies* inland. Due to the steep topographical gradients the local climate has a strong impact on vegetation, which over short distances may vary from southern boreal to high arctic/alpine.

The weather is dominated by cyclonic low pressures and high precipitation. The length of the growing season in the lowlands ($\overline{T} \geq 5°C$) varies from 120 days in the northeast to 180 days in the south of the region. Snow covers the ground for 6–8 months of the year.

The bedrock (Fig. 6.1) varies from acid gneiss and arkose to basic marble and mica schist. The most acid areas are concentrated along the coast. The Holocene marine limit varies from 0 or negative (Vesterålen) to 150 m asl (Namdalen).

The palaeoecology of Finnmark is well documented by, for example, Hyvärinen (1975, 1976, and 1985), and Prentice (1981, 1982). For the rest of northern Norway there is a fairly tight network of ^{14}C-dated sequences. During recent years new material has been added to the picture presented here.

TYPE REGION TRANSECT N-mo–N-si, NAMDALEN, MIDDLE NORWAY
(B. Mørkved, S. F. Selvik, and H. Ramfjord)

Location and area: The Namdalen region (Fig. 6.22) covers an area bounded by *ca.* 64–65°N latitude and 11–14°E longitude and includes four different natural grographic regions (Abrahamsen *et al.* 1977). During the Late Weichselian the entire Namdalen was glaciated, an area of *ca.* 12000 km^2.

Altitude: 0–1200 m, mainly 0–800 m.

Climate: Mean January temperature 0 to –10°C (west–east), July 12–15°C. Precipitation 700–1200 mm yr^{-1}.

Geology: In the east Cambro-Silurian bedrock prevails, in the west mainly Precambrian granite and gneiss. The main Quaternary deposits are tills and glacio-fluvial deposits above the marine limit, with prominent marine terraces of clay, sand, and gravel in the lower valleys (especially the Namdalen valley itself).

Topography: An undulating landscape with fjords and islands in the west. The Namsen River domi-nates the landscape, trending north–south to Grong, then turning east–west. Between Grong and the watershed to the east, the Sanddøldalen valley continues the lower Namsen straight eastwards. In the eastern part of the district, large lakes and mires dominate. The relief increases rather continuously from *ca.* 150 m on the islands in the west to *ca.* 800 m at the watershed and then decreases eastwards to the Swedish border.

Population: Approximately 3 persons km^{-2}.

Vegetation: Picea forest dominates and reaches the coastal area at the –0.5°C January isoline. The timberline is formed by *Picea* and *Betula. Pinus* may dominate on the eastern glacio-fluvial sediments. *Betula* and *Alnus incana* are common along the rivers. *Ulmus glabra* occurs in limited stands east to Tunnsjøen *ca.* 400 m above sea level, and *Corylus avellana* occurs only in the west and in the central lowland. *Alnus glutinosa* and *Frangula alnus* occur sporadically.

Soils: Podzols predominate. *Ca.* 20% of the area below the forest limit consists of mires.

Land use: Forestry and farming, mostly hay production, but also some cereal cultivation (*Hordeum* and *Avena*).

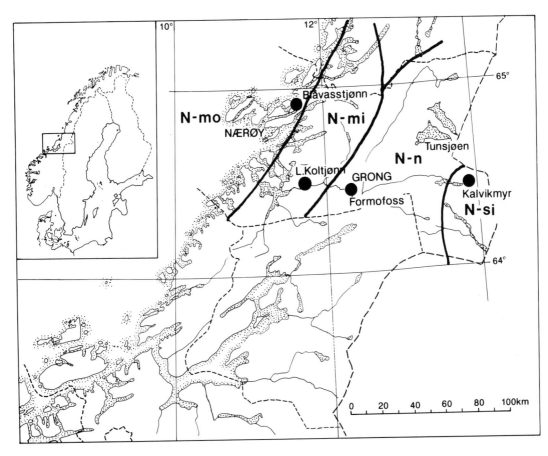

Fig. 6.22 Map of Namdalen with natural geographic regions and reference sites

Physical-geographical references: Birkeland (1958), Vorren, B. (1968), Bruun (1967), Bruun & Håland (1970), Bjørkvik (1976), Vorren, K.D. (1979a), Moen (1987, 1988).

Sub-region eastern Namdalen, Lierne: inland area

Altitude: 300–800 m.

Climate: Humid and continental to slightly oceanic. Annual precipitation *ca.* 700 mm. July mean temperature *ca.* 13°C. January mean temperature *ca.* −9.5°C.

Geology: Bedrock of Cambro-Silurian origin. The area was deglaciated about 9000 BP. Several traces of dead-ice lakes. Thick layers of till and glacio-fluvial deposits.

Topography: Undulating landscape with large bog areas and several large lakes. In the surrounding areas hills and peaks between 500 m and 1200 m altitude.

Population: 0.7 persons km^{-2}.

Vegetation: Coniferous forest dominated by spruce (*Picea abies*) and with some Scots pine (*Pinus sylvestris*) and birch (*Betula pubescens*). The timberline is at ca. 650 m altitude, formed by *Betula pubescens*.

Soils: Podzols dominate. Large areas of peat bogs.

Land use: Forestry, sheep farming, small-scale milk production, reindeer-hunting, and inland fishing.

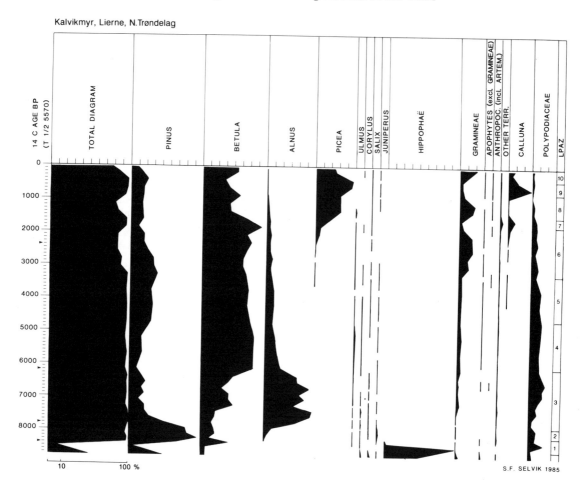

Fig. 6.23   Holocene pollen diagram from Kalvikmyr. Recalculated and redrawn after Selvik (1985). ^{14}C-datings marked with arrows

Reference site 9. Kalvikmyr (Selvik 1985)

Latitude 64°27′N, Longitude 13°52′E. Elevation 430 m. Age range 8700 BP to present. Ombrotrophic mire, core 2.05 m. Four radiocarbon dates. 10 local pollen-assemblage zones (paz) (Fig. 6.23).

Age	Local paz
1. 8700–8300 BP	*Hippophaë–Juniperus– Salix*
2. 8300–8000 BP	*Pinus–Juniperus*
3. 8000–6200 BP	*Alnus*
4. 6200–4700 BP	*Betula–Ulmus*
5. 4700–3300 BP	*Betula–Pinus*

Age	Local paz
6. 3300–1800 BP	*Betula–Pinus–* Gramineae
7. 1800–1500 BP	*Betula–Picea*
8. 1500–800 BP	*Picea* –Gramineae
9. 800–400 BP	*Picea–Calluna*
10. 400–0 BP	*Picea–Betula*–Gramineae

Sub-region Middle Namdalen—1. Eastern part: Grong

Altitude: 100–800 m.

Climate: Humid and continental to slightly oceanic.

Annual precipitation *ca.* 1200 mm. July mean temperature *ca.* 15°C, January *ca.* –4.5°C.

Geology: Bedrock of Precambrian origin. The area was deglaciated at *ca.* 9500 BP. Till, many terraces in glacio-fluvial material.

Topography: Plain at marine level, formed by a marginal moraine. Surrounded by slightly undulating landscape, terraces, and hills at 600–800 m altitude.

Population: 2.3 persons km^{-2}.

Vegetation: Coniferous forest dominated by *Picea abies* with some *Pinus sylvestris* and *Betula pubescens*. Timberline at *ca.* 450 m altitude, usually formed by *Picea*.

Soils: Podzols dominate. Several peat bogs.

Land use: Forestry, some grazing, hunting, and inland fishing.

Reference site 10. Formofoss (Selvik 1985)

Latitude 64°23′N, Longitude 12°21′E. Elevation 175 m. Age range probably 7000–0 BP. Ombrotrophic mire, core 2.40 m. Two radiocarbon dates. Six local pollen-assemblage zones (paz). (Fig. 6.24)

	Age	Local paz
1.	–5500 BP	*Alnus–Betula–Calluna*
2.	5500–3500 BP	*Pinus*
3.	3500–2300 BP	*Pinus*–Gramineae
4.	2300–2000 BP	*Pinus–Betula*
5.	2000–1300 BP	*Pinus–Picea*–Gramineae
6.	1300–0 BP	*Picea*–Gramineae

Sub-region Middle Namdalen—2. Western part: Overhalla

Altitude: 30–700 m.

Climate: Slightly oceanic. Annual precipitation *ca.* 1200 mm. July mean temperature *ca.* 15°C, January *ca.* –5°C.

Geology: Precambrian rocks, mostly granite and gneiss, and an area of calcareous-silicate-gneiss of great importance for the vegetation. Along the rivers are large marine terraces of clay, sand, and gravel. Marine limit at about 155 m asl.

Topography: The rivers are central in the landscape, with Namsen as the largest. Fluvial terraces are mostly cultivated but also contain some large bog systems. Otherwise the landscape is undulating with forested hills and sloping mires. The mountains reach 500–700 asl.

Population: *Ca.* 5 persons km^{-2}, mainly on farms but also in a few small villages.

Vegetation: *Picea abies* forests dominate the landscape. The timberline at about 400 m altitude is also formed by *Picea* and *Betula pubescens*. *Alnus incana* dominates along the rivers and in some south-facing hills. Some relics of *Corylus avellana* and *Ulmus glabra* exist. On the terraces there are large ombrotrophic mire-complexes. In the hill areas sloping mires prevail.

Soil: Podzol series dominate, but also brown forest soils occur.

Land use: Farming, mostly grass production, but also some cereal cultivation (*Hordeum*, *Avena*) and forestry.

Reference site 11. L. Koltjønn (B. Vorren 1969)

Latitude 64°31′N, Longitude 11°4′E. Elevation 159 m. Age range 8500–3000 BP. Lake sediments. Detritus mud. Core 5.30 m. Three radiocarbon dates. Five local pollen-assemblage zones (paz). (Fig. 6.25)

	Age	Local paz
1.	8500–8300 BP	*Salix–Juniperus–Hippophaë*
2.	8300–8000 BP	*Pinus*
3.	8000–5600 BP	*Alnus*
4.	5600–4500 BP	*Pinus–Ulmus–Corylus*
5.	4500–3000 BP	*Pinus–Betula*–Gramineae

Sub-region Western Namdalen, Nærøy: coastal area, fjord landscape

Altitude: 0–400 m asl.

Climate: Generally oceanic, but occasionally with low winter temperatures. Mean annual precipitation is about 1100–1200 mm, mainly occurring in the period September–March. Mean annual temperature is about 4°C, January mean *ca.* –1°C, and July

Formofoss, Grong, N.Trøndelag

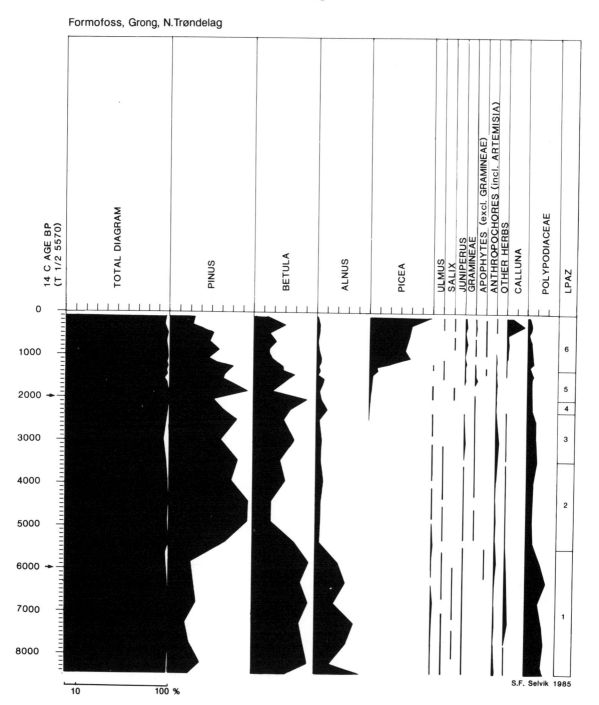

Fig. 6.24 Holocene pollen diagram from Formofoss. Recalculated and redrawn after Selvik (1985). ^{14}C-datings marked with arrows

mean about 14°C. Prevailing SE wind in winter and NW in summer.

Geology: Dominated by Precambrian rocks, viz. gneiss and granite (amphibole gneiss around the sites mentioned). No calcareous substratum of any importance occurs. The Younger Dryas ice-marginal deposits (11000–10000 BP) cross the area, and both western Namdalen sites considered here formed in the area glaciated during the Younger Dryas stadial.

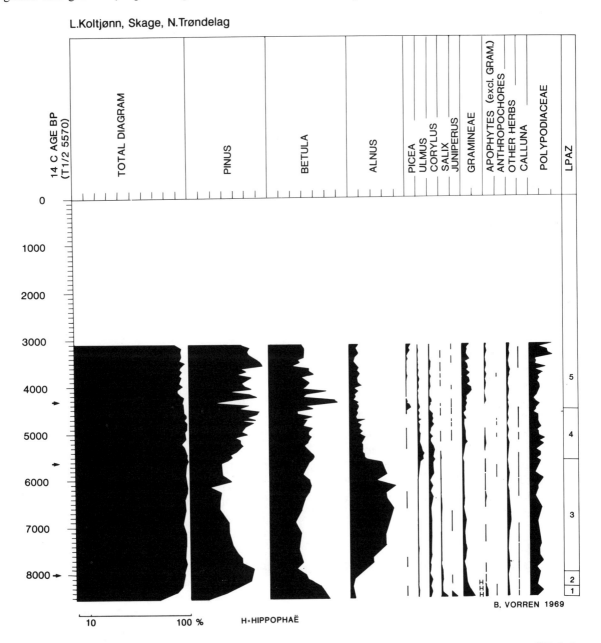

Fig. 6.25 Holocene pollen diagram from L. Koltjønn. Recalculated and redrawn after Vorren (1969). ¹⁴C-datings marked with arrows

Topography: Dominated by a complex fjord system, although the relief gives a rounded and even impression. Mountains seldom exceed 400 m.

Population: 1.7 persons km^{-2}.

Vegetation: Phytogeographically the area is included in the northern Coniferous Region of the Eurosiberian forest area. Predominating species are *Betula pubescens*, *Pinus sylvestris* and *Picea abies*, the latter frequently forming the local timberline.

Soils: Soligenous and ombrogenous mire types are common, due to the humid climate. The nutrient content is generally poor because of the non-calcareous bedrock.

Land use: Farming, mostly hay production. Small units, where the users frequently combine farming with other activities, such as forestry and fishing.

Reference sites 12. Blåvasstjønn and Gorrtjønn
(Ramfjord 1979, 1982)

Blåvasstjønn, Nærøy: Latitude 64°54′N, Longitude 11°38′E. Age range: 9000–0 BP. Elevation 92 m. Lake sediments. Gyttja. Isolated from sea since *ca.* 8950 BP. Core 4.40 m. Four radiocarbon dates. Gorrtjønn, Nærøy: Latitude 64°55′N, Longitude 11°38′E. Elevation 66 m. Lake sediments. Gyttja. Isolated from sea since *ca.* 8680 BP. Core 4.65 m. Three radiocarbon dates.

Stratigraphy: Both basins have a layer of algal gyttja above the marine clay, followed by quite homogeneous fine-detritus gyttja throughout the rest of the series. The recalculated and redrawn pollen diagram from Blåvasstjønn (Fig. 6.26) illustrates the western Namdalen vegetation development. Six local pollen assemblage zones (paz) in both diagrams (Fig. 6.26).

Age	Local paz
1. 9000–8500 BP	*Hippophaë –Salix–Betula–* Gramineae
2. 8500–8000 BP	*Betula–Pinus*
3. 8000–6000 BP	*Alnus–Betula–Corylus*
4. 6000–5000 BP	*Alnus–Ulmus*
5. 5000–2500 BP	*Pinus*
6. 2500–0 BP	*Picea–Pinus*–Gramineae

Synthesis of the Holocene vegetational development in the whole Namdalen region (Fig. 6.27)

Common patterns

(1) The early-Holocene pollen assemblages are dominated by *Hippophaë rhamnoides*, *Juniperus*, *Salix*, Gramineae, and *Betula*. The zones are metachronous within the time interval 9000–8300 BP.
(2) *Pinus* expanded between 8500 and 8300 BP.
(3) *Alnus* expanded between 8200 and 8000 BP.
(4) Lithological changes indicate a transition to a period with more humid climate *ca.* 6300 BP.
(5) *Ulmus* expanded between 6000 and 5800 BP.
(6) *Corylus* probably existed in western Namdalen from about 8000 BP. In middle Namdalen it occurred before or together with *Ulmus* from 6000 BP.
(7) *Picea* existed in the area from about 2500 BP. However, there may have occurred Sub-boreal *Picea* enclaves. *Picea* expanded at different periods, mainly around 2300, 2000, 1300, and 600 BP.

Special Patterns

(1) In western Namdalen *Betula* immigrated *ca.* 9500 BP (cf. diagram Nedre Lisetjønn, Ramfjord 1982).
(2) In western Namdalen an early immigration of *Ulmus* and *Corylus ca.* 7500 BP cannot be excluded. A possible climatic depression *ca.* 6300 BP may have caused its retreat before a new expansion *ca.* 6000–5800 BP.
(3) In western Namdalen the earliest records of agricultural activity may be dated at *ca.* 2600 BP, according to present knowledge. Generally in Namdalen agricultural expansion occurred between 2000 and 1700 BP.
(4) *Picea* established *ca.* 2300 BP in eastern Namdalen, but in central Namdalen not until 600 BP. The West-Namdalen *Picea* occurrence constitutes one of the most maritime in Norway at present.

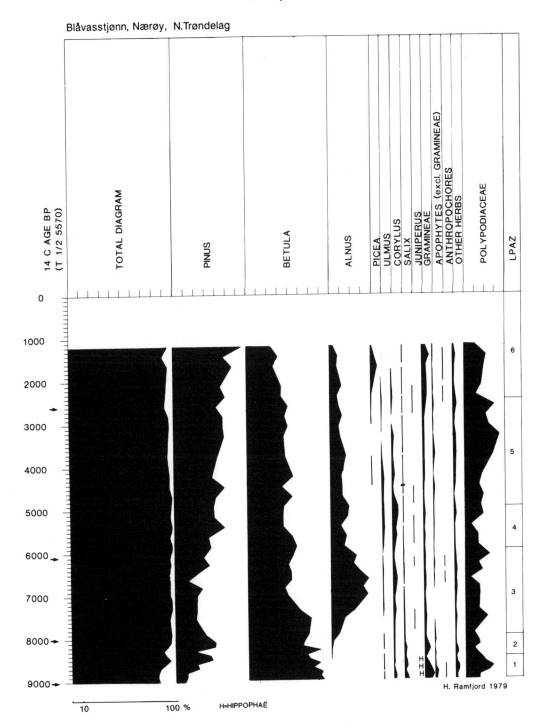

Blåvasstjønn, Nærøy, N.Trøndelag

H=HIPPOPHAË

Fig. 6.26 Holocene pollen diagram from Blåvasstjønn. Recalculated and redrawn after Ramfjord (1979, 1982). ^{14}C-datings marked with arrows

TYPE REGION N-mo   NAMDALEN, MIDDLE NORWAY

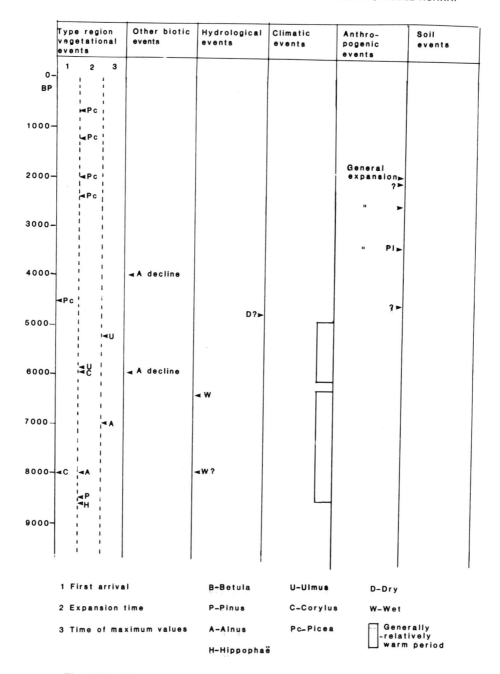

Fig. 6.27 Holocene event stratigraphy for type region N-mo, Namdalen

TYPE REGIONS N-u, N-v, N-w, N-x AND N-z
LOFOTEN, VESTERÅLEN AND SOUTHERN
TROMS (K.-D. Vorren, T. Alm, S. Fimreite, B.
Mørkved, and E. Nilssen).

Area: Latitude 68°00′–70°15′N, and Longitude
13°00′–19°30′E. *ca.* 30000 km².(Fig. 6.28).

Altitude: 0–1700 m.

Climate: July mean temperature between 11.3 and
14.2°C at sea level. Coldest month between 1.3 and
–9.0°C, and annual mean 3–4 to –0.1°C. Precipitation *ca.* 750 mm at the extreme coast, 100–1500 mm
along the coastal mountain range, and 600–300 mm
yr⁻¹ in the interior, eastern area.

Geology: Variations from acid coastal gneiss to
calcic Cambro-Silurian rocks in the fjord and valley
district.

Topography: Peaty strandflat in the west and
wooded Quaternary deposits in river valleys in the
east. Mountains range from high and steep to low
and hilly, increasing generally in elevation from *ca.*
300 m in the west to 1200–1700 m asl in the east.

Population: 6.3 km⁻² for the entire area, 16.1 km⁻² in

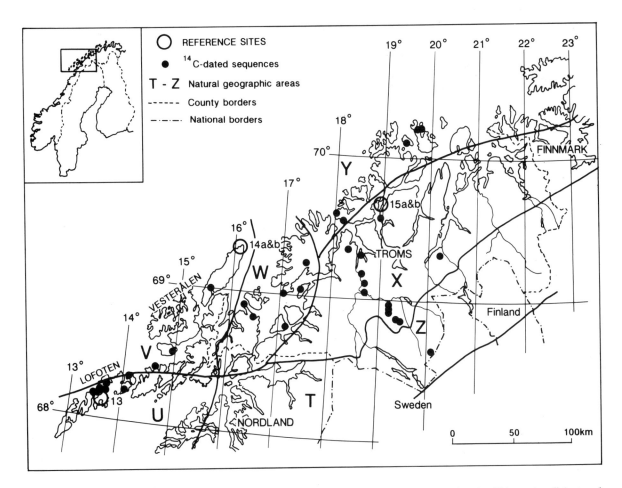

Fig. 6.28 Map of the reference area N III, Lofoten–Vesterålen–southern Troms, covering the "bio-regions" (natural
geographic areas) N-t, N-u, N-v, N-w, N-x, and N-z. The position of the most important biostratigraphical sites of the
reference region and the reference sites N13–15 have been marked

Tjernet,Tromsø

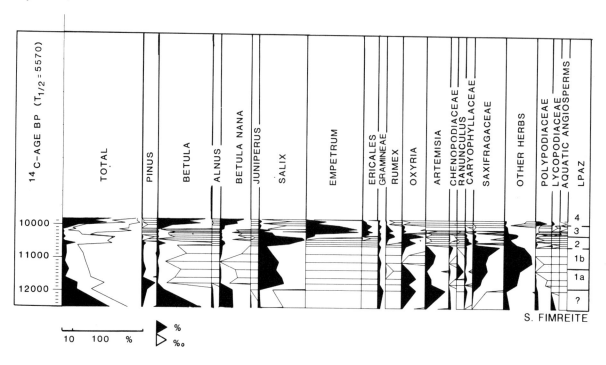

Fig. 6.29 Late-Glacial pollen diagram from Tjernet, Tromsø, close to reference site N15, reconstructed on the basis of material from Fimreite (1980)

the Vesterålen plus Lofoton. More than 50% live in urban areas.

Vegetation: Anthropogenically induced and controlled heaths, grasslands, and woodland in the west and along the fjords. "Virgin" mixed pine forests in the valleys in the east. Although *Alnus incana* is lacking in most of Vesterålen and Lofoten, it is rather abundant in the fjord and valley districts. The *Betula* tree line runs at 200 m asl at the outer cost, and at 500–700 m asl in the interior.

Soils: Mostly podzols, but also grey–brown forest soil (cambisol) in rich *Alnus* stands (Vorren & Vorren 1984). "Arctic brown" soils in the richest mountains. Peatland most extensive on the standflat of Andøya.

Land uses: Grass production, <1% fields, mostly potatoes.

Physical-geographical references: Abrahamsen (1977), Benum (1958), Bruun (1967), Undås (1967), Andersen (1968), Møller & Sollid (1972, 1973), Bergström (1973), Landmark (1973), Eurola & Vorren (1980), Statistisk Årbok (Statistical Yearbook) (1981), Fjelland *et al.* (1983), Møller (1984, 1986, 1987, 1989), Sigmond *et al.* (1984), Meteorlogisk Årbok (Meteorological annual, any year).

Palaeoecological references: Sonesson (1968), Hyvärinen (1975, 1976), Fjellberg (1978), Foged (1974), Vorren (1978, 1979, 1983, 1986), Fimreite (1980), Eronen & Hyvärinen (1982), Moe (1983), Nilssen (1983,1988), Stabell (1985), Vorren & Alm (1985), Alm (1986, 1993, in press), Vorren & Moe (1986), Mørkved (1987, 1990), Nilssen & Vorren (1987, 1991), T. Vorren *et al.* (1988), Vorren *et al.* (1990), Alm & Birks (1991), Alm & Willassen (1993), Alm & Vorren (in prep.).

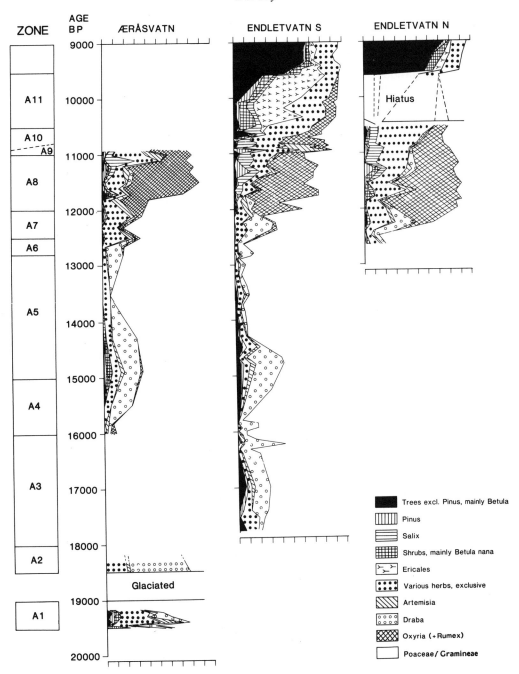

Fig. 6.30 The Late-Glacial (Late Weichselian) vegetation development on northern Andøya (reference site N14), an effort to correlate three different sequences from respectively Nedre Æråsvatn, Endletvatn south, and Endletvatn north total diagrams

Late Weichselian (20 000–10 000 BP)

Reference site 15 (a).Tjernet (Fimereite 1980)

Latitude 69°44′N. Longitude 18°57′E. Elevation 101 m. Age range 12 400–9800 BP. 0.5 ha. Type region N-x. (Fimreite 1980) (Fig. 6.29). Three radiocarbon dates. Four local pollen assemblage zones (paz).

Age	Local paz
12 400–11 690 BP	Redeposited?
1. 11 690–10 580 BP	*Dryas–Artemisia–Saxifraga oppositifolia*
	(a) *Dryas–Artemisia–Oxyria* subzone
	(b) *Dryas–Artemisia* subzone
2. 10 580–10 270 BP	*Oxyria–Salix*
	(a) *Oxyria–Salix–Polygonum viviparum* subzone
	(b) *Oxyria–Salix–Rumex–Empetrum* subzone
3. 10 270–10 000 BP	*Empetrum*–Ericales–*Rumex*
4. 10 000–9800 BP	*Betula–Empetrum*–Filicales

Reference sites 14 (a). Endletvatn and Æråsvatn
(Vorren 1978, Vorren *et al.* 1988, Alm 1986, 1993, in press, Alm & Birks 1991, Alm & Vorren in prep.)

Latitude 69° 16′N, Longitude 16° 05′E. Elevations 36 and 35 m. Type area N-v. Age range *ca.* 19 500–9500 BP with continuation in the Holocene. Lake size of Endletvatn 35 ha, but divided into two subbasins by a high rock threshold. Reference site sub basin is about 20 ha. Core *ca.* 7 m. Start of lacustrine sedimentation dated at *ca.* 18 500 BP. Lake size of Nedre Æråsvatn is *ca.* 12 ha, and the analysed late Weichselian sequence is 5.6 m. A total of 35 radiocarbon dates from the three sequences are considered in Fig. 6.30. 11 pollen-assemblage zones (paz).

	Age	Local paz
A1	19 500–19 000 BP	Gramineae–*Artemisia*–herbs, Endletavn-ice–covered
–	19 000–18 500 BP	Glacial advance over Æråsvatn

	Age	Local paz
A2	18 500–18 000 BP	*Draba*
A3	18 000–16 000 BP	Gramineae–*Draba* 1
A4	16 000–15 000 BP	Gramineae–*Draba*–other herbs
A5	15 000–12,800 BP	Gramineae–*Draba* 2
A6	12 800–12 500 BP	Gramineae–*Draba* 3
A7	12 500–12 000 BP	Gramineae–*Draba–Oxyria*
A8	12 000–11 000 BP	*Oxyria–Salix*–Gramineae
A9	11 000–?10 900 BP	*Artemisia–Oxyria*–Gramineae
A10	?10 900–10 500 BP	*Oxyria–Salix*–Gramineae
A11	10 500–9500 BP	Ericales–Gramineae–various herbs

General Patterns

(1) A succession of Gramineae-dominated vegetation, with an *Oxyria*-phase, divided by an *Artemisia*-maximum, to a more or less pronounced *Empetrum*-phase at the transition to Preboreal, seems to be general. The succession is obviously metachronous: The *Artemisia*-phase has been dated at 11 200–10 900 BP on Andøya (Fig. 6.30), *ca.* 12 000–10 600 at Tromsø (Fig. 6.29), and 11 200–10 450/10 300 BP in eastern Finnmark (Prentice 1981, 1982; ages inter- and extrapolated). The *Artemisia*-phase is also characterized by *Saxifraga oppositifolia* and Chenopodiaceae and contains sparse pollen of *Ephedra, Helianthemum,* and *Hippohaë.*

(2) *Salix* occurs rather synchronously around 12 000 BP, although it comprises different time intervals and sites. *Betula nana* occurrence is probably synchronous with *Salix.*

Special patterns

(1) According to ¹⁴C-dates, the vegetation development in eastern Finnmark is metachronous, as compared with Troms and northern Nordland.

(2) Tjernet, in a calcareous rock environment, exhibits high proportions of *Dryas* and *Saxi-*

fraga oppositifolia during the *Artemisia*-rich paz (11690–10580 BP).

Holocene (10000 BP–present)

Reference site 15 (b). Prestvannet
(Fimreite 1980) (Figs 6.31 and 6.32)

Latitude 69°44′N, Longitude 18°57′E. Elevation 95.5 m. Age range >10000–3420 BP. Original lake size 1.1 ha (now dammed). Type region N-x. Eight radiocarbon dates. Four pollen-assemblage zones (paz).

	Age	Local paz
?	Not dated oppositifolia	Salix–Saxifraga
?	Not dated	*Dryas–Artemisia– Saxifraga oppositifolia*
1.	9950–9810 BP	*Betula–Empetrum–* Filicales
2.	9810–6790 BP	*Betula*–Gramineae– *Filipendula*
		(a) *Betula*–Gramineae– *Filipendula–Salix* subzone
		(b) *Betula*–Gramineae– *Filipendula–Pinus* subzone
		(c) *Betula–Pinus* subzone
3.	6790–3420 BP	*Betula–Alnus*
		(a) *Betula–Alnus* subzone
		(b) *Betula–Alnus*– Cyperaceae subzone

Hiatus slightly above 3420 BP.

Reference site 14 (b). Endletvatn
(Alm 1986, Alm & Vorren in prep.) (Fig. 6.33)

Latitude 69°16′N, Longitude 16°05′E. Elevation 36 m. Age range 9600–0 BP. Type region N-v. Lake size 35 ha, but divided into two subbasins by a high rock threshold. Reference subbasin is about 20 ha. Holocene sequence 3.90 m thick. Ten radiocarbon dates. Five pollen-assemblage zones (paz).

	Age	Local paz
A12	9500–7000 BP	*Betula–Pinus* 1
A13	7000–5250 BP	*Betula–Pinus–Alnus*
A14	5250–2800 BP	*Betula–Pinus* 2
A15	2800–1700 BP	*Betula–Pinus*– Gramineae
A16	1700–0 BP	*Betula–Pinus– Gymnocarpium*– Gramineae

Reference site 13. Dønvoldmyra
(Figs 6.34 and 6.35)

Latitude 68°08′N, Longitude 13°35′E. Elevation 10 m. Age range 8600–0 BP. Bog area *ca.* 12 ha. Type region N-u. Six local pollen-assemblage zones (paz). Six radiocarbon dates.

	Age	Local paz
1.	8600–2600 BP	*Betula–Pinus*–Filicales
		(a) *Betula–Pinus– Filipendula*–Filicales– Gramineae subzone (8600–4600 BP)
		(b) *Betula–Pinus* subzone (4600–3300 BP)
		(c) *Betula–Filipendula*– Gramineae subzone (3300–2600 BP)
2.	2600–500 BP	*Betula–Pinus*–Gramineae– Filicales
3.	500–0 BP	*Betula–Pinus*–Filicales– Gramineae

Humification intervals

8600–6300 BP	Warm and moist, with probably an increase of moisture and fall of temperature 7500–7300 BP
6300–4350 BP	Probably warm and optimal
4350–2950 BP	Intermediate temperature, but increasing humidity to 3100 BP, the last 150 radiocarbon years drier
2950–0 BP	Cool-humid, except for less humid periods

PRESTVANNET, TROMSØ, N.NORWAY
("Relative-diagram")

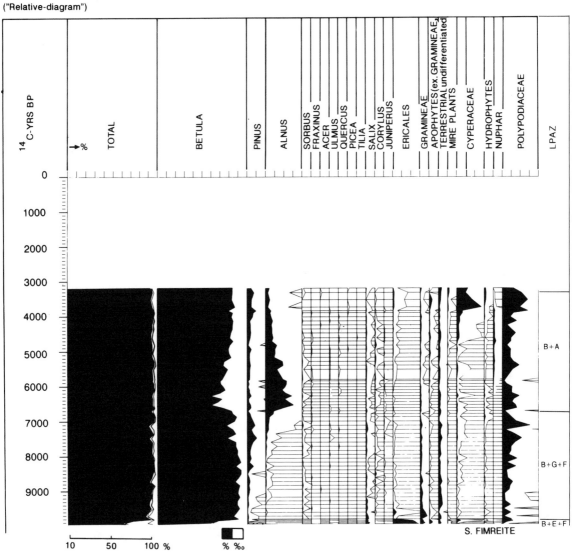

Fig. 6.31 Holocene pollen diagram from Prestvannet, Tromsø (reference site N15), based on data from Fimreite (1980), relative values. Note hiatus above *ca.* 3400 BP

between 2950 and 2700, and 2000 and 1800 BP

General vegetation patterns

(1) Most of the diagrams may be divided into the following successively conditioned meta-chronous zones: (a) a lower tree *Betula* zone, b) a *Betula–Pinus* zone, (c) a *Betula–Pinus–Alnus* zone, and (d) an upper *Betula* zone. In the northeast (b) and (c) may overlap.

(2) South of the Arctic Circle the zone transition

PRESTVANNET, TROMSØ, N.NORWAY
("Absolute-diagram")

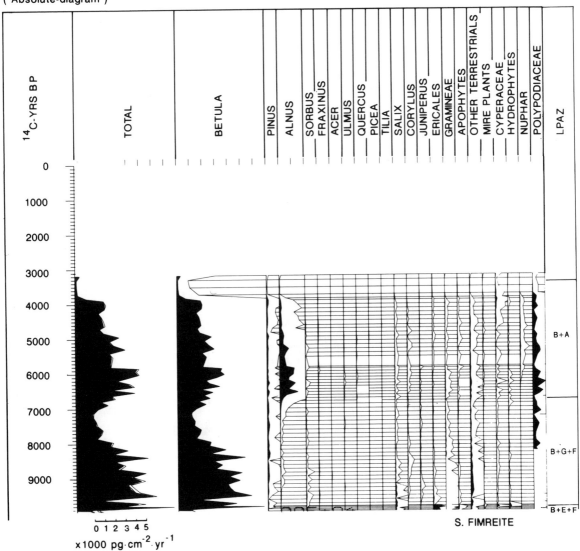

Fig. 6.32 Holocene pollen diagram from Prestvannet, Tromsø (reference site N15), based on data from Fimreite (1980), influx values

Endletvatn, Andøya

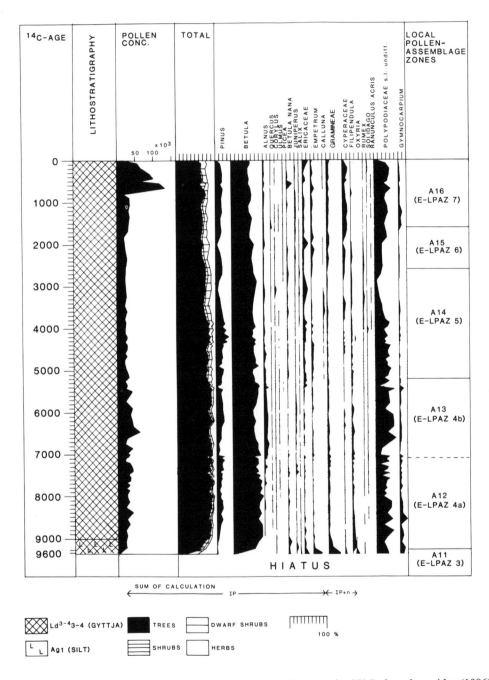

Fig. 6.33 Holocene pollen diagram from Endletvatn, Andøya (reference site N14), based on Alm (1986); relative values

DØNVOLDMYRA, VESTVÅGØY, N.NORWAY
("Relative-diagram")

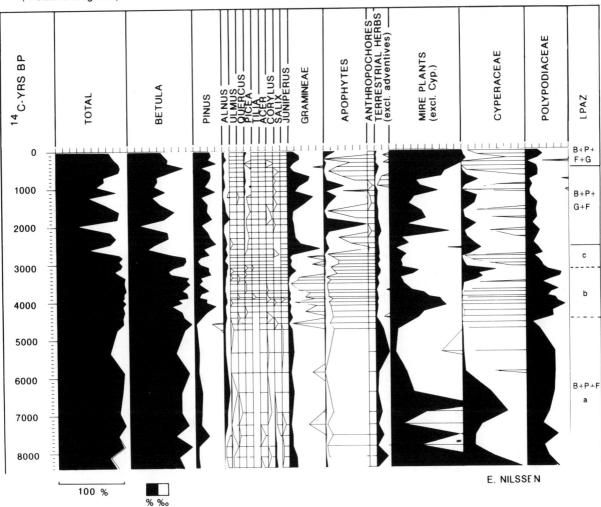

Fig. 6.34 Holocene pollen diagram from Dønvoldmyra, Lofoten (reference site N13), based on Nilssen (1983); relative values

area (b/c) contains some *Ulmus* and *Corylus*, and the upper *Betula* zone is replaced by a *Pinus–Betula–Picea* zone.

(3) The successively conditioned metachronous *Empetrum* heaths prevailed until *Betula* trees finally immigrated 10000–9200 BP (**8700 BP** in western Lofoten, zone N-u), depending on glacier retreat, migration speed, and climatic conditions.

The information in Fig. 6.36 and other diagrams constitute the basis for a brief survey of the forest history in northern Norway.

Betula pubescens (common tree birch; isochrones in map Fig. 6.37) probably immigrated from the east *ca.* 10000 BP or maybe earlier: evidence indicates presence about 10500 BP in the northern part of Vesterålen. Permanent enclaves were established in climatically favourable areas in the inner fjord districts of Troms and Finnmark, in front of the inland glaciers. A contemporary migration from the south along the coast into the southernmost part (Helgeland) may have occurred. The major spread of *Betula* trees occurred about 9700–9000 BP in most of the district. The final establishment in the latest deglaciated and the most maritime areas took place about 8600 BP.

Pinus sylvestris (Scots pine; isochrones in map Fig. 6.37) may have followed separate immigration

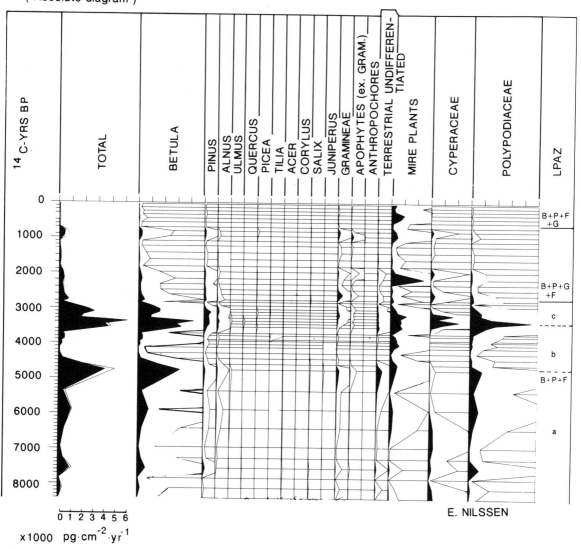

Fig. 6.35 Holocene pollen diagram from Dønvoldmyra, Lofoten (reference site N13), based on Nilssen (1983); influx values

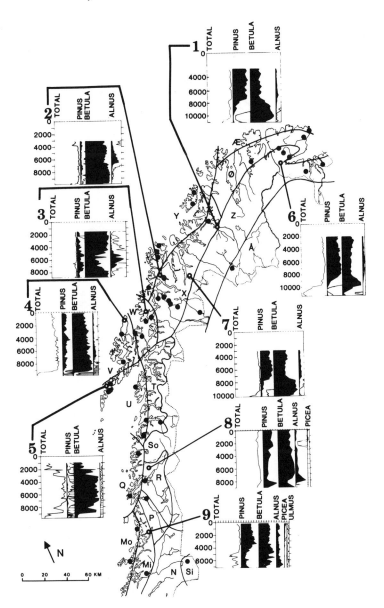

Fig. 6.36 Sites with selected survey diagrams (open circles) from North Norway, and other sites (dots) with published and unpublished pollen diagrams with radiocarbon datings, reflecting status for 1990. The selected pollen diagrams have been constructed so that the vertical axis represents number of years BP. The horizontal axis represents percentages of a total including trees and terrestrial herbs. The "Total" diagram shows the proportions of AP vs. terrestrial herbs in a percentage scale. Black: percentage values; white: permille values. Diagrams have been extracted from: 1: Alta (Hyvärinen 1985), 2: Tromsø (Fimreite 1980), 3: Senja (Vorren & Alm 1985), 4: Andøya (Alm 1986, Alm & Vorren in prep.), Vestvågøy (Nilssen 1982), 6: Varangerbotn, Bruvatnet (Hyvärinen 1975), 7: Skibotn (Eronen & Hyvärinen 1982), Rana, at Langvatn (Alm, unpublished) 9: Tosen (Mørkved, unpublished)

routes from east and south, but the major invasion came from the east during the period 8500–7500 BP. A second expansion primarily in the maritime districts occurred between 4600 and 4200 (3500?) BP. The major retreat from marginal positions along the coast and in the mountains was between 3000 and 2500 BP. In the southern fjord districts, however, the pine expanded at the cost of *Betula* during the late Subboreal and the early Subatlantic.

Alnus incana (grey alder; isochrones in map Fig. 6.37) is an eastern immigrant. It probably reached the northeast during the Boreal chronozone (9000–8000 BP) but expanded in three steps during the Atlantic: 8000 BP, 7500 BP, and 6500–6000 BP. It never became important along the outer coast. It decreased about 5000–4500 BP and further especially in the marginal positions along the coast and in the mountains about 4000–3500 BP, probably as a

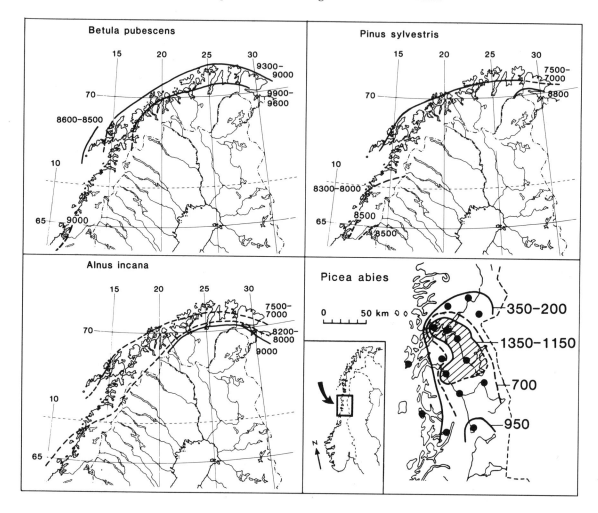

Fig. 6.37 Isochrones, or expansion limits towards the west and north at different ages BP for *Betula pubescens, Pinus sylvestris, Alnus incana,* and *Picea abies* in its main area in North Norway

combined result of human impact and climatic change. Its last major decrease happened between 3000 and 2500 BP.

Ulmus glabra (elm) probably had its greatest extension as a regional tree between 6000 and 5000 BP, north to the Arctic Circle.

Picea abies (spruce; cf. isochrone map Fig. 6.37) immigrated from the southeast to enclaves in the lower valleys and fjord heads in the southern part of the district, probably during the early Subatlantic.

The main expansion, however, occurred around 1350–1150 BP. Later two more expansions followed: *ca.* 700 and 350–200 BP (Mørkved 1987, 1990).

Agricultural influence

Man-made pasture (?) vegetation was locally created as early as *ca.* 5500 BP (Bøstad: Vorren 1979b). Small-scale *Hordeum* cultivation was introduced between 4000 and 3400 BP. The main developmental stage of pastures was between 2900 and 2400 BP, along the west coast. After a decrease of agricultural activity around 2100–1800 BP, a very rapid expansion of pastures and man-made heaths followed *ca.* 1800–1600 BP. In the northernmost districts (Finnmark) there may have been an agricultural "innovation" about 1200 BP, but the main expansion was in late Medieval time.

Palaeoecological events within the transect N-u/N-v—N-z (Fig. 6.38)

The litho-, chrono-, and biostratigraphy of several longer and shorter peat and lacustrine sequences (cf. Fig. 6.28) constitute the basis for the synthesis.

General patterns

(1) The cold intervals up to *ca.* 12800 BP are dominated by a Gramineae–*Draba*–rich vegetation. The warmer periods (thermomers) are characterized by a higher pollen influx and occurrence of different low-Arctic pollen taxa and distantly transported tree pollen.

(2) After *ca.* 12000 BP, *Oxyria* indicates a shift towards a more humid climate. An *Artemisia*-peak around 11000 BP is interpreted as a brief period of dry and winter-cold climate.

(3) The vegetational development expressed mainly through the changing of the forest tree composition of *Betula pubescens*, *Alnus incana*, and *Pinus sylvestris* is representative for central northern Norway.

(4) Man influenced forests/woodlands from *ca.* 4000–3700 BP. Apophytic grasslands expanded between 3000 and 2000 BP. Especially intensive heath- and grassland expansion from 1800–1600 BP.

(5) During the anthropogenic expansion there seems to have been an increased paludification and podzolization.

Unique patterns

(1) The coastal rim of the reference area was deglaciated early, with rather well-developed thermomers around 20000 to 19000 and 16000 to 15000 BP.

(2) The Bølling and Allerød chronozones do not exhibit thermophilous pollen taxa, except for a single *Hippophaë* pollen grain in the middle of the Allerød.

(3) The gradual increase in number of pollen taxa and pollen influx during the Younger Dryas indicates a rather gradual improvement from an Arctic steppe or desert climate to a humid montane-region climate during this chronozone. Low-alpine heath conditions may have occurred around 10500 BP, and even enclaves of a subarctic *Betula pubescens* woodland on favourable places.

(4) Final establishment of *Betula pubescens* woodland happened during the middle and later part of the Preboreal chronozone.

(5) A special problem along the extreme Atlantic coast (Vesterålen, western Lofoten) is the Tapes Sea oscillation, with a rise of sea level and submergence of land between the present 0- and 9-m level. Local climato-stratigraphical deviations have been recorded in the 7000–5500 BP interval.

(6) *Pinus* was at its westernmost expansion limit during 4600–4000 BP. An upper *Pinus* stump/root layer at Elgsnes was dated at *ca.* 2900 BP, which coincides with an agricultural expansion.

(7) Peat humification analyses indicate probable regional so-called recurrence surfaces (RY) around 8000–7850 (?), 5650–4850 (?), 4600–4400, 4150–4100, 3850–3750, 3500–3350, 3150–3000, 2950–2750, 2650–2500, 2400, 2200–2050, 1950–1900, 1750–1600, 1450–1300, 1200–1000, 950–850, 750–500, 400–200 (see Fig. 6.11: Hydrological events). These RYs seem to be related to contemporary humification oscillations in Danish bogs (Nilssen & Vorren 1991). The RYs around 2900–2750, 2650–2500, 1750–1600, and 1200–1000 coincide in part with agricultural expansions (cf. Vorren *et al.* 1990). Removal of forest in connection with land clearings causes a rise in the groundwater table, which may influence at least smaller bogs. Such a groundwater rise may exert influence on the peat humification. However, it is reasonable to believe that the humification changes from dark to light are basically induced by climatic changes.

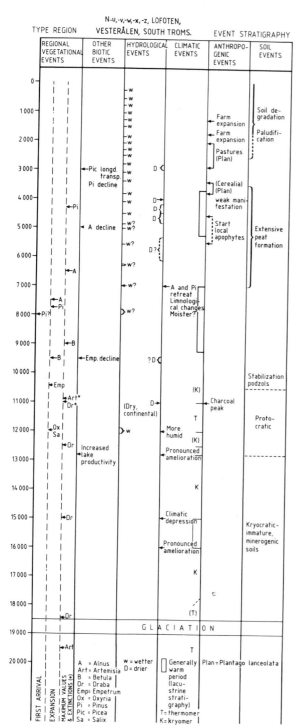

Fig. 6.38 Regional synthesis, stratigraphical events for reference area region N III (Lofoten, Vesterålen and southern Troms), covering the bio-regions N-u, N-v, N-w, N-x, and N-z. Holocene hydrological events are based on bog humification stratigraphy. Time scale to the left: ^{14}C years BP

REFERENCES

Aaheim, R. 1983: Forest and agricultural areas. *In Nasjonalatlas for Norge* Map 8.2.1. (Landbruksdepartementet). Statens kartverk, Hønefoss.

Abrahamsen, J., Jacobsen, N.K., Dahl, E., Kalliola, R., Wilborg, L. & Påhlsson, L. 1977: Naturgeografisk regioninndeling av Norden. *Nordisk utredningsserie B 34*, 137 pp.

Abrahamsen, N., Nybakken, S.E., Svanå, T.G. and Sørensen, R. (in press): Palaeomagnetism and other stratigraphic investigations in Lake Årungen, southeastern Norway. *Norsk Geologisk Tidsskrift.*

Alm, T. 1986: *Einletvatn og Æråsvatn, Andøya, Nordlanden palynologisk undersøkelse av utviklingen fra ca. 20000 BP til idag.* Cand. real. thesis, University of Tromsø, 230 pp.

Alm, T. 1993: Øvre Æråsvatn–palynostratigraphy of a 22,000 to 10,000 BP lacustrine record on Andøya, northern Norway. *Boreas 22*, 171–188.

Alm, T. (in press): Palynostratigraphy of Nedre Æråsvatn, Andøya, northern Norway, *ca.* 20000 to 10000 BP, with special reference to green algae, turbellarian cocoons and other lacustrine microfossils. *Review of Palaeobotany and Palynology.*

Alm, T. & Birks, H. 1991: Late Weichselian flora and vegetation of northern Andøya–macrofossil (seed and fruit) evidence from Nedre Æråsvatn. *Nordic Journal of Botany 11*, 465–476.

Alm, T. & Willassen, E. 1993: Late Weichselian chironomidae (Diptera) stratigraphy of Nedre Æråsvatn, Andøya, northern Norway. *Hydrobiologia 264*, 21–32.

Alm, T. & Vorren, K.-D. (in prep.): Late Weichselian and Holocene palynostratigraphy of lake Endletvatn, Andøya, northern Norway, *ca.* 18,000 BP to present. *Boreas.*

Andersen, B.G. 1968: Glacial geology of western Troms, North Norway. *Norges Geologiske Undersøkelser 320*, 1–74.

Andersen, B.G. 1980: The deglaciation of Norway after 10 000 BP. *Boreas 9*, 211–216.

Anundsen, K. 1985: Changes in shore level and ice-front position in Late Weichsel and Holocene, southern Norway. *Norsk Geografisk Tidsskrift 39*, 205–225.

Aune, B. 1993a: Klima. *In Nasjonalatlas for Norge.* Statens kartverk, Hønefoss, (English summary).

Aune, B. 1993b: Monthly temperatures of Norway. Map 1:7 mill. *In Nasjonalatlas for Norge.* Map 3.1.6. Statens kartverk, Hønefoss.

Aune, B. 1993c: Annual temperature. *In Nasjonalatlas for Norge.* Map 3.1.5. Statens kartverk, Hønefoss.

Aune, B. 1993d: Seasons of the year and growth season. In *Nasjonalatlas for Norge.* Map 3.1.5. Statens kartverk, Hønefoss.

Bang-Andersen, S., Prøsch-Danielsen, L. & Thomsen, H. 1986: Spennende oppdagelser på bunnen av en myr. *Frá haug ok heidni 11*, 112–121.

Benum, P. 1958: The flora of Troms Fylke. *Tromsø Museums Skrifter 6*, 402 pp + maps

Bergström, E. 1973: Den prerecenta lokalglaciationens utbredningshistoria inom Skanderna. *Naturgeografiska institutionen, Stockholms universitet. Forskningsrapport 16*, 1–214.

Birkeland, T. 1958: Geological and petrological investigations in Northern Trøndelag, Western Norway. *Norsk Geologisk Tidsskrift 38*, 327–420.

Birks, H.H. & Paus, Aa., 1991: *Osmunda regalis* in the early Holocene of western Norway. *Nordic Journal of Botany 11*, 635–640.

Bjørkvik, E. 1976: Trøndelag gjennom tusenåra. *In* Evensberget, S. & Mykland, R. (eds) *Bygd og by i Norge 3*, 91–131. Gyldendal, Oslo.

Bruun, I. 1962: *The air temperature in Norway 1931–60.* Det Norske Meterologiske Institutt, Oslo, 54 pp.

Bruun, I. 1967: *Climatological summaries for Norway. Standard normals 1931–60 of the air temperature in Norway.* Det Norske Meteorologiske Institutt, Oslo. I–LI + 270 pp.

Bruun, I. & Håland, L. 1970: *Climatological summaries for Norway. Standard normals 1931–60 of number of days with various weather phenomena.* Det Norske Meteorologiske Institutt, Oslo.

Caseldine, C.J. 1984, Pollen analysis of a buried arctic–alpine brown soil from Vestre Memurubreen, Jotunheimen, Norway: evidence for postglacial high-altitude vegetation change. *Arctic and Alpine Research 16*, 423–430.

Caseldine, C.J. & Matthews, J.A. 1985: ^{14}C dating of palaeo-soils, pollen analysis and landscape change: studies from the low- and mid-alpine belts of southern Norway. *In* Boardman, (ed.) *Soils and Quaternary landscape evolution*, 87–116. Wiley, Chichester.

Dahl, E. 1990. History of the Scandinavian alpine flora. *In* Gjærevoll, O. (ed.) *Maps of Distribution of Norwegian vascular plants. II. Alpine plants.* Tapir Publishers, Trondheim. 126 pp, 37 maps.

Dahl, E., Elven, R., Moen, A. & Skogen, A. 1986: Vegetasjonsregionkart over Norge 1:1,5 mill. *In Nasjonalatlas for Norge.* Map 4.1.1. Statens kartverk, Hønefoss.

Danielsen, A. 1970: Pollen-analytical Late-Quaternary studies in the Ra district of Østfold, southeastern Norway. *Universitetet i Bergen Matematisk–naturvitenskapelig Serie 14(1)*, 1–146.

DNMI, Det norske meteorologiske institutt 1991a: *Foreløbige temperatur og nedbørsnormaler for perioden 1961–1990.* Dataprint.

DNMI, Det norske meteorologiske institutt 1991b: *Nedbørsnormaler 1931–60.* Dataprint.

Eide, F. G. & Paus, Aa. 1982: Vegetasjonshistoriske undersøkelser på Kårstø, Tysvær kommune, Rogaland. *Botanical Institute, University of Bergen Report 23*, 45 pp.

Eronen, M. & Hyvärinen, H. 1982: Subfossil pine dates and pollen diagrams from northern Fennoscandia. *Geologiska Föreningens i Stockholm Förhandlingar 103*, 437–445.

Eurola, S. & Vorren, K.-D. 1980: Mire zones and sections in North Fennoscandia. *Aquilo, Series Botanica 17*, 39–56.

Fægri, K. 1960: Coast plants. *In* Berg, R., Fægri, K. & Gjærevoll, O. (eds) *Maps of distribution of Norwegian plants. I. University of Bergen Skr. 26*, 131 pp. 54 maps.

Farbregd, O. 1980: Arkeologi nordafjells. *Årbok for Trøndelag 1979*, 52–86.

Fimreite, S. 1980: *Vegetasjonshistoriske og palaeolimnologiske undersøkelser i Tromsø, Nord-Norge, fra Sen Weichsel og Holocen*. Cand. real. thesis, University of Tromsø, 179 pp.

Fjelland, M., Elven, R. & Johansen, V. 1983: Havstrand i Troms, botaniske verneverdier. *Miljøverndepartementet, Oslo. Rapport T-551*, 1–291.

Fjellberg, A. 1978: Fragments of a Middle Weichselian Fauna on Andøya, north Norway. *Boreas 7*, 39.

Foged, N. 1978: Diatoms from Middle and Late Weichselian and the Late Flandrian period on Andøya, north Norway. *Boreas 7*, 41–47.

Førland, E. 1993: Annual precipitation. *In Nasjonalatlas for Norge*. Map 3.1.1. Statens kartverk, Hønefoss.

Gjærevoll, O. 1950: Vegetasjonen på Gudfjelløyas sørberg, Røyrvik i Namdalen. *Blyttia 8*, 115–124.

Gjærevoll, O. 1973: *Plantegeografi*. Universitetsforlaget, Trondheim. 186 pp.

Gjærevoll, O. 1976: Floraen i Trøndelag. *In* Evensberget, S. & Mykland, R. (eds) *Bygd og by i Norge 3*, 56–73. Gyldendal, Oslo.

Gjærevoll, O. 1990: Alpine plants. *In* Berg, R., Fægri, K. & Gjærevoll, O. (eds) *Maps of distribution of Norwegian vascular plants. II.* Tapir Publishers, Trondheim. 126 pp., 37 maps.

Gjærevoll, O. 1992: *Plantegeografi*. Tapir publishers, Trondheim. 200 pp.

Göttlich, K., Hornburg, P., König, D., Schwaar, J. & Vorren, K.-D. 1983: Untersuchungen an einem Palsen mit Kieselgurschichten bei Kautokeino, Nord-Norwegen. *Norsk Geografisk Tidsskrift. 37*, 1–31.

Hafsten, U. 1956: Pollen-analytical investigations on the late Quaternary development in the inner Oslofjord area. *Universitetet i Bergen. Naturvitenskapelig rekke 8*, 1–161.

Hafsten, U. 1983: Shore-level changes in South Norway during the last 13000 years, traced by biostratigraphical methods and radiometric datings. *Norsk Geografisk Tidsskrift 37*, 63–79.

Hafsten, U. 1985: The immigration and spread of spruce forest in Norway, traced by biostratigraphical studies and radiocarbon datings. A preliminary report. *Norsk Geografisk Tidsskrift 39*, 99–108.

Hafsten, U. 1987: Vegetation, climate and evolution of the cultural landscape in Trøndelag, Central Norway, after the last ice age. *Norsk Geografisk Tidsskrift 41*, 101–120.

Hafsten, U., Henningsmoen, K.E. and Høeg, H.I. 1979: Innvandringen av gran til Norge. *In* Nydal, R., Westin, S., Hafsten, U. & Gulliksen, S. (eds) *Fortiden i søkelyset*, 171–198. NAVF, Strindheimstrykk, Trondheim.

Helland, A. 1909: Topografisk–Statistisk beskrivelse over nordre Trondhjems amt 2. *Norges Land og Folk 17*. Aschehoug & Co., Kristiania. 1099 pp.

Helldal, B. 1975: Klimaet i Ås. *In* Semb, G. (ed.) *Jorda i Ås*, 2–16. Landbruksforlaget, Oslo.

Henningsmoen, K. 1980: Trekk fra floraen i Vestfold. *In* Møller, V. (ed.) *Vestfold—Bygd og by i Norge*, 163–175. Gyldendal Norske Forlag, Oslo.

Høeg, H.I. 1982: Vegetational development from about 12000–6000 B.P. in the counties Agder and Telemark, South Norway. *Norsk Geografisk Tidsskrift 36*, 211–224.

Holtedahl, O. 1960: Geology of Norway. *Norges Geologiske Undersøkelser 208*, 540 pp.

Holtedahl, O. & Dons, J.A. 1960: Geologisk kart over Norge. Berggrunnskart 1:1 000 000. *Norges Geologiske Undersøkelser 208*.

Holten, J. 1978: Verneverdige edellauvskoger i Trøndelag. *Kongelige Norske Videnskabers Selskap Museet Rapport, Botanisk Serie 4*, 199 pp.

Holten, J. 1983: Flora- og vegetasjonsundersøkelser i nedbørfeltene for Sanddøla og Luru i Nord-Trøndelag. *Kongelige Norske Videnskabers Selskap Museet Rapport, Botanisk Serie 2*, 148 pp.

Hyvärinen, H. 1975: Absolute and relative pollen digrams from northernmost Fennoscandia. *Fennia 142*, 23 pp.

Hyvärinen, H. 1976: Flandrian pollen deposition rates and tree-line history in northern Fennoscandia. *Boreas 5*, 163–175.

Hyvärinen, H. 1985: Holocene pollen history of the Alta area, an isolated pine forest north of the general pine forest region in Fennoscandia. *Ecologia Mediterranea 11*, 69–71.

Indrelid, S. 1986: *Fangstfolk og bønder i fjellet. Bidrag til Hardangerviddas historie 8500–2500 år før nåtid*. Dr. thesis, University of Bergen. 448 pp.

Indrelid, S. & Moe, D. 1982: Februk på Hardangervidda i yngre steinalder. *Viking 46*, 36–71.

Johannessen, T.W. 1960: *Climatological summaries for Norway. Monthly frequencies of concurrent wind forces and wind directions in Norway*. Det Norske Meteorologiske Institutt, Oslo, I–XIX + 295 pp.

Johannessen, T.W. & Håland, L. 1969: *Climatological summaries for Norway. Standard Normals 1931–60 of monthly wind summaries for Norway*. Det Norske Meteorologiske Institutt, Oslo, 217 pp.

Johansen, O.-I., Henningsmoen, K.E. & Sollid, J.L. 1985: Deglasiasjons forløpet på Tingvollhalvøya og tilgrensende områder, Nordvestlandet, i lys av vegetasjonsutviklingen. *Norsk Geografisk Tidsskrift 39*, 155–174.

Kaland, P.E. 1984: Holocene shore displacement and shorelines in Hordaland, western Norway. *Boreas 13*, 203–242.

Kaland, P.E. 1986: The origin and management of Norwegian coastal heaths as reflected by pollen analysis. *In* Behre, K.A. (ed) *Anthropogenic indicators in pollen diagrams*, 19–36. A.A. Balkema, Rotterdam.

Kjemperud, A. 1982: *Late Weichselian and Holocene shoreline displacement in parts of Trøndelag, Central Norway.* Dr. Thesis, Department of Geology, University of Oslo. 191 pp.

Klemsdal, T. & Sjulsen O.E. 1992: Landforms. *In*: *Nasjonalatlas for Norge*. Map 2.1.2. Statens kartverk, Hønefoss.

Kollung, S. 1967: Geologiske undersøkelser i sørlige Helgeland og nordlige Namdal. *Norges Geologiske Undersøkelser 254*.

Kvamme, M. 1984: Vegetasjonshistoriske undersøkelser. *In* Meyer, O.B. (ed.) Breheimen Stryn. Konsesjonsavgjørende botaniske undersøkelser. *Botanical Institute, University of Bergen Report 34*, 238–275.

Kvamme, M. 1993: Holocene forest limit fluctuations and glacier development in the mountains of southern Norway, and their relevance to climate history. *Paläoklimaforschung 9, Special Issue: European Palaeoclimate and Man 4*, 99–113. Stuttgart.

Landmark, K. 1973: Beskrivelse til de geologiske kart "Tromsø" og "Målselv". II. Kaledonske bergarter. *Tromsø Museums Skrifter 15*, 1–263.

Landsskogtakseringen 1964–76: Nord-Trøndelag. *Norsk Institutt for Skogforskning, Landsskog-takseringen*, Ås, 226 pp.

Lid, J. 1959: The vascular plants of Hardangervidda, a mountain plateau of southern Norway. *Nytt Magasin for botanikk 7*, 61–128.

Lid, J. 1985: *Norsk, svensk og finsk flora.* Det norske samlaget, Oslo, 837 pp.

Lindblom, I. 1984: Former for økologisk tilpasning i Mesolitikum, Østfold. *Universitetets Oldsaksamling Årbok 1983/1984*, 43–86.

Lye, K.A. 1975: Survey of the main plant communities on Hardangervidda. *In* Wielgolaski F.E. (ed.) *Fennoscandian tundra ecosystems. I, 68–73.* Springer, Berlin. 366 pp.

Lye, K.A. & Lauritzen, E.M. 1975: Effect of grazing in alpine vegetation on vegetation. *Norwegian Journal of Botany 22(1)*, 7–13.

Mamen, H.C. (ed.) 1981: *Akershus—Bygd og by i Norge.* Gyldendal Norske forlag, Oslo. 576 pp.

Mangerud, J., Andersen, S.T., Berglund, B.E. & Donner, J.J. 1974: Quaternary stratigraphy of Norden, a proposal for terminology and classification. *Boreas 3*, 109–128.

Meteorologiske årbok (Meteorological annual). Det Norske Meteorologiske Institutt, Oslo.

Moe, D. 1970, The Post-Glacial immigration of *Picea abies* into Fennoscandia. *Botaniska Notiser 123*, 61–66.

Moe, D. 1973: Studies in the Holocene vegetation development on Hardangervidda, southern Norway. I. *Norwegian Archaeological Review 6*, 67–73.

Moe, D. 1974: Tilveksten av enkelte myrer på Hardangervidda. *In* "Hardangervidda, natur, kulturhistorie og samfunnsliv" (Ed: Utvalget for samordnet planlegging av Hardangervidda) 76–79. *NOU 30 B*.

Moe, D. 1978: *Studier over vegetasjonsutviklingen gjennom Holocen på Hardangervidda, Sør-Norge. II.* Part of Dr. thesis, University of Bergen. 97 pp.

Moe, D. 1979a: Tregrense-fluktuasjoner på Hardangervidda etter siste istid. *In* Nydal, R., Westin, S., Hafsten, U. & Gulliksen, S. (eds) *Fortiden i søkelyset*, 170–208. Strindheimtrykk, Trondheim.

Moe, D. 1979b: Studies on the first bog development in Lofoten and Vesterålen, North-Norway. *Norsk Geografisk Tidsskrift 33*, 1–5.

Moe, D. 1983: Studies in the vegetation history of Vestvågøy, Lofoten, North-Norway. *Tromura, naturvitenskap 39*, 28 pp. Tromsø.

Moe, D. 1984: The Late Quaternary history of *Rhamnus frangula* in Norway. *Nordic Journal of Botany 4*, 655–660.

Moe, D. 1991: Hustad, Arstad and Naurstad. A vegetational study of three farms in Salten, North Norway. *Norsk Geografisk Tidsskrift 45*, 11–24.

Moe, D. 1994: Climatic variations in western Norway during the last 13,000 years. A review. *Geologija, Lithuania 17*, 159–165

Moe, D. & Odland, A. 1992: The influence of the temperature climate on the vertical distribution of *Alnus incana* (Betulaceae) through the Holocene in Norway. *Acta Botanica Fennica 144*, 35–49.

Moe, D., Indrelid, S. & Kjos-Hansen, O. 1978: A study of environment and early man in the south Norwegian highlands. *Norwegian Archaeological Review 11*, 73–83.

Moe, D., Indrelid, S. & Fasteland, A. 1988: The Halne area, Hardangervidda. Use of a high mountain area during 5000 years: An interdisciplinary case-study. *In* Birks, H.H., Birks, H.J.B., Kaland, P.E. & Moe, D. (eds) *The cultural landscape – past, present and future*, 429–444. Cambridge University Press, Cambridge.

Moen, A. 1987: The regional vegetation of Norway, that of Central Norway in particular. *Norsk Geografisk Tidsskrift. 41*, 179–225.

Moen, A. 1988: Vegetasjonsregioner i Midt-Norge.—Fins " limes norrlandicus"? *Blyttia 46*, 53–64.

Møller, J. 1984: Holocene shore line displacement at Nappstraumen, Lofoten, North Norway. *Norsk Geologisk Tidsskrift 64*, 1–5.

Møller, J. 1985: Coastal caves and their relation to early postglacial shore levels in Lofoten and Vesterålen, North Norway. *Norges Geologiske Undersøkelser 400*, 51–65.

Møller, J. 1986: Holocene transgression maximum about

6000 years B.P. at Ramså, Vesterålen, North Norway. *Norsk Geografisk Tidsskrift 40*, 77–84.

Møller, J. 1987: Shoreline relation and prehistoric settlements in northern Norway. *Norsk Geografisk Tidsskrift 41*, 45–60.

Møller, J. 1989: Geometric simulation and mapping of Holocene relative sea level changes in northern Norway. *Journal of Coastal Research 5*, 403–417.

Møller, J. & Sollid, J.L. 1972: Deglaciation chronology of Lofoten—Vesterålen—Ofoten, North Norway. *Norsk Geografisk Tidssrift 26*, 101–133.

Møller, J. & Sollid, J.L. 1973: Geomorfologisk kart over Lofoten og Vesterålen. *Norsk Geografisk Tidsskrift 27*, 195–205.

Møller, V. (ed.) 1980: *Vestfold—Bygd og by i Norge*. 461 pp.

Mørkved, B. 1987: Granskogens historie i Nord-Norge. *In* Sveli, A. (ed.) *Skogbruk i Nord-Norge*, 24–31. Mosjøen.

Mørkved, B. 1990: Namdalsskogens 10.000-årige historie. *In* Hjulstad, O (ed.) *Skogrike Namdal. I*, 13–25. Namsos.

Nedbørnormaler 1983. *Det Norske Meteorologiske Institutt, Oslo*. 14 pp.

Nesje, A. & Dahl, S.O. 1991: Holocene glacier variations of Blåisen, Hardangerjøkulen, Central Southern Norway. *Quaternary Research 35*, 25–40.

Nesje, A. & Dahl, S.O. 1992: Holocene brevariasjoner ved Blåisen, Hardangerjøkulen. *Geonytt*, 5–6.

Nesje, A. & Kvamme, M. 1991: Holocene glacier and climate variations in western Norway: Evidence for early Holocene glacier demise and multiple Neoglacial events. *Geology 19*, 610–612.

Nesje A., Kvamme, M. & Rye, N. 1989: Neoglacial gelifluction in the Jostedalsbreen region, Western Norway: Evidence from dated buried palaeopodsols. *Earth Processes and Landforms 14*, 259–270.

Nilssen, E. 1983: *Klima- og vegetasjonshistoriske undersøkelser i Lofoten*. Cand. real. thesis, University of Tromsø.167 pp.

Nilssen, E. 1988: Development of the cultural landscape in the Lofoten area, North Norway. *In* Birks, H.H., Birks, H.J.B., Kaland, P.E. & Moe, D. (eds) *The cultural landscape—past, present and future*, 369–380. Cambridge University Press, Cambridge.

Nilssen, E. & Vorren, K.-D. 1987: Skogstrærnes innvandringshistorie i Nord-Norge, *In* Sveli, A. (ed.) *Skogbruk i Nord-Norge*, 11–23. Mosjøen.

Nilssen, E. & Vorren, K.-D. 1991: Peat humification and climate history. *Norsk Geologisk Tidsskrift 71*, 215–217.

Nordahl-Olsen, T. 1990: Ski Kvartærgeologisk kart 1914III–M1:50000, including description. *Norwegian Geological Survey, Skrifter 95*, 1–33.

Nordgård, A. 1972: Jordbruk i kontraksjon og spesialisering. *Ad Novas—Norwegian Geographical Studies 10*, 1–132.

Nybakken, S.E. 1985: Sedimentologi og diagenese i glasimarine, marine og lakustrine leiravsetninger,

Årungen, Sørøst Norge. *Agricultural University of Norway, Department of Geology, Report 22*, 1–220 (PhD thesis).

Oele, E., Schüttenhelm, R.T.E. & Wiggers, A.J. 1979: The Quaternary history of the North Sea. Acta Univ. Upsaliensis Symp. *Univ. Upsal. Annuum Quingentesimum Celebrantis 2*, 248 pp.

Øy, N.E. (ed.) 1978: *Østfold—Bygd og by i Norge*. 475pp.

Påhlsson, L. 1984: *Naturgeografisk regioninndelning av Norden*. Nordiska Ministerrådet. Berlingska, Arlöv. 289 pp.

Paus, Aa. 1982: *Paleo-økologiske undersøkelser på Frøya, sør-Trøndelag. Den vegetasjonshistoriske utviklingen fra senistiden og fram til idag*. Cand real thesis, Department of Botany, University of Trondheim. 234 pp.

Paus, Aa. 1988: Late Weichselian vegetation, climate, and floral migration at Sandvikvatn, North Rogaland, southwestern Norway, *Boreas 17*, 113–139.

Paus, Aa. 1989a: Late Weichselian vegetation, climate, and floral migration at Liastemmen, North Rogaland, southwestern Norway. *Journal of Quaternary Sciences 4*, 223–242.

Paus, Aa. 1989b: Late Weichselian vegetation, climate, and floral migration at Eigebakken, South Rogaland, southwestern Norway. *Review of Palaeobotany and Palynology 61*, 177–203.

Paus, 1990: Late Weichselian and early Holocene vegetation, climate, and floral migration at Uts-ira, North-Rogaland, southwestern Norway. *Norsk Geologisk Tidsskrift 70*, 135–152.

Pedersen, F. 1953: Unreduced mean temperatures for January – July for Norway. *Universitetet i Bergen naturvitensk. rekke 7*, 7 pp.

Prentice, H. 1981: A Late Weichselian and early Flandrian pollen diagram from Østervatnet, Varanger peninsula, NE Norway. *Boreas 10*, 53–70.

Prentice, H. 1982: Late Weichselian and early Flandrian vegetational history of Varanger peninsula, northeast Norway. *Boreas 11*, 187–208.

Ramfjord, H. 1979: *Vegetasjons-og klimahistorie gjennom de siste 9000 år i Nærøy, Nord-Trøndelag*. Cand. real. thesis, Universitetet i Trondheim. 108 pp.

Ramfjord, H. 1982: On the late Weichselian and Flandrian shoreline displacement in Nærøy, Nord-Trøndelag. *Norsk Geografisk Tidsskrift 62*, 191–205.

Rønnevig, H.K. 1971: *Kvartærgeologi på ytre del av Haugesundhalvøya*. Thesis,University of Bergen.

Røsberg, I. 1982: Karplanteflora og vegetasjon på Kårstø og Ognøy, Tysvær og Bokn kommuner i Rogaland. *Botanical Institute, University of Bergen, Report 22 (2)*, 155 pp.

Rosenfeld, H.J. 1978: Israndavsetninger i området Vestby—Ski. *Report 6*, 1–21.

Selsing, L. 1972: Undersøgelse af floraen i de brenære dele af Memurudalen, Jotunheimen 9.–23.juli 1972. *Botanical Museum, University of Bergen Report*, 25 pp.

Selvik, S.F. 1985: *Paleoøkologiske undersøkelser i Nord-Trøndelag.* Med hovedvekt på granskogens innvandring og etablering. Cand. real. thesis, Universitetet i Trondheim. 140 pp.

Selvik, S.F. & Stenvik, L.F. 1983: Arkeologiske registreringer og pollenanalytiske undersøkelser i Sanddølavassdraget, N-Trøndelag. *Kongelige Norske Videnskabers Selskap Musset Rapport, Arkeologisk Serie, 2*, 67 pp.

Sigmond, E., Gustavsson, M. & Roberts, D. 1984: *Berggrunnskart over Norge 1:1 million.* (Geological map of Norway 1:1 million). Norges Geologiske Undersøkelser.

Sigmond, E.M.O. 1985: Brukerveiledning til berggrunnskart over Norge (User's guide for bedrock map of Norway). *In Nasjonalatlas for Norge.* Norges geografiske oppmåling, 38 pp. (English text).

Simonsen, A. 1980, Vertikale variasjoner i Holocen pollensedimentasjon i Ulvik, Hardanger. *Ams-Varia (Stavanger) 8*, 68 pp.

Sjörs, H. 1967: *Nordisk växtgeografi.* 2nd edn. Stockholm 240 pp.

Skarland, P. 1974: *Oversikt over fjellgrunnen i Namdal.* Namdal Skogselskap, Namsos, 132 pp.

Sollid, J.L. 1976: Kvartærgeologisk kart over Nord-Trøndelag og Fosen. En foreløpig melding. *Norsk Geografisk Tidsskrift 30*, 25.

Sollid, J.L. & Sørbel, L. 1981: Kvartærgeologisk verneverdige områder i Midt-Norge. *Rapport T-524, Avdeling for naturvern og friluftsliv*, Miljøverndepartementet, 207 pp.

Sollid, J.L. & Sørbel, L. 1982: Kort beskrivelse til glasialgeologisk kart over Midt-Norge 1:500 000. *Norsk Geografisk Tidsskrift 36*, 225–232.

Sonesson, M. 1968: Pollen zones at Abisko, Torne Lappmark, Sweden. *Botaniska Notiser 121*, 491–500.

Sørensen, R. 1990a, Hva myrer og tjern forteller. *In* Ski–Quaternary geological map 1915 III, Scale 1:50 000, incl. description. *Norwegian Geological Survey, Skrifter. 95*, 14–16.

Sørensen, R. 1990b: Om grunnlaget for bosetning i søndre Follo—fra steinalderen til jernalder. *Follominne 28*, 184–194.

Sørensen, R., Lie, K.T. & Nybakken, S.E. 1990: *Kvartærgeologisk kart 1814 II—M 1:50,000.* Norwegian Geological Survey.

Stabell, B. 1985: Development of the diatom flora in Prestvannet, Tromsø, northern Norway. *Norsk Geologisk Tidsskrift 65*, 179–186.

Statistisk Årbok (Statistical Yearbook) 1981: Central Bureau of Statistics, Oslo.

Stenvik, L.F. 1980: Sanddøldalen—grenda som svartedauen la øde. Årbok for Namdalen, 7–19.

Svanå, T.G. 1983: *Biostratigrafiske undersøkelser (foraminiferer og pollen) i kvartære sedimenter fra Årungen (Akershus).* Thesis, Department of Geology, University of Oslo, 159 pp.

Tallantire, P.A. 1974: The palaeohistory of the grey alder (*Alnus incana* (L.) Moench.) and black alder (*A. glutinosa* (L.) Gaertn.) in Fennoscandia. *New Phytologist 73*, 529–546.

Thomsen, H. 1986: Spennende oppdagelser på bunnen av en myr. *Frå haug ok heidni 11(4)*, 112–121.

Throndsen, L. (ed.) 1977: *Buskerud—Bygd og by i Norge.* Gyldendal Norske forlag, Oslo. 414 pp.

Thoresen, M.K. 1990: Surficial materials. *In Nasjonalatlas for Norge.* Map 2.3.7. Statens kartverk, Hønefoss.

Undås, I. 1967: *Om maksimum av siste innlandsis i ytre Vesterålen.* Eides boktrykkeri, Bergen. 31 pp.

Vorren, B. 1968: Vegetasjonen i Overhalla. *In* Groven, G. (ed.) *Overhalla bygdebok V*, 1–24. Namsos.

Vorren, B. 1969: *Jordbrukshistorie, vegetasjons- og klimahistorie i Skage i Overhalla, Namdalen.* Cand. real. thesis, University of Trondheim, 95 pp.

Vorren, K.-D. 1978: Late and Middle Weichselian stratigraphy of Andøya, north Norway. *Boreas 7*, 19–38.

Vorren, K.-D. 1979a: Die Moorvegetation in Namdalen, Mittel-Norwegen. Eine Untersuchung mit besonderer Berücksichtigung des ozeanischen Gradienten der südborealen Hochmoorvegetation. *Tromura 8*, 102 pp. Tromsø.

Vorren, K.-D. 1979b: Anthropogenic influence on the natural vegetation in coastal North Norway during the Holocene. Development of farming and pastures. *Norwegian Archaeological Review 12 (1)*, 1–21.

Vorren, K.-D. 1983: Den eldste korndyrking i det nordlige Norge, *In* Sandnes, J., Kjelland, A. & Østlie, I. (eds) *Folk og ressurser i nord*, 11–46. Trondheim.

Vorren, K.-D. & Alm, T. 1985: An attempt at synthesizing the Holocene biostratigraphy of a "type area" in northern Norway by means of recommended methods for zonataion and comparison of biostratigraphical data. *Ecologia Mediterranea 11*, 53–64.

Vorren K.-D. & Moe D. 1986: The Early Holocene climate and sea-level changes in Lofoten and Vesterålen, North-Norway. *Norsk Geologisk Tidsskrift 66 (2)*, 135–143.

Vorren, K.-D. & Vorren, B. 1984: Gulsymre (*Anemone ranunculoides*) i Troms. *Polarflokken 8 (2)*, 88–100. (Tromsø).

Vorren, K.-D., Nilssen, E. & Mørkved, B. 1990: Age and agricultural history of the "-stadir" farms of North and Central Norway. *Norsk Geografisk Tidsskrift 44*, 79–102.

Vorren, T. 1979: Weichselian ice movements, sediments and stratigraphy on Hardangervidda, South Norway. *Norges Geologiske Undersøkelse 350*, 117pp.

Vorren, T., Vorren, K.-D., Alm, T., Gulliksen, S. & Løvlie, R. 1988: The last deglaciation (20,000 to 11 000 B.P.) on Andøya, Northern Norway. *Boreas 17*, 41–77.

7

Denmark

S.Th. Andersen, B. Aaby and B. Odgaard

INTRODUCTION
(S.Th. Andersen)

Before 1970 no radiocarbon-dated pollen diagrams covering the entire Holocene were available from Denmark. Since that time palaeoecological research in Denmark has aimed at providing such diagrams for the whole of Denmark. Four diagrams representing the type regions DK-a and DK-b are included in the regional syntheses (Fig. 7.1). Work is in progress at seven other sites in eastern and northern Denmark. In the near future considerably more extensive results will therefore be available. Besides the preparation of these regional pollen diagrams, research on local sites and soil horizons in burial mounds is in progress.

Radiocarbon-dated pollen diagrams are lacking from type region DK-c, the island of Bornholm. As for the rest of Denmark, coverage on a Scandinavian scale seems reasonable. However, recent experiences have demonstrated that anthropogenic influences on vegetation have varied even within distances of tens of kilometres. A much denser network of pollen diagrams is therefore needed. Besides Bornholm, conspicuous gaps occur in northeast and northwest Zealand, the islands of Falster, Lolland, and Funen, and parts of eastern, northern, and western Jutland. Coverage of these gaps is hampered by the scarcity of radiocarbon-datable deposits, as many lakes in these areas contain calcareous sediments. Recent developments in radiocarbon dating may, however, help to remedy this deficiency.

Detailed pollen analyses, macrofossil and sediment analyses, and radiocarbon dates are available for the reference sites that are bogs. The sections from lake deposits will to some extent be supplemented with diatom analyses.

The two Danish type regions are compared in detail in the regional syntheses. The main results discussed are similarities in trends in the early Holocene pollen diagrams despite differences in the frequency of certain trees, the great differences between eastern and western Denmark in the history of human occupation, and common patterns and variations in human occupational history within eastern Denmark.

TYPE REGION DK-a, WESTERN JUTLAND
(B. Odgaard)

Altitude: 0–111 m.

Climate: Mean January 0°C, July 16.5°C. Precipitation 750 mm yr^{-1}. *Ca.* 160 wet days yr^{-1}. Suboceanic.

Geology: Saalian till and Weichselian fluvioglacial sand (see Fig. 7.2).

Topography: Gently undulating hill areas surrounded by flat sandur plains. Dunes and marshes along the coast.

Palaeoecological Events During the Last 15 000 Years: Regional Syntheses of Palaeoecological Studies of Lakes and Mires in Europe.
Edited by B.E. Berglund, H.J.B. Birks, M. Ralska-Jasiewiczowa and H.E. Wright. © 1996 John Wiley & Sons Ltd.

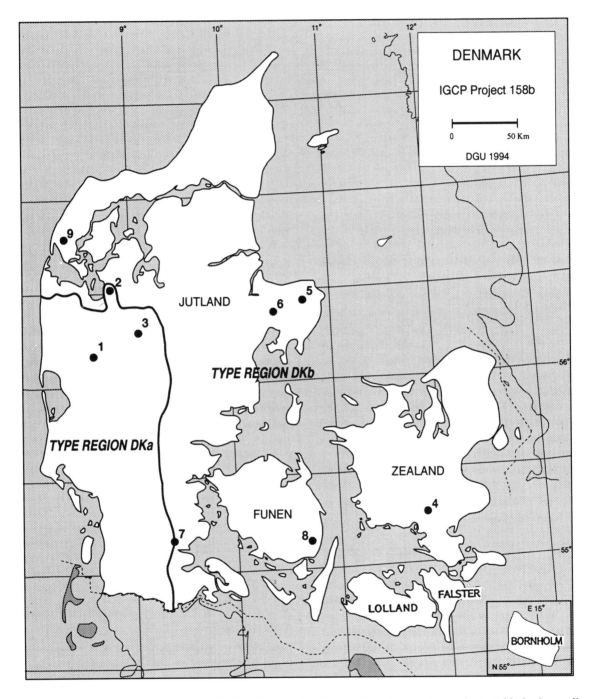

Fig. 7.1 Type regions in Denmark outside Bornholm and reference sites; the numbers refer to table in Appendix. Reference site 7 (Abkær bog) is referred to region DK-b. DK-a is western Jutland, DK-b is eastern Denmark outside Bornholm

Population: Approx. 60 people km⁻², mainly in villages and smaller towns.

Vegetation: Approx. 18% of the area is uncultivated. Most wetlands are dominated by ombrotrophic mires and oligotrophic fens. A few mesotrophic and oligotrophic lakes. Remnants of former heath areas are present. Forests form approx. 10% of the area and are dominated by planted *Picea*. The deciduous-forest vegetation consists mainly of *Quercus, Betula, Populus tremula,* and *Alnus glutinosa. Fagus sylvatica* is present.

Soils: Podzols, oligotrophic brown earths, and peat.

Land use: Approx. 65% agricultural area, 10% forest, uncultivated 18%, urban areas 7%. Barley, grass, and green fodder in rotation are the main crops.

Reference site 1. Lake Solsø (Odgaard 1994)

Latitude 56°08′N, Longitude 8°37′E. Elevation 40 m. Age range 10000–0 BP. Lake. Seven local pollen-assemblage zones (paz). Regional pollen-assemblage zones have been defined by Odgaard (1994). (Fig. 7.3)

S-1	–9000 BP	*Betula–Pinus–*Gramineae
S-2	9000–8000 BP	*Corylus–Pinus–Betula–Ulmus*
S-3	8000–5000 BP	*Corylus–Betula–Alnus–Quercus–Ulmus*
S-4	5000–4000 BP	*Calluna–Betula–Alnus–Quercus*
S-5	4000–2200 BP	*Calluna–*Gramineae*–Fagus*
S-6	2200–1200 BP	*Calluna–*Gramineae*–Fagus–Secale*
S-7	1200–0 BP	*Calluna–*Gramineae*–Fagus–Secale–Rumex*

General patterns

(1) The early Holocene pollen assemblages are dominated first by grasses, *Betula*, and *Pinus*, followed by *Corylus* as *Betula* and grasses decrease. *Ulmus* and later *Quercus* and *Alnus* expand.

(2) Only a weak decline in the *Ulmus* curve is seen at *ca.* 5000 BP.

(3) The *Calluna* curve increases in the early Subboreal and remains high.

(4) Single cereal pollen grains are found in the Subboreal, and their frequency increases at approx. 2200 BP, when *Secale* pollen first appears.

(5) *Rumex acetosella* expands after 1500 BP, followed by *Secale* at *ca.* 1200 BP.

(6) *Fagus* has never been a frequent pollen type.

TYPE REGION DK-b, EASTERN DENMARK OUTSIDE BORNHOLM
(B. Aaby)

Altitude: 0–173 m.

Climate: Mean January 0°C, July 16–17°C. Precipitation 600 mm yr⁻¹. *Ca.* 160 wet days yr⁻¹. Suboceanic.

Geology: Weichselian till, fluvioglacial sand, and Holocene marine sand and clay (see Fig 7.2).

Topography: Hilly landscape intersected by valleys. Former marine plateaux present in northern Jutland. Dunes along the coast in some areas.

Population: Approx. 200 people km⁻² on Zealand and Funen. Eastern Jutland 100 people km⁻². More than 1.5 million people live in and near the capital Copenhagen. Most people in towns.

Vegetation: Most land is intensively cultivated, and only about 10% of the area is not in use. Wetland is mainly confined to valley systems, former marine plains, and depressions in the landscape. Most lakes are eutrophic. Only a few small heath areas are present in eastern Denmark but occur more frequently in northern Jutland. Forests occupy approx. 10% of the area. *Picea* is commonly planted, mainly on acid soils, whereas deciduous trees are found on more fertile soils. *Fagus sylvatica* is the dominant natural tree in the region, but most other nemoral tree species are present too. *Acer campestre, Tilia platyphyllos, Ulmus laevis,* and other thermophilous tree species grow only in the southern part of the region.

Soils: Brown earths dominate, with some podzols and peat.

Land use: Approx. 70% agricultural area, 11%

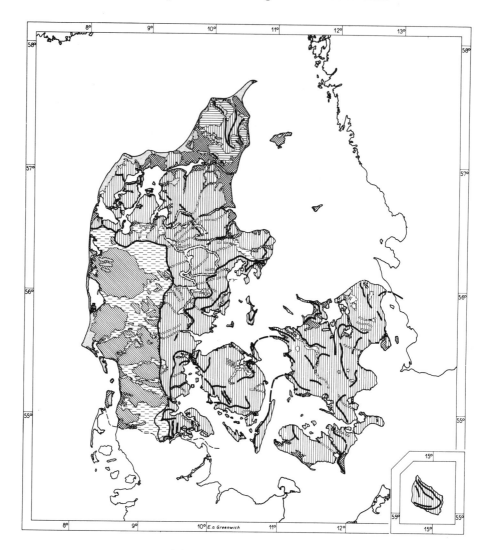

GEOLOGICAL MAP OF DENMARK
MAIN FEATURES OF THE QUATERNARY LANDSCAPE

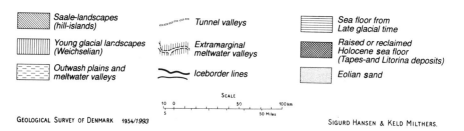

Saale-landscapes (hill-islands)	Tunnel valleys	Sea floor from Late glacial time
Young glacial landscapes (Weichselian)	Extramarginal meltwater valleys	Raised or reclaimed Holocene sea floor (Tapes-and Litorina deposits)
Outwash plains and meltwater valleys	Iceborder lines	Eolian sand

SCALE

GEOLOGICAL SURVEY OF DENMARK   1954/1993

SIGURD HANSEN & KELD MILTHERS.

Fig. 7.2 Geological map of Denmark. Western Jutland (type region DK-a) is a natural geological and geomorphological unit, delimited from the rest of Jutland by the main stationary line (iceborder) of the Weichselian glaciation

219

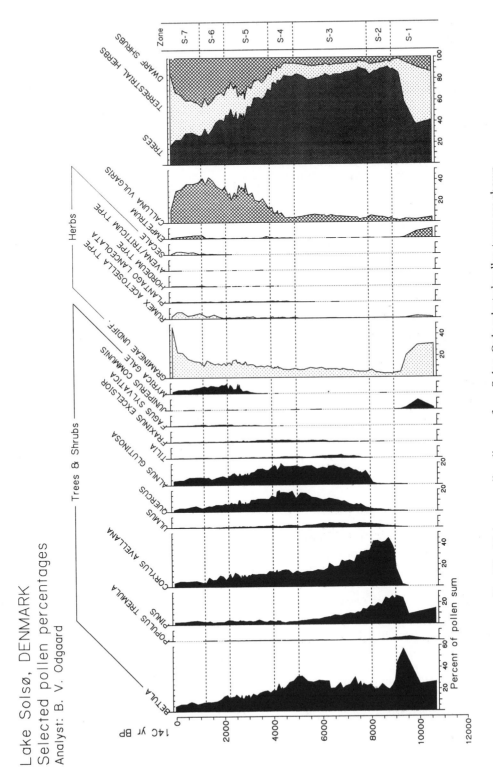

Fig. 7.3 Holocene pollen diagram from Solsø. Only selected pollen types are shown

forest, uncultivated 10%, urban areas 9%. Barley, wheat, grass, and green fodder are the main crops.

Reference site 4. Holmegaard Bog
(Andersen *et al.* 1983, Aaby 1986)

Latitude 55°18′N, Longitude 11°48′E. Elevation 31 m. Age range 10000–0 BP. The early Holocene sediments cannot be radiocarbon dated. Ombrotrophic mire. 12 local pollen-assemblage zones (paz). (Fig. 7.4)

–	–	*Betula–Pinus*
–	–	*Corylus–Pinus–Betula*
–	–	*Corylus*
–	–	*Corylus–Alnus–Ulmus–*
		Quercus–Tilia
7200–4800 BP		*Corylus–Alnus–Quercus–*
		Tilia–Ulmus
4800–4200 BP		*Corylus–Alnus–Betula*
4200–3100 BP		*Corylus–Alnus–Quercus–*
		Ulmus–Tilia
3100–2700 BP		*Corylus–Alnus–Tilia–Fagus*
2700–2300 BP		*Fagus–Alnus*
2300–1400 BP		*Alnus–Fagus–Corylus–*
		Gramineae
1400–750 BP		*Fagus*
750–0 BP		*Fagus–Alnus*–Gramineae

Reference site 5. Fuglsø Bog (Aaby 1985)

Latitude 56°29′N, Longitude 10°35′E. Elevation 31 m. Age range 6000–400 BP. Ombrotrophic mire. Eight local pollen-assemblage zones (paz). (Fig. 7.5)

–5050 BP	*Corylus–Alnus–Quercus–*
	Ulmus–Tilia
5050–4400 BP	*Corylus–Alnus–Betula*
4400–3800 BP	*Corylus–Alnus–Quercus–*
	Tilia
3800–3100 BP	*Corylus–Alnus–Tilia–*
	Gramineae
3100–2750 BP	*Corylus–Alnus–Tilia–Fagus*
2750–1800 BP	*Fagus–Alnus*–Gramineae
1800–750 BP	*Fagus–Alnus*
750–400 BP	*Fagus*–Gramineae

Reference site 7. Abkær Bog (Aaby 1988)

Latitude 55°11′N, Longitude 9°21′E. Elevation 50 m. Age range 9000–0 BP. Ombrotrophic mire. 11 local pollen-assemblage zones (paz). (Fig. 7.6)

–8000 BP	*Corylus–Pinus*
8000–7000 BP	*Corylus–Alnus–Pinus–Ulmus–*
	Tilia
7000–5000 BP	*Corylus–Alnus–Quercus–*
	Tilia–Ulmus
5000–4000 BP	*Corylus–Alnus–Betula*
4000–3100 BP	*Corylus–Alnus–Quercus–*
	Gramineae
3100–2800 BP	*Alnus–Corylus–Quercus–Fagus*
2800–1500 BP	*Alnus–Corylus*–Gramineae–
	Fagus
1500–850 BP	*Fagus–Alnus*
850–600 BP	*Fagus*–Gramineae
600–100 BP	*Fagus*–Gramineae–*Calluna*
100–0 BP	*Fagus*–Gramineae–*Picea*

The pollen diagrams

Three pollen diagrams from ombrotrophic mires in type region DK-b are presented. Local pollen-assemblage zones are given. Type-region pollen-assemblage zones have already been defined for eastern Denmark by Andersen (1978), based on the Jessen zone system (Jessen 1935).

Zone names are according to Andersen (1978), except for the youngest type-region paz (Fig. 7.7).

Type-region pollen-assemblage zones for DK-b:

Betula–Pinus paz. *Betula* dominant, *Pinus* increasing. Upper zone border at the *Corylus* expansion.

Corylus–Pinus–Betula paz. *Corylus* dominant, *Betula* first decreasing, followed by *Pinus*. Upper zone border at the *Alnus* expansion.

Corylus–Quercus–Ulmus–Tilia–Alnus paz. *Corylus* decreasing, *Quercus* , *Ulmus*, *Tilia*, and *Alnus* increasing. Upper zone border at an increase in *Quercus*.

Corylus–Alnus–Quercus–Ulmus–Tilia paz. *Corylus* and *Alnus* dominant. *Quercus*, *Alnus*, *Tilia* lower. Upper zone border at a decrease in *Ulmus*.

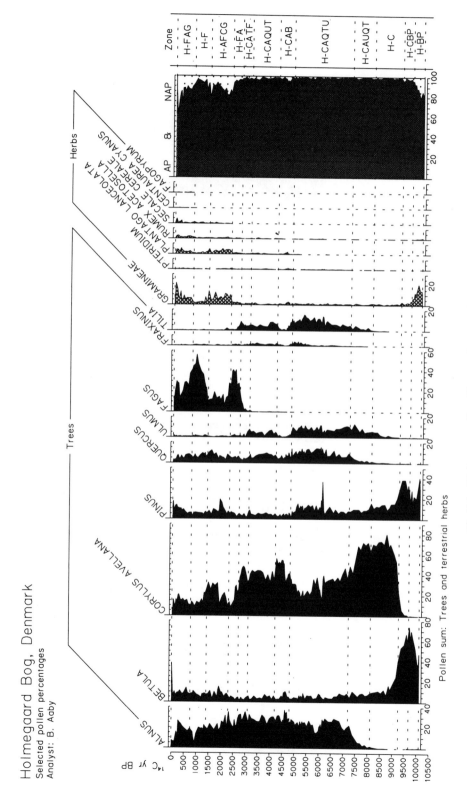

Fig.7.4 Holocene pollen diagram from Holmegaard Bog

Holmegaard Bog, Denmark
Selected pollen percentages
Analyst: B. Aaby

Pollen sum: Trees and terrestrial herbs

222

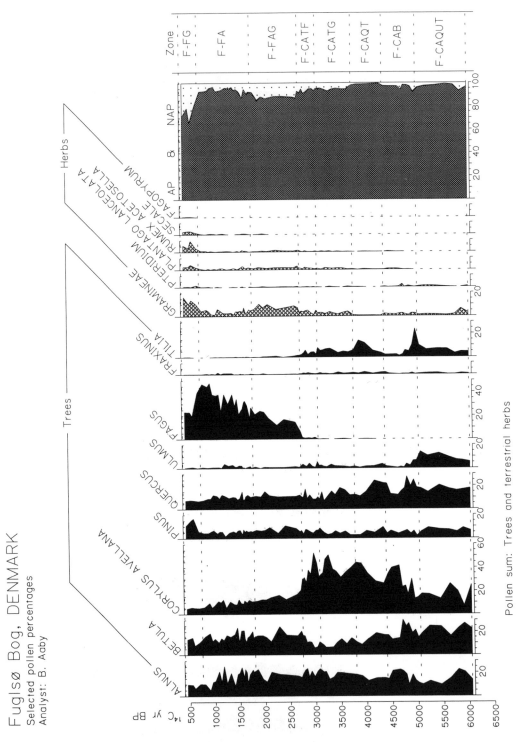

Fuglsø Bog, DENMARK
Selected pollen percentages
Analyst: B. Aaby

Pollen sum: Trees and terrestrial herbs

Fig. 7.5 Holocene pollen diagram from Fuglsø Bog

223

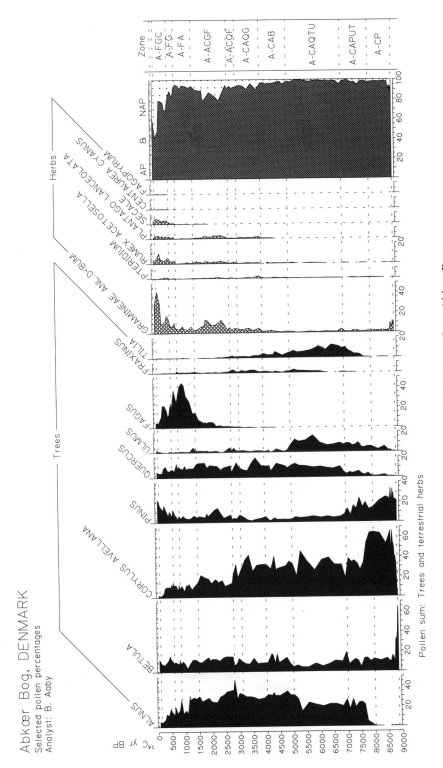

Fig. 7.6 Holocene pollen diagram from Abkær Bog

224

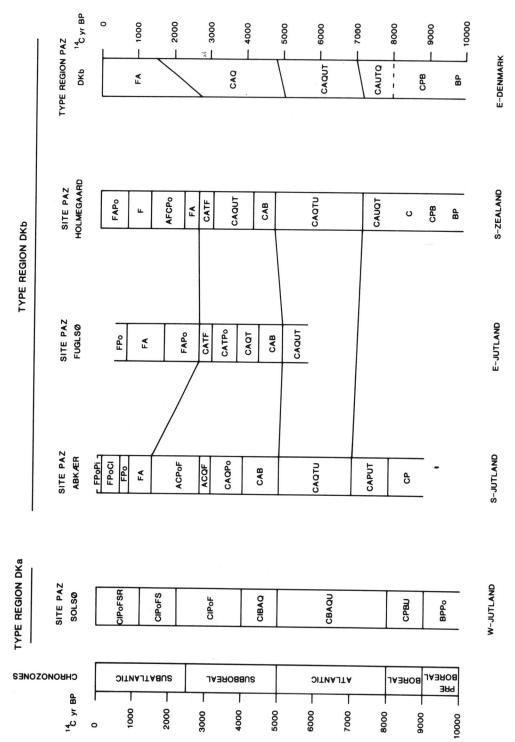

Fig. 7.7 Time–space correlation of type–region pollen–assemblage zones. Only radiocarbon–dated zone borders are mentioned. In particular, the zone border CAQ/FA is clearly asynchronous. A = *Alnus*, B = *Betula*, C = *Corylus*, Cl = *Calluna*, F = *Fagus*, P = *Pinus*, Pi = *Picea*, Po = *Gramineae*, Q = *Quercus*, R = *Rumex*, S = *Secale*, T = *Tilia*, U = *Ulmus*

Corylus–Alnus–Quercus paz. *Corylus* and *Alnus* dominant, *Quercus* lower. Upper zone border at the *Fagus* expansion.

Fagus–Alnus paz. *Fagus* dominant.

The *Fagus–Corylus–Quercus–Alnus* paz given by Andersen (1978) has been redefined because the Jessen zone border VIII/IX was based on lithostratigraphy as well as on pollen assemblages. The zone border of the *Fagus–Alnus* paz is generally placed at a higher level than the Jessen zone border VIII/IX, because *Fagus* is a dominant pollen type.

Common patterns

(1) The early Holocene pollen assemblages are dominated first by *Betula* followed by *Pinus* and *Corylus*. Later *Corylus* and *Alnus* are dominant pollen types. *Ulmus, Quercus,* and *Tilia* are more or less evenly represented.
(2) *Quercus* expansion at 7000–7200 BP.
(3) *Fraxinus excelsior* expansion in late Atlantic.
(4) *Betula* maximum and *Alnus glutinosa* expansion in early Subboreal.
(5) *Corylus* decline in late Subboreal.
(6) Continuous *Fagus sylvatica* pollen curve and increase to 1–2 % of AP about 3100 BP.
(7) *Fagus sylvatica* maximum in middle Subatlantic.
(8) First *Secale cereale* pollen 2200–2100 BP.

Unique patterns

(1) Different expansion time for *Corylus* in southern and northern parts of the region 9300–8800 BP.
(2) The *Ulmus* decline *ca.* 5000 BP is probably asynchronous within a range of 200 ^{14}C-years.
(3) The early Subboreal *Corylus* expansion is not obvious in all regional pollen diagrams (e.g. Abkær).
(4) *Ulmus* is frequent in middle Subboreal in the Holmegaard diagram, which also shows a "second" *Ulmus* decline.

PALAEOECOLOGICAL PATTERNS OF EVENTS (B. Aaby)

The pollen diagrams

Common patterns

(1) The early Holocene tree-pollen assemblages show similar trends.
(2) First *Secale cereale* pollen *ca.* 2200–2100 BP.

Unique patterns

(1) *Betula* more frequent and *Ulmus* and *Tilia* less frequent at Solsø than in eastern Denmark in the Holocene.
(2) *Calluna vulgaris* is a dominant pollen type in Subboreal lake sediments only in the Solsø region.
(3) A Subboreal *Corylus* decline is not found at Solsø.
(4) The *Fagus sylvatica* pollen curve remains low only at Solsø.

The landscape (Fig. 7.8)

Differences in landscape patterns between DK-a and DK-b

(1) The forest structure was more open in the Atlantic at Solsø than in eastern Denmark.
(2) Heathland was common at Solsø from the early Subboreal and from early Medieval time on oligotrophic soils in eastern Denmark. It is uncertain if heathland was common also in the early Subatlantic in some areas in eastern Denmark.
(3) The landscape remained open from the Subboreal only at Solsø.
(4) Glades were present in early Subboreal in eastern Denmark. Tree vegetation dominated the landscape during the following two millennia in most areas.
(5) The landscape changed its character in eastern Denmark in late Subboreal or early Subatlantic and became dominated by fields and pastures, but arboreal vegetation was still well established in most areas.

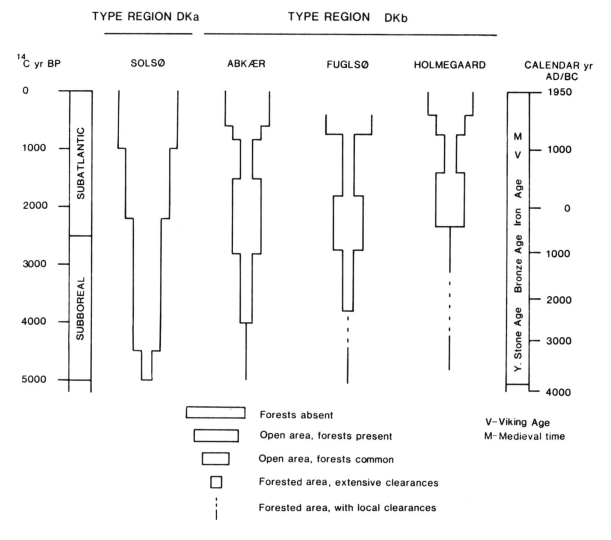

Fig. 7.8 Schematic interpretation of landscape openness based on the frequency of pollen types (including grasses) from dry-soil habitats

(6) Tree expansion at *ca.* 1800–1400 BP was only in eastern Denmark.

(7) New forest clearances took place in eastern Denmark *ca.* 1000–800 BP.

Anthropogenic events (Fig. 7.8)

Common pattern between DK-a and DK-b

(1) Increased human activity in the early Subboreal.

Unique patterns

(1) Periods of cultural cessation are not registered at Solsø.

(2) Cultural expansion was initiated at different times in type regions DK-a and Dk-b.

Hydrological events in type regions DK-a and DK-b

No hydrological events related to groundwater variations are registered in the present investigations.

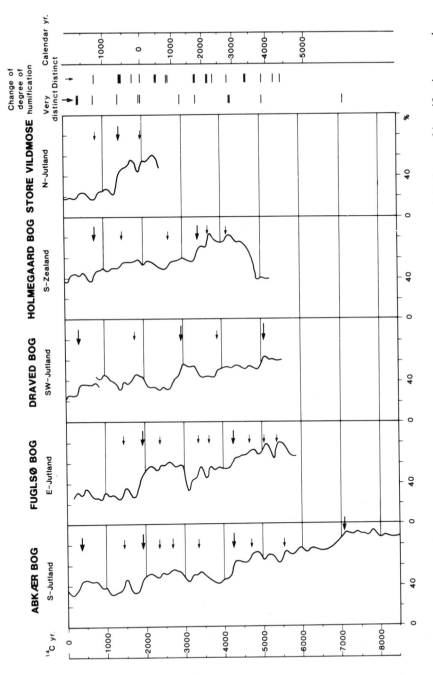

Fig. 7.9 Humification curves from ombrotrophic mires in Denmark. The curves show main trends in degree of humification as only one value per century is given. Only changes toward accumulation of less decomposed peat are considered to be caused by variations in mire hydrology. Arrows indicate distinct and very distinct humification changes. Draved Bog is located in type region Dk-a; the other mires are from type region DK-b

TYPE REGION   DKa WESTERN JUTLAND   EVENT STRATIGRAPHY

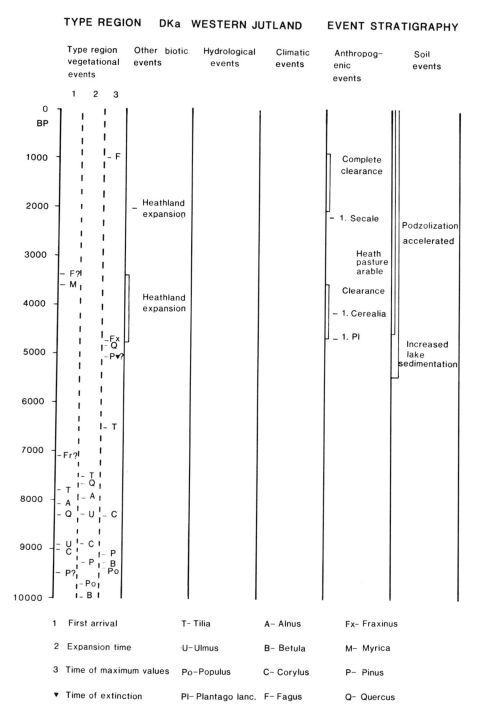

Fig. 7.10 Event stratigraphy for western Jutland, Investigations have demonstrated a close relationship between soil development, vegetation, and human activity (Iversen 1964, 1969, Andersen 1979, 1984, Aaby 1983, Odgaard 1985a, 1985b)

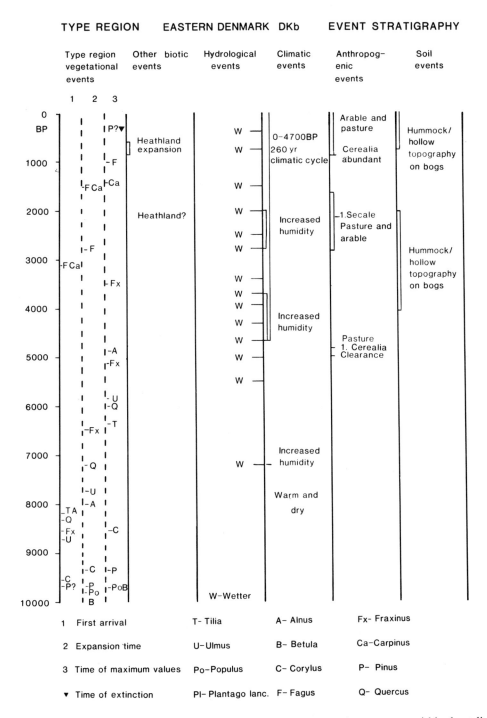

TYPE REGION   EASTERN DENMARK DKb   EVENT STRATIGRAPHY

1	First arrival	T- Tilia	A– Alnus	Fx- Fraxinus
2	Expansion time	U- Ulmus	B– Betula	Ca- Carpinus
3	Time of maximum values	Po- Populus	C– Corylus	P– Pinus
▼	Time of extinction	Pl- Plantago lanc.	F- Fagus	Q– Quercus

Fig. 7.11 Event stratigraphy for eastern Denmark, Podzolization has been asynchronous even within short distances in eastern Denmark. Leaching of nutrients and soil acidification have probably accelerated in type region DK-a since early Subboreal, when heathland originated

Local hydrological changes in ombrotrophic mires have been studied mainly from peat stratigraphy and macrofossil analysis.

(1) Several distinct and very distinct changes in peat humification are seen in all curves on Fig. 7.9. These changes are thought to reflect local hydrological events during the past 8000 radiocarbon years.

(2) Contemporary changes in peat humification are found in several bogs, indicating that these hydrological events must be controlled by a common external factor, namely climate.

(3) Increasing mire wetness is demonstrated from two or more bogs at approx. 350 BP, 750 BP, 1500 BP, 2000 BP, 2500–2800 BP, 3400 BP, 3700–3900 BP, 4300 BP, 4700 BP, 5000 BP, and 5400–5500 BP.

Climatic events in type regions DK-a and DK-b

The local hydrological changes in ombrotrophic mires (Fig. 7.9) are caused by climate. Unfortunately it cannot be stated whether the increased humidity is attributable either to higher precipitation, lower temperature, or a combination of the two. While this is true for a specific event, both precipitation and temperature are thought to have changed for a longer period of time. Drier climatic phases have occurred, but they are difficult to detect from mire stratigraphy.

Long-terms trends in climate

Temperatures must have been favourable in late Boreal and early Atlantic time to form the strongly decomposed ombrotrophic peat at Abkær, and the climate was probably rather continental. Humification values gradually decrease during middle and late Atlantic, which may reflect increasing humidity. Temperatures were probably rather high but decreasing. No overall climatic changes seem to be reflected in the humification curves until about 4700–3700 BP, when they show distinct and long-lasting humification minima. The climate probably was fluctuating and unstable in the early Subboreal, and changes in precipitation as well as temperature

are assumed for the climatic deterioration. The climate was probably rather stable during the following millennia, but fluctuations are seen from the distinct humification minima. Climatic changes are registered in several bogs in early Subatlantic time, at the onset of another climatic deterioration. Probably the overall climatic shift was gradual, including periods of both climatic amelioration and deterioration. The Subatlantic climate regime has lasted until present. The humification values are generally lower at Draved than at the other sites investigated, perhaps reflecting a more oceanic palaeoclimate in the Draved region, as it is today.

Short-term trends in climate

Stratigraphic investigations in combination with humification measurements and a detailed chronology have demonstrated short-term (*ca.* 260 yr) cyclic variations in the climate over the past 5500 calendar years (Aaby 1976, 1978).

Soil events in type regions DK-a and DK-b

Investigations have demonstrated a close relationship between soil development, vegetation, and human activity (Iversen 1964, 1969, Andersen 1979, 1984, Aaby 1983, Odgaard 1985a, 1985b). Podzolization has been asynchronous even within short distances in eastern Denmark. Leaching of nutrients and soil acidification have probably accelerated in type region DK-a since early Subboreal, when heathland originated.

REFERENCES

Aaby, B. 1976: Cyclic variations in climate over the past 5500 yrs. reflected in raised bogs. *Nature 263*, 281–284.

Aaby, B. 1978: Cyclic changes in climate during 5500 yrs. reflected in Danish raised bogs. *In*: Frydendahl, K. (ed.) *Proceedings of the Nordic symposium on climate changes and related problems, 1978*. Danish Meteorological Institute, Climatological Papers 4.

Aaby, B. 1983: Forest development, soil genesis and human activity illustrated by pollen and hypha analysis of two neighbouring podzols in Draved Forest, Denmark. *Danmarks Geologiske Undersøgelse 2. Series 114*, 1–114.

Aaby, B. 1985: Norddjurslands landskabsudvikling gennem 700 år belyst ved pollenanalyse og bestemmelse af støvindhold i højmosetørv. With an English summary. *Antikvariske Studier 7*, 60–84. Fredningsstyrelsen, København.

Aaby, B. 1986: Trees as anthropogenic indicators in regional pollen diagrams from eastern Denmark. *In* Behre, K.-E. (ed.) *Anthropogenic indicators in pollen diagrams*, 73–93. A. A. Balkema, Rotterdam.

Aaby, B. 1988: The cultural landscape as reflected in percentage and influx diagrams from two ombrotrophic mires. *In* Birks, H.H., Birks, H.J.B.,Kaland, P.E., Moe, D. (eds) *The cultural landscape—past, present and future*, 209–228. Cambridge University Press, Cambridge.

Andersen, S.T. 1970: The relative pollen productivity of North European trees and correction factors for tree pollen spectra. *Danmarks Geologiske Undersøgelse 2. Series 96*, 1–99.

Andersen, S.T. 1978: Local and regional vegetational development in eastern Denmark in the Holocene. *Danmarks Geologiske Undersøgelse, Årbog 1976*, 5–28.

Andersen, S.T. 1979: Brown earth and podzol: Soil genesis illuminated by microfossil analysis. *Boreas 8*, 59–73.

Andersen, S.T. 1980: The relative pollen productivity of the common forest trees in the early Holocene in Denmark. *Danmarks Geologiske Undersøgelse, Årbog 1979*, 5–19.

Andersen, S.T. 1984: Forests at Løvenholm Djursland, Denmark, at present and in the past. *Det Kongelige Danske Videnskabernes Selskab. Biologiske Skrifter 24(1)*, 1–208.

Andersen, S.T., Aaby, B. & Odgaard, B.V. 1983: Environment and man. Current studies in vegetational history at the Geological Survey of Denmark. *Journal of Danish Archaeology 2*, 184–196.

Iversen, J. 1964: Retrogressive vegetational succession in the Postglacial. *Journal of Ecology 52* (suppl.), 59–70.

Iversen, J. 1969: Retrogressive development of a forest ecosystem demonstrated by pollen diagrams from fossil mor. *Oikos, Supplement 12*, 35–49.

Jessen, K. 1935: Archaeological dating in the history of North Jutlands vegetation. *Acta Archaeologica 5*, 185–214.

Odgaard, B. 1985a: Kulturlandskabets historie i Vestjylland. Foreløbige resultater af nye pollenanalytiske undersøgelser. With an English summary. *Fortidsminder, Antikvariske Studier 7*, 48–59. Fredningsstyrelsen, København.

Odgaard, B. 1985b: A pollen analytical investigation of a Bronze Age and Pre-Roman Iron Age soil profile from Grøntoft, western Jutland. *Journal of Danish Archaeology 4*, 121–128.

Odgaard, B.V. 1994: The Holocene vegetation history of northern West Jutland, Denmark. *Opera Botanica 123*, 1–171.

Rasmussen, P. in press: Vegetation and land-use through 6000 years in Southeast Funen, Denmark, as reflected in a regional pollen diagram from Lake Gudme. *In* Nielsen, P.O. *et al.* (eds) The Gudme-Lundeborg Archaeological Project.

8

Sweden

B.E. Berglund, G. Digerfeldt, R. Engelmark,
M.-J. Gaillard, S. Karlsson, U. Miller and J. Risberg

INTRODUCTION
(B.E. Berglund, G. Digerfeldt)

The bedrock of Sweden is dominated by Archaean granites and gneisses (Fig. 8.1). Some limited occurrences of Palaeozoic sandstone, limestone, and shale are found in the southern and central part of the country, e.g. the islands Öland and Gotland. The bedrock of southernmost Sweden is partially formed by Mesozoic limestone and sandstone. The mountain range of northern Sweden, formed during the Caledonian orogenesis, is dominated by metamorphic Palaeozoic bedrock.

Large areas of Sweden are situated below the highest former coastline (Fig. 8.2). These areas are occupied by clayey or silty–sandy sediments or are dominated by bare bedrock. The areas above the highest coastline are largely covered by silty–sandy till. Occurrences of clayey till are found mainly in the areas of sedimentary bedrock. Peatland occupies 10% of the country.

The altitude of southern Sweden is mostly below 200 m, but a central upland reaches 200–350 m (Fig. 8.3). The altitude of central and northern Sweden increases successively from the Baltic coast towards the mountains. Most of the inland area is at an altitude of 200–500 m. The highest part of the mountain range reaches about 2000 m.

Sweden has a cold-temperate and humid climate. The mean July temperature decreases from 16°C in the south to 12°C in the north, and the mean January temperature from 0 to –15°C. The annual precipitation amounts to 500–700 mm in most of the country. However, precipitation reaching 700–1000 mm is found in rather large areas in western Sweden and along the mountain range.

Phytogeographically the southernmost part of the country belongs to the nemoral zone, dominated by *Fagus* and *Carpinus* together with *Ulmus*, *Tilia*, *Quercus*, and *Fraxinus*. However, most of southern Sweden falls within the boreo-nemoral zone, characterized by the dominance of *Picea* and *Pinus* but also by limited occurrences of the nemoral trees within particularly favourable areas. The boundary between the nemoral and boreo-nemoral zones is defined by the southern limit of *Picea*. In the boreal zone *Picea* and *Pinus* together with *Betula* prevail. The highest parts of the mountain range belong to the alpine zone.

The terrestrial soils vary depending on bedrock, subsoils, climate, and vegetation. Within the nemoral forest zone brown earths are dominant. Within the boreo-nemoral zone there is a mosaic of brown earths and podzols with the former more common in valleys with broad-leaved trees. Within the boreal zone podzols are most frequent. In the mountains alpine solifluction soils are common. Calcareous areas of southern Sweden have rendzina soils, e.g. on the islands Öland and Gotland. Further details of geology, topography, climate, and biogeo-

Palaeoecological Events During the Last 15 000 Years: Regional Syntheses of Palaeoecological Studies of Lakes and Mires in Europe.
Edited by B.E. Berglund, H.J.B. Birks, M. Ralska-Jasiewiczowa and H.E. Wright. © 1996 John Wiley & Sons Ltd.

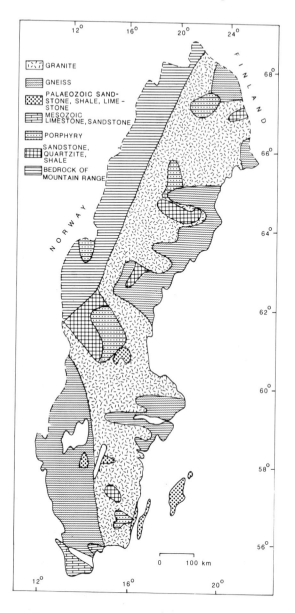

Fig. 8.1   Simplified bedrock map of Sweden

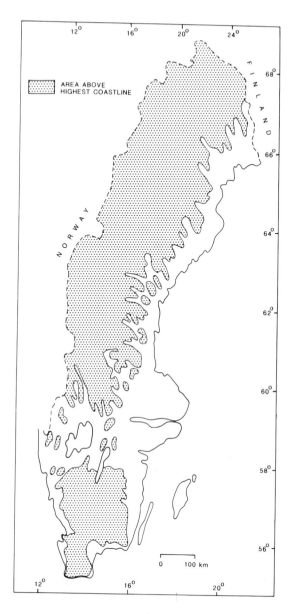

Fig. 8.2   Map of the highest coastline in Sweden

graphy are given in the national atlas for Sweden (Sveriges Nationalatlas).

Type regions and available palaeoecological data

Within the scope of IGCP 158 B, the country has been divided into 19 type regions (Fig. 8.4). The

division is based partly on the zonation of the modern vegetation, partly on geologic and geomorphic characteristics. The subdivision follows mainly the physicogeographic description of Norden by Nordiska ministerrådet (1984).

The glaciated landscape of Sweden has an abundance of suitable sites for palaeoecological studies,

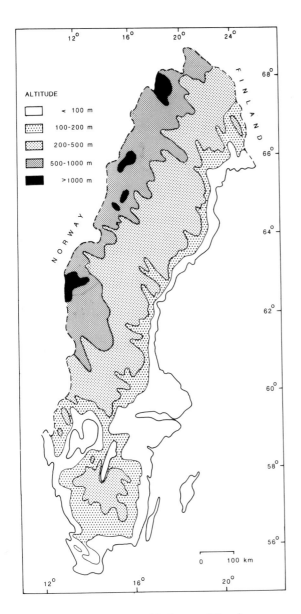

Fig. 8.3 Topographical map of Sweden

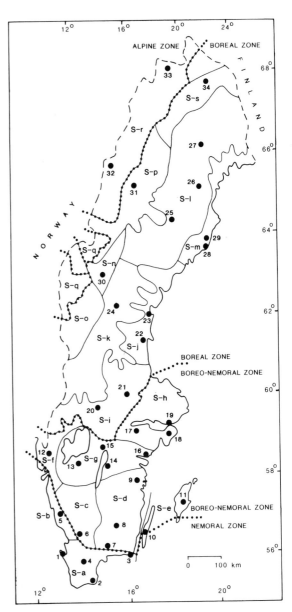

Fig. 8.4 Map of type regions and reference sites in Sweden. The boundaries between the biotic zones are indicated

lakes as well as bogs. Pollen diagrams covering Holocene time, and for southern Sweden also Late Weichselian time, are available from all type regions, but the time resolution and dating quality are less good for some regions in central Sweden. On the

other hand this is a rather uniform boreal forest area. In the table of reference sites (see Appendix) we have selected sites studied with modern techniques and where the chronology is based on a reasonable number of ^{14}C dates (or on pollen-stratigraphical correlations with dated profiles nearby). Some of these sites also include other palaeoecological analyses, such as geochemistry, diatoms, plant macrofossils, palaeohydrological studies, etc. The regional synthesis presented here together forms a transect from south to north.

(1) South Sweden

S-a   Plains and coast of Skåne and Blekinge
S-d   Eastern upland of Småland
S-g   Central plains of Västergötland and Östergötland

Together these three syntheses are meant to represent a S–N transect from the nemoral zone to the northern part of the boreo-nemoral zone.

(2) South-Central Sweden

S-h   Fissure-valley landscape of Södermanland and Uppland

This is a synthesis for the eastern part of the Central Swedish Lowland (below the highest coastline), within the northern part of the boreo-nemoral zone.

(3) North Sweden

S-m   Coastal plain of Bothnian Gulf
S-l   Upland plain of Lappland
S-p   Premontane area of Lappland
S-r   Northern mountain area of Lappland

Together these four syntheses are meant to represent a SE–NW transect from the coastal part of the boreal zone to the alpine zone of the mountain range—The Ume River valley.

SOUTH SWEDEN

TYPE REGION S-a, PLAINS AND COAST OF SKÅNE AND BLEKINGE
(B.E. Berglund, G. Digerfeldt, M.-J. Gaillard)

The southernmost provinces of Sweden, Skåne (W part), and Blekinge (E part) are well studied

concerning the Late Weichselian and Holocene palaeoecology. Two pollen-analytical monographs may be mentioned as bases for modern studies, the pioneer work by Nilsson (1935) about the vegetation history of Skåne, with 34 pollen sites, and the succeeding work by Berglund (1966a, 1966b) about the vegetation history in eastern Blekinge, with 14 pollen sites. The late-glacial environment has been the focus in some doctoral theses, from the Kullen peninsula in NW (Lemdahl 1988, Liedberg Jönsson 1988) and Blekinge (Björck 1979). Human impact has been studied and interpreted at several sites but particularly in the Ystad area in southernmost Skåne, where a multidisciplinary research team performed a detailed analysis (Berglund 1991). Vegetation history has been combined with palaeohydrological studies of several lake sites in Skåne (Digerfeldt 1971, 1974, 1988, Gaillard 1984). Bog-stratigraphical studies have involved interpretations of humification changes, e.g. at the classical site Ageröds Mosse (Nilsson 1964).

In this synthesis we present one Late Weichselian site and two Holocene sites. The Holocene diagrams represent characteristic features in this part of the country.

Altitude: 0–200 m.

Climate: Mean January −1 to −2°C, July 16–17°C. Precipitation 500–600 mm yr^{-1}. 140–160 wet days yr^{-1}. Cold-temperate, humid.

Geology and topography: Cambro-Silurian and Cretaceous sedimentary bedrocks dominate in the southwestern part of the region, in other parts Archaean gneisses and granites. Clayey till dominates in hummocky southwestern part, other parts characterized by silty–sandy till, higher and topographically more broken.

Shore displacement: This is an isostatic uplift region, with the highest shoreline from the time just after deglaciation at *ca.* 60 m in NW Skåne and in Blekinge, and *ca.* 20 m in S Skåne. The Holocene shoreline formed during the Littorina transgression is at the level of 5–10 m.

Population: 100–120 people km^{-2} in southwestern part, 40–50 people km^{-2} in other parts.

Vegetation: Existing forests characterized by *Fagus* and *Carpinus*, together with *Quercus, Ulmus, Tilia,* and *Fraxinus*. Forests of *Pinus* are common in northern part of region. Southern limit of *Picea* running along northern region boundary. Eutrophic lakes and fens characterize southern part, other parts meso- to oligotrophic. Ombrotrophic bogs mainly in northeastern part.

Soils: Brown soils in areas of sedimentary bedrock, in other parts podzolized brown soils and podzols.

Land use: Agricultural land in southwestern part occupies 60–80%, forest 10–20%, and peatland 0–5%. In other parts agricultural land is 30–40%, forest 40–50%, and peatland 5–10%.

Reference site 1. Håkulls Mosse (Berglund 1971, Berglund & Malmer 1971, Berglund & Persson unpubl., Lemdahl 1988, Liedberg Jönsson 1988)

Latitude 56°17′N, Longitude 12°31′E. Elevation 125 m. Age range *ca.* 12500–9300 BP. This is an ancient lake filled in by mid-Holocene time. Today it is a bog. Three main cores (A5, X4, B8:6) have been analysed. Here core A5 is presented. Seven regional pollen-assemblage zones (paz) with nine subzones are defined, but in core A5 the oldest zone is missing. The chronology is based on 20 ^{14}C dates in A5 (Berglund and Ralska-Jasiewiczowa 1986). (Fig. 8.5)

Age	Regional paz	Subzones
>12500 BP	Cyperaceae– Artemisia– Gramineae	
?		Salix– Artemisia
?		Artemisia– Gramineae
12500–12200 BP	Salix–Hippo- phaë–Cyper- aceae	

Age	Regional paz	Subzones
12200–12000 BP	Gramineae– Artemisia– Dryas– Ononis	
12000–11050 BP	Betula–Salix– Gramineae– Juniperus	
12000–11600 BP		Salix– Gramineae
11600–11200 BP		Empetrum
11200–11050 BP		Juniperus
11050–10200 BP	Gramineae– Cyperaceae– Gymnocarpi- um–Artemisia	
11050–10600 BP		Artemisia
10600–10200 BP		Betula nana– Empetrum
10200–10000 BP	Juniperus	
10000–9300 BP	Betula–Pinus	
10000–9700 BP		Betula– Juniperus
9700–9300 BP		Betula– Pinus

The stratigraphy at a nearby site, Björkeröds Mosse at the elevation of 75 m, has also been studied (references as above). Pollen analysis and ^{14}C dates indicate an early deglaciation. Revegetation starts *ca.* 14000 BP. From both sites records of pollen, plant macrofossils, and insects are available. A revision of earlier studies is under way (Berglund, Lemdahl & Persson).

Reference site 2. Lake Krageholmssjön
(Gaillard 1984, Regnéll 1989)

Latitude 55°30′N, Longitude 13°45′E. Elevation 43 m. Age range 10000–0 BP. Lake. Two main cores have been studied. Here we present the results from a core in the SW bay of the lake (Gaillard 1984). Four regional pollen-assemblage zones (paz) with 11 subzones. The chronology is based on pollen-analytical correlations with the ^{14}C-dated sequence for Ageröds Mosse (Nilsson 1964). (Fig. 8.6)

238

HÅKULLS MOSSE A5, Sweden
Selected Pollen Percentages
Analyst: T. Persson

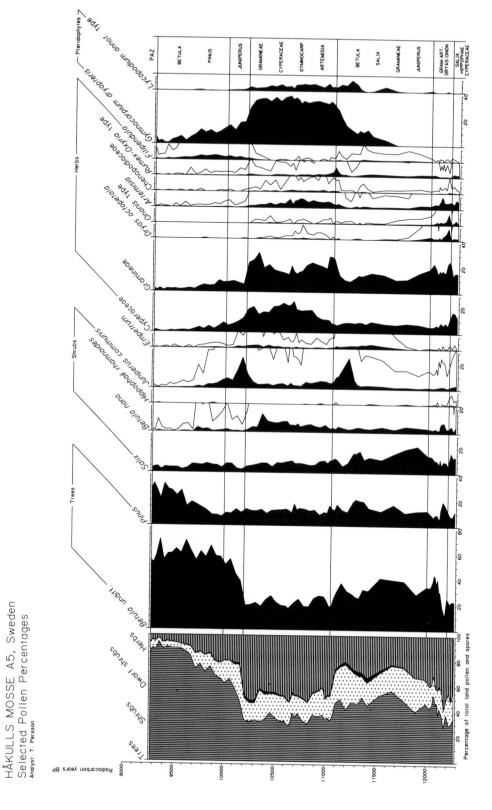

Fig. 8.5 Late Weichselian pollen diagram from Håkulls Mosse

239

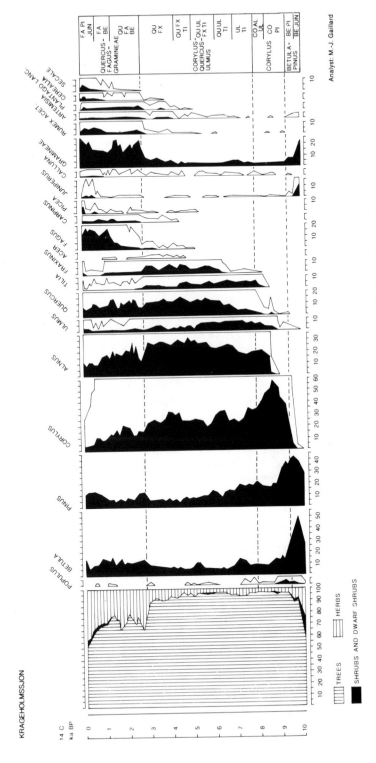

Fig. 8.6 Holocene pollen diagram from Lake Krageholmssjön

Age	Regional paz	Subzones
10000–9400 BP	Betula–Pinus	
10000–9800 BP		Betula–Juniperus
9800–9400 BP		Betula–Pinus
9400–7900 BP	Corylus	
9400–8400 BP		Corylus–Pinus
8400–7900 BP		Corylus–Alnus–Ulmus
7900–2700 BP	Corylus–Quercus–Ulmus	
7900–7000 BP		Ulmus–Tilia
7000–6200 BP		Quercus–Ulmus–Tilia
6200–5100 BP		Quercus–Ulmus–Fraxinus–Tilia
5100–4300 BP		Quercus–Fraxinus–Tilia
4300–2700 BP		Quercus–Fraxinus
2700–0 BP	Quercus–Fagus–Gramineae	
2700–1400 BP		Quercus–Fagus–Betula
1400–700 BP		Fagus–Betula
700–0 BP		Fagus–Pinus–Juniperus

Reference site 3. Lake Färskesjön
(Berglund 1966b).

Latitude 56°10′N, Longitude 15°52′E. Elevation 14 m. Age range 10800–0 BP. Lake. Five regional pollen-assemblage zones (paz) (cf. Birks and Berglund 1979) with four subzones. The chronology is based on pollen-analytical correlations with the ^{14}C-dated sequences for Hallarums Mosse and Store Mosse (Berglund 1966b). (Fig. 8.7)

Age	Regional paz	Subzone
10800–10000 BP	Pinus–Gramineae–Artemisia	

Age	Regional paz	Subzones
10000–9200 BP	Betula–Pinus	
9200–8100 BP	Pinus–Betula–Corylus	
8100–2300 BP	Pinus–Betula–Alnus	
8100–5000 BP		Quercus–Ulmus
5000–2300 BP		Quercus
2300–0 BP	Betula–Fagus–Calluna	
2300–800 BP		Quercus–Calluna
800–0 BP		Fagus–Juniperus

The reconstructed event stratigraphy of the type region is presented in Fig. 8.8.

TYPE REGION S-d, EASTERN UPLAND OF SMÅLAND (G. Digerfeldt)

Five modern pollen diagrams, all with a well established ^{14}C chronology, are available from this type region, viz. from Lake Trummen (Digerfeldt 1972), Lake Immeln (Digerfeldt 1974, 1975a), Lake Striern, and Lake Vån (Göransson 1977), and from the peat bog Dags Mosse (Göransson 1986, 1989). Lake Trummen was selected as reference site for the region, since the investigation of this lake includes not only regional vegetation history but also palaeolimnology and palaeohydrology.

All these pollen diagrams deal with the Holocene vegetation history, and human impact on the landscape development is given special attention. Regarding the Late Weichselian development some pollen diagrams have been presented in connection with an investigation of the deglaciation of the southern part of the region (Björck and Möller 1987).

As with Lake Trummen, the investigation of Lake Immeln also includes a palaeolimnological reconstruction based on diatom and sediment-chemical analyses. A special study of Holocene lake-level changes has been presented from Lake Växjösjön (Digerfeldt 1975b).

241

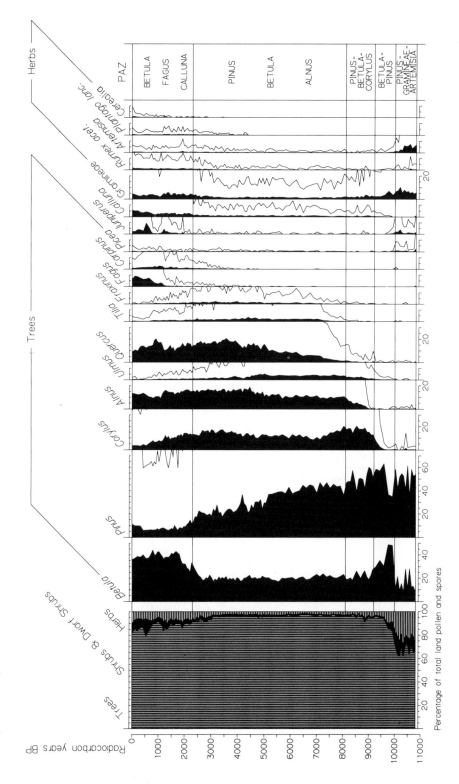

Fig. 8.7 Holocene pollen diagram from Lake Färskesjön

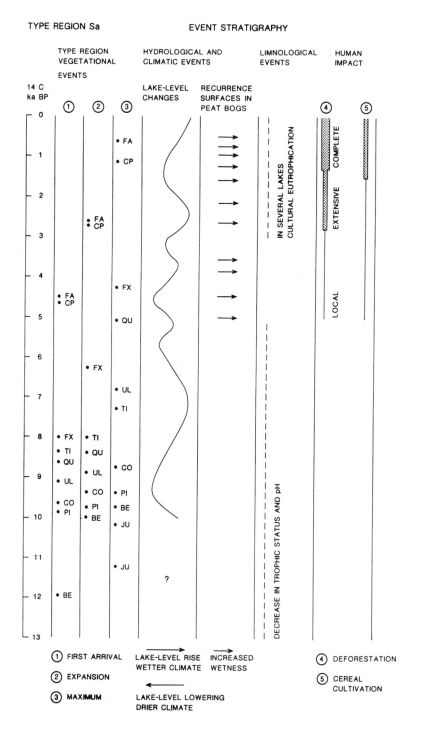

Fig. 8.8 Event stratigraphy for type region S-a. BE=*Betula,* PI=*Pinus,* CO=*Corylus,* UL=*Ulmus,* QU=*Quercus,* TI=*Tilia,* FX=*Fraxinus,* CP=*Carpinus,* FA=*Fagus,* PC=*Picea,* JU=*Juniperus,* AL = *Alnus*

Altitude: 0–350 m.

Climate: Mean January −2 to −3°C, July 15–16°C. Precipitation 500–600 mm yr^{-1}. 140–170 wet days yr^{-1}. Cold-temperate, humid.

Geology and topography: Archaean granites, Quaternary deposits dominated by sandy–silty till. Southern part of region characterized by gently undulating hills, northern part higher and topographically more broken.

Population: 20–25 people km^{-2}, mainly in towns and villages.

Vegetation: Forests of *Picea* and *Pinus* dominate, with *Betula* most common broad-leaved tree. Groves and stands of *Quercus*, *Ulmus*, *Tilia*, and *Fraxinus* mainly around farms and in previous meadows. Groves of *Fagus* locally around larger lakes. Northern limit of *Carpinus* running through southern part of region. Peatland dominated by ombrotrophic bogs, existing fens poor or moderately poor in character. Lakes usually oligotrophic with more or less humic water.

Soils: Podzols, locally podzolized brown soils.

Land use: Agricultural land occupies 10–15%, forest 50–60%, peatland 10–20%.

Reference site 8. Lake Trummen (Digerfeldt 1972)

Latitude 56°52′N, Longitude 14°50′E. Elevation 161 m. Age range 10000–0 BP. Lake. Six regional pollen-assemblage zones (paz) with 13 subzones. The chronology is based on 30 ^{14}C dates. (Fig. 8.9)

Age	Regional paz	Subzones
10000–9600 BP	*Betula*	
10000–9900 BP		*Betula–Juniperus*
9900–9600 BP		*Betula–Pinus*
9600–8500 BP	*Pinus–Betula*	
9600–9300 BP		*Pinus*
9300–8500 BP		*Pinus–Corylus*
8500–7100 BP	*Pinus–Ulmus*	
8500–8000 BP		*Pinus–Alnus*

Age	Regional paz	Subzones
8000–7100 BP		*Ulmus–Pinus*
7100–5300 BP	*Pinus–Quercus*	
7100–6700 BP		*Quercus–Ulmus*
6700–5300 BP		*Quercus–Tilia–Ulmus*
5300–1400 BP	*Betula–Quercus*	
5300–4100 BP		*Quercus–Tilia*
4100–2400 BP		*Quercus*
2400–1400 BP		*Quercus–Fagus*
1400–0 BP	*Betula–Picea–Juniperus*	
1400–900 BP		*Picea–Fagus–Quercus*
900–0 BP		*Picea–Pinus*

The reconstructed event stratigraphy for the type region is presented in Fig. 8.10.

TYPE REGION S-g, CENTRAL PLAINS OF VÄSTERGÖTLAND AND ÖSTERGÖTLAND
(G. Digerfeldt)

Three modern and useful pollen diagrams have been presented from this type region, viz. from Lake Flarken (Digerfeldt 1977), Lake Långa Getsjön (Florin 1977) and Lake Skyttasjön (Påhlsson 1985). The diagram from Lake Flarken, which has good time resolution and a well-established ^{14}C chronology, was selected as reference pollen diagram for the type region.

The diagrams from Lake Flarken and Lake Skyttasjön represent only the Holocene vegetation history. However, in the Långa Getsjön diagram part of the Late Weichselian history is also included. Some useful Late Weichselian pollen diagrams have been presented from Mt Hunneberg and the area around Mt Billingen in connection with investigations of the deglaciation chronology and the Late Weichselian shore displacement (Björck and Digerfeldt 1982, 1986).

The investigation of Lake Flarken includes besides the regional vegetation history also the palaeolimnology of the lake, which is reconstructed from diatom and sediment-chemical analyses. Hu-

244

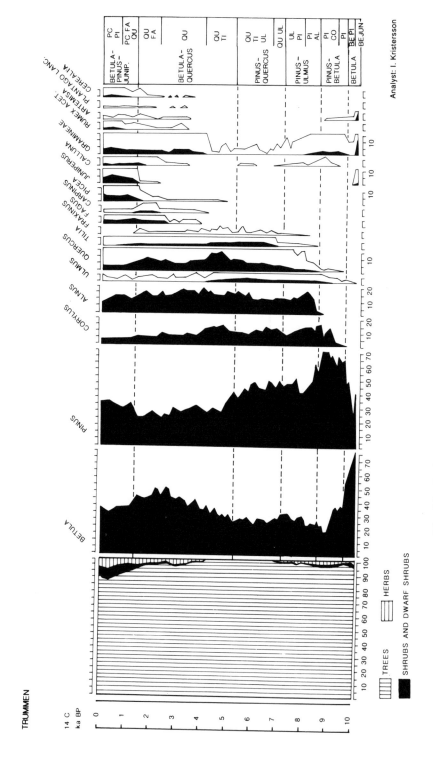

Fig. 8.9 Holocene pollen diagram from Lake Trummen

Analyst: I. Kristersson

man impact on the landscape development during late Holocene is described in the investigations of Lake Flarken and Lake Skyttasjön, but also in some other pollen-analytical studies from the region (Fries 1958, Digerfeldt & Welinder 1985).

Altitude: 0–200 m.

Climate: Mean January −2 to −3°C, July 15–16°C. Precipitation 500–600 mm yr⁻¹. 130–140 wet days yr⁻¹. Cold-temperate, humid.

Geology and topography: Archaean gneisses and granites, locally Cambro-Silurian sedimentary bedrock (mesas). Quaternary deposits dominated by clayey or sandy sediment plains, surrounded by hummocky till areas.

Population: 40–50 people km⁻², mainly in towns and villages.

Vegetation: Forests of *Picea* and *Pinus* dominate, *Betula* most common broad-leaved tree. Groves and stands of *Quercus, Ulmus, Tilia,* and *Fraxinus* around farms and locally in hilly areas of Cambro-Silurian bedrock. Northern limit of *Fagus* and *Carpinus* running through the southern part of the region. Existing lakes and fens are eutrophic in character. Ombrotrophic bogs in surrounding till areas.

Soils: Brown soils dominate, podzolized brown soils and podzols in till areas.

Land use: Agricultural land occupies 50–60%, forest 30–35%, peatland 5%.

Reference site 13. Lake Flarken (Digerfeldt 1977)

Latitude 58°35′N, Longitude 13°40′E. Elevation 109 m. Age range 10000–0 BP. Lake. Five regional pollen-assemblage zones (paz) with 10 subzones. The chronology is based on 13 ¹⁴C dates. (Fig. 8.11)

Age	Regional paz	Subzones
10000–9300 BP	*Betula*	
10000–9700 BP		*Betula–Juniperus*
9700–9300 BP		*Betula–Pinus*
9300–8600 BP	*Pinus–Betula–Corylus*	
8600–7200 BP	*Betula–Ulmus*	

Age	Regional paz	Subzones
8600–8300 BP		*Pinus–Alnus*
8300–7700 BP		*Ulmus–Betula*
7700–7200 BP		*Ulmus–Quercus*
7200–2200 BP	*Betula–Quercus*	
7200–5200 BP		*Quercus–Tilia–Ulmus*
5200–3800 BP		*Quercus–Tilia*
3800–2200 BP		*Quercus*
2200–0 BP	*Betula–Picea–Juniperus*	
2200–1100 BP		*Picea–Quercus*
1100–0 BP		*Picea–Pinus*

The reconstructed event stratigraphy for the type region is presented in Fig. 8.12.

PALAEOECOLOGICAL PATTERNS AND EVENTS IN TYPE REGIONS S-a, S-d, AND S-g
(B.E. Berglund and G. Digerfeldt)

Vegetational events

Late-glacial

The late-glacial history of the vegetation is here represented only by the site Håkulls Mosse in region S-a, because this site is situated in the early deglaciated southwest. However, the revegetation of South Sweden follows the same pattern all over regions S-a, S-d, and S-g according to studies at a large number of sites (cf. Berglund *et al.* 1994).

(1) Before 12500 BP a pioneer vegetation with tundra and steppe elements colonized the area, forming an arctic desert.

(2) Between 12500 and 12000 BP a shrub tundra with xeric elements characterized the landscape. Among shrubs are *Betula nana* and possibly also *Betula pubescens, Salix,* and *Hippophaë.*

(3) Between 12000 and 11000 BP a woodland tundra developed with *Betula pubescens, Salix,* and *Pinus silvestris* (in SE Sweden). A regression of woodlands started about 11200 BP.

TYPE REGION Sd   EVENT STRATIGRAPHY

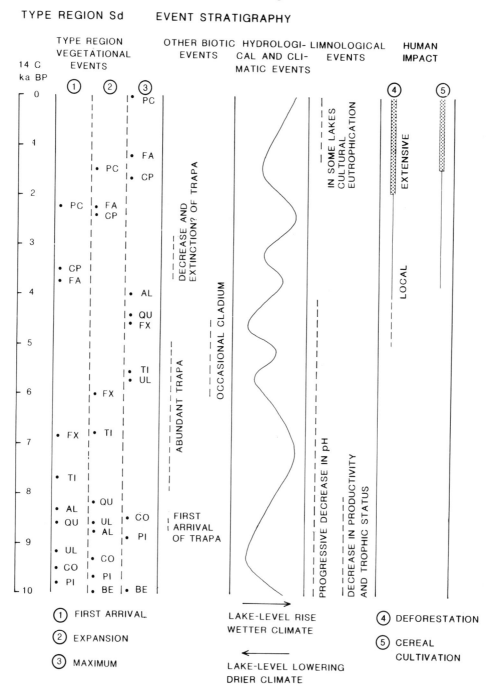

Fig. 8.10 Event stratigraphy for type region S-d. Vegetation abbreviations as for Fig. 8.8

247

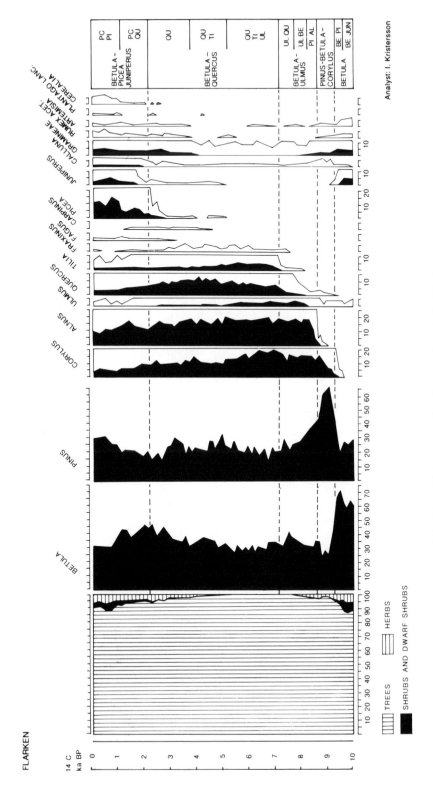

Fig. 8.11 Holocene pollen diagram from Lake Flarken

TYPE REGION Sg   EVENT STRATIGRAPHY

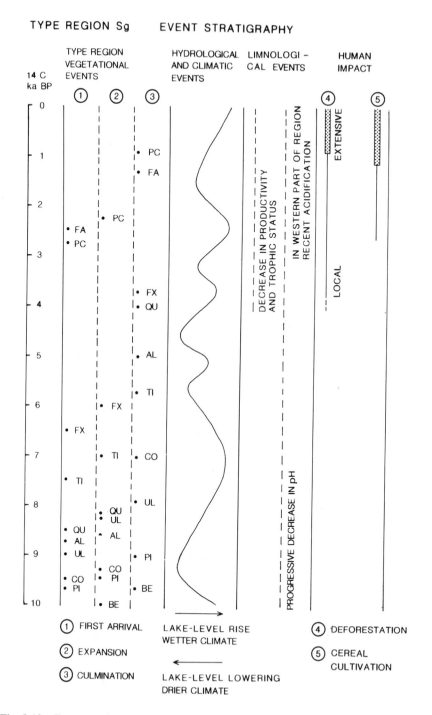

Fig. 8.12   Event stratigraphy for type region S-g. Vegetation abbreviations as for Fig. 8.8

(4) Between 11000 and 10200 BP a grass-shrub tundra without trees prevailed. Ferns like *Gymnocarpium* are common. Dwarf-shrubs like *Betula nana* and *Empetrum* expand after 10500.

(5) The period 10200–10000 BP is a shrub phase with *Juniperus* and *Betula*. This is the period of expanding woodlands before the Holocene dominance of forests.

Holocene

The regional pollen-assemblage zones and subzones distinguished in type regions S-a, S-d, and S-g are compared in Fig. 8.13. The arrival, expansion, and culmination of the forest trees represented are correlated in Fig. 8.14.

The order of arrival of the trees is the same at all reference sites. However, *Carpinus* never reached Lake Flarken, and *Picea* did not reach Lake Krageholmssjön.

Regarding the time of arrival, *Betula, Pinus, Corylus, Ulmus,* and *Quercus* reach all the reference sites at about the same time. The arrival of *Tilia, Fraxinus, Fagus,* and *Carpinus* is more or less delayed northwards. *Picea*, which immigrated from the north, arrived earlier at Lake Flarken than at Lake Trummen.

The order of culmination of the trees is the same at Lake Flarken and Lake Trummen. However, at Lake Krageholmssjön the order of culmination of *Ulmus, Quercus, Tilia,* and *Fraxinus* is somewhat different.

At all reference sites the early Holocene pollen assemblages are dominated by *Betula* and *Pinus*. The middle Holocene is characterized by the successive arrival and expansion of the nemoral forest trees. However, only at Lake Krageholmssjön do *Ulmus, Quercus, Tilia,* and *Fraxinus* dominate the pollen assemblages. At Lake Trummen and Lake Flarken *Betula* and *Pinus* remain dominant throughout the Holocene.

At Lake Krageholmssjön the late Holocene pollen assemblages are characterized by the culmination of *Fagus* and *Carpinus*, at Lake Trummen and Lake Flarken by the arrival and expansion of *Picea*.

Hydrological and climatic events

The late glacial climatic changes have been reconstructed on the basis of deglaciation patterns (Björck *et al.* 1988), vegetational history (Berglund *et al.* 1984), and insect-fauna changes (Lemdahl 1988). A synthesis was compiled by Berglund & Rapp (1988) and Berglund *et al.* (1994). Late-glacial lake-level changes are not studied in a regional perspective.

The following phases may be distinguished:

(1) 14000–12500 BP. High arctic with low temperatures, rather dry, strong winds.
(2) 12500–12000 BP. Subarctic to boreal, more continental character, dry and windy conditions particularly 12200–12000 BP.
(3) 12000–11000 BP. Subarctic and rather humid, alternating dry and humid periods. Cooling starts about 11200 BP.
(4) 11000–10500 BP. Arctic, possibly humid, strong winds.
(5) 10500–10200 BP. Subarctic with increasing temperature.
(6) 10200–9500 BP. Warm-temperate, gradually drier.

For the Holocene the climate reconstruction is mainly based on palaeohydrological studies. Systematical investigations of Holocene lake-level changes have been carried out in several lakes in South Sweden. Recorded lake-level changes are summarized in Fig. 8.15.

A distinct lowering in lake level culminated at about 9500–9200 BP, indicating a major period of drier climate in the early Holocene. After a succeeding period of increased humidity, recorded by rising and relatively higher lake levels, another major period of increased dryness began at about 6800–6500 BP. In contrast to the period in the early Holocene, the climate was not constantly drier, but a number of demonstrated lake-level fluctuations suggest a fluctuating climate. From the reconstruction in Lake Bysjön, dryness culminated at about 4900–4600 BP, and the major period lasted until about 2900–2600 BP. The successively increased humidity during the late Holocene resulted in a general rise in lake level.

Anthropogenic events

The human influence on the environment in South Sweden, associated with the settlement history, has been described and discussed by Berglund (1969, 1991), Digerfeldt & Welinder (1988), Regnéll (1989), Gaillard *et al.* (1991) and others. Human impact has been thoroughly treated also in the original descriptions of the reference sites.

Diagrams demonstrating human impact on vegetation at the reference sites are presented in Fig. 8.16. Lake Krageholmssjön represents a rather

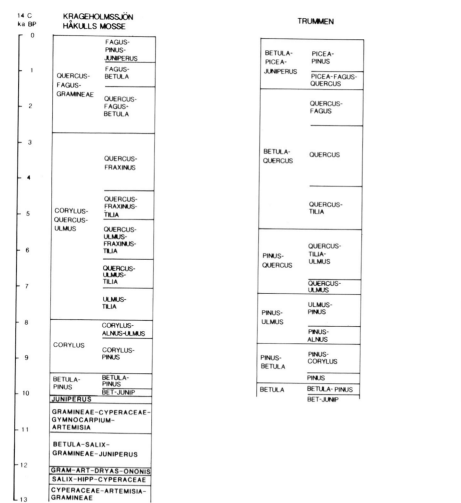

Fig. 8.13   Comparison of the Holocene regional pollen-assemblage zones and subzones at Lake Krageholmssjön, Lake Trummen, and Lake Flarken. Late Weichselian zones at Håkulls Mosse are also shown

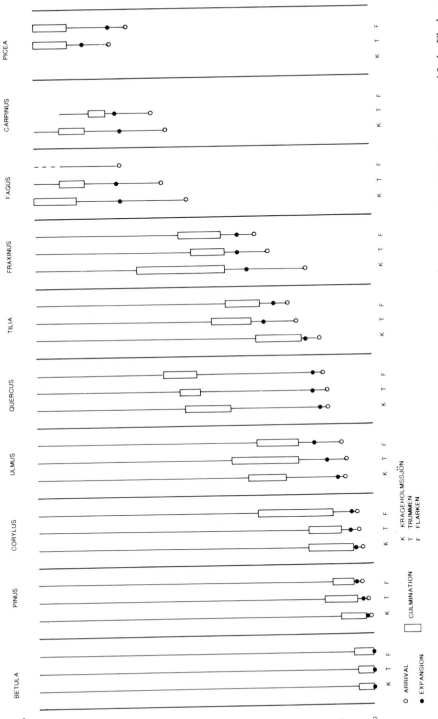

Fig. 8.14 Comparison of the arrival, expansion, and culmination of forest trees at Lake Krageholmssjön, Lake Trummen, and Lake Flarken

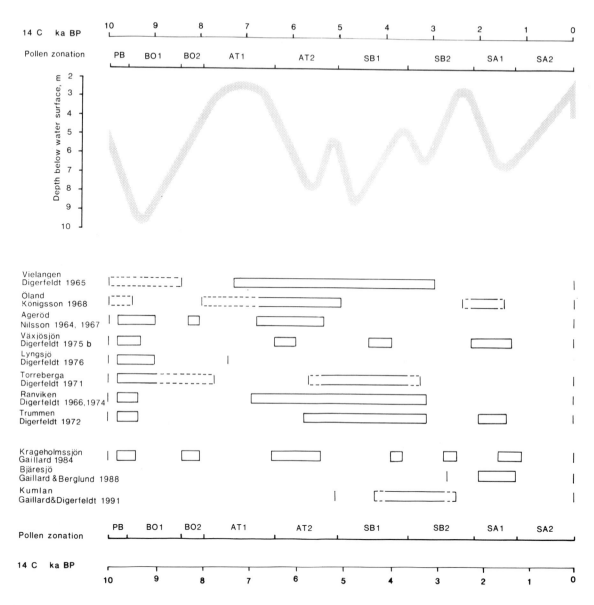

Fig. 8.15 *Above*: Quantitative reconstruction of Holocene lake-level changes in Lake Bysjön, based on recorded changes in sediment limit. *Below*: Lake-level changes recorded in lakes in South Sweden. The markings indicate periods of lowering. The left boundary of a marking corresponds to the beginning of lowering, and the right boundary to the end. In some cases (broken line) the lowering is not directly recorded but is indicated by a stratigraphical hiatus

intensely cultivated lowland area with introduction of agriculture about 5000 BP and large-scale deforestation already from about 3000 BP. Lake Trummen and Lake Flarken represent marginal settlements in upland areas with poorer soils, where human impact has been rather slight. The cultural landscape development may be summarized in the following way:

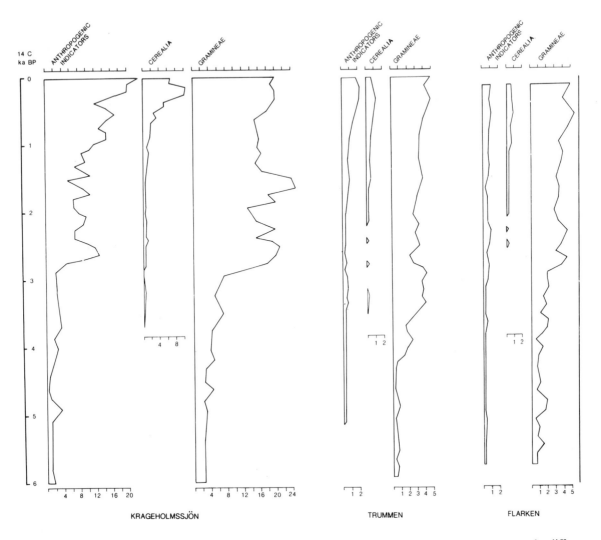

Fig. 8.16 Human-impact diagrams for Lake Krageholmssjön, Lake Trummen, and Lake Flarken. Note the different scale used for the Krageholmssjön diagram

Age	Coast area in Skåne, S-a	Upland area in Småland, S-d	Age	Coast area in Skåne, S-a	Upland area in Småland, S-d
5000 BP	Introduction of agriculture with small-scale arable fields and wood-pastures	Occasional, discontinuous small scale farming	3000 BP	Distinctly expanding agriculture with pastures causing deforestation	Slight expansion of agriculture with small-scale arable fields and wood-pastures
4000 BP	Increased pasturing	Introduction of agriculture	1300 BP	Expanding agriculture, particularly arable	Distinctly expanding agriculture,

Age	Coast area in Skåne, S-a	Upland area in Småland, S-d
	fields, second deforestation	causing deforestation
200–100 BP	Maximum deforestation, dry and wet meadows transformed to arable land	As above, followed by reforestation during the last 100 years

SOUTH-CENTRAL SWEDEN

TYPE REGION S-h, FISSURE-VALLEY LANDSCAPE OF SÖDERMANLAND AND UPPLAND (S. Karlsson, J. Risberg & U. Miller)

The provinces of Uppland and Södermanland in the central eastern part of Sweden are, to some extent, well studied concerning Holocene palaeoecology. There are *ca.* 35 published pollen investigations from the area since the first was carried out by Granlund (1931). Most of the pollen-analytical investigations cover the last 5000 years, for the major part of the area was submerged until the Subboreal (<50 m). The history of the Baltic Sea and shore displacement are decisive factors for the regional landscape development. Granlund (1928) was also the pioneer of shore displacement studies in the Stockholm region.

Modern palaeoecological studies on the history of lakes and mires in the Stockholm region have been carried out by Fries (1962), by Digerfeldt *et al.* (1980), and in connection with archaeological excavations by Miller (1973, 1982), Miller & Robertsson (1981, 1982), Robertsson (1987), and Robertsson *et al.* (1992). From the western part of the area pioneer work was carried out concerning vegetation history and shore displacement by M.-B. and S. Florin during the 1940s–1960s, in connection with archaeological excavations (e.g. M.-B. Florin 1945, 1957, 1958, S. Florin 1944, 1948, 1963). Later, the vegetation history and human impact from the western part have been described by Damell & Påhlsson (1982) and Hammar (1986, 1989). In the southern part Late-glacial and Preboreal pioneer vegetation in the Kolmården area was investigated by M.-B.

Florin (1969, 1977) and Olsson & Florin (1980). A comprehensive work on the Holocene environmental history in the southeastern Mälaren valley (central part of the area) has been compiled by Miller & Hedin (1988).

In this synthesis we present four Holocene sites that are representative for the central eastern part of the country.

Altitude: 0–120 m.

Climate: Mean January temperature –2.9°C (Stockholm) and –4.2°C (Uppsala), mean July temperature 17.8°C (Stockholm) and 17.3°C (Uppsala). Annual mean precipitation is 555 mm yr^{-1}. 160–165 wet days yr^{-1}. Cold-temperate, humid.

Geology and topography: The bedrock is composed of Archaean gneisses and granites, with minor occurrences of dolerite and limestone. Several fault and fissure systems cross the area, resulting in an uneven topography. The Quaternary deposits consist of till at higher altitudes and fine-grained soils (silt and clay) in the valleys. Glacifluvial deposits with esker ridges cross the landscape in a NNW–SSE direction. The region is situated below the highest coastline (*ca.* 150 m asl in the Stockholm area). This has resulted in wave-washed top layers.

Population: Mean density for Uppland *ca.* 90 and for Södermanland *ca.* 115 inhabitants km^{-2}.

Vegetation: Eastern Svealand belongs to the boreonemoral zone. Central eastern Svealand, including the Mälar valley and the Uppsala plain, is to a large extent cultivated. Areas with exposed bedrock are dominated by *Pinus silvestris*. The valley systems that are not cultivated are covered by *Picea abies* or deciduous trees, mainly *Quercus robur, Betula pubescens,* and *B. pendula*. Large mires are rare. The northern part, south of "Limes norrlandicus", mostly consists of low-lying flat areas dominated by coniferous forests, mainly *Picea abies*. Mires are abundant. In the Baltic archipelago a zonation of the vegetation from the outer skerries can be followed westwards: a) exposed bedrock sometimes with a diverse vegetation of herbs, b) in Uppland, a maritime *Betula*-forest, c) exposed bedrock dominated by *Pinus sylvestris*.

Soils: Podzols, peats, and more or less stable brown earths (cambisols).

Land use: Agriculture, industrial parks, towns, recreation areas with summer cottages and nature reserves.

Reference site 16. Lake Långa Getsjön
(Florin 1969, 1977).

Latitude 59°14′N, Longitude 14°43′E. Elevation 120 m. Age range 10000–0 BP (AP diagram), 4300–0 BP (AP + NAP diagram). Dystrophic lake. Five regional pollen-assemblages zones (paz) with 10 regional subzones. The chronology is based on 17 radiocarbon dates. (Fig. 8.17)

Age	Regional paz	Subzones
10000–9000 BP	Pinus–Betula	
10000–9500 BP		Betula
9500–9000 BP		Corylus–Alnus
9000–7000 BP	Pinus	
9000–8000 BP		Corylus–Alnus–Ulmus
8000–7000 BP		Ulmus–Quercus
7000–5000 BP	Pinus–Betula–Alnus	
7000–6500 BP		Ulmus–Tilia
6500–5000 BP		Tilia–Quercus
5000–2000 BP	Betula–Pinus	
5000–3500 BP		Quercus–Tilia
3500–2000 BP		Quercus–Corylus
2000–0 BP	Pinus–Picea	
2000–750 BP		Picea–Betula
750–0 BP		Picea–Juniperus

Reference site 18. Lake Ådran
(Risberg 1988, 1989, 1991, Risberg & Karlsson1989, Sandgren & Risberg 1990, Sandgren *et al.* 1990).

Latitude 59°10′N, Longitude 18°01′E. Elevation 45 m. Age range 10000–0 BP. Oligotrophic lake. Four regional pollen-assemblage zones (paz) with seven regional subzones. The chronology is based on five radiocarbon dates. (Fig. 8.18)

Age	Regional paz	Subzones
10000–7500 BP	Pinus–Betula	
10000–9500 BP		Betula
9500–9000 BP		Pinus–Corylus
9000–7500 BP		Alnus–Corylus–Ulmus
7500–4800 BP	Betula–Pinus–Alnus	
7500–6000 BP		Ulmus–Alnus
6000–4800 BP		Tilia–Quercus
4800–2000 BP	Pinus–Betula	
2000–0 BP	Pinus–Picea	
2000–1000 BP		Picea–Betula
1000–0 BP		Picea–Juniperus

Reference site 17. Lake Borsöknasjön
(Hammar 1986, 1989).

Latitude 59°20′N, Longitude 16°26′E. Elevation 24 m. Age range 4600–0 BP. Lake. Two regional pollen-assemblage zones (paz) with four regional subzones. The chronology is based on seven radiocarbon dates. (Fig. 8.19)

Age	Regional paz	Subzones
4600–2000 BP	Betula–Pinus	
4600–3500 BP		Quercus–Tilia
3500–2000 BP		Quercus–Corylus
2000–0 BP	Pinus–Picea	
2000–1000 BP		Quercus–Juniperus
1000–0 BP		Picea–Juniperus

256

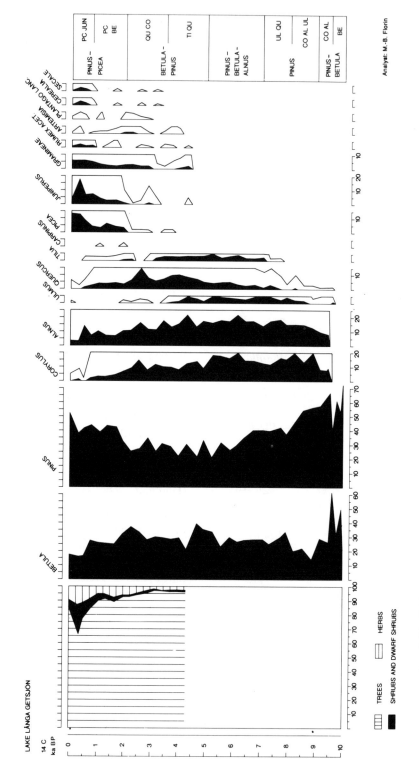

Fig. 8.17 Holocene pollen diagram for Lake Långa Getsjön, Södermanland. Only selected pollen types are shown

257

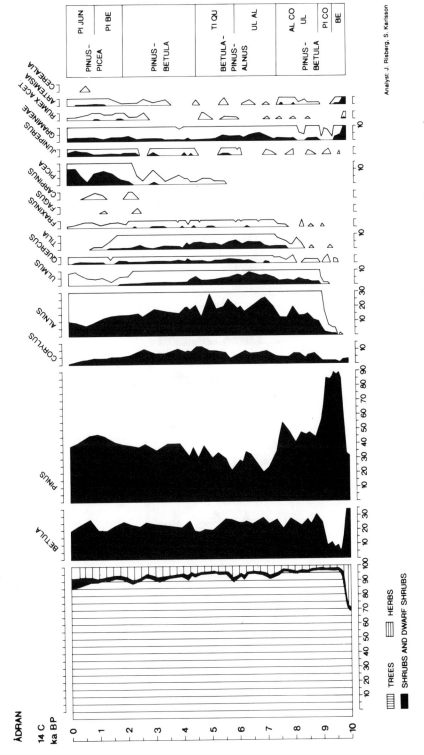

Fig. 8.18 Holocene pollen diagram for Lake Adran, Södermanland. Only selected pollen types are shown

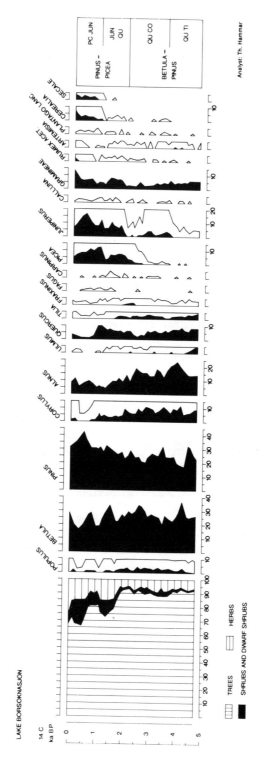

Fig. 8.19 Holocene pollen diagram for Lake Borsöknasjön. Only selected pollen types are shown

Reference site 19. Lake Laduviken

(Digerfeldt *et al.* 1980, Miller & Robertsson 1982).

Latitude 59°21′N, Longitude 18°04′E. Elevation 0.5 m. Age range 2700–0 BP. Eutrophic lake. Two regional pollen-assemblages zones (paz). The chronology is based on 11 radiocarbon dates. (Fig. 8.20)

Age	Regional paz
2700–1500 BP	*Betula–Pinus–Quercus*
1500–0 BP	*Pinus–Picea–Juniperus*

PALAEOECOLOGICAL PATTERNS AND EVENTS IN TYPE REGIONS S-h

The pollen-assemblage zones and subzones distinguished from the reference sites are compared in Fig. 8.22. The arrival, expansion, and culmination of the forest trees represented are correlated in Fig. 8.23. The vegetational patterns are based on pollen analyses from the four reference sites combined with a number of sites in eastern Södermanland and Uppland (Granlund 1932, M.-B. Florin 1945, 1957, Järnefors 1958, Fries 1962, 1969, S. Florin 1963, Königsson 1969, Persson 1973, 1981, Miller & Robertsson 1981, Damell & Påhlsson 1982, Ranheden 1989, Robertsson & Persson 1989).

The order of arrival of the forest trees is the same at all four reference sites. *Pinus*, *Betula*, and *Corylus* were the first trees to colonize the skerries emerging from the sea. A few hundred years later *Ulmus* and *Quercus* reached the region. *Tilia* and *Fraxinus* were the last nemoral trees to immigrate. *Picea*, which immigrated from the north, has its expansion contemporary with the earliest finds of *Fagus* and *Carpinus*.

The order of culmination of the forest trees is about the same in all reference sites. The minor differences for *Quercus* and *Tilia* in Lake Långa Getsjön and Lake Ådran are probably due to local conditions.

At all the sites the early Holocene pollen-assemblage zones are dominated by *Betula* and *Pinus*. The middle Holocene is characterized by the successive arrival and expansion of the nemoral forest trees. In all the pollen diagrams from the reference sites, however, *Betula* and *Pinus* remain dominant throughout the Holocene. The late Holocene is characterized by the arrival and expansion of *Picea* and minor occurrences of *Fagus* and *Carpinus*.

The Baltic stages

Eastern Svealand has undergone large areal changes during the Holocene. Land uplift and different lake and sea stages of the Baltic have affected the region (Brunnberg *et al.* 1985, Risberg *et al.* 1991).

The drainage of the Baltic Ice Lake occurred approximately when the Weichselian ice front passed the central parts of the type region at Stockholm (Brunnberg 1988).

The Yoldia Sea stage is represented by a reddish varved clay with finds of the mollusc *Portlandia* (*Yoldia*) *arctica* (Brunnberg & Miller 1990, Brunnberg & Possnert 1992). The highest parts of Södertörn (Tornberget, 110 m) emerged about 10000 BP (Fig. 8.24).

The boreal freshwater stage of the Baltic, the Ancylus Lake, started *ca.* 9400 BP. The Ancylus transgression is recorded as a retardation in the regressive shore displacement because of the intense isostatic uplift.

The slightly brackish initial phase of the Littorina Sea stage, started *ca.* 8200 BP. This transition phase, between the Ancylus and Littorina stages, is called the Mastogloia Sea stage, after a typical diatom flora, developing in the shallow littoral zone, where the water was brackish. Current studies have shown also that the plankton flora of the uppermost water column had a characteristic composition during the Mastogloia Sea stage (Witkowski & Miller 1995).

The proper marine–brackish water stage, called the Littorina Sea, was caused by the rise of sea level and transgression of the coastal land areas during the Atlantic climatic optimum. The transgression (L-1) culminated *ca.* 7000 BP (Littorina maximum *ca.* >55 <60 m asl in the central parts of Eastern Svealand) having an amplitude of *ca.* 6 m.

Since that time a mainly regressive trend to the present situation has been in progress. Several minor sea-level fluctuations that took place during the Littorina stage, L-2, L-3, and L-4 and the following postlittorina stage, mainly due to eustatic sea-level changes (Fig. 8.5), have been identified in the type region (Miller & Hedin 1988).

During the five last millennia the salinity content of the Baltic waters has decreased from approx. 20 ‰ to 5 ‰.

260

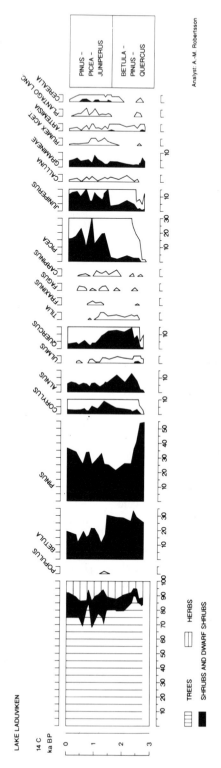

Fig. 8.20 Holocene pollen diagram for Lake Laduviken. Only selected pollen types are shown

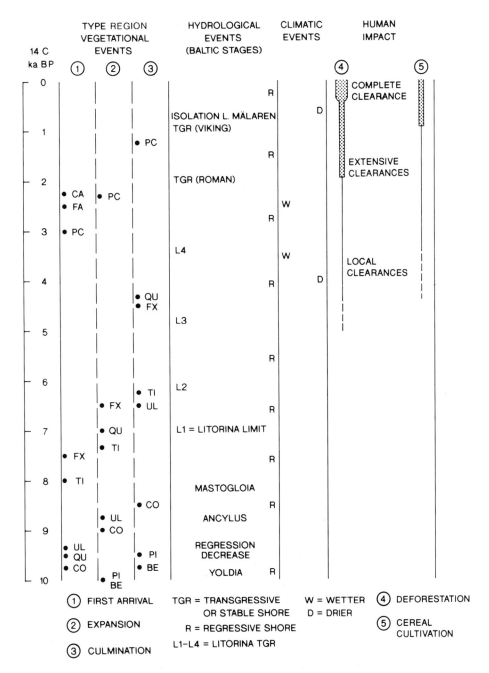

Fig. 8.21 Type region S-h. Event stratigraphy. Vegetation abbreviations as for Fig. 8.8

262

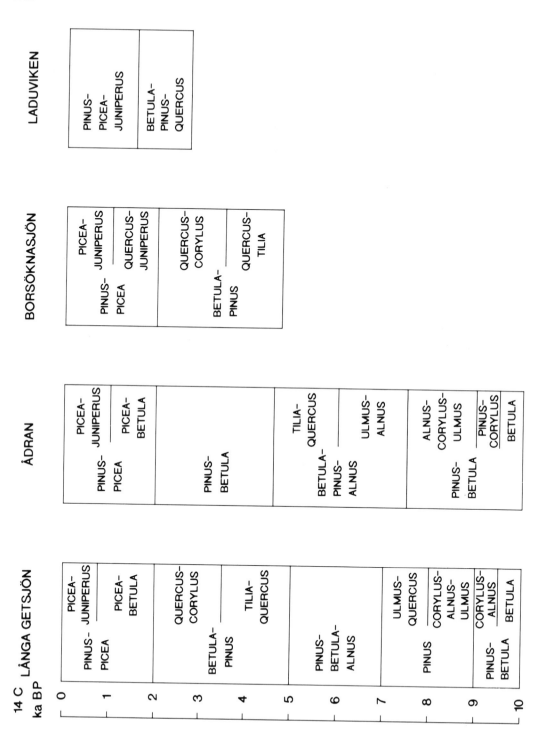

Fig. 8.22 Comparison of the Holocene regional pollen-assemblage zones and subzones in type region S-h, for Lake Långa Getsjön, Lake Ådran, Lake Borsöknasjön, and Lake Laduviken

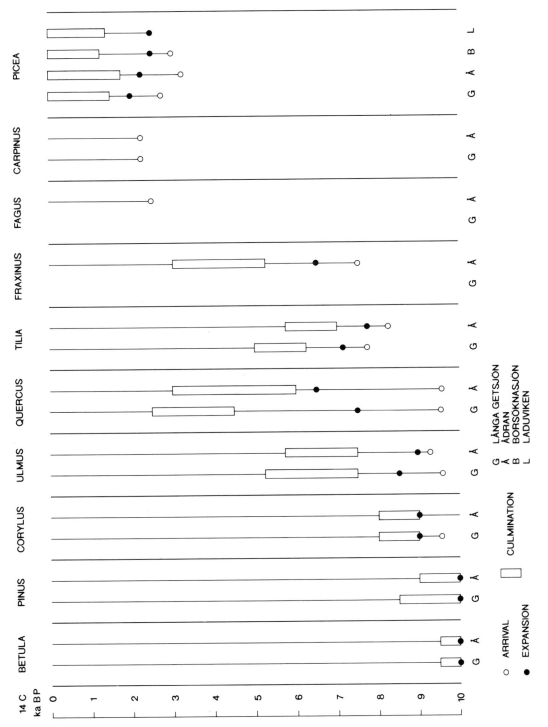

Fig. 8.23 Comparison of the arrival, expansion, and culmination of different forest trees in type region S-h for Lake Långa Getsjön, Lake Ådran, Lake Borsöknasjön, and Lake Laduviken

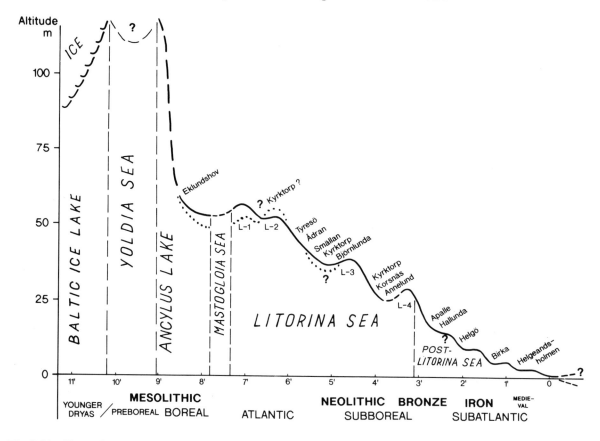

Fig. 8.24 Shore displacement curve and Baltic Sea stages in the Stockholm region, based on studies of lake sediment (Risberg *et al.* 1991)

The location of coastal settlements in the area has been greatly influenced by changes in relative sea level and the rate of shore displacement. The earliest coastal dwellings of Mesolithic age (*ca.* 8000 BP) were situated in the Södertörn peninsula south of Stockholm at 50–65 m asl.

The isolation of Lake Mälaren from the Baltic during Medieval times coincided with the foundation of Stockholm in the 13th century AD (Miller & Robertsson 1981, 1982).

Anthropogenic events

Diagrams demonstrating human impact on vegetation at the reference sites are presented in Fig. 8.25. A synthesis of human impact is also presented in Fig. 8.21.

Lake Borsöknasjön and Lake Laduviken represent rather intensely cultivated areas, where the regional vegetation has become distinctly affected by human activities from *ca.* 3500 BP (L. Borsöknasjön), and *ca.* 2500 BP (L. Laduviken). Lake Långa Getsjön and Lake Ådran represent marginal settlement areas on poorer soils, where human impact has been rather insignificant (at least until the Viking age).

Human influence on the environment in eastern Svealand, associated with the settlement history, has been described and discussed by, for example, M.-B. Florin (1958), Fries (1962), Königsson (1969), Digerfeldt *et al.* (1980), Damell & Påhlsson (1982), Miller & Robertsson (1982), Hammar (1986, 1989), and Ranheden (1989).

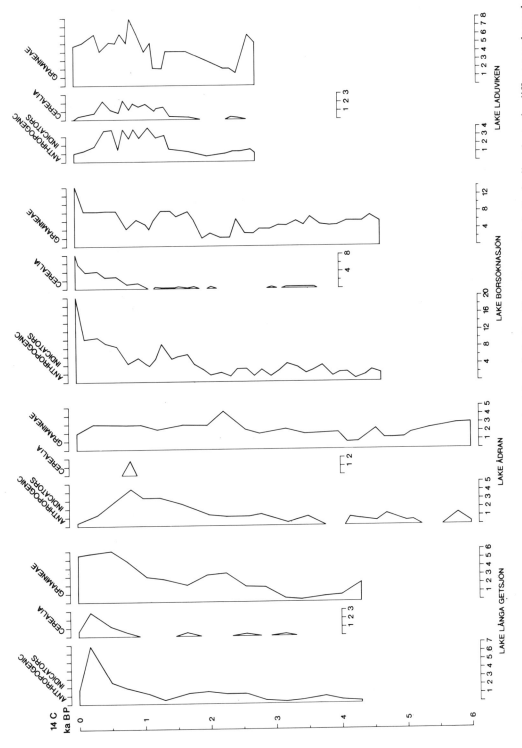

Fig. 8.25 Human impact diagrams for Låke Långa Getsjön. Lake Adran, Lake Borsöknasjön, and Lake Laduviken (note the different scale used for the diagram for Lake Borsöknasjon)

NORTH SWEDEN
(R. Engelmark)

Ume River is one of the seven big rivers crossing the regions described. The sources of Ume River are in the Caledonian mountains around the latitude 66°, and its outlet in the Bothnian Bay is 350 km to the southeast. The bedrock of the area is dominated by Archaean granite (Revsunds- and Linagranites) and in the coastal area veined gneiss. In the west the Caledonian mountains are formed of schist and greenstone. The limit of overthrusts reaches the region S-p.

The area is covered by till. Sorted sediments, mainly sand and silt, are abundant only in the coastal area. The bedrock, poor in nutrients, and the humid climate favour leaching and podzolisation. In the mountains (> 1000 m) frost-affected raw mineral soils predominate. The plateaux between the river valleys are covered with extensive fens and mixed mires, in some areas exceeding 60% of the total land area. Ombrotrophic peat bogs reach the southern part of S-m, and palsa bogs are restricted to the S-r-region.

The vegetation is dominated by mixed forests of *Pinus* and *Picea*. Besides the alpine birch belt the deciduous trees are limited to shores and secondary forests. The composition of the forests of the area is mapped by Malmström (1949).

The Bothnian coastal areas as well as the western mountain areas have a more maritime climatic regime than the continental inland areas. The physical-geographical description of the transect is presented in Table 8.1.

TYPE REGION S-m, COASTAL PLAIN OF BOTHNIAN GULF
(R. Engelmark)

The vegetational history and the human impact of the Bothnian landscape is fairly well studied, at least in the southern part. The later part of the Holocene is best represented, for the region has been influenced by substantial land upheaval, highest shoreline being 260 m asl. Only a few modern ^{14}C-dated pollen diagrams are published from the northern part of the region; two peat bogs, Bjurselet (Königsson 1970) and Gummark (Engelmark 1979)

and one lake, Fisktjärnen (Segerström 1990a). From the southern part pollen diagrams are published from peat stratigraphies (Engelmark 1976, Segerström *et al.* 1994) and lake sediments: Hamptjärn (Tolonen 1972), Prästsjön (Engelmark 1976), and Kassjön (Segerström 1990). The Kassjön diagram has a good varve chronology. The chronology of the other sites is based on ^{14}C-dating.

Two pollen profiles from sites at different elevation were selected to represent the Holocene vegetational development. The earlier Holocene periods of the region are only recorded from some hilly areas that are not representative for the main area, so they are not included as reference sites.

Reference site 28. Lake Prästsjön
(Engelmark 1976, Renberg 1976)

Latitude 63°50′ N, Longitude 20°10′ E. Elevation 32 m. Age range 3000–0 BP. The chronology is based on three ^{14}C-dates. (Fig. 8.26)

Age	Regional paz	Subzones
>2000 BP	*Betula*	
>3000 BP		*Betula–Alnus*
3000–2000 BP		*Betula–Picea*
2000–0 BP	*Picea*	

Reference site 29. Hömyren fen
(Engelmark, unpublished)

Latitude 64°00′N, Longitude 20°05′E. Elevation 100 m. Age range 6500–0 BP. A fen in Late Holocene and a lake in Middle Holocene. The chronology is based on four ^{14}C-dates. (Fig. 8.27)

Age	Regional paz	Subzones
6500–2000 BP	*Betula*	
6500–5500 BP		*Betula–Alnus–Pinus*
5500–4000 BP		*Betula–Alnus–Ulmus*
4000–3000 BP		*Betula–Alnus*
3000–2000 BP		*Betula–Picea*
2000–0 BP	*Picea*	
2000–1000 BP		*Picea–Pinus*
1000–0 BP		*Picea* Gramineae

267

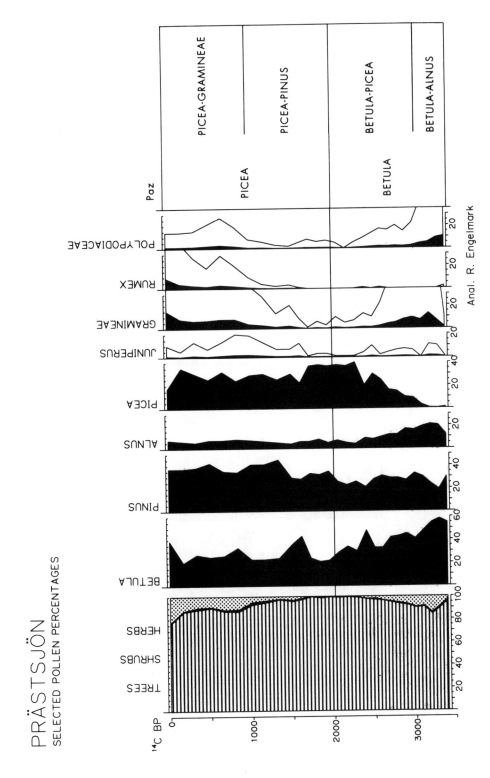

Fig. 8.26 Reference pollen diagram for Lake Prästsjön

268

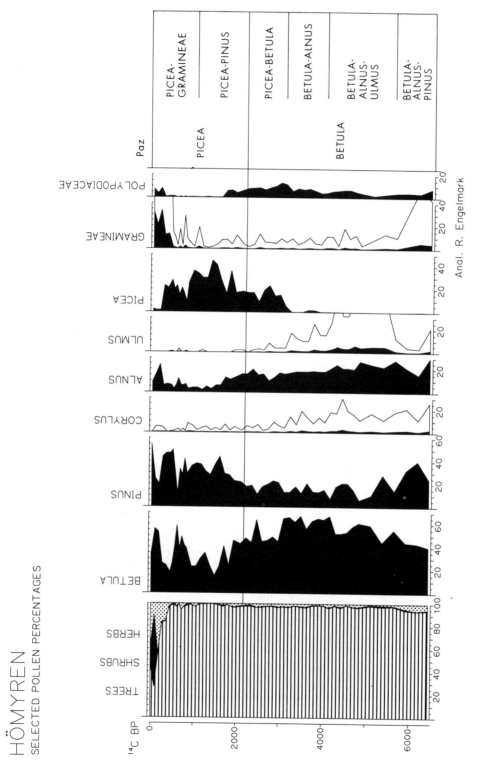

Fig. 8.27 Reference pollen diagram for Hömyren

Table 8.1 Regional physical-geographical data of the transect S-m, S-l, S-p, and S-i, the Ume River Valley

Region	S-m	S-l	S-p	S-r
Geology	Veined gneiss	Granite (Revsunds and Lina granite)	Granite, sparagmites, and quartzite	Caledonian schists and greenstone
Altitude (m)	0–250	200–500	400–900	500–2000
Soil parent material	Silty and sandy sediments	Till	Till	Till
Soils	Podzol	Podzol	Podzol	Podzol
Peat cover(%)	20–30	30–40	25	10
Precipitation (mm)	500	500	500–700	500–2000
Isotherm July (°C)	15–16	14–15	12–14	<12
Isotherm Jan.(°C)	−5 to −10	−10 to −12	−12 to −14	−11 to −13
Mean annual temp. (°C)	3	1–0	−1	1
Days >6°C	130	120	100	<100
Cultivated land (%)	5–10	1–5	<1	0

TYPE REGION S-l, UPLAND PLAIN OF LAPLAND

Very few modern pollen diagrams are published. From the northern part near River Luleälven are two varved lakes, Kroktjärnen and Strömbacka-tjärnen (Segerström 1990b) covering the last 3000 years. From the Byske River valley Robertsson & Miller (1985) have published pollen diagrams. In the south of the region along River Ångermanälven pollen stratigraphies are published from one peat bog in Hälla (Huttonen & Tolonen 1972) and from

two lakes, Håltjärn (Segerström 1982) and Stortjärn (Wallin 1983, 1986a)

Reference site 25. Hästlidmyren
(Engelmark, unpublished)

Latitude 64°30′N, Longitude 18°45′E. Elevation 220 m. Age range 8500–0 BP. There is only one ^{14}C-dating, and the chronology is based on pollen events dated in other diagrams within the region. (Fig. 8.28)

Age	Regional paz	Subzones
8500–8000 BP	*Pinus–Hippophaë*	
8000–2000 BP	*Pinus*	
8000–6000 BP		*Pinus–Alnus–Betula*
6000–4000 BP		*Betula–Pinus*
4000–2000 BP		*Pinus–Betula*
2000–0 BP	*Picea*	

TYPE REGION S-p, PREMONTANE AREA OF LAPLAND

A region mainly restricted to South and Central Lapland. Already by 1943 Erdtman had published peat and mor humus pollen stratigraphies when studying the pedogenesies of the podzols of the Storuman area. Bradshaw & Zackrisson (1990) have published a pollen stratigraphy from the northern part covering 2000 years. In the upper valley of the River Ångermanälven pollen stratigraphies from three peatbogs are published (Wallin 1980, 1983, 1986b). The ^{14}C-dated diagrams cover the Holocene of the area.

Reference site 31. Strömsundsmyren
(Engelmark, unpublished)

Latitude 65°20′ N, Longitude 16°35′ E. Elevation 350 m. Age range 8500–0 BP. The chronology is based on pollen events dated in other pollen diagrams. (Fig. 8.29)

Age	Regional paz	Subzones
8500–8000 BP	*Pinus–Hippophaë*	
8000–2000 BP	*Pinus*	
8000–6000 BP		*Pinus–Alnus–Betula*
6000–4000 BP		*Betula–Pinus*
4000–2000 BP		*Pinus–Betula*
2000–0 BP	*Picea*	

TYPE REGION S-r, NORTHERN MOUNTAIN AREA OF LAPLAND

Only from the very north of the region are any ^{14}C-dated pollen diagrams published (Sonesson 1968, 1974, Küttel 1984) and the vegetational development seems to differ in many respects from that in the south of the region.

Reference site 32. Peat bog at Hemavan
(Engelmark unpublished)

Latitude 65°50′ N, Longitude 15°00′ E. Elevation 450 m. Age range 8500–0 BP. The chronology is based on three ^{14}C-datings. (Fig. 8.30.)

Age	Regional paz	Subzones
8500–8000 BP	*Pinus–Hippophaë*	
8000–5500 BP	*Pinus–Alnus–Betula*	
5500–0 BP	*Betula*	
5500–2000 BP		*Betula–Pinus*
2000–0 BP		*Betula–Picea*

PALAEOECOLOGICAL PATTERNS AND EVENTS IN TYPE REGIONS S-m, S-l, S-p, AND S-r

Vegetational events

A time–space correlation diagram of the regional pollen-assemblages zones along the transect is presented in Fig. 8.31. The arrival, expansion and culmination of the represented forest trees are presented in Fig. 8.32.

Pinus and *Betula* arrived immediately after ice retreat at all reference sites. That is also true concerning *Corylus* in the S-m region. *Alnus* immigrated at the same time at all reference sites. The spread of *Picea* on the other hand is delayed westwards. The Holocene pollen assemblages at all reference sites are dominated by *Betula* and *Pinus*. Nemoral trees are only present in the coastal plain diagrams, and only *Corylus* and *Ulmus* are in sufficient frequency to indicate their actual presence. Subfossil hazelnuts are found in peat bogs in the southern part of region

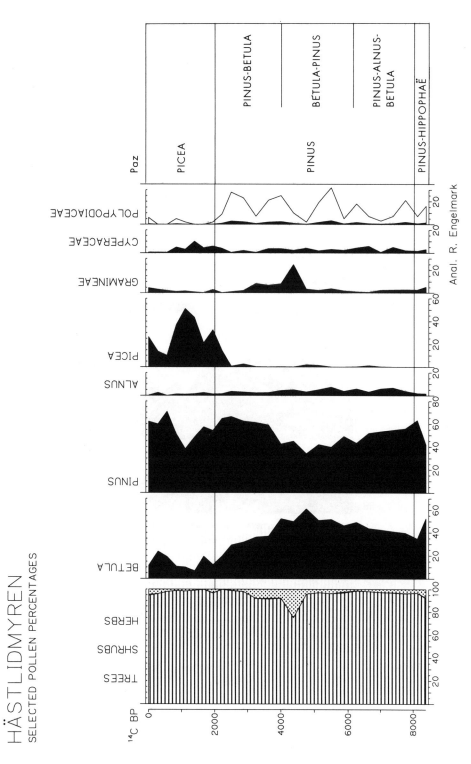

Fig. 8.28 Reference pollen diagram for Hästlidmyren

HÄSTLIDMYREN
SELECTED POLLEN PERCENTAGES

Anal. R. Engelmark

Paz

PICEA

PINUS

PINUS-HIPPOPHAË

PINUS-BETULA

BETULA-PINUS

PINUS-ALNUS-
BETULA

POLYPODIACEAE

CYPERACEAE

GRAMINEAE

PICEA

ALNUS

PINUS

BETULA

HERBS

SHRUBS

TREES

^{14}C BP

272

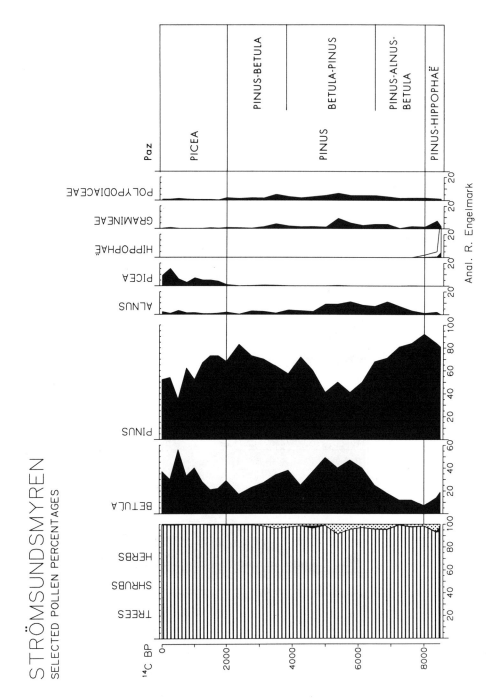

Fig. 8.29 Reference pollen diagram for Strömsundsmyren

273

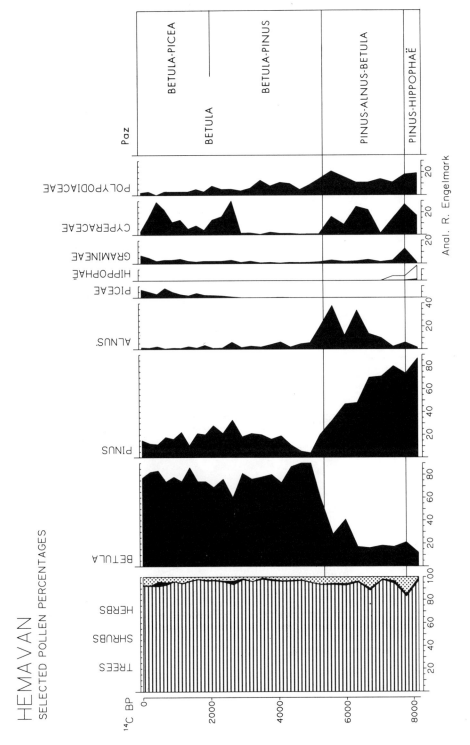

Fig. 8.30 Reference pollen diagram for Hemavan.

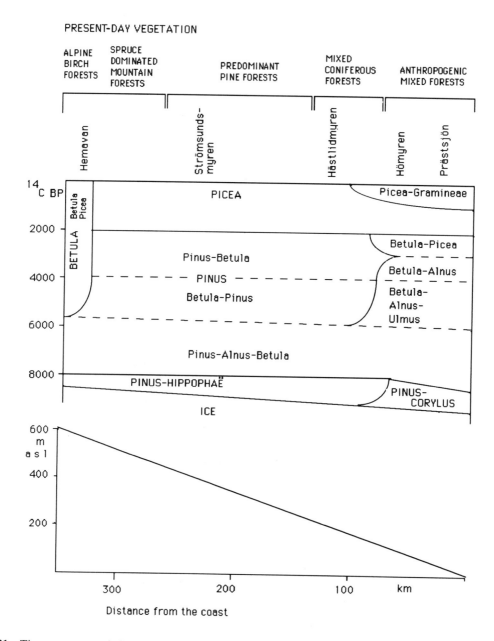

Fig. 8.31 Time–space correlation diagram of regional pollen-assemblage zones and sub zones along the Ume River valley transect

VEGETATIONAL EVENTS

Type Regions

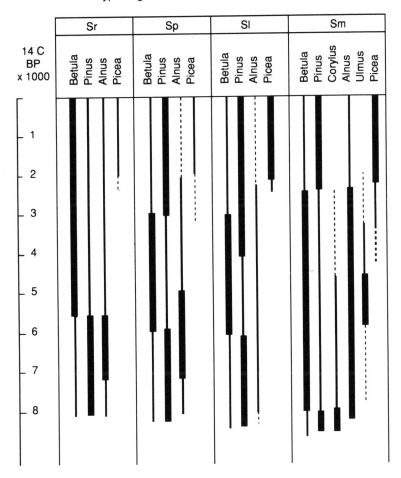

Fig. 8.32 Comparison of the arrival, expansion, and culmination of forest trees along the Ume River valley transect

S-m (Andersson 1902). The early Holocene is substantially dominated by *Pinus*. That is also valid for the mountain diagram (S-r), indicating that the tree-limit was *Pinus* at least in the south part of S-r, and ^{14}C-dated subfossil pine-stumps from above the present pine-limit in the southern Scandes fall in the early Holocene (cf. Kullman 1976, 1980).

In the middle Holocene *Betula* reached its culmination in all regions, and in the coastal region (S-m) also *Ulmus. Alnus* hàs its culmination during middle Holocene in all regions, but *Alnus* is too local to draw certain regional conclusions. The *Betula* culmination indicates the establishment at the tree-limit in the mountains in region S-r.

The late Holocene is dominated by *Pinus* or *Picea*, depending on local environmental factors and local forest history. In the region S-r, however, *Betula* continuously forms the tree-limit.

ANTHROPOGENIC EVENTS

Type Regions

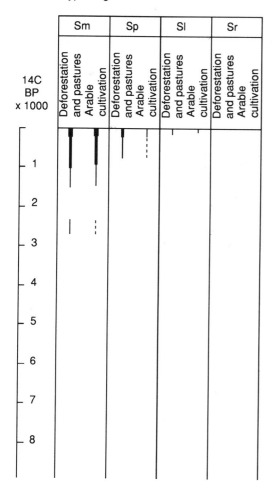

Fig. 8.33 Comparison of the human impact along the Ume River valley transect

Anthropogenic events

Pollen evidences of deforestation and grassland expansion and arable farming are diagramatically presented for the type regions in Fig. 8.33. The deforestation caused by modern industrial forestry is not considered here.

The regions S-l, S-p, and S-r were colonized by farmers very late. As late as 1750 AD only 100 settlements were recorded from the whole of Lappland (Bylund 1956). The impact of reindeer herding is not normally recorded in pollen diagrams except in very local ones (Aronsson 1991).

REFERENCES

Almquist-Jacobson, H. 1994: Interaction of Holocene climate, water balance, vegetation, fire, and the cultural land-use in the Swedish Borderland. *Lundqua Thesis 30*, 1–82.

Andersson, G. 1902: Hasseln i Sverige fordom och nu. *Sveriges Geologiska Undersökning Ca 3*, 1–168.

Aronsson, K.-Å. 1991: Forest reindeer herding A.D. 1–1800. *Archaeology and Environment 10*, 1–125.

Berglund, B.E. 1966a: Late-Quaternary vegetation in eastern Blekinge, southeastern Sweden. I. Late-Glacial time. *Opera Botanica 12(1)*, 1–180.

Berglund, B.E. 1966b: Late-Quaternary vegetation in eastern Blekinge, southeastern Sweden. II. Post-Glacial time. *Opera Botanica 12(2)*, 1–190.

Berglund, B.E. 1969: Vegetation and human influence in South Scandinavia during prehistoric time. *Oikos Suppl. 12*, 9–28.

Berglund, B.E. 1971: Late-Glacial stratigraphy and chronology in South Sweden in the light of biostratigraphic studies on Mt. Kullen, Scania. *Geologiska Föreningens i Stockholm Förhandlingar 93*, 11–45.

Berglund· B.E.(ed.) 1991 The cultural landscape during 6000 years in southern Sweden—The Ystad project. *Ecological Bulletins 41*, 1–495.

Berglund, B.E. & Malmer, N. 1971: Soil conditions and Late-Glacial stratigraphy. *Geologiska Föreningens i Stockholm Förhandlingar 93*, 575–586.

Berglund, B.E. & Ralska-Jasiewiczowa, M. 1986: Pollen analysis and pollen diagrams. *In* Berglund, B.E. (ed) *Handbook of Holocene palaeoecology and palaeohydrology*, 455–484. Wiley & Sons, Chichester.

Berglund, B.E. & Rapp, A. 1988: Geomorphology, climate and vegetation in NW Scania, Sweden, during the Late Weichselian. *Geographia Polonica 55*, 13–35.

Berglund, B.E., Lemdahl, G., Liedberg-Jönsson, B. & Persson, Th. 1984: Biotic response to climatic changes during the time span 13,000–10,000 BP— a case study from SW Sweden. *In* Mörner, N.-A. & Karlén, W. (eds) *Climatic changes on a yearly to millenial basis*, 25–35. Reidel Publishing Company, New York.

Berglund, B.E., Björck, S., Lemdahl, G., Bergsten, H., Nordberg, K. & Kolstrup, E. 1994: Late Weichselian environmental change in southern Sweden and Denmark. *Journal of Quaternary Science 9*, 127–132.

Birks, H.J.B. & Berglund, B.E. 1979: Holocene pollen stratigraphy of southern Sweden: a reappraisal using numerical methods. *Boreas 8*, 257–279.

Björck, S. 1979: Late Weichselian stratigraphy of Blekinge, SE Sweden, and water level changes in the Baltic Ice Lake. *Lundqua Thesis 7*, 1–248.

Björck, S. & Digerfeldt, G. 1982: Late Weichselian shore displacement at Hunneberg, southern Sweden, indicating complex uplift. *Geologiska Föreningens i Stockholm Förhandlingar 104*, 132–155.

Björck, S. & Digerfeldt, G. 1986: Late Weichselian–Early Holocene shore displacement west of Mt. Billingen, within the Middle Swedish end-moraine zone. *Boreas 15*, 1–18.

Björck, S. & Möller, P. 1987: Late Weichselian environment history in southeastern Sweden during the deglaciation of the Scandinavian ice-sheet. *Quaternary Research 28*, 1–37.

Björck, S., Berglund, B.E. & Digerfeldt, G. 1988: New aspects on the deglaciation chronology of South Sweden. *Geographia Polonica 55*, 37–49.

Bradshaw, R. & Zackrisson, O. 1990: A two thousand year history of a northern Swedish boreal forest stand. *Journal of Vegetation Science 1*, 519–528.

Brunnberg, L. 1988: Diatoms in glacial clay in the Södertörn area, Stockholm region. *In* Miller, U. & Robertsson , A.-M. (eds) *Proceedings of Nordic Diatomist Meeting. Stockholm, June 10–12, 1987.* University of Stockholm, Department of Quaternary Research. Report 12, 7–9.

Brunnberg, L. & Miller, U. 1990: Macro- and microfossils in late glacial clay from the Stockholm region, indicating brackish environment 10,400 years ago. *In* Lundqvist, J. & Saarnisto, M. (eds) *Termination of the Pleistocene. Field conference Norway–Sweden–Finland May 9–16, 1990.* IGCP Project 253. Geological Survey of Finland OPAS–Guide 31, 102–104.

Brunnberg, L. & Possnert, G. 1992: Radiocarbon dating of the Goti–Finiglacial boundary of the Swedish time scale. *Boreas 21*, 89–96.

Brunnberg, L., Miller, U. & Risberg, J. 1985: Project Eastern Svealand: Development of the Holocene landscape. *Iskos 5*, 85–91.

Bylund, E. 1956: Koloniseringen av Pite Lappmark t. o. m. 1867. *Geographica 30*, 1–448.

Damell, D. & Påhlsson, I. 1982: Kulturlandskapsutvecklingen i Borsöknaområdet, Eskilstuna. *Riksantikvarieämbetet och Statens Historiska Muséer. Rapport RAÄ 1982* (4), 27pp.

Digerfeldt, G. 1965: Vielången och Farlången. En utvecklingshistorisk insjöundersökning. *Skånes Natur 52*, 162–183.

Digerfeldt, G. 1966: Utvecklingshistoriska och limnologiska observationer i Ranviken av sjön Immeln. *Botaniska Notiser 119*, 216–232.

Digerfeldt, G. 1971: The Post-Glacial development of the ancient lake at Torreberga, Scania, South Sweden. *Geologiska Föreningens i Stockholm Förhandlingar 93*, 601–624.

Digerfeldt, G. 1972: The Post-Glacial development of Lake Trummen. Regional vegetation history, water-level changes, and palaeolimnology. *Folia Limnologica Scandinavica 16*, 1–96.

Digerfeldt, G. 1974: The Post-Glacial development of the Ranviken bay, Lake Immeln. I. The history of the regional vegetation, and II. The water-level changes. *Geologiska Föreningens i Stockholm Förhandlingar 96*, 1–32.

Digerfeldt, G. 1975a: The Post-Glacial development of Ranviken Bay in Lake Immeln. III. Palaeolimnology. *Geologiska Föreningens i Stockholm Förhandlingar 97*, 13–28.

Digerfeldt, G. 1975b: Post-Glacial water-level changes in Lake Växjösjön, central southern Sweden. *Geologiska Föreningens i Stockholm Förhandlingar 97*, 167–173.

Digerfeldt, G. 1976: A Pre-Boreal water-level change in Lake Lyngsjö, central Halland. *Geologiska Föreningens i Stockholm Förhandlingar 98*, 329–336.

Digerfeldt, G. 1977: The Flandrian development of Lake Flarken. Regional vegetation history and palaeolimnology. *Lundqua Report 13*, 1–101.

Digerfeldt, G. 1982: The Holocene development of Lake Sämbosjön. I. The regional vegetation history. *Lundqua Report 23*, 1–24.

Digerfeldt, G. 1988: Reconstruction and regional correlation of Holocene lake-level fluctuations in Lake Bysjön, South Sweden. *Boreas 17*, 165–182.

Digerfeldt, G. 1993: The Holocene paleolimnology of Lake Sämbosjön, Southwestern Sweden. *Journal of Paleolimnology 8*, 189–210.

Digerfeldt, G. & Welinder, S. 1985: An example of the establishment of the Bronze Age cultural landscape in SW Scandinavia. *Norwegian Archaeological Review 18*, 106–114.

Digerfeldt, G. & Welinder, S. 1988: The prehistoric cultural landscape in South-west Sweden. *Acta Archaeologica 58*, 127–136.

Digerfeldt, G., Håkansson, H. & Persson, Th. 1980: Palaeoecological studies of the recent development of the Stockholm lakes Långsjön, Lillsjön and Laduviken. *Lundqua Report 20*, 1–66.

Engelmark, R. 1976: The vegetational history of the Umeå area during the past 4000 years. *Early Norrland 9*, 75–111.

Engelmark, R. 1978: The comparative vegetational history of inland and coastal sites in Medelpad, N Sweden, during the Iron Age. *Early Norrland 11*, 25–62.

Engelmark, R. 1979: The paleoenvironment. *In* Broadbent, N. (ed.) Coastal resources and settlement stability. *Aun 3*, 158–173.

Engelmark, R. & Wallin, J.-E. 1985: Pollen analytical evidence for Iron Age agriculture in Hälsingland, Central Sweden. *Archaeology and Environment 4*, 353–366.

Erdtman, G. 1943: Pollenspektra från svenska växtsamhällen jämte pollenanalytiska markstudier i södra

Lappland. *Geologiska Föreningens i Stockholm Förhandlingar 65*, 37–66.

Florin, M.-B. 1945: Skärgårdstall och "strandskog" i västra Södermanlands pollendiagram. *Geologiska Föreningens i Stockholm Förhandlingar 67*, 511–533.

Florin, M.-B. 1957: Insjöstudier i Mellansverige. Mikrovegetation och pollenregn i vikar av Östersjöbäckenet och insjöar från preboreal tid till nutid. *Acta Phytogeographica Suecica 38*, 29 pp.

Florin, M-B. 1958: Pollen-analytical evidence of prehistoric agriculture at Mogetorp neolithic settlement, Sweden. *In* Florin, S. (ed.) Vråkulturen. *Kungliga Vitterhets Historie och Antikvitets Akademien*, 223–247.

Florin, M.-B. 1969: Late-glacial and Pre-boreal vegetation in Central Sweden. I. Records of pollen species. *Svensk Botanisk Tidskrift 63*, 143–187.

Florin, M.-B. 1977: Late-glacial and Pre-boreal vegetation in Southern Central Sweden. II. Pollen, spore and diatom analyses. *Striae 5*, 1–60.

Florin, S. 1944: Havsstrandens förskjutning och bebyggelseutveckling i östra Mellansverige under senkvartär tid, del 1. *Geologiska Föreningens i Stockholm Förhandlingar 66*, 551–634.

Florin, S. 1948: Kustförskjutningen och bebyggelseutveckling i östra Mellansverige under senkvartär tid, del 2. *Geologiska Föreningens i Stockholm Förhandlingar 70*, 17–196.

Florin, S. 1963: Land och vatten i forntiden. *The Institute of Quaternary Geology, University of Uppsala. Octavio Series 22*, 47–84.

Fries, M. 1958: Vegetationsutveckling och odlingshistoria i Varnhemstrakten. *Acta Phytogeographica Suecica 39*, 1–63.

Fries, M. 1962: Studies of the sediment and the vegetational history in the Ösbysjö basin north of Stockholm. *Oikos 13 (1)*, 76–96.

Fries, M. 1969: Sedimentproppar och pollendiagram från sjön Erken, östra Mellansverige. *Geologiska Föreningens i Stockholm Förhandlingar 91*, 353–365.

Gaillard, M.-J. 1984: A palaeohydrological study of Krageholmssjön (Scania, South Sweden). Regional vegetation history and water-level changes. *Lundqua Report 25*, 1–40.

Gaillard M.-J. & Berglund, B.E. 1988: Land-use history during the last 2700 years in the area of Bjäresjö, southern Sweden. *In* Birks, H.H., Birks, H.J.B., Kaland, P.E. & Moe, D. (eds) *The cultural landscape—past, present and future*, 409–428. Cambridge University Press, Cambridge.

Gaillard, M.-J. & Digerfeldt, G. 1990: Palaeohydrological studies and their contribution to palaeoecological and palaeoclimatic reconstructions. *Ecological Bulletins 41*, 275–282.

Gaillard, M.-J., Dearing, J.A., El-Daoushy, F., Enell, M. & Håkansson, H. 1991: A multidisciplinary study of Lake Bjäresjö (S Sweden): land-use history, soil ero-

sion, lake trophy and lake-level fluctuations during the last 3000 years. *Hydrobiologia 214*, 107–114.

Göransson, H. 1977: The Flandrian vegetational history of southern Östergötland. *Lundqua Thesis 3*, 1–148.

Göransson, H. 1986: Neolithic man and the forest environment around Alvastra Pile Dwelling in western Östergötland. *Theses and Papers in North-European Archaeology 20*, 1–90.

Göransson, H. 1989: Dags mosse—Östergötlands förhistoriska kalender. *Svensk Botanisk Tidskrift 83*, 371–408.

Granlund, E. 1928: Landhöjningen i Stockholmstrakten efter människans invandring. *Geologiska Föreningens i Stockholm Förhandlingar 50*, 207–232.

Granlund, E. 1931: Kungshamnsmossens utvecklingshistoria jämte pollenanalytiska åldersbestämningar i Uppland. *Sveriges Geologiska Undersökning C 368*, 51 pp.

Granlund, E. 1932: De svenska högmossarnas geologi. *Sveriges Geologiska Undersökning C 373*, 193 pp.

Hammar, Th. 1986: The development of the cultural landscape in the Hyndevad area, SW of Eskilstuna, southern middle Sweden. *In* Königsson, L.-K. (ed.) Nordic Late Quaternary biology and ecology. *Striae 24*, 172–176.

Hammar, Th. 1989: The development of the cultural landscape in the Borsökna area, SW of Eskilstuna, NW Södermanland, central Sweden. *In* Königsson L.K. (ed.) Dona Candolino. *Striae 25*, 27–52.

Hemmendorf, O. & Påhlsson, I. 1986: Storsjöbygdens vegetations-och kulturlandskaputveckling. *Riksantikvarieämbetet och Statens Historiska Museer, Rapport RAÄ 1986(1)*, 1–21.

Huttonen, P. & Tolonen, M. 1972: Pollen-analytical studies of prehistoric agriculture in northern Ångermanland. *Early Norrland 1*, 9–34.

Järnefors, B. 1958: Beskrivning till jordartskarta över Uppsalatrakten. *Sveriges Geologiska Undersökning Ba 15*, 46 pp.

Königsson, L.-K. 1968: The Holocene history of the Great Alvar of Öland. *Acta Phytogeographica Suecica 55*, 1–172.

Königsson, L.-K. 1969: Sju Riddares träsk. *Geologiska Föreningens i Stockholm Förhandlingar 91*, 366–373.

Königsson, L.-K. 1970: Traces of Neolithic human influence upon the landscape development at the Bjurselet settlement, Västerbotten, northern Sweden. *Kungliga Skytteanska Samfundets Handlingar 7*, 13–30.

Kullman, L. 1976: Recent trädgränsdynamik i V Härjedalen. *Svensk Botanisk Tidskrift 70*, 107–137.

Kullman, L. 1980: Radiocarbon dating of subfossil Scots pine (*Pinus sylvestris* L.) in the southern Swedish Scandes. *Boreas 9*, 101–106.

Küttel, M. 1984: Vuolep Allakasjaure eine pollenanalytische Studie zur Vegetationsgeschichte der Tundra in Nordschweden. Festschrift Max Welten. *Dissertationes Botanicae 72*, 191–212.

Lemdahl, G. 1988: Palaeoclimatic and palaeoecological studies based on subfossil insects from Late Weichselian sediments in southern Sweden. *Lundqua Thesis 22*, 1–12.

Liedberg Jönsson, B. 1988: The Late Weichselian macrofossil flora in western Skåne, southern Sweden. *Lundqua Thesis 24*, 1–12.

Malmström, C. 1949: Karta över trädslagsfördelningen år 1940 inom Västerbottens läns lappmarker. *Meddelanden från Statens Skogsförsöksanstalt 37.*

Miller, U. 1971: Diatom floras in the interglacial sediments at Leveäniemi. *Sveriges Geologiska Undersökning C 658*, 104–163.

Miller, U. 1973: Belägg för en subboreal transgression i Stockholms-trakten. *University of Lund, Department of Quaternary Geology, Report 3*, 96–104.

Miller, U. 1982: Shore displacement and coastal dwelling in the Stockholm region during the past 5000 years. *Annales Academiae Scientiarum Fennicae A. III. 134*, 185–211.

Miller, U. & Hedin, K. 1988: The Holocene development of landscape and environment in the south-eastern Mälaren valley, with special reference to Helgö. *Excavations at Helgö XI. Kungliga Vitterhets Historie och Antikvitets Akademien Stockholm*, 72 pp.

Miller, U. & Robertsson, A.-M. 1981: Current biostratigraphical studies connected with archaeological excavations in the Stockholm region. *In* Königsson, L.-K. & Paabo, K. (eds) Florilegium Florinis Dedicatum. *Striae 14*, 167–173.

Miller, U. & Robertsson, A.-M. 1982: The Helgeandsholmen excavation: An outline of biostratigraphical studies to document shore displacement and vegetational changes. *Second Nordic Conference on the Application of Scientific Methods in Archaeology, PACT Journal 7*, 311–328.

Nilsson, T. 1935: Die pollenanalytische Zonengliederung der spät- und postglazialen Bildungen Schonens. *Geologiska Föreningens i Stockholm Förhandlingar 57*, 385–562.

Nilsson, T. 1964: Entwicklungsgeschichtliche Studien in Ageröds mosse, Schonen. *Lunds universitets årsskrift N.F. 59 (7)*, 1–52.

Nilsson, T. 1967. Pollenanalytische Datierung mesolitischer Siedlungen im Randgebiet des Ageröds mosse im mittleren Schonen. *Acta Universitatis Lundensis II. 16*, 1–80.

Olsson, I.U. & Florin, M.-B. 1980: Radiocarbon dating of dy and peat in the Getsjö area, Kolmården, Sweden, to determine the rational limit of *Picea*. *Boreas 9*, 289–305.

Påhlsson, I. 1977: A standard pollen diagram from the Lojsta area of central Gotland. *Striae 3*, 1–40.

Påhlsson, I. 1985: Vegetations-och kulturlandskapsutveckling i Tiveden—en pollenanalytisk undersökning från Skyttasjön, norra Västergötland. *Skaraborgsnatur 22*, 48–57.

Persson, Ch. 1973: Indications of a Litorina transgression in the Nyköping area. *Sveriges Geologiska Undersökning C 680*, 23 pp.

Persson, Ch. 1981: Three peat deposits in south-eastern Södermanland, Sweden. *Geologiska Föreningens i Stockholm Förhandlingar 103*, 91–103.

Ranheden, H. 1989: Barknåre och Lingnåre. Human impact and vegetational development in an area of subrecent land uplift. *Striae 33*, (Thesis) 78 pp.

Regnéll, J. 1989: Vegetation and land use during 6000 years. Palaeoecology of the cultural landscape at two lake sites in southern Skåne, Sweden. *Lundqua Thesis 27*, 1–62.

Renberg, I. 1976: Palaeolimnological investigations in Lake Prästsjön. *Early Norrland 9*, 113–160.

Renberg, I. 1978: Palaeolimnology and varve counts of the annually laminated sediment of Lake Rudetjärn, Northern Sweden. *Early Norrland 11*, 63–91.

Risberg, J. 1988: The diatom stratigraphy of Lake Ådran basin, Södertörn, central eastern Svealand, Sweden. A preliminary report. *In* Miller, U. & Robertsson, A.-M. (eds) Proceedings of Nordic Diatomist Meeting, Stockholm, June 10–12, 1987. *University of Stockholm, Department of Quaternary Research, Report 12*, 69–76.

Risberg, J. 1989: Grain-size distribution of sediments: a comparison between pipette and sedigraph analysis. *Geologiska Föreningens i Stockholm Förhandlingar 111*, 247–250.

Risberg, J. 1991: Palaeoenvironment and sea level changes during the early Holocene on the Södertörn peninsula, Södermanland, eastern Sweden. *University of Stockholm, Department of Quaternary Research, Report 20*, 27 pp.

Risberg, J. & Karlsson, S. 1989: Pollen stratigraphy in a sediment core from Lake Ådran, Södertörn, central eastern Svealand, Sweden. *University of Stockholm, Department of Quaternary Research, Report 14*, 1–27.

Risberg, J., Miller, U. & Brunnberg, L. 1991: Deglaciation, Holocene shore displacement and coastal settlements in eastern Svealand, Sweden. *Quaternary International 9*, 33–37.

Robertsson, A.-M. 1971: Pollen-analytical investigation of the Leveäniemi sediments. *Sveriges Geologiska Undersökning C 658*, 82–97.

Robertsson, A.-M. 1987: Fornlämningarna 13 och 69. Hallunda, Botkyrka socken, Södermanland. Arkeologisk undersökning 1969–1971. Del IV: Naturvetenskapliga rapporter och analyser. *Riksantikvarieämbetet och Statens Historiska Museer. Rapport UV 1987 (4)*, 37–51.

Robertsson, A.M. & Miller, U. 1985: Garaselet—biostratigraphical studies of human impact during different periods of settlement from the Mesolithic to Medieval times. *Iskos 5*, 127–140.

Robertsson, A.-M. & Persson, Ch. 1989: Biostratigraphical studies of three mires in northern Uppland, Sweden. *Sveriges Geologiska Undersökning C 821*, 19 pp.

Robertsson, A.-M., Karlsson, S., and Aronsson, M. 1992: Fatburssjöns 3000-åriga historia—från bronsålder till 1700-tal. *In* Bergendahl, C. (ed.) *Fatburen 3000 år. Från en vik i skärgården till Bofills båge*, 8–23 Sigma, Stockholm.

Sandgren, P. & Risberg, J. 1990: Magnetic mineralogy of the sediments in Lake Ådran, eastern Sweden, and an interpretation of early Holocene water level changes. *Boreas 19*, 57–68.

Sandgren, P., Risberg, J. & Thompson, R. 1990: Magnetic susceptibility in sediment records of Lake Ådran, eastern Sweden: correlation among cores and interpretation. *Journal of Paleolimnology 3*, 129–141.

Segerström, U. 1982: Pollenanalytiska belägg för tidig medeltida odling vid Åsele tätort, Västerbotten. *Riksantikvarieämbetet Rapport*, 24 pp.

Segerström, U. 1990a: The Post-Glacial history of vegetation and agriculture in the Luleälv River valley. *Archaeology and Environment 7*, 1–80.

Segerström, U. 1990b: *The vegetational and agricultural history of a northern Swedish catchment, studied by analyses of varved lake sediments.* Umeå University, Dept. of Ecological Botany, Dissertation, 33 pp.

Segerström, U., Bradshaw, R., Hörnberg, G. & Bohlin, E. 1994: Disturbance history of a swamp forest refuge in northern Sweden. *Biological Conservation 68*, 189–196.

Sonesson, M. 1968: Pollen zones at Abisko, Torne Lappmark, Sweden. *Botaniska Notiser 121*, 491–500.

Sonesson, M. 1974: Late Quaternary forest development of the Torneträsk area, North Sweden. 2. Pollen-analytical evidence. *Oikos 25*, 288–306.

Thelaus, M. 1989: Late Quaternary vegetation history and paleohydrology of the Sandsjön-Årshult area, southwestern Sweden. *Lundqua Thesis 26*, 1–77.

Tolonen, K. 1972: On the palaeoecology of the Hamptjärn basin, 1. *Early Norrland 1*, 42–52.

Wallin J.-E. 1980: Vegetationsutvecklingen efter istiden vid Vojmsjöns utlopp, Vilhelmina socken, Lapland. En undersökning baserad på pollenanalys. *Riksantikvarieämbetet Rapport*, 16 pp.

Wallin, J.-E. 1983: Vegetationsutvecklingen under 8200 år inom Ångermanälvens övre nederbördsområde—en pollenanalytisk studie. *Riksantikvarieämbetet Rapport*, 27 pp.

Wallin, J.-E. 1986a: Naturgeografisk och paleobotanisk undersökning vid Vojmsjön. Studier i Norrländsk Forntid 2. *Acta Bothniensia Occidentalis 8*, 9–19.

Wallin, J.-E. 1986b: Vegetationshistorisk och naturgeografisk undersökning vid Stalon, Vilhelmina. Studier i Norrländsk Forntid 2. *Acta Bothniensia Occidentalis 8*, 20–32.

Witkowski, A. & Miller, U. 1995: Diatom flora of the Mastogloia Sea—a key to present and future environmental changes in the Baltic Sea. *Proceedings of the 13th Diatom Symposium, Italy.* Biopress (in press).

9

Finland

Y. VASARI, G. GLÜCKERT, S. HICKS, H. HYVÄRINEN,
H. SIMOLA and I. VUORELA

INTRODUCTION

Finland in eastern Fennoscandia is a part of the ancient Precambrian shield (Figs 9.1 and 9.2), with gradual gradients of climatic and vegetational patterns from south to north (Fig. 9.4). It is geographically rather homogeneous, and the different regions represent only gradual changes. Land uplift since the withdrawal of the continental ice has played a major role in the Late Quaternary history of Finland, and the initiation of the vegetational history is thus different in different parts of the country.

Type-region syntheses are presented for the following regions (Fig. 9.3).

SF-a	The Åland Islands and SW Archipelago
SF-b	The coastal area of SW Finland
SF-c	Coastal South Ostrobothnia
SF-d	The Lake District of Finland (W, Central Finland)
SF-d and SF-f	The Lake District of Finland (E, North Karelia)
SF-e and SF-h	North Ostrobothnia and South Lapland
SF-i	Kuusamo
SF-k and SF-l	Northeast Fennoscandia
Nz and Nae	Northeast Fennoscandia

No syntheses are made for the other Finnish type regions.

Regions SF-a and SF-b both belong to the hemiboreal (or boreo-nemoral) vegetation zone (Fig. 9.4). Climate is mild, slightly oceanic, and indifferent (Fig. 9.4). Limestone bedrock or calcareous glacial drift have locally created patches of rich vegetation. The broad-leaved deciduous trees (*Acer, Corylus, Fraxinus, Quercus, Tilia, Ulmus*) are regular components in the forest vegetation. Region SF-a comprises the southwestern archipelago of Finland, region SF-b the oak region on the mainland. Both were uplifted out of the Baltic Sea relatively late. The share of mires is very small (Fig. 9.5).

Regions SF-c and SF-d together comprise the southern boreal vegetation zone of Finland (Figs 9.3 and 9.4). Region SF-c includes the geologically relatively young coastal plains formed mostly of marine clay. Region SF-d is the Lake District of Finland. The proportion of mires is less than the average one for Finland (Fig. 9.5).

The three southern regions SF-a, SF-b, and SF-c are the main cereal-producing areas of Finland. Regions SF-e and SF-f together comprise the middle boreal vegetation zone in Finland (Figs 9.3 and 9.4). Region SF-e is geologically the younger and consists largely of postglacial marine clays. The terrain there is characteristically flat and largely covered by peatland (Fig. 9.5). Region SF-f has a more varied topography; the bedrock consists largely of common metamorphic rocks, resulting in large vegetational and floristic variation. The climate is indifferent and

Palaeoecological Events During the Last 15 000 Years: Regional Syntheses of Palaeoecological Studies of Lakes and Mires in Europe.
Edited by B.E. Berglund, H.J.B. Birks, M. Ralska-Jasiewiczowa and H.E. Wright. © 1996 John Wiley & Sons Ltd.

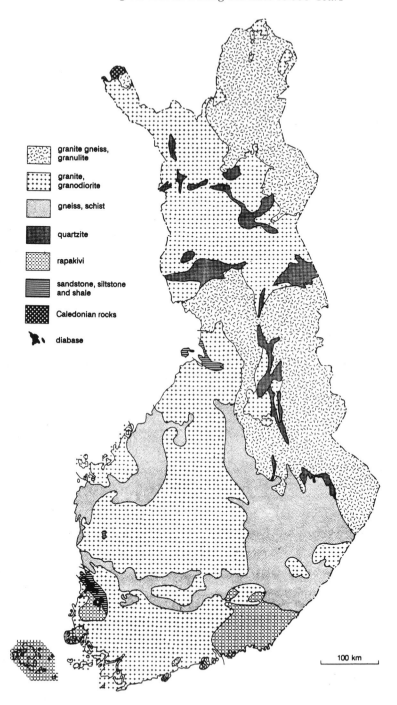

Fig. 9.1   Bedrock in Finland (Teemakarttoja Suomesta 1989)

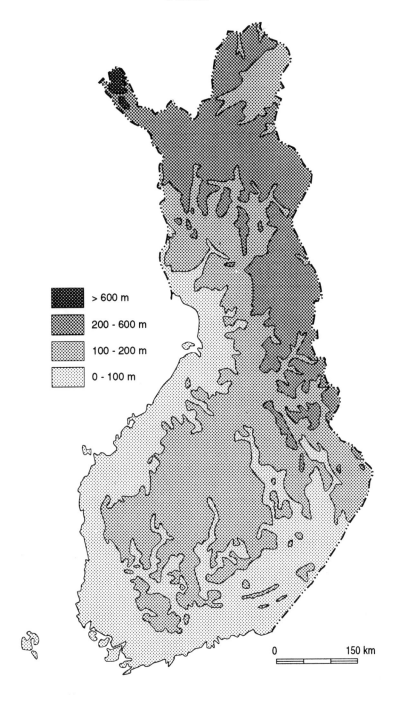

Fig. 9.2 Absolute altitude in Finland (metres above sea level)

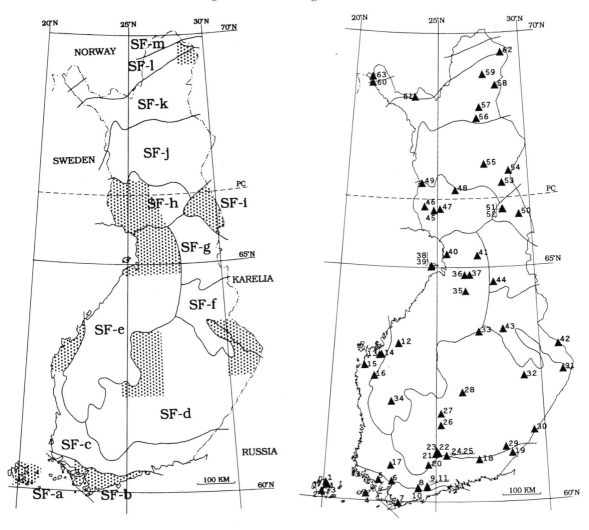

Fig. 9.3 IGCP 158 B type regions (left) and reference sites (right) in Finland. Shading indicates areas from which regional syntheses are presented. SF-a = SW Archipelago, SF-b = Oak region of SW Finland, SF-c = S and W coastal plains, SF-d = Lake District of Finland, SF-e = Ostrobothnia, SF-f = N Karelia–S Kainuu, SF-g = N Kainuu, SF-h = S Lapland, SF-i = Kuusamo, SF-j = Middle Lapland, SF-k = Outa Lapland, SF-l = Birch Lapland, SF-m = Fell Lapland. Reference sites are listed in the Appendix

slightly continental.

All the remaining regions belong to the northern boreal vegetation zone (Figs 9.3 and 9.4). Regions SF-g and SF-h are in many ways intermediate areas between southern and northern Finland. Region SF-h, North Ostrobothnia, is geologically younger and has been largely subaquatic, and region SF-g has been mainly supraaquatic. The coastal plains in SF-h are largely paludified (Fig. 9.5).

The small region SF-i (Kuusamo) is well delimited from its surroundings. It has a rugged topography founded mainly on metamorphic folded rocks. The bedrock is quite rich in calcium, and accordingly the area is floristically rich and famous for its luxurious vegetation types. The climate is slightly oceanic. The region has a long Quaternary history.

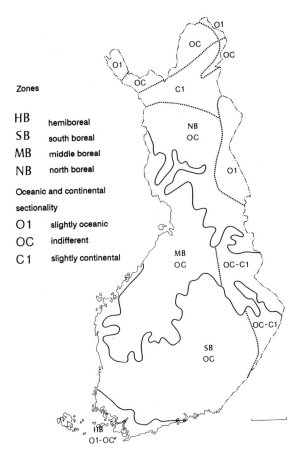

Fig. 9.4 Forest vegetation zones and their sections in Finland (according to Ahti *et al.* 1968)

The relatively flat region SF-j, Middle Lapland, is characterized by large wet aapamires and forests. The climate is indifferent as to continentality–oceanicity. Region SF-k comprises the northern-most area of coniferous forests in Finland. Climatically it is slightly continental.

Regions SF-l and SF-m comprise the area of subalpine birch forests and of orohemiarctic fjell-heath vegetation in Finland. Palsa mires exist locally. Climatically Sf-l is indifferent, whereas SF-m is slightly oceanic. Both these regions are characterized by fjell-landscape. The Kiolen mountains of Caledonian age penetrate into northwestern Finland, causing locally considerable floristic richness.

TYPE REGION SF-a, THE ÅLAND ISLANDS AND SOUTHWEST ARCHIPELAGO
(Gunnar Glückert)

Altitude: 0–130 m above sea level. High rocky areas in the north.

Climate: Cool maritime. Mean July temperature 16°C, February –4°C, annual 5°C, precipitation 550 mm yr^{-1}.

Geology: Rapakivi granites of Precambrian age, covered with glacial and postglacial deposits.

Topography: High rock areas in north part, lower undulating hills in other parts, and flat clay-covered valleys. Relief 0–130 m.

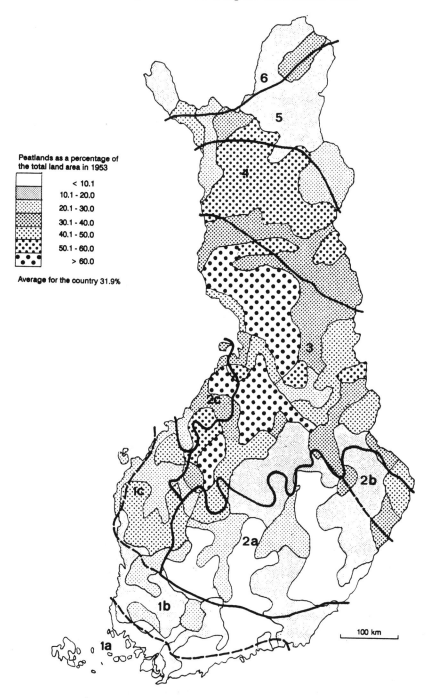

Fig. 9.5 The regional division (Eurola *et al.* 1984) and distribution (*Atlas of Finland* 1960) of mires in Finland. 1. Concentric kermi bogs: (a) plateau bogs of the archipelago; (b) concentric kermi bogs of southern Finland (these present a morphology and vegetation intermediate between the plateau bogs and the following shield-like kermi bogs); (c) shield-like kermi bogs of Satakunta and Ostrobothnia. 2. Eccentric kermi bogs and *Sphagnum fuscum* bogs: (a) of Lake Finland, (b) of northern Karelia, and (c) of the coastal areas of the Bothnian Bay. 3. Southern aapa mires of Ostrobothnia. 4. Northern aapa mires of Peräpohjola. 5. Northern aapa mires of Forest Lapland. 6. Palsa mires (chiefly aapa mires with palsas) of Fjeld Lapland

Land uplift: Recent uplift 0.5–0.6 mm yr⁻¹. The northern part emerged 9000–6000 BP (Yoldia, Ancylus), the lower areas since 6000 during the Litorina Sea. In the archipelago now about 6500 isles.

Population: *Ca.* 22000 people in an area of 1500 km², mainly in small villages and farms. The capital Mariehamn has 15000 inhabitants.

Vegetation: Barren pine and spruce forests in rocky areas, broad-leaved forests (with *Betula*) or cultivated fields in lower clayey areas. Small mires, mainly in depressions.

Soils: Acid clayey soils (podzol), partly fertile earths in low-lying agricultural areas, infertile earths in rocky areas with glacial deposits.

Human history: Signs of human influence since 5000 BP, of agriculture since 4000 BP; between about 2500–1600 BP no signs of agriculture.

Land use: About 80% rocks and uncultivated forests, 10% arable areas with agriculture in central and southern parts, 10% lakes, mires, and villages.

The type area SF-a belongs to the coastal area of SW Finland in the northernmost part of the oak region of hemiboreal mixed forests. For the present synthesis 27 pollen diagrams with 32 radiocarbon dates published by Fries (1961, 1963), Glückert (1978, 1989) and Sarmaja-Korjonen *et al.* (1991) have been used. The tree-arrival and tree-presence diagram (Fig. 9.8) is based on eight representative sites from the main island in Åland and from the island of Brändö:

- Kasmossen, Saltvik (time span 6000–0 BP) (Glückert 1978)
- Tjärnbergen, Saltvik (5000–0 BP) (Glückert 1978) (Fig. 9.7)
- Tjärnan, Saltvik (3500–0 BP) (Glückert 1978)
- Kolmilaträsk, Saltvik (Sarmaja-Korjonen *et al.* 1991)
- Kvarnträsk, Finström (Sarmaja-Korjonen *et al.* 1991)
- Lillträsk, Geta (4000–0 BP) (Glückert 1978)
- Degermossa, Brändö (3500–0 BP) (Glückert 1978)
- Dalkarbyträsk, Jomala (3000–0 BP) (Fries 1961)

The tree-arrival and tree-presence diagram is based

on few ¹⁴C dates and often on a pollen sum of only 150–200 AP, so the vegetational events can be only presented generally. Only a few mires and lakes have useful sediments for pollen analysis.

The vegetational history of the individual sites has been influenced by the rate of emergence of the land from the Baltic. The biostratigraphic sites lie at quite a low altitude (below 60 m above sea level), so postglacial events older than 6500 BP cannot be studied in the Åland area. Thus the time 6500 BP has been used for the beginning for the tree-arrival and tree-presence studies. The time span is thus 6500–0 BP.

According to Fries (1961, 1963) the first sign of human influence is from 5000 BP (occasional settlement), and the beginning of agriculture from 4000 BP. Between 2500 and 1600 BP no clear signs of agriculture have been recorded from Åland. According to Glückert (1978) the first sign of agriculture is from about 2500 BP, and no sign can be detected between 2500 and 1500 BP. From 1600 BP there is clear evidence of continuous agriculture and human settlement in Åland (Sarmaja-Korjonen *et al.* 1991).

Reference site 1. Tjärnbergen

The basin of Tjärnbergen (Glückert 1978) has been chosen for a reference pollen diagram for the type region SF-a (Åland) (Figs 9.6 and 9.7).

The site is a pond surrounded by rocks and bordered by a narrow mire. For pollen-assemblage zones (paz), see Fig. 9.7; for tree-arrival and tree-presence, mean tree-composition, the invasion of *Picea*, and event stratigraphy, see Figs 9.8 and 9.9.

Betula–Pinus–Alnus–Corylus–Ulmus (6000–4000 BP)
Time before isolation of the basin from the Baltic. High values of *Betula, Pinus, Alnus, Ulmus,* and *Quercus*. High values of *Ulmus, Quercus,* and *Tilia* (QM). *Fraxinus* appears about 6000 BP but is quite uncommon. The amount of *Quercetum Mixtum* (QM) decreases about 4000 BP. The time span of 6000–4000 BP is not presented in its typical form, because of the low altitude of the basin.

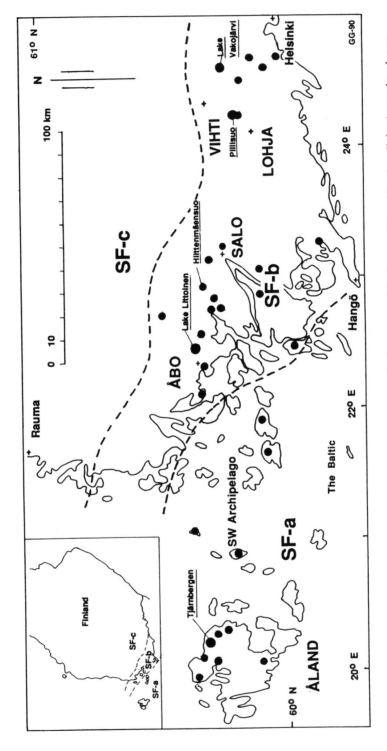

Fig. 9.6   Type regions SF-a and SF-b and reference sites (large dots) mentioned in the text. Other sites (small dots) are also shown

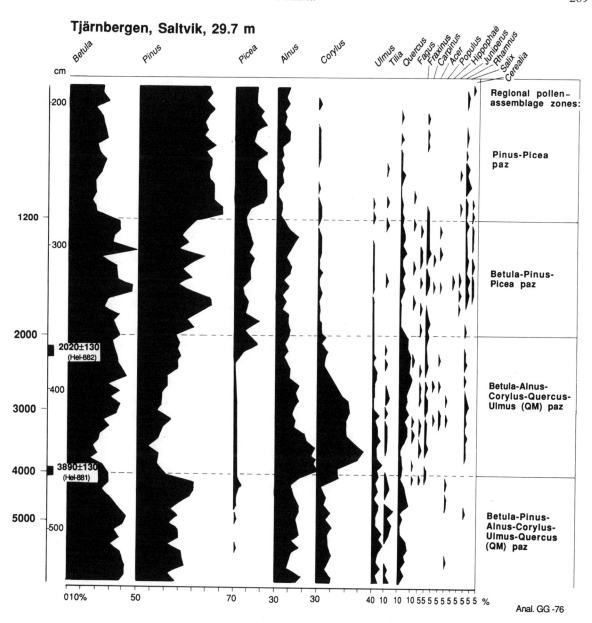

Fig. 9.7 Pollen diagram with selected taxa for Tjärnbergen bog, Saltvik, Åland (Glückert 1978)

Betula–Alnus–Corylus–Quercus–Ulmus(4000–2000 BP)

The time during and just after the isolation of the basin. Rich shore vegetation, with locally high values especially of *Alnus* and *Corylus* (up to 30%/AP). Maximum values of *Quercus* (3–10%/AP), a remarkable decrease of *Alnus, Corylus*, and QM trees about 2500 BP, and a minimum of *Pinus, Carpinus,* and

Fagus appear about 4000 BP.

Betula–Pinus–Picea (2000–1200 BP)

Betula and *Pinus* increase. *Picea* expanded in Åland, and QM trees decreased about 2000 BP.

Pinus–Picea (1200–0 BP)

Alnus, Betula and QM trees still decrease, *Pinus* and *Picea* increase. *Carpinus* and *Fagus* disappear about 600 BP.

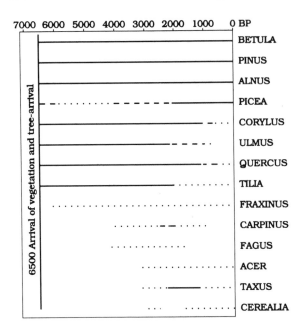

Fig. 9.8 Tree-arrival and tree-presence diagram for Åland. Full line indicates frequent occurrence, broken line sparse occurrence, dotted line rare or absent

Anthropogenic, hydrologic, climatic, and soil events

The Åland Islands are rich in archaeological finds, but the pollen evidence of human presence is consistently slight throughout. Signs of human influence have been recorded since about 5000 BP (e.g. Dreijer 1941; Meinander 1958, 1960), and of agriculture since about 4000 BP at the end of the Stone Age (Fries 1961, 1963; Glückert 1978). Between 2500 and 1600 BP there is no evidence of continuous agriculture on the island (Fig. 9.9).

The hydrologic events on the island are intimately related to the history of the Baltic, the best documented point being the limit of the Litorina Sea around 7000 BP. The continuing land uplift has exerted an influence, with the time of isolation of the basin becoming younger and younger as one approaches the coast. A part of the flat land has formed a suitable substrate for mire development. As the climate became wetter and cooler at about 2500 BP it caused a decrease of deciduous trees and a strong increase of peat growth in the mire develop-

ment. At the same time the soils and waters became more acid.

TYPE REGION SF-b, COASTAL AREA OF SOUTHWEST FINLAND (Gunnar Glückert)

The type region SF-b in the coastal area of SW Finland belongs to the oak region of SW Finland, a part of the hemiboreal mixed forests (Fig. 9.4). It is divided into two subregions with somewhat different palaeoecological data: the Åbo–Salo area (coastal) and the Lohja–Vihti area (inland) (Fig. 9.6).

Altitude: 0–160 m above sea level.

Climate: Cool maritime (coast), cool maritime/continental (inland). Mean temperature: July 17°C, February −5 to −7°C, annual 4–5°C. Precipitation 550–600 mm yr^{-1} (coast), 600–700 mm yr^{-1} (inland).

Geology: Mica schist, gneiss, and granite of Precambrian age, covered partly with glacial sediments and postglacial deposits (clay, peat) on the coast.

Land uplift: Recently 0.4–0.6 mm yr^{-1}. The highlying areas emerged 10000–7000 BP (Yoldia, Ancylus), the low areas later (Litorina).

Population: About 500000 people in an area of 15000 km^2, mainly in towns, villages, and farms. Åbo has 160000 inhabitants, the Helsinki region (= Helsinki, Espoo, Kauniainen, Vantaa) *ca*. 850000.

Vegetation: Mixed forests with *Betula, Pinus, Alnus, Picea*, and some broad-leaved trees locally (*Quercus, Ulmus, Tilia, Fraxinus, Acer*, and *Corylus*). Inland some areas have many small mires.

Soils: Acid clayey soils in the cultivated areas, infertile soils in rocky areas with glacial deposits.

Human history: Signs of human influence (local settlement) from the Neolithic since 6000 BP, of agriculture since 2000 BP.

Land use: 70% uncultivated forests (rocks, glacial deposits with forests), 20% arable areas, 10% lakes, mires, towns, and villages.

In the type region SF-b more than 100 pollen diagrams with about 100 radiocarbon dates are available for this synthesis. The pollen-analytical

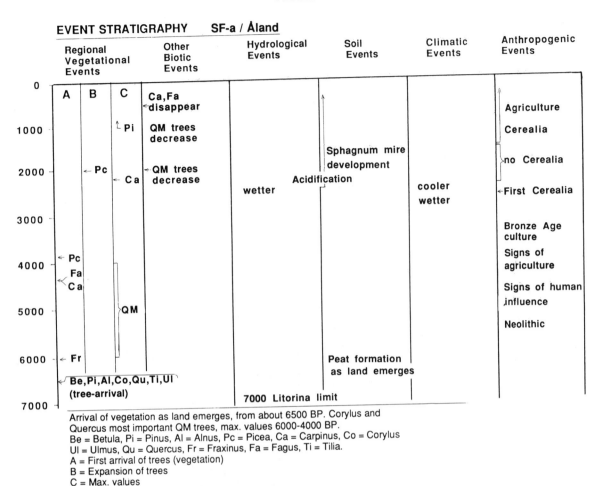

Arrival of vegetation as land emerges, from about 6500 BP. Corylus and
Quercus most important QM trees, max. values 6000-4000 BP.
Be = Betula, Pi = Pinus, Al = Alnus, Pc = Picea, Ca = Carpinus, Co = Corylus
Ul = Ulmus, Qu = Quercus, Fr = Fraxinus, Fa = Fagus, Ti = Tilia.
A = First arrival of trees (vegetation)
B = Expansion of trees
C = Max. values

Fig. 9.9   Event stratigraphy for Åland

results of 12 sites are used for this regional synthesis
(Fig. 9.6). Eight sites are from the Åbo–Salo
subregion, viz. Lake Littoinen (Littoistenjärvi)
in Kaarina and Lieto near Åbo (Fig. 9.10), Raho-
lansuo in Aura (Aurola 1938; Glückert 1976), Sand-
brinksmossen in Dragsfjärd (Glückert 1976), Vohtenkellarinsuo (Glückert 1976, 1977; Vuorela 1983),
Meltolansuo and Nummensuo (Glückert 1976) in
Paimio, Hiittenmäensuo in Halikko (Fig. 9.11), and
Santamäensuo in Salo (Tolonen, M. 1985). From
the Lohja–Vihti subregion pollen-analytical data
are used from four sites, viz. Nälköönsuo (Sauramo
1958; Glückert 1970; Tolonen, K. & Ruuhijärvi
1976) and Pillisuo in Lohja (Fig. 9.14) and from

Kaitlampi (Glückert 1979) and Vakojärvi in Vihti
(Fig. 9.15). Information concerning tree-arrival and
tree-presence (Figs 9.12 and 9.17) is based not only
on diagrams mentioned above but also on the papers
by Aartolahti (1966), Salonen *et al.* (1981), Räsänen
& Salonen (1983), and Ristaniemi & Glückert
(1988).

The time span is 9500–0 BP for the Åbo–Salo
subregion, and 10500–0 BP for the Lohja–Vihti area
(Fig. 9.6).

The rate of deglaciation and emergence of the
land from the Baltic has influenced the vegetational
history of the individual sites (cf. e.g. Sauramo
1958).

292

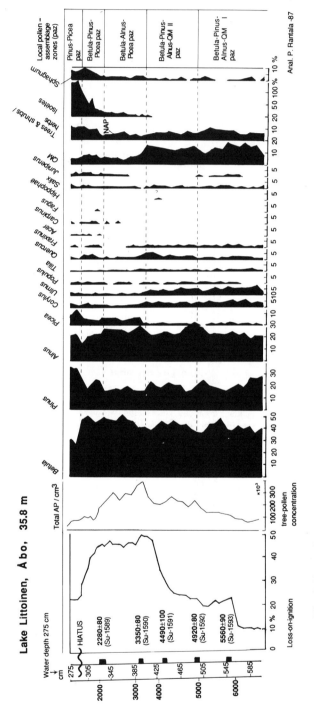

Fig. 9.10 Pollen diagram with selected taxa for Lake Littoinen, Åbo (Glückert *et al.* 1992)

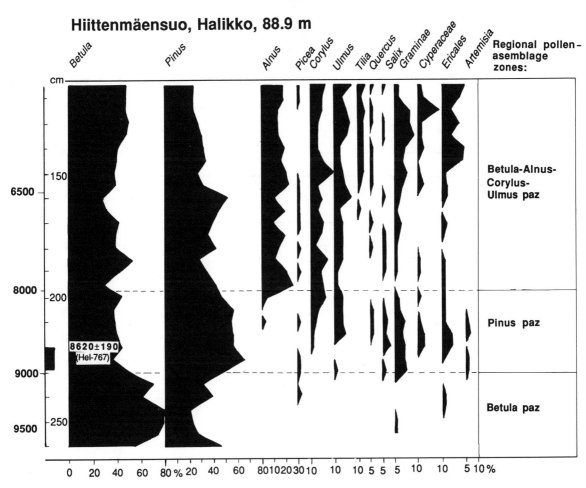

Fig. 9.11 Pollen diagram with selected taxa for Hiittenmäensuo bog, Halikko (Glückert 1976, M. Tolonen 1985)

The first signs of vegetation in the Åbo–Salo area are from about 9500 BP. Signs of an early appearance of *Picea* about 9000 and 8000 BP can be noticed in the Lohja–Vihti area. Between 9000 and 8000 BP an extremely dry climate caused a low ground-water level in southern Finland and very slow sedimentation (about 0.1 mm yr^{-1}) in lake basins. The first sign of human influence is from about 6000–5000 BP north of Åbo (Lieto, Paattinen, Maaria) (Kivikoski 1961), and of the beginning of agriculture about 2000–1500 BP. In the Lohja–Vihti subregion signs of human influence are younger, about 4000 BP, and that of agriculture about 1500–1000 BP.

Coastal area of SW Finland, the Åbo-Salo subregion

Reference site 5. Lake Littoinen

The basin of Lake Littoinen (Littoistenjärvi) has been chosen for a reference pollen diagram for a relatively young basin in the type region SF-b, the Åbo–Salo subregion (Glückert *et al.* 1992) (Figs 9.6 and 9.10).

The lake is about 1.5 km² in area and only 1–3 m deep, and it lies in a barren rocky area about 6 km from the town of Åbo. For pollen-assemblage zones (paz), see Fig. 9.10, for tree-arrival and tree-presence, Fig. 9.12, and for event stratigraphy, Fig. 9.13.

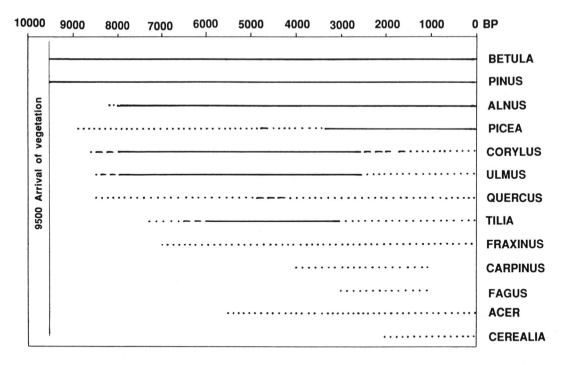

Fig. 9.12   Tree-arrival and tree-presence diagram for the Åbo–Salo subregion. See Fig. 9.8 for symbols

Before the Atlantic, *Corylus*, *Ulmus*, and *Quercus* appeared in the Turku area about 8500 BP. *Tilia* became common about 6500 BP. *Fraxinus* appeared about 7000 BP. The time span of QM was 8000–3000 BP.

*Betula–Alnus–Pinus–*QM I (6000–5000 BP)
Time before and after the isolation of the basin of Lake Littoinen from the Baltic. High values of *Betula, Alnus, Pinus, Corylus,* and *Ulmus,* some *Tilia* and *Quercus* (QM).

*Betula–Alnus–Pinus–*QM II (5000–3500 BP)
Rich vegetation of deciduous trees, especially *Betula, Alnus, Corylus,* and *Ulmus,* some *Quercus.*

Betula–Alnus–Picea (3500–2000 BP)
Forests of pine and spruce became common in the area, while the broad-leaved trees decreased. *Betula* and *Alnus* common, some *Corylus* and *Quercus*. The rapid invasion of *Picea* took place in the Åbo area about 3300 BP. *Acer* and *Carpinus* appeared. A noticeable peat (mire) formation began at about 2500 BP.

Betula–Pinus–Picea (2000–1200 BP)
Pine and spruce forests dominated in the area, *Alnus* still common, also some *Corylus* and *Quercus*. Also *Fagus* appeared. Peat formation continued.

Pinus–Picea (1200–0 BP)
Pine and spruce still increased. *Betula* and *Alnus* and the sparse QM representatives still decreased.

Reference site 6. Hiittenmäensuo

The bog basin of Hiittenmäensuo in Halikko has been chosen as a reference pollen diagram for an old basin of the type region SF-b, the Åbo–Salo subregion (Glückert 1976; Fig. 9.6). The bog lies in a depression of a high-lying rocky area. Time span 9500–0 BP. For pollen-assemblage zones (paz), see Fig. 9.11, for tree-arrival and tree-presence, Fig. 9.12, and for event stratigraphy, Fig. 9.13.

Betula (9500–9000 BP)
After the ice retreat at about 9800 BP, the vegetation arrived in the high-lying areas at about 9500 BP.

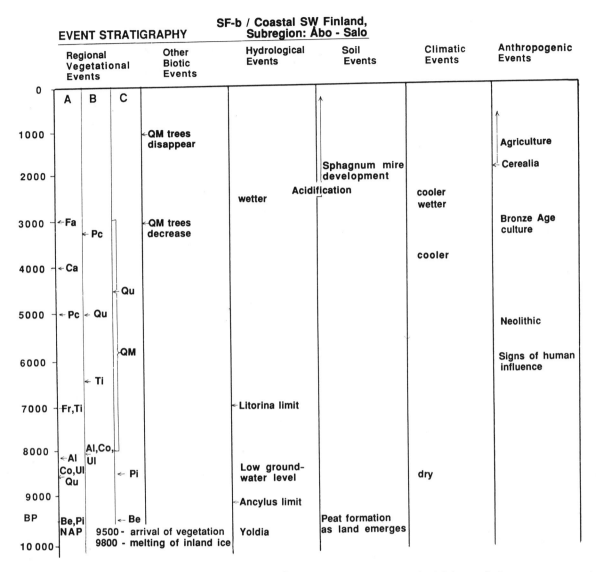

Fig. 9.13 Event stratigraphy for the Åbo–Salo subregion. See Fig. 9.9 for symbols

A typical *Betula* maximum characterizes the pollen spectra. No pollen of QM were identified. NAP records the first vegetation in the area. The early appearance of *Picea* can be noticed.

Pinus (9000–8000 BP)
The *Pinus* paz is typical for southern Finland, and some pollen of *Corylus, Ulmus,* and *Quercus* occurs (1–6%/AP each).

Betula–Alnus–Corylus–Ulmus (8000–0 BP)

The paz begins with the rise of *Alnus* and is characterized by *Betula* and a maximum of QM. The amount of *Corylus* and *Ulmus* each rises to 7–10%. The amount of *Pinus* decreases.

Anthropogenic, hydrologic, climatic, and soil events

In SW Finland, the area of Åbo and Salo is rich in archaeological finds. Signs of human influence have been recorded for more than 6000 BP (e.g. Kivikoski

1961). Pollen evidence for agriculture can be noticed since about 2000 BP (e.g. Vuorela 1983).

The hydrologic events on the Åbo–Salo subregion are related to the history of the Baltic, and the best documented point is the Litorina limit about 7000 BP. The continuing land uplift has exerted an influence, with the time of isolation of the basins becoming younger and younger as one approaches the coast. During the Boreal the ground-water level was low because of a dry climate in southern Finland. The Atlantic is characterized by warm and wet climate, with rich vegetation of deciduous trees and high groundwater level.

As the climate became cooler and wetter about 2500 BP, a part of the flat areas formed a suitable substrate for the mire development that strongly increased. *Sphagnum* played an increasing part in the mire development from 2500 BP onwards. As the pine and spruce forests as well as the mire development increased, the soils and waters became more acid. Also the limnological development of Lake Littoinen indicates a eutrophic phase during the Atlantic, changing into a dystrophic stage about 4000 BP (Fig. 9.13).

Inland area of SW Finland, the Lohja–Vihti subregion

Reference sites 10 and 9. Pillisuo and Lake Vakojärvi

The Pillisuo bog in Lohja (Glückert 1970) and the basin of Lake Vakojärvi in Vihti (Donner 1971, 1972) have been chosen for reference pollen diagrams of the type region SF-b, the Lohja–Vihti subregion (Fig. 9.6). The bog lies in a depression between the 1st Salpausselkä moraine and bedrock. Relative and pollen-concentration diagrams are presented (Figs 9.14–9.16).

For pollen-assemblage zones (paz), see Figs 9.14–9.16, for tree-arrival and tree-presence Fig. 9.17, and event stratigraphy Fig. 9.18.

Artemisia (10500–10000 BP)

The initial *Artemisia* paz is only recorded at high-lying sites in the Lohja and Vihti area that emerged just after deglaciation. Tundra vegetation with *Betula*, rich in NAP, e.g. *Artemisia*, Ericales and Gramineae, indicate the first vegetation in the area, following the retreating and melting ice margin

and land emergence (Figs 9.14–9.18). The end of the *Artemisia* paz is represented in several basins within the Lohja–Vihti subregion. Examples of old basins with tundra vegetation are Pillisuo in Lohja, Lakiassuo in Vihti (Glückert 1970), and Kaitalampi in Vihti, time span 10500–0 BP (Glückert 1979).

Betula (10000–9000 BP)

A distinct decrease of NAP at the beginning of the *Betula* paz. The pioneer birch woodland followed the retreating ice. *Betula* dominated (70–90%/AP), *Pinus* rare. First appearance of *Picea*, *Corylus*, and *Ulmus* about 9500 BP.

Pinus (9000–8000 BP)

Pinus dominated at *Betula*'s expense. *Corylus* and *Ulmus* became common, about 8500 BP, 3–5%/AP each. *Alnus* and *Quercus* appeared at the end of the paz, about 8200 BP.

Betula–Alnus–Corylus–Ulmus (8000–3500 BP)

The Atlantic is characterized by abundance of deciduous trees, not only *Betula* and *Alnus* but also *Corylus* and *Ulmus* at least 5–10%/AP each. The pollen curve of *Alnus* rose rapidly to its maximum rate, about 20%. *Quercus* became common, and *Tilia* and *Fraxinus* appeared about 7000 BP. *Tilia* became common about 6500 BP. *Acer* is very rare about 5000–3000 BP. The proportion of deciduous trees was about 7–18%/AP in total. The total tree pollen concentration ranges between 200000 and 900000 grains cm⁻³. The QM maximum was 8500–3000 BP. The Subboreal invasion of *Picea* took place about 4000 BP.

Betula–Pinus–Picea (3500–0 BP)

As the climate became cooler from 4000 BP onwards, the amount of broad-leaved trees noticeably decreased. During the Subboreal, coniferous trees forced the shade-sensitive trees out of the forests. *Tilia* and *Fraxinus* disappeared about 2000 BP, *Corylus*, *Ulmus*, and *Quercus* about 1500–1000 BP. Some pollen of *Carpinus* was found about 3500–1000 BP and *Fagus* about 2500–1000 BP.

Anthropogenic, hydrologic, climatic, and soil events

The Lohja–Vihti subregion is quite rich in archaeological finds, and signs of human presence have been noticed since 5000 BP. Pollen of Cerealia is found since

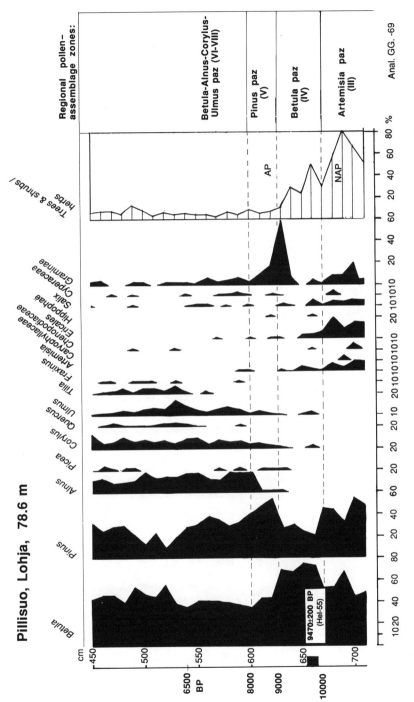

Fig. 9.14   Pollen diagram with selected taxa for Pillisuo bog, Lohja (Glückert 1970)

298

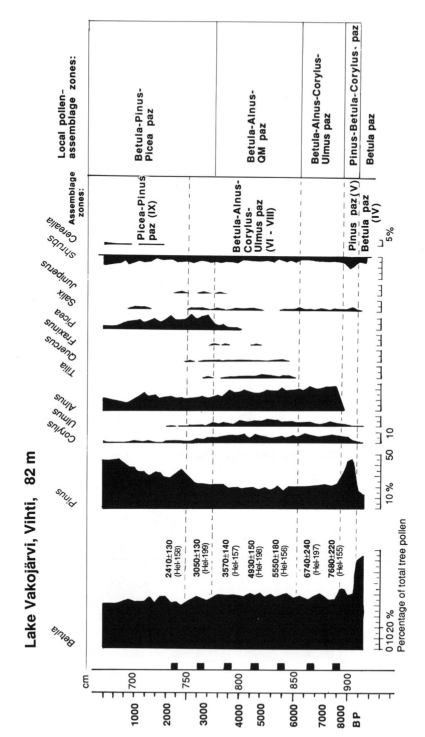

Fig. 9.15 Relative pollen diagram with selected taxa for Lake Vakojärvi, Vihti (modified from Donner 1972)

299

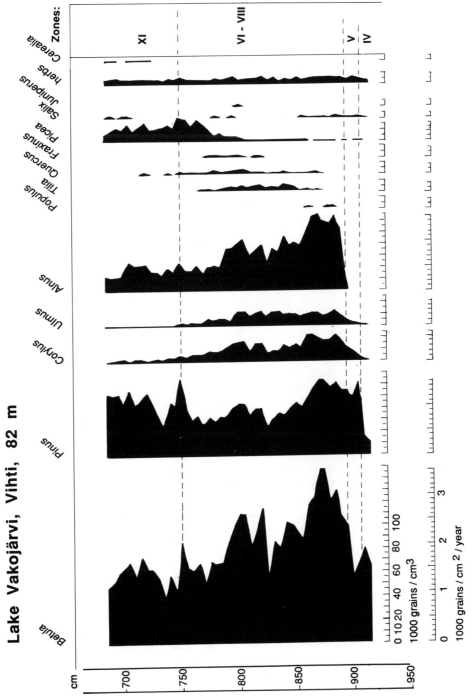

Fig. 9.16 Pollen concentration diagram with selected taxa for Lake Vakojärvi (modified from Donner 1972)

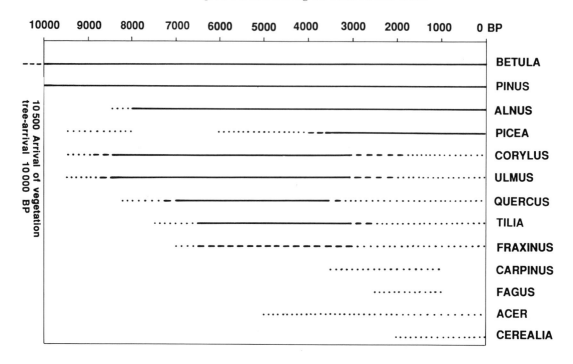

Fig. 9.17 Tree-arrival and tree-presence diagram for the Lohja–Vihti subregion. See Fig. 9.8 for symbols

2000 BP, but indicators of agriculture and human presence are distinctly rarer than in the Åbo–Salo area.

The hydrologic events are related to the history of the ancient Baltic, the best documented points being the Ancylus limit about 9000 BP and the Litorina limit about 7000 BP. During the Boreal the extremely dry climate caused a low groundwater level and, in general, a very scanty sedimentation in the basins. During the Atlantic and later the groundwater was higher. As the climate became wetter and cooler from about 4000 BP onwards, the groundwater level rose. The waters and soils became acid as a result of the increase of pine and spruce forests and the strong increase of peat growth in the mires, especially from 2500 BP onwards. *Sphagnum* played an increasing part in the mire development (Fig. 9.18).

TYPE REGION SF-c, COASTAL SOUTH OSTROBOTHNIA (I. Vuorela)

The synthesis region covers the northwestern part of the type region SF-c, the coastal plains of Ostro-

bothnia between latitudes 62° and 64°N, an area with a recent uplift rate of approximately 8.5 mm yr⁻¹.

The topography of the type region is flat, the gently undulating plains being composed of Litorina clay deposited during the last 5000 years.

The pollen-analytical results of five reference sites have been used for the regional synthesis: (1) Kaluneva in Jurva commune, (2) Inmossen in Malax, (3) Lintunemossen in Vörå, (4) Marjenemossen in Vörå, and (5) Nääverpakanneva in Purmo (Fig. 9.19).

According to the pollen data and ^{14}C determinations *Picea* arrived in the region shortly after 4000 BP. The final reduction of broad-leaved deciduous trees (*Corylus*, *Quercus*, *Tilia*) accompanied the more advanced agricultural practices around AD 1200–1400. This phase was preceded by a retrogressive cultural phase following an advanced Roman Iron Age settlement.

Altitude: 0–50 m (SE part of the area 50–100 m).

Climate: Mean annual temperature (1921–1950) 4°C, July 16°C, February –7°C. Precipitation 500–550 mm yr⁻¹. The number of days in the year with at

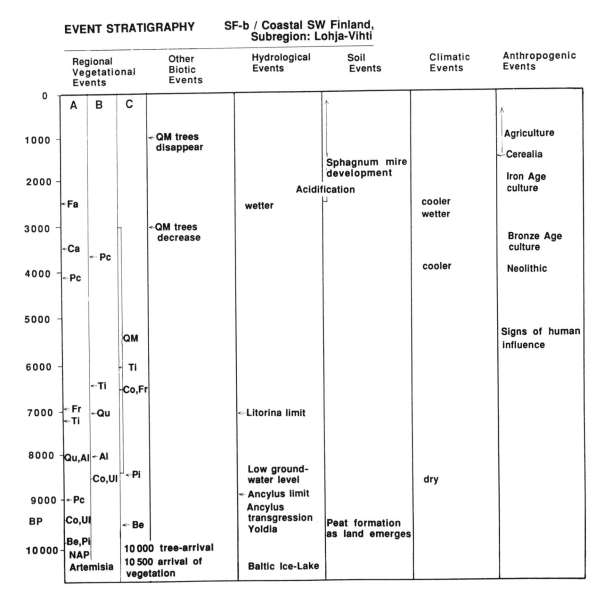

Fig. 9.18 Event stratigraphy for the Lohja–Vihti subregion. See Fig. 9.9 for symbols

least 1 mm precipitation: 90–100. Snow cover: 10–15 December to approximately 20 April. Snow depth 30–40 cm.

Geology: Northern part of the area: Precambrian acid plutonic rocks. Southern part: Precambrian metamorphic rocks, mainly migmatite. SE corner of the area: granodiorite and quartz diorite. The rocks are covered by Quaternary minerogenic deposits, mainly till, with sand and clay in river valleys. Recent land uplift rate 8.5 mm yr^{-1}.

Topography: Mainly flat or very gently undulating plains (relief <10 m) on marine clays of the Litorina Sea. No lakes; several rivers flowing E–W to the Gulf of Bothnia.

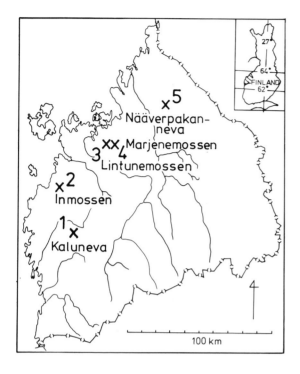

Fig. 9.19   Type region SF-c and the reference sites

Population: 150 registered farmsteads in 1540s. Present population density (mainly concentrated in the towns, villages, and along the rivers) 10–20 km⁻². Around Vaasa: 20–50 inhabitants km⁻². Urban population: Vaasa 60000–70000, Uusikaarlepyy 5000, Kaskinen and Kristiinankaupunki 2000 each.

Vegetation: The western part of the type region belongs to the southern boreal spruce forests, while the easternmost part is included in the middle boreal spruce forests (Ozenda et al. 1979). The area is crossed by the present northern limit of *Tilia cordata*. Other native trees are *Alnus incana, A. glutinosa, Betula pubescens, B. pendula, Pinus sylvestris*, and *Picea abies. Picea*-dominant forests cover 50–60% of the total area of productive forest land (mainly on the shore district), *Pinus*-dominant forest correspondingly 30–40%. Predominant forest type is *Myrtillus*-type (MT). Predominant type of peatland is *Pinus* peat bog (Ilvessalo 1960). Extensive peatlands, 20–30% of the total land area; in SE parts 40–50% of the land area (Fig. 9.5). Of the total area of peatlands 40–50% are *Pinus* peat bogs, 30–

40% are *Picea* and *Betula* peat bogs, and 20–30% are treeless *Sphagnum* bogs (Ilvessalo 1960).

Soils: Acid, in places abundantly sulphate rich clays and till. Minor areas of nonfertile glaciofluvial sand. Weak podzolization. Extensive mires.

Land use: In 1953 (30-) 50–60% of total land area was productive forest land. The share of cultivated land was 20–30 %. The most common cultivated crop was formerly wheat, nowadays predominantly oats and hay.

Reference site 16. Kaluneva, Jurva
(Hyvärinen, R. unpubl.) (Fig. 9.20)

Mire. Two regional pollen-assemblage zones (paz) with four local subzones.

Ka-1	3800–500 BP	QM–*Picea*
	3800–3000 BP	*Betula*–NAP
	3000–1600 BP	*Picea–Pinus*
	1600–1000 BP	*Alnus*–NAP
	1000–500 BP	*Pinus*
Ka-2	500–0 BP	*Pinus*–NAP– Ericaceae

Reference site 15. Inmossen, Malax
(Miettinen, M. 1986, Vuorela 1986, Miettinen, M. & Vuorela 1988).

Mire. Three local pollen-assemblage zones (paz). The stratigraphy of the earliest paz is strongly affected by redeposition caused by floods.

1700–1250 BP	Herbs–Cerealia
1250–	Reforestation
	Recent cultivation

Reference site 13. Lintunemossen, Vörå
(Vuorela 1987a and b, Miettinen, M. & Vuorela 1988; see also Tolonen, K. *et al.* 1979).

Mire. Four local pollen-assemblage zones (paz).

Li-1	1500–1200 (?) BP	*Pinus*–NAP
Li-2	1200–1100 BP	*Betula*–Ericaceae
Li-3	1100–1000 BP	*Betula*–QM
Li-4	1000–0 BP	*Pinus*–Ericaceae

303

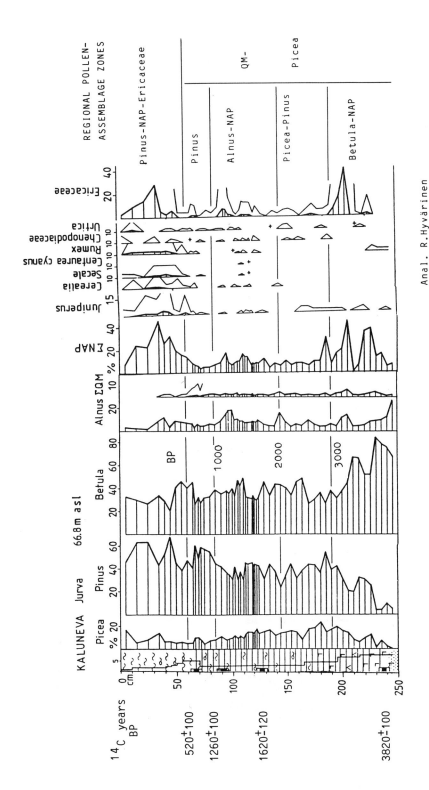

Fig. 9.20 Pollen diagram with selected taxa for Kaluneva Bog, Jurva (R. Hyvärinen 1987, unpubl.). See Fig. 9.24 for stratigraphical explanations

304

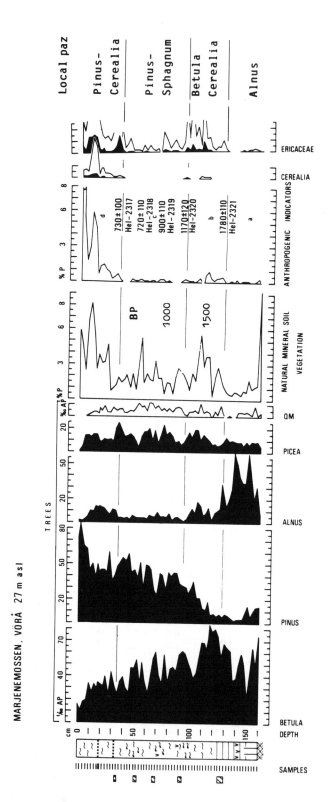

Fig. 9.21 Pollen diagram with selected taxa for Marjenemossen, Vörå (Miettinen & Vuorela 1988). See Fig. 9.24 for stratigraphical explanations

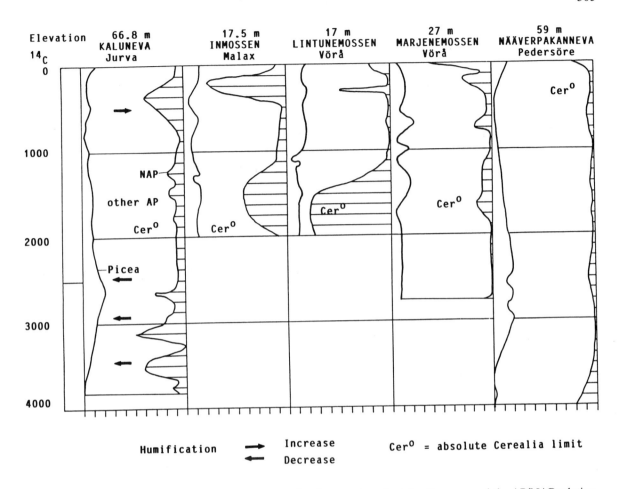

Fig. 9.22 Pollen diagram showing the relative *Picea* pollen frequencies of total pollen sum and the AP/NAP relation, drawn on a chronological scale. The humification trend is determined for the Kaluneva diagram

Reference site 14. Marjenemosse, Vörå (Vuorela 1987a and b,Miettinen, M.& Vuorela 1988) (Fig. 9.21).

Mire. Four local pollen-assemblage zones (paz).

Ma-1	2800–1800 BP	*Alnus* (sea shore)
Ma-2	1800–1100 BP	*Betula*–Cerealia
Ma-3	1100–730 BP	*Pinus–Sphagnum* (reforestation)
Ma-4	730–0 BP	*Pinus*–Cerealia

Reference site 12. Nääverpakanneva, Purmo (Vuorela 1982)

Mire. Three local pollen-assemblage zones (paz).

Np-1	4000 (?)–3250 BP	*Betula–Alnus*
	4000–3600 BP	NAP

	3600–3250 BP	Ericaceae
Np-2	3250–1000 BP	*Pinus–Picea*
Np-3	1000–0 BP	*Pinus*–NAP–Ericaceae

The pollen diagrams

The synthesis is based on five pollen diagrams from relatively young ombrotrophic *Sphagnum* bogs represented here by that from Kaluneva in Jurva (Fig. 9.20) and from Marjenemossen in Vörå (Fig. 9.21). For Kaluneva, which extends to the Subboreal chronozone (Mangerud *et al.* 1974), regional pollen-assemblage zones are proposed; for Marjenemossen, local pollen-assemblage zones are

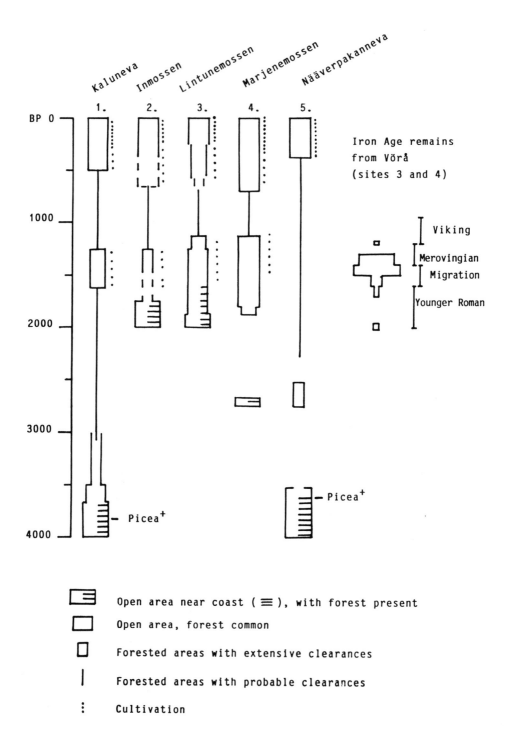

Fig. 9.23 Schematic interpretation of landscape openness in South Ostrobothnia (SF-c) based on the pollen dat represented in the diagrams for five sites

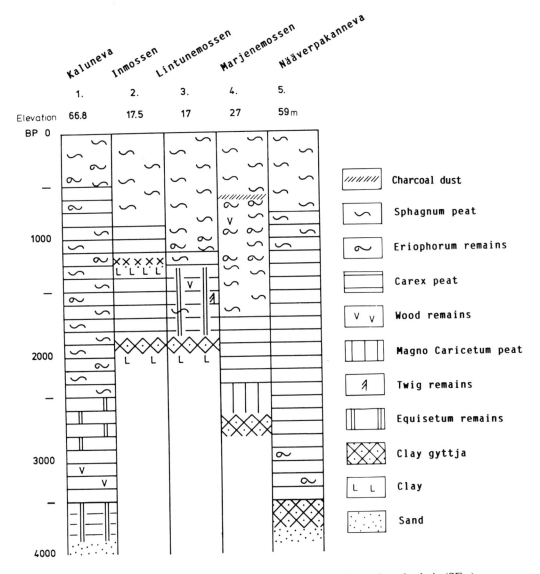

Fig. 9.24 The stratigraphy of the reference sites in South Ostrobothnia (SF-c)

recorded. In all diagrams the natural vegetational succession can be followed from the strongly uplifted shoreline through the shore meadow and *Alnus* and *Picea* forest stages up to the present-day *Pinus*-dominated *Sphagnum* peat bogs (cf. Kujala 1926).

Palaeoecological patterns and events

Common patterns

Strong NAP pollen maxima, reflecting shore mead-

ows, are clearly detectable in all diagrams (Fig. 9.22).

The invasion of *Picea* into the area took place around 4000 BP. The local pollen-assemblage zones b and d in the diagrams from Vörå (Fig. 9.21) clearly represent fluctuations in human impact. A prehistoric agricultural phase dating to the Roman Iron Age and withdrawal phase in the period of *c.* 1200–800 BP are apparent in the diagrams from most reference sites. See, however, also Wallin & Segerström (1994).

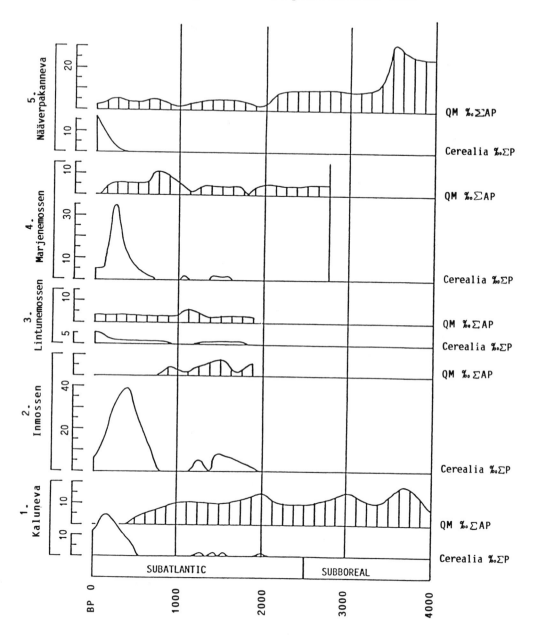

Fig. 9.25   Relative frequencies of Cerealia and QM pollen data at the reference sites of South Ostrobothnia (SF-c)

The present-day *Sphagnum* peat bogs in the reference area mainly developed during the last 1000
years. They were preceded by *Picea–Alnus–Betula–*
QM-dominated forests (Kujala 1926).

Relatively high pollen frequencies of Ericaceae,
mineral soil herbs, and anthropogenic indicators

occur during the period starting 500–800 BP.

The landscape (Fig. 9.23)

Because of the higher elevation of sites 1 and 5, the
landscape there naturally developed from sea-shore

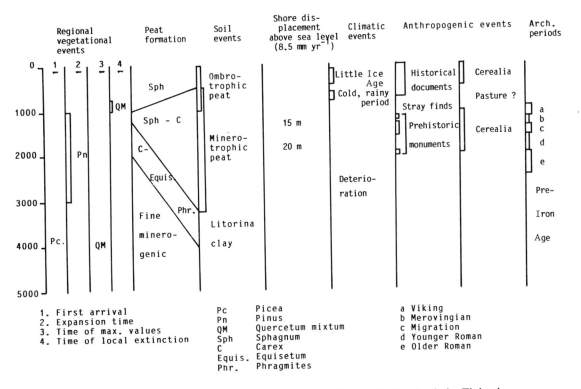

Fig. 9.26 Summary diagram for the type region SF-c, South Ostrobothnia, Finland

meadows into dense forests. Human activity later (around 1600–1200 BP) caused clearances, which were mostly reforested during 1200–800 BP. This was also the time when the growth of *Sphagnum* peat strongly increased and the present-day *Pinus*-dominated bogs originated (Fig. 9.24).

Hydrological events in type region SF-c
(Figs 9.22 and 9.24)

The relative peat humification was determined only at Kaluneva (site 1). The decreasing values up to 500 BP reflect increasing hydrological inputs during the Subatlantic chronozone. The same phenomenon is emphasized at sites 2–4 during the reforestation period of 1200–800 BP, probably resulting also in agricultural retrogression. The increasing moisture is reflected by stratigraphic and palynological evidence, as well as by the rapid growth of *Sphagnum* peat.

Short-term trends in climate

According to the pollen data no major changes in the mean temperature occurred during the late Subboreal and Subatlantic chronozones. A major decrease around 2000 BP is reflected in the QM data, and the final decrease of the broad-leaved deciduous trees coincides with the start of agricultural activities (Fig. 9.25).

Soil events (Fig. 9.26)

The coastal location of type region SF-c is reflected in the pedology of the region. The terrestrial sediments on the Litorina clay consist of *Carex–(Equisetum–Eriophorum)* peat in the early phase, later developing into *Sphagnum* peat, mainly around 1000 BP.

TYPE REGION SF-d (W), THE LAKE DISTRICT OF FINLAND (CENTRAL FINLAND)
(Yrjö Vasari and Leena Koivula)

Between about latitude 61°30′–63°30′N and longitude 24°–27°E, this synthesis region covers a part of the northwestern Lake District (SF-d) on the eastern side of the water divide between the Gulf of Bothnia and the Gulf of Finland. It corresponds with the administrative province of Central Finland.

Climate: Mean January temperature –8°C to –10°C, mean July 16–15°C, mean annual 3–2°C. Precipitation 600 mm yr^{-1}.

Geology: Precambrian granites prevail in the area (Fig. 9.1). A cover of till is more or less continuous; drumlins are characteristic in part of the region. The Jyväskylä–Jämsä end moraine (*ca.* 9800–9700 BP) runs across the region (Ristaniemi 1987, Fig. 2). Glaciofluvial material occurs in the form of eskers oriented NW–SE. Clay deposits are found in the northwestern part of the Päijänne basin. Peatlands cover 11–30% of the total land area (Fig. 9.5).

Topography and altitude: Hilly moraine lowland (80–250 m above sea level) with rather low (30–100 m) relief. The lakes cover a very large part of the region. The biggest lake is Päijänne (78 m above sea level).

Population: 5–20 people km^{-2}. Jyväskylä, with about 50000 inhabitants, is the biggest city and the capital of the administrative province.

Vegetation: The region belongs to the southern boreal forest, close to the boundary with the middle boreal forest (Fig. 9.4). Forests cover more than 80% of the total land area. Pine and spruce are almost equally important forest trees (31–50%) of the total area of productive forest land; the corresponding share of birch-dominated forests is about 11–30%. *Tilia cordata* has scattered occurrences, and *Corylus avellana* and *Ulmus glabra* occur in the southern part of the region.

Mires cover less than 30% of the total land area; they are mainly forested raised bogs (Fig. 9.5). Spruce and hardwood mires are also very common.

Soils: Predominantly podzols and acid peats.

Land use: Cultivated land and settlement covers about 5–20% of the area. Hay, oats, and barley are the most important crops.

Holocene history and reference sites: The region became free of ice about 9800–9400 BP. The deglaciation took place mainly from southeast to northwest. Most of the area was covered by the Baltic Sea; supra-aquatic areas occur mainly in the western part of the region (Fig. 9.27). The lakes were isolated from Ancylus Lake between about 8900 and 8300 BP. Until 6000 BP the waters from the region flowed northwestwards into the Gulf of Bothnia, across the present water-divide (Sauramo 1958; Aario 1965; Ristaniemi 1987).

Most pollen diagrams published from the region have been used for dating the shoreline displacement. The present regional synthesis is based on the pollen material studied by Koskinen (1983), Koivula (1987), Koivula *et al.* (1994), Ristaniemi (1987) and Vuorela (1993a; Vuorela *et al.* 1993).

Reference site 28. Mäyrälampi, Hankasalmi
(Koivula 1987; Koivula *et al.* 1994) (Fig. 9.28)

Lake. Four local pollen-assemblage zones (paz) with three local subzones.

HMä-1	>8480 BP	*Betula*
HMä-2	8480–8260 BP	*Pinus–Betula*
HMä-3	8260–5330 BP	*Betula–Pinus–Alnus*
HMä-4	5330–0 BP	*Pinus–Picea*
-4a	5330–2100 BP	*Pinus–Picea–Alnus* subzone
-4b	2100–580 BP	*Pinus–Picea–NAP* subzone
-4c	580–0 BP	*Pinus–Picea–Secale* subzone

Reference site 27. Aholammi Jämsä
(Koivula 1987; Koivula *et al.* 1994) (Fig. 9.29)

Lake. Three local pollen-assemblage zones (paz) with four local subzones.

Aho-1	>8030 BP	*Pinus–Betula*
Aho-2	8030–5440 BP	*Betula–Pinus–Alnus*
Aho-3	5440–0 BP	*Picea–Betula–Pinus*

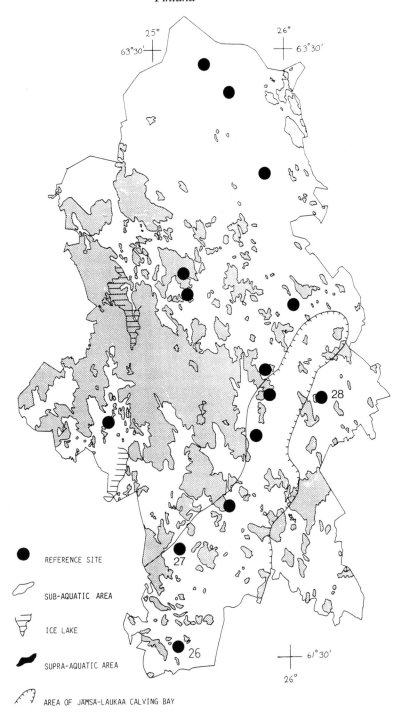

Fig. 9.27 Type region SF-d (W), the province of Central Finland within the Lake District of Finland. Based upon Enclosure Map II in Ristaniemi (1987). Reference sites indicated with numbers (see Appendix), dots without numbers indicate radiocarbon-dated AP diagrams of Koskinen (1983), Koivula (1987), Koivula *et al.* (1994), Ristaniemi (1987), M. Tolonen (1990) and Vuorela (1993a; Vuorela *et al.* 1993)

Mäyrälampi, Hankasalmi

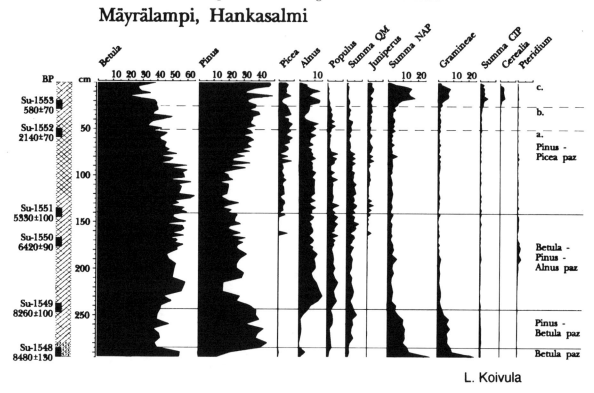

L. Koivula

Fig. 9.28   Pollen diagram with selected taxa for Mäyrälampi (Hankasalmi commune). (Modified from original in Koivula 1987). CIP indicates Culture Pollen Indicators. Analysed by L. Koivula

Aholammi, Jämsä

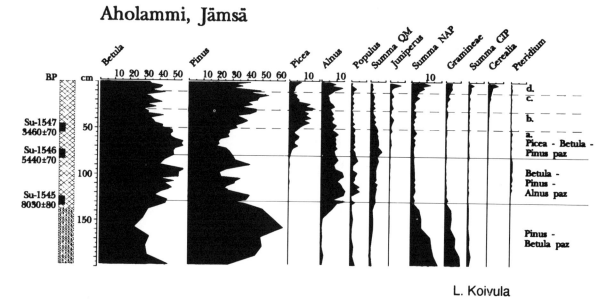

L. Koivula

Fig. 9.29   Pollen diagram with selected taxa for Aholammi (Jämsä commune). (Modified from original in Koivula 1987.) Analysed by L. Koivula

-3a	5440–3460 BP	*Betula–Picea–Pinus* subzone
-3b	3460–1300 BP	*Picea–Pinus–*NAP subzone
-3c	1300–700 BP	*Pinus–Picea–Secale* subzone
-3d	700–0 BP	*Pinus–Betula–*Gramineae subzone

Palaeoecological patterns and events

Common patterns (Figs 9.30 and 9.31)

No signs of treeless pioneer vegetation can be detected in the pollen diagrams from this region. Even birch forests, which supposedly must have spread to the area as the first forests, can be seen in only three radiocarbon-dated pollen diagrams from this region (Iso Pirttijärvi in Laukaa commune (Ristaniemi 1987), Linnasuo in Kuhmoinen commune (M. Tolonen 1990) and Mäyrälampi in Hankasalmi commune).

The oldest information concerning the local vegetation is from the Preboreal in Laukaa, where birch forests began to be replaced by pine forests about 9370 BP. *Pinus* has been a major forest tree from the very beginning according to most of the pollen diagrams, either co-dominant with *Betula* or as the sole dominant. The phase of pine dominance lasted until about 8100 BP.

Mixed pine–deciduous forest, which besides tree birches and *Alnus* included also *Corylus* and *Ulmus*, dominated the region between about 8100 and 5000 BP.

From 3000 BP the deciduous trees *Corylus, Ulmus,* and *Tilia* became rare, with some relic localities. *Fraxinus* already disappeared from the region between 6000 and 5000 BP.

The invasion of spruce began about 6400 BP, at first in the form of some local stands only. The general spread of spruce occurred about 5000 BP (5330–4810 BP), when spruce became a major component in the forests.

Human influence (Fig. 9.32)

The region is a part of the vast wilderness area that until recently was believed to have been settled permanently only after the 1550s. Recent pollen analyses by Leena Koivula (1987), Koivula *et al.* (1994), M. Tolonen (1990) and Irmeli Vuorela

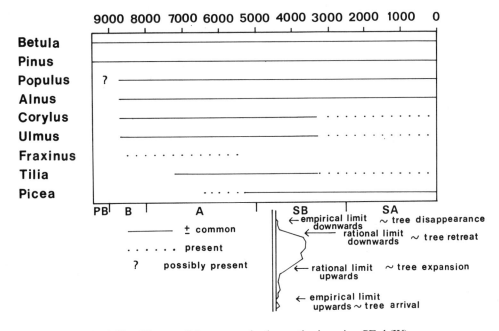

Fig. 9.30 History of forest trees in the synthesis region SF-d (W)

EVENT STRATIGRAPHY:

Type Region SF-d: The Lake District of Finland.

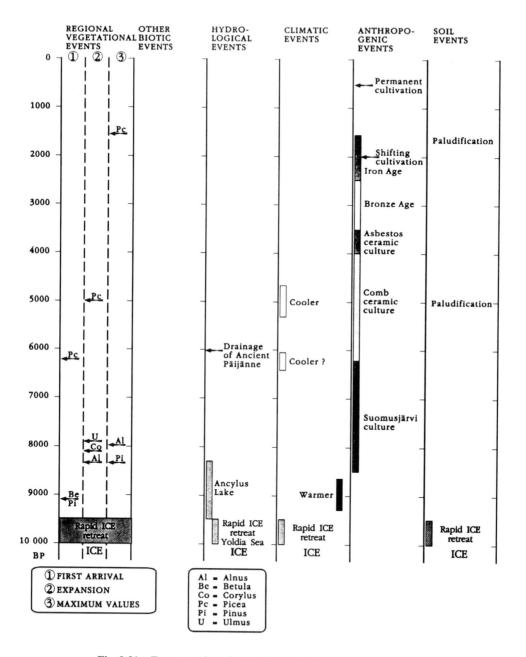

Fig. 9.31   Event stratigraphy for the synthesis region SF-d (W)

(1993a; Vuorela *et al.* 1993), however, have shed more light on the settlement history of the region.

Possible signs of cultural disturbance of the natural vegetation are visible in the pollen diagrams as early as the Stone Age (5300–4100 BP). Such signs are the occurrence of, for example, *Urtica, Rumex*

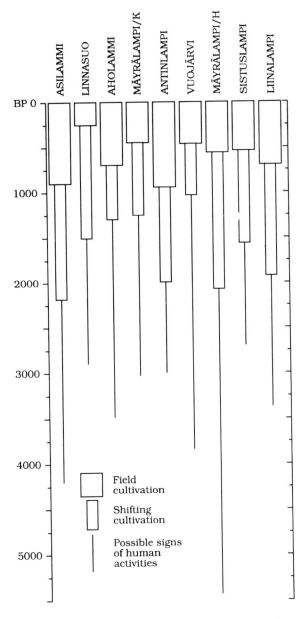

Fig. 9.32 Human impact in the synthesis region SF-d (W)

acetosa, Plantago major, and *Artemisia* and also an increase of *Juniperus.* Signs such as these became more regular between about 2000 and 1200 BP. By about 2000 BP indications of cultivation, including even *Secale* and other Cerealia pollen, became so regular as to indicate shifting (slash-and-burn) cultivation. A marked increase in cultivation indicators occurs between 900 and 400 BP, signifying the beginning of permanent agricultural settlement in the area. A tendency towards later agriculture can be detected from south to north.

Hydrologic, climatic, and soil events

Parts of the region were submerged by the waters of the Baltic basin until between 9000 and 8000 BP, when the Ancylus Lake receded from there. The vast lake system of Päijänne drained northwestwards into the Gulf of Bothnia until it was diverted through the Salpausselkä moraines 6000 BP. This event led to the formation of the present River Kymi, flowing southwards into the Gulf of Finland, exposing vast clay regions in the upper (northwestern) parts of the ancient Päijänne basin.

The present summary (Fig. 9.31), based on lake sediments, does not give any information concerning the history of mires or soils.

TYPE REGIONS SF-d (E) and SF-f, NORTH KARELIA (Heikki Simola, Pirjo Uimonen-Simola and Jukka Vuorinen)

Between about latitude 61°45′–63°55′N and longitude 28°15′–31°35′E this synthesis region covers the easternmost corner of Finland. It corresponds with the administrative province of North Karelia (Fig. 9.33A).

Climate: January mean temperature −10.5°C. July mean 16.7°, mean annual 3.0–1.5°C, annual rainfall 600 mm, snow cover 5 months, length of growing season 140–160 days, effective temperature 950–1150 ddu (sum of day-degrees above +5°C). The most continental area of Finland (Laaksonen 1979).

Geology: In north and east mainly crystalline bedrock (Prekarelidic gneisses and granites), in

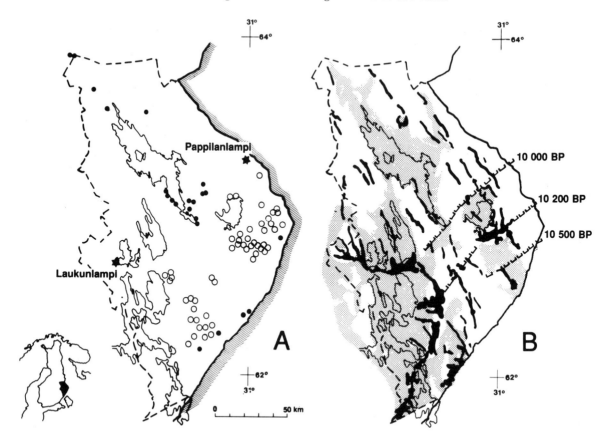

Fig. 9.33 The province of North Karelia in East Finland. A. Large lakes and stratigraphic study sites. Black dots = lakes, open circles = mires. B. Glaciofluvial formations (black), ice-margin positions (with dates; bracket line) and maximal extent of sub-aquatic areas (hatched)

south and west metamorphic sedimentary rocks (Karelidic schists). Bedrock largely covered by till. Three large ice-marginal formations accumulated by the Finnish Lake District and North Karelian ice lobes: Salpausselkä I (10500 BP, Salpausselkä II–Selkäkangas (10200 BP) and Jaamankangas–Uimaharju (10000 BP). Proglacial clays mainly in the largest lake basins (Fig. 9.33B) (Hyvärinen, H. 1966, 1971, 1973; Repo & Tynni 1967).

Altitude: In the south mainly 75–150 m asl, in the north generally 100–250 m, with the highest hills rising to over 300 m.

Topography: Four large and innumerable small lakes (16% of the total area). Greatest relief about 200 m. The northern border of North Karelia coincides with the Maanselkä water-divide (160+ m).

Population: 180000, of which 45% is in the four towns. The capital Joensuu has about 47000 inhabitants. Marginal areas largely unpopulated, rural populations declining.

Vegetation: Coniferous forests, lowlands extensively paludified, up to 15% of land area. The boundary between the south and middle boreal forest zones is the divide between SF-d and SF-f; in this area it roughly coincides with the border between the southern raised bogs and the northern aapa mires (Ahti *et al.* 1968; Eurola *et al.* 1984; Jelina 1985; Tolonen, K. & Ruuhijärvi 1976).

Soils: Predominantly podzols and acid peat.

Land-use history: Sporadic slash-and-burn agriculture since AD 900 (e.g. Grönlund & Asikainen 1992; Grönlund *et al.* 1992; Grönlund 1995); agricultural expansion since 1500 (predominantly confined to eastern part of SF-d); about 6% of land presently cultivated, mainly dairy farming. Recently, extensive manipulation and exploitation of forests and peatlands for timber and fuel peat.

Reference site 42. Lake Pappilanlampi (Fig. 9.34)

An oligotrophic, polyhumic forest lake, maximum depth 3.3 m. The lake is surrounded chiefly by *Pinus* forests of the *Empetrum–Vaccinium*-type; hill-slope forests nearby are of the *Vaccinium myrtillus*-type dominated by *Picea abies*. Pollen stratigraphy of the site (Fig. 9.34), controlled by 20 ^{14}C dates, is described in detail by Vuorinen and Tolonen, K. (1975) and its diatom stratigraphy by Tolonen, K. (1967, p. 343).

The following local pollen-assemblage zones (paz) can be discerned (modified from Vuorinen and Tolonen, K. 1975; uncorrected ^{14}C ages):

pre-8600 BP | NAP zone; non-arboreal pollen, particularly Ericaceae, making up to 30–50% of total pollen

8600–8400 BP | *Betula* zone; *Betula* up to 90%; spores of *Lycopodium* and Polypodiaceae abound

8400–5300 BP | *Pinus–Betula* zone (divided by Vuorinen and Tolonen into a lower *Pinus* and an upper *Pinus–Betula* zone). Zone begins with a sharp rise in *Pinus*; the continuous curves of *Alnus, Corylus,* and *Ulmus* start in the lower half of this zone

5300 BP–present | *Pinus–Picea* zone; beginning with abrupt increase of *Picea*; within the zone the proportion of *Pinus* gradually increases and the species of the mixed oak forest decline

Reference site 32. Lake Laukunlampi

A recently eutrophicated clear-water kettle-hole lake, water depth 27 m, meromictic (Simola *et al.* 1984). Interpretation rather problematic because of

complexities related to the water-level changes of Lake Saimaa, and inconsistencies of the ^{14}C dates (partly controlled by varves). A core 610 cm long was collected from the deepest part of the lake. Six ^{14}C dates (Hel-2345–2350) were obtained for the profile. The lowest of these, 498–508 cm, was dated to 9980±200 (Hel-2345). Below 498 cm the sediment is silt and sand with low pollen concentrations.

Three local pollen-assemblage zones can be discerned:

10000–9500? BP | *Pinus*–Gramineae zone, 500–450 cm; *Pinus* in excess of 50%; Gramineae 10–20%. Obviously the site at that time was mainly surrounded by open water, so the assemblage mainly reflects long-range transport of pollen, hence the high proportion of pine

9500?–5500 BP | *Betula–Pinus* zone, 450–300 cm; *Betula* 40–50%, *Pinus* 30%. *Alnus* appears at the base of this zone, gaining a 10 to 15% proportion

5500 BP–present | *Betula–Pinus–Picea* zone, from 300 cm upwards. Appearance of spruce at the beginning of the zone. Within this zone, the proportion of birch gradually decreases and that of pine increases

Palaeolimnological analyses (sediment characteristics, unpublished diatom and *Cladocera* analyses of H.S. and P.U.S.) have led to the following preliminary interpretation of the lake's development:

498 cm: | Shift from silt to silty gyttja, coinciding with a *Fragilaria* maximum, may indicate isolation of the basin from Yoldia Sea

490–430 cm: | The most eutrophic phase of the lake: *Stephanodiscus* sp. and *Bosmina longirostris* typical for this period that could reflect

LAKE PAPPILANLAMPI

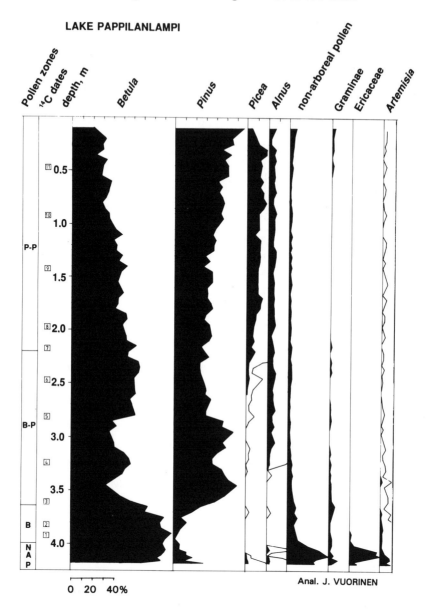

Fig. 9.34   Pollen diagram with selected taxa for Lake Pappilanlampi. For more details, see Vuorinen & Tolonen (1975)

reestablished contact with the then transgressive Lake Saimaa (Saarnisto 1970); this event could include inwash of old organic material, possible explanation for the anomalous ^{14}C date Hel-2347 (462–469 cm; 10200±190 BP).

430–60 cm: Slow development towards oligotrophy; the diatom community characterized by alternation of *Cyclotella comta* and *C. kuetzingiana;* eventual (120–70 cm) scarcity of plankton diatoms and abundance of Chrysophyte scales

NORTH KARELIA, EVENT STRATIGRAPHY

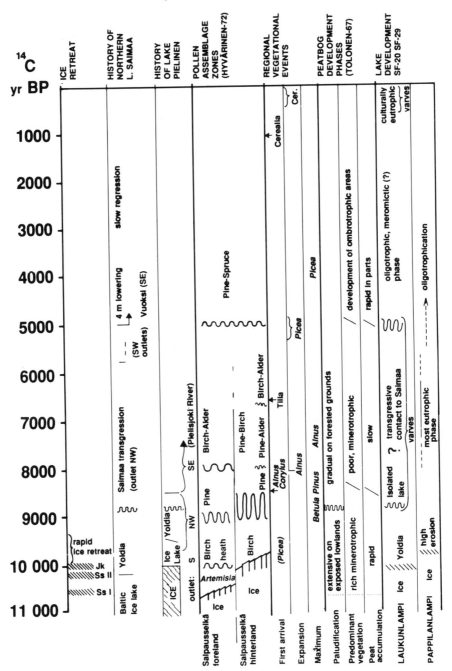

Fig. 9.35 Event stratigraphy for North Karelia (SF-d (E) and SF-f). Compiled from various sources

60–0 cm: Eutrophication of the lake by human
             agency from *ca.* AD 1800 onwards
             (fibre-plant retting; more recently
             local waste waters)

Regional pollen assemblage zones

For compilation of the event stratigraphy (Fig.
9.35), in addition to the sites described above, numerous other published pollen diagrams (e.g.
Tolonen, K. 1963, 1967; Hyvärinen, H. 1972;
Vuorinen 1978) have been taken into account.

The generalized regional pollen-assemblage zonation (paz) for North Karelia is as follows:

Before 10000 BP	*Artemisia*; encountered in the foreland of the Salpausselkä moraines (Hyvärinen, H. 1972)
10000–9000/8500 BP	*Betula*
9000/8500–8000 BP	*Pinus*
8000–5500/5000 BP	*Betula–Pinus–Alnus*
5500/5000 BP–present	*Pinus–Picea*

Palaeoecological patterns and events

Hydrological and pedological changes

The low-lying parts of the area (eastern SF-d) have
been much affected by water-level changes in the
large lakes. The main events are summarized in
the event stratigraphy chart (Fig. 9.35). Accounts of
paludification history (mainly according to
Tolonen, K. 1967) and developmental history of the
study lakes are also included.

Human influence

Human impact on the natural environment has been
generally weak in this area. However, slash-and-
burn agriculture, starting during the Late Iron Age,
peaking in the period AD 1500–1900, had a pro-
found influence on the forest structure in quite large
areas, although not in the vicinity of either of the
reference sites (these are both surrounded by rather
poor forests). Locally, near villages, fibre-plant
retting eutrophicated small lakes, often quite dra-
matically (Grönlund *et al.* 1986). During the 20th

century, especially since the 1950s, both the modern
forestry technology and the urge to exploit peatlands (for forestry and for fuel peat production) have
led to major changes in the natural landscape of
North Karelia.

TYPE REGIONS SF-e and SF-h, NORTH OSTROBOTHNIA AND SOUTH LAPLAND
(Sheila Hicks)

Altitude: 0–280 m. Southern part 0–150 m, northernmost area 100–280 m (Fig. 9.36).

Climate: Mean February temperature –10°C, July
16°C, mean annual 2.5–0.5°C. Precipitation 550 mm
yr^{-1}, 90 wet days yr^{-1}. Continental.

Geology: Precambrian granite, gneiss, phyllite, mica
schist, and quartzite with smaller amounts of meta-
basalt, granite, and gabbro and one area of siltstone
in the south. Bedrock almost entirely covered with
till, sands, and gravel, with some clay in the north.

Topography: Almost flat, particularly near the coast
but rising inland, crossed by a regular series of rivers.
More hilly in the north with an increasing number of
discrete summits. Extensive areas of mire.

Population: 296400 people in an area of 26000
km^2, predominantly along the coast and in the river
valleys.

Vegetation: The region lies within the area covered
by the middle boreal spruce forests (Ozenda *et al.*
1979). Between 45 and 65% of the land is forested.
Pinus covers 15–45% in the north and 45–75% in the
south, *Picea* 25–55% in the north and 5–35 % in the
south, and *Betula ca.* 25% throughout (Salminen
1973). Much of the forest is on the aapa mires, which
cover >60% of the land surface (Fig. 9.37).

Soils: Peat and acid podzols, some with an iron pan.

Land use: Cultivated land occupies <10% of the land
surface and is situated along the river banks or at the
coast. The main crops are barley and hay. Forestry is
important. Some of the peat areas are cut to fuel a
power station.

NORTHERN OSTROBOTHNIA and SOUTHERN LAPLAND

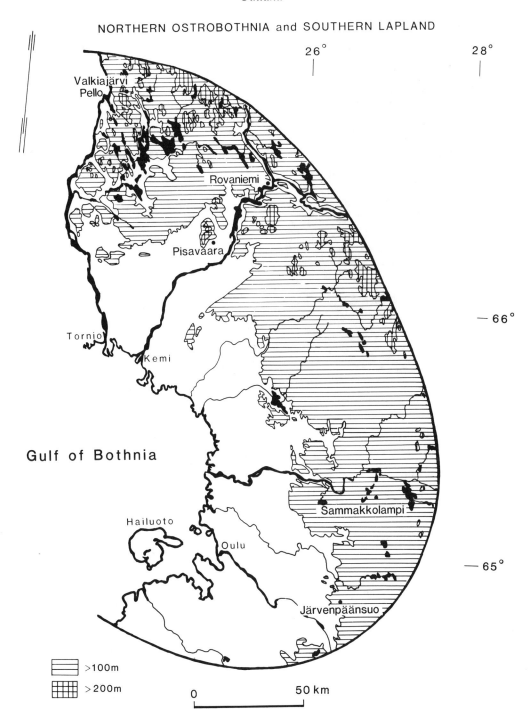

Fig. 9.36 Map of synthesis area Northern Ostrobothnia and Southern Lapland (SF-e and SF-h) showing the degree of altitudinal variation and the location of the four reference sites

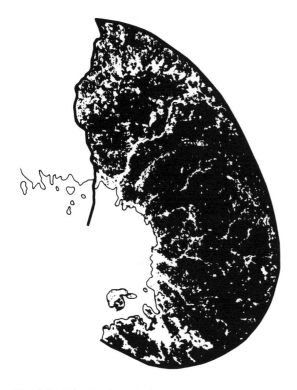

Fig. 9.37 Distribution of mires (*Atlas of Finland*, 1988)

Reference site 49. Valkiajärvi
(9260–2500 BP) (Saarnisto 1981) (Fig. 9.38)

Lake. The pre-isolation sediments are older than 9260 BP. Three local pollen-assemblage zones (paz). (Fig. 9.38)

pre 9260–8100 BP	*Betula*
8100–3500 BP	*Betula–Pinus–Alnus*
3500–2500 BP	*Pinus–Betula–Picea*

Reference site 47. Pisavaara (Kuprujänkä)
(5400–0 BP) (Juola-Helle 1982) (Fig. 9.39)

Mire. Two local pollen-assemblage zones (paz). (Fig. 9.39)

5400–3100 BP	*Betula–Pinus–Alnus*
3100–0 BP	*Pinus–Picea*

Reference site 41. Sammakkolampi
(8500–0 BP) (Haapalahti 1982) (Fig.9. 40)

Small lake. The radiocarbon dates indicate a con-

stant rate of sedimentation and are consistent with each other but seem to be all some 1000 years too old. Two local pollen-assemblage zones (paz). (Fig. 9.40)

8500–3800 BP	*Pinus–Betula*
3800–0 BP	*Pinus–Picea*

Reference site 37. Järvenpäänsuo
(>7500–0 BP) (Holappa 1976) (Fig. 9.41)

Mire. The pre-isolation sediments are older than 7330 BP. Three local pollen-assemblage zones (paz). (Fig. 9.41)

pre-7500 BP	*Betula–Alnus* (NB Holappa himself places the boundary at 6890 BP)
7500–3760 BP	*Pinus–Betula*
3760–0 BP	*Pinus–Picea*

The pollen diagrams

Four reference pollen diagrams combine to cover the whole region, and for each the local pollen-assemblage zones are given. Four regional pollen-assemblage zones are recognized, of which the oldest, the *Betula* paz, is represented at only two sites.

Betula paz.
Betula is overwhelmingly dominant, and NAP values are at their highest.

Pinus–Betula–Alnus paz.
Pinus values rise to equal or exceed those of *Betula*, although at some sites the situation may be reversed in the second half of the zone. *Alnus* is the only other tree pollen type to be continuously represented, and it can achieve relatively high values.

Pinus-Picea paz.
Pinus values remain high or increase, and *Picea* rises to become a characteristic component of the pollen assemblage. *Alnus* values commonly decrease.

The boundary between the *Betula* paz and the *Pinus–Betula–Alnus* paz falls within the time range 8100–7500 BP, the *Betula* paz representing the pioneer *Betula* woodland that followed the retreating ice. However, because the major part of this region

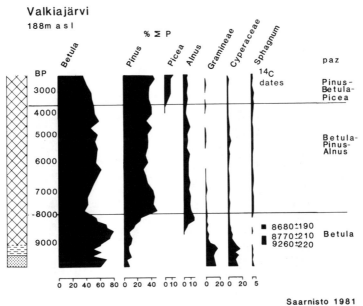

Fig. 9.38 Pollen diagram with selected taxa for lake Valkiajärvi

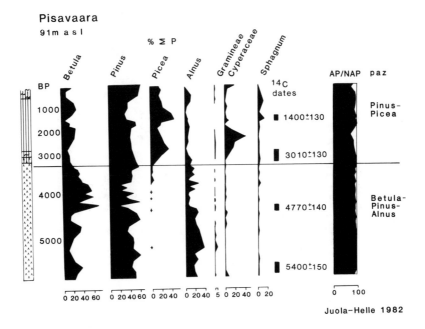

Fig. 9.39 Pollen diagram with selected taxa for mire Pisavaara (Kuprujänkä)

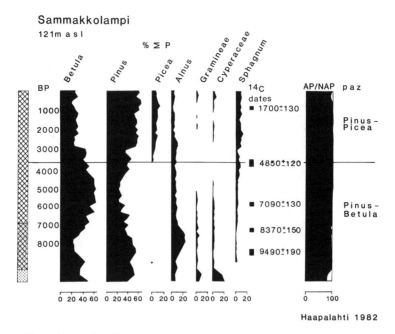

Fig. 9.40 Pollen diagram with selected taxa for lake Sammakkolampi

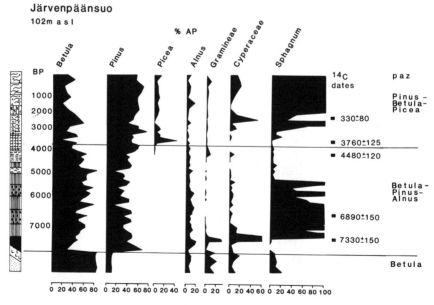

Fig. 9.41 Pollen diagram with selected taxa for mire Järvenpäänsuo. Although this is based on %AP (the raw data are not available for % ΣP calculations), the NAP species are predominantly local mire ones and are not particularly abundant, so that the overall regional picture remains unchanged. The very high values for *Sphagnum* are also a result of this calculation as %AP rather than %AP + *Sphagnum*

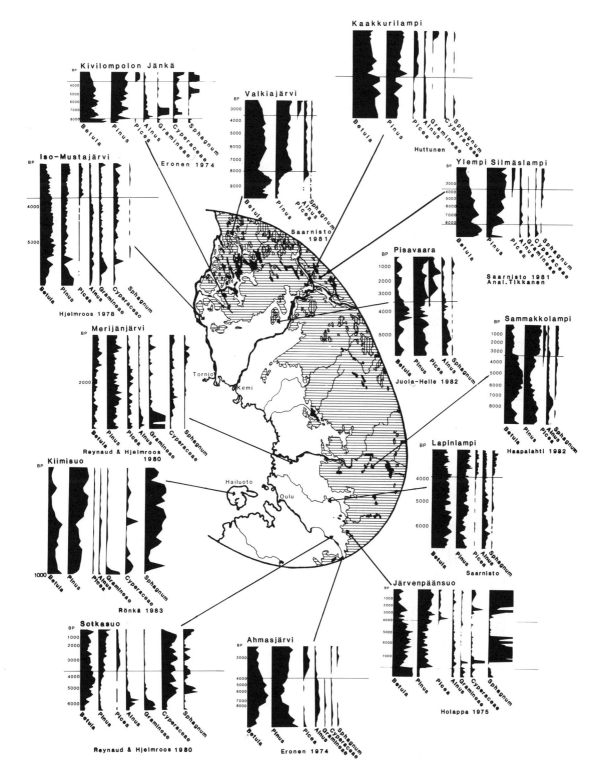

Fig. 9.42 Composite diagram with pollen silhouettes showing vegetation development for different parts of the region and different altitudes

326

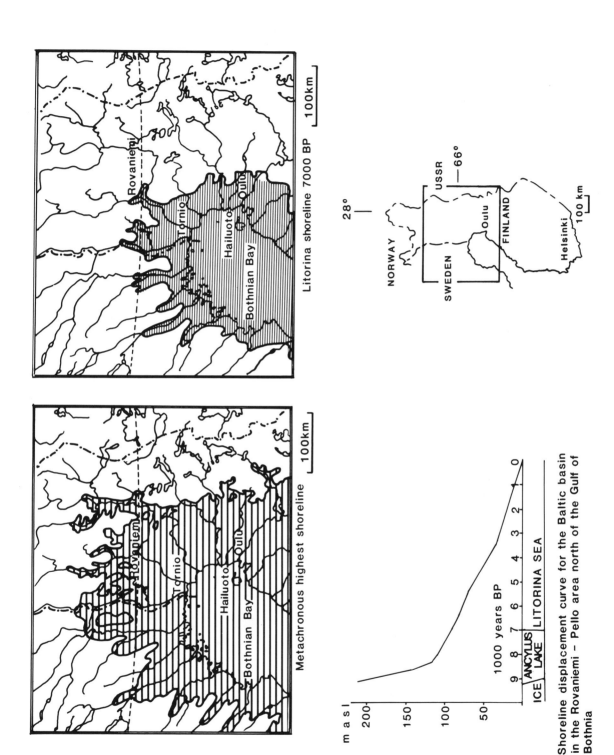

Metacronous highest shoreline

⊢—⊣ 100km

Litorina shoreline 7000 BP

⊢—⊣ 100km

28°

66°

NORWAY

SWEDEN

USSR

FINLAND

Oulu

Helsinki

⊢—⊣ 100km

m a s l

200

150

100

50

0   1   2   3   4   5   6   7   8   9

1000 years BP

ICE | ANCYLUS LAKE | LITORINA SEA

Shoreline displacement curve for the Baltic basin
in the Rovaniemi – Pello area north of the Gulf of
Bothnia

Fig. 9.43 Illustration of the rate of land uplift and the extent of the sub-aquatic area at different points in time (after Saarnisto 1981)

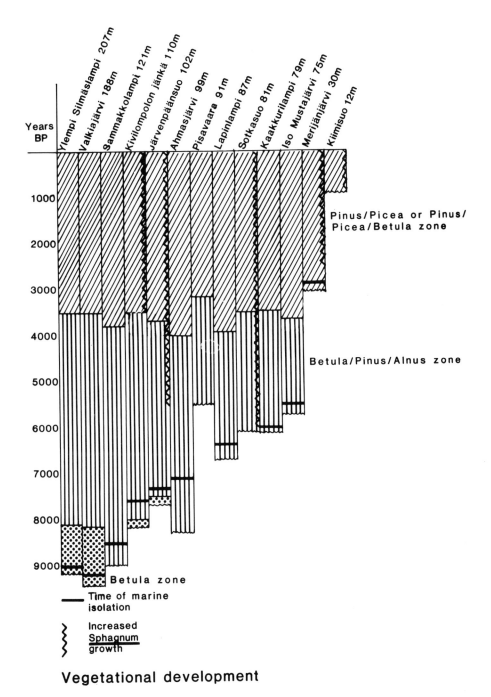

Vegetational development

Fig. 9.44 Comparison of vegetational development for the sites illustrated in Fig. 9.42

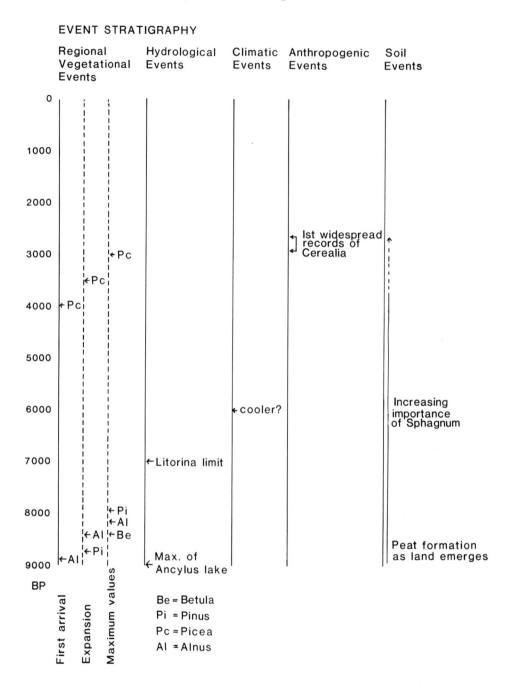

Fig. 9.45   Event stratigraphy for the type regions SF-e and SF-h–Northern Ostrobothnia and Southern Lapland

remained sub-aquatic following deglaciation, this initial *Betula* zone is only recorded at those sites with an elevation of over 100 m, which became supra-aquatic during this time range or shortly after. By the time the remaining sites became isolated, or their sediments began to accumulate, a *Pinus–Betula* woodland as represented by the *Pinus–Betula–Alnus* paz was already established. *Picea* arrived in the region around 4000 BP in the east and south and around 3500 BP in the west and north. In the most westerly sites, which remained sub-aquatic the longest, sediment did not start to accumulate until after *Picea* was well established in the forest.

This pattern of regional pollen-assemblage zones is illustrated in Figure 9.42, in which the four reference sites are supplemented with diagrams from other sites at differing elevations. Figure 9.43 shows the way in which the rate of emergence of the land from the Baltic has influenced the vegetational history of the individual sites. It is likely that the abundance of *Alnus* is related more to the availability of damp conditions around the sampling sites than to the general forest successions. The vegetational development with respect to time and site elevation is summarized in Figure 9.44.

Palaeoecological events

Anthropogenic events

Archaeological evidence confirms that man has occupied the different shorelines of the Baltic and the various river valleys from earliest times, for the Kemi River valley in particular and to a lesser degree also the Tornio River valley are rich in finds of artifacts. However, the pollen evidence for human presence is consistently slight throughout. Even the present-day agricultural situation is recorded by only very low values of the relevant indicator pollen types. This is primarily due to the high pollen production of the boreal forest, which still remains the dominant vegetation type, and the filtering effect of the trees themselves. What evidence is available suggests that agriculture was first widely practised after 3000 BP, there being Cerealia pollen evidence from a number of sites between 3000 and 2500 BP. Earlier and more sporadic evidence exists, but its exact dating is still open to debate.

Hydrological, climatic, and soil events (Fig. 9.45)

The hydrological events of the region are intimately related to the history of the Baltic, the best documented point being the limit of the Litorina Sea at around 7000 BP (Fig. 9.43). The continuing and rapid land uplift has clearly exerted a strong influence, with the time of isolation of the various lakes becoming younger and younger as one approaches the coast (Fig. 9.44). Similarly much of the relatively flat land thus revealed has formed a suitable substrate for mire development. Climatic conditions have always favoured peat growth, but *Sphagnum* plays an increasing part in mire development from 6000–5000 BP onwards (Fig. 9.44). The extreme of this general development trend is seen on the island of Hailuoto, which itself has existed for less than 2000 years and where peat development on the low-lying silty areas dates from 1000 BP at the earliest.

TYPE REGION SF-i, KUUSAMO (Yrjö Vasari)

Between about latitude 65°30′–66°30′N and longitude 27°–30°E, this synthesis region covers the northern part of the original type region SF-g N Kainuu–Kuusamo. The material presented originates from two communes, Kuusamo and Posio (Fig. 9.46). The major part of the area lies on the eastern side of the main water divide (Maanselkä) between the Baltic Sea and the White Sea.

Climate: The climate is thermally continental, with annual amplitude of mean temperatures from −12°C (January) to 11°C (July), mean annual ±0°C, but hydrologically oceanic, with annual precipitation of 520–560 mm (cf. also Koutaniemi 1983). The length of the vegetative period (5° to 5°) is 130 days.

Geology: The Precambrian bedrock consists of acid granite and granite–gneiss traversed by a belt of generally basic Karelidic quartzites, schists, and other metamorphic rocks (Fig. 9.1). The dominant Quaternary deposits are till (commonly in the form of drumlins) and peat (30–40% of the area).

Topography and altitude: Rugged hilly country with relief of 50–100 (to 200 m). The general altitude of the region is 200–300 m above sea level (Fig. 9.2). The highest hills rise to almost 500 m. As a watershed area, the region is very rich in lakes and rivers.

Population: 2–3 people km^{-2}. The main village of Kuusamo has about 10000 inhabitants.

Vegetation: The region belongs to the zone of northern boreal coniferous forest (Fig. 9.4). The dominant trees are *Pinus sylvestris* (50–60%), *Picea abies* (25–40%), *Betula pendula*, and *B. pubescens*. The highest hill tops above *ca.* 430 m are non-forested, although always with some stunted trees. The region is well known for the richness of arctic–alpine and eastern elements in the flora. Mires form an essential element in the vegetation, covering about 30–40% of the area (Fig. 9.5). Mires on hill slopes are characteristic for the region.

Soils: Dominantly podzols and minerogenic peats.

Land use: The area of cultivated land is only about 2% of the total area (mainly hay and barley). The economy of the region is based on forestry, animal husbandry (cattle and reindeer), and tourism.

Vegetational history and reference sites: The vegetational history of the region has been studied intensively through pollen analyses, both with and without radiocarbon dates. The pollen-analytical results of eight stratigraphic sites have been used for the regional synthesis (Fig. 9.46: Rytisuo, Heino 1987; Purkuputaansuo, L. Miettinen 1985; Kolmiloukkonen & Keski-Pohjassuo, A. Huttunen 1987; Kangerjoki, Hicks 1975, 1985; Liippasuo, Seppälä & Koutaniemi 1985; Perä-Puikkonen, Heikkinen & Kurimo 1977; Maanselänsuo, Vasari 1965a). One diagram is given here, viz. Kolmiloukkonen (lake) (Fig. 9.47).

Reference site 51. Kolmiloukkonen
(Huttunen, A. 1987)

Lake. Three local pollen-assemblage zones (paz) with five local subzones (Fig. 9.47).

KLL-1	9310–8400 BP	*Betula*
KLL-2	8400–5000 BP	*Pinus–Betula*
2a	8400–7200 BP	*Pinus–Alnus* subzone
-2b	7200–5000 BP	*Betula–Alnus* subzone

KLL-3	5000–0 BP	*Pinus–Picea*
-3a	5000–3800 BP	*Pinus–Betula–Picea* subzone
-3b	3800–1200 BP	*Picea–Pinus* subzone
-3c	1200–0 BP	*Picea–Betula–Pinus* subzone

The material is essentially augmented by numerous other pollen sequences and a considerable number of macrofossil analyses, which have been particularly helpful in following the developmental history of the lakes (e.g. Vasari 1962, 1967, Vasari *et al.* 1963; Hicks 1974, 1975, 1985; Koutaniemi 1979; Huttunen, A. & Huttunen, R. 1989; Hyvärinen, H. *et al.* 1990).

Palaeoecological patterns and events

Common patterns (Figs 9.48 and 9.49)

The region became ice-free around 9500 BP during a very rapid east–west retreat of the ice margin, presumably in the form of a wide dead-ice zone. The first periglacial vegetation was of a steppe–tundra character. Some tree *Betula* and *Pinus* were also present from the very beginning.

This short-lived pioneer vegetation was followed by birch forests, similarly of rather short duration. About 8500 BP *Pinus* became a dominant component in the forests. Mixed *Pinus–Betula* forests (with *Alnus incana* since *ca.* 8500 BP) dominated the region until about 5000–4000 BP, when *Picea abies* became a major component in the forests.

The immigration of *Picea* occurred in the form of local advances, and by about 5000 BP *Picea* was common almost everywhere.

Peat formation had begun already by 9000 BP and became very active around 5000 BP, when the mires on hill slopes began to form on a large scale (Vasari 1965b; Huttunen, A. 1987).

The climatic improvement was a very rapid phenomenon, and the period of climatic optimum with high humidity and high temperature is best placed between 7000 and 5000 BP. At about 5000 BP a very marked climatic deterioration began. The "Little Ice Age" (about AD 1300–1870) was a marked cold and moist period in the region and led to high

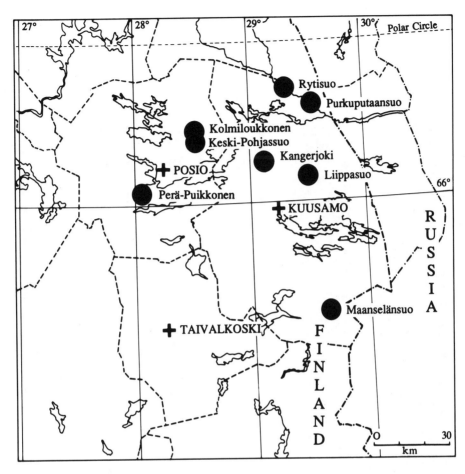

Fig. 9.46 Synthesis region SF-i–Kuusamo and the stratigraphic sites

groundwater levels and floods in the Oulanka River (Koutaniemi 1979; Koutaniemi & Ronkainen 1983; Koutaniemi & Sillanpää 1985).

The lake development shows a very clear pattern: initially, the lakes had a high electrolyte concentration. A period of rich aquatic vegetation is roughly 8500–7000 BP. After that a general process of acidification began.

The southernmost site, Maanselänsuo, shows a pattern rather different from the rest of the sites (Vasari 1965a). *Pinus* evidently became common there earlier than usual elsewhere, and the immigration of *Picea* took place very early (almost 7000 BP).

Human influence

A major cultural change in the region took place about 300 years ago, when the previously dominant hunter–fisher–reindeer culture of the Lapps was replaced by permanent Finnish agricultural settlement. The first agriculture involved slash-and-burn cultivation until late in the 19th century. This form of cultivation affected the forests very profoundly; practically all the better forest sites were burnt at least once. Mires were commonly used for hay-making until the 1950s. The human influence left only negligible signs in the pollen diagrams (Hicks 1985; Vasari & Väänänen 1986).

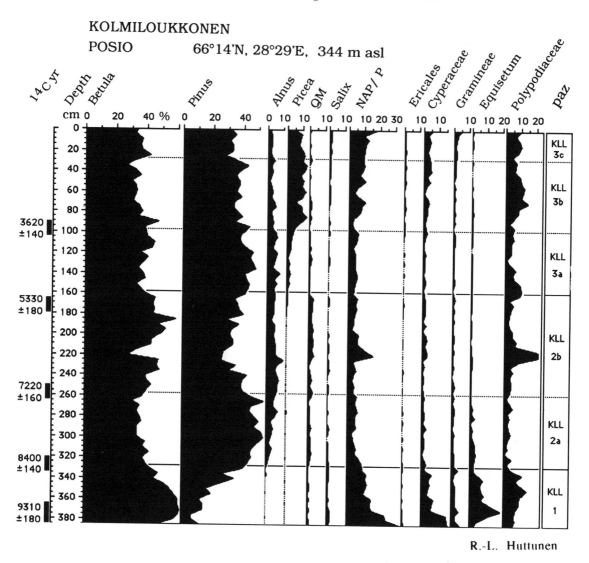

Fig. 9.47   Pollen diagram with selected taxa for lake Kolmiloukkonen (Huttunen 1987)

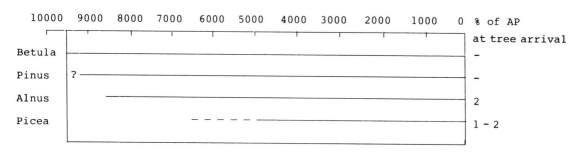

Fig. 9.48   History of forest trees in the synthesis region SF-i–Kuusamo

EVENT STRATIGRAPHY:

Type Region SF-i: Kuusamo.

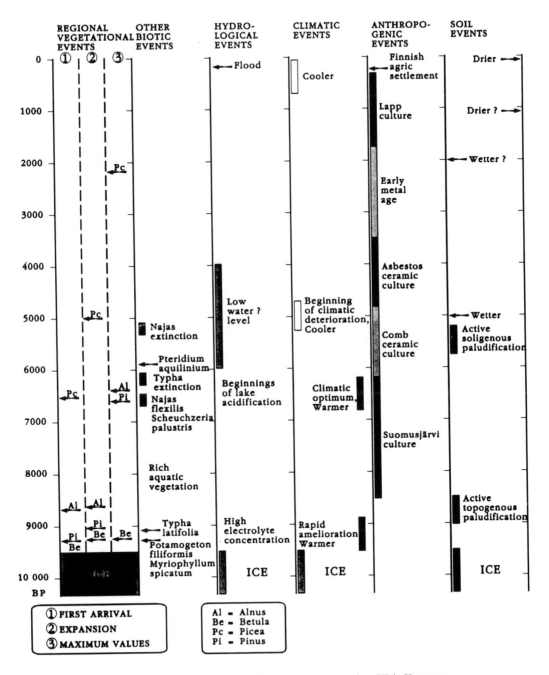

Fig. 9.49   Event stratigraphy for the synthesis region SF-i–Kuusamo

334

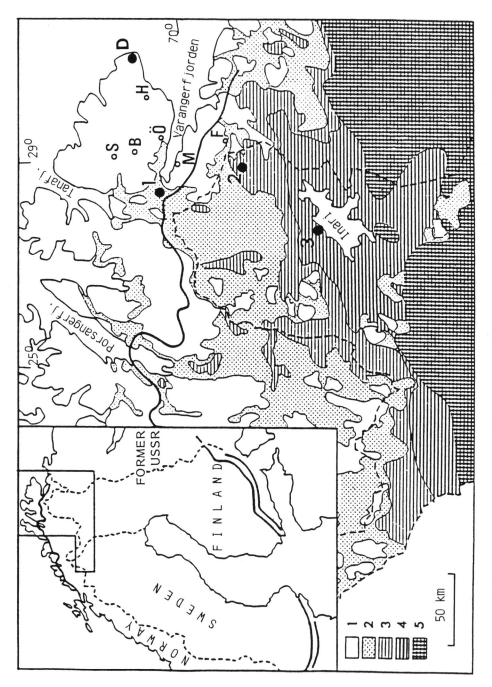

Fig. 9.50 Map of northern Fennoscandia showing the principal vegetational regions: (1) treeless, (2) mountain birch (*Betula pubescens* ssp. *tortuosa*) scrub and woodland, (3) birch (mainly mountain birch) and pine woods, (4) pine forest, (5) pine and spruce forest. Reference sites for Northeast Fennoscandia: D = Domsvatnet, 1 = Bruvatnet, 2 = Suovalampi, 3 = Akuvaara. Additional sites: S = Stjernevatnet, B = Bergebyvatnet, Ö = Östervatnet, M = Mordvatnet, F = Faerdesmyra. The heavy lines mark the Younger Dryas end moraines

TYPE REGIONS SF-k, SF-l, N-z and N-ae, NORTHEAST FENNOSCANDIA
(Hannu Hyvärinen)

Extending from northern Finnish Lapland to the Varanger Peninsula, this region shows marked climatic and vegetational gradients from the relatively continental southern interior to the more oceanic coast in the north.

Climate: Mean January temperature between −13°C (south interior) and −6°C (coast), mean July between 13°C and 9°C, mean annual −1 to 0°C; precipitation 400 mm yr^{-1} (outer coast up to 600 mm).

Geology: Precambrian granite–gneiss and granulite prevail in the south. Near Varangerfjorden they dip under Eocambrian sediments (sandstones, tillites) of the Sparagmite Group. There is a more or less continuous cover of till, except for rocky summits and steep slopes; locally extensive fluvioglacial and fluvial sands and gravels are found; Younger Dryas (Main Sub-Stage) moraines cross the area near Varangerfjorden.

Topography and altitude: Hilly and hummocky moraine lowland (under 200 m) around Lake Inari; north of Inari basin rounded hilly landscape (200–400 m); around Varangerfjorden a hilly plateau (300–600 m) descends, often with steep slopes, to a narrow coastal lowland or beach.

Population: 2–3 people km^{-2}, dominantly coastal; very sparse in the interior (<2 km^{-2}), mainly small lake- and river-side settlements.

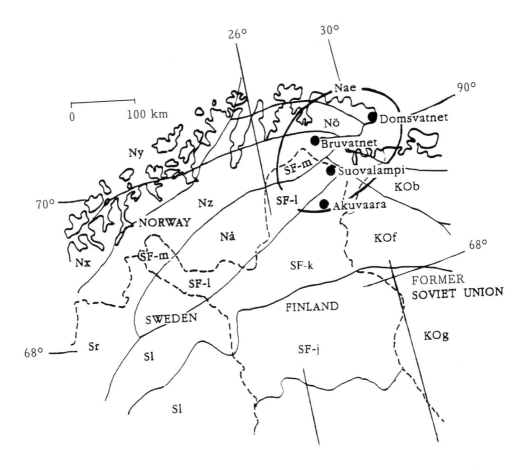

Fig. 9.51 Synthesis region Northeast Fennoscandia, with type regions and reference sites

Vegetation: The southern part is occupied by pine forests of open lichen–woodland type; mountain birch woodland and scrub prevails north of (and above) the pine forest limit, north of the Inari basin. Hill tops and outer coasts are treeless; the altitude of treeline (birch) ranges from sea level to 300 m in the southern interior.

Soils: Dominantly podzols, acid peats, and heathlands.

Land use: Reindeer grazing, some sheep grazing on the coast, limited forestry in the south.

The synthesis region of northeast Fennoscandia includes parts of six type regions, four of which are represented by reference sites. The sites belong to different vegetation zones, from pine forest to treeless landscapes in the north, and together they form a transect across the present limits of pine and birch (Fig. 9.50). The sequence of ice retreat was from north-northeast to south-southwest, so that the two northern sites (north of the Tromsö–Lyngen moraine) were deglaciated in late-glacial times and the two southern sites during the early Holocene.

Reference site 59. Akuvaara (Figs 9.51 and 9.52; Hyvärinen, H. 1975)

Situated within SF-k. Age range 9500–0 BP (lowest ^{14}C age: 8840 BP). Lake 400 by 100 m. Three local pollen assemblage-zones (paz) with two subzones.

Aku 1	–7500 BP	*Betula*
1a	–9000 BP	Ericales subzone
1b	9000–7500 BP	*Lycopodium* subzone
Aku 2	7500–3200 BP	*Pinus–Betula–Alnus*
Aku 3	3200–0 BP	*Pinus–Betula–Picea*

Reference site 62. Suovalampi (Figs 9.54 and 9.55; Hyvärinen, H. 1975)

Situated within SF-l. Age range 9500–0 BP. Lake 800 by 200 m. Three local pollen-assemblage zones (paz) with two subzones.

Suo 1	–7500 BP	*Betula*
1a	–9000 BP	Ericales subzone
1b	9000–7500 BP	*Lycopodium* subzone

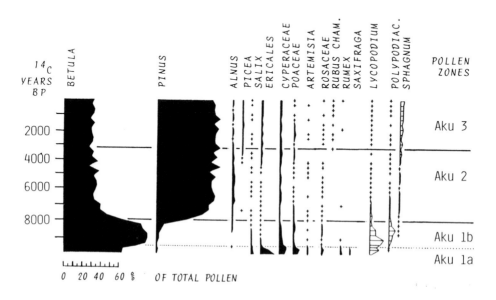

Fig. 9.52 Pollen diagram with selected taxa for lake Akuvaara

EVENT STRATIGRAPHY:

Type Region SF-k: Ref. site Akuvaara.

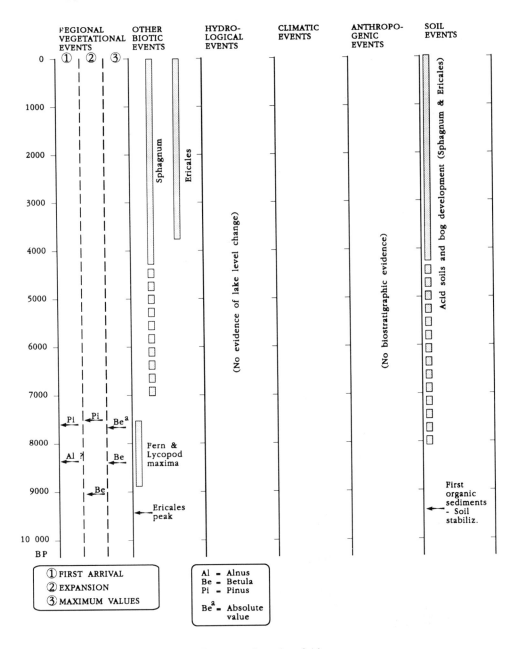

Fig. 9.53 Event stratigraphy of Akuvaara

Suo 2   7500–3200 BP   *Pinus–Betula–*
                          Alnus
Suo 3   3200–0 BP   *Pinus–Betula–*
                          Picea

Reference site N16. Bruvatnet (Figs 9.56 and 9.57; Hyvärinen, H. 1975)

Latitude 70°11′N, Longitude 28°25′E. Elevation 119m. This site within Norway, type region N-z. Age range 10500(10280)–0 BP. Lake 1000 by 600 m. Four local pollen-assemblage zones (paz) with two subzones.

Br 1   –10200 BP   *Artemisia*
Br 2   10200–8500 BP   *Betula*
   2a   10200–9500 BP   Ericales subzone
   2b   9500–8500 BP   *Lycopodium* subzone
Br 3   8500–3200 BP   *Pinus–Betula–Alnus*
Br 4   3200–0 BP   *Betula–Pinus–Picea*

Reference site N17. Domsvatnet (Figs 9.58 and 9.59; Hyvärinen, H. 1976)

Latitude 70°19′N, Longitude 32°02′E. Elevation

120 m. This site within Norway, type region N-ae. Lake 600 by 150 m. Age range 9500(8570)–0 BP. Three local pollen-assemblage zones (paz) with two subzones.

D 1   –8000 BP   *Betula*
   1a   –9200 BP   Herb–Graminid–
                          Ericales subzone
   1b   9200–8000 BP   *Lycopodium*
                          subzone
D 2   8000–3500 BP   *Betula–Pinus–*
                          Alnus
D 3   3500–0 BP   *Betula–Pinus–*
                          Picea

Additional sites: Östervatnet (Varanger Peninsula – Prentice 1981); Bergebyvatnet, Holmfjellvatnet, Stjernevatnet (Varanger Peninsula–Prentice 1982); Mordvatnet (South Varanger–Donner *et al.* 1977); Faerdesmyra (South Varanger–Vorren 1972).

Common and differential patterns between sites

Pollen zones and major patterns in the vegetational development of Northeast Fennoscandia are sum-

SUOVALAMPI

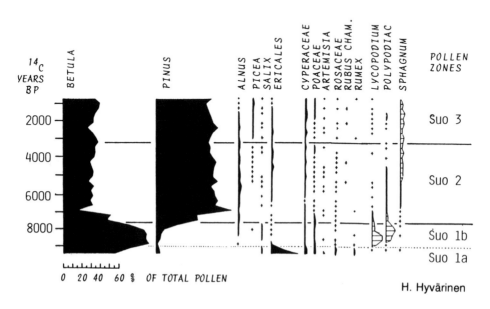

Fig. 9.54   Pollen diagram with selected taxa for lake Suovalampi

EVENT STRATIGRAPHY:

Type Region SF-l: Ref. site Suovalampi.

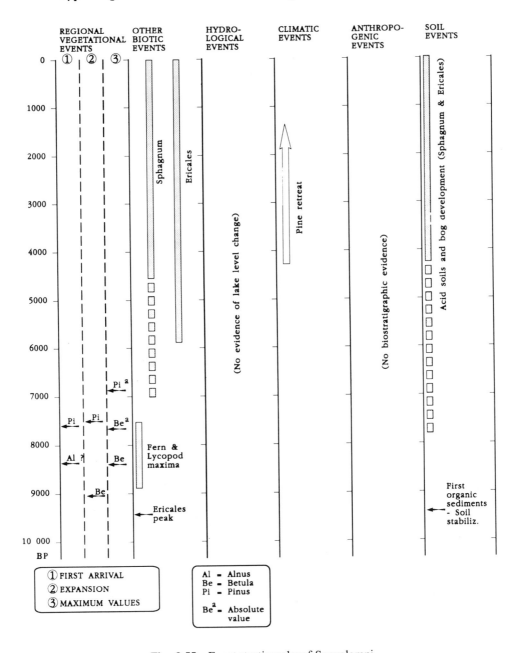

Fig. 9.55 Event stratigraphy of Suovalampi

marized in the correlation table (Fig. 9.60).

Common patterns
Retreat of tree-lines is documented since 5000–4500 BP. Continuous presence of *Picea* pollen (long-distance transport) since 3200 BP.

Differential patterns
Early Holocene spread of *Betula* is delayed inland, following deglaciation. Likewise, the spread of *Pinus* is time-transgressive, being earlier at inner Varangerfjorden (Bruvatnet, Mordvatnet) than farther inland and west; this suggests a relatively early migration north of *Pinus* via an eastern route. Domsvatnet (and probably all Varanger Peninsula) is outside the former range of *Pinus*.

Human impact
During historic times, selective use of pine by man, along with sheep grazing on the coast, must have contributed to the decline of pine and birch at least locally. However, the present pollen diagrams only record regional (climatic) trends of tree-line retreat

starting nearly 5000 years ago, and any anthropogenic changes of later date are totally masked by these main trends.

CONCLUSIONS

Figure 9.61 summarizes in a simplified manner the general vegetational history of Finland on the basis of the regional syntheses.

- NAP-rich pioneer vegetation is met with only on the southern and eastern side of the Salpausselkä terminal moraines.
- *Betula* dominated the first forests everywhere in supra-aquatic areas, with the exception of the north west part of the Finnish Lake Plateau (SF-d W), where *Pinus* was an important forest tree from the very beginning, together with *Betula*.
- In southern Finland, in the hemiboreal and southern boreal zones, a short-lived *Betula–Pinus* phase is found between the *Betula* phase and the mesocratic mixed forests.

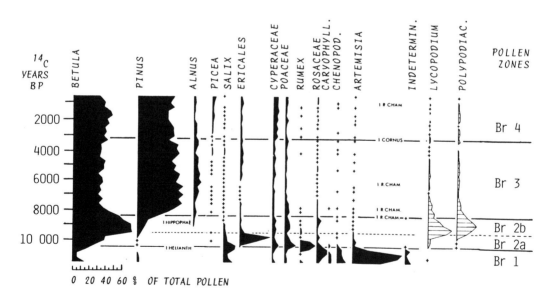

H. Hyvärinen

Fig. 9.56 Pollen diagram with selected taxa for lake Bruvatnet

EVENT STRATIGRAPHY:

Type Region Nz: Ref. site Bruvatnet.

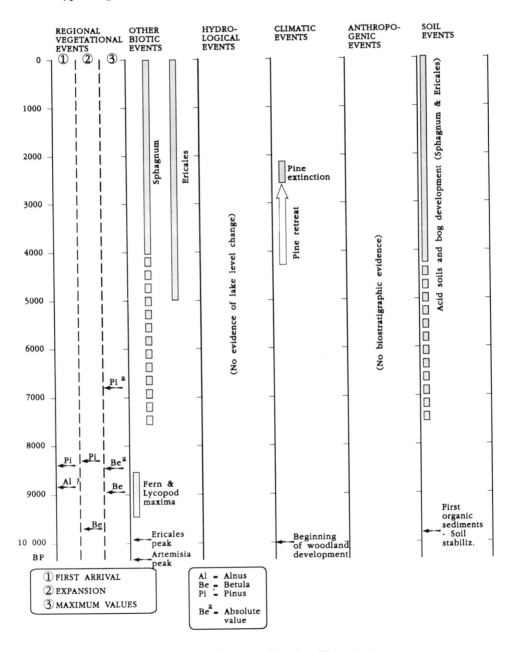

Fig. 9.57 Event stratigraphy of Bruvatnet

EVENT STRATIGRAPHY:

Type Region Nae: Ref. site Domsvatnet.

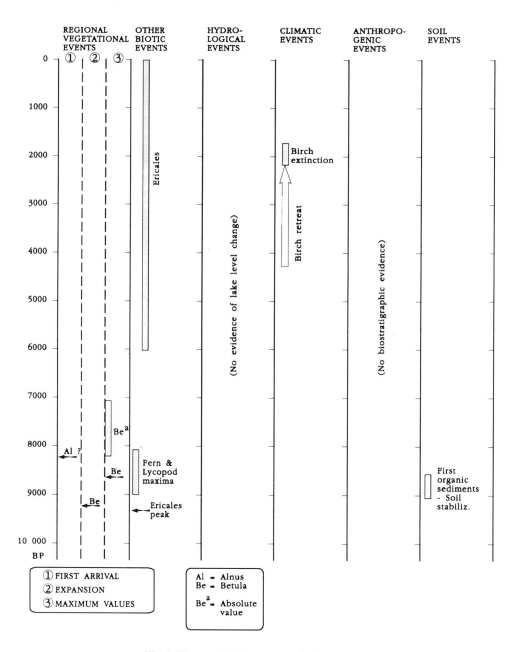

Fig. 9.58 Event stratigraphy of Domsvatnet

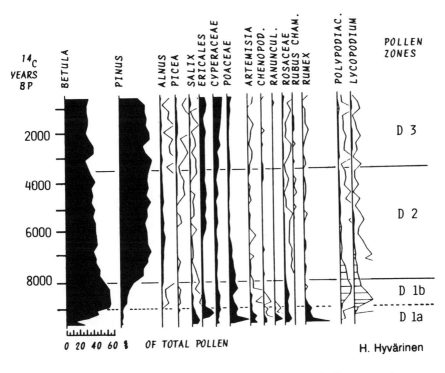

DOMSVATNET

0 20 40 60 % OF TOTAL POLLEN

H. Hyvärinen

Fig. 9.59 Pollen diagram with selected taxa for lake Domsvatnet

– The end of the *Betula* zone is dated in SE Finland to 9000–8500 BP. Northwards, around the present boundary between coniferous forests and subalpine birch forests in east Lapland (= SF-k/l), it is 7500 BP. Such a pattern has been described by Donner (1971) and Hyvärinen, H. (1972).

The forest vegetation of the mesocratic stage is remarkably uniform, showing only gradual differences in a mainly S–N direction. Beginning about 8000–7500 BP mixed coniferous–deciduous forests dominate. This phase, however, ends at greatly different times in different parts of the country.

– The SW archipelago (SF-a) and the coastal oak region (SF-b), together forming the present hemiboreal zone, clearly deviated from the rest of the country as far as vegetational development was concerned. There the role of the broad-leaved deciduous trees (mainly *Corylus, Ulmus, Quercus,*

and *Tilia*) was quite considerable, and such a situation also lasted later than elsewhere, until about 2000 BP (SW Archipelago)–3000 BP (mainland).

– The Ostrobothnian coastland also saw a continuation of QM richness (*Corylus, Quercus,* and *Tilia*) until the last millennium.

– In the Finnish Lake Plateau the broad-leaved deciduous trees (*Corylus, Ulmus, Fraxinus, Tilia*) were quite rare in the forests, even during the Hypsithermal. The spread of *Picea* ended the period of forests rich in deciduous trees there somewhat prior to 5000 BP.

– In northern Ostrobothnia–Kuusamo (SF-h and SF-i) broad-leaved deciduous trees were not present in the mesocratic forests, but their role was taken by *Alnus*. The phase of mixed coniferous–deciduous forests ends around 5000–4000 BP, in the east somewhat earlier than in the west.

344

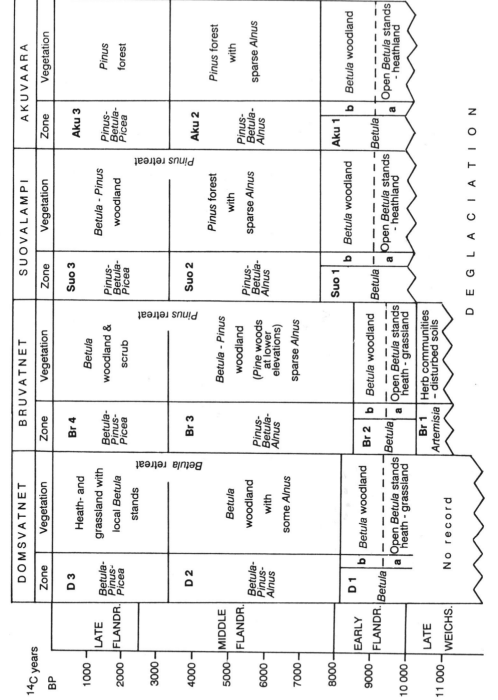

Fig. 9.60 Correlation table for pollen zones and vegetational events in Northeast Fennoscandia

Fig. 9.61 Time correlation of regional pollen-assemblage zones in Finland. The breadth of the columns is roughly proportional to the width of the respective zones on the land. Be = *Betula*, Pi = *Pinus*. Vertical scale in 10^3 yr BP, emphasized by shaded patterns

– In Lapland the phase of mixed forests continues up to 3200 BP.

The spread of *Picea* marked the beginning of the last phase of the natural history of forests.

– In the hemiboreal zone the onset of this new phase was delayed from east to west (3000–2000 BP) (cf. also Tolonen, K. & Ruuhijärvi 1976).
– In northern Lapland (SF-l and SF-m) *Picea* never became really important in the forests.
– The earliest appearance of the signs of human culture are met with in SW Finland between 5000 and 4000 BP. First Cerealia pollen there occurs before 2000 BP. Cerealia finds are clearly younger in the coastal oak zone, somewhat more than 1000 years old.
– In coastal Ostrobothnia first clearances can be dated to 3000 BP, and cultivation there began about 1500 BP. In northern Ostrobothnia the first clear signs of agriculture similarly date from 3000 BP. In the Lake Plateau the first weak signs of human influence generally appear slightly prior to 3000 BP, and they become more common only after 1500 BP. The slash-and-burn cultivation that started there in the Late Iron Age peaked in AD 1500–1900 (see also Simola *et al.* 1988; Grönlund 1995).
– Farther east and north the signs of human cultivation are negligible.

The present zonal division of Finland is visible throughout the period under consideration. A clear south–north (and, in some respects, east–west) gradient is present throughout. Human influence has been present in the southern, central, and northwestern parts of the country ever since the Neolithic. The real agricultural conquest of the main part of the country took place during the last two millennia only.

REFERENCES

Aario, R. 1965: Development of ancient Lake Päijänne and the history of the surrounding Forests. *Annales Academiae Scientiarum Fennicae, Series A III. Geologica–Geographica 81*, 1–191.

Aartolahti, T. 1966: Über die Einwanderung der Fichte in Finnland. *Annales Botanici Fennici 3*, 368–379.

Ahti, T., Hämet-Ahti, L. & Jalas, J. 1968: Vegetation zones and their sections in northwestern Europe. *Annales Botanici Fennici 5*, 169–211.

Atlas of Finland 1960. The Geographical Society of Finland and Department of Geography, Helsinki University, Helsinki.

Atlas of Finland 1988. Finnish Geographical Society, Helsinki.

Aurola, E. 1938: Die postglaziale Entwicklung des südwestlichen Finnlands. *Bulletin de la Commission géologique de Finlande 121*, 166 pp.

Battarbee, R., Cronberg, G. & Lowry, S. 1980: Observations on the occurrence of scales and bristles of *Mallomonas* spp. (Chrysophyceae) in the micro-laminated sediments of a small lake in Finnish North Karelia. *Hydrobiologia 1971*, 225–232.

Bondestam, K., Lemdahl, G., Eskonen, K., Vasari, A. & Vasari, Y. 1994: Younger Dryas and Preboreal in Salpausselkä foreland, Finnish Karelia. *Dissertationes Botanicae 234*, 161–206.

Donner, J.J. 1971: Towards a stratigraphical division of the Finnish Quaternary. *Societas scientiarum Fennica, Commentationes Physico-Mathematicae 41*, 281–305.

Donner, J.J. 1972: Pollen frequencies in the Flandrian sediments of Lake Vakojärvi, South-Finland. *Commentationes Biologicae, Societas Scientiarum Fennica 53*, 19 pp.

Donner, J., Eronen, M. & Jungner, H. 1977: The dating of Holocene relative sea-level changes in Finnmark, North Norway. *Norsk Geografisk Tidsskrift 31*, 103–128.

Donner, J.J., Alhonen, P., Eronen, M., Jungner, H. & Vuorela, I. 1978: Biostratigraphy and radiocarbon dating of the Holocene lake sediments of Työtjärvi and the peats in the adjoining bog Varrassuo west of Lahti in southern Finland. *Annales Botanici Fennici 15*, 258–280.

Dreijer, M. 1941: Ålands äldsta bebyggelse. *Finskt Museum 47/1940*, 66 pp.

Eronen, M. 1974: The history of the Litorina sea and associated Holocene events. *Commentationes Physico-Mathematicae, Societas Scientiarum Fennica 44*, 195 pp.

Eronen, M. & Hyvärinen, H. 1982: Subfossil pine dates and pollen diagrams from northern Fennoscandia. *Geologiska Föreningens i Stockholm Förhandlingar 103*, 437–445.

Eurola, S., Hicks, S. & Kaakinen, E. 1984: Key to Finnish mire types. *In* Moore, P.D. (ed.) *European mires*, 11–117. Academic Press, London.

Fries, M. 1961: Pollenanalytiskt bidrag till vegetations- och odlingshistoria på Åland. *Finskt Museum 68*, 5–20.

Fries, M. 1963: Pollenanalyser från Åland. *Åländsk Odling 24*, 102–125.

Glückert, G. 1970: Vorzeitliche Uferentwicklung am Ersten Salpausselkä in Lohja, Südfinnland. *Annales Universitatis Turkuensis, Series A. II. Biologica–Geographica–Geologica 45*, 116 pp.

Glückert, G. 1976: Post-Glacial shore-level displacement of the Baltic in SW Finland. *Annales Academiae Scientiarum Fennicae. Series A. III. Geologica–Geographica 118*, 92 pp.

Glückert, G. 1977: Itämeren rannansiirtymisestä Turussa ja sen lähiympäristössä. Abstract: Post-Glacial shore-level displacement of the Baltic at Turku, SW Finland. *Publications of the Department of Quaternary Geology, University of Turku 21*, 36 pp.

Glückert, G. 1978: Östersjöns postglaciala strandförskjutning och skogens historia på Åland. *Publications of the Department of Quaternary Geology, University of Turku 24*, 106 pp.

Glückert, G. 1979: Itämeren ja metsien historia Salpausselkävyöhykkeessä Uudenmaan länsiosassa. Summary: Shore-level displacement of the Baltic and the history of vegetation in the Salpausselkä belt, western Uusimaa, South Finland. *Publications of the Department of Quaternary Geology, University of Turku 39*, 77 pp.

Glückert, G. 1989: Itämeren rannansiirtyminen, metsien, asutuksen ja viljelyn historia Ahvenanmaan Kumlingen saarella. Abstract: Shore-level displacement of the Baltic, and the development of forests, settlement and agriculture on the island of Kumlinge, the Åland Islands, SW Finland. *Publications of the Department of Quaternary Geology, University of Turku 61*, 10 pp.

Glückert, G., Illmer, K., Kankainen, T., Rantala, P. & Räsänen, M. 1992: Littoistenjärven ympäristön kasvillisuuden kehitys jääkauden jälkeen ja järven luonnollinen happamoituminen. Abstract: Vegetational history in the surroundings of Lake Littoinen and its natural dystrofication. *Publications of the Department of Quaternary Geology, University of Turku 75*, 27 pp.

Grönlund, E. 1995: A palaeoecological study of land-use history in East Finland. University of Joensuu. *Publications in Sciences 31*, 44 pp. + App.

Grönlund, E. & Asikainen, E. 1992: Reflections of slash-and-burn cultivation cycles in a varved sediment of Lake Pitkälampi (North Karelia, Finland). *In* Proceedings of Vth Nordic conference on the application of scientific methods in archaeology. *Laborativ Arkeologi 6*, 43–48.

Grönlund, E., Simola, H. & Huttunen, P. 1986: Paleolimnological reflections of fiber-plant retting in the sediment of a small clear-water lake. *Hydrobiologia 143*, 423–431.

Grönlund, E., Kivinen, L. & Simola, H. 1992: Pollen-analytical evidence for Bronze-age cultivation in Eastern Finland. *In* Proceedings of Vth Nordic conference on the application of scientific methods in archaeology. *Laborativ Arkeologi 6*, 37–42.

Haapalahti, R. 1982: *Pudasjärven Sammakkolammen ja sen ympäristön jääkauden jälkeisestä kehityksestä.* Unpublished thesis, Department of Botany, University of Oulu.

Heikkinen, O. & Kurimo, H. 1977: The postglacial history of Kitkajärvi, North-Eastern Finland, as indicated by trend-surface analysis and radio-carbon dating. *Fennia 153*, 1–32.

Heino, J. 1987: *Oulangan (Ks) Rytisuon nykykasvillisuudesta ja paleoekologiasta.* Unpublished MSc thesis, Department of Botany, University of Oulu.

Hicks, S. 1974: A method of using pollen rain values to provide a time-scale for pollen diagrams from peat deposits. *Memoranda Societatis pro Fauna et Flora Fennica 49*, 21–33.

Hicks, S. 1975: Variations in pollen frequency in a bog at Kangerjoki, N.E. Finland during the Flandrian. *Commentationes Biologicae, Societas Scientiarum Fennica 80*, 28 pp.

Hicks, S. 1985: Problems and possibilities in correlating historical/archaeological and pollen-analytical evidence in a northern boreal environment: an example from Kuusamo, Finland. *Fennoscandia archaeologica II*, 51–84.

Hicks, S. 1988: Siitepölytodisteita Hailuodon varhaisesta asutuksesta (Palynological evidence for the occupation of Hailuoto). *In* Julku, K. & Satokangas, R. (eds) *Hailuodon Keskiaika. Studia Historica Septentrionalia 15*, 35–88.

Hicks, S. 1992: Modern pollen deposition and its use in interpreting the occupation history of island Hailuoto, Finland. *Journal of Vegetation History and Archaeobotany 1*, 75–86.

Hjelmroos, M. 1978: Den äldsta bosättningen i Tornedalen. En paleoekologisk undersökning. *University of Lund, Department of Quaternary Geology, Report 16.*

Holappa, K. 1976: *Utajärven Järvenpäänsuon kehityksestä ja stratigrafiasta.* Unpublished thesis, Department of Botany, University of Oulu.

Huttunen, A. 1987: *Kasvillisuuden kehitys Riistunturin alueella.* Unpublished Lic.Phil. thesis, Department of Botany, University of Oulu.

Huttunen, A. 1990: Vegetation and palaeoecology of a bog complex in S Finland. *Aquilo, Series Botanica 28*, 27–37.

Huttunen, A. & Huttunen R. 1989: Kuusamon Valtavaaran alueen kasvillisuuden historia. *Oulun yliopiston Oulangan biologisen aseman monisteita 13*, 19–24.

Huttunen, P. 1980. Early land use, especially the slash-and-burn cultivation in the commune of Lammi, southern Finland, interpreted mainly using pollen and charcoal analyses. *Acta Botanica Fennica 113*, 1–45.

Huttunen, P. & Tolonen, K. 1977: Human influence in the history of Lake Lovojärvi, S. Finland. *Finskt Museum 1975, Finska fornminnesföreningen*, 68–105.

Hyvärinen, H. 1966: Studies on the late-Quaternary history of Pielis-Karelia, eastern Finland. *Commentationes Biologiae, Societas Scientiarum Fennica 29*, 72 pp.

Hyvärinen, H. 1971: Two Late Weichselian stratigraphical sites from the eastern foreland of the Salpausselkäs in Finland. *Commentationes Biologicae, Societas Scientiauum Fennica 40*, 1–12.

Hyvärinen, H. 1972: Flandrian regional pollen assemblage zones in eastern Finland. *Commentationes Biologicae, Societas Scientiarum Fennica 59*, 1–25.

Hyvärinen, H. 1973: The deglaciation history of eastern Fennoscandia—recent data from Finland. *Boreas 2*, 85–102.

Hyvärinen, H. 1975: Absolute and relative pollen diagrams from northernmost Fennoscandia. *Fennia 142*, 1–23.

Hyvärinen, H. 1976: Flandrian pollen deposition rates and treeline history in northern Fennoscandia. *Boreas 5*, 163–175.

Hyvärinen, H. 1993: Holocene pine and birch limits near Kilpisjärvi, Western Finnish Lapland: pollen stratigraphical evidence. *In* Frenzel, B. (ed.) *Oscillations of the alpine and polar tree limits in the Holocene*, 19–27. Gustav Fischer Verlag, Stuttgart.

Hyvärinen, H. & Alhonen, P. 1994: Holocene lake level changes in the Fennoscandian tree-line region, western Finnish Lapland: diatom and cladoceran evidence. *The Holocene 4*, 249–256.

Hyvärinen, H., Martma, T. & Punning, J.-M. 1990: Stable isotope and pollen stratigraphy of a Holocene lake marl section from NE Finland. *Boreas 19*, 17–24.

Hyvärinen, R. unpubl.: Jurvan alueelta tehdyt siitepölyanalyyttiset tutkimukset vuosina 1984–87. *Report December 1987. National Board of Antiquities, Prehistorical Section.*

Ikonen, L. 1993: Holocene development and peat growth of the raised bog Pesänsuo in southwestern Finland. *Geological Survey of Finland Bulletin 370*, 58 pp.

Ilvessalo, Y. 1960: Suomen metsät kartakkeiden valossa. (The forests in Finland in the light of maps; Die Wälder Finnlands im Licht von Karten). *Communicationes Instituti Forestalis Fenniae 52*, 70 pp. +30 maps.

Jelina, G. 1985: The history of vegetation in Eastern Karelia (USSR) during the Holocene. *Aquilo, Series. Botanica 22*, 1–36.

Juola-Helle, M. 1982: *Pisavaaran (PP) Kuprujängän nykykasvillisuudesta sekä alueen jääkauden jälkeisestä kasvillisuudesta.* Unpublished M. Sc. thesis, Department of Botany, University of Oulu.

Kankainen, T. 1993: Appendix 1: Radiocarbon analyses of Pesänsuo, a raised bog in southwestern Finland. *Geological Survey of Finland Bulletin 370*, 1–15.

Kivikoski, E. 1961: *Suomen esihistoria (Suomen historia I).* Porvoo, Helsinki. 310 pp.

Koivula, L. 1987: *Keski-Suomen viljelyhistoriaa: Siitepölytutkimus viidestä Keski-Suomen kunnasta.* Unpublished Lic.Phil. thesis, Department of Biology, University of Jyväskylä.

Koivula, L., Raatikainen, M., Kankainen, T. & Vasari, Y. 1994: Ihmisen vaikutus luontoon Keski-Suomessa siitepöytutkimuksen valossa. Summary: Human influence upon nature in the province of Central Finland. *Jyväskylän ylidpisto Historian laitos. Suomen historian julkaisuja 21*, 1–93

Koponen, M. & Nuorteva, M. 1973: Über subfossile Waldinsekten aus dem Moor Piilonsuo in Südfinnland. *Acta Entomologica Fennica 29*, 1–84.

Korhola, A. 1992: The Early Holocene hydrosere in a small acid hill-top basin studied using crustacean sedimentary remains. *Journal of Paleolimnology 7*, 1–22.

Korhola, A.A. & Tikkanen, M.J. 1991: Holocene development and early extreme acidification in a small hill-top lake in southern Finland. *Boreas 20*, 333–356.

Korhola, A. & Tikkanen, M. 1992: The lateglacial—early postglacial transition in the pollen stratigraphy of lake Pieni Majaslampi, Espoo, Southern Finland. *In* Grönlund, E. (ed.) *The first meeting of Finnish palaeobotanists; state of the art in Finland—May 2–4, 1990. University of Joensuu, Publications of Karelian Institute 102*, 85–98.

Koskinen, E. 1983: *Kuusivaltaisesta kasvillisuudesta ja sen historiasta Saarijärven Pyhän-Häkin kansallispuistossa.* Unpublished MSc thesis, Department of Botany, University of Oulu.

Koutaniemi, L. 1979: Late-glacial and post-glacial development of the valleys of the Oulanka river basin, northeastern Finland. *Fennia 157*, 13–73.

Koutaniemi, L. 1983: Climatic characteristics of the Kuusamo uplands. *Oulanka Reports 3*, 3–29.

Koutaniemi, L. & Ronkainen, R. 1983: Palaeocurrents from 5000 and 1600–1500 BP in the main rivers of the Oulanka basin, northeastern Finland. *Publicationes instituti geographici universitatis Ouluensis 85*, 145–158.

Koutaniemi, L. & Sillanpää, A. 1985: Hukkunut mänty—Niskalammen turvekerrostumien avaintodistaja (A relict Scots pine as a key to the origin of submerged peat deposits in Niskalampi). *Terra 97*, 153–163.

Kujala, V. 1926: Untersuchungen über die Waldvegetation in Süd- und Mittelfinnland. I. Zur Kenntnis des ökologisch-biologischen Charakters der Pflanzenarten unter spezieller Berücksichtigung der Bildung von Pflanzenvereinen. A. Gefässpflanzen. *Communicationes ex Instituto questionum forestalium Finlandiae 10*, 1–140.

Kukkonen, E. & Tynni, R. 1972: Sediment core from lake Lovojärvi, a former meromictic lake (Lammi, South Finland). The development of Lovojärvi on the basis of its diatoms. *Aqua Fennica 1972*, 70–80.

Laaksonen, K. 1979: Effective temperature sums and

durations of the vegetative period in Fennoscandia (1921–1950). *Fennia 157*, 171–197.

Lappalainen, E. 1970: Über die spätquartäre Entwicklung der Flussufermoore Mittel-Lapplands. *Bulletin de la Commission Géologique de Finlande 244*, 1–79.

Mäkelä, E. 1992: *Holoseeniset metsärajatapahtumat Jierstivaaran alueella. Siitepölyanalyyttinen tutkimus. (Holocene tree-line history in the area of Jierstivaara. Pollen analytical study).* Unpublished MSc thesis, Department of Geology, Division of Geology and Palaeontology, University of Helsinki (in Finnish).

Mangerud, J., Andersen, S.Th., Berglund, B.E. & Donner, J.J. 1974: Quaternary stratigraphy of Norden, a proposal for terminology and classification. *Boreas 3*, 109–128.

Meinander, C.F. 1958: Stenåldersundersökningar på Åland 1957. *Åländsk odling 19*, 102–103.

Meinander, C.F. 1960: Arkeologi, 84–87. I text till *Atlas över Skärgårds-Finland* jämte kartbl. nr. 16. Nordenskiöld-Samfundet i Finland, Helsingfors.

Miettinen, L. 1985: *Kuusamon Purkuputaan lähdesuon kasvillisuudesta ja paleoekologiasta.* Unpublished M.Sc. thesis, Department of Botany, University of Oulu.

Miettinen, M. 1986: Cultural palaeoecology in Malax, southern Ostrobothnia: Archaeological background. *In* Königsson, L.-K. (ed.) Nordic Late Quaternary biology and ecology. *Sriae 24*, 161–164.

Miettinen, M. & Vuorela, I. 1988: Archaeological and palynological studies of the agricultural history of Vörå and Malax, Southern Ostrobothnia. *Fennoscandia Archaeologica V*, 47–68.

Nykänen, J. 1984: *Rautavaaran Hamusenjärven ja sen lähiympäristön paleoekologiasta.* Unpublished MSc thesis, Department of Botany, University of Oulu.

Ozenda, P., Noirfalise, A., Tomaselli, R. & Trautmann, W. 1979: Vegetation map (scale 1:3,000,000) of the Council of Europe member States. *Council of Europe Nature and Environment Series 16*, 1–99 + 3 maps.

Prentice, H. 1981: A late Weichselian and early Flandrian pollendiagram from Östervatnet, Varanger Peninsula, NE Norway. *Boreas 10*, 52–70.

Prentice, H. 1982: Late Weichselian and early Flandrian vegetational history of Varanger Peninsula, northeast Norway. *Boreas 11*, 178–208.

Räsänen, M. & Salonen, V.-P. 1983: Turun Kakskerranjärven tila ja sen kehitys. Abstract: The trophic status and its evolution on lake Kakskerranjärvi, Turku. *Publications of the Department of Quaternary Geology, University of Turku 50*, 38 pp.

Repo, R. & Tynni, R. 1967: Zur spät- und postglazialen Entwicklung im Ostteil des ersten Salpausselkä. *Comptes Rendus de la Sociéte géologique de Finlande 39*, 133–159.

Reynaud, C. & Hjelmroos, M. 1980: Pollen analysis and radiocarbon dating of human activity within the natural forest vegetation of the Pohjanmaa region (N. Finland). *Candollea 35*, 257–304.

Ristaniemi, O. 1987: Itämeren korkein ranta ja Ancylusraja sekä Muinais-Päijänne Keski-Suomessa. (The highest shore and Ancylus limit of the Baltic Sea and the ancient Lake Päijänne in Central Finland). *Annales Universitatis Turkuensis, Series C 59*, 1–102.

Ristaniemi, O. & Glückert, G. 1988: Ancylus- ja Litorinatransgressiot Lounais-Suomessa. *In* Tutkimuksia geologian alalta. *Annales Universitatis Turkuensis, Series C 67*, 129–145.

Rönkä, A. 1983: *Hailuodon Kiimisuon paleoekologiasta.* Unpublished M.Sc. thesis, Department of Botany, University of Oulu.

Ruohomäki, A.-M. 1983: *Hanhilammen (Iisalmi) ja sen ympäristön kasvillisuuden historia.* Unpublished thesis, Department of Botany, University of Oulu.

Saarnisto, M. 1970: The Late Weichselian and Flandrian history of the Saimaa lake complex. *Commentationes Physico-Mathematicae, Societas Scientiarum Fennica 37*, 1–107.

Saarnisto, M. 1981: Holocene emergence history and stratigraphy in the area north of the Gulf of Bothnia. *Annales Academiae Scientiarum Fennicae Series A. Geologia 130*, 42 pp.

Saarnisto, M., Huttunen, P. & Tolonen, K. 1977: Annual lamination of sediments in Lake Lovojärvi, Southern Finland, during the past 600 years. *Annales Botanici Fennici 14*, 35–45.

Salminen, S. 1973: Reliability of the results from the fifth national forest inventory and a presentation of an output mapping technique. *Communicationes Instituti Forestalis Fenniae 78*, 64 pp.

Salonen, V.-P., Ikäheimo, M. & Luoto, J. 1981: Rautakautisen ja historiallisen asutuksen ilmeneminen paleontologisin ja arkeologisin keinoin Piikkiön Kuoppajärven ympäristössä Lounais-Suomessa. Abstract: The activity of man in the Kuoppajärvi area in Piikkiö, SW Finland, in the light of palaeontology and archaeology. *Publications of the Department of Quaternary Geology, University of Turku 44*, 23 pp.

Sarmaja-Korjonen, K., Vasari, Y. & Hæggström, C-A. 1991: Taxus baccata and influence of Iron Age man on the vegetation in Åland, SW Finland. *Annales Botanici Fennici 28*, 143–159.

Sauramo, M. 1958: Die Geschichte der Ostsee. *Annales Academiae Scientiarum Fennicae, Series A III. Geologica–Geographica 51*, 522 pp.

Seppälä, M. & Koutaniemi, L. 1985: Formation of a string and pool topography as expressed by morphology, stratigraphy and current processes on a mire in Kuusamo, Finland. *Boreas 14*, 287–309.

Simola, H., Huttunen, P. & Meriläinen, J. 1984: Varvedated eutrophication history of a small lake. *Verhandlungen der internationalen Vereinigung für Limnologie 22*, 1404–1408.

Simola, H., Grönlund, E. & Uimonen-Simola, P. 1988: Etelä-Savon asutushistorian paleoekologinen tutkimus. (Paleoecological investigation of the history of agriculture in the province of South Savo, Finland). *University of Joensuu, Publications of Karelian Institute 84*, 1–55.

Sorsa, P. 1964: Über das Spätglazial in Salla, Nordfinnland. *Annales Botanici Fennici 1*, 97–103.

Sorsa, P. 1965: Pollenanalytische Untersuchungen zur spätquartären Vegetations- und Klimaentwicklung im östlichen Nordfinnland. *Annales Botanici Fennici 2*, 301–413.

Stén, C.-G. 1993. Appendix 2: Macrofossils of the raised bog Pesänsuo in southwestern Finland. *Geological Survey of Finland Bulletin 370*, 1–6.

Teemakarttoja Suomesta 1989. *Helsingin yliopiston maantieteen laitoksen opetusmonisteita 34*. Department of Geography, University of Helsinki (stencil).

Tikkanen, M. & Korhola, A. 1993: Divergent successions in two adjacent rocky basins in southern Finland: a physiographic and palaeoecological evaluation. *Annales Academiae Scientiarum Fennicae Series A III. Geologica–Geographica 157*, 26 pp.

Tolonen, K. 1963: Über die Entwicklung eines nordkarelischen Moores im Lichte der C[14]-Datierung. Das Moor Puohtiinsuo in Ilomantsi (Ost-Finland). *Archivum Societatis Zoologicae Botanicae Fennicae "Vanamo" 18*, 41–57.

Tolonen, K. 1967: Über die Entwicklung der Moore im Finnischen Nordkarelien. *Annales Botanici Fennici 4*, 219–416.

Tolonen, K. 1987: Natural history of raised bogs and forest vegetation in the Lammi area, southern Finland studied by stratigraphical methods. *Annales Academiae Scientiarum Fennicae Series A III. Geologica–Geographica 144*, 46 pp.

Tolonen, K. & Ruuhijärvi, R. 1976: Standard pollen diagrams from the Salpausselkä region of Southern Finland. *Annales Botanici Fennici 13*, 155–196.

Tolonen, K. & Tolonen, M. 1988: Synchronous pollen changes and traditional land use in South Finland, studied from three adjacent sites: A lake, a bog and a forest soil. *In* Lang, G. & Schlüchter, Ch. (eds) *Lake, mire and river environment during the last 15,000 years*, 83–97. A.A. Balkema, Rotterdam/Brookfield.

Tolonen, K., Siiriäinen, A. & Hirviluoto, A.-L. 1979: Iron Age cultivation in SW Finland. *Finskt Museum 1976*, 1–66.

Tolonen, M. 1978a: Palaeoecology of annually laminated sediments in Lake Ahvenainen, S. Finland. I. Pollen and charcoal analyses and their relation to human impact. *Annales Botanici Fennici 15*, 177–208.

Tolonen, M. 1978b: Palaeoecology of annually laminated sediments in Lake Ahvenainen, S. Finland. II. Comparison of dating methods. *Annales Botanici Fennici 15*, 208–222.

Tolonen, M. 1978c: Palaeoecology of annually laminated sediments in Lake Ahvenainen, S. Finland. III. Human influence in the lake development. *Annales Botanici Fennici 15*, 223–240.

Tolonen, M. 1985: Palaeoecological reconstruction of vegetation in a prehistoric settlement area, Salo, SW Finland. *Annales Botanici Fennici 22*, 101–116.

Tolonen, M. 1990: Appendix 5. Pollen-analytical evidence of ancient human action in the hillfort area of Kuhmoinen, Central Finland. *In* Taavitsainen, J.-P. 1990; Ancient hillforts of Finland. Problems of analysis, chronology and interpretation with special reference to the hillfort of Kuhmoinen. *Suomen Muinaismuis-toyhdistyksen Aikakauskirja 94*, 247–264.

Vasari, Y. 1962: A study of the vegetational history of the Kuusamo district (North East Finland) during the Late-quaternary period. *Annales Botanici Societatis "Vanamo" 33*, 1–140 + Appendix.

Vasari, Y. 1965a: Studies on the vegetational history of the Kuusamo district (North East Finland) during the Late-quaternary period. III. Maanselänsuo, a Lateglacial site in Kuusamo. *Annales Botanici Fennici 2*, 219–235.

Vasari, Y. 1965b: Studies on the vegetational history of the Kuusamo district (North East Finland) during the Late-quaternary period. IV. The age and origin of some present-day vegetation types. *Annales Botanici Fennici 2*, 248–273:

Vasari, Y. 1967: New additions to the sub-fossil flora of the Kuusamo district, North East Finland. *Aquilo, Series Botanica 6*, 71–83.

Vasari, Y. & Väänänen, K. 1986: Stratigraphical indications of the former use of wetlands. *In* Behre, K.-E. (ed.) *Anthropogenic indicators in pollen diagrams*, 65–71. A.A. Balkema, Rotterdam.

Vasari, Y., Vasari, A. & Koli, L. 1963: Purkuputaanlampi, a calcareous mud series from Kuusamo, North East Finland. *Archivum Societatis Zoologicae Botanicae Fennicae "Vanamo" 18*, 96–104.

Vasari, Y., Tonkov, S., Vasari, A. & Nikolova, A. (unpubl.): Early Holocene vegetation history in North-Eastern Finland in the light of a re-investigation of Aapalampi in Salla. (manuscript).

Vorren, K.-D. 1972: Stratigraphical investigations of a palsa bog in northern Norway. *Astarte 5*, 39–71.

Vuorela, I. 1982: Vad berättar pollenanalysen om människoaktiviteten på Hundbacka. *Bottnisk kontakt I, Skrifter från Örnsköldsviks museum I*, 21–24.

Vuorela, I. 1983: Vohtenkellarinsuo, A bog in Paimio, SW Finland with a cultural origin. *Bulletin of the Geological Society of Finland 55*, 57–66.

Vuorela, I. 1986: Cultural palaeoecology in Malax, southern Ostrobothnia: pollen analysis. *In* L.-K. Königsson (ed.) *Nordic Late Quaternary biology and ecology, Striae 24*, 165–168.

Vuorela, I. 1987a: Pollenanalyser från Vörå mossar. *I Rågens rike. Vörå årspublikation 1987*, 56–58.

Vuorela, I. 1987b: Type region: coastal S. Ostrobothnia, Finland. *IGCP 158. Palaeohydrological changes in the temperate zone in the last 15,000 years.* Symposium at Höör, Sweden, 18–26 May 1987, Posters: 135–138.

Vuorela, I. 1990a: Pollenanalytiska studier. *In* Zilliacus, K. (ed.) *"Finska Skären". Studier i Åboländsk kulturhistoria utgivna av Konstsamfundet till dess 50-årsjubileum 1990*, 115–132.

Vuorela, I. 1990b: Siitepölyanalyyttinen tutkimus Puolangan kunnan kasvillisuuden kehityksestä. *Geological Survey of Finland, Raport/Ympäristötutki mukset*, 1–19.

Vuorela, I. 1993a: Vuojärven kerrostumat Laukaan historian arkistoina. *In* Hänninen, H. (ed.) *Kohisevien koskien Laukaa*, 287–293. Gummerus Kirjapaino Oy, Jyväskylä.

Vuorela, I. 1993b: Luonnon ja kulttuurimaiseman kehitys Taipalsaaressa. *Geological Survey of Finland, Report P 34.4.108*, 1–46.

Vuorela, I. 1995: Palynological evidence of the Stone Age settlement in southern Finland. *In* Autio, S. (ed.) *Geological Survey of Finland, Current Research. Geological Survey of Finland,Special Paper 20*, 139–143.

Vuorela, I. (in press): Pollen evidence of Stone Age and Early Metal Age settlement in Taipalsaari, southern Karelia, eastern Finland. *Fennoscandia Archaeologica.*

Vuorela, I. & Aalto, M. 1982: Palaeobotanical investigations at a Neolithic dwelling site in Southern Finland, with special reference to Trapa natans. *Annales Botanici Fennici 19*, 81–92.

Vuorela, I. & Kankainen, T. 1991: Siitepölyanalyyttinen tutkimus asutuksen vaikutuksesta kasvillisuuteen Puolangan kunnan Kotilan kylässä. *Geological Survey of Finland, Report P 34.4.100*, 1–21.

Vuorela, I., Uutela, A., Saarnisto, M., Ilmasti, M. & Kankainen, T. 1993: Vuojärven ja Antinlammen kerrostumat Laukaan asutus- ja luonnonhistorian arkistoina. *Geological Survey of Finland, Report P 34.4.108*, 1–91.

Vuorinen, J. 1978: The influence of prior land use on the sediments of a small lake. *Polskie Archiwum Hydrobiologii 25*, 443–451.

Vuorinen, J. & Tolonen, K. 1975: Flandrian pollen deposition in lake Pappilanlampi, eastern Finland. *Publications of the University of Joensuu (BII) 3*, 1–12.

Wallin, J.-E. & Segerström, U. 1994: Natural resources and agriculture during the Iron Age in Ostrobothnia, western Finland, investigated by pollen analysis. *Journal of Vegetation History and Archaeobotany 3*, 89–105.

10

Russian Karelia

G.A. Elina and L.V. Filimonova

INTRODUCTION

Russian Karelia lies in the eastern part of the Baltic Shield, between Finland in the west and the White Sea in the east. Its northern boundary lies at latitude 66°39′N, the southern at 60°41′N. The area of the republic is 173000 km² (Karelian ASSR 1956).

The Baltic Shield is composed of Archean and Proterozoic gneiss, granite, schist, diabase, and other crystalline rocks. The modern relief of Karelia depends on the form of the bedrock surface, but its final shape has been affected by Quaternary glaciation. The peneplains and the system of shield fractures account for the general pattern of orography and hydrography. Lukashov (1976) distinguishes three relief levels, which descend by steps from the west to the White Sea in the east. The upper level reaches 250–300 m above sea level, and denudational-tectonic forms prevail. The middle level ranges from 150 to 250 m and is characterized by the alternation of ridges of denudational-tectonic genesis and moraines with kames. The lower level (0–150 m asl) is represented by submarine and lacustrine plains, which are confined to the White Sea and Onega and Ladoga lakes.

Twelve landscape–geobotanic regions are distinguished in Karelia (Yakovlev & Voronova 1959, Ramenskaya & Shubin 1975, Elina *et al.* 1984).

North taiga:

K-a   Northwest middle mountain with Piceetum + empetroso–cladinosum and Betuletum (open)

K-b   Northwest lake with Pinetum hylocomiosum and P. sphagnosum

K-c   Pribelomorskaya lowland with Pinetum cladinoso–hylocomiosum and ombrotrophic bogs

K-d   North Karelian upland with Piceeto–Pinetum hylocomiosum

K-e   Onega–White Sea watershed with Piceeto–Pinetum hylocomiosum and Piceeto–Pinetum sphagnosum and aapa mires

Middle taiga:

K-f   Yanisjöki–Suojöki watershed with Pinetum hylocomiosum

K-g   Suna–Shuya watershed with Piceeto–Pinetum cladinoso–hylocomiosum and aapa mires

K-h   Povenets Bay–Vyg watershed with Pineto–Piceetum sphagnoso–hylocomiosum with ombrotrophic bogs

K-i   Priladozhskaya lowland with Piceeto–Pinetum hylocomiosum with oligotrophic and mesotrophic mires

K-j   Northern Zaonezhje with Piceetum hylocomiosum, Pinetum cladinosum and P. hylocomiosum

K-k   Onega–Ladoga watershed with Piceetum nemori–herbosum and mesotrophic mires

K-l   Vodlozerskaya hollow with Pinetum hylocomiosum and mesotrophic mires

Presented here is a detailed description of type regions K-c, K-g, and K-j.

Palaeoecological Events During the Last 15 000 Years: Regional Syntheses of Palaeoecological Studies of Lakes and Mires in Europe.
Edited by B.E. Berglund, H.J.B. Birks, M. Ralska-Jasiewiczowa and H.E. Wright.   © 1996 John Wiley & Sons Ltd.

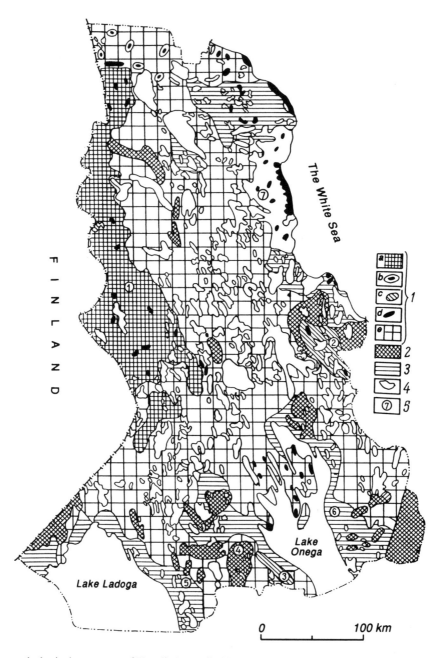

Fig. 10.1 Geomorphological structures of Karelia (compiled by G.T. Lak on the materials of A.D. Lukashov, V.A Ilyin, I.M. Ekman, G.S Biske and G.T. Lak). 1–denudational-tectonic type of relief: a) large elevations, b) individual massives, c) ridges, d) crystalline foundation, e) dissected surface of crystalline foundations overridden by moraine cover of different thickness; 2 – glacial type of relief; 3 – glacio-marine and glacio-lacustrine type of relief; 4 – mires; 5 – landscape regions: 1) West Karelian upland, 2) Vetrenny Poyas, 3) Shoksha ridge, 4) Olonets upland, 5) Ladoga plain, 6) East Onega plain, 7) Pribelomorskaya plain

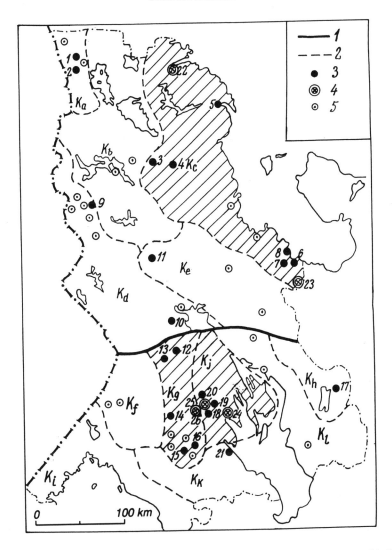

Fig. 10.2 Landscape–geobotanic zonation of Karelia and location of profiles provided with palynological and radiocarbon data. 1–The heavy line separates the north-taiga from the middle-taiga. 2–Dashed lines bound the geobotanic regions. Reference profiles (see the text): 3–documented within Project 158 B in 1986; 4–studied in 1985–1987; 5–complete and typical

TYPE REGION K-c, PRIBELOMORSKAYA LOWLAND

The Pribelomorskaya lowland is heterogeneous in its geologic, geomorphic, and hydrogeologic characteristics and it is divided into the western and eastern regions.

The western part of the region

Altitude: 100–150 m above sea level.

Climate: Mean January temperature –12°C, July 11°C, annual 0.6°C. Precipitation 450 mm yr⁻¹ (Romanov 1961).

Geology: Archean rocks (granite, granodiorite,

gneiss, amphibolite) are covered with a blanket of Quaternary sediments, mainly sand, loamy sand, and clay of glacial-lake genesis (Biske 1959).

Topography: A hilly plain with orographic elements generally oriented to the northwest. Linear ridges and parallel depressions are occupied by lakes and mires.

Population: Less than 1 person km^{-2}.

Vegetation: North taiga: Pinetum empetroso–myrtilloso–hylocomiosum and ledoso–sphagnosum forests –55% of the area; Piceetum polytrichosum and sphagnosum–13%; peatlands–35% (25–50%). Aapa mire herb–*Sphagnum* on ridge–hollow and ridge–pool eutrophic–mesotrophic (minerotrophic) bogs.

Soils: Ferrous-humic illuvial and ferrous illuvial podzols, and peat.

Land use: Practically no cultivated land.

Reference site 4. Shombashuo

Latitude 65°05′N, Longitude 33°03′E. Elevation 97 m. Age range 11500–0 BP. Minerotrophic mire. Pollen diagram (see Fig. 6 in Eelina 1981). Nine pollen assemblage zones (paz).

1.	–11000 BP (AL)	Scattered forests of *Betula pubescens*; tundra of *Betula nana–Salix* – Ericaceae
2.	11000–10150 BP(DR$_3$)	Tundra of *Betula nana–Salix–* Ericaceae
3.	10150–9200 BP(PB)	Forest tundra of *Betula pubescens–* Polypodiaceae
4.	9200–8500 BP(BO$_1$)	North-taiga forests of *Betula pubescens–* Bryales
5.	8500–7800 BP(BO$_2$)	North-taiga forests of *Betula pubescens–Pinus*
6.	7800–6500 BP(AT$_1$)	Middle-taiga forests of *Pinus–Betula pubescens*
7.	6500–4800 BP(AT$_2$)	South-taiga forests of *Betula pubescens–Pinus* with *Ulmus*
8.	4800–2500 BP(SB)	North-taiga forests of *Betula pubescens–Pinus* and *Picea–Pinus*
9.	2500–0 BP(SA)	North-taiga forests of *Pinus* and *Picea–Pinus*

The eastern part of the type region K-c

Altitude: 0–100 m above sea level.

Climate: Mean January temperature –10° to –11°C, July 14°C, annual 0.5°C. Precipitation 400–450 mm yr^{-1}.

Geology: Proterozoic crystalline rocks (mainly gneiss and schist) are covered with late- and postglacial marine clays or glacial-lake clays, loams and sands.

Topography: A slightly undulating plain has a slope towards the White Sea; the southern part is terraced. Late- and postglacial terraces are separated by abrasion ledges and are at different altitudes, from 7 to 27 m. The northern part of the lowland has a more dissected relief, with ridges composed of bedrock with west–east orientation.

Population: Less than 1 person km^{-2} .

Vegetation: North-taiga Pinetum hylocomiosum (*Empetrum, Vaccinium myrtillus, V. vitis-idaea*), Pinetum empetroso-cladinosum and Pinetum sphagnosum forests make up 28–30% of the area; along river valleys are Piceetum hylocomiosum (*Empetrum, Vaccinium vitis-idaea, V. myrtillus*) and Piceetum uligini-herbosum, 10–12%. The degree of peat formation is 70%. Mires on late-glacial and Boreal terraces have liverwort–lichen–*Sphagnum* ridge–pool oligotrophic (ombrotrophic) cover of 60%; on Atlantic and Sub-boreal terrace *Sphagnum* ridge–hollow oligotrophic (ombrotrophic) mires cover 10%.

Soils: Ferrous illuvial podzol and peat.

Land use: Practically no cultivated land.

Reference site 6. Zarutskoe (Fig. 10.3)

Latitude 63°50′N, Longitude 36°20′E. Elevation 20 m. Age range 8500–0 BP. Ombrotrophic mire. Pollen diagram (see Fig. 10.3 in Jelina 1981). Five pollen assemblage zones (paz).

1.	8500–7800 BP (BO$_2$)	North-taiga forests of *Pinus–Betula pubescens–*Polypodiaceae
2.	7800–6500 BP (AT$_1$)	Middle-taiga forests of *Betula pubescens–Pinus*
3.	6500–4800 BP (AT$_2$)	South-taiga forests of *Betula pubescens–Pinus* and *Pinus–Picea* with *Ulmus*
4.	4800–2500 BP (SB)	Middle-taiga forests of *Picea–Pinus* and *Picea* with *Corylus*
5.	2500–0 BP (SA)	North-taiga forests of *Betula–Pinus* and *Picea–Pinus*

TYPE REGION K-j, NORTHERN ZAONEZHYE

Altitude: 33 (Onega Lake level) to 80 m above sea level.

Climate: Mean January temperature −11.4°C, July 13°C, annual 1.6°C. Precipitation 570 mm yr^{-1}.

Geology: Proterozoic crystalline rocks are separated by joints into individual blocks. A northwest and submeridional trend is generally characteristic of structural elements.

Topography: Ridges are covered with a thin layer of Quaternary sediments or have crystalline rocks exposed. Moraines and glacial lake plains.

Population: 1–1.5 persons km^{-2}.

Vegetation: Middle-taiga forests of Piceetum myrtilloso–hylocomiosum and Oxalidoso–hylocomiosum with *Tilia*–about 40% of the area; Piceetum polytrichosum–16%; Pinetum cladinosum and polytrichosum–14%. Secondary Betuletum and Populetum forests–20%. Peat formations–10–15%. Herb–*Sphagnum* ridge–hollow and ridge–pool eutrophic–mesotrophic (minerotrophic) and herb–Bryales mires.

Soils: Young sandy and crushed stony podzol, mid- and modal podzol loamy sand and loam.

Land use: Secondary birch forests in felled areas; agricultural lands–about 2% of the area.

Reference site 18. Gotnavolok (Fig. 10.4)

Latitude 62°10′N, Longitude 33°45′E. Elevation 88 m. Age range 11500–0 BP. Ombrotrophic mire. Pollen diagram (Elina 1981, Elina & Filimonova 1987). Eight pollen assemblage zones (paz).

1.	11500–11000 BP (AL)	Scattered forests of *Betula pubescens*, occasionally *Picea*; *Alnus incana*; tundra of *Betula nana*; periglacial complexes of *Artemisia–*Chenopodiaceae
2.	11000–10150 BP (DR$_3$)	Tundra of *Betula nana– Lycopodium dubium–Diphasiastrum alpinum*; periglacial complexes of *Artemisia–*Chenopodiaceae
3.	8500–7800 BP (BO$_2$)	Middle-taiga forests of *Pinus–Betula pubescens*
4.	7800–6500 BP (AT$_1$)	South-taiga forests of *Pinus–Betula pubescens* with *Ulmus*
5.	6500–4800 BP (AT$_2$)	South-taiga forests of *Pinus–Betula pubescens, Picea–Pinus* with *Ulmus* and *Corylus*

358

Zarutskoe, KARELIA
Selected pollen percentages
Analyst: G. A. Elina

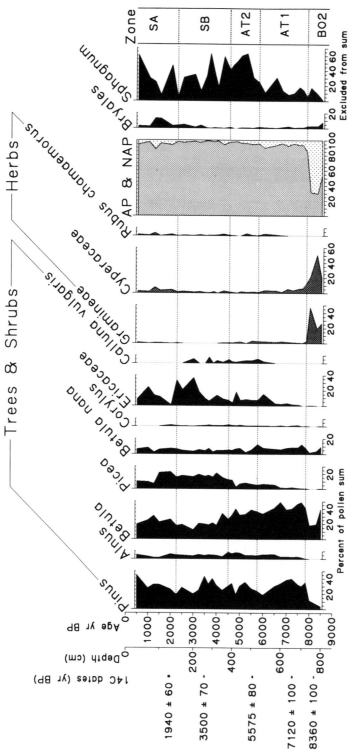

Fig. 10.3 Pollen–spore diagram of Zarutskoe mire

359

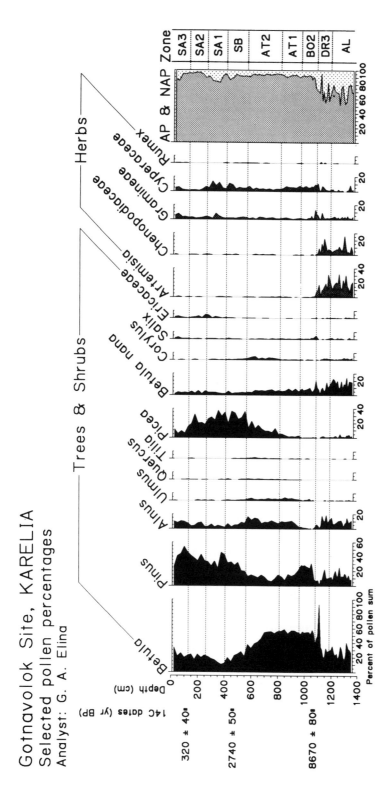

Fig. 10.4 Pollen–spore diagram of Gotnavolok mire

6. 4800–2500 BP (SB)   South-taiga forests of
                          Pinus and *Picea*
7. 2500–1200 BP (SA₁)   Middle-taiga forests of
                          Pinus and *Picea*
8. 1200–0 BP (SA₂)   Middle-taiga forests of
                          Picea and *Picea–*
                          Pinus

TYPE REGION K-g, SUNA–SHUYA WATERSHED

Altitude: 123–133 m above sea level.

Climate: Mean January temperature –11.4°C, July 13°C, annual 1.6°C. Precipitation 570 mm yr⁻¹.

Geology: Proterozoic crystalline rocks.

Topography: A rolling moraine plain of glacial origin. Ridges and depressions are orientated to the northwest. Relief of ridges 10–15 m.

Population: 1–1.5 persons km⁻².

Vegetation: Middle-taiga forests of Pinetum cladinosum and P. hylocomiosum–45%; Pineto–Piceetum hylocomiosum and Piceetum polytrichosum–25%. Degree of peat formation–30–35%; herb–*Sphagnum* ridge–hollow mesotrophic and oligotrophic mires.

Soils: Sandy and loamy podzols.

Land use: Secondary *Betula* and *Populus tremula* forests–35%. Cultivated land–less than 1%.

Reference site 14. Bezdonnoe (Fig. 10.5)

Latitude 62°00′N, Longitude 32°30′E. Elevation 123 m (mire), 133 m (ridges). Profile age range *ca.* 10000–0 BP. Minerotrophic mire. Pollen diagram (Elina 1981). Seven pollen assemblage zones (paz).

1. –9200 BP (PB)   North-taiga forests of
                          Betula pubescens,
                          tundra of *Betula*
                          nana–Salix with
                          Lycopodium dubium
                          and *Diphasiastrum*
                          tristachium

2. 9200–7800 BP (BO)   Middle-taiga forests of
                          Betula pubescens–
                          Pinus
3. 7800–6500 BP (AT₁)   South-taiga forests of
                          Betula pubescens–
                          Pinus with *Ulmus*
                          and *Corylus; Alnus*
                          incana.
4. 6500–4800 BP (AT₂)   South-taiga forests of
                          Betula pubescens–
                          *Pinus–*Polypodiaceae;
                          Picea–Pinus with
                          Ulmus, Quercus and
                          Corylus; Alnus incana,
                          Alnus glutinosa
5. 4800–2500 BP (SB)   South-taiga forests of
                          Betula–Pinus and
                          Pinus–Picea with
                          nemoral herbs
6. 2500–1500 BP (SA₁)   Middle-taiga forests of
                          Pinus and *Pinus–*
                          Picea
7. 1500–0 BP (SA₂)   Middle-taiga forests of
                          Pinus and *Picea–Pinus*

PALAEOECOLOGICAL PATTERNS AND EVENTS

Appearance and spread of arboreal species during the Holocene

Migration of arboreal species after glacier retreat had already begun in the Alleröd and continued at different rates depending on temperature and precipitation. For much of Karelia *Betula pubescens* was a pioneer species and had already formed open forest–tundra formations in the PB period. These bordered on north-taiga *Betula* forests as far as Lat. 63°N. Expansion of coniferous (*Pinus*) forests coincided with a significant rise of temperature and increased precipitation in the mid-Boreal period (see Fig. 10 in Jelina 1985). In the BO period middle-taiga forests became predominant, and only in the northern part of the area (64–66°N) were forests of north-taiga pattern (Figs 10.6–10.8).

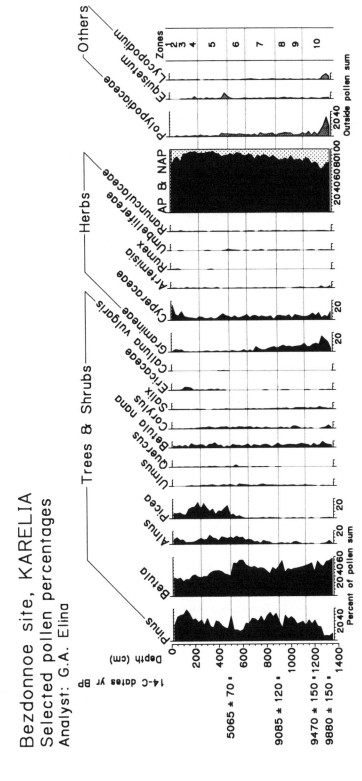

Fig. 10.5 Pollen–spore diagram of Bezdonnoe mire

TYPE REGION K-c   PRIBELOMORSKAYA LOWLAND, SITE 6

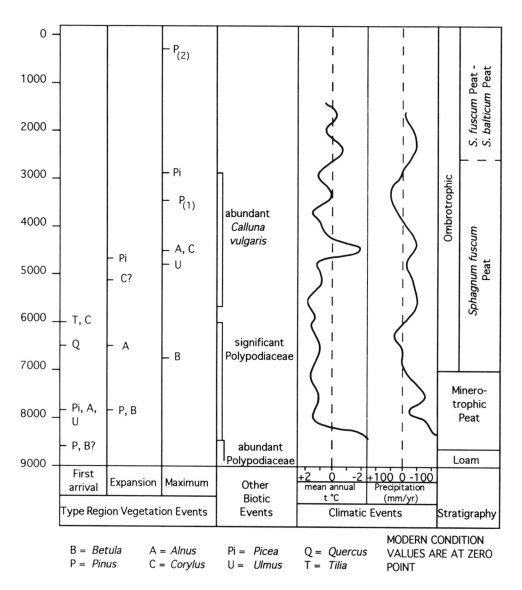

Fig. 10.6   Event stratigraphy for the Zarutskoe site in Pribelomorskaya Lowland

With increased temperature and humidity in AT (annual temperature 4°C and precipitation 100–150 mm year^{-1} higher than in late BO), *Pinus* and *Betula–Pinus* tall-herb and green-moss south-taiga forests began to spread in Karelia. In the early half of the AT$_1$ period *Ulmus* appeared in the forests, in

late AT$_1$ *Quercus*, and in early AT$_2$ *Tilia* (Elina 1981). In the AT$_2$ broad-leaved coniferous (subtaiga) forests began to spread in southern Karelia within 60–62°30′ N.

In Karelia migration of *Picea* occurred in a southeast–northwest direction. In the southeastern part

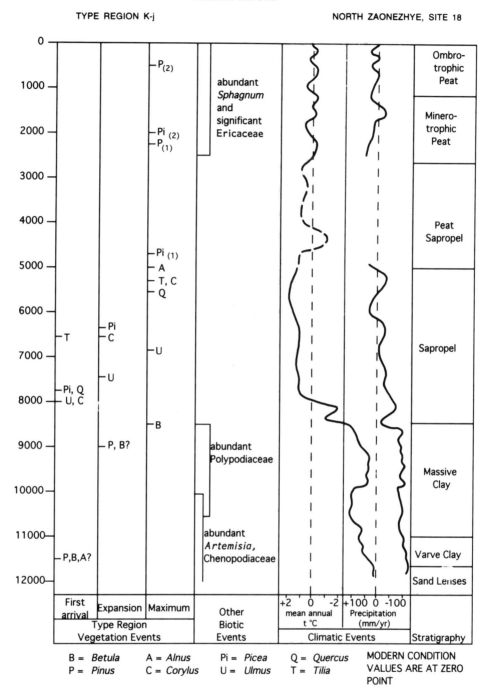

Fig. 10.7 Event stratigraphy for the Gotnavolok site in North Zaonezhye

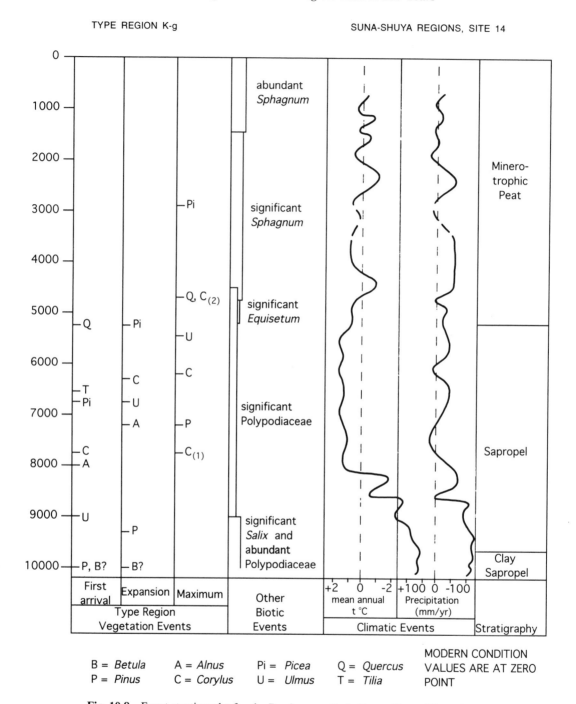

Fig. 10.8 Event stratigraphy for the Bezdonnoe site in Suna–Shuya Watershed

of the area it was already recorded by 8000 BP, on the shore of Onega Bay of the White Sea 7000 BP, and in the northwest of Karelia 5000 BP. *Picea* became a dominant species in the southeast *ca.* 6000 BP, and in interior northern Karelia only by 4000 BP. The expansion of *Picea* forests is connected with a general increase in erosion and increased fertility of loamy and loamy-sand soils (major habitats of *Picea*). The latter was the result of intensification of nutrient and energy cycling in the warm and wet AT period. *Picea* forests, which looked like south-taiga in AT and particularly in SB and early SA, are united and turned into middle-taiga. In the SA_2 period wide expansion of mires and soil acidification resulted in a replacement of *Picea* forests by *Pinus* in most of Karelia.

Palaeoclimate

The Karelian climate significantly changed during the Holocene. In the periods characterized by a rise of temperature, summer temperature increased more significantly, whereas in the periods when it was getting colder winter temperature decreased. Thus if in the early Holocene the summer temperature was 4°C lower than the present values, then January temperature was 10–12°C lower and annual temperature 6–7°C lower. With a rise of temperatures registered *ca.* 5500 BP, January temperature was 2–3°C higher, July 3–4°C higher, and annual 2–2.5°C higher than at present. Since the early Holocene a consistent rise of temperature was observed until 10000 BP, after which a new stage of temperature decrease began but with temperatures somewhat higher than in the early Holocene. Then again they increased until 8500 BP, and at approximately 8300–8200 BP a sharp fall of temperature was registered to levels higher than at the previous stage. Then for a long period temperatures were higher than the present-day values. The next sharp fall of temperature was observed *ca.* 4500 BP. Two other pronounced colder periods occurred *ca.* 2500 BP and 600–700 BP (see Figs 10.6–10.8).

Precipitation also changed. The early Holocene was characterized by the driest periods: the annual precipitation was 200 mm lower than at present. For a long period of time (until 8500 BP or later) precipitation changed within 50 mm, generally correlated with rises and falls of temperature: in the first case it increased, and in the second it decreased. At the turning-points in the temperature regime the amount of precipitation always decreased, and between them it increased.

Palaeohydrology

During the Holocene the palaeohydrological regime in the Karelian area was determined both by the ratio between warmth and humidity and by the fluctuations in the level of large water-bodies: the White Sea, Onega and Ladoga Lakes, and smaller lakes.

The entire low-elevation area in northern Karelia was affected by the White Sea. The hydrological regime of the territory was especially influenced by a drop in the water level of postglacial lakes, which took place *ca.* 9000 BP, i.e. simultaneously with the Boreal regression of the sea. In the latter half of the BO period a break in sedimentation was recorded in many mires, or the rates of sedimentation were very low and often interrupted. That period could be called "dry", because moisture was insufficient for normal mire development.

The "wet" period coincided with active expansion of mires during the AT, with a rapid increase in peat and transgression of the White Sea. Another "dry" period is registered *ca.* 4000 BP in the peat layer of corresponding age. It was the result not only of the White Sea regression in the SB period but also of a significant rise of temperature. As a result evaporation increased and erosion decreased.

In the SA period peat bogs developed via the "wet" cycle in interior Karelia.

The stratigraphy of lake and mire sediments from Onega Lake basin also shows "dry" and "wet" periods. Their comparison with the same periods from the mire profiles of the White Sea basin indicates that an increase of decrease of humidity was synchronous in the north and south.

REFERENCES

Biske, G.S. 1959: *Quaternary sediments and geomorphology of Karelia.* 305 pp. Petrozavodsk (in Russian).

Elina, G.A. 1981: *Principles and methods for reconstruction and mapping the Holocene vegetation.* 159 pp. Leningrad (in Russian).

Elina, G.A. & Filimonova, L. 1987: Late-glacial vegetation on the territory of Karelia. Palaeohydrology of temperate zone. vol. 111—*Mires and lakes,* 53–69. "Valgus", Tallinn.

Elina, G.A. & Lak, G.C. 1989: Peat mires–indicators of natural-climatic processes in Holocene. *In Paleoclimates of late Glacial and Holocene,* 52–57 (in Russian).

Elina, G.A., Kuznetsov, O.L. & Maksimov, A.I. 1984: *Structural-functional organization and dynamics of mire ecosystems in Karelia.* 128 pp. Leningrad (in Russian).

Elina, G.A., Filimonova, L.V., Kuznetsov, O.L., Lukashov, A.D., Stoikina, N.V., Arslanov, G.V. & Tertichnaya, T.V. 1994: The effect of paleohydrology on the dynamics of mire vegetation and peat accumulation. *Botan. journ. 79*, 1, 53–69 (in Russian).

Filimonova, L.V. 1985: On palynological study of the mires of the morainic plain in middle Karelia. *In Questions of the ecology of plants, mire habitats and peat deposits,* 122–132. Petrozavodsk.

Jelina, G. (Elina) 1985: The history of vegetation in Eastern Karelia (USSR) during the Holocene. *Aquilo, Series Botanica 22,* 1–36.

Karelian ASSR 1956: M. 335 pp. (in Russian).

Lukashov, A.D. 1976: *Newest tectonic geology of Karelia.* 108 pp. Leningrad (in Russian).

Ramenskaya, M.L. & Shubin, B.I. 1975: Natural zonation in terms of the problems of forest regeneration. *In* Forest regeneration in the Karelian ASSR and Murmansk region, 180–198. Leningrad (in Russian).

Romanov, A.A. 1961: *On the Karelian climate.* 140 pp. Petrozavodsk (in Russian).

Yakovlev, F.S. & Voronova, V.S. 1959: *Forest types of Karelia and their natural zonation.* 190 pp. Petrozavodsk (in Russian).

11

Estonia

L. Saarse, H. Mäemets, R. Pirrus A.-M. Rôuk,
A. Sarv and E. Ilves

INTRODUCTION

Estonia offers a wide variety of landscapes, subdivided into Lower and Upper Estonia. The boundary between them is marked by the highest shore line of the Baltic Ice Lake. In northern, western, and central Estonia the bedrock is composed mainly of Cambrian clay and Ordovician and Silurian limestone, marl, and dolomite. In South Estonia is Middle Devonian sandstone, whereas Upper Devonian limestone occurs in the southeastern corner of Estonia (Fig. 11.1). The inland ice left behind undulating glacial relief (Fig. 11.2) with Quaternary deposits ranging up to 207 m thick.

Estonia belongs to the northern part of the nemoral forest subzone of the north-temperate zone. The boundary between the West and East Baltic geobotanical zones of L. Laasimer in general coincides with the Upper and Lower Estonia boundary. Because of their more calcareous soils and maritime climate West and Northwest Estonia are richer in species (especially Atlantic and Subboreal species) than the rest of the mainland. East and South Estonia are characterized by Eurasian continental species, including the Pontic and Pontosarmatic element.

Based on the geology, geomorphology, and vegetation, seven type regions are distinguished in Estonia (Fig. 11.3). Type-region syntheses are presented for the following regions

E-b   West Estonia
E-d   Central Estonian Watershed
E-e   Peipsi and Vôrtsjärv Lowlands
E-g   Upper Devonian Plateau

No syntheses have been made for the other Estonian type regions.

Region E-b (West Estonia) belongs to the West Baltic geobotanical zone. Climate is mild, slightly maritime. Limestone bedrock and calcareous till favoured the spread of the broad-leaved trees *Ulmus*, *Tilia*, *Quercus*, *Corylus*, *Acer*, and *Fraxinus*.

Region E-d (Central Estonian Watershed) together with North Estonia and Middle Devonian Plateau (not included) are the main cereal-producing areas of Estonia today. The topography is varied; bedrock mostly consists of Silurian carbonaceous rocks and is covered by a thick Quaternary mantle with fertile soils. The proportion of spruce forests is high, containing an admixture of *Betula*, *Pinus*, *Alnus*, *Acer*, and *Corylus*.

The relatively flat region E-e (Peipsi and Vôrtsjärv Lowlands) is characterized by large mires and forests. Climate is moderately continental. Podzols and peaty and gley soils are covered by coniferous forests with *Betula* stands.

Region E-g comprises the Upper Devonian bedrock outcrop area. Topography is undulating with a mosaic vegetation of mostly *Pinus*, *Picea*, and *Betula* forests. Climate is continental with prolonged snow cover.

Palaeoecological Events During the Last 15 000 Years: Regional Syntheses of Palaeoecological Studies of Lakes and Mires in Europe.
Edited by B.E. Berglund, H.J.B. Birks, M. Ralska-Jasiewiczowa and H.E. Wright.   © 1996 John Wiley & Sons Ltd.

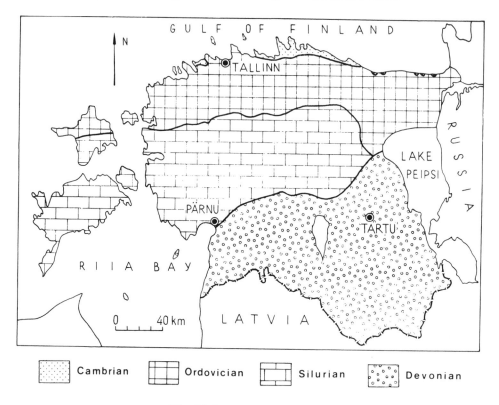

Fig. 11.1   Bedrock in Estonia

TYPE REGION E-b, WEST ESTONIA
(L. Saarse)

Saaremaa Island is in the area of Weichselian glaciation and was uncovered before 11200 BP.

Altitude: 0–54 m, mostly about 15 m.

Climate: Mean February temperature –4 to –5°C, July +16.5°C, precipitation 500–550 mm yr⁻¹. Maritime type.

Geology: Silurian calcareous bedrock covered by a thin Quaternary mantle, commonly less then 5 m thick but up to 25 m thick in endmoraines and 150 m in ancient valleys. Tills, fluvioglacial sands, limnoglacial varved clays, and marine sands are the most widespread drift deposits.

Topography: The Western Saaremaa Upland, 50 km long, 5–6 km wide, and with the highest point at 54 m, is situated above the Ancylus shore line (Fig. 11.4). Ancylus beach ridges, dunes, and escarpments

are widespread on its slopes. Below these on the marine plain are Limnea and Litorina Sea terraces and coastal formations. Along the modern sea-shore there are well-preserved Silurian klint cliffs, gentle depressions of lake–river systems, mires, and the unique meteorite crater of Lake Kaali.

Population: 32800 inhabitants on 2668 km². Sparsely populated areas with 13 inhabitants km⁻², and with the small town of Kuressaare and the town-type settlement of Orissaare. The density of the rural population is 7.7 persons km⁻².

Vegetation: Forest dominated by *Pinus, Betula, Alnus, Corylus*, and the broad-leaved trees makes up 44.3% of the region. Cultivated land (arable, grass-lands, gardens) accounts for 17.7%. Human impact on modern vegetation is moderate, mostly caused by land cultivation, limestone mining, gravel excavation, and tourism.

Soils: Coarse-grained soils on limestone, podzolic sandy soils on the marine plain, alvar rendzinas,

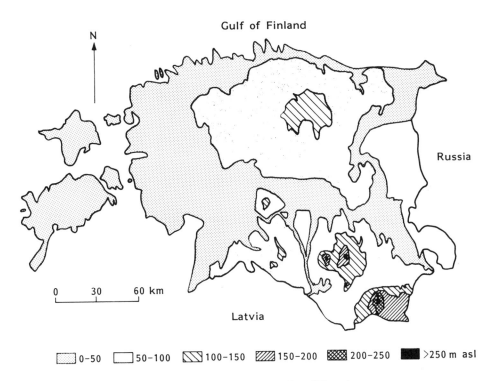

Fig. 11.2 Modern topography of Estonia

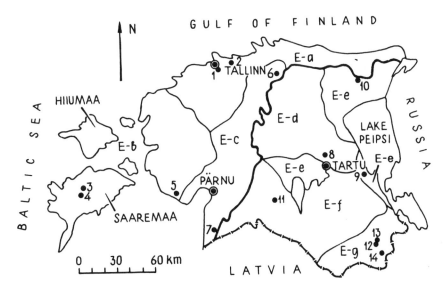

Fig. 11.3 IGCP 158 B type regions in Estonia (Saarse & Raukas 1984). Type regions: E-a–North Estonia, E-b– West Estonia, E-c–Intermediate Estonia, E-d–Central Estonian Watershed, E-e–Peipsi and Võrtsjärv Lowlands, E-f–Middle Devonian Plateau, E-g–Upper Devonian Plateau. Studied lakes and mires. 1–Ülemiste, 2–Maardu, 3–Pelisoo, 4–Karujärv, 5–Ermistu, 6–Viitna, 7–Nigula, 8–Raigastvere, 9–Saviku, 10–Kalina, 11–Päidre, 12–Vaskna, 13–Tuuljärv, 14–Kirikumäe

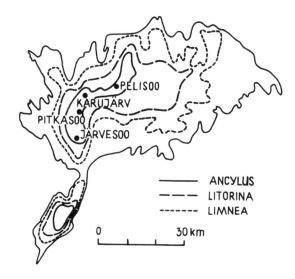

Fig. 11.4   Reference sites in type region E-b in Saaremaa Island, with Baltic shore lines

calcareous gley soils on weakly drained tills and varved clays, and slightly saline primeval soils on the modern neotectonically rising coast.

Land use: Fields, pastures, cultivated grasslands, woody meadows, town and rural settlements, fisheries, roads, forestry, recreation, tourism, peat cutting, limestone quarries, and gravel pits.

Archaeology: First settlement at the end of Mesolithic. Comparatively dense populations since the Late Neolithic.

Reference sites: Pelisoo, Karujärv, Järvesoo, Pitkasoo (Fig. 11.4).

Reference site 3. Pelisoo

Latitude 58°26′N, Longitude 22°13′E. Elevation 32.5–34.3 m. Age range 9000–0 BP. Raised bog on an overgrown lake. Three pollen diagrams, five radiocarbon dates (Saarse *et al.* 1990). 10 local pollen-assemblage zones (paz). (Fig. 11.5)

1.	?–8600 BP	*Pinus–Betula*
2.	8600–7700 BP	*Pinus–Betula–Ulmus*
3.	7700–5200 BP	*Ulmus–Alnus–Corylus*
4.	5200–4250 BP	*Tilia–Alnus–Pinus*
5.	4250–3750 BP	*Picea*

6.	3750–2800 BP	*Quercus–Tilia–Picea*
7.	2800–1800 BP	*Betula–Pinus–Alnus–Picea*
8.	1800–1000 BP	*Pinus–Betula–Picea–Alnus*
9.	1000–800 BP	*Picea–Pinus–Betula*
10.	800–0 BP	*Pinus–Betula*

The pollen diagram

The first decline in tree pollen is at the end of the Atlantic, about 5200 BP (Fig. 11.5), when *Pinus*, *Betula*, and *Corylus* diminish, and *Alnus*, *Tilia*, and Ericales rise. In the second half of the Early Subboreal the openness of the landscape decreased (Fig. 11.6), mostly due to the expansion of *Pinus*. The short-term forest regeneration was followed by a long-lasting deforestation (3750–2000 BP), caused by intensive human activities: cattle breeding, clearances, and primitive land cultivation. During that time *Picea* culminated, and *Pinus* and *Betula* then started to expand. Between 2900 and 2400 BP *Tilia* was cut down near Pelisoo. It reappeared between 2400 and 800 BP but has been absent since then.

Between 2000 and 600 BP the next forest expansion took place, first *Pinus* and *Betula* and then *Alnus* and *Corylus* coppices. Distinct forest clearance occurred 500–300 BP, accompanied by a rapid increase in *Secale cereale* pollen (up to 2%).

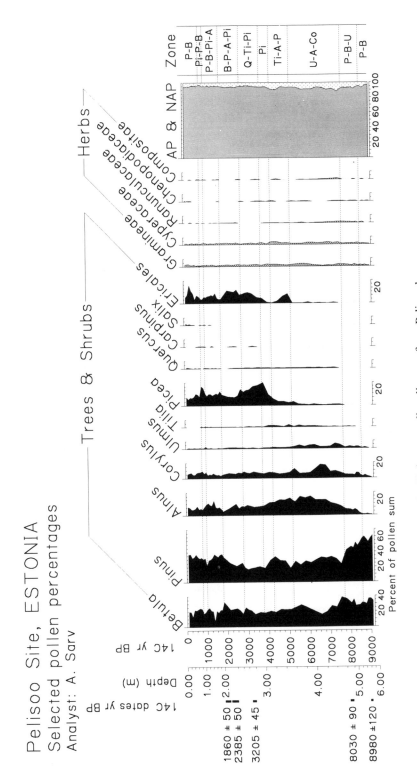

Fig. 11.5 Holocene pollen diagram from Pelisoo bog

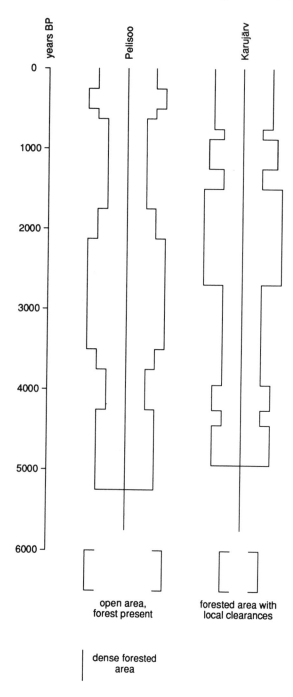

Fig. 11.6   Schematic interpretation of landscape openness
in type region E-b

On the basis of the Pelisoo diagram we can recognize a sharp *Ulmus* decline. *Tilia* declines at 5000 BP, with a new rise at *ca.* 4000 BP. The single *Carpinus betulus* grains, perhaps transported from a distance, were found at 3250 BP. *Fagus sylvatica* grains appear later, i.e. at 2000 BP.

Reference site 4. Karujärv

Latitude 58°22′N, Longitude 22°07′E. Located on the centre of the Western Saaremaa Upland (Fig. 11.4) at an elevation of 32 m. Age range 8000–0 BP. Eutrophic lake with area of 330 ha. Three pollen diagrams and 10 radiocarbon dates. 10 local pollen-assemblage zones (paz). (Fig. 11.7)

1.	?–7800 BP	*Pinus–Betula–Corylus*
2.	7800–7100 BP	*Ulmus–Pinus–Corylus–Alnus*
3.	7100–6600 BP	*Ulmus–Alnus–Corylus–Tilia*
4.	6600–5000 BP	*Tilia–Alnus–Corylus–Quercus*
5.	5000–4100 BP	*Quercus–Alnus–Picea–Ulmus–Tilia–Corylus*
6.	4100–3500 BP	*Picea–Betula–Alnus*
7.	3500–2600 BP	*Betula–Pinus–Alnus–Picea*
8.	2600–1700 BP	*Pinus–Betula–Picea–Alnus*
9.	1700–1100 BP	*Pinus–Betula–Picea*
10.	1100–0 BP	*Pinus–Betula*

The surroundings of Lake Karujärv were densely forested during the Late Boreal and Atlantic. The first deforestation, caused by Neolithic clearance, started about 5000 BP (Fig. 11.6), and lasted for about a thousand years. When this first interference lessened, abandoned areas were reforested, mostly during the late Neolithic and Bronze Age. The second deforestation started at the beginning of the Pre-Roman Iron Age (*ca.* 2600 BP), with short-term reforestation periods at 1500 and 900–800 BP, lasting up to modern times.

The first distinct *Ulmus* decline occurred about 5100 BP, the second one about 4000 BP. Between these a rather clear *Ulmus* rise is traceable in the Early Subboreal. The *Tilia* curve has the same trend as *Ulmus*, with its first decline a little earlier (5200 BP) and the second at 4300 BP. *Tilia* culminated 900–1000 years later than *Ulmus*.

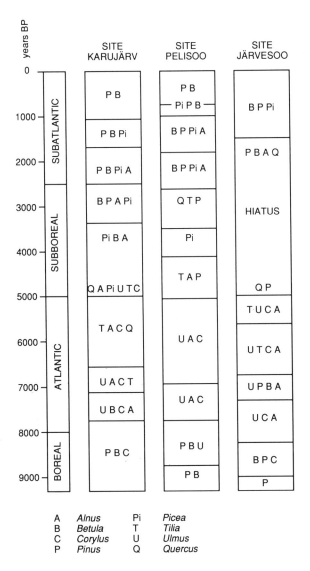

years BP

| | SITE KARUJÄRV | SITE PELISOO | SITE JÄRVESOO |

SUBATLANTIC
SUBBOREAL
ATLANTIC
BOREAL

SITE KARUJÄRV:
P B
P B Pi
P B Pi A
B P A Pi
Pi B A
Q A Pi U T C
T A C Q
U A C T
U B C A
P B C

SITE PELISOO:
P B
Pi P B
B P Pi A
B P Pi A
Q T P
Pi
T A P
U A C
U A C
P B U
P B

SITE JÄRVESOO:
B P Pi
P B A Q
HIATUS
Q P
T U C A
U T C A
U P B A
U C A
B P C
P

A	Alnus
B	Betula
C	Corylus
P	Pinus

Pi	Picea
T	Tilia
U	Ulmus
Q	Quercus

Fig. 11.7 Time-space correlation of Saaremaa pollen-assemblage zones. The additional site Järvesoo is situated on SW Saaremaa

dence, especially by the rapid increase in population (Eesti esiajalugu 1982).

The second deforestation, which started at the beginning of the Subatlantic in the Pre-Roman Iron Age, brought about a sharp decrease in *Ulmus, Tilia, Corylus,* and *Quercus* and a contemporaneous rise of *Betula, Alnus,* and *Pinus.* Such changes in plant communities were caused by climatic change and by slash-and-burn agricultural practice. *Betula* and *Alnus* were the first immigrants into the abandoned fields. According to Laasimer (1981), the Subatlantic maximum of *Picea* is also connected with human activities, as it coincides with the rapid erection of settlements and more advanced land cultivation. In the present-day vegetation the deciduous forests have remained only as small groves in maritime regions, where the conditions for *Picea* as the main competitor are not especially favourable (Kalda 1981).

Palaeoecological patterns and events

General patterns

Betula and *Pinus* immigrated into Saaremaa Island during the Late-Glacial, *Corylus* and *Ulmus* in the Preboreal, *Alnus* a little later about 9000 BP. *Ulmus* expanded at 9000 BP but culminated at 7500–5200 BP. Its first decline started *ca.* 5200–5000 BP. *Tilia* immigrated 8200 BP, expanded 7600 BP, culminated at 6500–5200 BP, diminished 5200–5000 BP, and disappeared in the Late Subatlantic. *Picea* immigrated at 8100 BP, expanded after 6600 BP, and culminated in the Late Subboreal and mid-Subatlantic. Single cereal pollen grains were found in the Subboreal. Very distinct *Secale* pollen appeared at 500–300 BP.

Hydrological events (Fig. 11.8)

On the basis of sediment lithology and deposition rates, diatom succession, abundant molluscan fauna (Männil 1963), and the aquatic pollen types, it may be concluded that the water level in the Saaremaa Island lakes was rather high at 7800, 6800, 5300, 4900, 1400 and 500 BP (Fig. 11.8).

During the first deforestation in the Early Subboreal the *Betula* and *Pinus* pollen amounts decreased remarkably, but *Ulmus* and *Tilia* had their second rise against the background of the *Quercus* culmination. In this region the most important factor responsible for the *Ulmus* decline seems to be human activities, as proven by archaeological evi-

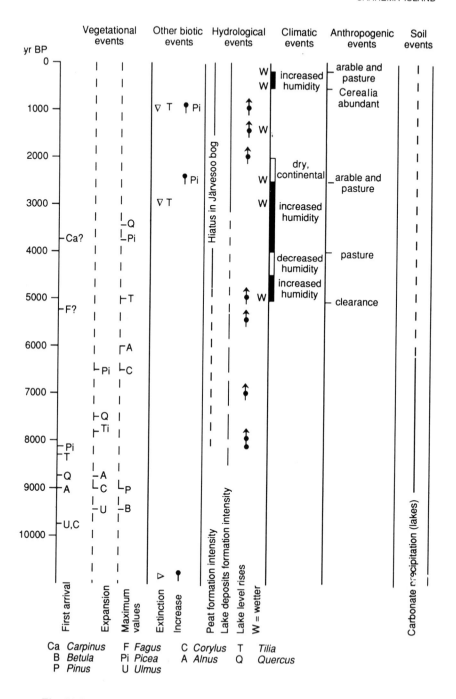

Fig. 11.8 Palaeoecological events in type region E-b (Saaremaa Island)

Archaeological events (Fig. 11.9)

Two Late Mesolithic and four Early Neolithic settlements are known on Saaremaa Island. In the Late Neolithic (4200–3500 BP) domestication of animals started, first of all wild boars, favoured by the simultaneous expansion of oak forests. Population increased rapidly, and since the end of the Neolithic primitive cattle breeding started and became the main focus of human activities during the Bronze Age (Eesti esiajalugu 1982). During the Late Bronze Age fortified settlements appeared to defend the peasants' surplus production. The climate must have been mild, as cattle stayed out over the winter. According to osteological studies, 75–80% of bones belong to domestic animals (Eesti esiajalugu 1982). The main crops were barley, oats, wheat, and flax.

During the Pre-Roman Iron Age (2600–1900 BP) the climate deteriorated, becoming more continental and drier and leading to decreased land use and diminished human influence. In the Roman Iron Age (1900–1500 BP), besides the farming on long-fallow land and burnt-over clearings (slash- and-burn agriculture), permanent fields were cultivated. After 1800 BP land cultivation decreased. The light and most fertile soils were impoverished and abandoned. In the middle Iron Age (1500–1100 BP) the density of population grew, as the yields of crops rapidly increased because of fertilization of permanent fields. During the Late Iron Age (1100–800 BP) agriculture made noticeable progress. All suitable arable lands were cultivated. Rye, barley, wheat, oats, and flax were the main crops, and a three-field rotation system was used.

Conclusions

Saaremaa Island serves as a special region for studying palaeoecological and anthropogenic events. Its

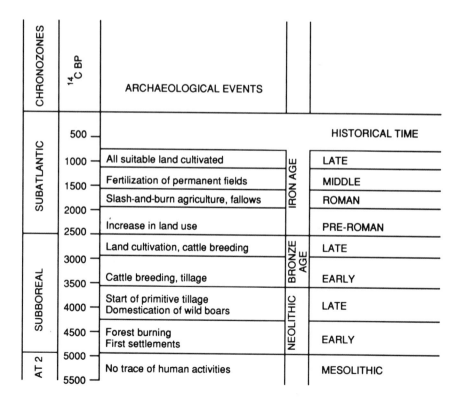

Fig. 11.9   Archaeological events on Saaremaa Island

modern climate is more maritime than that on the mainland, and it seems to have been so throughout the entire Holocene. Accumulation of peat and lacustrine deposits started here *ca.* 9200–9000 BP. The immigration, expansion, and culmination of different trees are now clarified, as well as re- and deforestation patterns. A good accordance of palaeoecological and archaeological events has resulted.

The first clearance started at the end of the Mesolithic with the appearance of permanent settlers. Bronze Age cattle breeding brought about new clearances. Later on, when the most suitable arable lands were impoverished and partly abandoned, they became reforested again, so the advanced agriculture did not cause permanent landscape openings (Fig. 11.6).

Concerning the climate, two drier periods at 4500–3800 and 2200–1700 BP were identified, with three more humid phases at 5000–4500, 3700–2700, and 500–300 BP. During the Late Bronze Age the climate must have been rather mild (the cattle stayed out all winter), but during the Pre-Roman Iron Age it deteriorated once more, as shown by diminished land cultivation, slow peat growth or complete hiatus in some bogs, and rapid expansion of *Betula* and *Pinus*.

Saaremaa Island is also important with its alvars and rare plant species (Kask & Laasimer 1987, Kukk 1987). Originally the alvars were also forested, but burnt-over clearings and grazing caused their deforestation (Laasimer 1981). Now, when the grazing has stopped, the alvars themselves have begun to grow over naturally.

TYPE REGION E-d, THE CENTRAL ESTONIAN WATERSHED
(R. Pirrus and A.-M. Rôuk)

Four main landscape types exist in the Central Estonian watershed type region E-d, the Vooremaa reference area (Saadjärv drumlin field) being palynologically studied in greatest detail. The other main landscapes are the Pandivere Uplands, Central Estonian moraine plain, and Türi drumlin field. They all belong to the same geobotanical region, so the development of the vegetation is to some extent similar, in spite of differences in geology and geomorphology. The characteristic features of the Vooremaa reference area are large drumlins in rather thick Quaternary deposits, fertile soils, elongated lakes in troughs between drumlins, and the domination of cultural landscapes.

Vooremaa reference area (Saadjärv drumlin field), east Estonia:

Altitude: 38–144 m.

Climate: Mean February temperature −6.5 to −7.5°C, July 16–17°C. Precipitation 600–650 mm yr^{-1}. Approximately 185 rainy days yr^{-1}. Transition from maritime to continental.

Geology: Weichselian till, limnoglacial clay and loam, fluvioglacial sand and gravel, Holocene gyttja and lacustrine marl, peat.

Topography: Large drumlins of NW–SE orientation, separated by elongated depressions filled with peat, lacustrine deposits or elongated lakes. A few moraine plains, kame areas, eskers, and moraine hillocks.

Population: Approx. 14 people km^{-2}, mainly small villages; one small town.

Vegetation: Most wetlands are dominated by fens fed by groundwater rich in minerals, partly by surface and flood water. Elongated eutrophic lakes. Remnants of former mixed forests are few and small. Forest occupying approx. 11% of the area, mainly in paludified depressions, are dominated by *Picea*, *Betula*, and *Alnus*. Also present are *Tilia*, *Ulmus*, *Quercus*, *Fraxinus*, and *Corylus*.

Soils: Brown soils, pseudopodzolic brown soils, and peat.

Land use: Approx. 65% agricultural area, 11% forest, 21% uncultivated, 1.6% lakes. Rye, wheat, barley, grass, and green fodder are the main crops.

Reference site 8. Lake Raigastvere
(Pirrus *et al.* 1987a)

Latitude 58°35′N, Longitude 26°39′E. Elevation 51.8 m. Age range 10000–0 BP. Lake. Holocene lake sediments: calcareous gyttja and clayey algal gyttja. 14 pollen-assemblage zones (paz).

1. 10000–9700 BP   *Betula–Salix*
2. 9700–9100 BP   *Betula*
3. 9100–8500 BP   *Betula–Pinus*
4. 8500–7850 BP   *Betula–Corylus–Ulmus–*
                   Alnus
5. 7850–6600 BP   *Ulmus–Corylus–Alnus*
6. 6600–5000 BP   *Tilia–Ulmus–Corylus–*
                   Quercus–Alnus
7. 5000–3800 BP   *Quercus–Alnus*
8. 3800–3500 BP   *Picea* paz I
9. 3500–3300 BP   *Betula–Alnus–Picea*
10. 3300–2800 BP   *Picea* paz II
11. 2800–1900 BP   *Betula–Alnus–Pinus*
12. 1900–1500 BP   *Picea* paz III
13. 1500–800 BP   *Betula–Pinus*
14. 800–0 BP   *Alnus–Betula–Pinus*

One pollen diagram from lake sediments is presented (Fig. 11.10). In addition to the Raigastvere reference site there are quite complete pollen diagrams for lacustrine sediments from the lakes of Soitsjärv (Pirrus & Rôuk 1979), Pikkjärv (Pirrus & Rôuk 1988), and Kuremaa (Pirrus *et al.* 1988) and from a kettle hole Siniallika (Pirrus *et al.* 1987b) occurring within the boundaries of Saadjärv drumlin field.

[14]C dates have been obtained on the sediments from the Raigastvere reference profile and the Kuremaa profile. The latter are not reliable because of hard-water effects. In the future the correlation of ^{14}C dates on peat will provide fewer sources of error. There are 11 pollen-assemblage zones for Vooremaa reference area.

(1) *Betula–Pinus–Salix* paz. *Betula* dominant. *Pinus* and *Salix* decreasing.
(2) *Betula* paz. *Betula* dominant. Upper zone border at the *Ulmus* arrival and an increase in *Pinus*.
(3) *Betula–Pinus* paz. *Betula* dominant. *Pinus* increase followed by *Pinus* maximum. *Alnus* arrival. Upper zone border at the *Corylus* and *Ulmus* expansion.
(4) *Betula–Corylus–Alnus–Ulmus* paz. *Betula* dominant, but lower. *Corylus, Ulmus,* and *Alnus*

expansion. *Tilia* arrival. Upper zone border at an increase in *Ulmus, Corylus,* and *Alnus* and a decrease in *Betula*.
(5) *Ulmus–Alnus–Corylus–Tilia* paz. *Ulmus, Alnus,* and *Corylus* dominant, *Tilia* increasing. *Betula* decreasing. Upper zone border at the distinct increase in Quercetum mixtum, including *Quercus*.
(6) *Tilia–Ulmus–Corylus–Alnus* paz. Quercetum mixtum and *Alnus* reach their culmination. Upper zone border at a decrease in *Ulmus* and *Corylus* and an increase in *Quercus*.
(7) *Quercus–Alnus* paz. *Quercus* and *Alnus* dominant. *Quercus* culmination. *Ulmus, Tilia, Corylus,* and *Alnus* lower. *Picea* expansion. Upper zone border at an increase in *Picea*.
(8) Lower *Picea* paz. *Picea* dominant. *Alnus* decreasing. Upper zone border at a decrease in *Picea* and an increase in *Betula*.
(9) *Betula–Pinus–Alnus* paz. *Betula* dominant. *Pinus* and *Alnus* slightly increasing. *Picea* decreasing. Upper zone border at an increase in *Picea*.
(10) Upper *Picea* paz. *Picea* dominant. *Betula* and *Alnus* decreasing. Upper zone border at a decrease in *Picea*.
(11) *Betula–Pinus–Alnus* paz *Betula* and *Pinus* increasing. *Betula* dominant.

Palaeoecological patterns and events

Common patterns

(1) The early-Holocene pollen-assemblages are dominated by *Betula*.
(2) Extremely low *Pinus* frequency in all Holocene pollen assemblages.
(3) In the Boreal *Pinus* increases but does not become dominant.
(4) High concentration of Quercetum mixtum in Atlantic as well as in late Boreal and early Subboreal.
(5) *Corylus* and *Ulmus* expansion at *ca.* 8500 BP followed by *Alnus*.
(6) *Quercus* expansion *ca.* 6500 BP.
(7) *Ulmus* and *Corylus* decline *ca.* 5000 BP.
(8) First Cerealia 2800–2000 BP.

378

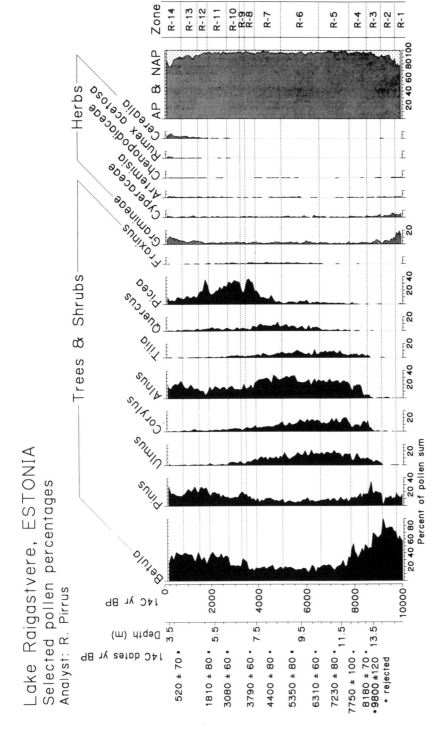

Fig. 11.10 Pollen diagram of Raigastvere reference site

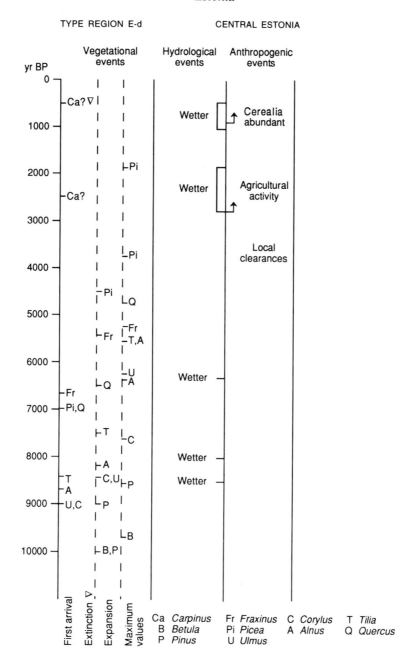

Fig. 11.11   Palaeoecological events in type region E-d, Central Estonian Watershed (Vooremaa reference area)

Unique patterns

(1) The borders of the Lower and Upper *Picea* pollen-assemblage zones to a certain extent may be asynchronous in different parts of the area.

(2) The Lower *Picea* paz is common for most sites of the Vooremaa area. In the middle of it the *Betula–Alnus–Picea* paz (3500–3300 BP) is characteristic of the Raigastvere reference site.

(3) Above *Picea* paz III we cannot separate two paz of Vooremaa area.

Anthropogenic events

Local forest clearance as a result of human impact is inferred at *ca.* 3500 BP (Fig. 11.11). In order to have more open landscapes the forest was burnt down in places, and as a result *Picea* was temporarily replaced by *Betula* and *Alnus*. After that *Picea* dominated again.

Signs of early agricultural activities are traceable *ca.* 2800 BP. Cerealia expansion and open-area increase show that a significant increase in human activities took place *ca.* 1000 BP. Within the boundaries of the Vooremaa area as a whole, cultural expansion seemed to have differed in time and extent.

Conclusions

The development of the vegetation with its specific characteristics in the Vooremaa area was closely related to the fertility of soils and the variability of landscape components. Of great significance were also the peculiar topographic and hydrological conditions.

Within the Central Estonian watershed area the pollen diagrams of the Vooremaa reference area have several features in common with those obtained for small lakes and mires in the Pandivere Uplands.

TYPE REGION E-e, PEIPSI LOWLAND
(L. Saarse, A. Sarv and E. Ilves)

The region became deglaciated before 12 000 BP, at the end of the Pandivere stage.

Altitude: 30–70 m, large areas about 30–40 m.

Climate: Mean February temperature –6.5 to –7.0°C, July +17°C, with extreme values –38 and +32°C. Precipitation average 600 mm yr^{-1}, with fluctuations between 350 and 800 mm yr^{-1}. Continental type.

Geology: Ordovician and Middle Devonian carbonate rocks and sandstone are covered by Quaternary deposits with a thickness up to 10–15 m in ancient valleys and 40–80 m in esker ridges and kame fields. Tills and fluvioglacial, limnoglacial, eolian, alluvial, and limnic deposits and terrestrial peat are the most common surface-sediment types.

Topography: Very flat, previous limnoglacial plain, with esker ridges and Illuka kame field north of Lake Peipsi.

Population: Sparsely populated, with less than 10 people km^{-2}. Rural settlements.

Vegetation: Highly paludified area, mainly forested with swamps and peatland forests. Within the forests *Pinus* forms 41%, *Betula* 35%, and *Picea* 17%. The total area of mixed forest is 42%. The western part of Peipsi basin on the Suur-Emajôgi River mouth is mostly paludified, with large natural grasslands on the floodplains (*Carex elata*, *Phalaris arundinacea*, and *Carex lasiocarpa* meadows; Laasimer 1964).

Hydrology: Two maxima in groundwater table are common–in April–May and October–November, with low levels in March and August.

Soils: Mostly podzols, half-hydromorphic, gley, and hydromorphic soils.

Land use: Forests, pastures, fields, meadows, swamps, raised bogs, forestry areas, peat cutting, oil-shale mines, gravel pits, rural settlements.

Archaeology: First settlements in Early Neolithic.

Reference sites: Studies on the chosen reference lakes Rääkjärv and Lahepera are in progress. This region is characterized by the mires Saviku and Kalina and by Lake Rääkjärv. The other sites studied in this region are Lake Ümarjärv and the mires Liivjärv, Alasooo, and Kurtna.

Reference site 9. Saviku

Latitude 58°26′N, Longitude 27°15′E. The average elevation is 30–31.3 m, maximum 34 m, below the highest shore line of the Baltic Ice Lake. The Saviku sequence is in the large mire complex at the Suur-Emajôgi River mouth, with a total area of more than 250 km². Rivers and streams divide the mire into several smaller units. It is a mostly minerogenic mire (155 km²), with smaller mesotrophic and ombrotrophic belts.

The bio- and chronostratigraphically studied sequence is located on the left bank of the Suur-Emajôgi River (Fig. 11.12). Peat deposits 4.90 m thick are underlain by gyttja 3.15 m thick. The transition between peat and gyttja lies at an elevation of 25.1 m and is dated to 5800 BP (Sarv & Ilves 1975). One pollen diagram, 13 radiocarbon dates. 11 local pollen-assemblage zones (paz). (Fig. 11.13)

1.	?–10200 BP	*Pinus–Betula–Salix*
2.	10200–9000 BP	*Betula*
3.	9000–7700 BP	*Pinus–Betula*
4.	7700–5800 BP	*Alnus–Ulmus–Corylus*
5.	5800–4700 BP	*Tilia–Ulmus–Corylus*
6.	4700–4000 BP	*Alnus–Quercus*
7.	4000–3000 BP	*Picea*
8.	3000–2300 BP	*Betula–Alnus*
9.	2300–1000 BP	*Picea–Pinus*
10.	1000–700 BP	*Betula–Pinus*
11.	700–0 BP	*Pinus*

In the preparation of the Saviku pollen diagram (Fig. 11.13) one radiocarbon date (7110±70 BP, TA-326) was rejected because it implied a hiatus between 7200 and 8000 BP, not otherwise indicated. Nevertheless, the upper limit of the *Pinus–Betula* paz. is still under question.

The first *Ulmus* decline is dated at 5200 BP, with a minimum at 4900 BP, and the second decline at 4200 BP. *Tilia* representation in the Saviku diagram (Fig. 11.13) is moderate without any sharp maximum. One well-expressed clearance occurred 2900–2200 BP, when the forest area rapidly decreased and was replaced by dry-soil herbs. The increase in *Betula* pollen percentages, the appearance of *Plantago lanceolata*, and the Akali settlement all point to human activities. The other features – wet-soil herbs diminishing, dry-soil herbs flourishing – suggest a decline in the groundwater level.

The Middle and Late Subatlantic periods are characterized by high peat growth and by expanding *Pinus* and *Picea* forests, with one sharp *Betula* peak and synchronous *Pinus* and *Picea* decreases about 900 BP (Fig.11.13). It was caused by short-term clearances or by large forest fires. The first Cerealia pollen appeared at 2200 BP (Fig.11.13).

Reference site 10. Kalina mire

Latitude 59°16′N, Longitude 27°16′E. Elevation 70 m, above the highest shore line of the Baltic Ice Lake. Mire area 1210 ha. The central ombrotrophic part is surrounded by meso- and minerotrophic mire belts. Average peat thickness is 3.1 m. In the biostratigraphically studied sequence the till at a depth of 3.02 m (from the mire surface) was covered by gyttja (3.02–2.75 m), reed-peat (2.75–2.50 m), woody-reed peat (2.50–2.40 m), reed-*Sphagnum* and woody-*Sphagnum* peat (2.40–1.95 m), and *Sphagnum fuscum* peat (1.95–0 m; Ilves & Sarv 1969). One pollen diagram,

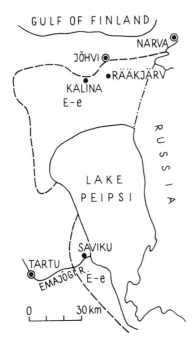

Fig. 11.12 Type region E-e in Peipsi Lowland, showing the reference sites

382

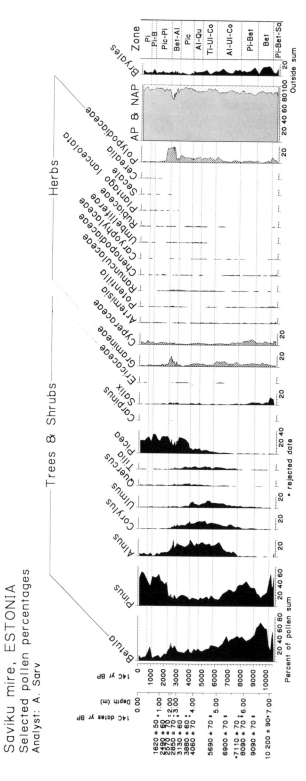

Fig. 11.13 Holocene pollen diagram of mire Saviku

12 radiocarbon dates. 11 local pollen assemblage zones (paz). (Fig. 11.14)

1.   ?–9100 BP   *Betula–Salix*
2. 9100–8000 BP   *Betula–Pinus*
3. 8000–6300 BP   *Alnus–Corylus–Ulmus*
4. 6300–5400 BP   *Ulmus–Alnus–Tilia*
5. 5400–4700 BP   *Tilia–Ulmus–Corylus*
6. 4700–3600 BP   *Alnus–Quercus*
7. 3600–3300 BP   *Picea*
8. 3300–2600 BP   *Betula*
9. 2600–1700 BP   *Picea–Pinus*
10. 1700–800 BP   *Betula*
11.   800–0 BP   *Betula–Pinus*

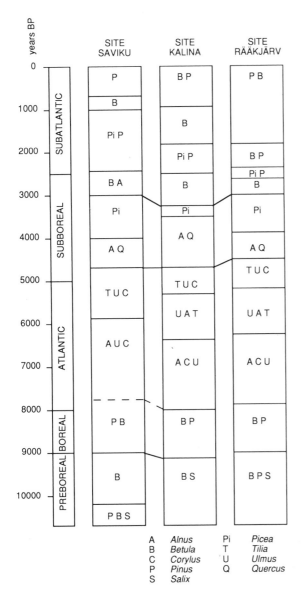

Fig. 11.14 Time-space correlation of type region E-e pollen-assemblage zones

Specific patterns in pollen diagram

(1) The first sharp decrease in the tree-pollen curve occurred about 6100 BP. It coincides with an expansion of *Picea*, an increase in Gramineae pollen, and a decrease in *Pinus* and *Corylus*. This level marks the transition from minerotrophic to mesotrophic mire.

(2) The first *Ulmus* decline is traceable at 6400 BP, minimum 6200 BP, the second decline about 5000 BP.

(3) Stepwise deforestation started about 1300 BP, accompanied by an *Artemisia* pollen increase.

(4) The peat accumulation rate was extremely low in the Early Subatlantic (0.13 mm yr^{-1}) and Late Subboreal (0.16 mm yr^{-1}; Ilves & Sarv 1969).

Palaeoecological patterns and events

General patterns

(1) *Betula* and *Pinus* immigrated into northeastern Estonia in Late-Glacial time, and *Ulmus*, *Corylus*, and *Alnus* in the Preboreal.

(2) Broad-leaved trees except *Quercus* started to expand in the Boreal. *Quercus* immigrated at the beginning of the Early Atlantic and expanded about 6000 BP.

(3) *Picea* immigrated at *ca.* 7000 BP and started to expand about 5700 BP, although its single pollen grains are present in Preboreal and Boreal spectra.

(4) The Boreal period in this region is represented by *Pinus–Betula* or *Betula–Pinus* paz (Fig. 11.14). In the Atlantic period three paz are most common, not two, as is shown in the local unified stratigraphical scheme for the Baltic region (Kajak *et al.* 1976).

TYPE REGION E-e   EAST AND NORTH ESTONIA

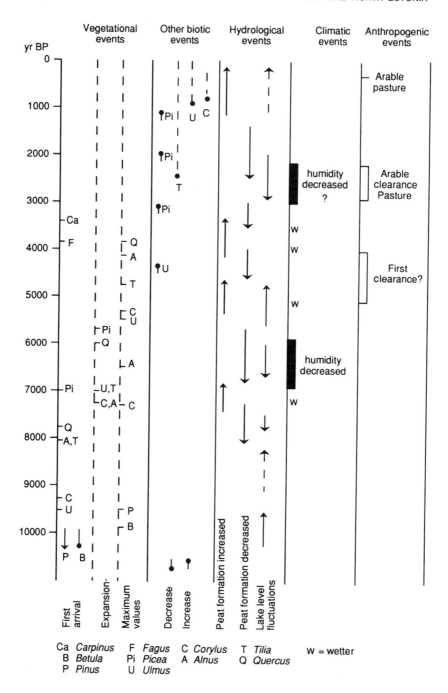

Fig. 11.15   Palaeoecological events in type region E-e (Peipsi Lowland)

Hydrological events (Fig. 11.15)

Based on lithology, sediment-accumulation rate, aquatic pollen, and levelling of the terrace fragments in the Illuka kame field, it was concluded that at the beginning of the Preboreal lake-level was rather high, with a tendency to decrease. It stayed low in the Boreal, rose a little at the turn of Boreal, and sank again in mid-Atlantic, at about 6500–6200 BP (Ilomets 1987, Saarse 1987). Between 6200 and 4800 BP the lake-level rose, as shown by the intensified paludification in these regions. The lake and mires closely related to large Lake Peipsi experienced a transgression from the mid-Atlantic onwards, proved also by archaeological evidence (Jaanits 1959). This rather continuous transgression at the Saviku site was interrupted at 2900–2200 BP by a standstill or lowering. The forest area rapidly de-creased, wet-soil herbs diminished and dry-soil herbs flourished, showing the lowering in ground-water table. The synchronous increases in *Betula* and *Ulmus* stands indicate a new human invasion and forest clearance at the beginning of the Iron Age, partly caused by the need for timber for iron smelting. The high peat-accumulation rate in the Early Subboreal contradicts the climate aridity and suggests that this water-level lowering had a local character.

Archaeological events (Fig. 11.16)

Two Neolithic settlements, Akali and Kullamägi, are known to have existed in the region studied. Akali settlement stretched over the relatively large area of 2 ha. Five cultural layers with different

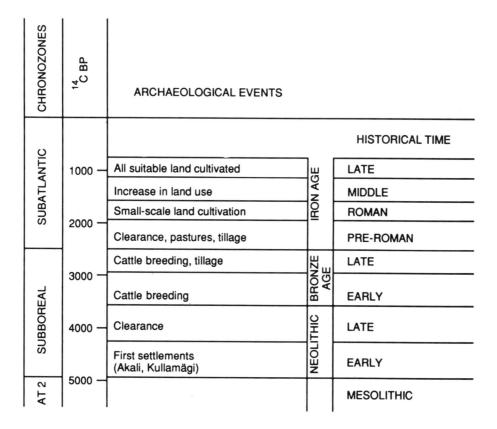

Fig. 11.16 Archaeological events in type region E-e (Peipsi Lowland)

ceramics have been studied, as well as stone and bone artefacts (Jaanits 1959, Moora *et al.* 1988). Akali was inhabited for about 2000 years, so it gives a good picture of the development of the tools and habits of its settlers.

The first Neolithic tribes who erected their settlement on the Akali River bank (1.5 km southeast of the Saviku site), were hunters and fishermen. According to osteological finds they hunted elk, wild boar, aurochs, beaver, deer, etc. (Eesti esiajalugu 1982). The water-level rise in the surroundings forced them to move upwards to small hummocks, and when the whole area was finally paludified at the end of the Bronze Age the settlement was abandoned at the beginning of Iron Age (Jaanits 1959). Now the lowest cultural layers are carpeted by a peat layer 2.5 m thick.

There is no clear evidence of the activities of the Akali settlers during the Early Bronze Age. Obviously they continued with fishing and hunting, but this by no means eliminates the start of cattle breeding. Due to intensive paludification this area stayed sparsely populated throughout the Pre-Roman and Roman Iron Age. In the Middle Iron Age, when tillage became the main means of living, the population density increased, and at the beginning of historical time the location of farmyards and rural settlements obtained their present features.

In general human impact during prehistoric time in the Peipsi Lowland was smaller than in North and West Estonia, for example, due to the wide distribution of hydromorphic and gley soils with poor drainage conditions.

Conclusions

Peipsi Lowland, with its large mires and wild forests, is a very specific region. Continental climate may have been the main reason why *Corylus*, *Ulmus*, and *Alnus* expanded here about 500 years later than in West Estonia, although by rough estimation the time of their culmination was the same.

Peat-formation intensified about 7000, 5000, 4000–3500, and 1200–0 BP, and decreased 8000–7000, 6700–5700, 4500–4000, 3500–3000, and 2500–1500 BP, reflecting the peculiarities in the

mire-system development as well as fluctuations in the hydrological regime. The wetter periods occurred 7000, 5000, 4000, and 3500 BP, and the more arid ones 6700–5800 and 2900–2200 BP.

The first clearance at the Saviku site took place 5000–4000 BP, the second 2900–2200 BP. The first Cerealia pollen was found in the deposits of the Late Subboreal (K. Kimmel unpubl. comm.) in Saviku at 2200 BP. *Plantago lanceolata* pollen appeared earlier, at 2900 BP.

TYPE REGION E-g, UPPER DEVONIAN PLATEAU (E. Ilves and H. Mäemets)

The region of Haanja Heights belongs to the area of Weichselian glaciation. It was deglaciated about 13500–13200 BP.

Altitude: 100–318 m, mostly above 180 m.

Climate: Mean January temperature −7°C, July +16.8°C, precipitation 740–800 mm yr^{-1}. Temperature amplitudes are continental, but elevation causes higher precipitation and the mosaic of hilly topography results in great variation in microclimatic conditions.

Geology: Upper Devonian sandstone, siltstone, and dolomite are covered by thick Quaternary mantle (50–180 m, Kajak 1968), formed during the different glaciations.

Topography: Highly undulating moraine relief, divided by valleys and depressions. The central highest part consists of morainic and fluvioglacial hills and plateaux, the latter often covered by limnoglacial clays. This area is surrounded by small-scale hills and hillocks, intersected by meltwater valleys (Arold 1979).

Population: Approximately 12 people km^{-2}.

Vegetation: The most characteristic forest types in the central part of Haanja are nemoral or boreal *Picea abies* forests. In marginal areas *Pinus* forests are dominant. Of the deciduous trees *Betula*, and *Alnus* and *Populus* are numerous. *Quercus*, *Fraxinus*, *Ulmus*, and *Acer* are less frequent, and *Tilia* is

387

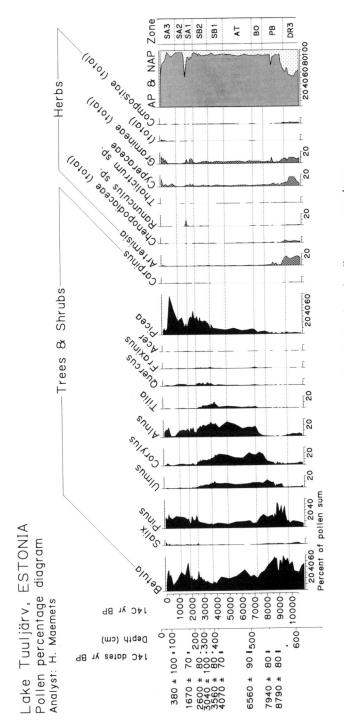

Fig. 11.17 Pollen diagram from Tuuljärv. Only selected pollen types are shown

present. During the last 30 years in the central part of Haanja the cultivated area has increased from 31% to 46%. Small, mostly eutrophic fens are numerous.

Soils: Podzolic and carbonaceous soils together with deluvial and wet deluvial soils dominate, suffering from erosion on steep slopes.

Land use: Approximately 40% forests, 27% arable land, 20% pastures. Small villages and rural settlements dominate. Forestry, gravel pits, tourism, traffic. Main crops: rye, barley, potatoes.

Archaeology: The Kääpa settlement (on the northern slope of Haanja Heights) with the first ceramic finds belongs to Early Neolithic. Villa and Tamula settlements with typical and late combed ware pottery. Bronze Age settlements are little known. Population increased since the Middle Iron Age, mainly on the marginal areas–Rôuge settlement (W slope of Haanja Heights).

Reference site 12. Lake Tuuljärv

Latitude 57°41′N, Longitude 27°08′E. Elevation 257 m above sea level, the highest situated lake in Estonia. Age range 10500–0 BP. One pollen diagram, nine radiocarbon dates. Nine local pollen-assemblage zones (paz).

1.	?–10000 BP	*Betula–Artemisia–*
		Chenopodiaceae–
		Helianthemum
2.	10000–8800 BP	*Betula–Pinus*
3.	8800–8000 BP	*Betula–Pinus–Ulmus*
4.	8000–4500 BP	*Alnus–Corylus–Ulmus*
5.	4500–3700 BP	*Tilia–Alnus–Ulmus*
6.	3700–2600 BP	*Quercus–Picea–Alnus*
7.	2600–1700 BP	*Picea*–nemoral herbs
8.	1700–400 BP	*Picea–Betula*
9.	400–0 BP	*Pinus–Betula*–Cerealia

The pollen diagram (Fig. 11.17)

Only one pollen diagram is selected from the few ¹⁴C-dated diagrams of Haanja Heights. General patterns are drawn on the basis of all pollen diagrams at our disposal. Type-region pollen assemblages have still

not been defined, as variability of the diagrams in the region is high. There is a great discrepancy between the pollen and ¹⁴C records. Commonly broad-leaved trees make their maximum in Late Atlantic, but according to the present dates their maximum appears in Early Subboreal.

Palaeoecological patterns and events

General patterns

In the Younger Dryas open areas dominated. *Betula, Juniperus, Salix, Hippophaë*, and *Populus* were represented, with *Pinus* and *Picea* presence. *Betula, Pinus*, and *Picea* immigrated to South Estonia during Late-Glacial, *Alnus, Ulmus* and *Corylus* in Preboreal. In the Preboreal the forest canopy was open in a number of areas.

There are no distinct single peaks of *Betula* in Preboreal or *Pinus* in the Boreal. Their percentages remain high at the beginning of the Atlantic. *Ulmus* immigrated in the Preboreal, and expanded at

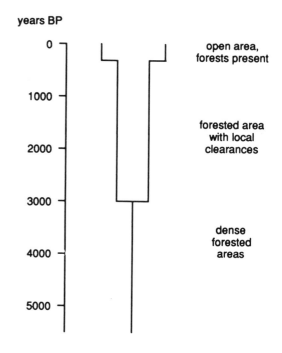

Fig. 11.18 Schematic interpretation of landscape openness at Tuuljärv

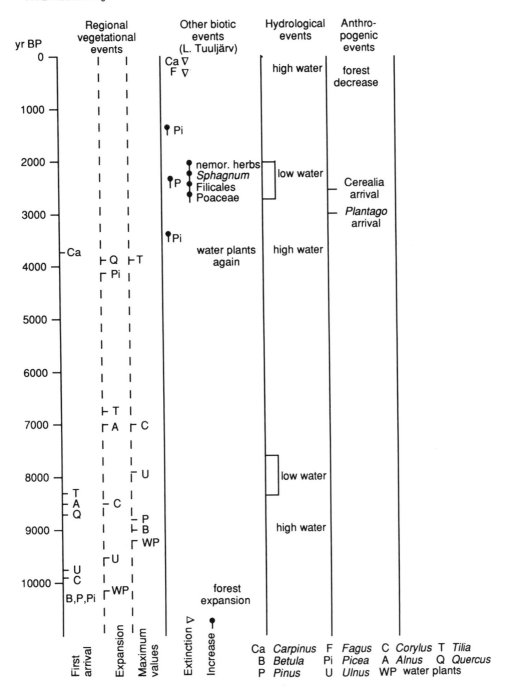

Fig. 11.19 Palaeoecological events in type region E-g (Upper Devonian Plateau)

the beginning of the Boreal without any clear culmination in the Early Atlantic. *Alnus* and *Tilia* expanded about 7000 BP, with *Tilia* culmination rather late, in the Early Subboreal. *Fagus sylvatica* and *Carpinus betulus* pollen is rare in the Atlantic but more regular in the Subboreal and Subatlantic. From the Atlantic onwards dense forest covered all the study area.

Pollen frequency of broad-leaved trees (excluding *Quercus*) and *Corylus* shows noticeable decreases between 2600 and 2000 BP.

The peaks of *Picea* are numerous, but it is difficult to use them in stratigraphy. There are remarkable percentages of *Picea* pollen already in the Atlantic period.

Weak anthropogenic indicators are traceable since the Early Subboreal. Strong human influence on the landscape began about 400 BP.

Unique patterns of Tuuljärv diagram

(1) High frequency of *Betula* pollen during the entire Holocene.
(2) Reflection of changes in coastal vegetation, especially in the period 2500–1700 BP, when many nemoral herbs are represented with high pollen frequencies.

Hydrological events (Fig. 11.19)

Hydrological changes have been studied on the basis of sedimentation rates and sediment lithology, together with frequencies of *Alnus*, aquatics, and other indicative taxa. It may be concluded that the water level in Tuuljärv was rather high during the Preboreal, first half of Boreal, and in the Late Atlantic. It decreased in the Late Boreal and stayed rather low in the Early Atlantic. The second noticeable lake-level lowering occurred in the Early Subatlantic.

Anthropogenic events

The first settlements were erected in Kääpa in the Early Neolithic and later on in Tamula, both north of the Haanja Heights. During the Neolithic the population seems to have been sparse, because of small lakes, clayey soils, highly undulated relief and dense forests, which did not favour fishing and

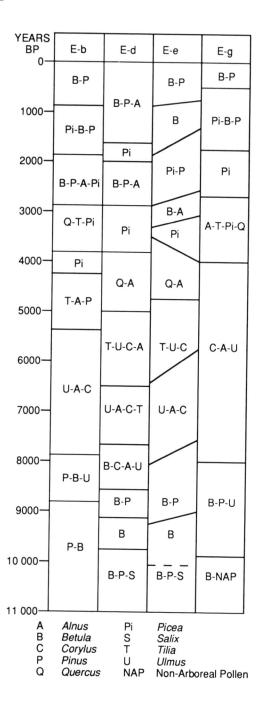

Fig. 11.20 Time correlation of regional pollen-assemblage zones in Estonia

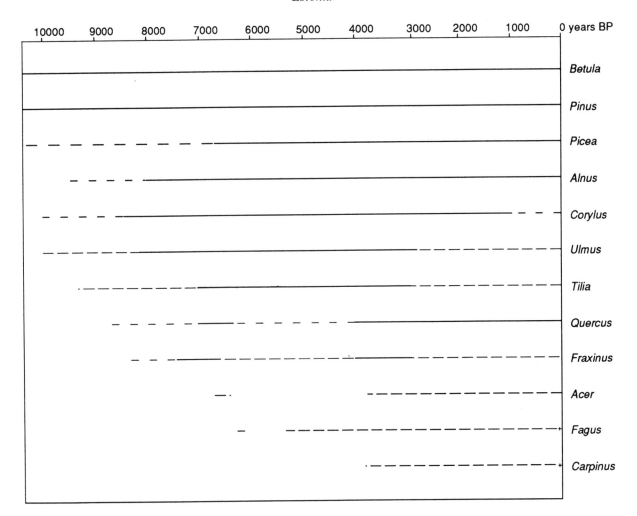

Fig. 11.21 Tree arrival and tree presence in South Estonia. Full line indicates frequent occurrence, broken line sparse occurrence, + extinct. Compiled by E. Mäemets

farming. Late Neolithic and Bronze Age archaeological evidence from South Estonia is rare (Eesti esiajalugu 1982), because this area was mainly used for hunting.

In the Roman Iron Age the population increased, and cultivation of fields expanded on the slopes of Haanja Heights. The most important evidence of the Roman period settlers is *tarand* graves. In the land cultivation slash- and-burn practice was in use. The fallow and three-field rotation were delayed in comparison with North and West Estonia. This is reflected also in the pollen diagrams, in which the

role of the anthropogenic indicators during the Bronze Age was weak. The first cereals and *Plantago lanceolata* appeared at the very beginning of the Pre-Roman Iron Age, about 2600 BP. In the central part of Haanja Heights only single archaeological finds are known.

Conclusions

The vegetation of the Haanja Heights differs from most of Estonia by its mosaic structure and by the prevalence of spruce forests in the central part.

Picea immigrated into South Estonia already in Late-Glacial and seems to have survived Preboreal and Boreal periods, being first suppressed by *Betula* and *Pinus*, later by *Corylus, Alnus,* and *Ulmus.* Expansion of spruce started in Atlantic.

Due to the sparse population, forest clearances are weakly expressed on the Tuuljärv diagram.

Lake-level was rather high in Preboreal and Early Boreal, followed by a low phase in Late Boreal. During the Atlantic it increased. Late Subboreal is characterized by an unstable lake level. New lowering occurred in the Early Subatlantic. Since the Middle Subatlantic lake-level has tended to rise.

CONCLUSIONS

Figures 11.20 and 11.21 give a simplified summary of the vegetational history based on the regional syntheses.

- *Betula* dominated in the preboreal forests everywhere.
- *Betula–Pinus* or *Pinus–Betula* forest phase followed, with *Ulmus* in the Upper Devonian bedrock area (E–g).
- The end of the *Betula* zone is dated to 8000–7700 BP, showing slight gradual difference in a S-N direction.
- The mesocratic stage lasted about 4000 years. The arrival and culmination of deciduous trees had a clear sequence: *Corylus + Alnus–Ulmus–Tilia–Quercus.*
- The first phase of the remarkable spread of *Picea* occurred around 4000 BP.
- The earliest traces of human activities are found on the archipelago about 5000–4000 BP.
- First cereal pollen in West Estonia and in the archipelago appeared *ca.* 4000 BP in Central Estonia 2800 BP.
- The first clearances started at the beginning of the Neolithic on Saaremaa and North Estonia, and in the Bronze Age on the other part of the mainland.

REFERENCES

Arold, I. 1979: Haanja kôrgustiku geomorfoloogiast ja maastikest. *In* Raukas, A. (ed.) *Eesti NSV saarkôr-gustike ja järvenôgude kujunemine,* 66–87. Valgus, Tallinn.

Eesti esiajalugu 1982: Jaanits, L., Laul, S., Lôugas, V. & Tônisson, E. (eds) Eesti Raamat, Tallinn.

Ilomets, M. 1987: Soode arengust Kurtna môhnastiku piirkonnas. *In* Ilomets, M.(ed.) *Kurtna Järvestiku looduslik seisund ja selle areng,* 47–54. Valgus, Tallinn.

Ilves, E. & Mäemets, H. 1987: Results of radiocarbon and palynological analyses of coastal deposits of lakes Tuuljärv and Vaskna. *In* Raukas, A. & Saarse, L. (eds) *Palaeohydrology of the temperate zone, III. Mires and lakes,* 108–130. Valgus, Tallinn.

Ilves, E. & Sarv, A. 1969: Die Stratigraphie und Chronologie der Organischen Ablagerungen im Hochmoor Kalina. *Eesti Teaduste Akadeemia Toimetised XVIII (4),* 377–384 (in Russian).

Jaanits, L. 1959: Neolithic and Early Iron Age settlers at the Suur-Emajôgi River mouth (Estonian SSR). 382 pp. Tallinn (in Russian).

Kajak, K. 1968: Kôrgustiku ja tema tuumiku vanus. *Eesti Loodus* 4, 200–202.

Kajak, K., Kessel, H., Liivrand, E., Pirrus, R., Raukas, A. & Sarv, A. 1976: Local stratigraphic scheme of the Quaternary deposits of Estonia. *In* Vaitekunas, P. & Gaigalas, A. (eds) *Stratigraphy of the Quaternary deposits of Baltic,* 4–42. Vilnius (in Russian).

Kalda, A. 1981: Human impact on the plant cover of Lahemaa National park. *In* Laasimer, L. (ed.) *Anthropogenous changes in the plant cover of Estonia,* 32–45. Tartu.

Kask, M. & Laasimer, L. 1987: The significance of species at the margins of their area in the Estonian flora. *In* Laasimer, L. & Kull, T. (eds) *The plant cover of the Estonian SSR. Flora, Vegetation and ecology,* 7–16. Valgus, Tallinn.

Kessel, H., Saarse, L., Vishnevskaja, E. & Sinisalu, R. 1986: Geological history of Lake Ülemiste (Tallinn, Estonian SSR). *Proceedings of the Academy of Sciences of the ESSR, Geology 35(1)* 35–41 (in Russian).

Kukk, Ü. 1987: Protection of rare plant species in the Estonian SSR. *In* Laasimer, L. & Kull, T. (eds) *The plant cover of the Estonian SSR. Flora, vegetation and ecology,* 17–21. Valgus, Tallinn.

Laasimer, L. 1964: Peipsi nôo taimkate ja selle kasutamise perspektiivid. *In Eesti Geograafia Seltsi Aastaraamat 1964,* 103–115. Tallinn.

Laasimer, L. 1981: Anthropogenic changes of plant communities and problems of conservation. *In* Laasimer, L. (ed.) *Anthropogenous changes in the plant cover of Estonia,* 18–31. Tartu.

Mäemets, A. 1977: *Eesti NSV järved ja nende kaitse.* 263 pp. Valgus, Tallinn.

Männil, R. 1963: The fresh-water lime deposits of Saaremaa, their malacofauna and age. *In Material of the geology of Upper Pleistocene and Holocene of Estonia,* 145–161. Tallinn (in Russian).

Moora, T., Ilomets, M. & Jaanits, L. 1988: Muistsetest loodusoludest Akali kiviaja asulakoha lähiümbruses. *In* Rôuk, A.-M. & Selirand, J. (eds) *Loodusteaduslikke meetodeid Eesti arheoloogias,* 26–38. Tallinn.

Pirrus, R. & Rôuk, A.-M. 1979: Uusi andmeid Soitsjärve nôo geoloogiast. *In* Raukas, A. (ed.) *Eesti NSV saarkôrgustike ja järvenôgude kujunemine.* 118–144. Tallinn.

Pirrus, R. & Rôuk, A.-M. 1988: Inimtegevuse kajastumisest Vooremaa soo- ja järvesetetes. *In* Rôuk, A.-M. & Selirand, J. (eds) *Loodusteaduslikke meetodeid Eesti arheoloogias,* 39–53. Tallinn.

Pirrus, R., Rôuk, A.-M. & Liiva, A. 1987a: Geology and stratigraphy of the reference site of Lake Raigastvere in Saadjärv drumlin field. *In* Raukas, A. & Saarse, L. (eds) *Palaeohydrology of the temperate zone, II. Lakes,* 101–122. Tallinn.

Pirrus, R., Rôuk, A.-M. & Koff, T. 1987b: Structure and development of the Siniallika kettle hole. *Proceedings of the Academy of Sciences of the ESSR, Geology 36 (1),* 1–5 (in Russian).

Pirrus, R.O., Rajamäe, R.A., Varvas, M.J., Rôuk, A.-M. & Liiva, A.A. 1988: Investigation of lake deposits of Saadjärv drumlin field by isotope methods. *In Isotopic studies of the Baltic and Belorus,* 175–185. Tallinn (in Russian).

Punning, J.-M., Ilomets, M., Koff, T. & Rajamäe, R. 1987: On the development of the Lake Ümarjärv (NE Estonia) in the Holocene. *In* Raukas, A. & Saarse, L. (eds) *Palaeohydrology of the temperate zone II. Lakes,* 123–136. Valgus, Tallinn.

Saarse, L. 1987: Kurtna järvestiku geoloogiline areng ja järvesetete koostis. *In* Ilomets, M. (ed.) *Kurtna järvestiku looduslik seisund ja selle areng,* 55–61. Valgus, Tallinn.

Saarse, L. & Raukas, A. 1984: Background to a multidisciplinary investigation of mires, lakes and rivers. *In* Punning J.-M. (ed.) *Estonia. Nature, man and economy,* 78–87. Tallinn.

Saarse, L., Sarv, A. & Karin, J. 1985: Sedimentation and geological development of lakes in the Illuka kame field. *Proceedings of the Academy of Sciences of the Estonian SSR. Geology 34(2),* 62–67 (in Russian).

Saarse, L., Vishnevskaya, E., Sarv, A. & Rajamäe, R. 1990: Evolution of the lakes of Saaremaa Island. *Proceedings of the Academy of Sciences of the ESSR, Biology, 39, (1),* 34–45 (in Russian).

Sarv, A. & Ilves, E. 1975: Über das Alter der Holozären Ablagerungen im Mündungsgebiet des Flusses Emajôgi (Saviku). *Proceedings of the Academy of Sciences of the ESSR, Chemistry & Geology 24,* 64–69 (in Russian).

Sarv, A. & Ilves, E. 1976: Stratigraphy and geochronology of Holocene lake and mire sediments from the southwestern part of Estonia. *In* Bartosh T. (ed.) *Palynology in the terrestrial and marine geological studies,* 47–59 (in Russian).

Veski, S. 1992: *The Holocene development of Lake Maardu and the vegetational History of North-Estonia.* Kvartärgeologiska avdelningen, Uppsala Universitet. Licentiatthesis, 41 pp. Uppsala.

12

Lithuania

M. KABAILIENÉ

INTRODUCTION

Lithuania is located on the eastern coast of the Baltic Sea. The whole territory is covered by Quaternary deposits as much as 300 m thick underlain by Cenozoic, Mesozoic, and Palaeozoic rocks. Crystalline Proterozoic and Archaean rocks, lying at a depth of 200–2300 m, make their basement. The surface of crystalline rocks slopes down from southeast Lithuania to the Baltic Sea.

During the Pleistocene, the Lithuanian area underwent several advances of glaciers. The topography is mainly formed by the last glaciation, although previous glaciations also were important. In the hilly areas the relief is caused by end moraines and kames. Central Lithuania has a low relief caused by ground moraine plains and glacial lake sediments and similarly southeast Lithuania has fluvioglacial and limnoglacial plains. In the maritime west the shore formations of the Baltic Sea are characteristic (Fig. 12.1).

Lithuania has about 4000 lakes, although those with area > 0.01 km^2 make up only 2500. They are concentrated on the uplands and in the South-Eastern Plain. The development of lake and bog depressions is mainly due to the action of a glacier and its meltwaters. The thickness of Late-Glacial and Holocene sediments ranges up to about 20 m and is about 6 m on average (Garunkštis 1975). In the main, the base of lake and bog formations is composed of Late-Glacial limnic and limnoglacial deposits. Bölling and Older Dryas deposits, without exception, are represented by terrigenous layers (clay, sand, mud, clayey loam); the admixture of organic matter is negligible. The deposits of Bölling and Older Dryas chronozones are rather homogeneous, and show no evidence of any distinct variation in their composition that might indicate changing physical and geographical conditions. Alleröd and Younger Dryas deposits have been subjected to palynological studies in hundreds of sections (Seibutis *et al.* 1960; Kabailiené 1967, 1990; Kunskas *et al.* 1975). Alleröd deposits are rich in organic remains and are often regarded as a stratigraphic reference horizon. Alleröd deposits are represented predominantly by peat, peaty sapropel, mud, and clay with abundant organic matter and remains of higher plants. They are overlain by lake deposits of Alleröd age that are usually poorer in plant remains. Double-layers like this are accounted for by frequent thermokarst phenomena in Alleröd time. Younger Dryas deposits are represented by silt, clay, and sand, except in some sections where they include sapropel and peat-like formations. The Holocene deposits are dominated by sapropel, peat, lacustrine lime, and mud.

According to physical and geographical characteristics Lithuania was divided into seven type regions (Fig. 12.2):

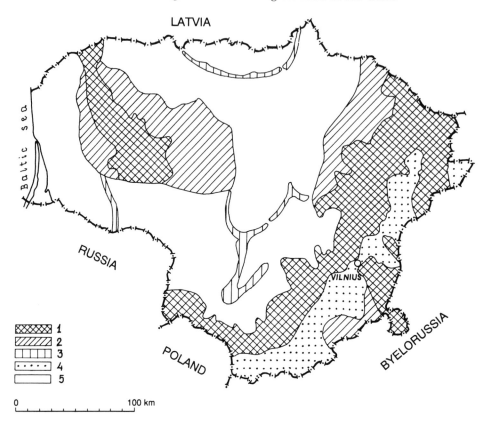

Fig. 12.1 Geomorphological subdivision of the Lithuanian landscape: 1–uplands, 2–plateau, 3–end morainic ridges, 4–plains, 5–lowlands

LT-a Baltic Coast Lowland
LT-b Žemaitija Upland
LT-c Central Lithuanian Plain
LT-d Baltic Ridge
LT-d¹ West Aukštaitija Plateau
LT-e South-Eastern Fluvioglacial Plain
LT-f Middle Pleistocene Upland

The climate is intermediate between the maritime climate of Western Europe and continental climate of Eastern Europe. Lithuania is in the forest zone of temperate climate and in the subzone of mixed forests. Pine forests make up 37.2% of total forest area in Lithuania, with 18.5% of spruce, 23.0% of birch, 7.5% of white alder, 5.7% of black alder, 4.8% of aspens, and 1.4% of oak and ash groves. Only the type region of the Baltic Ridge (LT-d) and reference site Bebrukas in this region is discussed here as no

valid radiocarbon-dated pollen diagrams are available from other sites and type regions (Šulija *et al.* 1967).

TYPE REGION LT-d, THE BALTIC RIDGE

Altitude: Up to 257 m.

Climate: Mean annual temperature 6°C, mean January –4.8°C, mean July 17.2°C. Precipitation 680 mm yr⁻¹.

Geology: Marginal morainic uplands of the last glaciation.

Topography: Hilly surface, rich in lakes and small bogs.

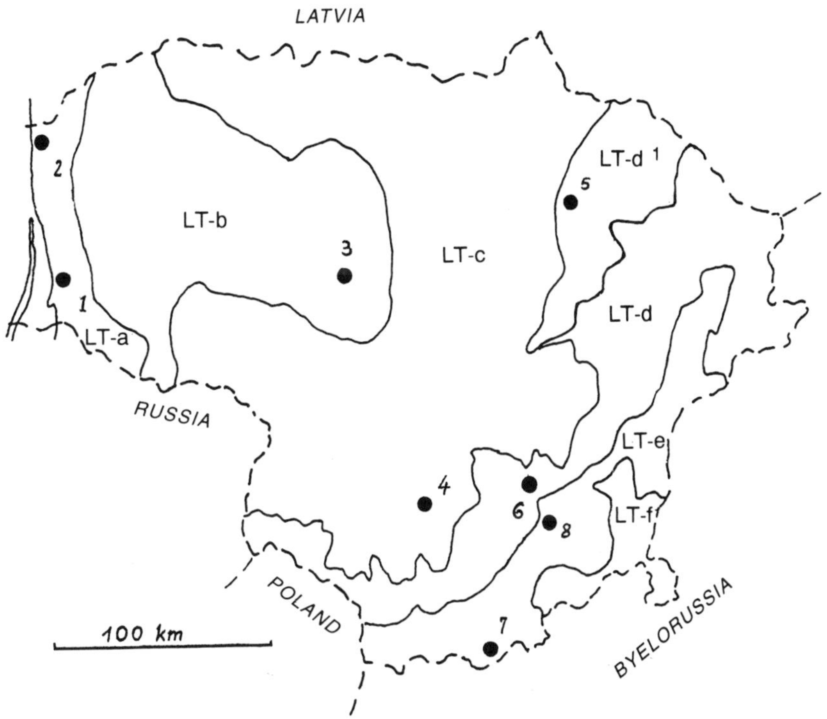

Fig. 12.2 Type regions and reference sites in Lithuania. Reference sites: 1. Šventelé, 2. Šventoji, 3. Tytuvénu tyrelis, 4. Žuvintas, 5. Šepeta, 6. Bebrukas, 7. Čepkeliai, 8. Glúkas (see Appendix)

Population: Approximately 52 persons km^{-2}.

Vegetation: Forests cover approx. 30% of the area, with *Pinus* prevailing, together with *Picea, Betula, Alnus*, and some *Quercus* and *Fraxinus*.

Soils: Sod-podzolic soils prevail.

Land use: Approx. 30% forest, 40–50% agricultural land, 10–20% uncultivated land, and 10% urban areas. Main crops are rye, wheat, potatoes, sugar beet, and grasses.

Reference site 6. Lake Bebrukas
(Kabailiené 1965, 1986, 1987a)

Latitude 54°35′N, Longitude 24°38′E. Elevation 160 m. Age range 11800–0 BP. Eleven local pollen-assemblage zones (paz).

B-1	–11800 BP	Herbs and shrubs; tundra and forest-tundra types
B-2	11800–11300 BP	*Pinus*
B-3	11300–10000 BP	Forest–tundra types–*Betula*
B-4	10000–9000 BP	*Betula*
B-5	9000–7800 BP	*Pinus–Betula*
B-6	7800–6800 BP	*Alnus–Ulmus*
B-7	6800–5400 BP	*Tilia–Ulmus–Corylus*
B-8	5400–4050 BP	*Picea–Quercus–Alnus*
B-9	4050–3250 BP	*Pinus–Betula*
B-10	3250–2000 BP	*Alnus–Quercus–Tilia*
B-11	2000–0 BP	*Pinus–Picea*–herbs

Only one pollen diagram is presented (Fig. 12.3).

General patterns

(1) High quantities of *Pinus* (generally about 35–60%, with maximum in Alleröd) and decrease of *Picea*, *Quercus, Alnus*, and *Corylus*.
(2) Rather low content of *Betula* in late Holocene

398

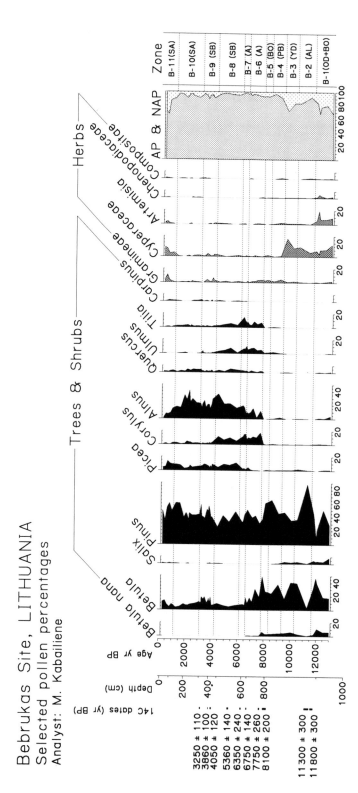

Fig. 12.3 The Late-Glacial and Holocene pollen diagram from Lake Bebrukas, Lithuania

with higher values in Boreal, Pre-Boreal, and Late-Glacial (to 48%).

(3) Culmination of *Ulmus* in the first half of the Atlantic. During the Holocene climatic optimum the broad-leaved species prevailed: *Tilia, Ulmus,* and *Quercus.* High values of *Corylus* in the Atlantic and first half of the Subboreal.

(4) Small quantities of *Carpinus* appear at the very end of the Atlantic.

(5) Cerealia is widespread in the Subatlantic.

Patterns of events for the Baltic Ridge type region

Pollen diagram characteristics

(1) High values for *Pinus.* Most obvious peak is in the Alleröd, slightly lower peak is in the Boreal, and still lower one in the upper part of the Subboreal and Subatlantic.

(2) *Betula* prevails in the Preboreal.

(3) Among the broad-leaved species *Tilia* dominates, appearing at the end of the Boreal with maximum in the second half of the Atlantic and considerable decrease in the middle of the Subboreal. *Ulmus* appears in the beginning of the boreal; its maximum coincides with that of *Tilia,* then follows a decrease in the first half of the Subboreal. *Quercus* appears in the Atlantic with low quantities.

(4) *Picea* has slight maxima in the early Subboreal and late Subatlantic.

(5) *Corylus* expands in the Atlantic with a considerable decrease in the middle of the Subboreal.

Landscape events

(1) In the Arctic period (Older Dryas and Bölling chronozones) tundra and forest–tundra landscape prevailed.

(2) During the Alleröd, thermokarst processes and regeneration of lake hollows expanded.

(3) In the Preboreal and the first half of the Boreal, forests were open.

(4) Beginning in the second half of the Boreal, forest landscapes spread.

(5) In the Subboreal, fields and pastures began to spread.

Anthropogenic events

(1) In the early Subboreal, meadow-pastures and cattle breeding began as well as cultivation of cereals.

(2) In the Subatlantic (especially in the second half), *Secale* cultivation increased.

Hydrological events

Hydrological changes have been studied mainly on the basis of diatom data, taking into account sediment composition and lake-hollow structure.

(1) In the Arctic period (Older Dryas and Bölling chronozones) lakes were shallow.

(2) Decrease in water level, formation of the third terrace, and intensive development of the river network were related to the Alleröd.

(3) In the Preboreal and first half of the Boreal, water level in the lakes was low.

(4) In the second half of the Boreal and Atlantic, lake level increased. Fluctuations in lake level were characteristic of the Atlantic.

(5) Lake level decreased greatly (by 2–3 m), and the second lake terrace was formed in the second half of the Subboreal.

(6) In the second half of the Subatlantic, lake level decreased and the first terrace was formed.

Climatic events

For climate reconstructions using pollen-analytical data, mathematical methods were applied (Kabailiené, 1982).

(1) In the Preboreal, climate was temperate but not humid. Annual temperature was *ca.* 3–4°C lower than today. Precipitation was lower.

(2) During the first half of the Boreal, it became warmer (annual temperature *ca.* 6°C), but climate was still dry. During the second half of the Boreal, it became warmer and more humid.

(3) In the Atlantic, annual temperature was higher than today by several degrees, and precipitation was higher too.

(4) In the Subboreal, as compared to the Atlantic period, precipitation was lower and annual temperature was close to today's (6°C).

(5) In the Subatlantic, as compared to the recent period, the climate was slightly warmer and humid.

Soil events

Soil acidification and leaching of nutrients occurred since the early Subboreal.

CONCLUSIONS

Pollen from the Late-Glacial and Holocene sediments in Lithuania were studied in more than 100 sections. However, few radiocarbon data are available.

Characteristic features of pollen diagrams in Lithuania are high values of *Pinus* with the highest peak in the Alleröd, a slightly lower peak in the Boreal, and the lowest peak in the upper part of the diagrams (beginning in the second half of the Subboreal). During the Boreal, *Pinus* was most widespread in the Baltic Coast Lowland and Žemaitija Upland. Later *Pinus* migrated eastwards, and from the end of the Boreal its highest quantities were found in southeast Lithuania, where sand soils are common.

Quantities of *Betula* were low during the entire Late-Glacial and Holocene, even during the Preboreal, when it was most widespread; sometimes the content of *Pinus* was higher than that of *Betula*. The lowest quantities of *Betula* pollen were in the Atlantic, and a slight increase of *Betula* occurred in the second half of the Subboreal.

Picea during the Late-Glacial (mainly in the first half of the Alleröd and Younger Dryas) is found in small quantities in several locations in Lithuania. During the early Holocene, pollen of *Picea* was scarce. Beginning in the second half of the Atlantic, its quantities increased. The Holocene peak of *Picea* in Lithuania was in the lower half of the Subboreal, and a slightly lower peak was in the upper half of the Subatlantic. During the Holocene, the highest and the lowest quantities of *Picea* were in the area of the Žemaitija Upland and southeast Lithuania respectively.

Alnus during the Late-Glacial and early Holocene was scarce, but from the end of the Boreal its quantities significantly increased; among tree species,

Alnus prevails only in the second half of the Atlantic and in the early Subboreal. The highest quantities of *Alnus* in Lithuania were found in the Baltic Coast Lowland.

Ulmus gradually began to expand in the Boreal, with maximum Holocene values in the second half of the Atlantic, except the area of the Baltic Coast Lowland, where this expansion began earlier, i.e. in the second half of the Boreal. Quantities of *Ulmus* decreased significantly in the entire area of Lithuania at the beginning of the Subboreal.

Tilia began to spread in the late Boreal with migration from southeast Lithuania. Culmination of values in Lithuania was in the second half of the Atlantic. Later *Tilia* gradually decreased, with the exception in the Central Lithuanian Plain and southeast Lithuania, where its quantities slightly increased in the Subatlantic.

Quercus in the area of Lithuania was found to appear in solitary sites from the Boreal. *Quercus* migrated from southwest. Its culmination in various areas differs with time: in the Baltic Coast Lowland during the second half of the Atlantic, in the Central Lithuanian Plain during the first half of the Sub-Boreal, and in south east Lithuania during the upper half of the Atlantic and the Subboreal.

Carpinus in Lithuania was scarce with maximum values in the Sub-Atlantic. The earliest signs of human influence in Lithuania is at the end of the Atlantic and beginning of the Subboreal, when meadow-pastures began to develop, pollen of *Triticum* is found in the sediments of this time. *Secale* appears later, i.e. in the Subatlantic.

REFERENCES

Basalykas, A. 1965: *The physical geography of Lithuania*, 2. Mintis, Vilnius, 495 pp. (in Lithuanian).

Garunkštis, A. A. 1975: *Sedimentation processes in the lakes of Lithuania*. Mokalas, Vilnius, 295 pp. (in Russian).

Kabailiené, M. 1959: Development of vegetation during the Late-Glacial and Holocene on the coastal zone of Lithuania and South Latvia. *Geographical annals 2*, 477–505 (in Lithuanian).

Kabailiené, M.V. 1965: Some questions on stratigraphy and palaeogeography of Holocene in SE of Lithuania. *Proceedings of Geology Institute (Vilnius) 2*, 302–335 (in Russian).

Kabailiené, M. 1968: Lacustrine and peaty sediments of lake Žuvintas and stratigraphy. *In* Šivickis P. (ed.) *The reservation of Žuvintas*, 40–49 Mintis, Vilnius. (in Russian).

Kabailiené M.V. 1971: On interpretation of Lithuanian Holocene pollen diagrams. *In* Neustadt, M.I. (ed.) *Palynology of Holocene*, 63–70. Moscow (in Russian).

Kabailiené, M. 1973: *Formation of pollen spectra and its interpretation methods: application to stratigraphy and forest history of Holocene in Lithuania.* Hab. dr. dissertation, 1, 390 pp. Vilnius (in Russian).

Kabailiené, M. 1979: *The essentials of applied palynology.* Mokslas, Vilnius, 147 pp. (in Lithuanian).

Kabailiené, M.V. 1986: Pollen analyses methods. *In* Kvasov, D.D., Davydova, N.N. & Rumiancev, V.A. (eds): *The main regularities of lakes origin and development methods for study of lake history*, 119–128. Leningrad (in Russian).

Kabailiené, M.V. 1987a: Pollen stratigraphy and correlation of the Late-Glacial and Holocene deposits in Lithuania using mathematical methods. *Transactions of the University of Lithuania Geology 8*, 123–135 (in Russian).

Kabailiené, M. 1987b: Correlation in diatom analysis data and stratigraphy of deposits in small lakes with the application of mathematical methods. *In* Kabailiene, M. (ed.). *Methods for investigations of lake deposits:* *palaeoecological and palaeoclimatological aspects, 91–98.* Vilnius.

Kabailiené, M. 1990: Holocene in Lithuania. 176 pp. Mokslas, Vilnius (in Lithuanian).

Kunskas, R. 1962: Main stages of Žuvintas mire development. *The Academy of Sciences of Lithuania, Institute of Geology and Geography, Scientific Reports 14* (2), 97–115 (in Lithuanian).

Kunskas, R.A., Vai ěviliené, B., Savukyniené, N.P. 1975: Main stages of development and stratigraphy of some lakes in Southern and Eastern Lithuania. *In* Klimkaité, J. (ed.). *Sedimentation in small lakes of Southern Lithuania*, 405–458. Vilnius (in Russian).

Rimantiené, R. 1979: *Šventoji, I.* 188 pp. Mokslas, Vilnius (in Lithuanian).

Savukyniené, N. 1976: Features of development of agriculture in the district of Čepkeliai. *Geographical Annals 14*, 169–175 (in Lithuanian).

Seibutis, A. Sudnikavičiené, F. 1960: On beginning of formation of peatbogs in Holocene on territory of Lithuania. *Geographical Annals 3*, 259–263 (in Lithuanian).

Šulija, K.S., Lujanas, V.J., Kibilda, Z.A., Banys, J.J. & Genutiené, I.K. 1967: Stratigraphy and chronology of lacustrine-peaty sediments in the hollow of lake Bebrukas. *Proceedings of Geology Institute (Vilnius) 5*, 231–239 (in Russian).

13

Poland

M. RALSKA-JASIEWICZOWA and M. LATAŁOWA

INTRODUCTION

The position of Poland within Europe is typically transitional, linking the peninsular western Europe with the continental eastern areas. The main patterns of the landscapes, however, are roughly latitudinal, with the natural barriers of the Baltic Sea to the north and the Carpathian and Sudety Mountains to the south.

The important expression of this specific position is the crossing of geographical gradients of related natural phenomena such as climate, like the N–S increase in mean July temperatures, the W–E increase of mean January temperatures, the N–S decrease in mean annual rainfall, and W–E increase in the duration of snow cover. It may result in the skewness of some derived variables, like mean annual temperature or length of the growing season (Fig. 13.1). The successive Quaternary glaciations developed oblique NW–SE boundaries, most distinctly in the case of the last cold stage.

The last three glaciations covered the entire lowland part of Poland with moraines and outwash, whereas the upland areas to the south, of very differentiated geological structure, were partly covered with loess. In the Carpathian Mountains, which are formed mostly of flysch sandstone and shale, and the Sudety Mountains, built of old metamorphic rocks, the Quaternary cover is thin and discontinuous (Fig. 13.2).

The entire area of Poland lies within the zone of nemoral forests except for its northeastern corner,

which belongs to the subboreal zone. The dominant lowland forest communities today are mixed *Pinus–Quercus* and *Pinus* forests on poorer and drier soils and mixed deciduous woodlands (mainly *Carpinus, Quercus, Tilia,* and in some areas also *Fagus*) on more fertile and humid soils. In the mountains the altitudinal zonation of vegetation below the timberline is expressed by the dominance of *Fagus–Abies* and *Abies–Picea* forests at lower elevations and *Picea* forests above. The natural vegetation is mostly destroyed by man; only 8.5% of Poland supports large complexes of natural plant communities (Faliński 1976). However, the maps of reconstructed natural vegetation indicate a distinct geographical differentiation of woodlands related to the influences of an oceanic climate in the northwest and of a continental climate to the east-southeast, as shown for instance by the distribution of different types of mixed deciduous forest (Fig. 13.3).

The subdivision of Poland into palaeoecological type regions (Fig. 13.4), following the IGCP Project 158 B recommendations, is based on the comparison of existing maps of climatic, vegetational, and geomorphic regions as well as on geologic and soils maps. It roughly approximates the geobotanical subdivision of Poland (Szafer & Zarzycki 1972) but is inevitably much simplified, especially in its second version (Ralska-Jasiewiczowa 1982, 1987), in which the two-rank system of regions and subregions is eliminated. This change emphasized still more the difference in defined regions between south and north. The 25 type regions are small and highly

Palaeoecological Events During the Last 15 000 Years: Regional Syntheses of Palaeoecological Studies of Lakes and Mires in Europe.
Edited by B.E. Berglund, H.J.B. Birks, M. Ralska-Jasiewiczowa and H.E. Wright. © 1996 John Wiley & Sons Ltd.

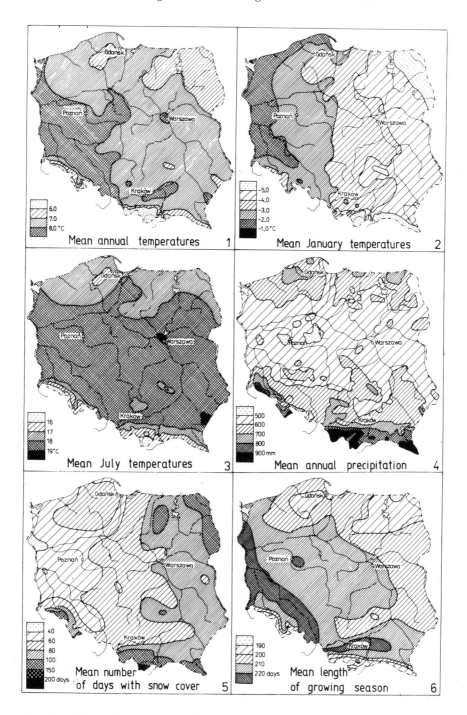

Fig. 13.1 Maps of some selected climatic variables in Poland (after Kondracki, 1978)

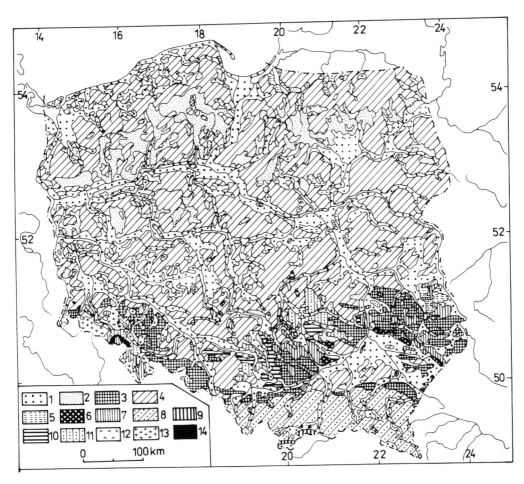

Fig. 13.2 Schematic geological map of Poland (after Birkenmajer in Szafer 1966; simplified). 1–gravel, sand, silt, and clay of fluvial (Holocene) and fluvioglacial (Pleistocene) accumulation, 2–outwash sand and gravel (Pleistocene), 3–loess and sandy loess (Pleistocene), 4–moraine till and boulder ramparts (Pleistocene), 5–gyttja, silt, and sand of freshwater origin (Pleistocene, Pliocene, Miocene), 6–limestone, sandstone, sand, and gypsum silt (Miocene marine facies), 7–marl, sandstone, and Cretaceous sand of Sudety, Cracow, and Lublin regions, 8–shale and sandstone of Carpathian flysch (Cretaceous, Paleogene), 9–dolomite, limestone, marl (Triassic, Jurassic, Cretaceous), 10–limestone, partly marl (Devonian, lower Carboniferous, Middle Triassic, Upper Jurassic, Upper Cretaceous), 11–shale, arkose, greywacke, shale, and conglomerate with coal beds, silt, clay, sand (Cambrian, Carboniferous, Permian, Middle Jurassic), 12–quartzite ranges of Holy Cross Mts (Cambrian), 13–granite, gneiss, syenite, gabbro, schist, quartzite, diabase (Precambrian, Cambrian, Ordovician, Silurian), 14–intrusive and extrusive rocks: melaphyre, porphyry, teschinite, andesite, basalt (Miocene, Pliocene)

varied in the mountainous and upland parts of Poland because of the differentiation of the landscapes, but they are much larger in the lowland areas. The maximum extent of the Vistulian (Weichselian) glaciation forms the main border between the denudation plains of middle Poland and the young morainic landscapes of north-northwest Poland. The mainly longitudinal subdivision of those two landscape zones into type regions expresses first of all the eastward increase of continental influences. The only exception is the northwesternmost area, where the roughly parallel pattern of type regions reflects the decreasing influence of the adjacent Baltic Sea.

The type regions are as follows:

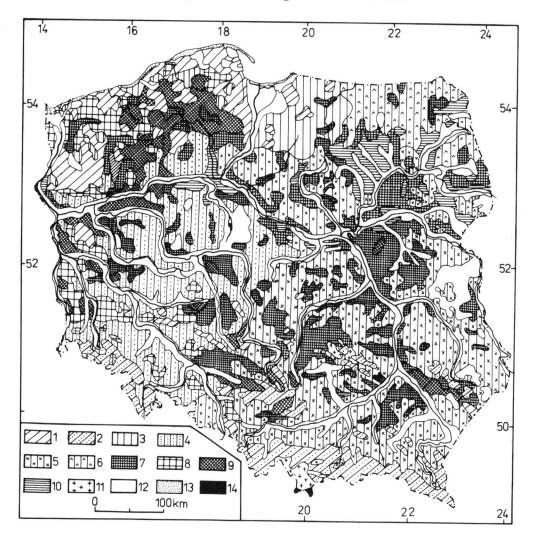

Fig. 13.3 Map of the potential natural vegetation of Poland (after Matuszkiewicz, in: *Geograficzny Atlas Świata* 1987, p. 75, simplified). 1–*Fagus* forests, lowland type, 2–*Fagus* forests, upland type and lower montane (Carpathian and Sudetian) forest zone type, 3–sub-oceanic fertile *Fagus–Quercus–Carpinus* forest, 4–middle-European *Quercus–Carpinus* forests, lowland and mountain-foreland types, 5–East-middle-European *Tilia–Quercus–Carpinus* forests (Minor Poland and central Poland), 6–East-middle-European subboreal *Tilia–Quercus–Carpinus* forests with *Picea*, 7–thermophilous *Quercus* woodlands and sub-continental mixed *Pinus–Quercus* woodlands with elements of mixed deciduous forests, 8–acidiphilous *Quercus* woodlands, 9–middle-European poor *Pinus* forests, with grass or moss–dwarf–shrub ground flora, 10–sub-continental *Pinus* forests, 11–*Picea* and *Abies* forests, 12–riverine and swamp forests, 13–complex of sea-coast vegetation, 14–high-mountain vegetation

P-a Inner West Carpathians
P-b West Beskidy Mts and forelands
P-c Low Beskidy Mts
P-d Jasło–Sanok Depression and Eastern
      Forelands

P-e Bieszczady Mts
P-f Sudety Mts
P-g Silesia Lowland
P-h Silesia–Cracow Upland
P-i Miechów Upland and Nida Basin

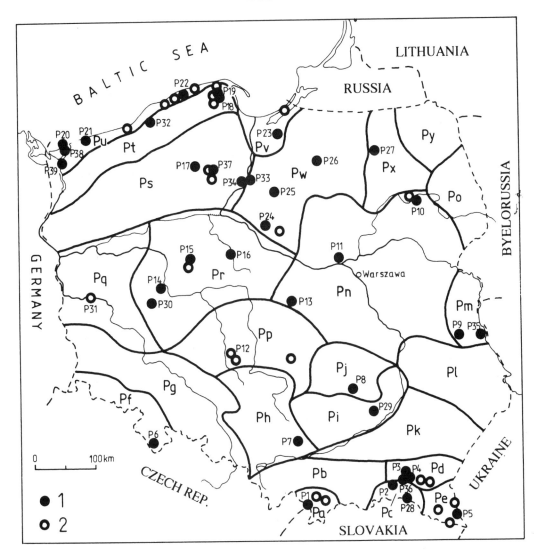

Fig. 13.4 Map showing the subdivision of Poland into the palaeoecological type regions and location of sites. 1–primary/secondary reference sites (numbered), 2–complementary sites (after Ralska-Jasiewiczowa 1987, corrected). The complete list of reference sites is in the Appendix

P-j   Holy Cross Mts
P-k   Sandomierz Basin
P-l   Lublin Upland and Roztocze Mts
P-m   Lublin Polesie
P-n   Masovia and Podlasie Lowlands
P-o   Białystok Upland and Biebrza Basin
P-p   Northern Marginal Uplands
P-q   Lubuskie Lake District
P-r   Poznań–Gniezno–Kujawy Lake Districts

P-s   West Pomeranian Lake Districts
P-t   Baltic Coastal Zone
P-u   Baltic Shore
P-v   Vistula Deltaic Area
P-w   Dobrzyń–Olsztyn Lake Districts
P-x   Masurian Lake District
P-y   Suwałki–Augustow Lake District

The synthesis is based on a total of 37 reference sites, investigated mostly within the framework of

IGCP Project 158 B but in some cases from other studies that approximated the Project requirements. Data from the additional sites, if providing any useful information, are also included (as indicated on the particular maps of synthesis regions). Still, the final coverage of Poland by the sites studied is incomplete; eight type regions have no sites, so they have had to be excluded from the synthesis. They are mostly in the marginal west-southwest, southeast, and northeast areas of the country.

Another problem arises from dating deficiencies, which were insuperable in cases of carbonate sediments. Eight of the total 37 sites have only one to four dates, and two sites are not dated at all.

Besides pollen analysis, the palaeoecological information is derived from the analyses of plant macrofossils (28 sites), diatoms (10 sites), Cladocera (5 sites), rhizopods (3 sites), molluscs (6 sites), and insects (4 sites) and also from more or less detailed chemical analyses (22 sites), including sedimentary plant pigments (4 sites) and stable isotopes (3 sites) (Ralska-Jasiewiczowa 1987).

The numerical zonation of pollen data, according to the ZONATION program of Gordon & Birks (1972) as adapted by A. Walanus, was performed at all sites, in order to give a consistent and uniform basis for defining local pollen-assemblage zones. In practice, however, its results were treated as an aid and not as the only criterion for pollen stratigraphy.

The synthesis is worked out in two steps: first, the information from the particular type regions is summarized in a uniform and concise way from the tables of event stratigraphy prepared by individual authors. These more restricted studies are grouped according to the general idea of synthesis regions into sections representing five large landscape units: mountains, uplands, mid-Polish lowlands, lake districts, and Baltic coastal zone (Ralska-Jasiewiczowa 1989).

These materials provide the basis for attempts at interregional correlations and for summarizing the data within the synthesis regions as defined above. However, many practical difficulties arise when a great deal of data worked out by different authors is put together, and these need some general comments.

The amount and quality of data available for the particular synthesis regions are different. In the

Baltic Coastal Zone and the mountains the dating standard is the best, and the distribution of sites (except in type region P-b) is quite satisfactory. In the lake districts the number and distribution of sites is good, but the dating at some sites is inadequate. From the upland areas only two dated sites are available, giving fragmentary knowledge about the region. In the Mid-Polish Lowlands both the site number and distribution and the dating level are insufficient. Consequently it is only possible to define type-region pollen-assemblage zones (paz) in two synthesis regions (the Carpathians and the Lake Districts); in the others, the interregional correlations are based on the local pazs. In this connection, at some sites it appeared useful to alter some of the original criteria for defining single pazs and to adapt their naming and position to the neighbouring sites for better correlation. We apologize to the authors for those changes.

The interregional synthesis of data for Poland was made by us. However, we feel that all the persons contributing with their original type-region studies are the co-authors of this paper, viz. Z. Balwierz, K. Bałaga, K. Binka, K. Harmata, M. Hjelmroos-Ericsson, E. Madeyska, B. Marciniak, G. Miotk-Szpiganowicz, B. Noryśkiewicz, A. Obidowicz, I. Okuniewska-Nowaczyk, K. Szczepanek, K. Szeroczyńska, K. Tobolski, and J. Zachowicz. We also received a lot of valuable help from co-operating specialists, such as S.W. Alexandrowicz, M. Pawlikowski, K. Różański, and others. Most of the radiocarbon datings were made by M.F. Pazdur[†] with some by S. Håkansson. All the computer work was done by A. Walanus. Figures were drawn by M. Jąkalska.

TYPE REGIONS P-a, P-b, P-c, P-d, AND P-e, POLISH CARPATHIAN MOUNTAINS

The Polish Carpathians cover the northern part of the West Carpathians with four type regions: Inner West Carpathians (P-a, Tatra Mts and Nowy Targ–Orawa Basin), Western Beskidy Mts (P-b), Low Beskidy Mts (P-c), Jasło–Sanok Depression with Eastern Forelands (P-d), and a small northwestern part of East Carpathians (P-e, Bieszczady Mts).

The synthesis does not include any data from the Tatra Mts (the highest southern-part of P-a region) nor from the Beskidy Mts (P-b). It is based on data from seven reference sites and seven complementary sites (Fig. 13.5).

Altitude: 230 m (P-d) –1348 m above sea level (excluding the Tatra Mts).

Climate: Altitudinal climatic zonation: warm-temperate climate up to 700 m, cool-temperate up to 1100 m, cool up to 1550 m, very cool up to 1850 m, and cold-temperate and cold above 1850 m, the continentality of climate increasing eastwards. The detailed data below do not cover zones above 1550 m: Mean January temperature −3 to −7°C, mean July 14.0 to 18.2°C (P-d), mean annual 4.0 to 8.1°C. Mean annual rainfall 700–1000 mm. Dominant westerly winds in western ranges, southwesterly in P-d region, and southerly in P-e region.

Geology: Cretaceous and Paleogene flysch sandstone and shale form most of the bedrock; crystalline rocks, Mesozoic limestone, and other sedimentary rocks occur only in the highest parts of the Tatra Mts. The Pleistocene is limited to gravel–clay covers and terraces, and the Holocene to thin layers of slopewash.

Topography: Young folded mountains culminating in the highest peaks in the Tatra Mts; lowlands (P-d) and subsidence depressions (P-a); topography very differentiated in different ranges.

Type regions P-c and P-d together form the lowest and narrowest parts of the Carpathians, with the highest peaks below 1000 m elevation. We call this the Carpathian Low.

Vegetation: Altitudinal zonation of vegetation: Foothill zone up to 450–600 m, originally occupied by mixed deciduous forests with dominant Tilio–Carpinetum association in different variants, nowadays mostly destroyed and occupied mainly by fields and meadows (Arrhenatheretalia and Molinietalia).

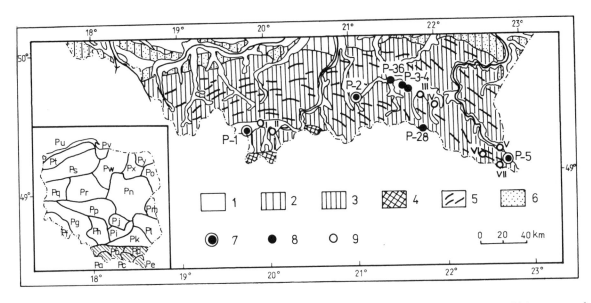

Fig. 13.5 Geomorphic map of the Polish Carpathian Mts synthesis region, showing location of sites. 1 – Pleistocene and Holocene river valleys and terraces, 2 – Neogene sedimentary covers, 3 – flysch folded structures, 4 – crystalline blocks and their sedimentary covers in the alpine zone, 5 – exposed ridges, 6 – loess, 7 – reference sites with the pollen diagrams included in this paper: P-1 Puścizna Rękowiańska, P-2 Szymbark, P-5 Tarnawa Wyżna, 8 – other reference sites: P-3 Tarnowiec, P-4 Roztoki, P-28 Jasiel, P-36 Jasło, 9 – complementary sites: I – Grel (Koperowa 1962), II – Bór na Czerwonem (Obidowicz 1978), III – Kępa (Gerlach *et al.* 1972), IV – Besko (Koperowa 1970), V – Zakole, VI – Smerek, VII – Wołosate (Ralska-Jasiewiczowa 1980)

Lower montane forest zone of Fagetum carpaticum and Abieti–Piceetum forests up to 1100–1250 m. Upper montane forest zone of Piceetum tatricum up to 1300–1500 m does not occur in the Bieszczady Mts. Pinetum mughi zone and subalpine and alpine meadow zones occur only in the Tatra Mts and are extremely rare at the highest peaks of the Beskidy Mts.

Soils: Leached to acidic brown soils generally prevail; deluvial podzols and pararendzinas are rare; mud and bog alluvial soils occur locally in the valley bottoms.

Archaeology: Penetration of Late Palaeolithic tribes during the Intrapleniglacial, Allerød, and Younger Dryas. Neolithic activities mostly in the marginal and foothill zones, earlier and more intensive in the eastern ranges; expansion of Corded Ware culture within the mountainous areas at the decline of Late Neolithic (pastoral economy dominant). Progress of colonization during the latest Bronze/Early Iron Age by the tribes of Lusatian culture. Growing evidence of occupation during the late phase of Roman influences. Since Late Medieval time progressive colonization in the whole area.

Reference site 1. Puścizna Rękowiańska
(Koperowa 1962, Obidowicz 1989, 1990)

Latitude 49°29′N, Longitude 19°49′E. Elevation 656 m. A mountain raised bog 280 ha in area, 7 m in maximum depth, today partly degraded by drainage and digging, situated at the bottom of the subsidence depression (Nowy Targ Basin) separating the Tatra Mts from the Outer Carpathians. Fills a fossil oxbow lake formed by dissection of alluvial fan of Czarny Dunajec River during the late-glacial. Age range 10000–0 BP. Ten local pollen-assemblage zones (paz), eight ^{14}C dates (Fig. 13.6). Obidowicz's investigations did not reach the late-glacial part of deposit studied by Koperowa (1962).

PR-1	(*ca.* 10000)–9000 BP	NAP–*Pinus*
PR-2	9000–8800 BP	*Betula*
PR-3	8800–8500 BP	*Ulmus*
PR-4	8500–7300 BP	*Corylus–Ulmus–Picea*

PR-5	7300–5000 BP	*Ulmus–Tilia–Quercus–Fraxinus*
PR-5a	7300–(6000) BP	*Corylus* subzone
PR-5b	(6000)–5000 BP	*Picea–Ulmus* subzone
PR-6	5000–3700 BP	*Picea*
PR-6a	5000–4500 BP	*Tilia–Fraxinus–Quercus* subzone
PR-6b	4500–3700 BP	*Carpinus* subzone
PR-7	3700–3000 BP	*Carpinus–Abies*
PR-8	3000–(1900) BP	*Abies–Fagus*
PR-9	(1900)–(700) BP	*Carpinus–Abies–Fagus*
PR-10	(700)–0 BP	NAP paz (with three local subzones illustrating development of colonization)

The data from two complementary sites are used in the synthesis for the P-a region: Grel, 600 m above sea level (Koperowa 1958, 1962); Bór na Czerwonem, 615 m (Obidowicz 1978).

Reference site 2. Szymbark
(Gil *et al.* 1974, Szczepanek 1989a)
Latitude 55°00′N, Longitude 21°06′E. Elevation 465 m. A mesotrophic mire formed in a depression behind the landslide ridge on the valley slope. Max. depth of deposit 5 m. Age range 8600–0 BP. Eight local pollen-assemblage zones (paz), five ^{14}C dates (Fig. 13.7).

Sz-1	(*ca.* 8700)–8300 BP	*Corylus–Ulmus*
Sz-2	8300–7700 BP	*Tilia–Ulmus–Corylus*
Sz-3	7700–(6000) BP	*Ulmus–Corylus–Fraxinus–Tilia*
Sz-4	(6000)–4500 BP	*Corylus–Tilia–Ulmus*
Sz-5	4500–3300 BP	*Picea–Tilia–Quercus*
Sz-6	3300–3000 BP	*Abies–Carpinus*
Sz-7	3000–(2100) BP	*Abies–Fagus*
Sz-8	(2100)–0 BP	*Alnus–Abies–Fagus*

The data from one other reference site (28, Jasiel, a mesotrophic mire, 675 m, SE part of P-c, Szczepanek 1987, 1989a) are used in the synthesis for type region P-c.

Puścizna Rękowiańska P-1 (650 m a.s.l.)

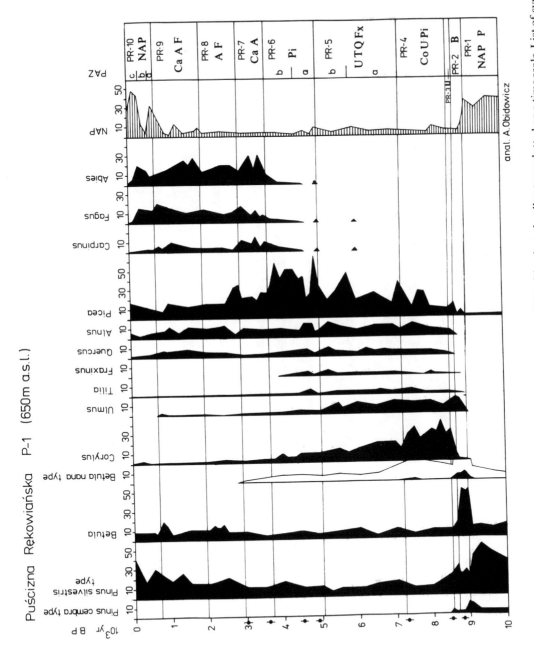

anal. A. Obidowicz

Fig. 13.6 Simplified pollen diagram from Puścizna Rękowiańska peat bog (P-1) with selected pollen taxa plotted on a time scale. List of symbols used for defining pollen types in all figures included in this paper: A – *Abies*, Al – *Alnus*, Art – *Artemisia*, B – *Betula*, Bn – *Betula nana*, Ca – *Carpinus*, Call – *Calluna*, Co – *Corylus*, Cy – *Cyperaceae*, Emp – *Empetrum*, F – *Fagus*, Fil – *Filipendula*, Fx – *Fraxinus*, Gr – *Gramineae*, H – *Hippophaë*, J – *Juniperus*, L – *Larix*, Nap – herbs, P – *Pinus*, Ps – *Pinus sylvestris* type, Pc – *Pinus cembra* type, Pi – *Picea*, Pl – *Plantago lanceolata*, Po – *Populus*, Pol – *Polypodiaceae*, Pte – *Pteridium*, Q – *Quercus*, S – *Salix*, Sec – *Secale*, Sph – *Sphagnum*, T – *Tilia*, U – *Ulmus*, Ur – *Urtica*

Szymbark P-2 (465m asl)

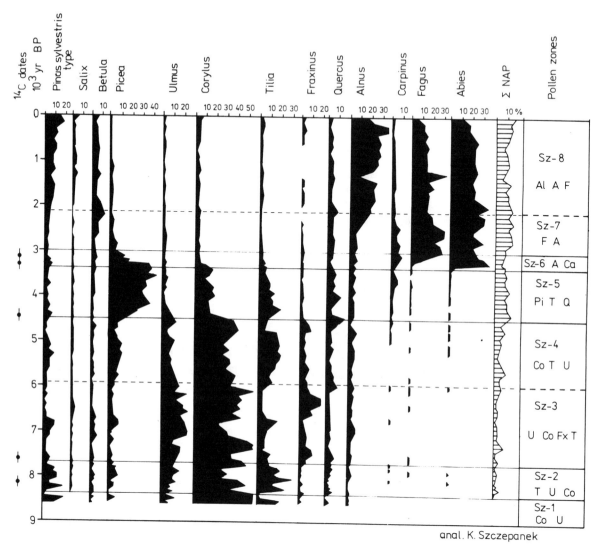

anal. K. Szczepanek

Fig. 13.7 Simplified pollen diagram from Szymbark mire (P-2) with selected pollen taxa, plotted on a time scale. For paz symbols see Fig. 13.6 caption

Reference site 3. Tarnowiec (Harmata 1987, 1989, Wójcik 1987, Alexandrowicz 1987)

Latitude 49°42′N, Longitude 21°37′E. Elevation 240 m. The late-glacial shallow basin filled with marl, overgrown by a mesotrophic mire at the beginning of the Holocene; recently drained and covered by a meadow. Age range *ca.* 12000–*ca.* 1000 BP. Ten local pollen-assemblage zones (paz), eight ^{14}C dates.

Tar-1	...–(11800) BP	NAP–*Juniperus*
Tar-2	(11800)–11450 BP	*Pinus cembra–Larix*
Tar-3	11450–(10700) BP	*Betula–Pinus–*
		Filipendula
Tar-4	(10700)–10250 BP	*Artemisia–*
		Chenopodiaceae–
		Betula
Tar-5	10250–... BP	*Pinus–Betula*
Tar-6	hiatus–5300 BP	*Alnus–Corylus–*

413

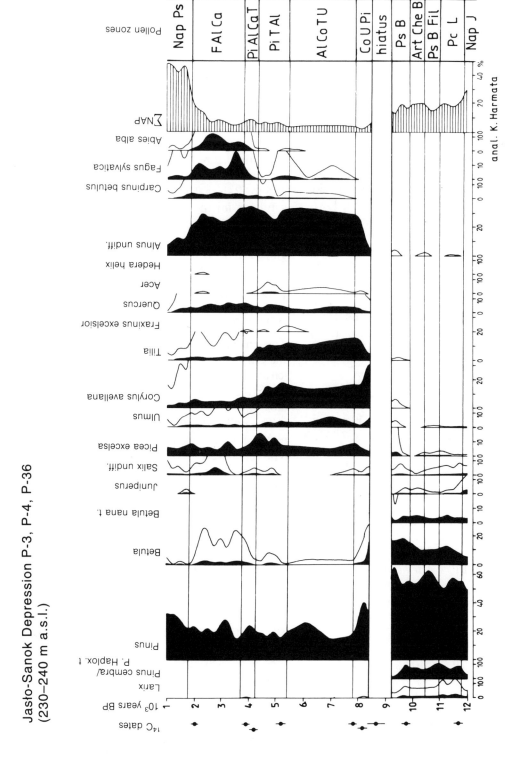

Jasło-Sanok Depression P-3, P-4, P-36
(230–240 m a.s.l.)

anal. K. Harmata

Fig. 13.8 Schematic smoothed pollen diagram compiled from sites Tarnowiec (P-3), Roztoki (P-4), and Jasło(p-36). For paz symbols see Fig. 13.6 caption

		Tilia–Ulmus
Tar-7	5300 – 4200 BP	*Picea–Tilia–Alnus*
Tar-8	4200 – 4000 BP	*Picea–Alnus–*
		Carpinus–Fagus
Tar-9	4000 – 1900 BP	*Fagus–Abies–*
		Carpinus
Tar-10	1900 – ... BP	NAP–*Pinus*

The data from site 3 were combined with those from two other reference sites located within a few kilometres distance and were used to construct the generalized pollen diagram (Fig. 13.8) and the regional pollen-assemblage zones for the type region P-d: site 4, Roztoki, 232 m elevation (Harmata 1987, 1989, Wójcik 1987, Różański *et al.* 1988) and site 36, Jasło, 229 m (Harmata unpubl.). Both are shallow late-glacial basins filled with marl and overgrown by a mire and then buried by river muds in the early Atlantic.

Additional information was also derived from two complementary sites of similar origin and stratigraphy: Besko at 288 m elevation (Koperowa 1970) and Kępa at 315 m (Gerlach *et al.* 1972) (Fig. 13.5).

Reference site 5. Tarnawa Wyżna

(Ralska-Jasiewiczowa 1980, 1989a)

Latitude 49°07′N, Longitude 22°50′E. Elevation 670 m. A mountain raised bog, 9 ha in area, 7.25 m in maximum depth, situated on the upper terrace of River San in its upper course. Age range *ca.* 11800–0 BP. Nine local pollen-assemblage zones (paz), nine ^{14}C dates (Fig. 13.9).

TW-1	*ca* 11800–10800 BP	*Pinus cembra–*
		Larix–Pinus (=P.
		sylvestris type)
TW-2	10800–9900 BP	*Artemisia–*
		Juniperus–Pinus
TW-3	9900–9500 BP	*Pinus cembra–*
		Pinus–Larix
TW-4	9500–8500 BP	*Ulmus–Picea–*
		Pinus
TW-5	8500–4400 BP	*Corylus–Ulmus–*
		Picea–Alnus

TW-5a	8500–(7000) BP	*Fraxinus* subzone
TW-5b	(7000)–4400 BP	*Picea*–Polypodia-
		ceae subzone
TW-6	4400–(3300) BP	*Alnus–Carpinus–*
		Fagus
TW-7	(3300)–285) BP	*Fagus*–NAP
TW-8	2850–(750) BP	*Fagus–Abies–*
		Carpinus
TW-9	(750)–0 BP	NAP

Data from three complementary sites (Smerek at 600 m elevation, Wołosate at 700 m, and Zakole at 585 m, Ralska-Jasiewiczowa 1980) are used in the synthesis for the type region P-e. The sequence of pollen-assemblage zones from Smerek illustrates the vegetational changes in the western part of the region (Fig. 13.10).

Palaeoecological patterns and events

General patterns of vegetational history

The vegetational history in the synthesis region of the Carpathian Mts is illustrated by three local pollen diagrams, one schematic pollen diagram, a correlation table (Fig. 13.10), and an event-stratigraphy table (Fig. 13.11).

The pollen diagrams (one in region P-a, two in region P-c, five in region P-e), are thought to be representative for the lower montane forest zones. A schematic smoothed pollen diagram, compiled from several site diagrams from the type region P-d, where no site with a continuous pollen sequence was found, is representative for the vegetation of foothill altitudes. Due to the high diversity of the area the sequences of pollen assemblage zones in Figure 13.10 cannot be treated as regional zones *sensu stricto*, although they also include data from complementary sites.

Two local paz sequences from the type region P-c (site 2 in the northwest part and site 28 in the southeast part of the region) can hardly be correlated with each other and reveal closer correlation with data from the adjacent regions P-a and P-e.

Fig. 13.9 Simplified pollen diagram from Tarnawa Wyżna peat bog (P-5), with selected pollen taxa, plotted on a time scale. For paz symbols see Fig. 13.6 caption

415

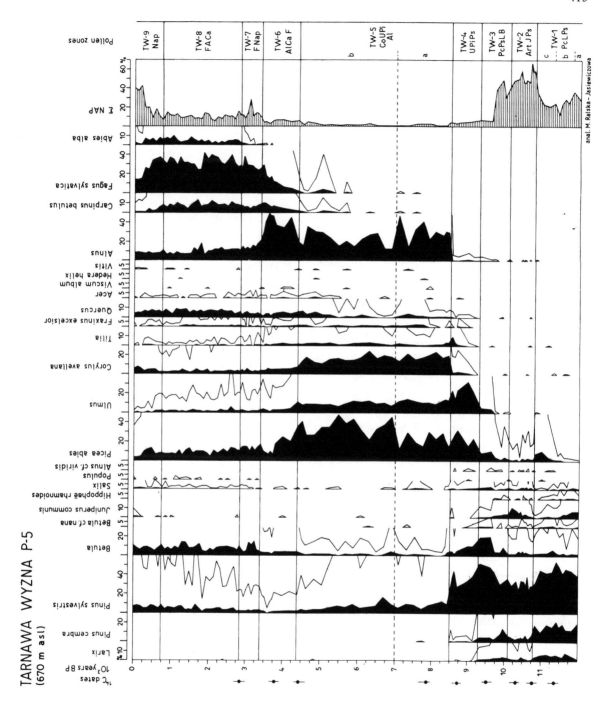

TARNAWA WYZNA P-5
(670 m asl)

anal. M. Ralska-Jasiewiczowa

As a consequence of such a differentiation, the only real correlation markers throughout the whole synthesis region are the immigration times for particular forest trees.

Late-glacial

Reliable information on the vegetational development between 12 000 and 10 000 BP is available only from regions P-d and P-e. The old pollen sequences from the Nowy Targ Basin (Koperowa 1958, 1962), interpreted as the record of the entire late-glacial since the Oldest Dryas, are clearly disturbed by input of rebedded pollen originating from flood deposits. However, they also give some useful information.

– The Allerød forests of the lower mountainous altitudes (no data from above 700 m) were mostly composed of conifers (*Pinus sylvestris, P. cembra, P. mugo, Larix,* and some *Picea*), *Betula* being generally of minor importance.
– Remnants of those forests survived the Younger Dryas cooling up to at least 650–700 m elevation (macrofossil evidence, P-a, P-d, P-e).

Holocene

– After a regeneration phase of the mixed conifer forests of late-glacial type, with a substantial contribution of *Betula* at the beginning of Holocene, their decline occurred generally between 9600 and 9000 BP.
– The formation of Holocene woodlands in different Carpathian ranges was closely connected with the migration routes and time of particular forest components (Ralska-Jasiewiczowa 1983). It generally progressed earlier in the eastern areas situated closer to the routes from the southeast around the Carpathian arc (recorded e.g. for *Ulmus* migration) or from the south via the Carpathian Low (e.g. *Tilia* migration).
– The mid-Holocene woodlands of low and middle altitudes were generally dominated by *Ulmus* (probably *U. glabra*) and *Corylus*, with other deciduous trees playing a more substantial role only in particular areas of the Carpathian Low (P-c, P-d regions). *Picea*, present in the entire synthesis region since the late-glacial, expanded after 9500–9000 BP on cool and humid valley floors, en-

croaching in places also on to mires (P-e). No information about its role in the forests of higher altitudes is discussed here. The contribution of *Picea* to the vegetation of P-c and P-d regions was of minor importance.

– The forest communities described above developed between 9500 and 8500 BP and remained rather stable until *ca.* (5300) 4700–4300 BP, with a locally increasing role of *Picea* in the younger part of this period.
– The late Holocene expansions of *Carpinus* and *Fagus* proceeded synchronously since 4500 BP (E)–3500 BP (W) and of *Abies* since 4000 BP (W)–2500 BP (E). They resulted in a complete transformation of Carpathian forests and gradual establishment of forest zones basically similar to those of today.

Unique patterns of vegetational history

Late-glacial

– The changes from Allerød to Younger Dryas vegetation in the low-lying area of the P-d region are poorly represented in the pollen record; they seem to express some reduction of conifer forest and the increased role of *Betula* and of *Artemisia*-dominated heliophyte communities.
– The continental east-Carpathian features of the Bieszczady (P-e) vegetation were expressed after the Allerød by the unusually high contribution of *Larix* in the conifer forests and by the appearance of *Alnus viridis* in the shrub/tall-herb communities.

Holocene

– The expansion of *Ulmus* and *Picea* was the earliest (from *ca.* 9500 BP) in the east (P-e). The phase of *Ulmus–Picea–Pinus* forests there preceded the development of deciduous woodlands by *ca.* 1000 years.
– *Tilia* (*cordata* type) expanded in the areas of the Carpathian Low (P-c, P-d) since at least *ca.* 9000 BP, playing there a substantial role in the forests until 4000–3300 BP, as one of the most distinctive lowland features of the vegetation in those regions.
– Two different patterns of *Alnus* history are re-

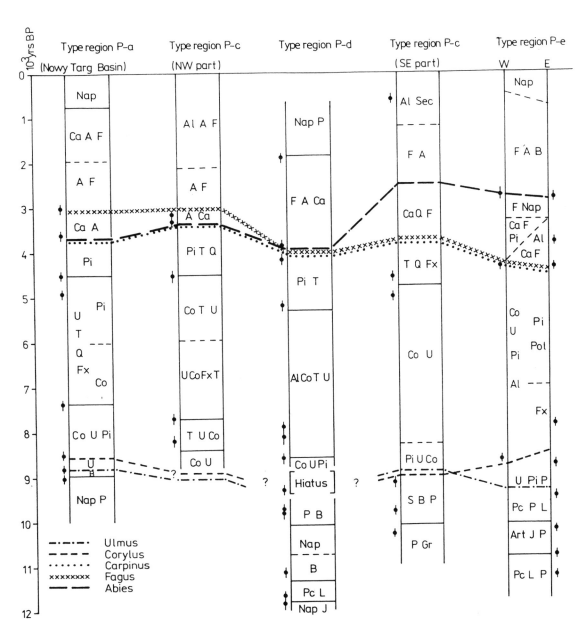

Fig. 13.10 Time–space correlation of type-region pollen–assemblage zones along the synthesis region of the Polish Carpathian Mts. The northwest and southeast parts of type region P-c (the Low Beskidy Mts) are treated separately. The main correlation lines follow the expansion times of some of the most important forest components. For paz symbols see Fig. 13.6 caption. Radiocarbon-dated borders are indicated with a solid line, and interpolated borders with a dashed line

418

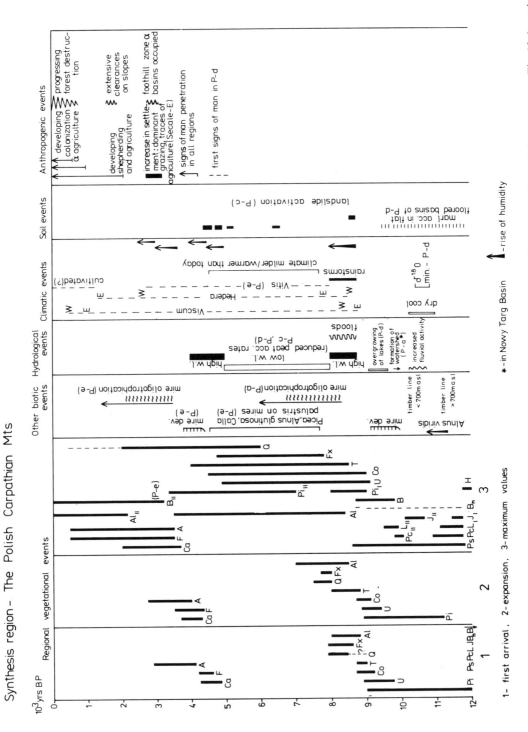

Fig. 13.11 Event stratigraphy for the Polish Carpathian Mts synthesis region. For symbols used in column of vegetation events see Fig. 13.6 caption

corded in the Carpathian profiles: (1) A rapid expansion since 8400–8000 BP, followed by high pollen values up to 40% or more until the late Holocene in the P-e and P-d regions, suggest the spread of *Alnus glutinosa* on mires, as confirmed by macrofossil evidence; (2) the very low pollen values throughout the mid-Holocene, with the rapid, clearly anthropogenic expansion in historical time, suggest the presence of *Alnus incana* only, as recorded in the P-c (Low Beskidy) region (sites 2, 28).

– The maximum spread of *Picea*, accompanying the decline of Atlantic mixed deciduous forests and preceding the expansion of *Carpinus* and *Fagus*, makes a pronounced phase between *ca.* 4500 and 3500 BP in the P-a and northwest part of the P-c regions and is earlier (5300–4000 BP) and less distinct in the P-d region. It does not appear in the southeast part of the P-c region, where *Picea* was always scarce. In the P-e region the maximum *Picea* development occurred much earlier (7000–4400 BP), its decline coinciding with the decline of *Ulmus–Corylus*-dominated woodlands, which was directly followed by the expansion of *Carpinus* and *Fagus*. Such a successional pattern suggests a different ecological role and azonal character of *Picea* communities in the Bieszczady Mts in the past, to that of today. *Picea* was there a peat-forming component on mires from the Preboreal, until they changed into ombrotrophic bogs between 4400 and 3300 BP.

– The formation of late-Holocene zonal forest communities proceeded in different ways, in eastern and western regions, depending on the directions and rates of tree migrations but also on the edaphic and climatic factors:

– *Carpinus* and *Fagus* woodlands started to develop in P-e and in the southeast part of P-c from *ca.* 4400–3800 BP, with *Fagus* becoming an absolutely dominant forest tree there from 3500–3000 BP, and *Abies* contributing as a minor forest component since *ca.* 2700–2200 BP only. In the low-lying region P-d, *Carpinus*, *Fagus*, and *Abies* expanded synchronously around 4000 BP, and in the P-a and northwest part of P-c a phase of *Carpinus* and *Abies* forests between 3600 (3300) and 3000 BP preceded the development of *Abies–Fagus* forests, with *Abies* being a major component.

Climatic events

No information about late-glacial climate is available from the regions P-a and P-c.

– A discrepancy exists between the stable-isotope and palaeobotanical data from the late-glacial profile at site 4 (region P-d); δ ^{18}O curve (Fig. 13.11) suggests a strong cooling covering the decline of Allerød and the early phase of Younger Dryas, a large temperature minimum at middle Younger Dryas, and a slow warming recorded from the late Younger Dryas; there is no corresponding record of drastic changes in plant communities.

– The late phase of the Younger Dryas was the driest in the eastern regions (P-d and P-e), but the temperatures were not severe enough to exterminate the temperate swamp communities of *Cladium mariscus*, *Typha latifolia*, and *Scirpus lacustris* ssp. *glaucus* at altitudes of *ca.* 700 m in region P-e.

– The climatic plant-indicators show warm summers (*Viscum*) since 8700 BP in P-e, 8300 BP in P-c (NW), and *ca.* 8200 BP in P-a, and mild winters (*Hedera*) since 7800 BP in P-e, 8400 BP in P-c (SE), and 8500 BP in P-a, clearly illustrating the W-E continentality gradient.

– The climate was distinctly warmer and milder than today in the region P-e, as suggested by the occurrence of *Vitis* (cf. *V. sylvestris*) and of swamp alderwood communities of lowland type with *Alnus glutinosa*, *Calla palustris*, *Carex elongata*, and others between *ca.* 8000 BP and 4500 BP at elevations of 600–700 m in the P-e region.

– The favourable conditions for mire development and oligotrophication recorded from *ca.* 4500 BP in the P-e region suggest a general rise of climatic humidity.

Hydrological events

No lakes exist today in the areas discussed. The hydrological events (Fig. 13.12) are mainly inferred from the peat stratigraphy.

– During the late-glacial in the region P-d numerous shallow water bodies accumulated marl at the bottoms of depressions and in the river valleys. In other regions, small waterpools and mires developing later into peat bogs were mostly connected with the river systems and subject to frequent and intensive hydrological processes.

- The lowering of water levels and reduction of fluvial activities during 9700–9200 BP resulted in

overgrowth of lakes in the Jasło–Sanok region and in stabilization of river terraces in the other regions, enabling continuous growth of mires. In the Nowy Targ Basin the watershed was formed then.

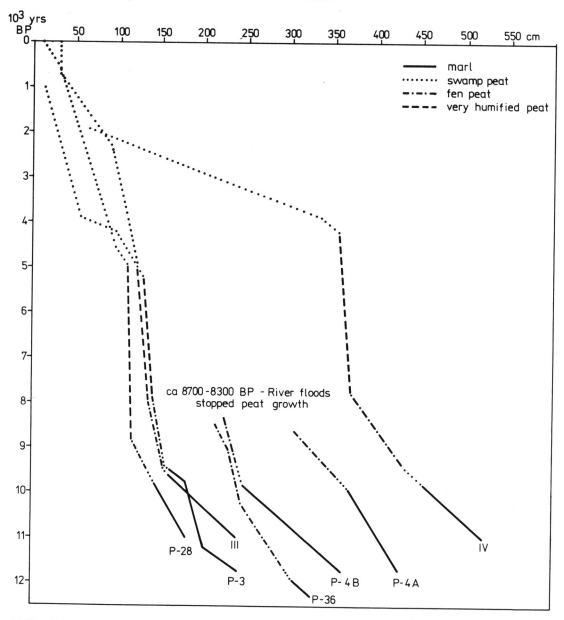

Fig. 13.12 Curves of sediment-accumulation rates in some mires from type regions P-c and P-d, showing generally rather high rates in the late-glacial basins accumulating marl, their gradual decrease after the lakelets are overgrown by swamp and fen at the beginning of the Holocene, overflooding of some mires around 8500 BP, a strong reduction of peat growth in the mid-Holocene, and an increase of accumulation rates in swamp peats forming since *ca.* 5000 BP

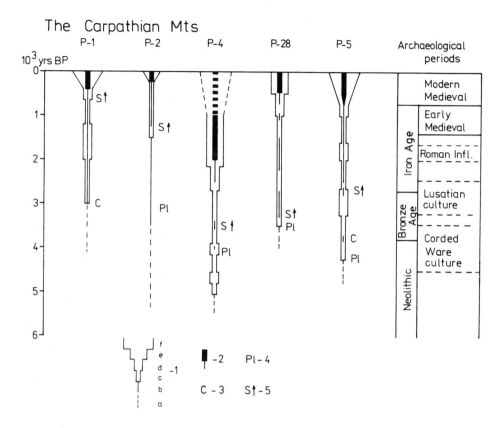

Fig. 13.13 Schematic interpretation of the intensity of human impact on vegetation, based on pollen frequencies from five sites representing the Carpathian Mts synthesis region. 1–human impact in general expressed in 6-grade scale: a–probable presence of man, b–slight local impact, c–regular clearings and land-use, d, e–increasing extent of clearings and land-use, f–large-scale deforestation and intensive economy. 2–slight to substantial indications of agriculture. 3–first appearance of cereals. 4–first appearance of *Plantago lanceolata*. 5–increase of *Secale* cultivation

– Increased precipitation around 8600–8300 BP is evidenced by (1) the series of floods depositing fluvial muds on the mires at river terraces in the regions P-e and P-d, (2) the activation of landslides in P-c, and (3) the transformation of minerogenic forested mires into ombrogenous bogs in P-a.

– Between *ca.* 8000 and 5000 BP most of the mires in P-c and P-d suffered from strong water deficit, reducing peat-accumulation rates to less than 0.1 mm yr^{-1}, although with no clear evidence of hiatuses (Ralska-Jasiewiczowa & Starkel 1988). In the bogs of P-a the reduction of peat-accumulation rates is recorded between 7400 and 4900 BP,

and in P-e the expansion of trees (mostly *Picea* on mires) occurs between 7000 and 4400 BP.

– The rise of water level after 5000 BP is evidenced by increasing accumulation rates and a change in peat-forming vegetation in mires of P-a, P-c, and P-d, landslide activation in P-c, development of new mires, and inundation and oligotrophication of older mires in P-e, recorded from *ca.* 4400 BP.

Anthropogenic events

– Traces of human activities (cattle breeding) appear first from *ca.* 5000 BP in the P-d area and then from *ca.* 4300–4000 BP in P-e (Fig. 13.13). They probably indicate the penetration of half-nomadic tribes of the Corded Ware culture earlier into the eastern parts of Carpathians than in the western ranges.

– Lusatian settlements during the Late Bronze/ Early Iron Ages are recorded from *ca.* 3300 BP in P-c (SE) and P-e, from *ca.* 3000 BP in P-a, and from 2700 BP in P-d. Evidence of agricultural practices appears more or less regularly, but the dominance of pastoral activities is obvious.

– The distinct settlement phases with continuous (P-a, P-e) or increasing (P-d) evidence of agriculture recorded since *ca.* 2000 BP correspond with the time of Roman influences.

– The local clearings during the above settlements particularly affected the lowest-lying *Carpinus*-dominated woodlands, and partly also *Fagus* forests higher on slopes (evidence from P-a and P-e). However, large-scale destruction of natural forests in the whole Polish Carpathians started in the Early Medieval/Medieval colonization since 1000– 700 BP.

– The low frequencies of cultural indicators in both sites from P-c may result from the location of sites on the slopes. However, the changes in tree-pollen curves (decline of *Carpinus* and *Fagus* and rise of *Alnus, Betula,* and *Pinus*) since *ca.* 2000 BP are obvious records of anthropogenic changes in forests.

TYPE REGION P-f, POLISH SUDETY MOUNTAINS

The Sudety Mts (Fig. 13.14) are old mountains of highly differentiated geological structure. Their history, including vegetational history developed independently from that of the Carpathian Mts.

In the geobotanical subdivision of Poland (Szafer & Zarzycki 1972) they represent a mountain subprovince of Europe different from the Carpathians, namely the Hercynic–Sudetic Subprovince, its main part belonging to Czech Republic. For these reasons, the only reference site investigated from this region (site 6, representing the area of Bystrzyckie Mts in the central Sudety) must be treated separately in the synthesis of Poland.

Altitude: 200–1600 m.

Climate: For the central-eastern Sudety:mean January temperatures –6 to –2°C, mean July temperatures 14–17.5°C, mean annual temperatures 3– 7.5°C. Mean annual rainfall 700–1350 mm. Western winds prevail, interrupted frequently by southwestern foehn winds.

Geology: Polygenetic structures consisting of Precambrian metamorphic rocks, secondarily folded Palaeozoic sediments, intrusive granite, melaphyre and porphyry, Tertiary basalt, Cretaceous sandstone, etc. (Fig. 13.14).

Topography: Horst mountains formed by tectonic dislocations in the Tertiary, with flat tops and steep slopes separated by subsidence or denudation depressions.

Vegetation: Compared with the Carpathians, it is much poorer, especially in mountain species, and the ranges of vegetational zones are 250–300 m lower. The foothill zone up to 400 (500) m elevation is occupied by Carpinion and Pino-Quercion forests. The lower montane forest zone up to 1000 m has a natural zonal community of Fagetum sudeticum but is now mostly formed by mixed *Picea* forests dominated by planted *Picea*; *Fagus* forests are of minor importance. The upper montane forest zone is occupied by Piceetum hercynicum. All zonal communities differ floristically from the corresponding communities of the Carpathians and are closer to the forests of western mountains. Numerous raised bogs occur in both montane forest zones. The subalpine and alpine zones occur only in the Western Sudety.

Soils: Variety of mountain soils connected with the diversity of substratum. Podzolic poor soils on acid granite and porphyry prevail. In the Kłodzko Basin rendzinas on loess.

Archaeology: Single Neolithic finds. The earliest regular colonization was in the Kłodzko Basin in the 10th century. Strongholds since the middle 13th century.

Reference site 6. Zieleniec (Madeyska 1989)

Latitude 50°21′N, Longitude 16°24′E. Elevation 750–760 m. A mountain raised bog with *Pinus uliginosa, ca.* 150 ha in area, maximum depth *ca.* 7 m.

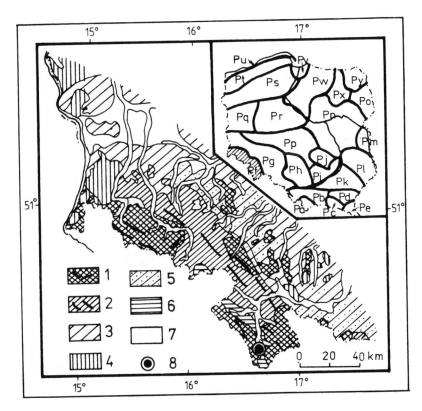

Fig. 13.14 Geomorphic map of the type region P-f (the Sudety Mts) showing the location of reference site. 1–crystalline and metamorphic blocks of Hercynian–Caledonian zone, 2–secondary folded structure, 3–Pleistocene denudation plains, 4–Neogene sedimentary covers, 5–Pre-Tertiary sedimentary covers, 6–loess, 7–Pleistocene and Holocene river valleys and terraces, 8–reference site P–6 Zieleniec

Age range *ca.* 9000–0 BP. Nine local pollen-assemblage zones (paz), six ^{14}C dates (Fig. 13.15).

ZL-1	9000–8500 BP	*Salix–Pinus–Betula–Ulmus*
ZL-2	8500–(8050) BP	*Corylus–Betula*
ZL-3	(8050)–(6150) BP	*Corylus–Picea–Ulmus–Tilia*
ZL-4	(6150)–5650 BP	*Corylus–Ulmus–Quercus*
ZL-5	5650–(4400) BP	*Corylus–Picea–Quercus–Ulmus*
ZL-6	(4400)–3750 BP	*Picea–Quercus–Fagus*
ZL-7	3750–(3300) BP	*Fagus–Carpinus–Alnus*
ZL-8	(3300)–(2300) BP	*Fagus–Abies*
ZL-9	(2300)–0 BP	*Fagus–Abies*–Cerealia

Palaeoecological patterns and events

General patterns of vegetational history (Fig. 13.16)

– The beginning of the Holocene is not recorded in the peat bog. The record starts *ca.* 9000 BP, when *Corylus* and *Ulmus* started to spread into the still-open *Pinus–Betula* forests with *Populus* and *Salix*.

– Between 8500 and 8100 BP the main forest components were *Betula* and *Corylus*, with a substantial contribution of *Ulmus*.

– Since *ca.* 8000 BP *Tilia*, *Quercus*, and *Fraxinus* contributed to the *Corylus*- and *Ulmus*-dominated forests of lower elevations, and *Picea* spread to higher elevations.

– Between 7500 and 7300 BP *Betula* retreated from the forests, and between 7200 and 6700 BP *Tilia* reached its maximum development.

Zieleniec   (750–760 m a s l)

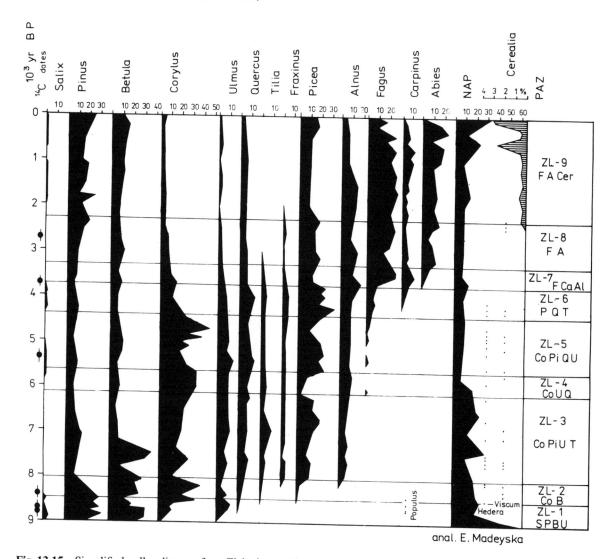

Fig. 13.15 Simplified pollen diagram from Zieleniec peat bog (P-6) with selected pollen taxa, plotted on a time scale. For paz symbols see Fig. 13.6 caption

– A more or less stable pattern of forest communities remained till *ca.* 4600 BP. Between 4500 and 4200 BP the dominant role of *Corylus* gradually declined, and *Fagus* started slowly to spread.

– The basic change of forest composition between 4000 and 3200 BP resulted from the complete retreat of Atlantic mixed deciduous forest and spread of *Fagus* and *Carpinus*, and between 3500 and 3200 BP also of *Abies*. Since that time the composition of forests remained basically unchanged, with human impact expressed by rise of NAP, cereals, and *Pinus* since *ca.* 2100 BP.

– The low contribution of *Alnus* until *ca* 3600 BP suggests the absence of *Alnus glutinosa* from the area.

425

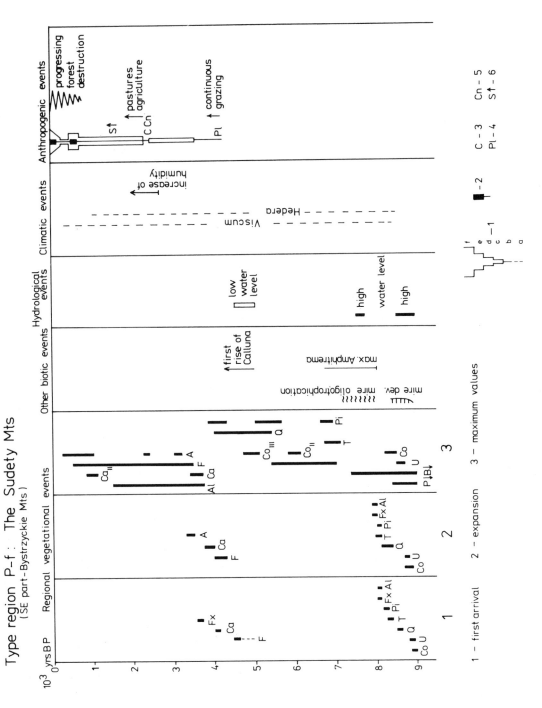

Fig. 13.16 Event stratigraphy for the type region P-f (Sudety Mts). 3– first appearance of cereals. 4– first appearance of *Plantago lanceolata*. 5– first appearance of *Centaurea cyanus*. 6–increase of *Secale* cultivation. For other symbols see Fig. 13.6 and 13.3 captions

Differences in vegetational patterns between the Carpathians and the Sudety Mts

The site 6 is situated in the higher part of the lower forest zone; most sites studied in the Carpathians are located in the lower part of the lower montane zone, some in the foothill zone.

- Nearly all forest components immigrating in the early Holocene expanded in the Sudety Mts later than in the Carpathians, e.g. the expansion of *Ulmus* was later by at least 800 years than in the eastern Carpathians (P-e), but only slightly later than in their western part (P-a). It spread here since *ca.* 8800 BP, synchronously with *Corylus*.
- The expansion of *Picea* was delayed by 1500 years (E Carpathians) to 1000 years (W Carpathians), because its first appearance in the Sudety area was only slightly before 8000 BP.
- The appearance of *Alnus* in the Sudety was 800–700 years later than in the Nowy Targ Basin (P-a).
- The *Corylus*-dominated communities reached their maximum development between *ca.* 6200 and 4600 BP; during that period in the Carpathians the role of *Corylus* gradually decreased (at the expense of expanding *Picea*?), as is especially distinct in the Nowy Targ Basin.
- *Fagus* appeared in the Sudety before 4000 BP, but its real expansion proceeded since *ca.* 3800–3700 BP. It was roughly synchronous with that of *Carpinus* and was followed by the spread of *Abies* since *ca.* 3500–3400 BP. *Fagus* remained dominant until *ca.* 1000 BP. This pattern is different from that of the West Carpathians (Nowy Targ Basin), where *Abies* expanded earlier than *Fagus* and remained dominant until recent times.

All these differences suggest independent forest development in the Sudety and West Carpathians, although at least one of the main migration routes (from SSW via the Moravian Gate) was common for both mountain areas.

Climatic events

Little climatic information can be derived from the existing data, beyond that suggested by the general development of vegetation.

- *Hedera* and *Viscum* pollen is recorded regularly from *ca.* 8500 to 4000 BP and then only sporadically about 3000 BP and 1000 BP (*Hedera*) and about 2500–2300 BP and 500 BP (*Viscum*). The end of regular record at *ca.* 4000 BP coincides with the decline of all Atlantic deciduous-tree taxa, indicating the change of dominant forest communities of the lower mountain elevations.
- The general increase in climatic humidity around 5000 BP and 2800–2500 BP is reflected in the increase in peat-accumulation rates.

Hydrological events

- High water levels in the mire are recorded by the presence of aquatics and by high pollen values of telmatophytes twice: in its initial phase around 9000 BP and again between 8000 and 7500 BP, accompanying its gradual change into a *Sphagnum* bog.
- Low water level is suggested by the spread of *Calluna* on the peat bog around 4800 BP (?).

Anthropogenic events

- Anthropogenic pollen indicators, including *Plantago lanceolata*, appear regularly from *ca.* 4200 BP, suggesting some land-use (probably grazing) within the region (Fig. 13.16).
- Some increase in pollen frequencies of cultural indicators, including cereals and *Centaurea cyanus* around 2100 BP, may express the appearance of agricultural practices at some distance from the peat bog, as confirmed by the first *Secale* pollen at *ca.* 1700 BP.
- Deforestation is recorded by changes in tree-pollen curves and by increases in pollen values of nitrophilous and acidiphilous weeds and other culture indicators since *ca.* 600 BP only.

TYPE REGIONS P-h, P-i, AND P-j, MIDDLE POLAND UPLANDS

Three type regions are discussed jointly as a "synthesis region": Silesia–Cracow Upland (P-h), Miechów Upland and Nida Basin (P-i), and Holy Cross Mts (P-j) (Fig. 13.17). Together they make an area differentiated with respect to the geology, geomorphology, and present vegetation. The palynological data show that differentiation of the vegetation was very clear in the past as well.

The most important questions concerning the area are the arrival time of some trees, the origin of xerothermic steppe communities, and the anthropogenic transformations of the landscape. However, the results we have so far obtained are still unsatisfactory and do not permit any comprehensive answers.

The lack of appropriate data is caused by the scarcity of sites suitable for palynological investigations. Even those profiles that were analysed are characterized by hiatuses and periodically by very low accumulation rates. Most of the data do not meet the requirements of modern pollen analysis because of the unsatisfactory sampling methods, contamination, very low sporomorph counting, or lack of radiocarbon dates. Thus the basis for this discussion is limited to only two radiocarbon-dated

sites: Wolbrom (7) and Słopiec (8). The other sites, including the undated site of Czajków (29), provide only minor additional information.

Altitude: 150–611 m above sea level.

Climate: Mean annual temperature 7.5°C, mean January −3.5°C, mean July 18°C, annual temperature amplitude over 22°C. Mean annual precipitation 600–800 mm.

Geology: Palaeozoic bedrock covered mainly by Triassic, Jurassic, and Cretaceous limestones, which are still subject to active karst processes. The highest part of the Holy Cross Mts is built of Cambrian quartzite. During the last glaciation a part of the area was covered by loess. Pleistocene sands overlie older formations in the depressions and river valleys.

Topography: An upland landscape of varied relief; the Holy Cross Mts represent the highest part of the area (max. 601 m elevation).

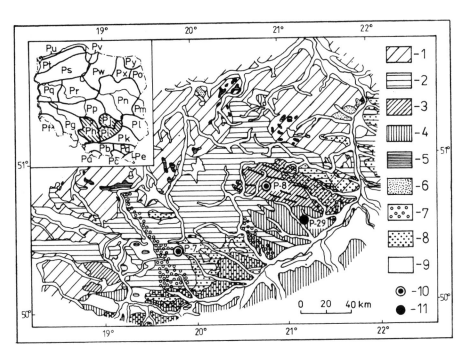

Fig. 13.17 Geomorphic map of the Middle Poland Uplands synthesis region showing location of sites. 1–Pleistocene denudation plains, 2–sedimentary covers of pre-Tertiary rocks, 3–secondary folded structure in the Hercynian–Caledonian zone, 4–Neogene sedimentary covers, 5–denudation remnants, 6–outwash plains, 7–karst sink-holes, 8–loess, 9–Pleistocene and Holocene river valleys and terraces, 10– reference sites, with the pollen diagrams included in this paper: P-7 Wolbrom, P-8 Słopiec, 11–other reference sites: P-29 Czajków

Vegetation: The area is intensively deforested. For a large part of the region the plant cover depends on loess, which is strongly eroded and exposes the bedrock. On such substrate xerothermic plant communities are present. The forests are composed of different communities in which *Quercus*, *Fagus*, *Carpinus*, and both *Tilia* species (*T. cordata* and *T. platyphyllos*) are important; at higher elevations *Abies* and *Picea* are also significant. A great peculiarity of the Holy Cross Mts is the high contribution of *Larix decidua* spp. *polonica* (Racib.) Domin to the forests. *Pinus* forests predominate on sandy soils in the extensive depressions and river valleys.

Soils: Rendzinas on limestone and gypsum, chernozems and podzolized soils on loess, brown soils and different kinds of acid soils on the sandy formations.

Archaeology: The oldest traces of human activity date back to the Palaeolithic. Neolithic and subsequent settlements were distributed unevenly, but in some areas they were very intensive. In the Holy Cross Mts there are imposing iron works dated at 1900–1500 BP.

Reference site 7. Wolbrom (Latałowa 1976, 1988a, 1989a, Latałowa & Nalepka 1987, Obidowicz 1976)

Latitude 50°23′N, Longitude 19°46′E. Elevation 375 m. Age range *ca.* 13000–5500 BP and *ca.* 2500–2300 BP. Large peat bog. Ten local pollen-assemblage zones (paz), nineteen ^{14}C dates (Fig. 13.18).

The sediments reveal a hiatus covering the younger Atlantic and the Subboreal periods. The absence of the upper deposits is a result of peat cutting and land reclamation. The pollen diagram presented in this synthesis (Fig. 13.18) is compiled from the three diagrams presented by Latałowa & Nalepka (1987).

Wol-1	>12000 BP	*Pinus–NAP*
Wol-2	?–12000 BP	*Pinus–Betula–NAP*
Wol-3	12000–11800 BP	*Betula–Juniperus*
Wol-4	11800–10800 BP	*Pinus–Betula*
Wol-5	10800–10000 BP	*Betula–Larix–*
		Juniperus–
		Artemisia

Wol-6	10000–9300 BP	*Pinus–Betula–*
		Filipendula–
		Polypodiaceae
Wol-7	9300–8800 BP	*Pinus–Picea–Ulmus–*
		Corylus
Wol-8	8800–7000 BP	*Corylus–Quercus–*
		Tilia–Alnus
Wol-9	7000–?	*Ulmus–Fraxinus–*
		Corylus–Sphagnum
Wol-10	^{14}C dates:	*Carpinus–Fagus–*
	2420±70 BP	*Abies–Sphagnum*
	2300±70 BP	

Reference site 8. Słopiec (Szczepanek 1961, 1982, 1989b)

Latitude 50°47′N, Longitude 20°47′E. Elevation 248 m. Age range 10280–0 BP. Medium-size mesotrophic mire, eight local pollen-assemblage zones (paz), seventeen ^{14}C dates. Between *ca.* 9000 and 3600 BP a very low peat-accumulation rate is recorded (Fig. 13.19).

Sł-1	10280–9900 BP	*Juniperus–Pinus*
Sł-2	9900–8900 BP	*Salix–Betula–Ulmus*
Sł-3	8900–7000 BP	*Corylus–Alnus*
Sł-4	7000–3600 BP	*Corylus–Quercus–Picea*
Sł-5	3600–2500 BP	*Alnus–Carpinus–*
		Fagus–Abies
Sł-6	2500–1100 BP	*Carpinus–Fagus–*
		*Abies–Pinus–*NAP
Sł-7	1100–400 BP	*Pinus*
Sł-8	400–0 BP	NAP–*Pinus*

The additional information concerning the type region P-i comes from the undated site Czajków 29 (Fig. 13.17) studied by Szczepanek (1971).

Palaeoecological patterns and events

General patterns of vegetational history

Two pollen diagrams (Fig. 13.18 and 13.19) are discussed in conjunction with some additional data. The time correlation of the local pollen-assemblage zones from these diagrams is presented on Figure 13.20. These data, however are not sufficient to define regional pollen-assemblage zones. The event-stratigraphy table (Fig. 13.21) summarises the most important information on environmental changes.

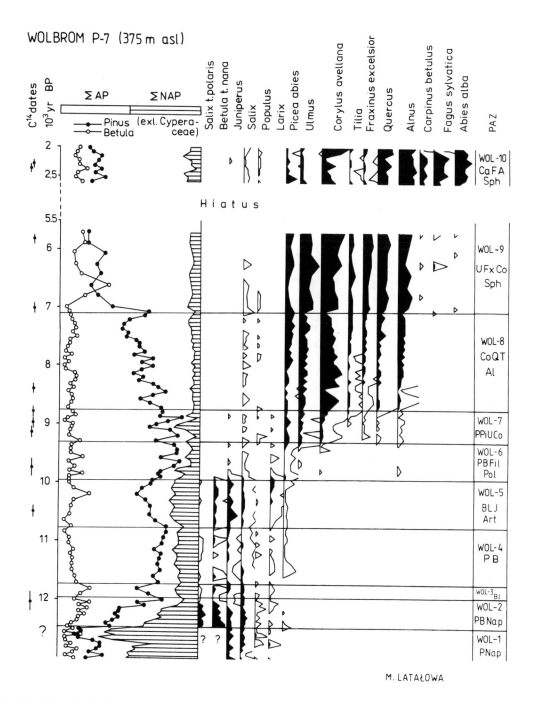

Fig. 13.18 Simplified pollen diagram from Wolbrom peat bog (P-7), with selected pollen taxa, plotted on a time scale. For paz symbols see Fig. 13.6 caption

Słopiec P-8 (248 m a.s.l.)

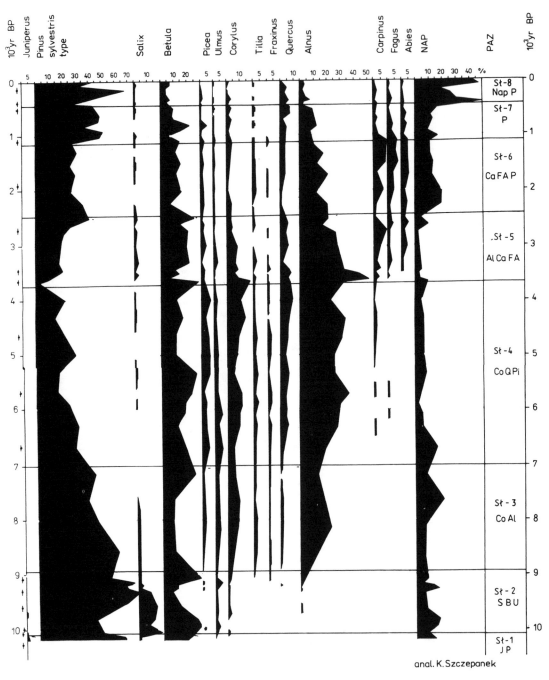

anal. K. Szczepanek

Fig. 13.19 Simplified pollen diagram from Słopiec mire (P-8), with selected pollen taxa, plotted on a time scale. For paz symbols see Fig. 13.6 caption

Late-glacial

Fully developed late-glacial sediments are represented by the pollen diagram from Wolbrom. The vegetational changes are described by the following pollen assemblage zones:

Pinus–NAP paz (Oldest Dryas): The vegetation consisted of treeless tundra.

Pinus–Betula–NAP paz (Bølling):
The expansion of *Betula* was followed by the development of woodland with a high proportion of *Pinus*. *Pinus cembra* was present.

Betula–Juniperus paz (Older Dryas):
It was probably a short-lived thinning of woodland vegetation and a slight expansion of *Juniperus*.

Pinus–Betula paz (Allerød):
Betula–Pinus and then *Pinus* woods became widespread; *Picea* and *Alnus* probably appeared.

Betula–Larix–Juniperus–Artemisia paz (Younger Dryas):
Woodland became more open; *Larix* was an important constituent in the forests; *Juniperus* and *Artemisia* expanded again.

Holocene

- From around 10000 BP the *Pinus* woods with *Betula* spread, *Picea* gradually became more important, and *Larix* was present.
- From *ca.* 9300 BP, *Picea*, *Ulmus*, and *Corylus* expanded; the expansion of *Ulmus* was probably much earlier in particular areas (10000–9700 BP?).
- From *ca.* 9000 (8800) BP the proportion of *Pinus* gradually declined, and deciduous woodland communities increased in importance; *Tilia*, *Fraxinus*, *Quercus*, and *Alnus* expanded.
- *ca.* 7000–3500 BP, the optimum development of mixed deciduous forests.
- From *ca.* 5000 BP, the probable expansion of *Carpinus* in the forest communities, and at *ca.* 3500 BP *Fagus* and *Abies*.
- From *ca.* 1000 BP, total deforestation began.

Unique patterns for type regions P-h, P-i, and P-j

- In P-i and probably P-j, *Ulmus* expanded earlier than *Corylus*, between 10000 and 9700 BP (?); in

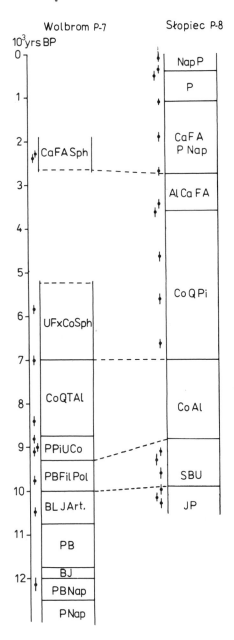

Fig. 13.20 Time–space correlation of local pollen-assemblage zones from the sites representing the Middle Poland Uplands synthesis region. For paz symbols see Fig. 13.6 caption

P-h *Ulmus* expanded somewhat later, about 9300 BP.
- In P-h the participation of *Picea* started much earlier and was stronger than in P-i and P-j.

Synthesis Region – Uplands of Middle Poland

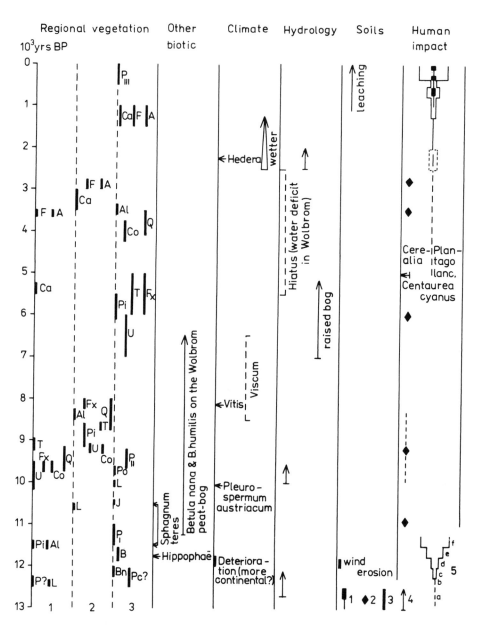

Fig. 13.21 Event stratigraphy for the Middle Poland Uplands synthesis region. 1–slight to substantial indications of agriculture, 2–extensive fires, 3–range of a phenomenon (generally), 4–rise of water level, 5–intensity of human impact expressed in a 6-grade scale. For other explanations see Fig. 13.6 and 13.16 captions

– In P-h and P-i there were better conditions for *Corylus* expansion and for mixed deciduous-forest development.
– The forest communities with *Carpinus* and those with *Fagus* and *Abies* were more widely distributed in P-h and P-i than in P-j.
– P-j represents the poorest edaphic conditions.

Hydrology of mires

The phases of the high ground-water level correspond with Bølling and beginning of Allerød, and they also occur later at *ca.* 10000–9000 BP, *ca.* 2500 BP, and *ca.* 480–370 BP (Słopiec–deforestation).

The phases of low ground-water level lasted from the middle Allerød to the Younger Dryas and from 9000 to 2500 BP, with the strongest lowering since *ca.* 7000 BP.

Human impact

Great incompatibility exists between the archaeological evidence of early intensive human activity in the loess uplands and the lack of palynological evidence for such intensive human impact.

Palynological data on human impact are few in the Wolbrom diagrams (P-h region) due to the lack of the uppermost parts of the profiles, and also due to the hiatus extending throughout the Subboreal. In all probability the charred layers occurring in the peat deposits in all Wolbrom profiles since the late-glacial are of anthropogenic origin. Worthy of notice also are the pollen grains of *Plantago lanceolata* coincident with the decrease in the *Ulmus* curve close to the radiocarbon date of 5850±70 BP. Pollen of *Plantago lanceolata*, *Secale*, *Triticum* type, *Centaurea cyanus*, and *Fagopyrum* found in the layer dated at 2420±70 and 2300±70 BP record the Iron Age settlement.

In the pollen diagrams from P-i and P-j regions (Szczepanek 1961, 1971, 1982) single pollen grains of Cerealia and *Plantago lanceolata* appear around 5000 BP; they are more frequent since *ca.* 3500 BP and since *ca.* 2700 BP, but typical settlement phases cannot be distinguished. The intensive human impact reflected by the anthropogenic indicators can be dated only to 1000 BP.

The reasons for the weakly marked settlement phases are probably the very low sediment-accumulation rate and the low pollen sums counted (Słopiec slightly above 200, Czajków 300–500). The frequent charcoal layers in the profiles and changes in the tree-pollen curves, however, suggest early and intensive human impact.

TYPE REGIONS P-n AND P-m, THE EASTERN PART OF THE MID-POLISH LOWLANDS

The synthesis region of the Mid-Polish Lowlands (Fig. 13.22) includes the territories situated between the maximum extent of the Vistulian ice sheet to the north-northwest and the margins of uplands to the south-southwest. The region is broadly opened to the east, adjoining the vast areas of Polesie and the Byelorussian plateau. The synthesis involves two type regions: the Masovia and Podlasie Lowlands (P-n) and the Lublin Polesie (P-m). No data were available from the third type region belonging here, the Białystok Upland and Biebrza Basin (P-o), although one of the reference sites discussed (10) lies on the border between P-m and P-o.

Altitude: 50–220 m asl.

Climate: Mean January temperatures –2 (SE) to –5°C (NE), mean July temperatures 19.6 (SE) to 17.5°C (NE), mean annual temperatures 8.4 (SE) to 6.5°C (NE). Mean annual rainfall 480 mm (SE and central lowlands) to 650 mm (NE). Generally western winds prevail. Growing season from 203 days in southeast part to maximum of 220 days in central lowlands.

Geology: A large tectonic Tertiary basin forms the base for the Pleistocene deposits in the central-western part of the area. In the southeastern part the pre-Quaternary surface is built of Jurassic and Cretaceous limestones and Tertiary sands and mudstones. Carboniferous coal beds occur along the Bug River valley. The Quaternary cover, originating mainly from the Middle Polish glaciation, was levelled by periglacial processes.

Topography: Generally flat areas mostly up to 150 m

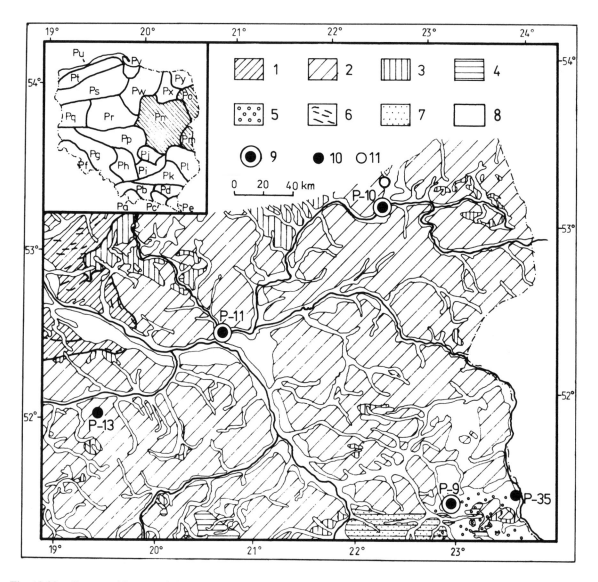

Fig. 13.22 Geomorphic map of the eastern part of the Mid-Polish Lowlands synthesis region, showing the location of sites. 1–ground moraines, 2– Pleistocene denudation plains, 3–outwash plains, 4– sedimentary covers of pre-Tertiary rocks, 5–karst sink-holes, 6–terminal moraines, 7–loess, 8– Pleistocene and Holocene river valleys and terraces, 9–reference sites with the pollen diagrams included in this paper: P-9 Lake Łukcze, P-10 Lake Maliszewskie, P-11 Lake Błędowo, 10–other reference sites: P-13 Witów, P-35 Krowie Bagno, 11–complementary site: Wizna (Żurek 1975, 1978)

above sea level incised by the broad ice-marginal valleys of pre-Vistulian origin, used today by the big rivers– Vistula, Biebrza, and partly Bug and Wieprz. The main features of the landscape are those of denudation, accumulation, and glacial outwash. In the southwest part (type region P-m) karst processes

of different age formed numerous sink-holes now filled by lakes and peat bogs.

Vegetation: *Pinus* forests in different variants, mixed *Pinus* forests (Pino–Quercetum), and riverside woodlands of Populetalia albae and Alnetalia glutinosae were dominant in the whole region, mixed deciduous forests being of minor importance. The fen and reedswamp communities were originally widespread in the broad river valleys. Today this part of Poland is the most heavily deforested and changed by human impact, except for some areas in the northeast and southeast parts. In the Lublin Polesie region (P-m), rich in lakes and mires, eutrophic limnophyte and fen communities are dominant; oligotrophic bogs and dystrophic lakes are less frequent.

Soils: Podzols and podzolic soils formed on sandy or sandy-clay subsoils prevail; brown soils are less common; mud and marsh soils occur along the river valleys and in the karst depressions of P-m.

Archaeology: Evidence of Late Palaeolithic and Mesolithic cultures in dune and river-terrace habitats; scarce Early Neolithic sites along the Vistula and Wieprz Rivers. Nearly the whole area inhabited since the Middle Neolithic; intensification of settlement connected with the development of Lusatian populations (3300–2300 BP); expansion of Przeworsk and late Wielbark cultures during the period of Roman influences; after cultural regression during the Migration period, cultural development has been continuous since Early Medieval times.

The synthesis is based on four reference sites, representing central (11) and northeastern (10) parts of the P-n type region and the southern part of P-m (9, 35). All sites are more or less inadequately radiocarbon-dated, so it appears unwise to present the pollen diagrams on a time scale. Moreover, the age estimates used in the event stratigraphy and correlation tables should be treated as approximate in places. Some information from site 13 (Witów, Wasylikowa 1964, 1978) (Fig.13.22) is also used.

Reference site 9. Lake Łukcze
(Bałaga 1982, 1989, 1990)

Latitude 51°30′N, Longitude 22°57′E. Elevation 163 m. Eutrophic lake without any inlet or outlet, marginally overgrown. 56.5 ha in area, 8.9 m max. depth. Origin explained as combined effects of karst phenomena and ground-ice melting. Age range *ca.* 12800–0 BP. Twelve local pollen-assemblage zones (paz), ten ^{14}C dates (Figs 13.23 and 13.24).

Ł-1	... –(12800) BP	*Artemisia–Salix*
Ł-2	(12800)–(11800) BP	*Betula–Salix*
Ł-3	(11800)–10900 BP	*Pinus–Betula*
Ł-4	10900–(10000) BP	*Pinus–Artemisia*
Ł-5	(10000)–(9400) BP	*Betula–Pinus– Ulmus*
Ł-6	(9400)–9100 BP	*Betula–Ulmus– Corylus*
Ł-7	9100–(8400) BP	*Corylus–Ulmus*
Ł-8	(8400)–7800 BP	*Alnus–Quercus*
Ł-9	7800–(5000) BP	*Ulmus–Quercus– Corylus-Tilia– Fraxinus*
Ł-10	(5000)–(3800) BP	*Pinus–Quercus*
Ł-11	(3800)–1000 BP	*Pinus–Quercus– Carpinus-Fagus*
Ł-12	1000–0 BP	NAP

The paz sequence is based on data from four pollen diagrams, and the ages in parentheses are interpolated from the dates in four profiles. Data from another site were additionally used for the synthesis: site 35, Krowie Bagno (Fig. 13.22), southeast part of P-m region, 166 m elevation, an extinct lake overgrown during the Subboreal, lying within the largest fen area in the region (36.4 km²) (Bałaga *et al.* 1980/81).

Reference site 11. Lake Błędowo (Binka *et al.* 1988, 1991, Binka & Szeroczyńska 1989)

Latitude 52°32′N, Longitude 20°40′E. Elevation 78 m. Eutrophic lake of a regularly circular shape, *ca.* 10 ha in area, 5.6 m in max. depth, *ca.* 250 m distant from the Wkra River channel and *ca.* 100 km south of the maximum extent of the Vistulian ice sheet. Lake basin formed in Pliocene clays. Age range *ca.* 11800–0 BP. Eight local pollen-assemblage zones (paz), three ^{14}C dates (Fig. 13.25). Most of the ages of paz boundaries are calculated tentatively from the sedimentation rates, supported by few ^{14}C dates. The correctness of this calculation has lately been supported by the correlation with the pollen diagram of a varved sediment sequence from Lake Gościąż covering *ca.* 12750 cal. years (Ralska-Jasiewiczowa *et al.* (eds.) in prep.). The lake is located *ca.* 90 km to the east of the Błędowo site.

436

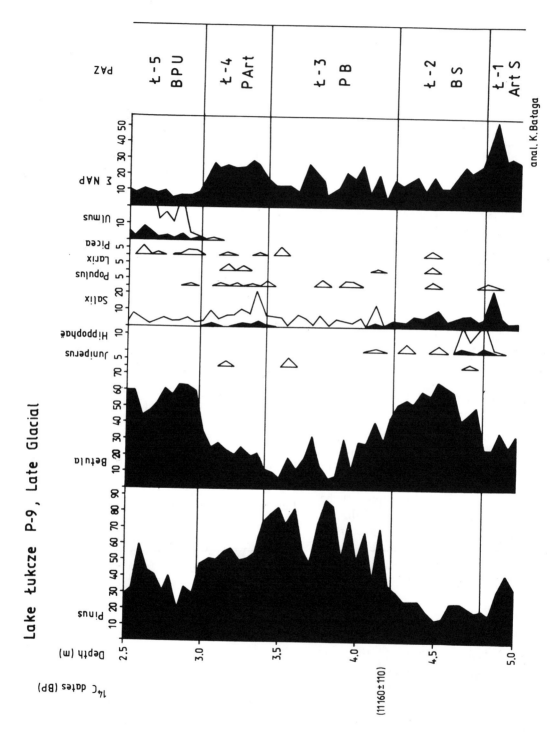

Fig. 13.23 Simplified pollen diagram from Lake Łukcze (P-9), profile I, late-glacial part, with selected pollen taxa, plotted on a depth scale. The radiocarbon date in parentheses is transferred from the profile III. For paz symbols see Fig. 13.6 caption

Lake Łukcze P-9, Holocene

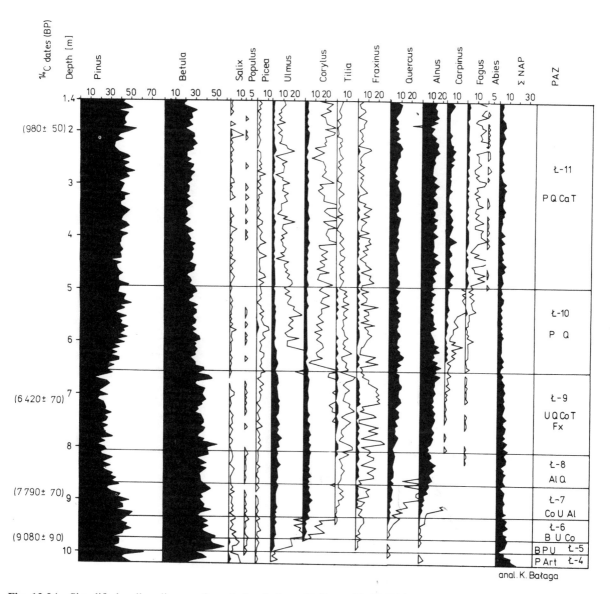

Fig. 13.24 Simplified pollen diagram from Lake Łukcze (P-9), profile II, Holocene part, with selected pollen taxa, plotted on a depth scale. The radiocarbon dates in parentheses are transferred from the profiles III and IV. For paz symbols see Fig. 13.6 caption

B-1	(*ca.* 11800)–11000 BP	*Pinus–Betula*		*Ulmus–Quercus*
B-2	11000–(10200) BP	*Juniperus*–NAP	B-5 (7000)–(5500) BP	*Ulmus–Fraxinus–Quercus*
B-3	(10200)–(9000) BP	*Betula–Pinus*	B-6 (5500)–(3100) BP	*Quercus–Corylus*
B-3a	(10200)–(9900) BP	*Populus* subzone	B-7 (3100)–(900) BP	*Carpinus–Betula–*
B-3b	(9900)–(9000) BP	*Ulmus* subzone		*Quercus*
B-4	(9000)–(7000) BP	*Corylus–Alnus–*	B-8 (900)–0 BP	NAP–*Pinus*

Lake Błędowo P-11

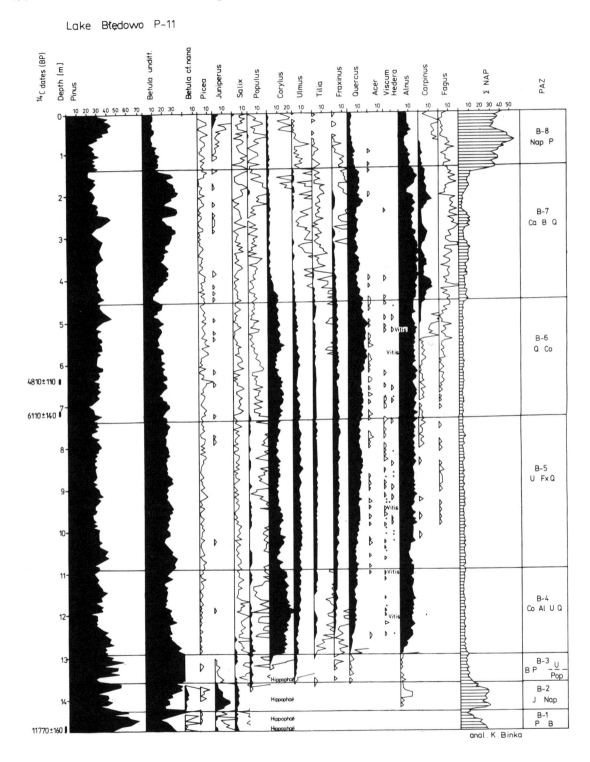

anal. K. Binka

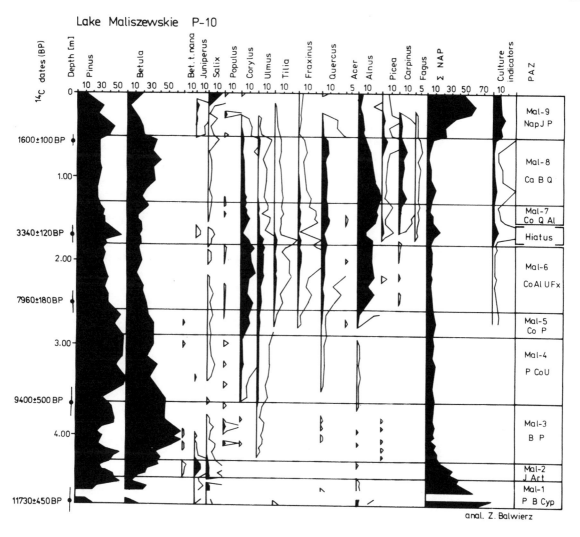

Fig. 13.26 Simplified pollen diagram from Lake Maliszewskie (P-10), with selected pollen taxa, plotted on a depth scale. Hiatus is suggested by the authors of this paper. For paz symbols see Fig. 13.6 caption

Reference site 10. Lake Maliszewskie
(Żurek 1978, Balwierz & Żurek 1987, 1989)

Latitude 53°10′N, Longitude 22°30′E, Elevation 104.1 m. The lake is located within the complex of Wizna peatlands (*ca.* 9000 ha), and occupies the wide ice-marginal valley formed during the Middle Polish glaciation and used today by the Biebrza River close to its outlet to the Narew River. The lake is eutrophic, *ca.* 80 ha in area, max. depth less than 1 m, currently being rapidly overgrown, partly by thick mats of floating vegetation, invaded in places by shrubs and trees. The profile studied was collected in the marginal part of the lake (max. sedi-

Fig. 13.25 Simplified pollen diagram from Lake Błędowo (P-11), with selected pollen taxa, plotted on a depth scale. For paz symbols see Fig. 13.6 caption

ment depth in the lake centre–22.5 m). Age range *ca.* 11800–0 BP. Nine local pollen-assemblage zones (paz), five ^{14}C dates (Fig. 13.26). The paz ages in parentheses are calculated from the ^{14}C dates in the same profile.

Mal-1	11750–(10800) BP	*Pinus–Betula–*
		Cyperaceae
Mal-2	(10800)–(10000) BP	*Juniperus–Artemisia*
Mal-3	(10000)–9400 BP	*Betula–Pinus*
Mal-4	9400–(8500) BP	*Pinus–Corylus–*
		Ulmus
Mal-5	(8500)–8000 BP	*Corylus–Pinus*
Mal-6	8000– ... hiatus ?*	*Corylus–Alnus–*
		Ulmus–Fraxinus
Mal-7	...–(2800) BP	*Corylus–Quercus–*
		Alnus
Mal-8	(2800)–1600 BP	*Carpinus–Betula–*
		Quercus
Mal-9	1600–0 BP	NAP–*Juniperus–*
		Pinus

*The hiatus is assumed by the authors of this paper.

Palaeoecological patterns and events

General patterns

The pollen-analytical information from three sites which represent totally different environments geographically distant from each other and inadequately radiocarbon-dated gives no grounds for the definition of regional pollen-assemblage zones. The comparison of the local pollen-assemblage zones (Fig. 13.27) suggests the following conclusions.

Late-glacial

Only one site (9) records the full succession of late-glacial vegetation with the following sequence:

– *Artemisia*-dominated communities with willow shrubs (Art–S paz, Oldest Dryas).
– Expansion of *Betula* and formation of open *Betula–Salix* copses initiated by a phase of *Hippophaë* shrubs (B–S paz, Bølling).
– Expansion of *Pinus* and development of *Pinus–Betula* forests (P–B paz, Allerød). This and succeeding stages of vegetational changes are recorded at all three sites.
– Some recession of woodlands and retrogressive

expansion of *Artemisia*–Chenopodiaceae-dominated vegetation with *Salix* and *Betula nana* shrubs and varied contribution of *Juniperus* (Younger Dryas).

Holocene

– Expansion of *Betula*–dominated *Betula-Pinus* forests, with the early appearance of *Ulmus* migrating from the southeast.
– Gradual expansion of *Corylus* in the understorey of forests 9900–9100 BP and of *Alnus* at moist habitats, slow spread of *Quercus*, and appearance of other deciduous trees.
– Optimum development of mixed deciduous forests between 8000 and 7000 BP (hiatus at site 11).
– Slow recession of *Ulmus*, *Tilia*, and *Fraxinus*; optimum development of *Quercus* forests with *Pinus* or *Corylus* 5500–5000 BP; first human impact.
– Expansion of *Carpinus* between 3800 and 3100 BP; formation of woodlands of Querco–Carpinetum and Pino–Quercetum type and of secondary *Betula* woods, with increasing human impact.
– Large-scale destruction of forests by man.

Unique patterns

– A very early expansion of *Ulmus* (around 10000 BP) in the southeastern P-m region, delayed towards the northern part of P-n (site 10) until *ca.* 9400 BP.
– Expansion of *Corylus* in the western part of P-n (site 11), delayed and synchronous with the spread of *Quercus* and *Alnus ca.* 9000 BP.
– Optimum development of mixed deciduous forests earlier in P-m (since *ca.* 8000–7800 BP) and later in the western part of P-n (since 7200–7000 BP); the insignificant role of *Corylus* and *Tilia* in the Atlantic forests at site 9; much more substantial contribution especially of *Corylus* and *Fraxinus* at site 11.
– Dominant role of *Pinus–Quercus* forests in P-m since *ca.* 5000 BP.
– Metachronous immigration and expansion of *Carpinus* between *ca.* 4000 BP (P-m) and 2700 BP (P-n, northern part).
– Some contribution of *Fagus* to the forests of P-m since *ca.* 3800 BP and of *Picea* in the northern part of P-n since *ca.* 1600 BP.

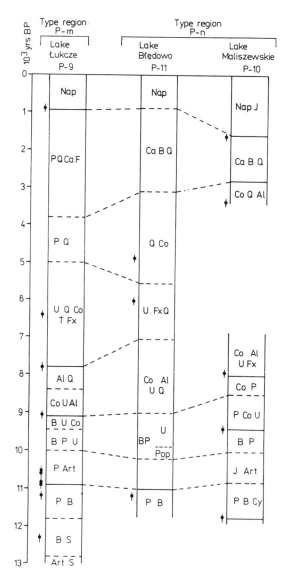

Fig. 13.27 Time–space correlation of local pollen-assemblage zones from the sites representing the Eastern Mid-Polish Lowlands synthesis region. The radiocarbon-dated borders are indicated with a solid line, the borders interpolated or calculated from the accumulation rates with a dashed line. For paz symbols see Fig. 13.6 caption. A hiatus at P-10 is suggested by the authors of this paper

Climatic events

– According to the general vegetational development, including the occurrence of climatic plant indicators at site 13 (Fig. 13.28) (Wasylikowa 1964), the mean July temperatures in the western part of the synthesis region during the late-glacial temperate oscillations are assumed to be rather high (Bølling 15°C, Allerød more than 16°C), and their decreases during the coolings rather strong (Older Dryas 10–12°C, Younger Dryas 12°C) and accompanied by dry strong winds (activation of aeolian processes).

– In the eastern part of the synthesis region (site 9) no cooling separates Bølling from Allerød at all, although a drier climate at that time is indicated. The climate of Younger Dryas there was dry and continental but not very cool, at least in its younger phase, as evidenced by the reappearance of temperate plant indicators.

– There is a trace of a Youngest Dryas oscillation (*sensu* Behre 1978) during the Preboreal, as recorded at site 11 by rises of NAP, *Salix*, and *Populus* values.

– Indicators of the climatic optimum (*Viscum* and *Hedera*) appear regularly between 8500 (8200) and *ca.* 1000 (1500) BP and are accompanied by sporadic *Vitis* pollen between *ca.* 8000 and 4500 BP in the western part of P-m at site 11 (see also P-e and P-h regions!); in the eastern sites (9, 10) *Hedera* is nearly absent, and *Viscum* appears at site 9 since *ca.* 7500 BP.

Hydrological events

The information on hydrological events (Fig. 13.29) is too heterogeneous to be summarized in the event stratigraphy: at site 11 it comes from a lake of unknown origin located in the terrace system of Wkra River, but with no natural flow connection to it; at site 10 it comes from a lake lying within the mire complex accompanying the Biebrza River, with a subsurficial inflow and outflow; sites 9 and 35 are partly overgrown lakes in the karst area.

The data from site 13 (after Wasylikowa 1964, 1978, and unpubl.) indicate lake-level increases around 12500–12300 BP, *ca.* 11300–10900 BP, and 10000–9000 BP and low water levels around 12000–11500 BP and 10500–10000 BP. The lake was overgrown by a mire around 9000 BP, its growth being periodically disturbed by the activation of aeolian

Synthesis region: Eastern Mid-Polish Lowlands

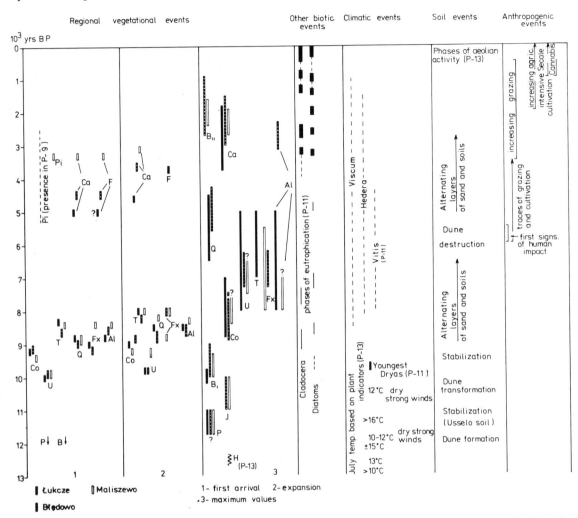

Fig. 13.28   Event stratigraphy for the Eastern Mid-Polish Lowlands synthesis region. For symbols used in the column of vegetational events see Fig. 13.6 caption

processes. It was finally buried by a dune between 3500 and 3000 BP.

The lakes at sites 11 and 10 show rises of water level from *ca.* 11000–10500 BP and a generally high level until *ca.* 8000 BP, followed at site 11 by a lowering between 8000 and 7000 BP and then a period recording river floods until *ca.* 4500 BP, corresponding roughly to a hiatus at site 10, where a new rise of water level is recorded at *ca.* 3500 BP. The continuous tendency to shallowing begins at

both sites *ca.* 1700 BP. At sites in the karst area the rises of water level are recorded between 12000 and 11500 BP and before 10000 BP, with shallowing/ overgrowing of lakes 3700–3200 BP.

Anthropogenic events

– The uncertain traces of Early Neolithic man are noted in the southeastern (site 9) and central (site 11) parts of the synthesis region at 6000–5500 BP

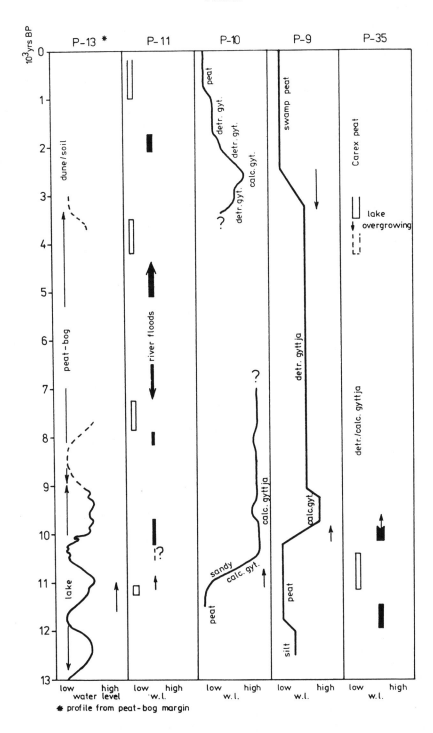

Fig. 13.29 Correlation of data on hydrological events from five sites representing the Eastern Mid-Polish Lowlands synthesis region. P-13, P-11, P-10, P-9, P-35 – after the individual authors. At P-10 a hiatus is suggested by the authors of this paper. Open blocks – low water levels; black blocks – high water levels; arrows – rises and declines of water level

(Fig. 13.30), but more reliable evidence of Neolithic activities, including indicators of grazing, agriculture, and local clearings, is recorded only at site 9 since *ca.* 4500 BP.

- The Late site Bronze/Early Iron Age expansion of Lusatian culture is evidenced at site 9 since *ca.* 3500 BP (slightly too old because of interpolation error?), and at site 11 since *ca.* 3200 BP. From the differences in the evidence it can be concluded that settlements at site 9 were located within the region but at some distance from the lake. At site 11 the settlements close to the lake are indicated by clearing cycles that affected distinctly local forests (ratio of *Carpinus–Quercus* to *Betula* alternating with higher values of cultural indicators).

- The decline in human colonization at *ca.* 2300 BP is recorded at sites 9 and 11 synchronously.

- A distinct settlement phase corresponding to the period of Roman influences is recorded at site 11 only. Changes express extensive deforestation and developed agriculture.

- Since *ca.* 1500 BP the continuously advancing colonization and deforestation is evidenced at all

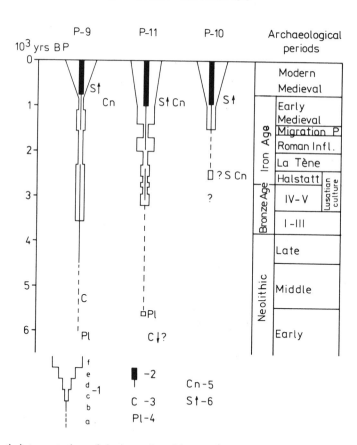

Fig. 13.30 Schematic interpretation of the intensity of human impact on vegetation based on pollen frequencies from three sites representing the Eastern Mid-Polish Lowlands synthesis region. 1–human impact in general, expressed in 6-grade scale: a–probable presence of man, b–evidenced slight local impact, c–regular clearings and land-use, d,e–increasing extent of clearings and land-use, f–large-scale or total deforestation and intensive economy, 2–slight to substantial indications of agriculture, 3–first appearance of cereals, 4–first appearance of *Plantago lanceolata*, 5–cultivation of *Cannabis*, 6–increase of *Secale* cultivation

three sites (9, 10, 11), with a temporary decline around 1100–800 BP recorded at site 9 only.

TYPE REGIONS P-r, P-s, P-w, AND P-x, LAKE DISTRICTS

The synthesis region of the Lake Districts covers the areas of northern and west-central Poland glaciated during the Vistulian glaciation, lying between the maximum extents of its Poznań, Leszno, and Pomeranian stages to the south and the Baltic Coastal Zone to the north (Fig. 13.31). It includes four type regions: Poznań–Gniezno–Kujawy Lake Districts (P-r), eastern part of West Pomeranian Lake District (P-s), Dobrzyń–Olsztyn Lake Districts (P-w), and Masurian Great Lakes District (P-x). The extreme northeast and southwest regions of the Lake Districts (P-y, P-q with the adjoining W part of P-s) are not included because of lack of data.

Altitude: 450–312 m above sea level.

Climate: Mean January temperatures –1 (SW) to –4.5°C (NE), mean July 16–18°C (19°C locally in SW), mean annual 8.1–7.6°C (SW) to below 6.0°C (NE). Mean annual rainfall: 500 mm (SW–W) to 650 mm (NE). Western winds prevalent, strong southwesterly and southeasterly winds frequent in northeast regions, which are the coldest part of Poland.

Geology: Miocene clays and brown coal facies locally outcropping in southwest and west parts of area, Oligocene glauconite sands in northwest part, Cretaceous marls and limestones and Mio/Pliocene silts, sands, and clays widespread in the whole region; in the eastern part a tectonic depression of Cretaceous–Oligocene age filled with the Miocene deposits mostly eroded by glacial activities; the Quaternary cover of tills and sands 60 to more than 200 m thick.

Topography: Young moraine landscape, formed mostly by the three last Vistulian stages; the parallel W–E moraine ridges and hills, dissected by perpendicular or oblique systems of channel lakes; eskers, kames, and glacial kettles; large outwash plains in the northwest and southeast parts of the region; a rich lake-system, including the two largest lakes of Poland, follows the N–S depression in the eastern part. The watershed of the Vistula and Pregola Rivers runs there.

Vegetation: The natural vegetation was composed of mixed deciduous woodlands (Carpinion betuli) prevailing on brown soils, with the contribution of *Picea* in the northeast, of *Fagus* in the northwest, and of various types of *Pinus* and mixed *Pinus–Quercus* forests (Dicrano–Pinion, Pino–Quercion) occurring on podzols and degraded brown soils. Numerous mires support rich fen and peat-bog communities of all types. Limnic/telmatic communities occur in lakes, and xerothermic vegetation on steep escarpments.

A significant participation of northern and boreal species, including *Picea abies*, distinguishes the eastern part of the Lake Districts (regions P-x, P-y) from the rest of the synthesis region. The distribution limits of many plants, including important trees like *Fagus sylvatica*, run through the P-w region, which is a typical transitional zone for climate and vegetation.

Soils: Brown soils, degraded brown soils, and podzols originating from tills and silts, and podzols originating from fluvioglacial sands; black soils mostly in the Kujawy area; gley and marsh soils in depressions.

Archaeology: The first human penetrations at the decline of Palaeolithic; inhabited since Mesolithic; Early Neolithic (Band Ceramic) cultural centre in eastern part of P-r region, but otherwise strong connections with the Neolithic of peri-Baltic areas; active development of Lusatian culture during the Late Bronze and Early Iron Ages with a network of strong-holds and with the tribal borderline along the Great Lakes Valley (P-x). The whole area was well populated during the time of Roman influences. After a set-back connected with the population decline during the Migration period the large-scale destruction of forests continued in Early Medieval time.

The synthesis is based on data from 12 reference sites and several additional sites.

446

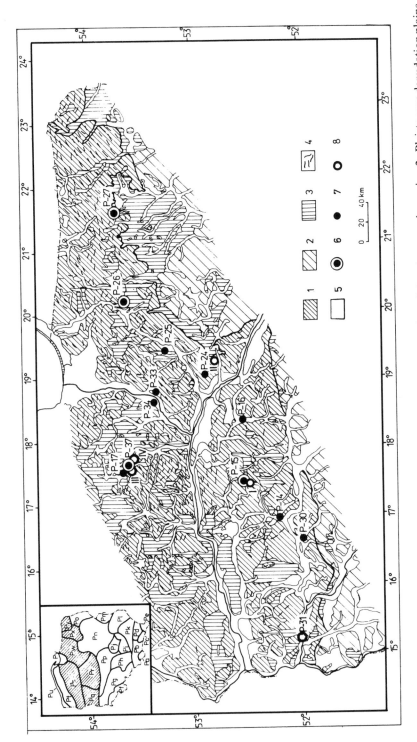

Fig. 13.31 Geomorphic map of the Lake Districts synthesis region showing the location of sites. 1 – ground moraine, 2 – Pleistocene denudation plains, 3 – outwash plains, 4 – terminal moraines, 5 – Pleistocene and Holocene river valleys and terraces, 6 – sites with the pollen diagrams included in this paper: P-15 Lake Skrzetuszewskie, P-26 Woryty, P-27 Lake Mikołajki, P-37 Lake Mały Suszek, 7 – other reference sites: P-14 Lake Skrzynka, P-16 Lake Gopło, P-17 Lake Wielkie Gacno, P-24 Lake Steklin, P-25 Lake Strażym, P-30 Wonieść, P-33 Lake Rudnickie Małe, P-34 Fletnowo, P-31 Pomorsko, I – Dziekanowice (Litt 1988), II – Lake Mielno (Kępczyński 1960), III – Suszek (Miotk-Szpiganowicz 1992), 8 – complementary sites: IV – Kęsowo (Miotk-Szpiganowicz 1992)

Reference site 15. Lake Skrzetuszewskie
(Tobolski & Okuniewska-Nowaczyk 1989)

Latitude 52°33′N, Longitude 17°23′E. Elevation 109.1 m. A small eutrophic lake *ca.* 3 ha in area, in a glacial-channel lake-system, located in deforested agricultural landscape. Age range 9000–0 BP. Seven local pollen-assemblage zones (paz), nine ^{14}C dates (Fig. 13.32).

Sk-1	...–8750 BP	*Pinus–Betula*
Sk-2	8750–8050 BP	*Corylus–Pinus*
Sk-3	8050–(6550) BP	*Alnus–Corylus*
Sk-4	(6550)–5150 BP	*Ulmus–Tilia–*
		Quercus–Fraxinus
Sk-5	5150–3350 BP	*Corylus–Quercus*
Sk-6	3350–1550 BP	*Carpinus–Betula*
Sk-7	1550–0 BP	NAP–*Betula*

Reference site 14. Lake Skrzynka (Fig. 13.31)

65.5 m elevation, a mesotrophic partly overgrown lake in a glacial channel, 3 ha in area, 3.1 m max. water depth, situated in the forests of the Wielkopolski (Great Poland) National Park (Offierska 1978, Offierska & Okuniewska 1986, Okuniewska-Nowaczyk 1987, Tobolski & Okuniewska-Nowaczyk 1989).

Reference site 16. Lake Gopło (Fig. 13.31)

77.0 m elevation, an eutrophic channel lake 25 km long, 16.5 m max. water depth, 15.25 m max. sediment depth (Jankowska 1980).

The information on the late-glacial vegetation in the P-r region was completed with data from an additional site–an overgrown lake at Dziekanowice near Lednogóra (Fig. 13.31) (Litt 1988). The sequence of pollen-assemblage zones from these four sites was correlated to define regional assemblage zones for the type region P-r (Fig. 13.37).

Reference site 37. Mały Suszek
(Miotk-Szpiganowicz 1989, 1992)

Latitude 53°43′N, Longitude 17°46′E. Elevation 115 m. A small dystrophic lake surrounded by bogs within the area of Bory Tucholskie (Tuchola Forests). Age range *ca.* 12000–*ca.* 250 BP. Eleven local

pollen-assemblage zones (paz), ten ^{14}C dates (Fig. 13.33).

MS-1	(*ca.* 12000)–11800 BP	*Pinus*–Cyperaceae
MS-2	11800–(11150) BP	*Pinus–Betula–*
		Gramineae
MS-3	(11150)–(10100) BP	*Juniperus–*
		Gramineae–
		Cyperaceae–
		Artemisia
MS-4	(10100)–(9750) BP	*Betula–Pinus*
MS-5	(9750)–8950 BP	*Pinus–Betula–*
		Corylus
MS-6	8950–7300 BP	*Pinus–Betula–*
		Corylus–Alnus–
		Ulmus
MS-7	7300–5500 BP	*Pinus–Betula–*
		Ulmus–
		Quercus–Tilia
MS-8	5500–3700 BP	*Pinus–Betula–*
		Corylus–
		Quercus–Alnus
MS-9	3700–2200 BP	*Pinus–Betula–*
		Alnus–Carpinus
MS-10	2200–(750) BP	*Pinus–Betula–*
		Alnus–
		Carpinus–Fagus
MS-11	(750)–(250) BP	*Pinus*–NAP

One additional reference site was used in the synthesis for the type region P-s: Lake Wielkie Gacno (17), (Fig. 13.31) (Hjelmroos 1981), 130 m elevation, an oligotrophic "Lobelia" lake, 13 ha in area, being a part of a glacial channel, max. water depth 6.1 m, max. sediment depth 8 m, situated in the area of poor outwash sandy soils, surrounded by dunes and *Pinus* forests (Hjelmroos-Ericsson 1981) of Tuchola Forests complex. The data from sites 17 and 37 were used for the definition of the regional pollen-assemblage zones for the western part of type region P-s: Bory Tucholskie (Tuchola Forests region Fig. 13.37).

Reference site 25. Lake Strażym (Noryśkiewicz 1987, Niewiarowski 1987, Lankauf 1987, Różański 1987, Boińska 1987, Błędzki 1987)

Latitude 53°20′N, Longitude 19°27′E. Elevation 71 m. Eutrophic lake occupying a part of a subglacial

Lake Skrzetuszewskie   P-15

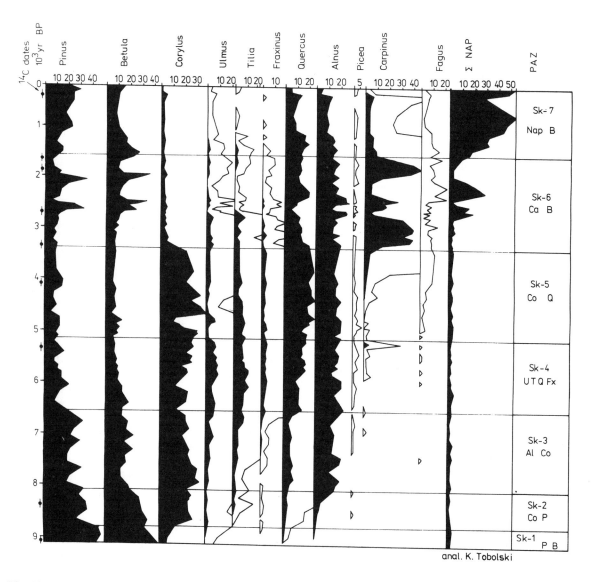

anal. K. Tobolski

Fig. 13.32   Simplified pollen diagrams from Lake Skrzetuszewskie (P-15), with selected pollen taxa, plotted on a time scale. For paz symbols see Fig. 13.6 caption

channel *ca.* 2 km long, lake area 73.4 ha, max. water depth 9 m, max. sediment depth 10.60 m. Age range *ca.* 12000–0 BP. Ten local pollen-assemblage zones (paz), seven ¹⁴C dates, all except the basal date coming from lake-margin profiles, with no continuous sediment sequences. The paz dates transferred to the main profile are indicated in parentheses.

Sm-1	11000–	*Pinus–Betula–* Cyperaceae
Sm-2		*Juniperus– Artemisia*

Lake Mały Suszek P-37

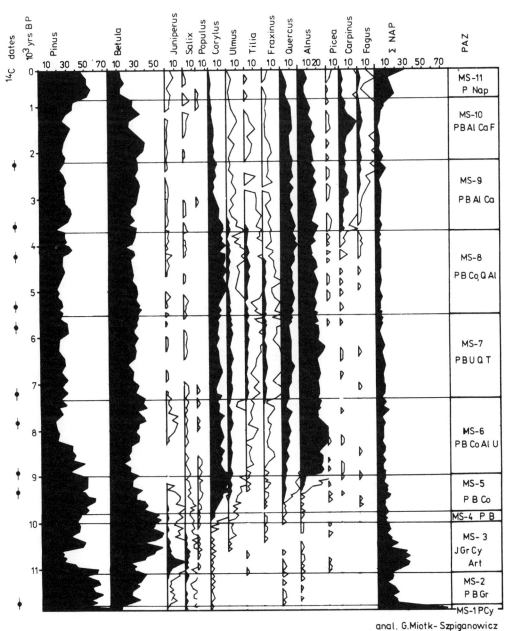

anal. G.Miotk- Szpiganowicz

Fig. 13.33 Simplified pollen diagram from Lake Mały Suszek (P-37), with selected pollen taxa, plotted on a time scale.
For paz symbols see Fig. 13.6 caption

Sm-3	...–(*ca.*9700) BP	*Pinus–Betula*
Sm-4	(*ca.*9700) BP– ...	*Pinus–Betula–*
		Corylus
Sm-5	...–(5000) BP	*Corylus–Alnus–*
		Quercus–
		Ulmus–Tilia
Sm-6	(5000)–(*ca.* 3500) BP	*Quercus–*
		Corylus
Sm-7	(*ca.*3500) BP– ...	*Pinus–Carpinus*
Sm-8		*Carpinus–*
		Quercus–
		Alnus
Sm-9		*Betula–Pinus*
Sm-10		*Pinus*–NAP

Reference site 26. Woryty (Cieśla *et al.* 1978, Marciniak 1979, Dąbrowski 1981, Pawlikowski *et al.* 1982, Szeroczyńska 1985, Noryśkiewicz & Ralska-Jasiewiczowa 1989)

Latitude 53°45′N, Longitude 20°12′E. Elevation 105 m. An ancient overgrown lake 30.5 ha in area, max. sediment depth 10.25 m, evidence of inflow and outflow in the past. The site is situated in the centre of a dwelling area of Lusatian culture population (Late Bronze Age); remnants of earlier and later settlements found too. The site was the subject of interdisciplinary palaeoecological studies, including geology, chemistry, mineralogy, Cladocera, Rhizopoda, Diatomae, plant macrofossils, and pollen. Age range *ca.* 11800–200 BP. Eleven local pollen-assemblage zones (paz), eighteen ^{14}C dates (Fig. 13.34).

W-1	(*ca.*11800)–11300 BP	*Pinus–Betula*
W-2	11300–11050 BP	*Pinus–Juniperus*
W-3	11050–(10100) BP	NAP–*Juniperus–*
		Salix
W-4	(10100)–(8900) BP	*Betula–Pinus–*
		Populus
W-5	(8900)–8300 BP	*Corylus–Pinus–*
		Ulmus
W-6	8300–(6900) BP	*Alnus–Pinus–*
		Ulmus–Tilia
W-7	(6900)–5050 BP	*Ulmus–Tilia–*
		Quercus–Alnus
W-8	5050–3400 BP	*Corylus–Quercus–*
		Alnus
W-9	3400–2300 BP	NAP–*Betula–*

		Pinus–Carpinus
W-10	2300–900 BP	*Carpinus–Quercus–*
		Betula
W-10a	2300–2050 BP	*Carpinus–*
		Tilia–Fagus subzone
W-10b	2050–1800 BP	NAP subzone
W-10c	1800–900 BP	*Carpinus* subzone
W-11	900–(200) BP	NAP–*Pinus*

Reference site 24. Lake Steklin (Fig. 13.31)

Latitude 52°56′N, Longitude 19°0′E. Elevation 73.7 m, a narrow hardwater channel-lake 5 km long and 112 ha in area, incised into the moraine plateau (Noryśkiewicz 1982, Marciniak 1987).

Reference site 27. Lake Mikołajki
(Ralska-Jasiewiczowa 1966, 1989b and unpubl., Więckowski 1966, Marciniak 1973)

Latitude 53°46′N, Longitude 21°35′E. Elevation 116 m. Eutrophic lake filling part of a subglacial channel 38 km long, connected with the largest lake of Poland (Lake Śniardwy). Length of the lake 5 km, max. water depth 27.8 m, max. sediment depth 10.2 m, mostly calcareous gyttja. Age range *ca.* 13000–0 BP. Ten local pollen-assemblage zones (paz), five ^{14}C dates (Fig. 13.35). Dates are transferred from a profile collected in the N part of the lake, occupied by a mire until *ca.*9300 BP (pollen diagram Ralska-Jasiewiczowa unpubl.).

M-1	12000–11000 BP	*Hippophaë*–NAP
M-2	11000–10700 BP	*Betula–Pinus*–NAP
M-3	10700–10250 BP	NAP–*Juniperus–Pinus*
M-4	10250–9300 BP	*Betula–Pinus*
M-5	9300–... BP	*Corylus–Pinus–Ulmus*
M-6	?	*Ulmus–Alnus–Pinus–*
		Quercus
M-7	?	*Quercus–Alnus–Tilia*
M-8	?	*Corylus–Quercus–*
		Alnus
M-9	?	*Carpinus–Betula*–NAP
M-10	?	*Pinus–Picea–*
		Carpinus–NAP

Data from sites 24, 25, 26, and 27 were used for the definition of the regional pollen-assemblage zones for the type regions P-w and P-x (Fig. 13.37).

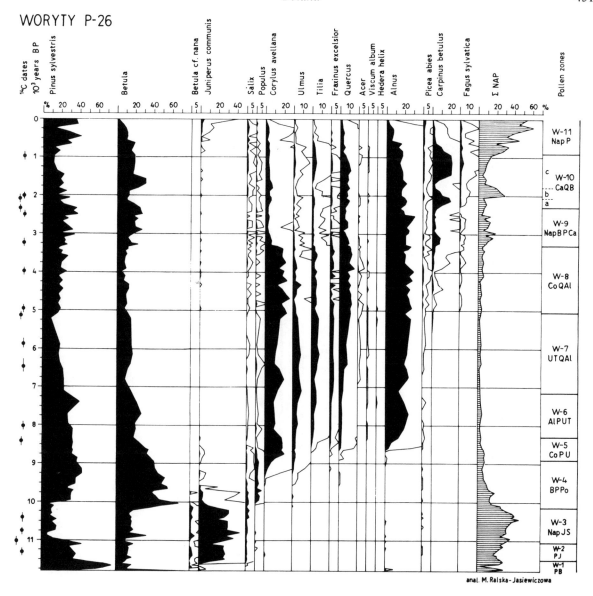

WORYTY P-26

anal. M. Ralska-Jasiewiczowa

Fig. 13.34 Simplified pollen diagram from Woryty (P-26), combined from two profiles–prof. 82 (Late-glacial, early Holocene) and prof. 80 (mid and late Holocene), with selected pollen taxa, plotted on a time scale. For paz symbols see Fig. 13.6 caption

Palaeoecological patterns and events

General patterns

Correlation of pollen-assemblage zones from ten sites located in the Lake Districts (Fig. 13.36) provide the basis for defining three sequences of re-gional assemblage zones representative for type regions P-r, P-s, and P-w + P-x, treated jointly because of the limited information available from P-x and general dating inadequacy (Fig. 13.37). Some other significant events are presented on Figure 13.38. Only three sites, located at opposite ends of the

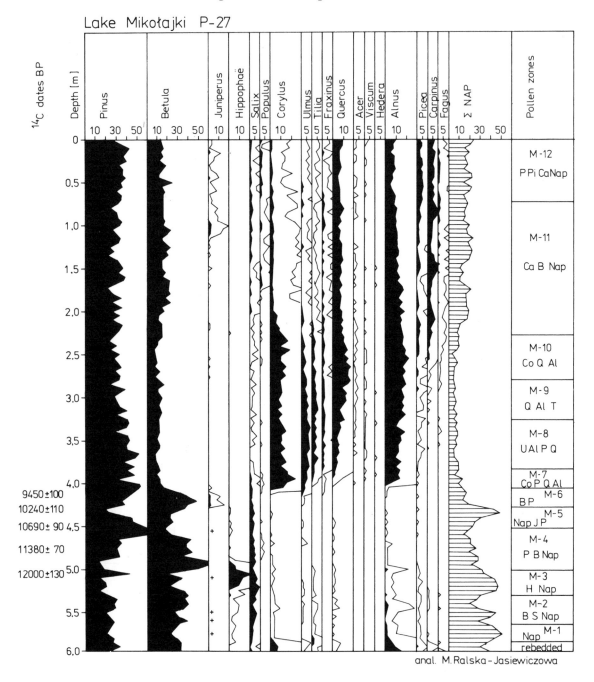

Fig. 13.35 Simplified pollen diagram from Lake Mikołajki (P-27), with selected pollen taxa, plotted on a depth scale. The radiocarbon dates are transferred from the profile 1984 (unpubl.). For paz symbols see Fig. 13.6 caption

synthesis region (27 in the northeast and 16 and Dziekanowice in the southwest) contain a full sequence of late-glacial events including the pre-Allerød sections. In spite of the distance they reveal some common features in the vegetational development:

- Initial phase of tundra-grasslands with dwarf *Salix* and *Dryas octopetala*; strong representation of calciphilous vegetation (Oldest Dryas).
- Expansion of *Betula* and formation of open *Betula* copses with high contribution of *Salix* and locally of *Hippophaë*; possible presence of *Pinus* in southwest areas (Bølling).
- Recession of *Betula* and short temporary spread of heliophyte vegetation with dominant *Hippophaë*, locally with abundant *Salix* or *Juniperus* shrubs (Older Dryas).

The next two stages are recorded at nearly all sites in the synthesis region:

- Development of *Betula* and/or *Pinus–Betula* forests (Allerød).
- Subsequent spread of heliophyte grasslands with a distinct domination of xeric (steppe-like?) elements and *Juniperus*; no return of calciphilous *Dryas*-tundra species; insignificant local contribution of *Empetrum* communities (Younger Dryas).

Correlation of regional pollen-assemblage zones (Fig. 13.37) revealed the main transitions between the subsequent stages in the vegetational development consistent in the whole synthesis region since the beginning of the Allerød.

The transitions indicate the beginnings of the following changes:

- First expansion of *Betula–Pinus* and *Pinus* woods.
- Recession of forests and spread of heliophilous herb vegetation with *Juniperus* scrub.
- Second expansion of *Pinus–Betula* woods.
- Expansion of *Corylus* within the *Pinus* woods; appearance of some deciduous trees (mostly *Ulmus*) in central and eastern regions.
- Expansion of *Alnus*; deciduous trees start to play locally a substantial role in the forests.
- Optimum development of mesophilous mixed deciduous forests.
- Optimum development of *Quercus–Corylus*

forests, recession of *Ulmus*, *Fraxinus*, and *Tilia*; first human impact.
- Expansion of *Carpinus* connected with the formation of anthropogenic *Betula* woods.
- Large-scale deforestation.

Particular common patterns in the Holocene

- Spread of *Alnus* (most probably *A. glutinosa*) on moist habitats was very rapid between 8800 and 8500 BP in the whole synthesis region (Fig. 13.38).
- The dominant role of *Pinus* in the forests lasted generally until *ca.* 7000 BP.
- Recession of mixed deciduous forests, in spite of their very differentiated developmental processes, began generally between 5500 and 5000 BP.
- Subsequent development of *Quercus*- and *Corylus*-dominated woodlands between *ca.* 5000 and 3500 BP depended not only on soil but also on changing light conditions (also by opening of forests by man).
- *Carpinus* started to spread in the whole region between 3700 and 3400 BP, but its further expansion depended most of all upon the processes of settlements.

Unique patterns

- Very abundant development of *Juniperus* in the Younger Dryas vegetation in the central lake districts (east part of P-s, P-w) commenced already in the late Allerød (around 11300 BP).
- Younger Dryas recession of forests was the shortest in the east (P-x, 10700–10250 BP).
- The earliest expansion of *Corylus*, since *ca.* 9900–9800 BP, is recorded in the east part of P-s and the adjacent southwest part of P-w regions.
- *Quercus* expansion started earliest in the P-s region since *ca.* 9200 BP.
- *Tilia* and *Fraxinus* expanded first in the eastern and central regions (P-x, P-w).
- Expansion of all deciduous trees in the southwest part of synthesis region (P-r) was generally delayed.
- The metachronous appearances of *Fagus* and *Picea* in the particular sites illustrate the different migrational processes at the marginal areas of

454

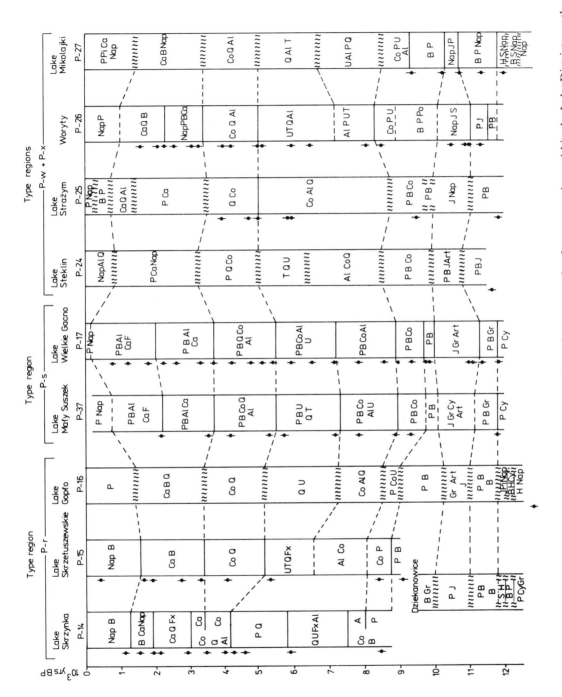

Fig. 13.36 Time–space correlation of local pollen-assemblage zones from sites representing four type regions within the Lake Districts synthesis region. The radiocarbon-dated borders are indicated by a solid line, the borders interpolated from the dates at the site by a dashed line; and the borders roughly calculated using interpolations from other sites and accumulation rates by S-shaped symbols. For paz symbols see Fig. 13.6 caption

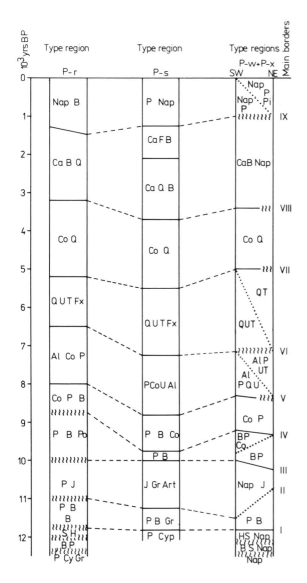

Fig. 13.37 Time–space correlation of regional pollen-assemblage zones within the Lake Districts synthesis region. The main correlation borders are indicated with the Roman numerals to the right; the subregional differences in pazs by a dotted line; SW–southwest; NE–northeast. For other explanations see Figs 13.6 and 13.36 captions

their distribution, *Fagus* appearing from 4000–3500 BP in the southwest part of synthesis region, from 3300–2800 BP in its central western part, at *ca.* 2000 BP in the central eastern part, never reaching the eastern (P-x) regions and never com-

mon in the whole region. *Picea* appeared sporadically between 6000 and 4000 BP in the central and eastern regions, making local small expansions between 4000 and 2000 BP, and expanded more substantially only in the east (P-x) from 2000 BP.

Hydrological events

The information on the hydrological changes (Figs 13.38 and 13.39) includes the geomorphologic, geologic, sedimentologic, and palaeoecologic evidence from lakes or extinct lakes located in most cases in the subglacial channels formed during the recession of the Vistulian ice sheet and filled with dead ice. The initial phase was normally a mire on the buried ice surface, formed by melting during the Allerød warming and followed by lake deepening. During the early stages the lakes developed in an individual way, depending largely on the local situation.

- The first lowering of water level was often connected with the Allerød/Younger Dryas transition.
- The second lowering resulted from the rapid warming and concomitant melting out of dead ice around 10000–9500 BP.
- A tendency for a water level rise is mostly observed from before 9000 BP, with local fluctuations between 8500 and 8000 BP. The sites from P-s (E) show another distinct lowering of water table then. A period of a rather stable high water table lasted until *ca.* 6800–6500 BP.
- The third lowering is recorded at all sites from *ca.* 6500 BP as a combined effect of successional processes, increasing lake productivity, high evapo-transpiration, and downcutting of river channels.
- The majority of sites record a rise of water level from *ca.* 5000 BP, at some sites (17, 37) registered as the highest Holocene water level, at other sites (25, 15) delayed until *ca.* 4000 BP.
- Later hydrological changes are too dependent on the local human activities to be discussed in terms of changing natural conditions.

Climatic events

- The Younger Dryas vegetation indicates a rather dry and continental but not particularly cold climate; at site 25, a slightly warmer middle phase is recorded (δ ^{18}O, Fig. 13.38).

456

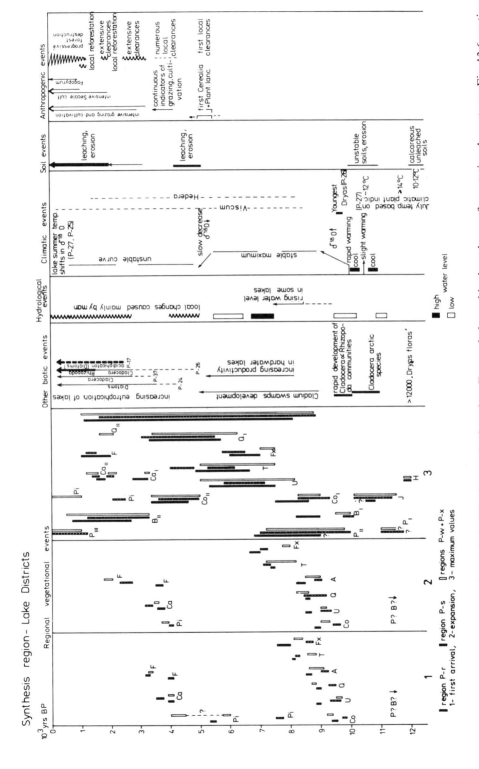

Fig. 13.38 Event stratigraphy for the Lake Districts synthesis region. For symbols used in the column of vegetational events see Fig. 13.6 caption

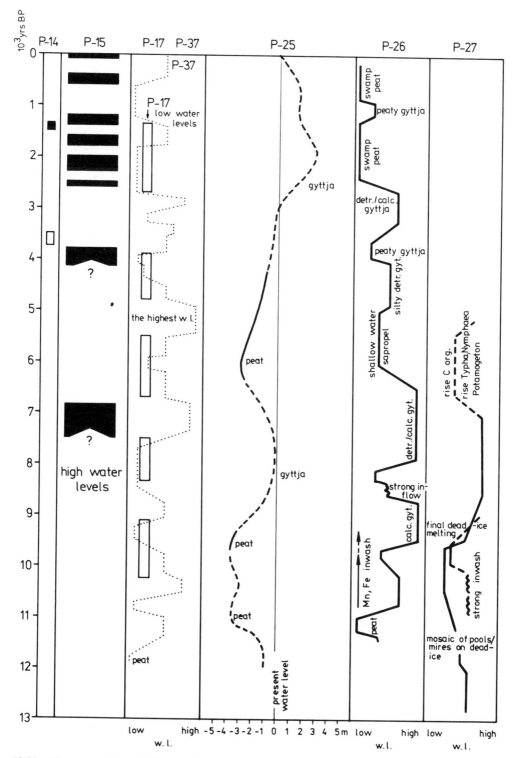

Fig. 13.39 Time correlation of data on hydrological events from seven sites representing the Lake Districts synthesis region, after the individual authors. Open blocks–low water levels; black blocks–high water levels (P-14, 15, 17). A rough estimation of low, intermediate, and high water levels is shown by a broken line (P-26, 27, 37) and more approximate quantification of changes by a smoothed curve (P-25)

– A climatic oscillation at around 9700 BP is suggested at site 26 by evidence for a temporary spread of heliophyte herbs followed by *Populus tremula* and *Betula*.

– The δ ^{18}O curves from sites 25 and 27 (Różański 1987, Różański *et al.* 1988) indicate rapid warming between *ca.* 10000 and *ca.* 9200 BP, still progressing slowly until *ca.* 8000 BP, stabilizing at maximum level until *ca.* 5100 (5500) BP, and then followed by a slow temperature decrease until middle Subatlantic, but with more and more unstable δ ^{18}O values.

– The appearance of climatic pollen indicators suggests warm summers (*Viscum*) from *ca.* 8500 BP synchronously over the entire synthesis region,

and then warm autumns and mild winters (*Hedera*) since 8300–7000 BP. *Hedera* disappears from pollen spectra around 1500 BP at the latest.

Anthropogenic events

The differentiation of the synthesis region in terms of the development of human settlements is well reflected in pollen diagrams.

– The earliest record of grazing (from *ca.* 5800 BP) and of agriculture (from *ca.* 5500 BP) comes from the area of the oldest Early Neolithic cultural centre around sites 16 and 15 (Fig. 13.40).

– Reliable evidence of Middle Neolithic activities appears in the western regions from before 5000

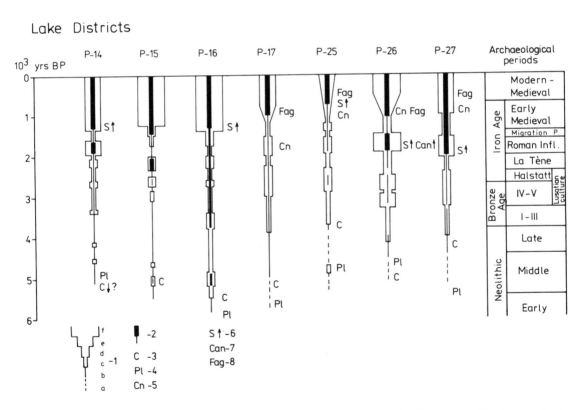

Fig. 13.40 Schematic interpretation of the intensity of human impact on vegetation based on pollen frequencies from seven sites representing the Lake Districts synthesis region. 1– human impact in general, expressed in 6-grade scale: a–probable presence of man, b–slight local impact, c–regular clearings and land-use, d,e – increasing extent of clearings and land-use, f–large-scale or total deforestation and intensive economy, 2–slight to substantial indications of agriculture, 3–first appearance of cereals, 4–first appearance of *Plantago lanceolata*, 5–first appearance of *Centaurea cyanus*, 6 –increase of *Secale* cultivation, 7–cultivation of *Cannabis*, 8–cultivation of *Fagopyrum*

BP, also in sandy areas of poor settlement (site 17), and in the eastern regions from *ca.* 5000 BP, with the record of agricultural practices locally delayed until *ca.* 4300 (site 27)–3800 BP (site 25).

- Forest thinning by Neolithic people might be one of the essential reasons for expansion of *Corylus* and abundant flowering of *Quercus* between *ca.* 5000 and 3500 BP (*Corylus–Quercus* paz).
- Development of the Lusatian population during the Late Bronze and Early Iron Ages, building of strongholds, and extensive clearings caused large-scale destruction of natural forests in the whole region from *ca.* 3200 BP.
- Decline of settlements during the La Tène period in the central and eastern regions (P-a, P-w, P-x) resulted in forest regeneration, with *Carpinus* dominance between *ca.* 2400 and 2000 BP. In southwestern areas (P-r) settlement was well developed then.
- Intensive development of agriculture during the period of Roman influences from *ca.* 2000 BP was connected with the common introduction of *Secale* and *Cannabis* into cultivation in central and eastern regions. A reduction of settlements is then noticed in western sites 15 and 16.
- The Migration period (*ca.* 1600–1300 BP) was a time of economic decline in western and central regions and of maximum development of *Carpinus*-dominated woodlands. No reduction in agriculture indicators is observed in the east (site 27).
- Early Medieval colonization is recorded from *ca.* 1400–1300 BP in the SW region and from *ca.* 1000 BP in the central and eastern regions. It was connected with the progressive destruction of forest cover and development of complex agriculture. The common cultivation of *Secale* is then recorded in southwest regions, and of *Fagopyrum* in the central-eastern regions.

TYPE REGIONS P-u, P-t, AND P-v, BALTIC COASTAL ZONE OF POLAND

The synthesis covers three type regions: Baltic Shore (P-u), Baltic Coastal Zone *s.str.* (P-t), and Vistula Deltaic Area (P-v) (acc. to Ralska-Jasiewiczowa 1987), which today and in the past were all influenced by the Baltic Sea (Fig. 13.41). This part of Poland was affected by the last glaciation the longest; the southern boundary of the area comprises the stretch of terminal moraine delimiting the extent of the Pomeranian stage.

The synthesis is based on palaeoecological data from seven reference sites and eight complementary sites. The new pollen diagrams and additional radiocarbon dates help to correct some of the earlier information.

Altitude: 1.8 m below sea level to 140 m above sea level.

Climate: Mean annual temperature 7–8°C, mean January –1°C, mean July 16.5–17.5°C, annual temperature amplitude 19°C. Mean annual precipitation 600–700 mm. Winds mainly from the west and northwest.

Geology: Quaternary ground-moraine plateau and hills are mostly formed of till and loamy sand; patches of outwash plains are of sand and gravel; valleys are filled mostly with peat.

Topography: The landscape was shaped by the Vistulian ice sheet and its meltwaters, and the Pleistocene plateau was dissected by deep ice-marginal valleys. Cliffs, dunes, and marsh plains are present along the Baltic shore. The Vistula delta surface descends below sea level to form a shallow depression.

Vegetation: Present-day vegetation is largely a result of human activities. Remnants of deciduous forests comprise mainly Luzulo pilosae–Fagetum and Fago–Quercetum – two types of acidophilous communities present on the ground moraine. On inland sandy soils mostly Leucobryo–Pinetum and Pino–Quercetum phytocoenoses occur, whereas along the Baltic shore the sandy soils are occupied by Empetro nigri–Pinetum phytocoenoses. Fen communities are of primary importance in the peat vegetation.

Soils: Podzols, pseudopodzols, acid brown soils, muds, and bog soils.

Archaeology: The area was inhabited from the Late Palaeolithic. Settlement was most intensive at the end of the Bronze Age, at the beginning of the Iron Age, and in the Early Middle Ages.

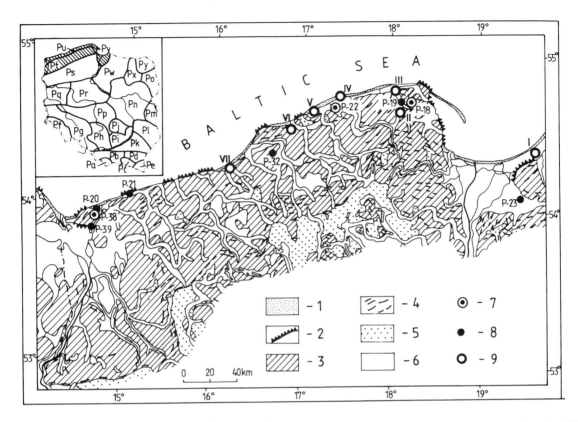

Fig. 13.41 Geomorphic map of the Baltic Coastal Zone synthesis region, showing the location of sites. 1–bars, 2–cliffs, 3–ground moraine, 4–terminal moraines, 5–outwash plains, 6–Pleistocene and Holocene river valleys and terraces, 7–reference sites with the pollen diagrams included in this paper: P-18 Darżlubie Forest, P-22 Kluki, P-38 Kołczewo, 8–other reference sites: P-19 Zarnowiec peat bog, P-20 Lake Racze, P-21 Niechorze, P-23 Lake Druzno, P-32 Zurawiec, P-39 Wolin II, 9–complementary sites: I–Vistula Lagoon (Zachowicz 1985), II–Orle (Latałowa 1988b), III–Lake Żarnowieckie (Latałowa 1982b, Bogaczewicz-Adamczak & Latałowa 1985), IV–Łebska Bar (Tobolski 1975), V–Lake Gardno (Bogaczewicz-Adamczak & Miotk 1985), VI–Ustka (Marsz & Tobolski 1993), VII–Lake Jamno (Dąbrowski *et al.* 1985, Przybyłowska-Lange 1979)

Reference site 38. Kołczewo (Latałowa 1989c, 1992)

Latitude 53°55′N, Longitude 14°40′E. Elevation *ca.* 15 m. Age range 12000–*ca.* 800 BP. Small peat bog, eleven local pollen-assemblage zones (paz) and six subzones, sixteen ^{14}C dates (Fig. 13.42). Very detailed pollen data especially for the last 4500 years. In the late-glacial it was a shallow water body, which became overgrown by mesotrophic moss communities at the start of the Holocene. Around 3000 BP, it was transformed into a raised bog. Since reclamation a *Pinus* wood has covered this site.

K-1	?–12000 BP	*Betula–Juniperus*–NAP
K-2	12000–11800 BP	*Betula–Hippophaë*

K-3	11800–11000 BP	*Pinus–Betula–Populus*
K-4	11000–10000 BP	*Juniperus–Betula nana– Artemisia–Empetrum*
K-5	10000–9200 BP	*Pinus–Betula–Populus– Urtica*
K-6	9200–8100 BP	*Pinus–Corylus–Ulmus*
K-7	8100–5800 BP	*Quercus–Ulmus– Tilia–Betula*
K-8	5800–2900 BP	*Quercus–Corylus– Tilia–Ulmus– Fraxinus*
K-8a	5800–3400 BP	*Corylus* subzone
K-8b	3400–2900 BP	*Pinus–Betula* subzone
K-9	2900–2200 BP	*Pinus–Fagus*–Gramineae

K-9a	2900–2700 BP	Gramineae–Cyperaceae–*Plantago lanceolata* subzone
K-9b	2700–2200 BP	*Pinus* subzone
K-10	2200–1000 BP	*Fagus–Pinus–Carpinus*
K-10a	2200–2100 BP	*Quercus–Fagus* subzone
K-10b	2100–1000 BP	*Fagus* subzone
K-11	1000–800 BP	*Pinus*–Gramineae–Cerealia–*Juniperus*

Reference site 22. Kluki (Tobolski 1982, 1987, Latałowa & Tobolski 1989)

Latitude 54°40′N, Longitude 17°19′E. Elevation 2.1 m. Age range *ca.* 10000–0 BP, twenty ^{14}C dates (Fig. 13.43). For better correlation with other pollen diagrams, two modifications were introduced to the ten paz originally described by Tobolski (1987). The site is situated within the large mire complex extending on the west and southwest shore of Łebsko Lake.

Kl-1	10000–9000 BP	*Betula–Pinus*
Kl-2	9000–8320 BP	*Pinus–Corylus*
Kl-3	8320–7200 BP	*Pinus–Ulmus–Alnus–Quercus*
Kl-4	7200–6500 BP	*Corylus–Alnus–Quercus–Fraxinus–Ulmus*
Kl-5	6500–5010 BP	*Pinus–Tilia*
Kl-6	5010–3920 BP	*Quercus–Corylus*
Kl-7	3920–1750 BP	*Carpinus*
Kl-7a	3920–2600 BP	*Quercus* subzone
Kl-7b	2600–1750 BP	*Pinus* subzone
Kl-8	1750–865 BP	*Fagus*
Kl-9	865–230 BP	*Alnus–Sphagnum*
Kl-10	230–0 BP	*Pinus*–Cerealia

Reference site 18. Darżlubie Forest (Latałowa 1982a, b, 1989b)

Latitude 54°42′N, Longitude 18°10′E. Elevation 40 m. Age range 10000–*ca.* 800 BP. Small peat bog, seven local pollen-assemblage zones (paz), four subzones, thirteen ^{14}C dates (Fig. 13.44).

The deposits are characterized by a great diversity of macrofossil assemblages and by varied accumulation rates. Between 6700 and 4000 BP the rate of accumulation was very low. The data from this time are certainly incomplete.

PD-2	?–10000 BP	*Juniperus–Pinus–Betula*
PD-3	10000–9100 BP	*Pinus–Betula*
PD-4	9100–7700 BP	*Corylus–Pinus*
PD-4a	9100–8200 BP	*Corylus* subzone
PD-4b	8200–7700 BP	*Alnus* subzone
PD-5	7700–4700 BP	*Tilia–Ulmus–Pinus*
PD-6	4700–2800 BP	*Quercus–Corylus*
PD-7	2800–1200 BP	*Quercus–Carpinus*
PD-7a	2800–1700 BP	*Betula–Pinus* subzone
PD-7b	1700–1200 BP	*Carpinus* subzone
PD-8	1200–*ca.*800 BP	*Pinus–Fagus–Juniperus*

Palaeoecological patterns and events

General patterns in the vegetational history

The area under discussion is highly varied, as reflected by the pollen diagrams. In consequence, in many cases the pollen-assemblage zones compared on Figure 13.45 are not equivalent from an ecological point of view, and correlations concern mainly the time and direction of vegetational changes. Regional pollen-assemblage zones cannot be defined, and only the main stages in the vegetational development will be characterized.

Late-glacial

The oldest well-documented sediments were found at Kołczewo (Latałowa unpubl.), Niechorze (Ralska-Jasiewiczowa & Rzętkowska 1987), and Ustka (Marsz & Tobolski 1993). In pre-Allerød time *Betula* forest probably dominated in the area. The Older Dryas climatic oscillation is recorded. The palaeobotanical data show that between *ca.* 11800 and 11000 BP (Allerød) forest communities with *Betula* and *Pinus* were present. The numerous pollen diagrams from the Younger Dryas (*ca.* 11000–10000 BP) illustrate the open landscape with *Betula* and *Pinus* (park tundra?), with the substantial participation of *Juniperus* thickets on dry, elevated habitats. In the valleys and along the lakes channels, moist tundra communities were dominant. *Empetrum* heath was broadly distributed.

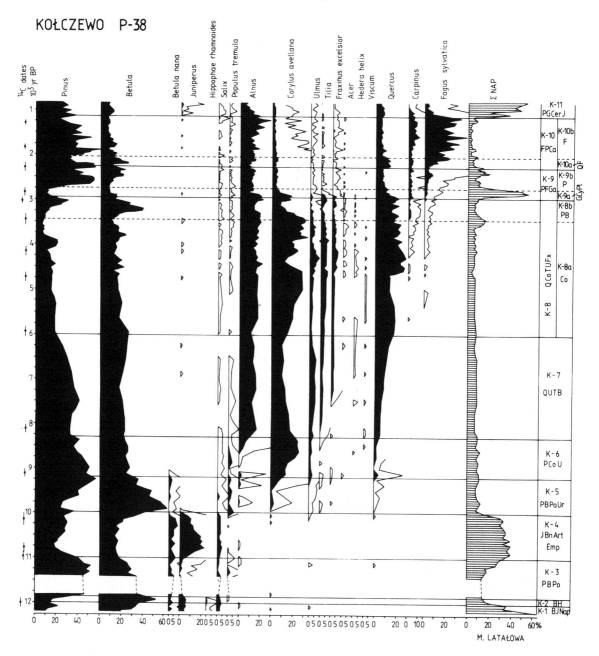

Fig. 13.42 Simplified pollen diagram from Kołczewo peat bog (P-38), with selected pollen taxa, plotted on a time scale. For paz symbols see Fig. 13.6 caption

Holocene

As shown on Figures 13.45 and 13.46, most of the vegetational changes began earlier in the western part of the area. At the beginning of the Holocene they resulted from the earlier immigration of deciduous trees and the early development of mesophilous forest communities. In the late Holocene metachroneity of the changes was connected mainly with

KLUKI - P 22

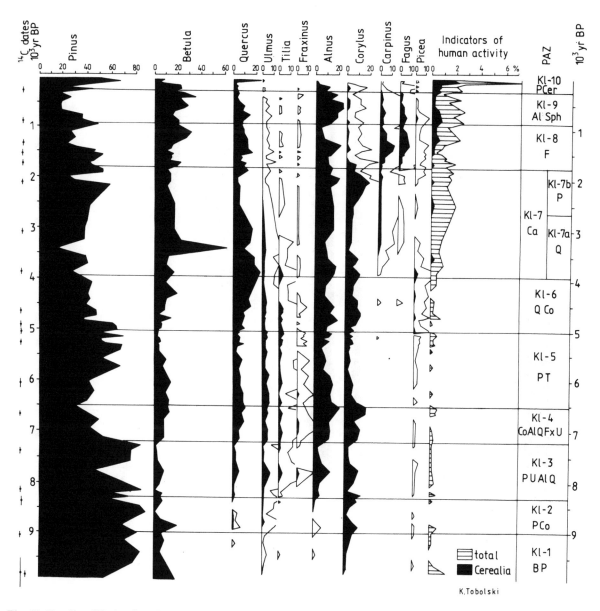

Fig. 13.43 Simplified pollen diagram from Kluki mire (P-22), with selected pollen taxa, plotted on a time scale. For paz symbols see Fig. 13.6 caption

the different intensity of human impact.

The main boundaries between pollen-assemblage zones I–VIII are placed at the start or at the intensification of the particular changes (Fig. 13.45). They are characterized as follows:

I First spread of the *Betula–Pinus* and *Pinus* forests

II First spread of *Corylus* in the understorey of *Pinus* forests

III First spread of the *Alnus* communities

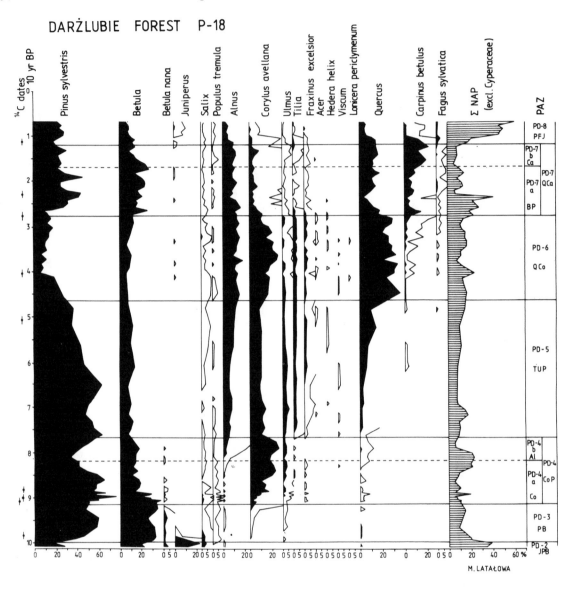

Fig. 13.44 Simplified pollen diagram from Darżlubie Forest mire (P-18), with selected pollen taxa, plotted on a time scale. For paz symbols see Fig. 13.6 caption

IV First spread of the mixed deciduous forests with *Quercus* and *Tilia* and increasing importance of *Ulmus*; gradual reduction of communities with *Pinus* and *Corylus*

 V Beginning of the optimum phase of the *Quercus* forests

VI Definite reduction of mixed deciduous forests

with *Quercus*, *Ulmus*, *Tilia*, and *Fraxinus*; increasing *Fagus* and *Carpinus*

VII Beginning of reforestation by communities with *Fagus* in the west, *Carpinus* in the east and both in the middle part of the area

VIII Beginning of total deforestation caused by the Early Medieval farming

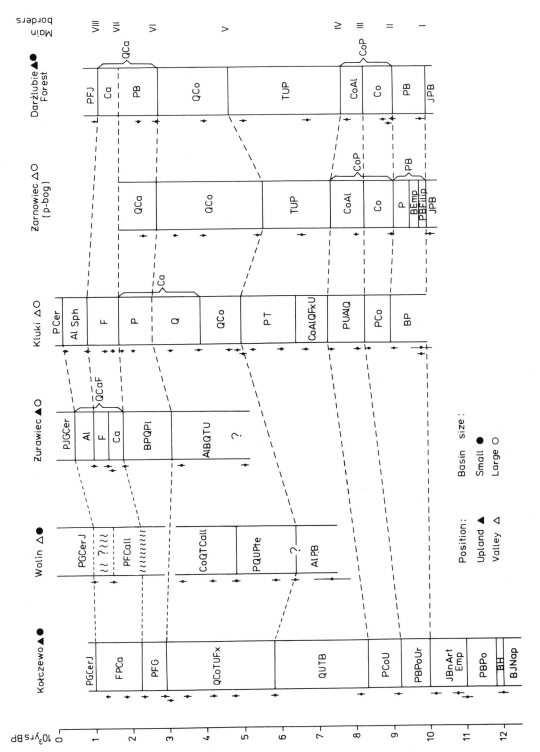

Fig. 13.45 Time–space correlation of local pollen- assemblage zones from sites representing the Baltic Coastal Zone synthesis region. The main correlation borders are indicated with the **Roman** numerals to the right. For paz symbols see Fig. 13.6 caption

Common patterns

- Presence of the rich "Dryas flora" and significant role of *Empetrum* in the Younger Dryas chronozone.
- Decrease of *Pinus* in the forest communities between 6500 and 6000 BP.
- Late optimum phase of *Quercus*-dominated forests, 4500–3000 BP.
- Definite decrease of *Tilia*, *Ulmus*, and *Fraxinus* in the woodland at about 2800–2500 BP.

Unique patterns

Two main factors have influenced the diversity in the vegetational history of the area:

(1) Differentiation depending on the geographical position:

- appearance of all the deciduous trees earlier at the western sites,
- the lack of a significant *Carpinus* phase on Wolin Island (west), and a prominent *Carpinus* phase in the eastern and middle parts of the area,
- *Fagus* migration from the west, with its optimum in the western and middle parts of the area; delay of the *Fagus* phase in the east by about 600 years.

(2) Differentiation depending on the physiographical conditions

- within the upland areas the culmination of *Juniperus* and higher *Pinus* representation in the Younger Dryas chronozone,
- in the extensive peat-filled valleys a much weaker culmination of *Juniperus* and a delay in forest expansion,
- substantial *Carpinus* and *Fagus* phases only on the uplands,
- heaths with *Calluna vulgaris* only in areas of poor sandy soils of southern Wolin Island (west),
- higher *Pinus* values in sites located in the valleys.

Climate

The climate of the area appears to have been always of a maritime character. In the Younger Dryas, which is here the best known part of the lateglacial, this character is marked by the considerable contribution of *Empetrum*.

In some of the profiles (Latałowa 1982a,b, 1988b, unpubl.) the Preboreal period (*ca.* 10000–9100 BP) seems to be tripartite, suggesting a climatic oscillation ("Youngest Dryas" *sensu* Behre 1978).

Pollen grains of the climatic bioindicators–*Viscum* and *Hedera*–appear around 8000 BP; *Hedera* appears first in the profiles from the west, and *ca.* 1000 years later in the eastern sites; *Viscum* occurs simultaneously in the diagrams from the western and eastern sites. The disappearance of pollen of both genera around 2500 BP cannot be unequivocally connected with the climate because of the strong anthropogenic changes at that time.

The palaeoecological data indicate favourable hydrological conditions for raised-bog development from about 1500 BP, which can be related to the increasing humidity of the climate.

Hydrology

Hydrological information is derived mainly from different kinds of mires, former lakes, and coastal lakes. The changes recorded in these localities were dependent on various factors, and the hydrological reaction was different among sites. The factor that significantly affected the hydrology of the whole area was fluctuations of the Baltic Sea level. The biostratigraphic studies of the peat deposits at Kluki (Tobolski 1987) reveal six transgression phases in the southern Baltic: three Littorina (1–3) and three post-Littorina (4–6) phases–Fig. 13.46. Only phase 3 exceeded the present level of the Baltic (by about 1 m). These results are generally in agreement with palaeolimnological information derived from the coastal lakes (Bogaczewicz-Adamczak & Miotk 1985, Dąbrowski *et al.* 1985, Przybyłowska-Lange 1976, 1979, Żachowicz 1985, Zachowicz *et al.* 1982, Zachowicz & Kępińska 1987) and Żarnowieckie Lake (Bogaczewicz-Adamczak & Latałowa 1985).

Soils

Soil acidification was accelerated by human influence relatively early within the area (*ca.* 5000–4500

Synthesis Region – Baltic Coastal Zone of Poland

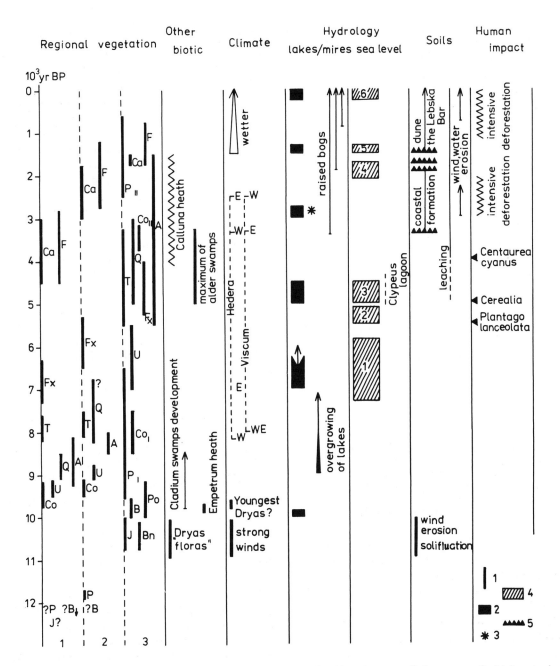

Fig. 13.46 Event stratigraphy for the Baltic Coastal Zone synthesis region. 1–range of phenomena, 2–high water level, 3–rise of water level caused by deforestation, 4–transgression phases, 5–horizons of fossil soils

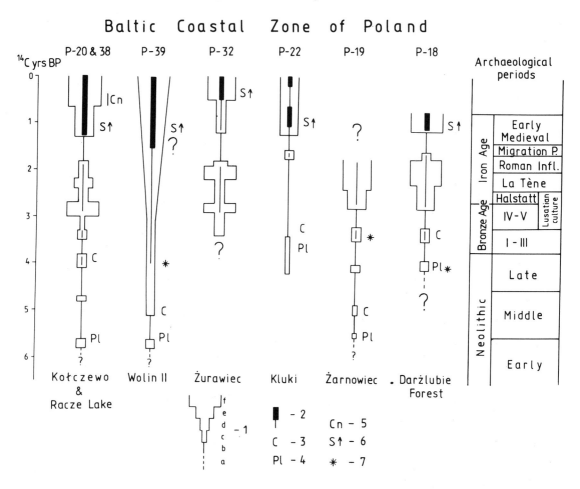

Fig. 13.47 Schematic interpretation of the intensity of human impact on vegetation, based on pollen frequencies from six sites representing the Baltic Coastal Zone synthesis region. 1–human impact in general, expressed in 6-grade scale: a–probable presence of man, b–slight local impact, c - regular clearings and land-use, d,e– increasing extent of clearings and land-use, f–large-scale or total deforestation and intensive economy, 2–slight to substantial indications of agriculture, 3– first appearance of cereals, 4–first appearance of *Plantago lanceolata*, 5–cultivation of *Cannabis*, 6–increasing cultivation of *Secale*, 7– first appearance of *Centaurea cyanus*

BP) because of the prevalent sandy formations. Wind and water erosion of soils was very strong in the intensively settled areas about 3000–2500 BP, and then in the Early Middle Ages.

The extensive dune area on the Łebska Bar was formed partly due to anthropogenic activity. Consecutive generations of dunes were formed after forest fires about 3400, 1900, and 1700 BP. The youngest dune-forming phase lasted from about 1500 years ago to the present day (Tobolski 1975).

Human impact

The earliest traces of agriculture in the pollen diagrams are from Early/Middle Neolithic (Fig. 13.47). In all localities, signs of forest clearance are accompanied by indications of grazing. The first pollen of Cerealia appear *ca.* 5500 BP. During the Bronze Age human influence was well marked but not continuous.

Because of the cultural diversity in the area, settlement phases at different sites do not quite coincide in the Bronze and Early Iron Ages. However, during the late phase of the Lusatian culture and the La Tène period, settlement was generally intensive. It was somewhat reduced during the Roman influence period (though not at Kluki), whereas during the Migration period the entire area went into economic decline. Settlement increased again between the 8th and 10th centuries AD; at that time total deforestation began. Only at one of the localities (39 Wolin II) does settlement seem to be continuous from *ca.* 5000 BP. At this site, on light sandy soils, *Calluna* heath started to develop between *ca.* 5000 and 4000 BP as a result of forest grazing and burning. It is characteristic of the whole area that cereal cultivation was of no great importance before the Early Middle Ages.

REFERENCES

Alexandrowicz, S.W. 1987: Malacofauna of the late-Vistulian and early Holocene lacustrine chalk from Roztoki near Jasło (Jasło–Sanok Depression). *Acta Palaeobotanica 27(1)*, 67–74.

Alexandrowicz, S.W. & Nowaczyk, B. 1983: Late-glacial and Holocene lake sediments at Pomorsko near Sulechów. *Questiones Geographicae 8*, 5–17.

Bałaga, K. 1982: Vegetational history of the Lake Łukcze environment (Lublin Polesie, E. Poland) during the Late Glacial and Holocene. *Acta Palaeobotanica 22(1)*, 7–22.

Bałaga, K. 1989: Type Region P-m: Lublin Polesie. *Acta Palaeobotanica 29(2)*, 69–73.

Bałaga, K. 1990: Development of Lake Łukcze and regional vegetation changes in the south-western part of Lublin Polesie during the last 13000 years. *Acta Palaeobotanica 30(1–2)*, 33–85.

Bałaga, K., Buraczyński, J. & Wojtanowicz, J. 1980/81: Budowa geologiczna i rozwój torfowiska Krowie Bagno (Polesie Lubelskie)—Geological structure and development of the Krowie Bagno Peatland (Polesie Lubelskie) (summary). *Annales Universitatis M. Curie-Skłodowska sect. B 35/36(4)*, 37–62.

Balwierz, Z. & Zurek, S. 1987: The late-glacial and holocene vegetational history and palaeohydrological changes at the Wizna site (Podlasie Lowland). *Acta Palaeobotanica 27(1)*, 121–136.

Balwierz, Z. & Zurek, S. 1989: Type Region P-n: Masovia and Podlasie Lowlands, NE part: Podlasie Lowland. *Acta Palaeobotanica 29(2)*, 65–68.

Behre, K.-E. 1978: Die Klimaschwankungen im europäischen Präboreal. *Pettermanns Geographische Mitteilungen 2*, 97–102.

Binka, K. & Szeroczyńska, K. 1989: Type Region P-n: Masovia and Podlasie Lowlands, W part: Masovia Lowland. *Acta Palaeobotanica 29(2)*, 59–63.

Binka, K., Madeyska, T., Marciniak, B., Szeroczyńska, K., Więckowski, K. & Wicik, B. 1988: Site Lake Błędowo. *In*: Starkel, L., Rutkowski, J., & Ralska-Jasiewiczowa, M. (eds) *Late glacial and Holocene Environmental changes. Vistula Basin*, 124–128. Publ. House Wydawnictwo Akademii Górniczo-Hutniczej, Cracow.

Binka, K., Cieśla, A., Łącka, B., Madeyska, T., Marciniak, B., Szeroczyńska, K. & Więckowski, K. 1991: The development of Błędowo Lake (Central Poland)—a palaeoecological study. *Studia Geologica Polonica 100*, 3–83

Błędzki, L.A. 1987: Cladoceran remains analysis in sediments of Lake Strażym (Brodnica Lake District). *Acta Palaeobotanica 27(1)*, 311–317.

Bogaczewicz-Adamczak, B. & Latałowa, M. 1985: The stratigraphic position of brackish-water bottom sediments from the Lake Żarnowiec (N. Poland). *Acta Palaeobotanica 25(1–2)*, 139–148.

Bogaczewicz-Adamczak, B. & Miotk, G. 1985: Z badań biostratygraficznych nad osadami z rejonu jeziora Gardno. From biostratigraphical studies of sediment from the region of Lake Gardno (summary). *Peribalticum III*, 79–96.

Boińska, U. 1987: Analysis of macrofossils in bottom deposits of Lake Strażym (Brodnica Lake District). *Acta Palaeobotanica 27(1)*, 305–310.

Cieśla, A., Ralska-Jasiewiczowa, M. & Stupnicka, E. 1978: Paleobotanical and geochemical investigations of the lacustrine deposits at Woryty near Olsztyn (NE Poland). *Polskie Archiwum Hydrobiologii 25(1/2)*, 61–73.

Dąbrowski, J. (ed.) 1981: Woryty—studium archeologiczno–przyrodnicze zespołu osadniczego kultury Łużyckiej. Woryty—an archaeological and naturalistic study of the settlement complex of Lusatian Culture (summary). *Polskie Badania Archeologiczne 20*, 3–256 .

Dąbrowski, M.J., Lubliner-Mianowska, K., Zachowicz, J. & Wypych, K. 1985: Z palinologii osadów Jeziora Jamno. Bottom sediments of lake Jamno in the light of palynological studies (summary). *Peribalticum III*, 37–52.

Drozdowski, E. 1974: Geneza Basenu Grudziądzkiego w świetle osadów i form glacjalnych. Genesis of the Grudziądz Basin in the light of its glacial forms and deposits (summary). *Prace Geograficzne 104*, 3–136.

Faliński, J. 1976: Antropogeniczne przeobrażenia roślinności Polski. Anthropogenic changes of the vegetation of Poland (summary). *Acta Agrobotanica 29(2)*, suppl., 275–290.

Geograficzny atlas świata, cz. I. 1987: Państwowe Przedsiębiorstwa Wydawnictw Kartograficznych im. E. Romera, Warszawa–Wrocław.

Gerlach, T., Koszarski, L., Koperowa, W. & Koster, E. 1972: Sediments lacustres postglaciaires dans la Depression de Jasło–Sanok. *Studia Geomorphologica Carpatho-Balcanica 6*, 37–61.

Gil, E., Gilot, E., Kotarba, A., Starkel, L. & Szczepanek, K. 1974: An early Holocene landslide in the Niski Beskid and its significance for paleogeographical reconstructions. *Studia Geomorphologica Carpatho-Balcanica 8*, 69–83.

Gordon, A.D. & Birks, H.J.B. 1972: Numerical methods in Quaternary palaeoecology. I. Zonation of pollen diagrams. *New Phytologist 71*, 961–979.

Harmata, K. 1987: Late-glacial and Holocene history of vegetation at Roztoki and Tarnowiec near Jasło (Jasło–Sanok Depression). *Acta Palaeobotanica 27(1)*, 43–65.

Harmata, K. 1989: Type Region P-d; Jasło–Sanok Depression. *Acta Palaeobotanica 29(2)*, 25–29.

Hjelmroos, M. 1981: The Post-Glacial development of Lake Wielkie Gacno, NW-Poland, the human impact on the natural vegetation recorded by means of pollen analysis and ¹⁴C dating. *Acta Palaeobotanica 21(2)*, 129–144.

Hjelmroos-Ericsson, M. 1981: *Holocene development of Lake Wielkie Gacno area, northwestern Poland.* Lundqua Thesis, 10. Department of Quaternary Geology, University of Lund.

Jankowska, B. 1980: Szata roślinna okolic Gopła w późnym glacjale i holocenie oraz wpływ osadnictwa na jej rozwój w świetle danych paleobotanicznych. The vegetation in the Gopło region in the Late Glacial and Holocene and the influence of settlement on its development (summary). *Przegląd Archeologiczny 27*, 5–41.

Kępczyński, K. 1960: Zespoły roślinne Jezior Skępskich i otaczających je łąk. Plant groups of the Lake District of Skępe and the surrounding peatbogs (summary). *Studia Societatis Scientiarum Toruniensis, suppl. 6*, 1–244.

Kępczyński, K. & Noryśkiewicz, B. 1968: Roślinność i historia torfowiska Fletnowo w pow. Grudziądzkim. Flora und Geschichte des Moores Fletnowo in dem Kreise Grudziądz (summary). *Zeszyty Naukowe UMK, Toruń 21/Biol. 9*, 49–95.

Kondracki, J. 1978: *Geografia fizyczna Polski.* 463 pp. PWN, Warszawa (in Polish only).

Koperowa, W. 1958: Późny glacjał z północnego podnóża Tatr w świetle analizy pyłkowej. A Late-Glacial pollen diagram at the north foot of Tatra Mountains (summary). *Monographiae Botanicae 7*, 107–133.

Koperowa, W. 1962: Późnoglacjalna i holoceńska historia roślinności Kotliny Nowotarskiej. The history of the Late-Glacial and Holocene vegetation in Nowy Targ Basin (summary). *Acta Palaeobotanica 2(3)*, 3–57.

Koperowa, W. 1970: Późnoglacjalna i holoceńska historia roślinności wschodniej części Dołów Jasielsko–Sanockich. Late-Glacial and Holocene history of the vegetation of the eastern part of the "Jasło-Sanok-

Doły", Flysch Carpathians (summary). *Acta Palaeobotanica 11(2)*, 3–42.

Lankauf, K.R. 1987: Results of physical and chemical studies on Lake Strażym deposits (Brodnica Lake District). *Acta Palaeobotanica 27(1)*, 269–276.

Latałowa, M. 1976: Diagram pyłkowy osadów późnoglacjalnych i holoceńskich z torfowiska w Wolbromiu. Pollen diagram of the Late-Glacial and Holocene peat deposits from Wolbrom (S. Poland) (summary). *Acta Palaeobotanica 17(1)*, 55–80.

Latałowa, M. 1982a: Major aspects of the vegetational history in the eastern Baltic coastal zone of Poland. *Acta Palaeobotanica 22(1)*, 47–63.

Latałowa, M. 1982b: Postglacial vegetational changes in the eastern Baltic coastal zone of Poland. *Acta Palaeobotanica 22(2)*, 179–249.

Latałowa, M. 1988a: The Late-glacial and early Holocene history of the vegetation in the Wolbrom area (Silesian–Cracovian Upland), S. Poland. *In*: Lang, G. & Schlüchter, Ch. (eds) *Lake, mire and river environments*, 9–22. Balkema, Rotterdam.

Latałowa, M. 1988b: A palaeobotanical study of the peatbog at Orle in the Reda–Łeba ice-marginal valley. *Folia Quaternaria 58*, 45–58.

Latałowa, M. 1989a: Type Region P-h: The Silesia–Cracow Upland. *Acta Palaeobotanica 29(2)*, 45–50.

Latałowa, M. 1989b: Type Region P-t: Baltic Coastal Zone. *Acta Palaeobotanica 29(2)*, 103–108.

Latałowa, M. 1989c: Type Region P-u: Baltic Shore, W part Wolin Island. *Acta Palaeobotanica 29(2)*, 115–120.

Latałowa, M. 1992: Man and vegetation in the pollen diagrams from Wolin Island (NW Poland). *Acta Palaeobotanica 32(1)*, 123–249.

Latałowa, M. & Nalepka, D. 1987: A study of the Late-glacial and Holocene vegetational history of the Wolbrom area (Silesian–Cracovian Upland). *Acta Palaeobotanica 27(1)*, 75–115.

Latałowa, M. & Tobolski, K. 1989: Type Region P-u: Baltic Shore. *Acta Palaeobotanica 29(2)*, 109–114.

Litt, Th. 1988: Untersuchungen zur spätglazialen Vegetations-entwicklung bei Dziekanowice (Umgebung Lednógora,Wielkopolska). *Acta Palaeobotanica 28(1–2)*, 49–60.

Madeyska, E. 1989: Type Region P-f: The Sudety Mts–Bystrzyckie Mts. *Acta Palaeobotanica 29 (2)*, 37–41.

Marciniak, B. 1973: Zastosowanie analizy diatomologicznej w stratigrafii osadów późnoglacjalnych jeziora Mikotajskiego. The application of the diatomological analysis in the stratigraphy of the late glacial deposits of the Mikołajki Lake (summary). *Studia Geologica Polonica 39*, 3–153.

Marciniak, B. 1979: Dominant diatoms from Late Glacial and Holocene lacustrine sediments in Northern Poland. *Nova Hedwigia 64*, 411–426.

Marciniak, B. 1987: Diatoms from the Holocene sediments of Lake Steklin (Dobrzyń Lake District). *Acta Palaeobotanica 27(1)*, 319–334.

Marsz, A. & Tobolski, K. 1993: Osady późnoglacjalne i holoceńskie w klifie między Ustką a ujściem Potoku Orzechowskiego. Late Glacial and Holocene deposits in the cliff between Ustka and Potok Orzechowski mouth (summary) *In*. Florek, W. (ed.) *Geologia i Geomorfologia Środkowego Pobrzeża i Południowego Bałtyku*, 201–250 WSP, Słupsk.

Miotk-Szpiganowicz, G. 1989: Type region P-s: W-Pomeranian Lake Districts: E part: Bory Tucholskie. *Acta Palaeobotanica 29(1)*, 81–84.

Miotk-Szpiganowicz, G. 1992: The history of the vegetation of Bory Tucholskie and the role of man in the light of palynological investigations. *Acta Palaeobotanica 32*, 39–122.

Niewiarowski, W. 1987: Development of Lake Strażym (Brodnica Lake District) during the late glacial and holocene. *Acta Palaeobotanica 27(1)*, 251–268.

Noryśkiewicz, B. 1982: Lake Steklin—a reference site for the Dobrzyń–Chełmno Lake District, N Poland—report on palaeoecological studies for the IGCP Project No. 158 B. *Acta Palaeobotanica 22(1)*, 65–83.

Noryśkiewicz, B. 1987: History of vegetation during the late glacial and holocene in the Brodnica Lake District in the light of pollen analysis of Lake Strażym deposits. *Acta Palaeobotanica 27(1)*, 283–304.

Noryśkiewicz, B. & Ralska-Jasiewiczowa, M. 1989: Type Region P-w: Dobrzyń–Olsztyn Lake Districts. *Acta Palaeobotanica 29(2)*, 85–93.

Obidowicz, A. 1976: Geneza i rozwój torfowiska w Wolbromiu. Genesis and development of the peat-bog at Wolbrom (S. Poland) (summary). *Acta Palaeobotanica 17(1)*, 45–54.

Obidowicz, A. 1978: Genese und Stratigraphie des Moores "Bór na Czerwonem" in Orawa–Nowy Targ Mulde. *Fragmenta Floristica et Geobotanica 24(3)*, 447–466.

Obidowicz, A. 1989: Type Region P-a: Inner West Carpathians–Nowy Targ Basin. *Acta Palaeobotanica 29(2)*, 11–15.

Obidowicz, A. 1990. Eine pollenanalythische und moorkundliche Studie zur Vegetationsgeschichte des Podhale-Gebietes (West-Karpaten). *Acta Palaeobotanica 30(1–2)*, 147–219.

Offierska, J. 1978: Pełzaki skorupkowe (Testacea) torfowiska otaczającego Jezioro Skrzynka w Wielkopolskim Parku Narodowym. Testacea of peatbog surrounding Skrzynka Lake in the National Park of Great Poland (summary). *Badania Fizjograficzne nad Polską Zachodnią 31, ser. C Zoologia*, 7–38.

Offierska, J. & Okuniewska, I. 1986: Analiza palinologiczna i fauny Testacea (Rhizopoda) z torfowiska przy Jeziorze Skrzynka w Wielkopolskim Parku Narodowym. Palynological study and analysis of Testacea (Rhizopoda) fauna from a peatbog close to Lake Skrzynka in the Great Poland National Park (summary). *Sprawozdania Polskiego Towarzystwa Przyjaciół Nauk 103/1984*, 146–148.

Okuniewska, I. 1985: Działalność antropogeniczna w okolicach Wonieścia w świetle analizy pyłkowej. Anthropogenic activity in the vicinity of Wonieść in the light of pollen analysis (summary). *Sprawozdania Polskiego Towarzystwa Przyjaciół Nauk 103/1984*, 144–146.

Okuniewska-Nowaczyk, I. 1987: Late-holocene history of the vegetation growing in the vicinity of Lake Skrzynka, the Greater Poland National Park, obtained from pollen analytical data. *Acta Palaeobotanica 27(1)*, 137–151.

Pawlikowski, M., Ralska-Jasiewiczowa, M., Schönborn, W., Stupnicka E. & Szeroczyńska, K. 1982: Woryty near Gietrzwałd, Olsztyn Lake District, NE-Poland—vegetational history and lake development during the last 12000 years. *Acta Palaeobotanica 22(1)*, 85–116.

Przybyłowska-Lange, W. 1976: Diatoms of lake deposits from the Polish Baltic Coast. I. Lake Druzno. *Acta Palaeobotanica 17(2)*, 35–74.

Przybyłowska-Lange, W. 1979: Diatoms of lake deposits from the Polish Baltic Coast. II. Lake Jamno. *Acta Palaeobotanica 20(2)*, 213–226.

Ralska-Jasiewiczowa, M. 1966: Osady denne Jeziora Mikołajskiego na Pojezierzu Mazurskim w świetle badań paleobotanicznych. Bottom sediments of the Mikołajki Lake Masurian Lake District in the light of palaeobotanical investigations (summary). *Acta Palaeobotanica 7(2)*, 1–118.

Ralska-Jasiewiczowa, M. 1980: *Late-glacial and Holocene vegetation of the Bieszczady Mts (Polish Eastern Carpathians)*. 202 pp. PWN, Warszawa–Kraków.

Ralska-Jasiewiczowa, M. 1982: Introductory remarks. *Acta Palaeobotanica 22(1)*, 3–6.

Ralska-Jasiewiczowa, M. 1983: Isopollen maps for Poland: 0–11000 years BP. *New Phytologist 49*, 133–175.

Ralska-Jasiewiczowa, M. 1987: Introductory remarks. *Acta Palaeobotanica 27(1)*, 3–8.

Ralska-Jasiewiczowa, M. (ed.) 1989: Environmental changes recorded in lakes and mires of Poland during the last 13000 years. III. *Acta Palaeobotanica 29(2)*, 1–120.

Ralska-Jasiewiczowa, M. 1989a: Type Region P-e: Bieszczady Mts. *Acta Palaeobotanica 29(2)*, 31–35.

Ralska-Jasiewiczowa, M. 1989b: Type Region P-x: Masurian Great Lakes District. *Acta Palaeobotanica 29(2)*, 95–100.

Ralska-Jasiewiczowa, M., Goslar, T., Madeyska, T. & Starkel, L. in prep. (1996): Lake Gościąż, central Poland – a monographic study. *Acta Palaeobotanica, Suppl. 2*.

Ralska-Jasiewiczowa, M. & Rzętkowska, A. 1987: Pollen and macrofossil stratigraphy of fossil lake sediments at Niechorze I, W. Baltic Coast. *Acta Palaeobotanica 27(1)*, 153–178.

Ralska-Jasiewiczowa, M. & Starkel, L. 1988: Record of the hydrological changes during the Holocene in the lake, mire and fluvial deposits of Poland. *Folia Quaternaria 57*, 91–127.

Różański, K. 1987: The ^{18}O and ^{13}C isotope investigations of carbonate sediments from the Lake Strażym (Brodnica Lake District). *Acta Palaeobotanica 27(1)*, 277–282.

Różański, K., Harmata, K., Noryśkiewicz, B., Ralska-Jasiewiczowa, M. & Wcisło, D. 1988: Palynological and isotope studies on carbonate sediments from some Polish lakes. Preliminary results. *In* Lang, G. & Schlüchter, Ch. (eds) *Lake, mire and river environments*, 41–49. Balkema, Rotterdam.

Szafer, W. (ed.) 1966: *The vegetation of Poland.* 738 pp. Pergamon Press, Oxford/PWN, Warszawa.

Szafer, W. & Zarzycki, K. (eds) 1972: *Szata roślinna Polski.* 615 pp. PWN, Warszawa (in Polish only).

Szczepanek, K. 1961: Późnoglacjalna i holoceńska historia roślinności Gór Świętokrzyskich. The history of the Late-Glacial and Holocene vegetation of the Holy Cross Mountains (summary). *Acta Palaeobotanica 2(2)*, 1–45.

Szczepanek, K. 1971: Kras staszowski w świetle badań paleobotanicznych. The Staszów karst in the light of palaeobotanical studies—South Poland (summary). *Acta Palaeobotanica 12(2)*, 60–140.

Szczepanek, K. 1982: Development of the peat-bog at Słopiec and the vegetational history of the Świętokrzyskie (Holy Cross) Mts, in the last 10000 years. *Acta Palaeobotanica 22(1)*, 117–130.

Szczepanek, K. 1987: Late-glacial and Holocene pollen diagram from Jasiel in the Low Beskid Mts (The Carpathians). *Acta Palaeobotanica 27(1)*, 9–26.

Szczepanek, K. 1989a: Type Region P-c: Low Beskidy Mts. *Acta Palaeobotanica 29(2)*, 17–23.

Szczepanek, K. 1989b: Type Region P-j Świętokrzyskie Mts (Holy Cross Mts). *Acta Palaeobotanica 29(2)*, 51–56.

Szeroczyńska, K. 1985: Cladocera jako wskaźnik ekologiczny w póżnoczwartorzędowych osadach jeziornych Polski północnej. Cladocera as ecological indicator in Late Quaternary lacustrine sediments in Northern Poland (summary). *Acta Palaeontologica Polonica 30(1–2)*, 3–69.

Tobolski, K. 1966: Poznoglacjalna i holocenska historia roslinosci na obszarze wydmowym w dolinie srodkowej Prosny. The Late Glacial and Holocene history of vegetation in the dune area of the middle Prosna valley (summary). *PTPN, Prace Komisji Biologicznej 32(1)*, 3–28.

Tobolski, K. 1975: Studium palinologiczne gleb kopalnych Mierzei Łebskiej w Słowińskim Parku Narodowym. Palynological study of fossil soils of the Łeba Bay Bar in the Słowiński National Park (summary). *PTPN, Prace Komisji Biologicznej 41*, 1–76.

Tobolski, K. 1982: Anthropogenic changes in vegetation of the Gardno–Łeba Lowland, N. Poland. Preliminary report. *Acta Palaeobotanica 22(1)*, 131–139.

Tobolski, K. 1987: Holocene vegetational development based on the Kluki reference site in the Gardno–Łeba plain. *Acta Palaeobotanica 28(1)*, 179–222.

Tobolski, K. & Okuniewska-Nowaczyk, I. 1989: Type Region P-r: Poznań-Gniezno–Kujawy Lake Districts. *Acta Palaeobotanica 29(2)*, 77–80.

Wasylikowa, K. 1964: Roślinność i klimat póżnego glacjału w Środkowej Polsce na podstawie badań w Witowie koło Łęczycy. Vegetation and climate of the Late-glacial in Central Poland based on investigations made at Witów near Łęczyca (summary). *Biuletyn Peryglacjalny 13*, 262–417.

Wasylikowa, K. 1978: Aneks–Roślinność stanowiska mezolitycznego w Witowie w okresie borealnym. The vegetation of the Mesolithic site in Witów during the Boreal period (summary). *Prace i materiały Muzeum Archeologicznego i Etnograficznego w Łodzi.* Seria Archeologii 25, 82–86.

Więckowski, K. 1966: Osady denne Jeziora Mikołajskiego. Bottom deposits of Lake Mikołajki (summary). *Prace Geograficzne 57*, 3–112.

Wójcik, A. 1987: Late-glacial lacustrine sediments from Roztoki and Tarnowiec near Jasło (Jasło–Sanok Depression). *Acta Palaeobotanica 27(1)*, 27–41.

Zachowicz, J. 1985: Z badań biostratygraficznych nad osadami Zalewu Wiślanego. From biostratigraphical studies of sediments from the Vistula Lagoon (summary). *Peribalticum III*, 97–111.

Zachowicz, J. & Kępińska, U. 1987: The palaeoecological development of Lake Druzno (Vistula deltaic area). *Acta Palaeobotanica 28(1)*, 227–249.

Zachowicz, J., Przybyłowska-Lange, W. & Nagler, J. 1982: The Late-Glacial and Holocene vegetational history of the Żuławy region, N. Poland. A biostratigraphic study of Lake Druzno sediments. *Acta Palaeobotanica 22(1)*, 141–161.

Żurek, S. 1975: Geneza zabagnienia pradoliny Biebrzy. Genesis of bog formation in the Biebrza Urstromtal (summary). *Prace Geograficzne 110*, 3–104.

Żurek, S. 1978: Development of the fossil Holocene lakes in the Biebrza ice-marginal valley against the background of the Maliszewskie lake sediments. *Polskie Archiwum Hydrobiologii 25(1/2)*, 491–498.

14

Czech and Slovak Republics

E. Rybníčková and K. Rybníček

In collaboration with H. Hüttemann (Innsbruck), J. Dolejšová (Nitra), V. Jankovská (Brno), E. Krippel † (Bratislava), M. Peichlová (Brno), and H. Svobodová (Brno, Průhonice)

INTRODUCTION

The Czech and Slovak Republics have an extraordinarily high environmental diversity. The landforms, rocks, climate, soils, and consequently the vegetation all change within short distances. The altitude ranges between 85 and 2655 m above sea level, and 70% of the territory consists of uplands and mountains (Fig. 14.1). The hydrologic network is very dense and includes the continental divide. The Bohemian Massif forms the geologically older western part of the territory, and the West Carpathians form the younger eastern part (Fig. 14.2). Great geomorphic differentiation of the surface supports the climatic and mesoclimatic diversity. The mean annual temperature ranges between −3.7°C and 10°C, and the annual mean precipitation between 450 and 1550 mm. The western part of the territory is suboceanic, the eastern subcontinental. Among the major soil types the chernozem and brown soils cover the lowlands. Podzol, illimerized soils, and brown forest soils prevail in the uplands, mountains, and foothills. Mountain humic podzols developed in the summit areas of the mountains. Among azonal soils rendzinas cover the limestone areas; alluvial flood loams, semigleys, and gley soils fill in the river and stream valleys.

This environmental diversity and the position of the territory between three major phytogeographic regions of Europe–the Central European, Pannonian, and Carpathian–also influence the composition, distribution, and development of the vegetation. Two fundamental works of Czech and Slovak phytosociologists must be mentioned, namely the geobotanical map of the Czech Republic (Mikyška 1968) and its Slovak counterpart (Michalko 1987). According to these maps the present vertical zonation of the major climax vegetation types is as in Table 14.1.

Most of the region has never been glaciated, so a certain continuity of vegetation and soil development is to be expected during the entire Quaternary. Czech and Slovak territories have never been "tabula rasa", where the forming of ecosystems would have initiated from the very beginning, as is true in the glaciated region north of the republics' borders. The past human impact is another important historical factor, forming the past and present stages of ecosystems and their vegetation. It started more than 7000 years BP in the lowlands and about (500) 700–900 years BP in the uplands and mountains. For further information on the present geomorphology, hydrology, geology, climate and soils see Vesecký *et al.* (1958), Götz (1966), Pelíšek (1966), Demek & Střída (1971), Czudek (1972), and Hejný & Slavík (1988).

Because of these environmental and historical factors, palaeoecological investigation can greatly assist in the understanding of many key problems of

Palaeoecological Events During the Last 15 000 Years: Regional Syntheses of Palaeoecological Studies of Lakes and Mires in Europe.
Edited by B.E. Berglund, H.J.B. Birks, M. Ralska-Jasiewiczowa and H.E. Wright. © 1996 John Wiley & Sons Ltd.

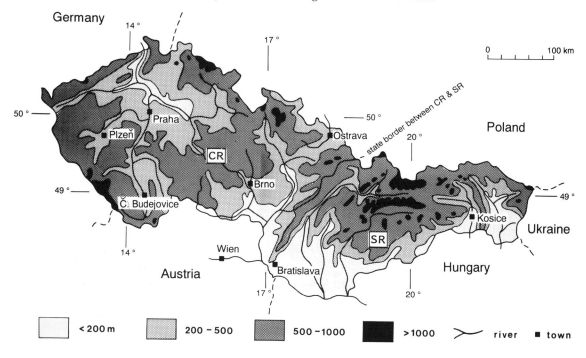

Fig. 14.1 Topographic and hyposometric map of Czech and Slovak Republics (CR, SR)

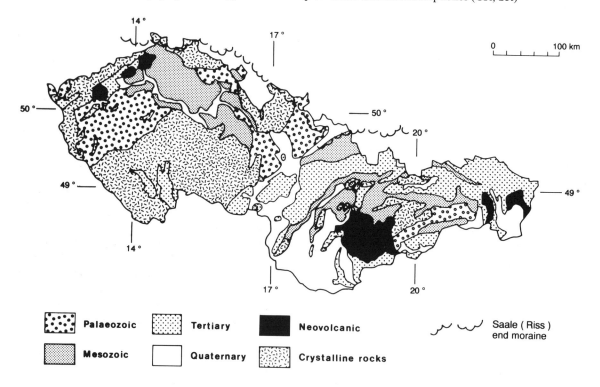

Fig. 14.2 Geological map of Czech and Slovak Republics. Very simplified

Table 14.1 Vertical zonation of major climax vegetation types

Vegetation belts	Upper limits (m)	
	Bohemian Massif	Carpathians
Oak, oak–hornbeam	(400)500	(500)–600
Beech, fir–beech with spruce	(400)500–900(1000)	(500)600–1100 (1300)
Spruce, fir–spruce	(900)1000–1300(1400)	(900)1300–1500(1600)
Pinus mugho	1300–1500	(1500)1600–1700(1900)
Treeless alpine	over (1000) 1500	over (1600) 1700(1900)

European palaeoecology and palaeogeography. On the other hand the same environmental diversity causes many methodological problems and complicates the interpretation of palaeoecological events.

In general, the Czech and Slovak national syntheses follow the recommended methodological and interpretive approaches of the IGCP 158 B (see Berglund 1986). Specific questions concerning the type-region and reference-site concepts, synchronization of biostratigraphic events, and some bioindication problems have been discussed and published earlier (Rybníček & Rybníčková 1991).

Type regions and reference sites

Fourteen type regions were delimited inside the former Czechoslovak territory, most of them extending into neighbouring countries (see Fig. 14.3, and the Appendix). No modern data with ^{14}C dates exist for eight type regions, at least for their CS parts, namely the West Bohemian Mountains (CS-a), the Silesian Lowlands (CS-e), the Danube Lowlands (CS-g), the outer West Carpathians (CS-h), the West Carpathian foothills of South Slovakia (CS-k), the Low Beskydy Mountains (CS-l), the Bieszczady/Vihorlat Mountains (CS-m) and the Tisza Lowlands (CS-n). Only nine reference sites have been chosen as examples to characterize the six remaining type regions and their subregions in this national synthesis. They represent a selection from 25 reference sites studied. However, all of them are used as data sources for drawing general conclusions. Complete data, if not published, are stored in the Czech Academy of Sciences in Brno as well as in the international project centres like the European Pollen Database (Arles).

TYPE REGION CS-b, CENTRAL BOHEMIAN BASIN

This includes several geographic units: the Bohemian Plateau, the České středohoří Mountains, the Mostecká pánev Basin, the Pražská plošina Uplands, and parts of some other neighbouring units.

Altitude: The altitude ranges between 130 and 250 m above sea level in the valleys, the mean altitude of the hilly landscape is about 400 m, and the highest point of the České středohoří Mountains reaches 837 m (Milešovka Mountain).

Climate: Warm and moderately warm, dry and moderately dry. Mean annual temperature 8.0–8.8°C, mean precipitation about 460 mm.

Geology: Cretaceous sediments (sandstone, marlstone, claystone) prevail, volcanics (phonolite, basalt) are common, and Paleozoic rocks can be found in marginal parts. Quaternary sediments cover the broad valleys (sands, flood-loams), but organic sediments are rare.

Topography: The axis of the region is formed by broad valleys of the Labe and Ohře Rivers. Volcanic hills occur in the České Středohoří Mountains, and sandstone cliffs constitute the northern and north-western margins. The remaining parts are marked by a slightly undulated hilly landscape.

Population: 200–300 km^{-2} (mean), or up to 1000 km^{-2} in urban agglomerations.

Vegetation: The present vegetation consists mostly of cropland, pine plantations, and remains of deciduous forests. Virgin vegetation includes subxero-

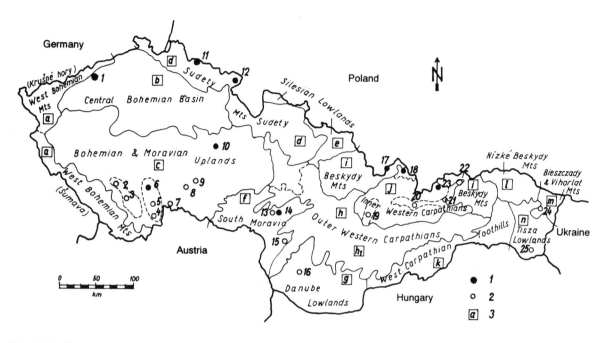

Fig. 14.3 Type regions (full lines) and subregions (dashed lines) of Czech and Slovak Republics and location of reference sites: 1. major reference sites described in text. 2. additional reference sites mentioned in text. For a list of reference sites see Appendix. 3. Type region letters

philous and thermophilous oak forests, xerothermic oak forests (*Quercus pubescens*), oak/hornbeam forests, and floodplain forests as the main types. Relict pine stands grow on sandstone cliffs. The flora is rich in species, especially on limestone and volcanics.

Soil types: Chernozems, regosols, rendzinas, illimerized soils, and brown forest soils.

Land use: Forest 15%, cultivated land 70%, permanent grassland (xerothermic) 5%, urbanized 10%, mires, water bodies, and other areas less than 1%.

Reference site 1. Komořanské jezero

One reference site from the northwest part of the region (Mostecká pánev Basin) is presented (Figs 14. 4 and 14.5). For full results see Jankovská (1983, 1984b, 1988b). Latitude 50°30′N, Longitude 13°32′E. Elevation 230 m. Age range 8300–250 BP. Ancient lake with aquatic sediments. Infilling from

the margins: tall sedges, reed swamps from 7500 BP, alder carr since 6000 BP. After drainage in AD 1831 the open water disappeared. The site has not existed at all since brown-coal mining started.

The pollen diagram reflects the vegetation of the Central Bohemian Basin and the thermophilic vegetation of the České středohoří Mountains. Pollen spectra, however, are influenced by the mountain vegetation of the nearby Krušné hory Mountains (*Fagus*, *Picea*, partly *Abies*). Early Holocene assemblages: *Pinus*, *Betula*, *Juniperus*, *Salix*, Gramineae, *Ephedra*, and heliophilous plants. After 8000 BP *Corylus avellana* and *Pinus* were dominant and QM species present. After 6000 BP *Picea*, *Fagus*, and *Abies* covered the mountain slopes and/or the northern slopes of the volcanic mountains, and mixed oak forests were present in the lowlands. At 3500 BP *Carpinus* invaded the mesophilous stands of the lowlands. *Alnus* formed important extralocal vegetation around the lake (cf. pollen over-representation). Human impact is apparent since the

Neolithic, and continuous agriculture developed since the Migration Period (1400–1500 BP).

TYPE REGION CS-c, THE BOHEMIAN AND MORAVIAN UPLANDS

The type region is the largest one in the territory of former Czechoslovakia. It is divided into two subregions. A) The Southbohemian basins include the Českobudějovická pánev and Třebo,ská pánev Basins. B) The upland subregion consists of the following orographic units: the Plzeř,ská pahorkatina Uplands, the Brdy Mountains, the Středočeská pahorkatina Uplands, the Českomoravská vrchovina Highlands, and the Drahanská vrchovina Highlands.

Altitudes: (250) 350–450 m above sea level in the basins, 450–650 m above sea level in the uplands, ridges over 650 m above sea level. The highest point is 862 m (Praha Mountain in the Brdy Mountains).

Climate: Moderately warm, moderately humid or humid. Mean annual temperatures 5–8°C, mean annual precipitation 550–800 mm.

Geology: Proterozoic crystalline rocks (granite, gneiss, granodiorite). In the east are Carboniferous culm facies (shale, greywacke, conglomerate), with Devonian limestones in the Moravian Karst of the Drahanská vrchovina Highlands. Senonian claystones, sands, Oligocene clays, and Quaternary sediments (sand, peat, flood-loams) in the Southbohemian basins.

Topography: Gently undulated upland peneplain with deep river valleys. Karst features and cliffs in the Moravian Karst. Flat relief in the basins.

Population: 40–60 km^{-2}.

Vegetation: Secondary agricultural communities and coniferous plantations (*Picea, Pinus*). Poor acidophilous flora except for limestone areas of the Karst. Virgin vegetation: mixed *Fagus–Abies* forests with *Picea, Abies* and *Picea* forests on humid soils, *Picea* (over 550 m) and *Alnus* stands on the waterlogged sites, and Carpinion communities in the foothills and valleys (in South Bohemia without *Carpinus betulus*).

Soil types: Illimerized and podzol soils, brown forest soils, gleys and pseudogleys; rendzinas in the Moravian Karst. Organogenic soils in Southbohemian basins.

Land use: Forest 35%, cultivated land (potatoes, rye, oat, wheat) including meadows 45–55%, urbanized 5%, water bodies about 2% (in South Bohemia to 10%), peatlands in South Bohemia to 15% (mostly cultivated).

Several reference sites were studied in the region, but only two of them are presented here: reference site CS 6 (Borkovice) is representative for Southbohemian Basins, and reference site CS 10 Kameničky for the uplands subregion. Further reference sites: for CS 2 Řežabinec see Rybníčková & Rybníček (1985), for CS 3 Zbudov see Rybníčková et al. (1975) and Rybníčková (1982), for CS 4 Červené blato see Jankovská (1980), for CS 5 Mokré louky see Jankovská (1987), and for CS 7 Bláto, CS 8 Řásná, and CS 9 Loučky see Rybníčková (1974) and Rybuíček & Rybníčková (1968).

Reference site 6. Borkovice

The pollen diagram (Figs. 14.6 and 14.7) comes from the Borkovice mire in the northwestern part of the Třebo,ská pánev Basin. For details see Jankovská (1980). Latitude 49°16′N, Longitude 14°36′E. Elevation 415 m. Age range 12200–0 radiocarbon years BP.

The mire started to grow in shallow pools with water from the outflows of artesian springs (*Potamogeton, Myriophyllum, Pediastrum*) in the Alleröd. At about 11000 BP swamp vegetation (*Carex, Equisetum*) replaced the aquatic stages. Minerotrophic transitional poor-fen (*Phragmites, Sphagnum*) covered the site during AT-1. At about 6000 BP bog communities (*Eriophorum vaginatum, Sphagnum*) prevailed, and ombrotrophic peat formed. Minerotrophic patches had persisted around the outflows of artesian waters (*Meesia triquetra, Calliergon trifarium*). *Pinus rotundata* (tree form) and *Ledum palustre* bog assemblage completed the succession during SA-2.

The Holocene development of vegetation has the usual sequence, but several specific features are

478

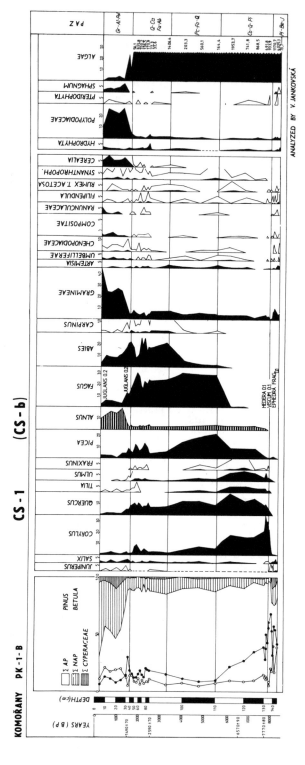

Fig. 14.4 Simplified pollen diagram from the reference site "Komořanské jezero" (CS 1) of the type region Central Bohemian Basin. For abbreviations used in local PAZ column see Fig. 14.5. *Alnus* pollen reduced 10 times

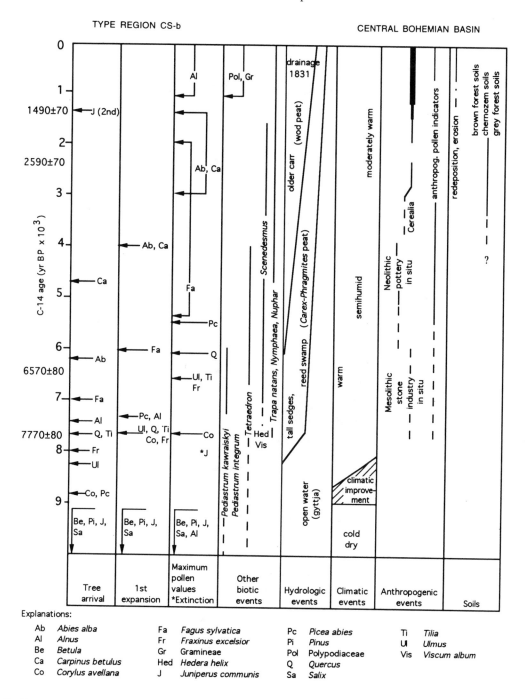

Fig. 14.5 Event stratigraphy table for the Central Bohemian Basin (CS-b), reference site "Komořanské jezero" (CS 1). After Jankovská (1983, 1984b, 1988b)

480

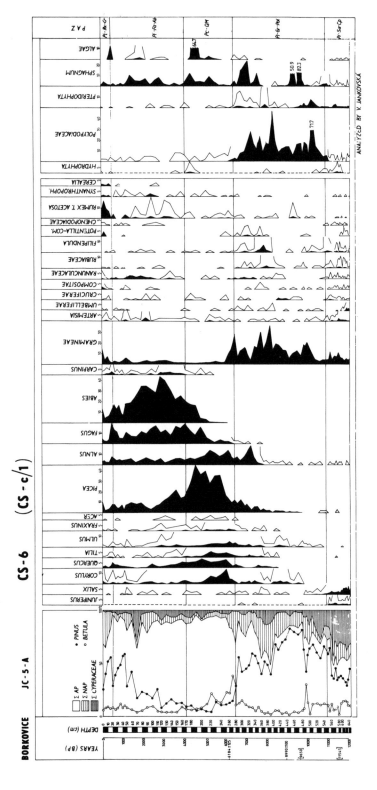

Fig. 14.6 Simplified pollen diagram from the reference site "Borkovice" (CS 6) of the type subregion Southbohemian Basins of the region Bohemian and Moravian Uplands (CS-c). For abbreviations see Fig. 14.7 and in addition: Cr – Cerealia, QM – Quercetum Mixtum, RX – *Rumex*

ANALYZED BY V. JANKOVSKÁ

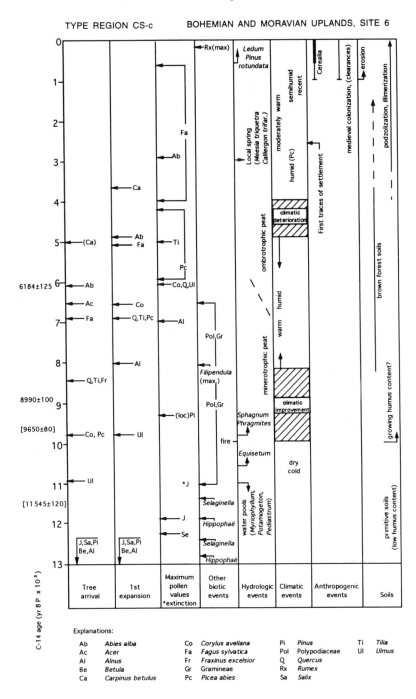

Fig. 14.7 Event stratigraphy table for the Southbohemian Basins of the Bohemian and Moravian Uplands, reference site "Borkovice" (CS 6). After Jankovská (1980)

observed. Low representation of *Corylus* is remarkable for BO and even AT. *Picea abies* seems to be a
dominant tree between 6000 and 4000 BP, and *Abies
alba* between 4000 and 1000 BP. The predominance
of coniferous trees at relatively low altitudes of the
Southbohemian Basins can be explained by higher
soil humidity and waterlogging of great areas. Acidophilic oak forest and acidophilic mixed beech
forests are reconstructed on dry-land soils since
6500 BP and 5000 BP respectively. Carpinion forests, usual at similar altitudes, did not develop at all
or did not develop as the main climax. *Carpinus
betulus* itself is absent in South Bohemia historically.

Reference site 10. Kameničky

The profile (Figs. 14.8 and 14.9) was exposed in a
spring mire at the village of Kameničky east of the
town of Hlinsko in northern part of the Českomoravská vrchovina Highlands. For full results of
investigation see Rybníčková & Rybníček (1988).
Latitude 49°44′N, Longitude 16°03′E. Elevation
625 m. Age range 11 500–0 years BP.

The accumulation of organogenic and organo-
minerogenic sediments started at the end of the AL.
There was open *Betula* carr with *Filipendula ulmaria*
dominating. A community of *Carex rostrata, C.
aquatilis, C. curta, Potentilla palustris*, and *Equisetum* prevailed in the PB and BO. Spruce carr covered the mire in the AT and remained until the SA-2.
Stands of low sedges (*Carex nigra, C. pilulifera,
Potentilla erecta, Peucedanum palustre, Eriophorum
angustifolium*), used for grazing, developed after the
deforestation and colonization in the 12th century.
The present vegetation is highly ruderalized.

Late-glacial and early Holocene pollen assemblages show the usual representation of *Pinus, Juniperus, Salix, Ephedra*, and Gramineae, with scattered pollen of *Corylus, Ulmus, Picea*, and *Alnus*. The
middle Holocene (AT, SB) was characterized by
Corylus, Picea abies, Ulmus, Tilia, and *Fraxinus.
Picea* pollen came partly from local sources. Since
4000 BP mixed *Fagus* forest with *Abies* and *Picea*
form the climax of the SA. *Abies* was dominant after
2500 BP. Intensive human impact (deforestation)
started about 750 BP. There is indication of grazing
activity since 500 BP (*Juniperus* reappearance).

TYPE REGION CS-d, SUDETY MOUNTAINS

The type region consists of the following major
orographic units: the Lužické hory Mountains, the
Jizerské hory and Krkonoše Mountains, the Orlické
hory Mountains, the Jeseníky Mountains, including
the Oderské vrchy Uplands and their foothills. The
region covers part of Poland.

Altitude: 350–700 m above sea level in the foothills,
600–1000 (1300) m in the mountains. Highest point
1602 m (Sněžka in the Krkonoše Mountains).

Climate: Cool and very humid in the mountains,
moderately warm and humid in the lower parts.
Mean annual temperature (0.2) 2–7 (8)°C, mean
annual precipitation (700) 800–1400 (1600) mm.

Geology: Proterozoic crystalline rocks prevail: granite, granodiorite, gneiss. Carboniferous rocks in the
Nízký Jeseník Mountains and the Oderské vrchy
Uplands. Permian and Cretaceous rocks in the foothills. Quaternary sediments (flood-loam, fluvial
sand, gravel) in the valleys, peat in the plateaux of
the Jizerské hory and Krkonoše Mountains. Geological conditions in general very complicated.

Topography: Mountain plateaux between mountain
ridges, block mountains. In the Krkonoše and Hrubý
Jeseník Mountains glacial cirques. Deep river valleys. Slightly undulated landscape in the foothills.

Population: 50–120 km^{-2} in the lower agricultural
areas, 10–40 km^{-2} in the forested mountains. Local
tourist overpopulation in mountain resorts.

Vegetation: *Picea* plantations are the main forest
type at present. Remnants of Carpinion communities in the foothills, subalpine vegetation (*Pinus
mugo* stands, dwarf-shrubs, subalpine meadows)
over *ca.* 1250–1300 m in the Krkonoše Mountains
and on the highest summits of the Jeseníky Mountains. Virgin vegetation: acidophilous oak and mesophilous Carpinion communities in the foothills (to
500 m), mixed *Fagus* forest (*Fagus, Picea, Abies*) in
the mountains to about 1100–1200 m, waterlogged
Picea forests on the mountain plateaux. *Pinus mugo*
stands exist over 1200 m and in the mires and even
lower in avalanche tracks and glacial cirques but are

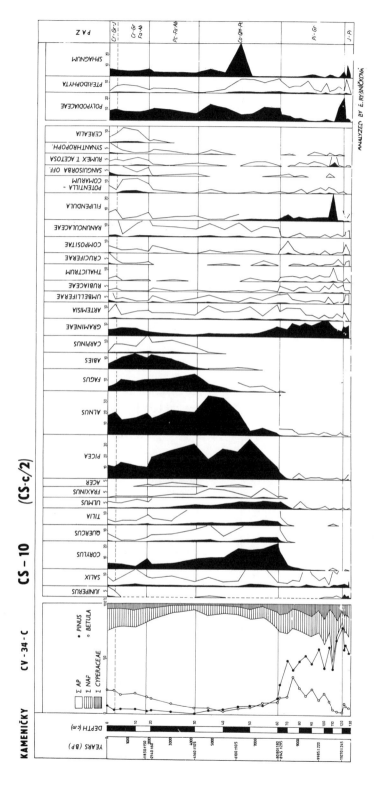

Fig. 14.8 Simplified pollen diagram from the reference site "Kameničky" (CS 10) of the type region Bohemian and Moravian Uplands (CS-c). For abbreviations see Fig. 14.9 and in addition; Cr – Cerealia, Gr – Gramineae

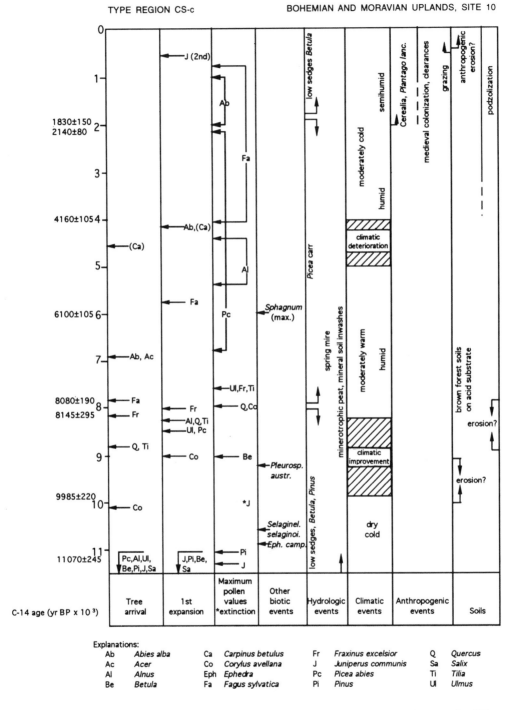

Fig. 14.9 Event stratigraphy table for the Bohemian and Moravian Uplands (CS-c), reference site "Kameničky" (CS 10). After Rybníčková & Rybníček (1988)

absent in the Jeseníky Mountains. Several relict and endemic plants in the Krkonoše Mountains and the Jeseníky Mountains, especially in their glacial cirques. Air pollution has caused catastrophic damage recently.

Soil types: Brown forest and podzol soils in the foothills and lower parts of the mountains, humic podzols in the mountain areas over 1000–1100 m. Organic soils in the Jizerské hory and Krkonoše Mountains. Screes in the glacial cirques and avalanche tracks.

Land use: Forest about 40–50% in the foothills, 70–80% in the mountains; mountains above the tree limit 5–15% (in the Krkonoše and Jeseníky Mountains only); cultivated land 20–50% (rye, oats, potatoes, wheat) in the foothills, less than 10–20% in the mountains (grazing places mainly); urbanized 5%; mires in the Jizerské hory and Krkonoše Mountains 1–5%.

Two reference sites from the Bohemian part of the region are presented. CS 11 Pančická louka Mire represents the highest ridges of the Sudety Mountains, and CS 12 Verněřovice reflects the conditions in the foothills and basins of the region. For the remaining part of the Sudety Mountains see the Polish reference site P 6 Zieleniec (750–760 m above sea level) by Madeyska (1989 and this volume).

Reference site 11. Pančická louka

The pollen diagram (Figs 14.10 and 14.11) comes from the Pančická louka mire in the western part of the Krkonoše Mountains. Full results of palaeoecological investigation were published by Hüttemann & Bortenschlager (1987). Additional information included by Rybníček (unpublished). Latitude 50°46′N, Longitude 15°32′E. Elevation 1325 m. Age range 7700–0 years BP.

The sediment accumulation began in the spring area of the Labe River with *Carex*–brown moss peat at the end of the BO and beginning of the AT-1. *Pinus* cf. *mugo* was present and formed a distinct wood layer between 6500 and 6000 BP and at about 5000 BP. Ombrotrophic *Sphagnum–Eriophorum vaginatum – Scirpus cespitosus* peat is intercalated between 2000 and 1500 BP.

The oldest pollen assemblage (end of the BO) indicates open grassland vegetation in the summit plateaux above 1200 m (?), with *Corylus* stands on the slopes below. *Pinus* (cf. *mugo* ?) occurred in mires. The *Pinus–Corylus*–QM pollen zone is dated to the AT. QM trees were distributed at lower altitudes. The *Picea abies* maximum corresponds to the AT-2 and lower part of the SB. *Abies alba–Fagus sylvatica* and *Picea abies* pollen zones cover the upper part of the SB and the whole SA-1. QM and *Carpinus* pollen are supposed to come from the foothills. According to the pollen diagram, *Fagus*, *Picea*, *Abies*, *Acer*, and *Fraxinus* could exist in krumholz form at the forest line, which was probably higher than today (over 1300 m) in SA-1. Human impact started about 600–700 years ago. In the summit areas it caused the decrease of the forest line (due to deforestation and summer grazing) and the spread of subalpine grasslands, heaths, and *Pinus mugo* stands (*Salix, Pinus* cf. *mugo*, *Calluna, Rumex*, Gramineae). The pollen diagram reflects also an introduction and extension of Medieval agriculture (cf. pollen of cereals) in the lower parts and foothills of the region.

Reference site 12. Verněřovice

The profile (Figs 14.12 and 14.13) was analysed by Peichlová (1979). It characterizes the lower altitudes of the region, namely the Broumovská kotlina Basin. Latitude 50°33′N, Longitude 16°21′E. Elevation 450 m. Age range (11800) 11500–0 years BP.

Combined valley and spring mire, maximum depth about 200 cm. Highly humified and drained secondary spruce forest and meadows cover the present surface. The minerogenic sedimentation started in oligotrophic *Carex–Sphagnum* communities in the AL. Organogenic sedimentation started under the transitional fen communities of *Betula, Phragmites*, and *Sphagnum* in the PB and BO. *Betula* carr developed in AT-1, *Betula* and *Picea abies* stands since AT-1. Humified organo-minerogenic sediments in the upper layers.

Pinus–Juniperus–Salix–Gramineae pollen zone corresponds to the late-glacial and early-Holocene periods. Representation of *Corylus, Picea abies, Alnus*, and *Acer* pollen is remarkable. The *Corylus–Tilia*–Gramineae–*Pinus* zone is dated to the BO.

486

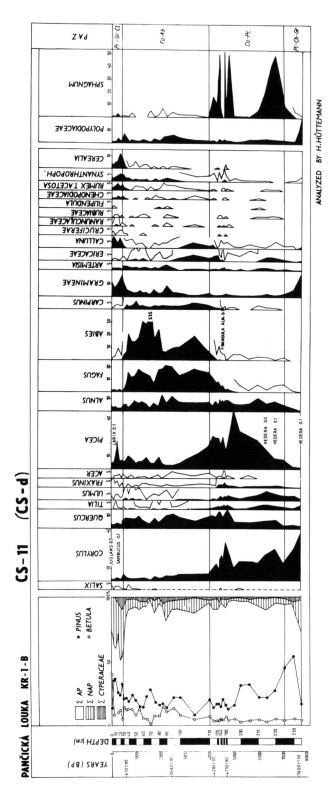

Fig. 14.10 Simplified pollen diagram from the reference site "Pančická louka" (CS 11) of the type region the Sudety Mountains (CS-d)–mountain subregion. For abbreviations see Fig. 14.11 and in addition: Gr – Gramineae

TYPE REGION CS-d   SUDETY MTS, SITE 11

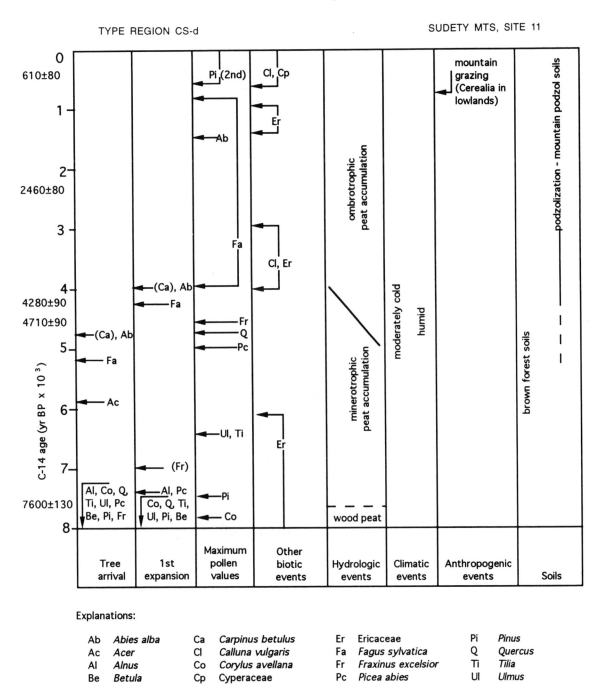

Fig. 14.11 Event stratigraphy table for the mountain parts of the Sudety Mountains type region (CS-d), reference site "Pančická louka" (CS 11). After Hüttemann & Bortenschlager (1987)

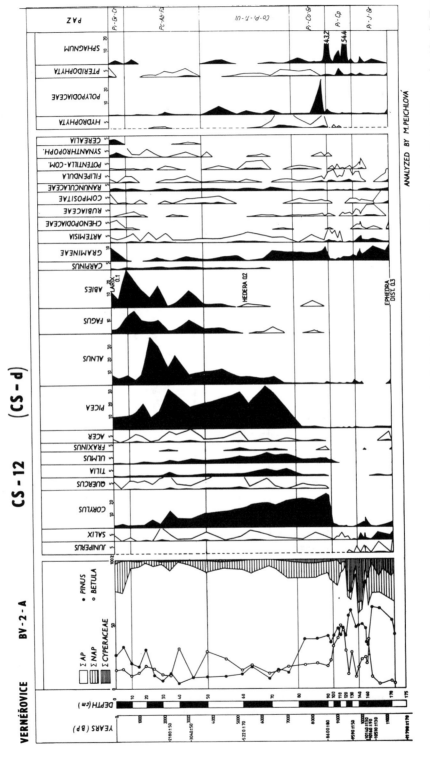

Fig. 14.12 Simplified pollen diagram from the reference site "Vernéřovice" (CS 12) of the lower parts of the type region Sudety Mountains (Cs-d). For abbreviations see Fig. 14.13 and in addition: Cr – Cerealia, Gr – Gramineae

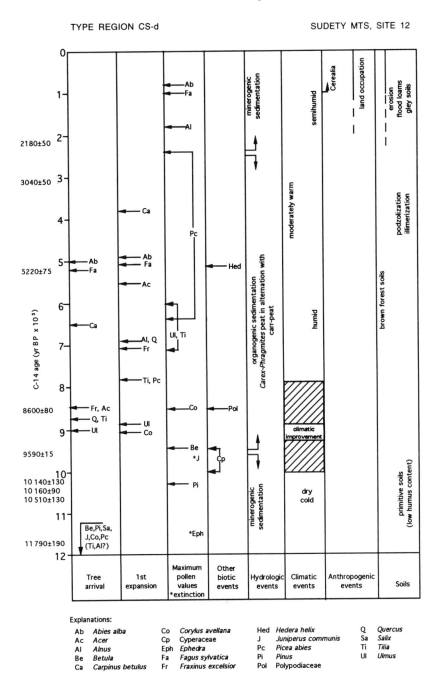

Fig. 14.13 Event stratigraphy table for the lower parts of the Sudety Mountains. type region (CS-d), reference site "Verněřovice" (CS 12). After Peichlová (1979)

Corylus–Tilia–Ulmus and *Picea* pollen zones characterize the middle Holocene periods (AT, SB). Low representation of *Quercus* is typical for the Broumovská kotlina Basin. *Abies–Fagus–Picea* pollen zone (*Abies* dominant) occurs since 3500 BP. The maximum of local alder pollen is dated to about 1500 BP. Human impact is indicated since about 750 BP (Medieval colonization), cf. pollen of Cerealia, synanthropic plants, decrease of AP. There is a continuous curve of *Carpinus* since 5500 BP (one of the oldest occurrences in the CS territory).

TYPE REGION CS-f, SOUTH MORAVIA

The region is a northwest extension of the Pannonian Lowlands, which has its centre in Hungary. It is connected with the Austrian part of the Lowlands and with southwest Slovakia. The Dyjskosvratecký úval Graben, Dolnomoravský úval Graben, and Záhorská nížina Lowlands with their neighbouring hills (including the Pavlovské vrchy Mountains) are the main orographic units of the region.

Altitudes: 150–300 (350) m above sea level. The highest point is 550 m in the Pavlovské vrchy Mountains.

Climate: Warm, dry, or moderately dry with mild winters. Mean annual temperature 8.5–9.5°C, mean annual precipitation 500–625 mm.

Geology: Miocene sediments (clay, marl, gravel, sand) covered with Quaternary loess and fluvial sediment (flood-loam) and in the southeast partly aeolian sands. Jurassic limestone in the Pavlovské vrchy Mountains.

Topography: Flat in broad valleys, slightly undulating in other parts. Limestone cliffs and karst features in the Pavlovské vrchy Mountains.

Population: About 200 km^{-2}.

Vegetation: Present landscape is deforested, and agricultural communities prevail. Xerothermic and thermophilic flora still persists in extreme habitats. Virgin vegetation: flood-plain forests in valleys, Carpinion or thermophilic oak forests in remaining parts of the region. *Quercus pubescens* communities on southern slopes. *Quercus–Pinus* forests on drift sands in the southeast part of the region.

Soil types: Alluvial soils, illimerized soils, chernozem soils (destroyed by wind and water erosion at present). Rendzinas and salt soils exceptional.

Land use: Forested land maximal 15% (*Pinus sylvestris, Robinia pseudacacia, Quercus, Carpinus*), cultivated land 80% (wheat, maize, vines, sugarbeet, vegetables, and fruits), urbanized about 5%. Other areas of negligible extent.

Three reference sites were studied in the region, and one of them CS 14 Vracov (Rybníčková & Rybníček 1972) is presented as representative here. For CS 13 Svatobořice-Mistřín see Svobodová (1989), for CS 15 Cerová-Lieskové see Krippel (1965).

Reference site 14. Vracov

The reference site (see Figs 14.14 and 14.15) is situated in the northeast part of the region near the towns of Kyjov and Bzenec. Latitude 48°58'N, Longitude 17°13'E. Elevation 190 m. Age range 12000–0 years BP.

Ancient lake. Algal gyttja of low humification in the late-glacial and early Holocene layers. Highly humified coarse-detritus gyttja with local intercalations of *Carex–Phragmites* peat in the middle Holocene layers. *Carex–Phragmites*–brown moss peats in the late Holocene. The infilling processes in the lake are described in detail by Rybníček (1983).

Pinus–Juniperus–Gramineae pollen zone corresponds to the late-glacial and early Holocene, with *Corylus–Quercus–Acer–Ulmus* and *Alnus* pollen present. *Corylus–Quercus–Ulmus–Tilia–Fraxinus* pollen zone appeared at the end of the BO and lasted for the whole middle Holocene. *Quercus–Carpinus* pollen zone characterizes the late Holocene. *Fagus* and *Abies* pollen come probably from the neighbouring marginal slopes of the West Carpathians. Traces of human impact (cereals, synanthropic plants, growing deforestation) since 6500 BP. Final deforestation and extension of agriculture at about 3000 BP.

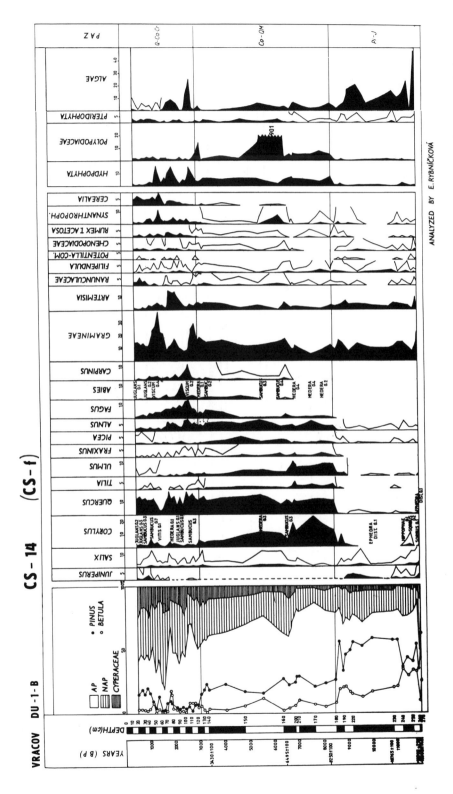

Fig. 14.14 Simplified pollen diagram from the reference site "Vracov" (CS 14) of the type region South Moravia (CS-f). For abbreviations see Fig. 14.15 and in addition: Cr – Cerealia, QM – Quercetum Mixtum

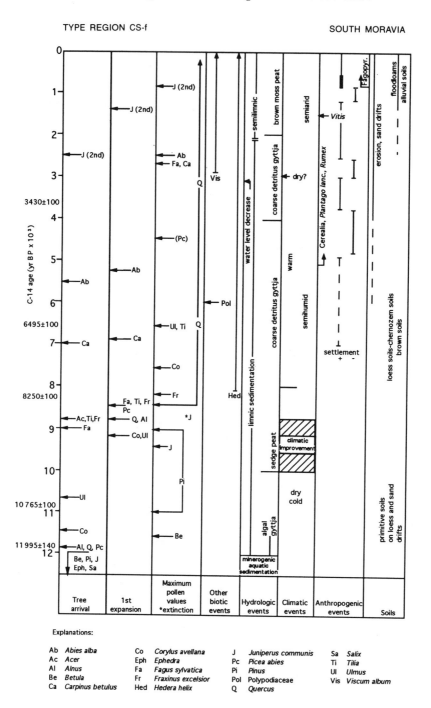

Fig. 14.15 Event stratigraphy table for the South Moravia (CS-f), reference site "Vracov" (CS 14). After Rybníčková & Rybníček (1972)

TYPE REGION CS-i, THE BESKYDY MOUNTAINS

About half of the type region belongs to Poland. The following main orographic units are included in the region from the CS side: the Moravskoslezské Beskydy Mountains, the Javorníky Mountains, and the Slovenské Beskydy Mountains in the western part and the Čerchov Mountains, Spišská Magura Mountains, and Levočské vrchy Mountains in its eastern part.

Altitudes: (300) 600–1000 (1500) m above sea level. The highest point is Babia hora Mountain (1725 m) in the Slovenské Beskydy Mountains.

Climate: Moderately warm and humid of highland character at lower altitudes, moderately cool and very humid or humid in the upper parts. Mean annual temperature 4–7°C, mean annual precipitation 750–900 mm (low altitude), 900–1400 (1550) mm (high altitude).

Geology: Palaeogene and Cretaceous flysch, sand- and claystone, with Cretaceous limestone klippe present. Quaternary sediments in valleys only (fluvial sands and loams), peat deposits very rare.

Topography: Several rounded mountain ridges separated and cut through by deep valleys. Foothills undulating, limestone cliffs in klippe areas.

Population: Mountain areas about 40 km⁻², foothills and valleys about 80 km⁻².

Vegetation: Spruce plantations prevail at present. Mountain pasture grasslands, agricultural communities in valleys only. Virgin vegetation: Carpinion-forests to about 600 m, mixed beech forests (*Abies alba, Picea abies*) to about 1100 m. Mountain spruce forests in limited areas 1100–1200 m only; *Pinus mugo* stands on Pilsko and Babia hora Mountains.

Soil types: Brown forest soils, mountain podzol soils, small areas of rendzinas on limestones. Grey forest soils in the Čerchov Mountains. Water erosion and landslides widespread in flysch areas.

Land use: Forest over 45% (*Picea abies* plantations, *Abies alba* still important, *Fagus* locally), cultivated land (rye, oats, potatoes, until recently *Fagopyrum*) less than 3–5%, pasture land about 15%, urbanized 3–5% (at lower altitudes). Mires small and rare.

One reference site CS 17 by Rybníčková and Rybníček (1989a) is available.

Reference site 17. Zlatnická dolina

The site is situated in the western part of the region in the Slovenské Beskydy Mountains. Simplified results are presented in Figures 14.16 and 14.17. Latitude 49°29′N, Longitude 19°16′E. Elevation 900 m. Age range 8200–0 years BP.

Small spring mire (about 6 ha) in the valley of the Zlatnický potok brook. Long minerogenic sedimentation phase since the BO to 4300 BP with *Carex, Equisetum*, and Polypodiaceae stands. Reed swamps open the organogenic phase of sedimentation between 3500 and 4000 BP (*Carex–Phragmites* peat). Poor *Carex*-fen with *Carex rostrata* and *Sphagnum* sect. *Cuspidata* between 2750 and 3500 BP. Dry period (?) with willow and birch stands between 2750 and 2500 BP (cf. corresponding wood layer). Extremely poor fen with *Eriophorum vaginatum, Carex* cf. *nigra, Carex curta, Sphagnum* cf. *recurvum*, and *S. medium. Picea abies* forms a transition to the ombrotrophic bog since about 2500 BP.

Corylus–Ulmus–Picea abies pollen zone represents the vegetation at the end of the BO. Middle Holocene vegetation is reflected by *Picea abies–*QM–*Corylus* pollen zone. Mountain mixed-deciduous forest (*Ulmus, Tilia, Fraxinus, Acer, Picea abies*, and *Corylus avellana*, and at lower altitudes also *Quercus*) was probably the main vegetational type of the AT and SB. *Picea* stands can be expected on northern slopes and in waterlogged habitats. In valleys are *Picea* and *Alnus* cf. *incana. Picea abies–Fagus sylvatica–Abies alba–Carpinus* pollen zone characterizes the late Holocene (SA-2). Mixed *Fagus–Abies* forests dominated, and *Abies–Picea* forests formed the second main type. Carpinion-communities in the lowest parts of the region. Human impact (the Wallachian colonization with its mountain grazing) is reflected by a decrease of climax tree pollen, increase of Gramineae, appearance of synanthropic pollen and re-occurrence of *Juni-*

494

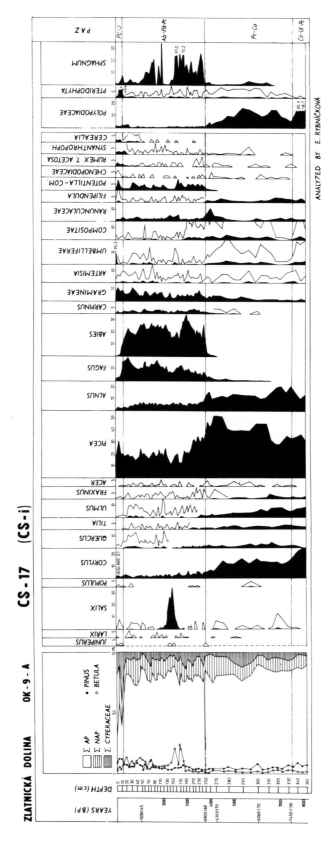

Fig. 14.16 Simplified pollen diagram from the reference site "Zlatnická dolina" (CS 17) of the type region the Beskydy Mountains (CS-i). For abbreviations see Fig. 14.17

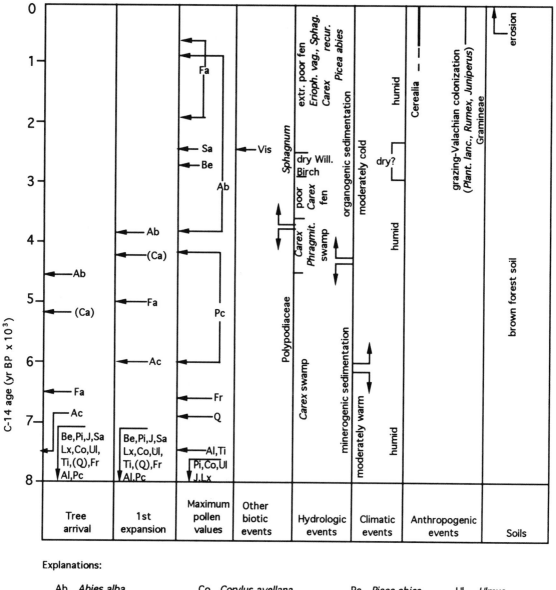

Fig. 14.17 Event stratigraphy table for the western part of the Beskydy Mountains type region (CS-i), reference site "Zlatnická dolina" (CS 17). After Rybníčkova & Rybníček (1989a)

perus at about 500 BP (AD 16th century). Traces of agricultural activity (cereals) in neighbouring lowlands and hills since 1000 BP.

TYPE REGION CS-j, THE INNER WEST CARPATHIANS

Two subregions were delimited within the region: A) The subregion of intermontane basins includes the Oravská kotlina, Turčianská kotlina, Liptovská kotlina, and Popradská kotlina Basins. B) The mountain subregion consists of the highest orographic units of the West Carpathians, namely the Vysoké Tatry Mountains, the Nízké Tatry Mountains, the Velká and Malá Fatra Mountains. The Oravská Magura and Choč Mountains are included.

Altitudes: (350) 500–900 m above sea level in the basins, 900–2300 m in the mountain parts. The highest point Gerlach 2655 m in the Vysoké Tatry Mountains. The region has the greatest topographic relief in the whole of the former Czechoslovakia.

Climate: Moderately cold, humid in the western and moderately humid in the eastern basins. Cold and very humid in the high mountains. Thermal inversion common in the intermontane basins represents a very important climatic feature. Mean annual temperature between 0 and 5°C, at the highest altitudes less than –3°C. Mean annual precipitation (700) 800–1400 (2000) mm.

Geology: Granite, granodiorite, Mesozoic limestones or dolomites, Palaeogene flysch facies (sandstones or claystones). Fluvioglacial sediments in the basins, travertine, scattered organogenic deposits.

Topography: Glacial mountain landscape (U-valleys, glacial cirques, moraines, glacial lakes) in the highest parts. In lower parts deeply dissected mountain relief with several ridges and limestone cliffs in the klippen zone. Intermontane basins flat or slightly undulating.

Population: About 100 km⁻² in the basin, less than 20 km⁻² in the mountains. Tourist overpopulation, especially in the Vysoké Tatry Mountains.

Vegetation: Agricultural communities (fields, pasture land) in the basins, secondary forest vegetation (*Picea, Pinus*) in small areas only. In the mountains mixed *Fagus* forests (*Picea abies, Fagus sylvatica, Abies alba*) or *Abies* forests (*Abies alba, Picea abies*) reaching to about 1100–1300 m. Most of the woodland is changed into spruce plantations. Mountain climax *Picea* forests over 900–1300 m in highest mountains only. *Pinus mugo* communities over 1500–1700 m, with *Pinus cembra* and *Larix* in the Vysoké Tatry Mountains. Alpine vegetation above the tree line in alpine meadows, on screes, rocks, avalanche tracks, etc. Virgin vegetation: coniferous forests (*Picea, Abies*) in the western basins, *Picea* and *Pinus* in the eastern basins, in other mountain parts forest belts similar to present. Relict *Pinus* forests on limestone cliffs. Flora rich in species, with several endemic plants. In the Poprad basin scattered xerothermic vegetation appears. Rich fen flora and vegetation in mires.

Soil types: Brown forest soils and podzols, mountain humic podzols, subalpine soils, rendzinas on limestone.

Land use: Basins: forests less than 10%, cultivated land about 80–90% (including pasture land 20–40%) with rye, wheat, potatoes, oats; urbanized about 5%, mires 1%. Mountain areas: forest 50–60%, cultivated land 20–30% (pasture land 10–20%), mountains above the tree limit 10–15%, urbanized 1%, mires occur exceptionally.

Six reference sites were studied in the region. Five of them are representative for the basins: CS 18 Bobrov (Rybníčková & Rybníček, 1989b; CS 19 Ivančiná (Krippel 1974); CS 20 Liptovský Ján (Dolejšová, unpublished); CS 21 Hozelec (Jankovská 1988a); CS 22 Sivár,a, (Jankovská 1984a; see also Obidowicz, 1989). One reference site, CS 23 Trojrohé pleso, should be representative for the highest West Carpathians (Hüttemann & Bortenschlager 1987). In this survey only two sites, one for each subregion, are presented.

Reference site 18. Bobrov

A summary of results of investigations by Rybníčkova & Rybníček is published here (see Figs

14.18 and 14.19). Publication of the full results is in preparation. Latitude 49°27′N, Longitude 19°34′E. Elevation 640 m. Age range 10800–0 BP.

The profile comes from a small calcitrophic spring mire (size about 3 ha) situated in the Oravská kotlina Basin. The mire development started with minerogenic sedimentation at the end of the late-glacial. Brown-moss fen vegetation formed brown-moss peat at the beginning of the early Holocene. Low sedge communities with scattered trees of *Betula* and *Picea abies*, forming a transitional fen peat, characterize a long period of mire growth between about 9800 and 4000 BP. Brown-moss fen communities covered the site from the SB until present. The mire ceased to exist after recent exploitation.

Pinus (including *Pinus cembra* and *P.* cf. *mugo*)–*Juniperus*–*Larix* pollen zone is dated to the late-glacial and the PB. Open mixed coniferous taiga-like forests were the main vegetation type in all intermontane basins of the West Carpathians (see also finds of seeds of *Pinus cembra* and *Larix* by Jankovská 1984a in the Poprad Basin in the same time). Heliophilic conifers (*Larix, Pinus cembra, Juniperus*) retreated at about 9500 BP. *Picea abies*–*Corylus*–*Ulmus* pollen zone is dated to the BO. *Picea*–*Abies* pollen zone dominated all remaining periods of the Holocene. *Alnus* became an important tree about 4000 BP. *Abies* and especially *Fagus* never played any great role in the lowest parts of all basins. Human impact started at about 500 BP in the Oravská kotlina Basin and is reflected by a decrease of tree pollen (*Picea, Alnus*), reappearance of *Juniperus* (Wallachian type of grazing), and increase of Gramineae, Cyperaceae, synanthrops, and cereal pollen. Intermontane basins were very probably isolated glacial refugia for *Picea abies* and *Alnus* cf. *incana*–see the continuous pollen curves.

Reference site 23. Trojrohé pleso

The subalpine mire at Trojrohé pleso Lake is situated at the tree limit in the eastern part of the Vysoké Tatry Mountains. For full results of investigations see Hüttemann & Bortenschlager (1987) and complementary information by Rybníček (unpublished). See Figs 14.20 and 14.21. Latitude 49°14′N, Longitude 20°15′E. Elevation 1650 m. Age range 6250–0 BP.

The mire originated in a small morainic depression. Minerotrophic peat accumulation (*Carex curta, C. nigra, Drepanocladus exannulatus, Sphagnum* sect. *Cuspidata, S.* sect. *Acutifolia*) was followed by ombrotrophic peat formation at about 4500 BP. *Eriophorum vaginatum, Empetrum hermaphroditum*, roots and twigs of Ericaceae, *Sphagnum compactum, S. medium, S. fuscum*, and *Pinus mugo* are the main components. The current Trojrohé pleso Lake is probably a result of surface peat erosion and frost and is in fact a large bog pool. No traces of lake sediments were observed in or under the peat layers.

The pollen diagram reveals a typical "mountain effect", which influences the composition of the pollen assemblage, with pollen from distant lowlands. When interpreting the diagram we must therefore try to distinguish pollen types of probable local and extralocal origin from those of distant transport. In the highest forest belts we can expect *Picea abies, Larix, Acer, Salix*, and, between 6200 and 5500 BP, possibly also *Corylus. Pinus* pollen is probably represented mostly by *Pinus mugo* from the habitats above the forest line, i.e. over about 1500 m. QM trees and *Carpinus* could never have grown so high on alpine soils in rocks and screes of the Vysoké Tatry Mountains, and their pollen represents the vegetation of the surrounding lowlands and hills. The *Picea*-*Fagus*-*Abies* pollen zone is dated between about 3500 and 750 BP, however, so even here we have to expect *Fagus* and *Abies* pollen transport from lower altitudes. Human activity started at about 500–600 BP. It lowered the forest line, and mountain pastures were established (cf. increase of Gramineae, *Rumex*, and other synanthrops). Pollen grains of cereals, however, certainly came from the lowlands.

TYPE REGION CS-m, BIESZCZADY MOUNTAINS AND VIHORLAT MOUNTAINS

Only a small part of the region lies in the CS territory (the Vihorlat Mountains and the Bukovské hory Mountains); most of it belongs to Ukraine (the Poloniny Mountains) and to Poland (the Bieszczady Mountains). The data from Hypka, a mire (CS 24) by Krippel (1971) can be used as representative for

498

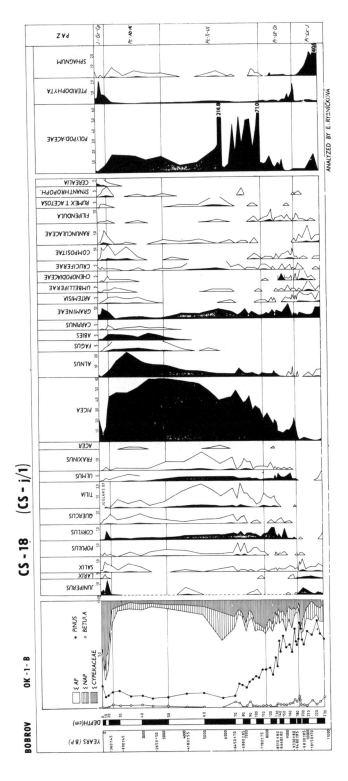

Fig. 14.18 Simplified pollen diagram from the reference site "Bobrov" (CS 18) of the type subregion Intermontane Basins of the Inner West Carpathians (CS-j). For abbreviations see Fig. 14.19 and in addition: Cp – Cyperaceae, Gr – Gramineae.

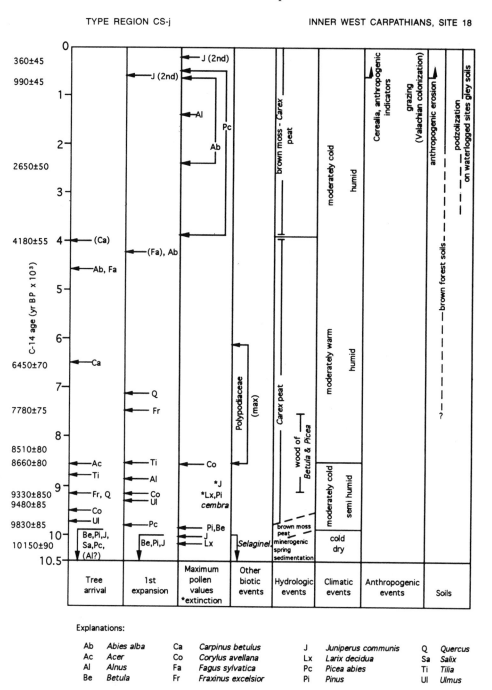

Fig. 14.19 Event stratigraphy table for the Intermontane Basins of the Inner West Carpathians (CS-j), reference site "Bobrov" (CS 18). After Rybníčkova & Rybníček (1989b)

Vihorlat Mountains, but correlation with radiocarbon-dated diagrams by Ralska-Jasiewiczowa (1980, 1989) from the Bieszczady Mountains is necessary.

TYPE REGION CS-n, THE TISZA LOWLANDS

The main part of the type region belongs to Hungary. One reference site (CS 25 Rad, no radiocarbon dating) was studied by Krippel (unpublished) from the Východoslovenská nížina Lowlands.

CONCLUSIONS

The results of palaeoecological and palaeogeobotanical investigations of 25 reference mire and (ancient) lake sites in the Czech and Slovak Republics included in the IGCP 158 B project are summarized from the following viewpoints: (l) tree immigration, tree spreading, and vegetational changes, (2) climatic and hydrologic changes, (3) soil development, and (4) anthropogenic changes in a landscape.

(l) Tree immigration never progressed in only one direction in CS territory. Because of the great environmental diversity of both countries several migration centres and routes are to be expected. This is the main reason for relatively long time ranges of tree arrivals, expansion, and maximum occurrences. The mean time spans of tree appearances are summarized in Figure 14.22. Consequently the development of past vegetation types was not at all synchronous over the whole territory.

(2) Great climatic variability as in the present Czech and Slovak territories must also have existed in the past. This caused difficulties in finding synchronous broad-scale climatic and hydrologic variation through palynological/palaeogeobotanical methods only. Thus the reflection of local (extralocal) climatic and hydrologic conditions prevails. From biological indications, general climatic improvement (rise of temperature and humidity) can be dated between 10000 and 8000 BP and climatic deterioration (fall of temperature first of all) between 4500 and 4000 BP. A

wide distribution of conifers (*Picea*, later also *Abies alba*) and *Alnus* especially in SB and SA also in lower altitudes indicates the existence of specific local hydrological conditions, including high air and soil humidity or even waterlogging of many areas, especially in the basins. Therefore we have to consider specific water regimes, i.e. the retardation of water outflow and consequent natural accumulation of water probably in all CS basins. Only one more or less synchronous short period of drought can be observed in three reference sites between about 2750 and 2500 BP, as reflected by the water-level decrease in the ancient lake at Vracov (CS 14) around 2600 BP, by the end of organogenic sedimentation in the profile Verněřovice (CS 12) between 2600 and 2500 BP, and by the drying up of the bog surface in the Zlatnická dolina (CS 17) between 2750 and 2500 BP (see Figs 14.13, 14.15 and 14.17). The statistical significance of the phenomenon, however, is weak and needs further investigation.

(3) Pedogenic changes can be reconstructed from bioindicators only. The existence of brown forest soils should be connected with the existence of mesophytic mixed deciduous forests (since the AT). Similarly, the beginning of podzolization, connected with higher relative and absolute humidity, can be correlated with the extension of *Picea* stands in the SB.

(4) The beginning of human impact is dated between the Neolithic (in lowlands) and 16–17th century in the Carpathian Mountains, when the present landscape was in general completed. Grazing and cultivation of arable fields were of similar importance in the lowlands, with grazing more extensive in the mountains. Irregular floods, soil erosion in the upper parts of river basins, and accumulation of thick layers of loam in broad floodplains are very important results of extensive deforestation during the medieval colonization. Secondary spreading of plants and shifts in limits between vertical vegetational belts are also results of deforestation and succession of anthropogenic habitats, revealing new specific mesoclimatic and local hydrologic conditions.

501

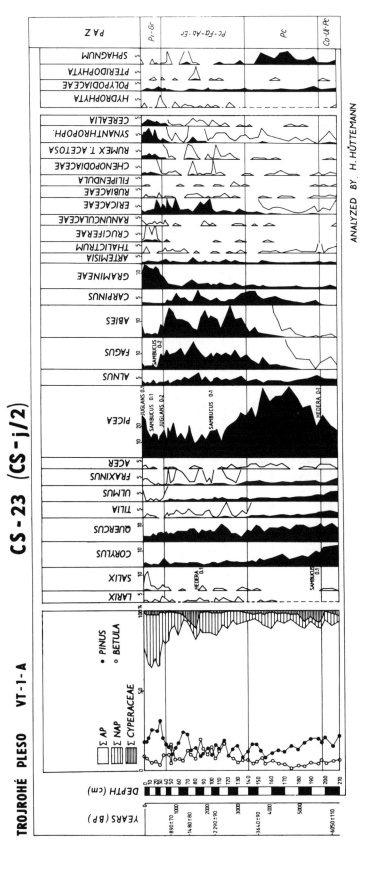

Fig. 14.20 Simplified pollen diagram from the reference site "Trojrohé pleso" (CS 23) of the mountain parts of the type region Inner West Carpathians (CS-j). For abbreviations see Fig. 14.21 and in addition: Er – Ericaceae, Gr – Gramineae

TYPE REGION CS-j   INNER WEST CARPATHIANS, SITE 23

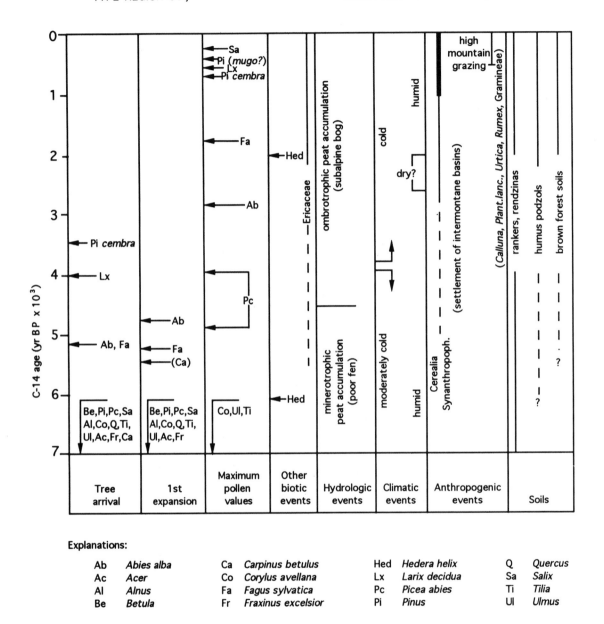

Explanations:

Ab	*Abies alba*	Ca	*Carpinus betulus*	Hed	*Hedera helix*	Q	*Quercus*
Ac	*Acer*	Co	*Corylus avellana*	Lx	*Larix decidua*	Sa	*Salix*
Al	*Alnus*	Fa	*Fagus sylvatica*	Pc	*Picea abies*	Ti	*Tilia*
Be	*Betula*	Fr	*Fraxinus excelsior*	Pi	*Pinus*	Ul	*Ulmus*

Fig 14.21   Event stratigraphy table for the mountain parts of the Inner West Carpathians (CS-j), reference site "Trojrohé pleso" (CS 23). After Hüttemann & Bortenschlager (1987)

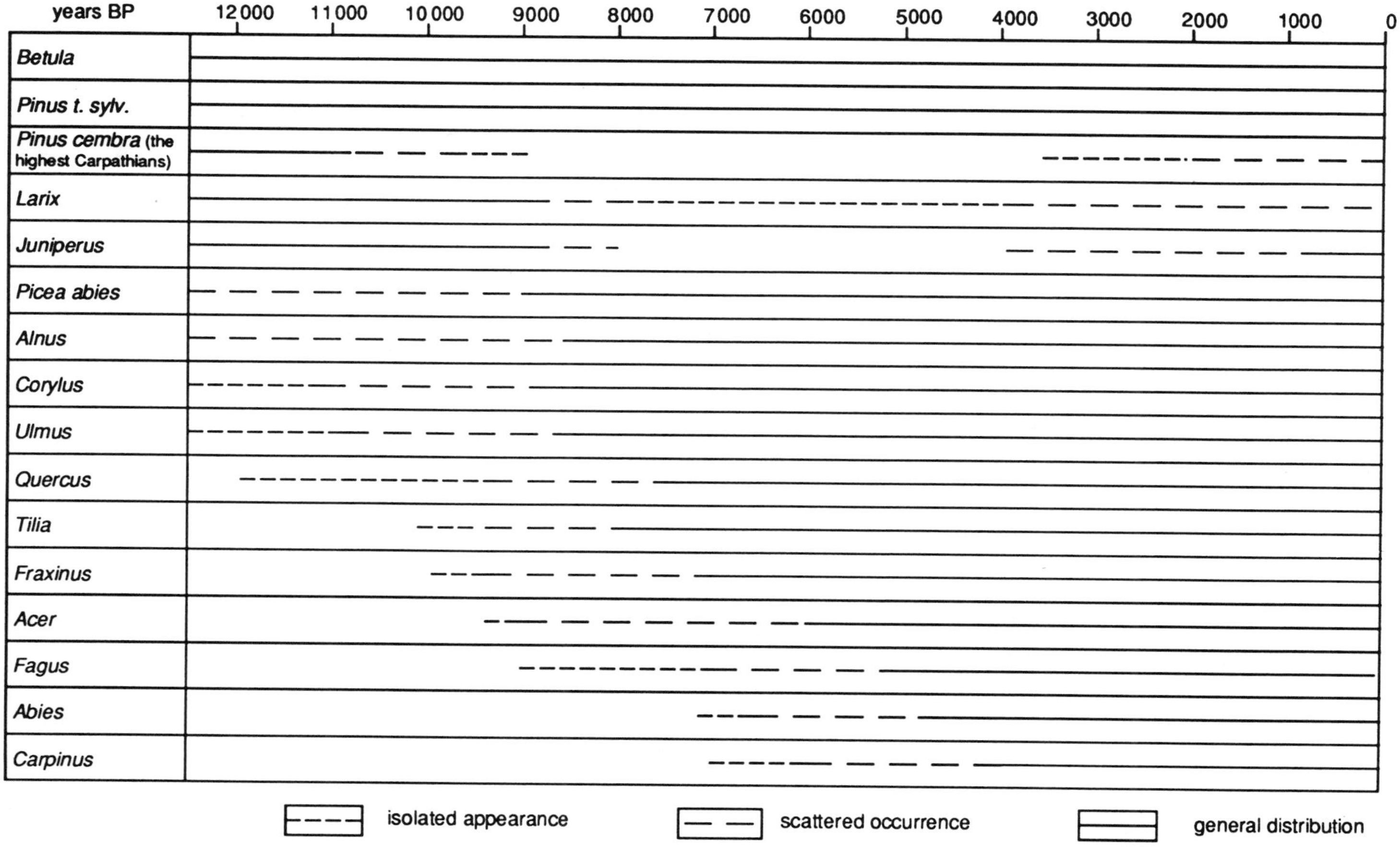

Fig. 14.22 Immigration and time distribution of major trees in the territory of Czech and Slovak Republics. Additional reference sites and other pollen diagrams have also been used for the construction of the histogram

ACKNOWLEDGEMENTS

We are indebted to the heads and staffs of the following C-14 Laboratories, who dated the profiles presented: Lund, Hannover, Wisconsin, Wien, Berlin. Our gratitude also goes to the international project leader Professor Björn E. Berglund (University of Lund) for his continuous interest, advice, and support.

REFERENCES

Berglund, B.E. (ed.) 1986: *Handbook of Holocene palaeoecology and palaeohydrology.* 869 pp. Wiley, Chichester.

Czudek, T. (ed.) 1972: Geomorfologické členění ČSR. *Studia Geographica 23,* 5–95. Brno.

Demek, J. & Střída, M. (eds) 1971: *Geography of Czechoslovakia.* 330 pp. Academia, Praha.

Götz, A. (ed.) 1966: *Atlas Československé socialistické republiky.* 58 sheets, Ústřední Správa Geodesie a Kartografie, Praha.

Hejný, S. & Slavík, B. (eds) 1988: *Květena České socialistické republiky.* 557 pp. Academia, Praha.

Hüttemann, H. & Bortenschlager, S. 1987: Beiträge zur Vegetationsgeschichte Tirols VI: Riesengebirge, Hohe Tatra–Zillertal, Kühtai. Ein Vergleich der postglazialen Vegetationsentwicklung und Waldgrenzeschwankungen. *Berichte des Naturwissenschaftlich-medizinischen Vereins in Innsbruck 74,* 81–112.

Jankovská, V. 1980: *Paläogeobotanische Rekonstruktion der Vegetationsentwicklung im Becken Třebo,ská pánev während des Spätglazials und Holozäns.* 151 pp. Vegetace ČSSR A 11. Academia, Praha.

Jankovská, V. 1983: Palynologische Forschung am ehemaligen Komořany See (Spätglacial bis Subatlantikum). *Věstník Ústředního Ústavu Geologického 58(2),* 99–107. Praha.

Jankovská, V. 1984a: Late Glacial finds of *Pinus cembra* L. in the Lubovnianská kotlina Basin. *Folia Geobotanica et Phytotaxonomica 19,* 323–325. Praha.

Jankovská V. 1984b: Radiokarbondatierung der Sedimente aus dem ehemaligen Komořany-See (NW Böhmen). *Věstník Ústředního Ústavu Geologického 59,* 325–326. Praha.

Jankovská, V. 1987: Entwicklung des Moores Mokré louky bei Třebo, im Postglazial. (Paläoökologische Studie). *Folia Geobotanica et Phytotaxonomica 22,* 199–216. Praha.

Jankovská, V. 1988a: A reconstruction of the Late-Glacial and Early-Holocene evolution of forest vegetation in the Poprad Basin. *Folia Geobotanica et Phytotaxonomica 23,* 303–319. Praha.

Jankovská, V. 1988b: Palynologische Erforschung archäologischer Proben aus dem Komořanské jezeroSee bei Most (NW Böhmen). *Folia Geobotanica et Phytotaxonomica 23,* 45–77. Praha.

Krippel, E. 1963: Postglaziale Entwicklung der Vegetation des nördlichen Teils der Donauebene. *Biológia 18,* 730–742. Bratislava.

Krippel, E. 1965: Postglaciálny vývoj lesov Záhorskej nížiny. (Historicko-geobotanická štúdia). *Biologické Práce 11(3),* 1–100. Bratislava.

Krippel, E. 1971: Postglaciálny vývoj vegetácie východného Slovenska. *Geografický Časopis 23,* 225–241. Bratislava.

Krippel, E. 1974: Rekonštrukcia rostlinnej pokrývky Turčianskej kotliny na základe pelovej analýzy. *Geografický Časopis 26,* 42–53. Bratislava.

Madeyska, E. 1989: Type region P-f: Sudetes Mts.–Bystrzyckie Mts. *Acta Palaeobotanica, 29(2),* 37–41. Warszawa.

Michalko, J. (ed.) 1987: *Geobotanical map of ČSR. Part 2 Slovak Socialist Republic.* 167 pp. Veda, Bratislava.

Mikyška, R. (ed.) 1968: *Geobotanische Karte der Tschechoslowakei. 1. Böhmische Länder.* 204 pp. Vegetace ČSSR A 2. Academia, Praha.

Obidowicz, A. (1989): Type region P-a: Inner West Carpathians–Nowy Targ Basin. *Acta Palaeobotanica 29(2):* 11–15. Warszawa.

Peichlová, M. 1979: *Historie vegetace Broumovska.* Dissertation, Botanical Institute of the Academy of Sciences, Průhonice, Brno, 122 pp. (manuscript).

Pelíšek, J. 1966: *Výšková půdní pásmitost střední Evropy.* 366 pp. Academia, Praha.

Ralska-Jasiewiczowa, M. 1980: *Late-Glacial and Holocene vegetation of the Bieszczady Mts (Polish Eastern Carpathians).* 202 pp. Warszawa.

Ralska-Jasiewiczowa, M. 1989: Type region P-e: The Bieszczady Mts. *Acta Palaeobotanica 29(2),* 31–35. Warszawa.

Rybníček, K. 1983: The environmental evolution and infilling process of a former lake near Vracov (Czechoslovakia). *Hydrobiologia 103,* 247–250. The Hague.

Rybníček, K. & Rybníčková, E. 1968: The history of flora and vegetation in the Bláto-mire in the Southeastern Bohemia, Czechoslovakia. *Folia Geobotanica et Phytotaxonomica 3,* 117–142. Praha.

Rybníček, K. & Rybníčková, E. 1987: Palaeogeobotanical evidences of middle Holocene stratigraphic hiatuses in Czechoslovakia and their explanation. *Folia Geobotanica et Phytotaxonomica 22,* 313–327, Praha.

Rybníček, K. & Rybníčková, E. 1991: Methodological comments on the Czechoslovak national syntheses of the IGCP-158 B. *Quaternary Studies in Poland 10,* 95–101. Pozna, .

Rybníčková, E. 1974: *Die Entwicklung der Vegetation und Flora im südlichen Teil der Böhmisch-Mährischen Höhe während des Spätglazials und Holozäns.* 163 pp. Vegetace ČSSR A 7. Academia, Praha.

Rybníčková, E. 1982: Absolute C-14 dates of the profile from the Zbudovská blata Marshes (southern Bohemia). *Folia Geobotanica et Phytotaxonomica 17,* 99–100. Praha.

Rybníčková, E. & Rybníček, K. 1972: Erste Ergebnisse paläogeobotanischer Untersuchungen des Moores bei Vracov, Südmähren. *Folia Geobotanica et Phytotaxonomica 7*, 285–308. Praha.

Rybníčková, E. & Rybníček, K. 1985: Palaeogeobotanical evaluation of the Holocene profile from the Řežabinec Fish-pond. *Folia Geobotanica et Phytotaxonomica 20*, 419–437. Praha.

Rybníčková, E. & Rybníček, K. 1988: Holocene palaeovegetation and palaeoenvironment of the Kameničská kotlina Basin (Czechoslovakia). *Folia Geobotanica et Phytotaxonomica 23*, 285–302, Praha.

Rybníčková, E. & Rybníček, K. 1989a: The vegetation development of the Slovenské Beskydy Mts. *Excursion Guide Book of the 12th International Meeting of European Quaternary Botanists, Czechoslovakia*, pp. 118–120, Brno.

Rybníčková, E. & Rybníček, K 1989b: The Holocene development of vegetation in the Oravská kotlina Basin. *Excursion Guide Book of the 12th International Meeting of European Quaternary Botanists, Czechoslovakia*, pp. 115–116, Brno.

Rybníčková, E., Rybníček, K. & Jankovská V. 1975: Palaeoecological investigations of buried peat profiles from the Zbudovská blata Marshes, south Bohemia. *Folia Geobotanica et Phytotaxonomica 10*, 157–178, Praha.

Svobodová, H. 1989: Rekonstrukce přírodního prostředí a osídlení v okolí Mistřína. (Palynologická studie). *Památky Archeologické 80*, 188–206. Praha.

Veselský, A. *et al.* (eds) 1958: *Atlas podnebí Československé Republiky—Klimaatlas der ČSR*. 100 map sheets. Ústřední Správa Geodesie a Kartografie, Praha.

15

Germany

K.-E. Behre, A. Brande, H. Küster and M. Rösch

INTRODUCTION
(K.-E. Behre)

There has been no national project for IGCP 158 B in Germany because all potential contributors were at that time involved in other long-running projects of their own. However, several colleagues have been asked to summarize the present knowledge of certain type regions and to arrange it as far as possible according to the IGCP guidelines. The contributors are H. Küster for the type region D-o, M. Rösch for D-n, D-l, and D-r, A. Brande for D-s and K.-E. Behre for D-d and introduction.

The contributions given below present only a few of the type regions within Germany (see Fig. 15.1) and do not represent the great variety of different natural regions, which extend northward from the Alps and their foreland across uplands and lowlands to the marine clay district bordering the North Sea. Great differences in the history of vegetation and climate as well as in the development of bogs and mires have occurred during the last 15000 years within Germany.

The following regional syntheses may give an impression of the main events that happened in this part of Europe. As these regions represent the most distant ones in Germany, i.e. the northwestern and eastern lowlands and the Alpine foreland in the south, these contributions may serve as reasonable examples for the whole country. All publications up to now are documented in the "Vegetationsgeschichtliche Kartierung Göttingen" (Beug & Henrion 1971).

Germany comprises a great variety of landscapes (see Fig. 15.2). In the southernmost part it shares the northern Limestone Alps with altitudes up to almost 3000 m asl. To the north is the Alpine Foreland with its young morainic (Würmian) landscape and then old morainic (pre-Würmian) and Tertiary sediments. North of the Danube River are many different upland areas, mainly consisting of Triassic and Jurassic sediments. Only a few crystalline mountain areas exist such as the Black Forest in the west and the Bavarian and the Bohemian Forest and the Erzgebirge in the east, the latter two representing the margins of the Bohemian Massif. The Central German Upland (Mittelgebirge) extends north as far as the Harz and the ridges south of Hannover and in Westphalia. Here also Cretaceous sediments are common. To the west is the wide tectonic graben of the Upper Rhine, which extends from Basel to Mainz and is characterized by a particularly warm climate. To the northwest the uplands of the Rheinisches Schiefergebirge, which stretches into Belgium and France, are formed by Palaeozoic rocks.

In several places in the upland areas Tertiary volcanic rocks are present. Most impressive is the huge massif of the Vogelsberg northeast of Frankfurt. Other areas, particularly the graben valleys of the Upper Rhine and the Wetterau, are covered by loess of different ages. The pattern of different upland soil types results in a mosaic of various biotic zones as well.

Quite different from the elevated central and southern parts of Germany are the North German Lowlands (Norddeutsche Tiefebene), a landscape

Palaeoecological Events During the Last 15000 Years: Regional Syntheses of Palaeoecological Studies of Lakes and Mires in Europe.
Edited by B.E. Berglund, H.J.B. Birks, M. Ralska-Jasiewiczowa and H.E. Wright. © 1996 John Wiley & Sons Ltd.

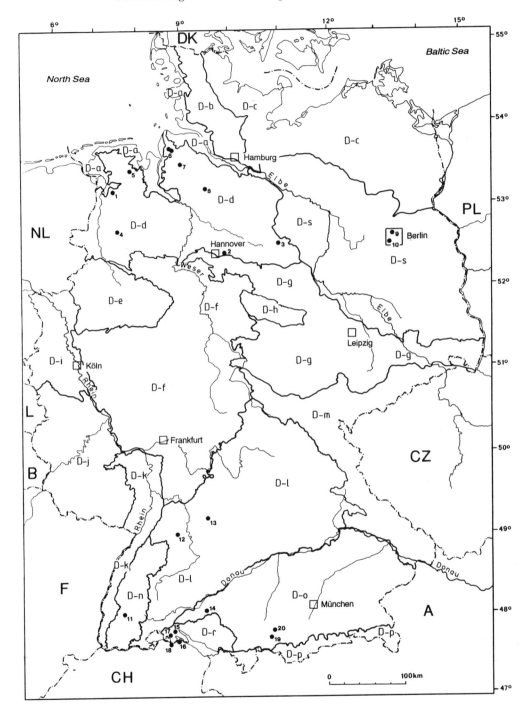

Fig. 15.1 IGCP type regions in Germany. Numbers of reference sites are according to the Appendix

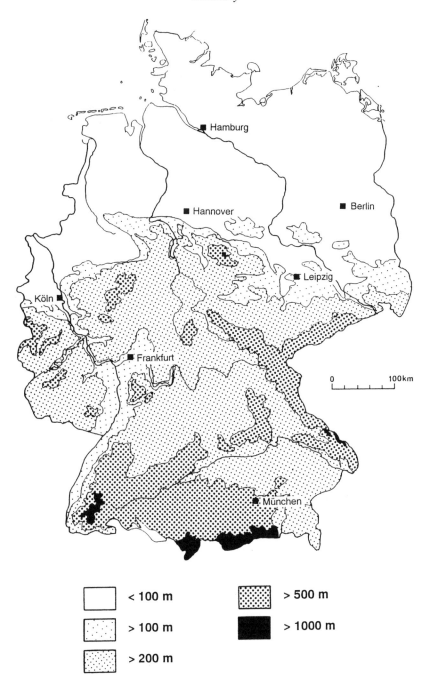

Fig. 15.2 Topographic map of Germany

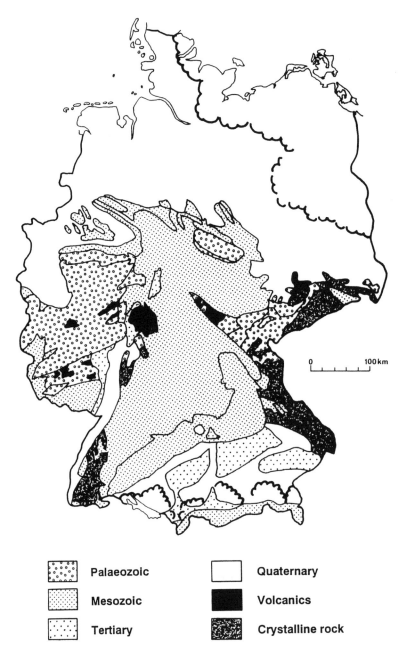

Fig. 15.3 Simplified geological map of Germany

that continues to the west as far as the Netherlands and northern Belgium and to the east as far as Poland, the Baltic countries, and Russia.

This North German Lowland consists of Quaternary deposits and is divided into two different regions. The first one is the old morainic landscape (the "Geest"), which was formed during the Elsterian and Saalian glaciations and, because of the absence of an ice cover during the last glaciation, has been evened out and leached during subsequent time. As most depressions were filled in, few lakes exist there. Poor acid soils and corresponding vegetation units prevail. In northwestern Germany large areas have been covered by raised bogs since late Atlantic times. In contrast to this, the second region of young morainic landscape is situated within the range of the last (Weichselian) glaciation, which did not reach beyond the River Elbe. This young morainic landscape is characterized by fresh geomorphologic forms with many lakes and good soils as well as outwash plains. The whole Pleistocene area is structured by a network of wide ice-marginal valleys, which generally run from east to west, eventually turning northwest to the North Sea.

The lowest parts of these wide valleys as well as the marginal areas along the North Sea around the German Bight are filled with marine and brackish Holocene deposits ("Marsch"), mainly resulting from the postglacial sea-level rise.

Several approaches have been made to subdivide Germany into natural and biotic regions. For the IGCP type regions we follow the map of Meynen *et al.* (1962) "Naturräumliche Gliederung Deutschlands".

TYPE REGION D-d, LOWER SAXONY OLD MORAINE LANDSCAPE (K.-E. Behre)

Germany north of the Central Uplands consists of three main types of landscape: the young moraine area, the area of the older moraines, and the Holocene along the coasts and rivers. Type region D-d comprises the area of the older moraines east and west of the Weser River in Niedersachsen. This lowland region was formed during the penultimate (Saale) glaciation by deposits brought from Scandinavia, and it remained ice-free during the last

(Weichselian) glaciation. During more than 100000 years elevations were levelled and depressions filled in. As a consequence only very few lakes exist at present, the soils are leached, and most of the area is dominated by poor acid sands, large parts of which were transformed into heathlands during the Middle Ages. Because of the oceanic climate, large parts of the region are covered by extensive raised bogs (now mostly cultivated), which were formed mainly during the last 5000 years. To the south are loess deposits (type region D-g).

No special projects were undertaken for the IGCP, so several sites investigated for various different aims are used for this compilation (see Fig. 15.4). Because of the scarcity of modern lakes, suitable lacustrine deposits are only available for the Lateglacial, whereas the younger periods are best represented in bog profiles.

Altitude: 0–170 m.

Climate: Mean January temperature –1.0 to 1.0°C, July 16–18°C; precipitation around 720 mm yr^{-1}, oceanic.

Geology: Saalian till, Saalian and Weichselian fluvioglacial deposits, Holocene raised bogs.

Topography: Most of the area more or less plain, in some parts hills or low ridges (Saalian end moraines) with gentle slopes.

Population: 100–150 people km^{-2}, mainly in smaller towns and villages.

Vegetation: Apart from the raised bogs, natural vegetation in most parts of the region was acid mixed *Quercus* forest (Quercion robori–petraeae) and *Quercus–Carpinus* forest (Querco–Carpinetum). Today forests cover about 15% of the region and consist mainly of planted *Picea* and *Pinus*. Some deciduous forest areas are formed of *Quercus, Fagus, Betula, Carpinus*, and *Fraxinus*. Small remnants of the former large raised bogs and heath areas are also present.

Soils: Podzol (in parts *Calluna* podzol), brown earth, pseudo-gley, and peat.

Land use: Mainly agricultural areas with the main crops rye, barley, maize, and potatoes, some pastures and meadows, 15% forest, and small uncultivated areas.

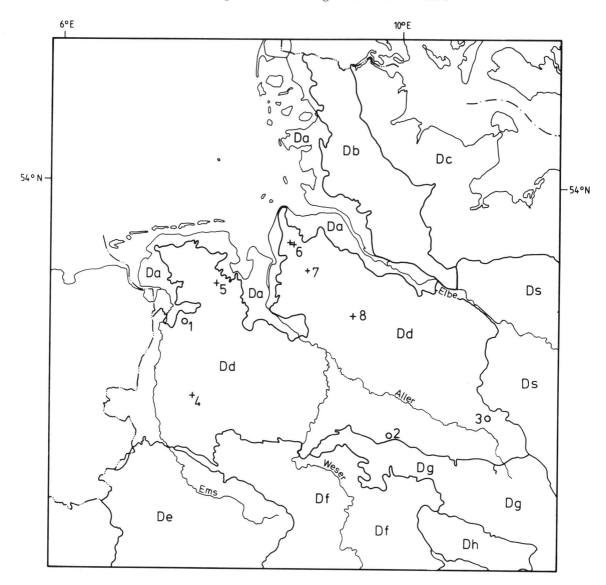

Fig. 15.4 Sites within the type region D-d: circles, late-glacial and early postglacial; crosses, middle and late postglacial

Reference sites 1–3

Type region D-d has no suitable pollen diagrams comprising the complete late- and postglacial sequences. Therefore the time-span has been divided into two parts: late-glacial to early postglacial and middle and younger postglacial.

For the late-glacial and early postglacial the following three sites, evenly distributed across the area, are used:

(1) Westrhauderfehn I in East Frisia (Latitude 53°7′N, Longitude 7°33′E; Behre 1966).
(2) Seckbruch SB 1 near Hannover (Latitude 52°24′N, Longitude 9°52′E; Dietz *et al.* 1958).
(3) Kaiserwinkel Ka$_{B1}$ (Latitude 52°31′N, Longitude 10°58′E; Golombek 1980).

Only site 1 is presented with a pollen diagram (Fig. 15.5).

Only a few radiocarbon dates exist for these sites

(one each from Westrhauderfehn and Kaiserwinkel, none from Seckbruch), so the dating of the zones relies on dates from other pollen diagrams that have characteristic marker horizons. The following pollen-assemblage zones (paz) can be distinguished:

–11800 BP	*Juniperus–Hippophaë* paz (corresponds to Older Dryas period). Not forested, NAP-dominated, shrubs invading (including *Betula nana*)
11800–10800 BP	1. *Betula–Pinus* paz (corresponds to Alleröd period). Rapid expansion of forests, in the first part with *Betula* prevailing, in the second part *Betula* and *Pinus* (boundary at the Laacher tuff where it is present).
10800–10200 BP	*Betula–Pinus* paz (corresponds to Younger Dryas period). Climatic deterioration to forest tundra with dwarf shrubs *Empetrum–Juniperus*-Gramineae
10200–10000 BP	2. *Betula–Pinus* paz (corresponds to Friesland oscillation, Preboreal a). Improvement of climate: reforestation
10000–*ca*. 9500 BP	*Betula–Empetrum*-Gramineae paz (corresponds to Rammelbeek phase or Preboreal b, Youngest Dryas period). Climatic setback to open forests: increase of shrubs and NAP
9500–9000 BP	3. *Betula–Pinus* paz (corresponds to Preboreal c). Final reforestation

Reference sites 4–8

The postglacial continuation of the vegetation history is represented by the following reference sites, also more or less evenly distributed across the type region D-d:

(4) Hahnenmoor in West Niedersachsen (Latitude 52°38′N, Longitude 7°38′E; Kramm 1978, Middeldorp 1984).

(5) Spolsener Moor, southwest of Wilhelmshaven (Latitude 53°23′N, Longitude 7°54′E; O'Connell 1986).

(6) Swienskuhle (Latitude 53°40′N, Longitude 8°43′E) and Flögelner Holz (Latitude 53°39′N, Longitude 8°46′E, northeast of Bremerhaven; Behre & Kucan 1986, 1994).

(7) Altes Moor WA 3, southeast of Bremerhaven (Latitude 53°26′N, Longitude 8°52′E; Dörfler 1989).

(8) Hohes Moor A, north of Rotenburg (Latitude 53°11′N, Longitude 9°27′E; Schneekloth 1963).

Only the profile Flögelner Holz is presented as a pollen diagram (Fig. 15.6).

For the postglacial period the following pollen-assemblage zones can be distinguished in the type region:

10200–9000 BP	see above (Preboreal)
9000–8700 BP	*Pinus–Betula* paz (Boreal, older part)
8700–8000 BP	*Corylus–Pinus* paz (Boreal, younger part)
8000–5200 BP	*Quercus–Alnus–Ulmus–Tilia* paz (Atlantic)
5200–4400 BP	*Quercus–Alnus–Tilia* paz (Subboreal, older part)
4400–2650 BP	*Quercus–Alnus–Fagus* paz (± Subboreal, younger part)
2650–1400 BP	*Quercus–Gramineae–Calluna* paz (most of Early Subatlantic)
1400–1200 BP	*Quercus–Fagus–Carpinus* paz (uppermost Early Subatlantic)
1200–0 BP	*Fagus–Calluna*-Gramineae-*Secale* paz (Late Subatlantic)

Palaeoecological patterns and events

Although the area is poor in lakes, most of the profiles that extend to the late-glacial start with limnic deposits. The open vegetation during the Older and Younger Dryas periods as described above normally led to the infilling of mainly clay and sand in the lakes, whereas during the Alleröd a remarkable increase of organic detritus occurred. Only in a

514

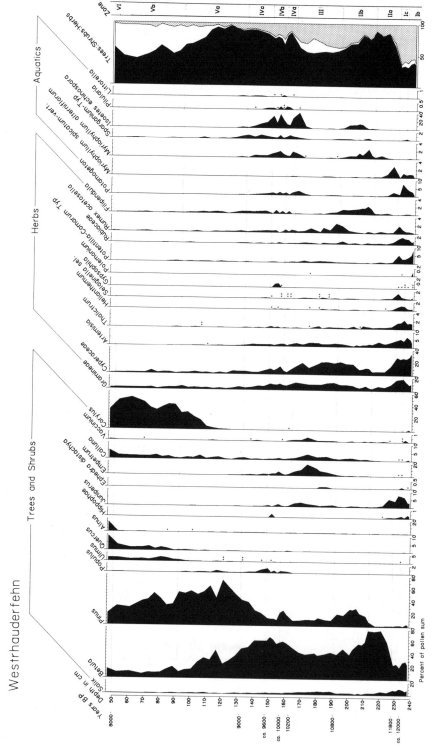

Fig. 15.5 Late-glacial and early Postglacial pollen diagram Westrhauderfehn from type region D-d. Only selected curves are given; for the complete diagram see Behre (1966). Basis = total terrestrial pollen. Analyst: K.-E. Behre

515

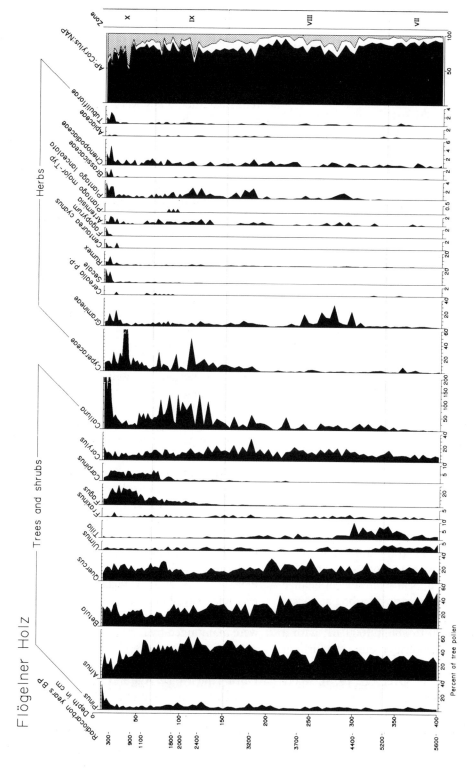

Fig. 15.6 Pollen diagram from the kettle-hole raised bog Flögelner Holz from type region D-d. Only selected curves are given; for the complete diagram see Behre & Kučan (1994). Basis = arboreal pollen; *Calluna* is omitted in the ratio AP–NAP. Analyst: D. Kučan

very few places did peat form as early as the Alleröd period.

Because of the regeneration of soils by the Weichselian periglacial climate, their nutrient content was improved, resulting in an adequate trophic state of the lakes. Even lake marl was occasionally deposited, e.g. at Seckbruch (site 2). At Westrhauderfehn (site 1) this eu- or mesotrophic state was short-lived only and was followed by deposits of an oligotrophic lake.

Hydrological changes

A good number of the Holocene profiles in the type region start in the Preboreal period, in most cases also with lacustrine deposits. The climatic and vegetational oscillations during the Preboreal in this (at that time) sensitive area have already been described. In most cases the transition from lake sediments to peat formation took place during the late-Preboreal period.

Raised bogs started to form at several places in the younger part of the Atlantic, in other areas during the Subboreal or even later. This resulted in extensive areas of treeless raised bogs, some of them covering more than 100 km². These expanded until their growth was halted by cultivation in modern times. In most parts of the type region D-d a lower layer of black and highly humified peat is covered by light and fresh *Sphagnum* peat. Both layers may have a thickness of several metres. The recurrence surface in between (earlier called "Grenzhorizont", now "Schwarz-Weißtorf-Kontakt" = SWK) indicates a change in water balance caused by increasing humidity. A major result of recent research is the evidence that this marked stratigraphic horizon is not synchronous even within a bog itself. The peat growth responds to the local hydrological change and may vary in different parts of a bog and even more among separate bogs. The dates of this SWK range from about 4000 to about 1000 BP, with a clear maximum in the centuries around the Birth of Christ. Some raised bogs also exist, in particular the large Gifhorner Moor (cf. Overbeck 1975), where several changes in humification occur, i.e. several recurrence surfaces in a section. In general the recurrence surfaces reflect climatic deteriorations in the younger postglacial.

Human impact

Evidence from several pollen diagrams shows that anthropogenic indicators generally start at about 5200 BP. The best indicator is given by the curve of *Plantago lanceolata*, which begins at this level at all reference sites. Only very few cereal-type pollen grains occur below this level.

Because of the generally low *Ulmus* values, the elm decline is not as sharp as in pollen diagrams from areas with better soils. There are also diagrams (e.g. several diagrams in Dörfler 1989) without any marked elm decline in spite of various proofs of human occupation. This relation favours the view that the elm decline has no anthropogenic origin but probably was caused by elm disease. On the other hand, a marked decline of *Tilia* in several diagrams can be attributed to human impact. This *Tilia* decline is not synchronous but normally occurs in the Neolithic, together with a "landnam-phase".

The decrease of *Tilia* on the better soils gives way to the first and slow expansion of *Fagus*. Also in later periods, during Pre-Roman Iron Age, Roman, and Migration periods, as well as during the Middle Ages, the main reason for the further expansion of *Fagus* is human interference in the forest by clearances. With the regeneration of the forest *Fagus* was able to compete much faster than in natural forests, where all sites were occupied by other trees, significantly delaying the expansion of *Fagus*.

The pollen diagrams show that in the type region D-d three main occupation phases led to severe changes of the vegetation, mainly destruction of the forests: middle/late Neolithic (from about 4400 BP), Roman Period, and Middle Ages.

Because of the poor soils in many parts of the Pleistocene geest area, regeneration of the forests was relatively slow, and grazing of the forests resulted in the formation of *Calluna* heaths, which were maintained by further grazing. The first small areas of such heaths can be dated back to Neolithic times (cf. Behre 1970, Behre & Kučan 1986, 1994). During the Bronze Age and Iron Age the heath area increased. Starting at about AD 1000 a very rapid expansion of the *Calluna* heaths took place. This resulted from a special form of agricultural economy, the plaggen technique, where the humus layer

TYPE REGION D–d Niedersächsisches Altmoränengebiet: EVENT STRATIGRAPHY

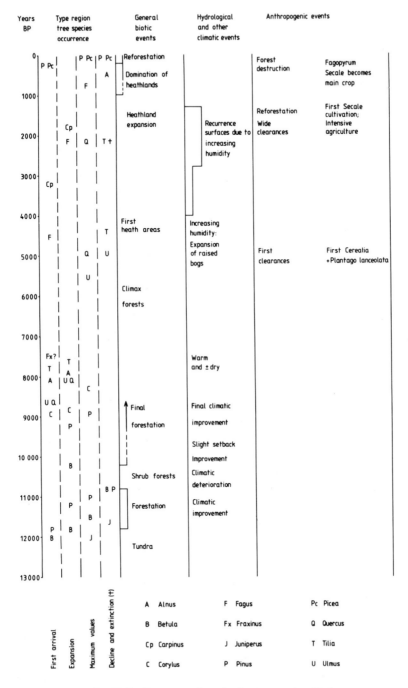

Fig. 15.7 Event stratigraphy for type region D-d

of the heath areas was cut and used as fertilizer for the fields (cf. Behre 1980). This resulted in a far-reaching destruction of the forest and a contemporary formation of heaths, which covered wide areas in the type region and had its maximum in modern times until the introduction of artificial fertilizers in the last century. The heath grounds were thus finally transformed into agricultural areas or afforested, mainly with introduced conifers such as *Pinus* and *Picea*.

TYPE REGION D-s, BERLIN (A. Brande)

The region D-s (Fig. 15.8) belongs to the Weichselian glaciated area of the North German lowlands and is a section of the older part of the young moraine landscape. The Warsaw–Berlin Urstromtal with the Spree River and the Havel lake–river system subdivide the region into four subregions, including a northern, southern and western ground moraine plateau. The region as a whole comprises the former West-Berlin, surrounded by the late GDR.

Altitude: 29–103 m, larger areas 32–60 m, excluding artificial hills made of rubble, etc.

Climate: Mean January –0.7°C, July 18°C, precipitation 580 mm yr⁻¹, 380 mm during vegetative period. Sub–continental within an oceanic-continental transition zone.

Geology: Weichselian fluvioglacial sand (Urstromtal); ice-marginal sand and gravel: marl, loam, and sand (ground moraine); Late Weichselian and Holocene sand, mud, and peat (14% of the region).

Topography: Flat Urstromtal terraces with sand dunes; gently undulating ground moraine plateaux; steep relief of dunes, kames, and end moraines; steep depressions in the lake–river system.

Population: 1.95 million (1988) on 480 km², large city area.

Vegetation: Urban and large city vegetation (spontaneous and planted); forests comprise 16% of the region, dominated by *Pinus sylvestris* and *Quercus* spp.; agriculture (arable, grass, and garden land) 7%; lakes, rivers, mires, ponds 7%. Different intensity of

actual human influence on vegetation (Böcker & Sukopp 1987).

Soils: Anthopogenic rendzinas on rubble, hortisols, etc; in the outskirts podzolized brown earths (forests); para-brown earth (Lessive) (formerly arable areas); oligo- to eutrophic fens (mires); more or less calcareous gyttja and sapropel (lakes, ponds). Different intensity of actual human influence on soils (Grenzius 1985).

Land use: Large city. Transport (17%, rail-, highways, etc.), forestry, recreation. Anthropogenic effects of extreme lake eutrophication, drying out of mires by groundwater depression, infill or sediment removal in ponds, etc.

Reference site 9 (Tegeler See) and reference site 10 (Pechsee)

Tegeler See

Latitude 52°35′N, Longitude 13°13′E. Elevation 31 m. Age range 11400–0 BP. Lake. 13 local pollen-assemblage zones (paz). Pollen diagram (Fig. 15.9) Subatlantic–Alleröd (Brande 1980, 1988b, and in prep.).

Pechsee

Latitude 52°29′N, Longitude13°13′E. Elevation 31 m. Age range 12500–11400 BP. Lake stage of oligotrophic mire. Three local pollen-assemblage zones (paz). Pollen diagram (Fig. 15.9) Bölling/ Older Dryas–Alleröd (Brande 1980, 1988b).

The 15 pollen-assemblage zones, derived from the reference sites as selected in the investigation area (Fig. 15.8), are of regional validity, excluding settlement areas (archaeological and historical sites). In Fig. 15.9 radiocarbon dates and calendar dates are given only for dating the transitions of the pollen-assemblage zones, obtained from peat at various sites in the region (Brande 1978/79, 1980, 1985, 1986a, Brande in Böcker *et al.* 1986, Böse & Brande 1986). The criteria for defining the pollen-assemblage zones (paz) are based on the pollen diagrams (Fig. 15.9) and the regional vegetational events (Fig. 15.10).

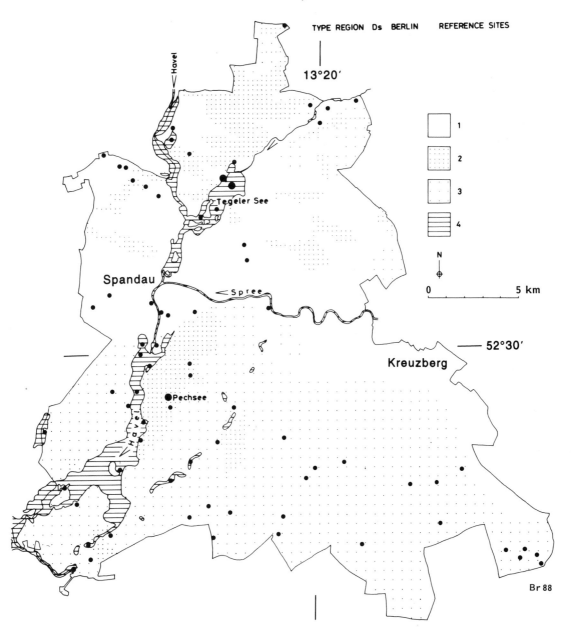

Fig. 15.8 Type region D-s, Berlin. Reference sites Pechsee and Tegeler See and location of other important palynological sites in the area of former Berlin (West), documented by Brande in Böcker *et al.* (1986), Brande (1988a), Brande & Hühn (1988), Brande *et al.* (1990), and unpublished. Landscape units: 1, Urstromtal sand terraces, wet to dry, 30–40 m elevation; 2, sand-dunes and ice-marginal sand and gravel, dry, 35–100 m elevation; 3, ground-moraine marl, loam, and sand, moist to dry, 35–60 m elevation; 4, recent lakes and rivers, 29–31 m elevation

1.	12500–11800 BP	*Betula–Hippophaë*
2.	11800–11500 BP	*Betula–Juniperus*
3.	11500–11000 BP	*Pinus–Betula*
4.	11000–10300 BP	*Pinus–Juniperus–* *Artemisia*
5.	10300–9000 BP	*Pinus–Betula–* *Thelypteris palustris*
6.	9000–8700 BP	*Pinus–Corylus–Ulmus*
7.	8700–8000 BP	*Pinus –Corylus– Quercus*
8.	8000–7500 BP	*Pinus–Corylus–Alnus*
9.	7500–6500 BP	*Pinus–Alnus–Tilia*
10.	6500–5000 BP	*Pinus–Fraxinus*
11.	5000–3700 BP	*Pinus–Plantago* *lanceolata–Cerealia*
12.	3700–2900 BP	*Pinus–Fagus–Corylus–* *Tilia*
13.	2900–2300 BP	*Pinus–Fagus*
14.	2300–900 BP	*Pinus–Fagus–* *Carpinus–Secale*
15.	900–0 BP	*Pinus–Secale–Rumex*

The pollen diagram

The pollen-assemblage zones partly coincide with chronozones. Although *Pinus* pollen was dominant from the younger Alleröd before the Laacher See tephra deposition, most woodland successional stages can be defined as being similar to adjacent regions poorer in *Pinus*, e.g. the Atlantic/Subboreal *Ulmus* decline. Accordingly the Subboreal/Subatlantic transition is characterized by the *Corylus* and *Tilia* decline, whereas the repeated *Corylus* expansion during the Atlantic and Subboreal chronozones is very weak or even absent. *Fagus sylvatica* and *Carpinus betulus* are present, confirmed by macroremains in the older Subatlantic chronozone and, in the case of *Fagus*, by high local pollen frequencies.

Pollen-assemblage zone transition 14/15 reveals the last extensive clearance (German period) and is found also in sites on areas still forested. Moreover, it coincides in the Urstromtal subregion with peat-stratigraphic criteria due to an artificial water-level rise by mill barrages, and in the ground-moraine subregion it coincides with historical data for large-scale landnams and deforestation. The radiocarbon age of the 14/15 transition (900 BP) includes a dendrochronologically corrected archaeological/historical date of about AD 1230 (Brande 1985, 1986a). However, a similarly defined pollen-assemblage-zone transition can be found in restricted areas of the region a few centuries older (Slavic period, Brande *et al.* 1987).

Palaeohydrological and palaeoecological events

The events shown in Fig. 15.10 are derived from many sites in the region (Fig. 15.8), including mires and ponds but excluding archaeological sites and locally confined events. Only a few events are reflected in the pollen diagram (Fig. 15.9). On the other hand, the reference site Tegeler See has been intensively studied by palaeolimnology, palynology, and ecological history (Brande 1980, 1988a,b, and in prep., Sukopp & Brande 1984/85, Böcker & Sukopp 1985, Bertzen 1987, Pachur & Röper 1987, Wolter 1992).

The water-level rise of the Havel–Spree lake–river system during the Atlantic chronozone is probably a distant effect of the Holocene North Sea transgression via the Elbe catchment area, which includes the region. The water level does not yet reach the level of flat mires, which during that period show severe reduction of growth or hiatus (Brande 1986b). This event in the non-ombrotrophic mire development accords with surrounding regions (Succow 1987).

At least since AD 1230 the lake level and corresponding groundwater table influenced the mire development in the sandy Urstromtal subregion, as documented by the Havel mill barrage, which caused increasing peat formation in hollows some distance from the lake–river system even in still forested areas. But already since the Subboreal chronozone peat formation increased, with some marginal paludification after the hiatus period. This is also true for the groundwater-independent kettle-holes

Fig. 15.9 Type region D-s Berlin. Reference site pollen diagram (percentage/time) Pechsee 3.8-3.95 m and Tegeler See 0–15.1 m, selected pollen types (from Brande 1980, 1988b, and in prep.). Analyst: A. Brande, G. Hinz

TYPE REGION D-s BERLIN   POLLEN DIAGRAM

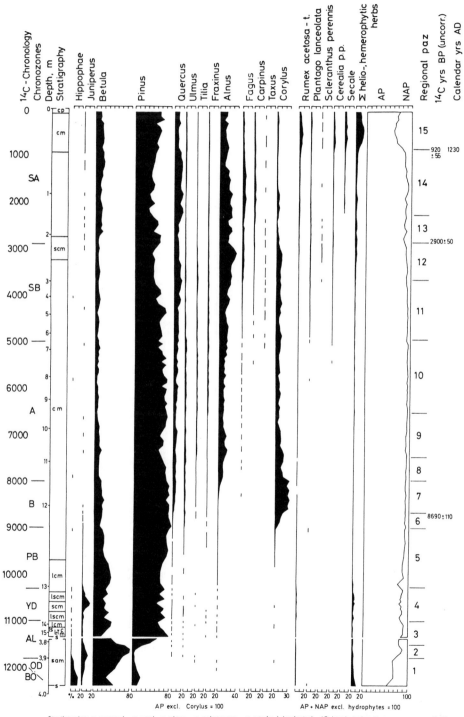

Stratigraphy: p sapropel, m mud, a algae, c calcareous, s sandy, l laminated, LT Laach tephra layer   Br 88

TYPE REGION Ds BERLIN   EVENT STRATIGRAPHY

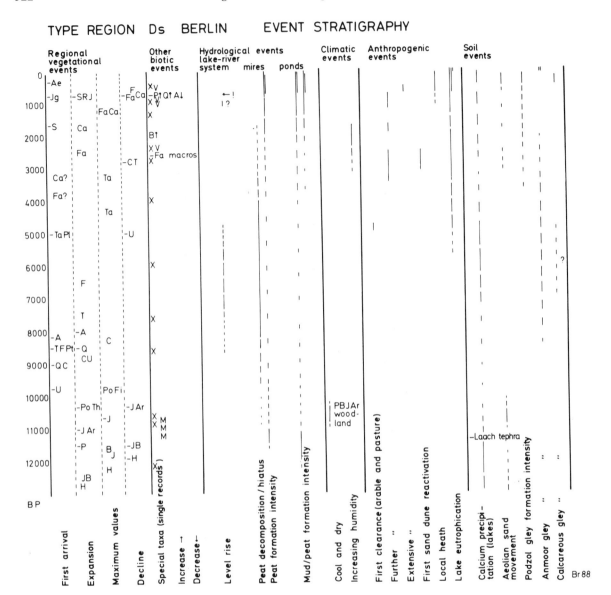

Fig. 15.10   Type region D-s Berlin. Event stratigraphy. A *Alnus*, Ae *Aesculus*, Ar *Artemisia*, B *Betula*, C *Corylus*, Ca *Carpinus*, F *Fraxinus*, Fa *Fagus*, Fi *Filipendula*, H *Hippophaë*, J *Juniperus*, Jg *Juglans*, M *Myriophyllum alterniflorum*, P *Pinus*, Pl *Plantago lanceolata*, Po *Populus*, Pt *Pteridium*, Q *Quercus*, R *Rumex acetosa* type, S *Secale*, T *Tilia*, Ta *Taxus*, Th *Thelypteris palustris*, V *Vitis*, X *Xanthium*

(ponds) on the ground-moraine plateaux, and therefore the increasing humidity of climate probably is one responsible factor. In this subregion organic sedimentation in the ponds also intensified since the 13th century AD because of more runoff water after large-scale clearances (Brande *et al.* 1990). Within the last 100 years in the Urstromtal subregion the mires dried out again, this time as a consequence of groundwater lowering caused by urban waterworks.

The climatic history (Brande 1990), derived from palynological data, is partly confirmed by oxygen-isotope analysis of the Tegeler See sediments (Pachur 1987, Wolter 1992) and is compared in some detail with northwest German regions.

In correspondence with the regional archaeological record the palynological traces of Neolithic land use go back in general only to the Atlantic/Subboreal chronozone transition or slightly earlier, linked with the first steps of anthropogenic lake eutrophication in the reference site.

Intensified land use is found in the Urstromtal subregion during the Late Bronze age (Subboreal chronozone, 11/12 transition) with the spread of *Scleranthus perennis* and the first indications of anthropogenic sand-dune reactivation since consolidation during the Younger Dryas/Preboreal transition. Large sand-dune movements took place between the 13th and 18th centuries AD until afforestation. Anthropogenic *Calluna* heathland also developed but not before Medieval times, thus dating back to the Slavic period. Typical natural *Calluna* stands are on the anmoor and podzol gley soils of dry mires, and mainly date from the Atlantic up to the older Subatlantic chronozones.

The progress of lake eutrophication is closely correlated to anthropogenic influence, increasing rapidly in the late 19th century AD during industrialization and drainage from waste-water disposal into the lake–river system, thus forming sapropel on calcareous mud (Fig. 15.9).

Changes in the composition of the remaining woodland on less fertile sandy soils (podzolized brown earth) depend on various types of land use since the 14/15 transition, favouring *Pinus sylvestris* and *Quercus* spp. at the expense of *Fagus, Carpinus, Fraxinus,* and *Alnus.*

Among the palaeofloristic results, the native status of *Xanthium* (Brande 1980, Opravil 1983) and *Vitis* has been confirmed as well as of *Trapa* and *Salvinia* by listing all single pollen finds in the region (Fig. 15.10).

During the stages of soil development calcium leaching of sandy soils came to an end already in the Late Weichselian at many sites, mostly in the Alleröd chronozone. Secondary calcium accumulation, forming calcareous gley soils, is probably a consequence of the mid-Holocene lake-level rise, at least of the artificial rise of lake and groundwater table in the upper Subatlantic chronozone. Late Weichselian soil leaching is indicated even in the ground-moraine subregion by the only occurrence of *Myriophyllum alterniflorum* at that time. Other processes of hydromorphic soil formation (podzol gley and anmoor gley) are also closely connected with the changes of groundwater table, runoff-water access, and their natural and anthropogenic nutrient content during the Holocene.

Patterns of events between regions

In terms of the woodland history, the region is part of the landscape farther south of the Baltic and west of the Elbe, where *Pinus* dominates in the tree pollen rain on prevailing anhydromorphic sandy soils, i.e. in the eastern parts of the German type regions with subcontinental climate. Slight climatic changes, based on very local mire and pollen stratigraphy (e.g. Müller 1971), are still contradictory (Brande 1985). The general trends in regional palaeohydrology, reflected by non-ombrotrophic peat growth and presented independently for the type region D-s, strongly confirm the results obtained from the surrounding regions of the Northeast German lowland, as discussed by Lange *et al.* (1978, 1986), Succow & Lange (1984), and Succow (1987). The precondition, however, is to take into account the local features of every site, e.g. groundwater table in relation to site elevation, runoff-water catchment area, intensity of earlier organic sedimentation in flat and small hollows, and compactibility and water content of sediments during dry periods.

The anthropogenic events discussed here are on a regional scale only, whereas results from archaeological sites reveal additional effects.

TYPE REGIONS D-n, D-l AND D-r, SOUTH-WEST GERMANY (M. Rösch)

The region discussed here (see Fig. 15.11) corresponds to the present-day German county of Baden-Württemberg. It covers an area of 35750 km² and had in 1983 9.3 million inhabitants (Borcherdt 1983). Several landscapes with widely varying geology, geomorphology, climatology, and biotic conditions can be distinguished (Huttenlocher 1972, Borcherdt 1983, Geyer & Gwinner 1986, see also Fig. 15.1). At the western border are the upper Rhine lowlands with Tertiary and Quaternary fluvial deposits, fertile soils, and a warm climate (annual average temperature more than 9°C), but without lakes or mires suitable for yielding reference pollen diagrams. It belongs to the type region D-k and will not be discussed further here.

Eastward the southern part of the region is covered by the Black Forest (type region D-n), highlands with an elevation up to 1500 m above sea level. Its northern part consists of Triassic sandstone, its southern part of granite and gneiss. These mountains were first colonized with permanent settlements in medieval times. From this type region D-n one reference site is presented.

The western part of type region D-l consists of several different landscapes: In the northernmost part of the region on the west are the hills of Odenwald and Spessart, moderate highlands with Triassic sandstones but without reference sites and, also without reference sites, the Mainfränkische Platten, lowlands with Keuper and Triassic limestones. In the central part of the region are the Neckar- and Tauber-Gäuplatten–lowlands with Keuper and Triassic limestones locally covered by loess, with a warm climate, early colonization, and dense population. These landscapes, very poor in lakes and mires, have, nevertheless, two reference sites in reference area a. To the south, in the eastern part of the region, are the highlands Keuperwaldberge and Schwäbische Alb, the first consisting of Keuper sandstone up to 500 m above sea level, the second of Jurassic limestone up to 1000 m above sea level, both without reference areas.

South of this area are the prealpine lowlands of Southwest Germany, between the Danube and Lake Constance (type region D-r). The Federsee basin in the Upper Swabian Prealpine lowlands is called reference area b, and the Lake Constance region reference area c. Both reference areas have several reference sites. The prealpine lowlands are covered by moraines of the Würmian or Rissian glaciation

or by sediments of the Tertiary Mollassic sea. The elevation of this landscape is between 400 (Lake Constance) and 700 m above sea level.

Due to intense and, in some landscapes, very early human colonization, no original natural vegetation remains in the type regions considered. For most of type region D-l and D-r the potential natural vegetation would be *Fagus sylvatica* forests, with a large element of *Quercus* and *Carpinus betulus* at the lowest altitudes and with a more or less large proportion of *Abies alba* at higher altitudes, especially in the Keuperwaldberge and Schwäbische Alb. In the type region D-n, Black forest, the potential natural vegetation would be mixed forests of *Abies alba* and *Fagus sylvatica*.

The history of human colonization starts in the 6th millennium BC in the loess-covered areas of upper Rhine lowlands and Neckar- and Tauber-Gäuplatten with the Linear-Pottery culture. The Neolithic cultures of the 5th millennium BC occupied the adjacent regions, for example the Prealpine lowlands, wherever they had more or less fertile soils. Even the highlands of the Schwäbische Alb were inhabited by man. The highlands of the Black Forest and Keuperwaldberge, where fertile soils are very rare or totally lacking, were first occupied by man in historical times.

TYPE REGION D-n, BLACK FOREST

Altitude: 500–1493 m.

Climate: Suboceanic. Mean annual temperature 3–9°C; mean January temperature −4 to 0°C; mean July temperature 11–18°C. Precipitation 800–2000 mm yr^{-1}.

Geology: Buntsandstein (Precambrian).

Topography: Mountains.

Population: 270 people km^{-2} on average, large regional differences.

Vegetation: Most of the area is cultivated. Small lakes in glacial cirques, as well as bogs. Forests are dominated by planted spruce and pine. Fir is abundant, as well as beech in the western part. Other deciduous trees are rare.

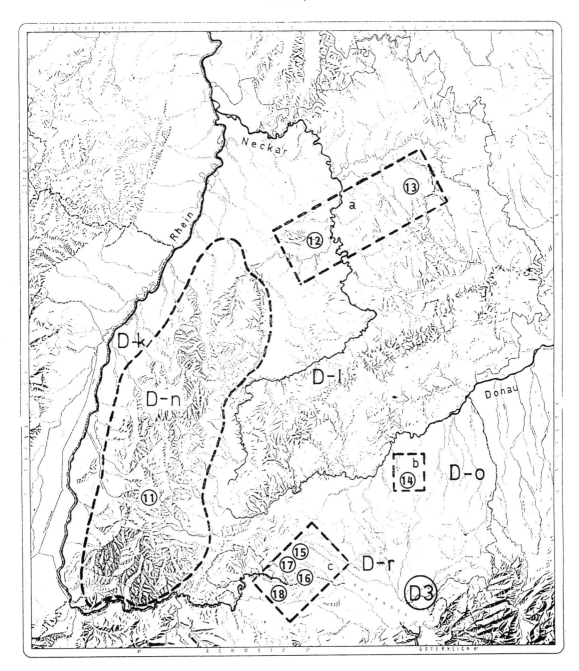

Fig.15.11 IGCP Project 158 B, type regions, reference areas, and reference sites in southwestern Germany. a, b, c are reference areas with the sites 11 Breitnau-Neuhof, 12 Sersheimer Moor, 13 Kupfermoor, 14 Buchau-Torfwerk, 15 Durchenbergried, 16 Hornstaad-Bodensee, 17 Feuenried, 18 Nussbaumer Seen

Soils: Oligotrophic brown earth, podzols, and peat.

Land use: Total area 35742 km². Urban areas 11.8%, agricultural area 49.1%, forest area 37.0%, and uncultivated area 2.1%.

Forest composition: Spruce 46.5%, beech 20.3%, fir 9.6%, pine 9.1%, and oak 5.5%.

Reference site 11. Breitnau-Neuhof (Rösch 1989)

Latitude 47°58'N, Longitude 8°12'E. Elevation 985 m, area 5 ha. Bog core 6.85 m long. Age range *ca.* 10000–1500 BP, 71 pollen samples, 15 ¹⁴C-dates. The reference site Breitnau-Neuhof is situated in the middle part of the Black Forest, 10 km north of its highest elevation, the Feldberg, 1493 m above sea level, and north of the Dreisam valley (Rösch 1989). A *Pinus uncinata* forest now grows on the surface of this drained bog. The core was investigated by pollen and plant macrofossil analyses and dated by radiocarbon analysis (15 samples). 15 local pollen-assemblage zones (paz).

1.	–9800 BP	*Pinus–NAP–Betula*
2.	9800–9400 BP	*Pinus*
3.	9400–9000 BP	*Pinus–Corylus*
4.	9000–8200 BP	*Corylus*
5.	8200–7200 BP	*Corylus–Quercus–Tilia–Ulmus*
6.	7200–6000 BP	*Quercus–Corylus–Tilia–Ulmus*
7.	6000–5400 BP	*Quercus–Tilia–Corylus–Ulmus*
8.	5400–5200 BP	*Corylus–Quercus–Abies–Ulmus–Tilia*
9.	5200–4600 BP	*Abies*
10.	4600–4000 BP	*Abies(–Fagus)*
11.	4000–3500 BP	*Abies–Fagus*
12.	3500–2500 BP	*Fagus–Abies*
13.	2500–1500 BP	*Fagus*
14.	1500–? BP	*Abies–Fagus–Corylus–Alnus*
15.	?–0 BP	*Pinus–Picea*

We can suppose that the increase of *Fagus sylvatica* and the decrease of *Abies alba* was induced by prehistoric man, who used these highlands seasonally for forest grazing. However, permanent settlement occurred first in historical times.

TYPE REGIONS D-l AND D-r

Altitude: 120–1000 m.

Climate: Suboceanic/subcontinental. Mean annual temperature 5–10°C; mean January temperature –4 to 0°C; mean July temperature 14–19°C. Precipitation 600–1100 mm yr⁻¹.

Geology: Keuper, Muschelkalk, Jura, Molasse, Rissian and Würmian till and fluvioglacial sediments.

Topography: Plain, hilly areas, and mountain areas.

Population: See type region D-n.

Vegetation: Most of the area is cultivated. Wetlands exist only along the rivers. The forests are dominated by planted spruce and pine. The deciduous tree vegetation consists mainly of beech, oak, hornbeam, alder, and ash.

Soils: Brown earth, rendzina, and para brown earth.

Land use: See type region D-n.

Forest composition: See type region D-n.

Reference area a, Neckar- and Tauber-Gäuplatten

Reference site 13. Kupfermoor (Smettan 1988)

Latitude 49°07'N, Longitude 9°43'E. Elevation 275 m, area 4 ha. Fen core 6.12 m long. Age range 8200–0 BP, 156 pollen samples, 34 ¹⁴C-dates, nine local pollen-assemblage zones (paz).

1.	–8200 BP	*Corylus*
2.	8200–7600 BP	*Corylus–Quercus–Ulmus*
3.	7600–6300 BP	*Quercus–Corylus– Ulmus–Tilia–Fraxinus*
4.	6300–5200 BP	*Corylus–Quercus–Tilia–Ulmus*
5.	5200–4500 BP	*Quercus–Corylus–Tilia*
6.	4500–4000 BP	*Quercus–Corylus–Fagus–Tilia*
7.	4000–3500 BP	*Fagus*
8.	3500–1500 BP	*Fagus–Corylus–Alnus–Betula–Quercus*
9.	1500–0 BP	*Betula–Fagus–Alnus–Quercus–NAP*

Reference site 12. Sersheimer Moor
(Smettan 1985)

Latitude 48°58′N, Longitude 9°03′E. Elevation 234 m, area 0.5 ha. Fen core 4.4 m long. Age range *ca.* 8500–0 BP, 16 ^{14}C-dates, nine local pollen-assemblage zones (paz).

1.	–8000 BP	*Pinus–Corylus*
2.	8000–7800 BP	*Corylus–Pinus–Quercus*
3.	7800–7400 BP	*Corylus–Quercus–Ulmus*
4.	7400–5700 BP	*Quercus–Corylus–Ulmus*
5.	5700–4500 BP	*Quercus–Corylus*
6.	4500–3900 BP	*Quercus–Corylus–Betula–Fagus*
7.	3900–2000 BP	*Fagus–Quercus–Corylus–Betula*
8.	2000–100 BP	*Quercus*
9.	100–0 BP	*Pinus–Quercus*

Reference area b, Federsee Basin

The reference area b, the Federsee Basin, is situated at the border between the Würmian and Rissian moraines. The recent level of the Federsee is 578 m above sea leavel. Nowadays it covers an area of about 2 km², and is surrounded by fens that cover the former lake basin, the area of which is about 25 km².

Reference site 14. Buchau-Torfwerk
(Liese-Kleiber 1991)

Latitude 48°03′N, Longitude 9°38′E. Elevation 584 m, area 15 km². Bog core 3.04 m long, consisting of calcareous gyttja, fen peat, and bog peat. Age range 8000–2300 BP, 175 pollen samples, 22 ^{14}C-dates, seven local pollen-assemblage zones (paz).

1.	–8000 BP	*Corylus*
2.	8000–7400 BP	*Corylus–Ulmus–Quercus*
3.	7400–5200 BP	*Betula–Corylus–Quercus–Ulmus*
4.	5200–4500 BP	*Betula–Corylus–Quercus–Alnus*
5.	4500–3000 BP	*Fagus*
6.	3000–2700 BP	*Alnus–Fagus–Corylus–Betula*
7.	2700–2300 BP	*Fagus–Alnus–Betula*

Reference area c, Lake Constance region

Reference area c, the Lake Constance region in the southernmost part of the Southwest German Prealpine lowlands, has several reference sites. The Nussbaumer Lakes are situated in Switzerland but belong to this reference area. These three lakes of glacial origin are at 434 m above sea level and cover an area of about 6 km². The ten cores investigated consist of lake marl, gyttja, and fen peat. The depth of the cores ranges from 2.5 to 12.5 m. The time span represented is from Late Würmian (Oldest Dryas) to Subatlantic (recent times). The cores were investigated by pollen and plant-macrofossil, stable-oxygen and chemical analysis and dated by radiocarbon analysis (15 dates, Rösch 1983, 1985a).

About 15 km north of the Nussbaumer Lakes is the site Feuenried, a small kettle-hole with a diameter of about 250 m. The surface is about 407 m above sea level. A 5.5 m core consisting of clay, gyttja, and fen peat was investigated by pollen and plant-macrofossil analysis and dated by radiocarbon analysis (33 samples). The deposits represent a time span from the Oldest Dryas to the Subatlantic (modern times). The time record is not complete. Gaps exist especially in the Bölling and Preboreal, and zones with very low peat accumulation rates occur in the Subboreal and Subatlantic (Rösch 1985b).

Seven kilometres northeast of this site is the Durchenbergried, a kettle-hole with a diameter of about 100 m. The last reference site Hornstaad/Bodensee is situated 10 km southeast from the Feuenried and 10 km south of Durchenbergried at the shores of Lake Constance. Only these last two sites are presented here.

Reference site 15. Durchenbergried
(Rösch 1986, 1990a)

Latitude 47°44′N, Longitude 8°59′E. Elevation 432 m, area 1.2 ha. Fen core 8.4 m long consisting of gyttja, bog peat, and fen peat. Age range 13000–0 BP, 480 pollen samples, 50 ^{14}C-dates, 31 local pollen-assemblage zones (paz).

1.	–13000 BP	NAP–*Betula*
2.	13000–12800 BP	*Juniperus–Betula*

3. 12800–12000 BP   *Betula*
4. 12000–11700 BP   *Betula–Pinus*
5. 11700–11000 BP   *Pinus–Betula*
6. 11000–10600 BP   *Pinus*
7. 10600–10200 BP   *Pinus–Betula*
8. 10200–10000 BP   *Betula–Pinus–Corylus*
9. 10000–9600 BP   *Corylus–Betula–Pinus*
10. 9600–9400 BP   *Corylus–Betula–Pinus–*
                        Quercus–Ulmus
11. 9400–8700 BP   *Corylus*
12. 8700–8400 BP   *Corylus–Quercus–Ulmus*
13. 8400–7400 BP   *Quercus–Corylus–*
                        Ulmus–Tilia
14. 7400–6300 BP   *Quercus–Corylus–*
                        Ulmus–Fraxinus–Tilia
15. 6300–6000 BP   *Quercus–Fagus–Corylus–*
                        Ulmus–Fraxinus
16. 6000–5200 BP   *Fagus–Quercus–Corylus–*
                        Ulmus
17. 5200–5000 BP   *Quercus–Corylus–*
                        Betula–Fagus–Alnus
18. 5000–4700 BP   *Corylus–Betula–Quercus–*
                        Alnus–Fagus
19. 4700–4500 BP   *Fagus*
20. 4500–4000 BP   *Corylus–Quercus–Betula–*
                        Alnus–Fagus
21. 4000–3600 BP   *Fagus–Betula–Alnus*
22. 3600–3000 BP   *Quercus(–Corylus–Alnus–*
                        Fagus)
23. 3000–2800 BP   *Betula–Corylus–Alnus*
24. 2800–2400 BP   *Quercus–Fagus–Betula–*
                        Corylus–Alnus
25. 2400–2200 BP   *Quercus–NAP*
26. 2200–2100 BP   *Betula*
27. 2100–1900 BP   *Quercus*
28. 1900–1400 BP   *Quercus–Betula–Fagus–*
                        Alnus
29. 1400–1000 BP   *Quercus–NAP*
30. 1000–400 BP   *Quercus–Pinus*
31. 400–0 BP   *Pinus–Picea–Quercus–*
                        Fagus

Reference site 16. Hornstaad-Bodensee
(Rösch 1992, 1993)

Latitude 47°40′N, Longitude 9°01′E. Elevation 395 m, area 50 km². Lake core 14 m long. Age range

13000–400 BP. 868 pollen samples, 25 ^{14}C-dates, 33 local pollen-assemblage zones (paz).

1.   –13000 BP   NAP–*Betula*
2. 13000–12800 BP   *Betula–Juniperus*
3. 12800–12000 BP   *Betula*
4. 12000–11700 BP   *Betula–Pinus*
5. 11700–11000 BP   *Pinus–Betula*
6. 11000–10600 BP   *Pinus*
7. 10600–10200 BP   *Pinus–Betula*
8. 10200–10000 BP   *Betula–Pinus–Corylus*
9. 10000–9600 BP   *Corylus–Pinus–Betula*
10. 9600–9400 BP   *Corylus–Betula–Pinus–*
                        Quercus–Ulmus
11. 9400–8700 BP   *Corylus*
12. 8700–8400 BP   *Corylus–Quercus–Ulmus*
13. 8400–7400 BP   *Quercus–Corylus–*
                        Ulmus–Tilia
14. 7400–6300 BP   *Quercus–Corylus–*
                        Ulmus–Fraxinus–Tilia
15. 6300–6000 BP   *Quercus–Fagus–Corylus–*
                        Alnus–Ulmus–Fraxinus
16. 6000–5200 BP   *Fagus–Corylus– Quercus–*
                        Alnus–Ulmus
17. 5200–5000 BP   *Corylus–Alnus–Quercus–*
                        Betula–Fagus
18. 5000–4700 BP   *Corylus–Alnus–Betula–*
                        Quercus
19. 4700–4500 BP   *Fagus–Alnus–Corylus*
20. 4500–4000 BP   *Corylus–Alnus– Betula–*
                        Fagus
21. 4000–3700 BP   *Fagus*
22. 3700–3100 BP   *Alnus–Corylus–Fagus-*
                        Quercus–Betula
23. 3100–2800 BP   *Fagus–Alnus–Betula–*
                        Corylus–Quercus
24. 2800–2400 BP   *Alnus–Betula–Corylus–*
                        Fagus–Quercus
25. 2400–2200 BP   *Quercus–Fagus–Alnus*
26. 2200–2100 BP   *Betula–Alnus*
27. 2100–1900 BP   *Quercus–Fagus–Alnus*
28. 1900–1600 BP   *Alnus–Betula–Corylus–*
                        Fagus–Quercus
29. 1600–1300 BP   *Fagus–Alnus–Quercus–*
                        Corylus–Carpinus
30. 1300–1200 BP   *Alnus*
31. 1200–800 BP   *Quercus–Fagus–*
                        Alnus–Corylus–NAP

| 32. | 800–600 BP | *Quercus–Pinus–Corylus*–NAP |
| 33. | 600–400 BP | *Pinus–Quercus*–NAP |

The pollen diagrams

Figure 15.11 shows the subregions and the reference sites. Pollen diagrams of some important sites are shown in Figures 15.12, 15.13, and 15.14. Local pollen-assemblage zones are given. The following regional pollen-assemblage zones have been defined for the Lake Constance region (Rösch 1990a).

Gramineae–Artemisia paz
Gramineae dominant, *Artemisia* subdominant. Upper zone border at the *Juniperus* expansion.

Juniperus paz
Juniperus dominant. Upper zone border at the expansion of *Betula* sect. Albae

Betula paz
Betula dominant. *Pinus* increasing. Upper zone border at the expansion of *Pinus*.

Betula–Pinus 1 paz
Betula and *Pinus* codominant. *Betula* decreasing, *Pinus* increasing. Upper border: *Pinus* becomes predominant.

Pinus paz
Pinus predominant. Upper border at the *Betula* expansion.

Betula–Pinus 2 paz
Betula and *Pinus* codominant. *Corylus* increasing. Upper zone border at the *Corylus* expansion.

Betula–Corylus–Pinus paz
Betula, *Corylus* and *Pinus* codominant. *Ulmus* and *Quercus* increasing. Upper zone border at the *Pinus* decrease.

Corylus paz
Corylus dominant, *Quercus*, *Ulmus*, *Tilia* increasing. Upper zone border: increase of *Fraxinus*.

Corylus–Mixed oak Forest 1 paz
Corylus and Mixed Oak Forest codominant, *Corylus* decreasing. Upper zone border: a distinct *Corylus* decrease.

Mixed Oak Forest paz
Mixed Oak Forest dominant, *Alnus* and *Fagus* increasing. Upper zone border: expansion of *Fagus*.

Mixed Oak Forest–*Fagus* paz
Mixed Oak Forest and *Fagus* codominant. Upper zone border: decrease of *Ulmus*, *Fagus* and *Tilia*.

Corylus–Mixed Oak Forest 2 paz
Corylus, *Quercus*, *Betula* and *Alnus* dominant. Upper zone border: increase of *Fagus*.

Fagus 1 paz
Fagus dominant. Upper zone border: increase of *Corylus*.

Corylus 2 paz
Corylus dominant (together with *Alnus* and/or *Quercus*. Upper zone border: increase of *Fagus*.

Fagus 2 paz
Fagus dominant. *Alnus*, *Corylus*, *Quercus*, *Betula* subdominant. Upper zone border: increase of *Betula*.

Quercus–Corylus paz
Corylus and/or *Quercus* dominant. Upper zone border: increase of *Betula*.

Betula–Corylus–Fagus–Quercus paz
Betula, *Corylus*, *Fagus* and *Quercus* dominant. Upper zone border: increase of *Quercus*.

Quercus 1 paz
Quercus (and *Betula*) dominant. NAP increasing. Upper zone border: Increase of *Betula*, *Alnus*, *Carpinus* and *Fagus*.

Quercus–Betula–Fagus–Carpinus paz
Fagus, *Alnus* and *Betula* dominant. Upper zone border: increase of *Quercus*.

Quercus 2 paz
Quercus dominant. *Pinus* and NAP increasing. Upper zone border: increase of *Juniperus*.

Pinus–Quercus paz
Pinus and *Quercus* dominant. Upper zone border: increase of *Picea*.

Pinus–Picea paz
Pinus dominant, *Picea* subdominant.

530

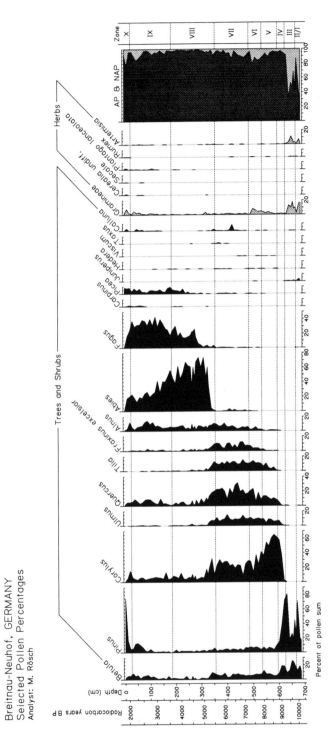

Fig. 15.12 Pollen diagram from Breitnau-Neuhof. Only selected pollen types are shown. Vertical scale is the corrected time scale (yr BC/AD).

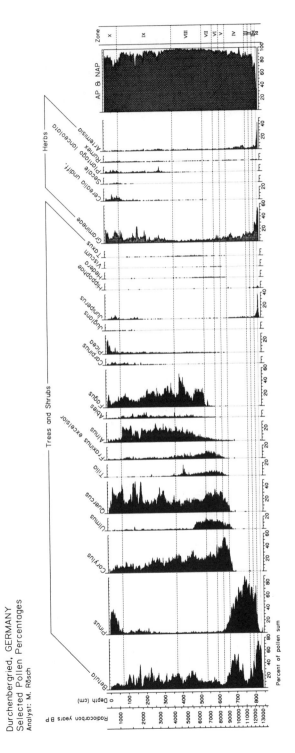

Fig. 15.13 Pollen diagram from Durchenbergried. Only selected pollen types are shown. Vertical scale is the corrected time (yr BC/AD)

532

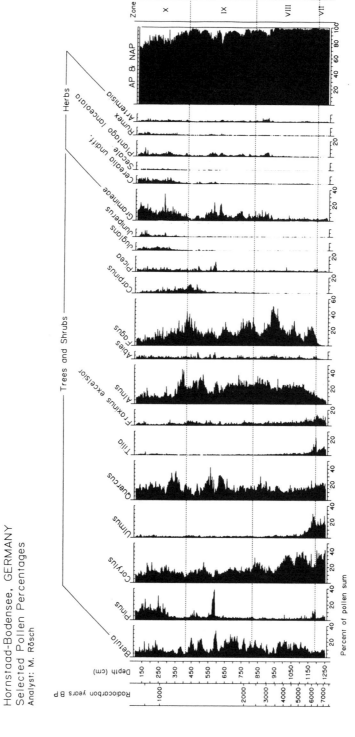

Fig. 15.14 Pollen diagram from Hornstaad–Bodensee. Only selected pollen types are shown. Vertical scale is the corrected time scale (yr BC/AD)

533

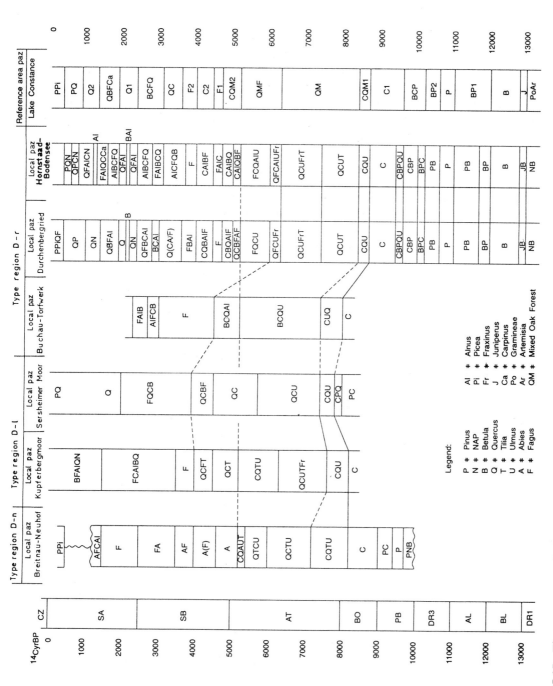

Fig. 15.15 Time–space correlation of local pollen-assemblage zones for type regions D-n, D-l, and D-r. Abbreviations: CZ Chronozones, SA Subatlantic, SB Subboreal, AT Atlantic, BO Boreal, PB Preboreal, DR3 Younger Dryas, AL Alleröd, BL Bölling, DR1 Oldest Dryas. These are chronozones according to Mangerud *et al.* (1974) modified *sensu* Welten (1982) (The Older Dryas is eliminated.)

PALAEOECOLOGICAL PATTERNS AND EVENTS IN TYPE REGIONS D-n, D-l AND D-r

Vegetation patterns

Common patterns

(1) The complete late Würmian is only represented in profiles from the Prealpine Lowlands.
(2) The late Würm and early-Holocene pollen assemblages show similar trends.
(3) The *Corylus* maximum is synchronous (about 8500 BP).
(4) The *Ulmus* decline at about 5200 BP is synchronous.
(5) The immigration of *Carpinus* at about 3000 BP is synchronous.
(6) First pollen of *Juglans, Castanea* and *Secale ca.* 2000 BP (single grains in some cases a little earlier).

Unique patterns

(1) Expansion of *Corylus* in D-n later than in D-l.
(2) Expansion of *Quercus, Ulmus,* and *Tilia* first at Lake Constance, later at Federsee and in the Neckar-Gäuplatten, latest in D-n .
(3) Expansion of *Fagus* at Lake Constance earlier (6000 BP) than in the other regions (4500 BP).
(4) Composition of Mixed Oak Forest:

  – *Tilia* more frequent in D-n than in D-l
  – *Quercus* most frequent in the Neckar Gäuplatten
  – *Ulmus* most frequent in the Prealpine Lowlands (D-r)

(5) Replacement of *Ulmus* and *Tilia* in D-n by *Abies*, in D-r by *Fagus, Corylus, Alnus, Betula*, in the Neckar-Gäuplatten by *Corylus* and *Quercus.*

The landscape

The succession from a landscape without any vegetation to pioneer vegetation, steppe and dwarf-shrub communities, and the following reforestation (*Juniperus–Hippophaë* shrubs, *Betula* forests, *Pinus* forests) can be recognized in many diagrams and seems to have been contemporary in large parts of the northern Prealpine Lowlands from the Rhine glacier to the Rhone glacier area.

At the end of the late Würmian *Betula* spread again and thermophilous tree species appeared. *Corylus* and Mixed Oak Forest species spread and pushed out *Pinus* and *Betula*.

In the Boreal *Corylus* predominated in the pollen diagrams.

In the early Atlantic *Tilia* spread and *Corylus* decreased. The first traces of *Fagus* and *Abies* occurred. The Mixed Oak Forest became dominant.

The last centuries of the late Atlantic were a period of great environmental change. In the Black Forest (D-n) the Mixed Oak Forest was completely replaced by *Abies*. In both type regions the final *Ulmus* decline took place. At Lake Constance a decline of *Fagus, Tilia* and *Fraxinus* also occurred, together with an increase of *Corylus* and *Betula*, whereas at Federsee *Fagus* increased and Mixed Oak Forest decreased.

In the middle of the Early Subboreal, *Fagus* increased in all regions, although the peaks in the Black Forest and the Neckar-Gäuplatten were rather moderate. At Lake Constance *Fagus* decreased a few centuries later and *Corylus* increased again.

At the transition from Early to Middle Subboreal at Lake Constance, *Corylus* and *Quercus* increased, and at Federsee *Corylus* and *Betula* expanded. In the Black Forest *Fagus* increased again, while in the Neckar-Gäuplatten the expansion of *Fagus* took place. Here, as well as in the Federsee area, *Fagus* mainly replaced *Quercus*. At Lake Constance the next *Fagus* peak is somewhat later. During this time the first traces of *Carpinus* occur. *Fagus* was replaced by *Quercus* or *Corylus* at Lake Constance to a high degree, at Federsee to a lesser one. In the Black Forest there was a further decrease of *Abies* and an increase of *Fagus*.

In the Early Subatlantic *Quercus* was most frequent at Lake Constance and in the Neckar-Gäuplatten. In the Federsee region *Fagus* remained most frequent, whereas in the Black Forest *Fagus* and *Abies* were most frequent.

In the Middle Subatlantic a re-establishment of *Fagus* forests, now with the inclusion of *Carpinus*, took place in D-l.

In the Late Subatlantic, after a time when *Quercus*

was very frequent, the natural forest components of the region, such as broad-leaved trees and *Abies*, were largely replaced by *Pinus* and *Picea*.

Anthropogenic events (see also Fig. 15.16)

Common patterns between D-n, D-r, and D-l

(1) Distinct clearances in Roman and early Medieval time.

Unique patterns

(1) Towards the end of the Middle Atlantic the first traces of Neolithic human impact can be recognized in the lowlands (reference areas a, b, c). The main features are cereal grains, a decline of *Tilia*, a first *Fagus* peak (up to 10%) at Lake Constance, and increased minerogenic input.
(2) At the transition from the Middle to the Late Atlantic, after additional forest clearances at Lake Constance, *Fagus* expanded. In the lowlands the *Ulmus* decline started.
(3) Towards the end of the late Atlantic in both type regions the final *Ulmus* decline took place. At Lake Constance there was also a decline of *Fagus, Tilia* and *Fraxinus*, together with an increase of *Corylus* and *Betula*, whereas at Federsee *Fagus* increased and Mixed Oak Forest decreased. This was the beginning of the Late Neolithic lake-shore dwellings. The changes mentioned above were caused by human occupation of the forests along the lake shores and by widespread changes of the forest composition by shifting cultivation, which is indicated not only by pollen analysis, but also by plant-macrofossil analysis of lake-shore dwellings (Rösch 1990b).
(4) In the Early Subboreal, distinct but not synchronous decreases in human impact can be observed.
(5) In the Middle Subboreal there were again decreases in human impact. In the Prealpine Lowlands they were more or less synchronous, but in the Neckar-Gäuplatten they were later.
(6) In the Late Subboreal, extensive but not synchronous clearances took place. Afterwards most sites show a reduction of human impact.

(7) In the Early Subatlantic, forest clearances became more extensive than before. Towards the end of the Early Subatlantic human impact decreased again.
(8) At the beginning of the Middle Subatlantic a strong human impact, caused by the Romans, can be observed in all regions. Afterwards, from the 3rd century AD, human impact became very low.
(9) Later on, in the early Medieval times, human impact became very strong again. The highest degree of deforestation was in the Medieval period (11th to 13th centuries AD), and in early Modern Times.

Hydrological events in type region D-r

Hydrological events related to groundwater variations were only observed as lake-level fluctuations in the lakes of the prealpine lowlands. The observations from Lake Constance (Rösch & Ostendorp 1988), the Federsee (Liese-Kleiber 1988), and the Nussbaumer Seen (Rösch 1983) are compiled in Figure 15.17. We distinguish rising and falling lake levels.

(1) In the late Würmian lake levels fell in general. More or less synchronous fluctuations can be observed in the Bölling and towards the end of the Younger Dryas.
(2) Some smaller Early and Middle Holocene lake-level fluctuations from Federsee and Nussbaumer Seen are known.
(3) Towards the middle of the Early Subboreal lake levels rose synchronously and later fell again.
(4) Another synchronous and rather distinct lake-level fluctuation occurred at the transition from the Middle to the Late Subboreal.

These general synchronous trends in lake-level fluctuation are noteworthy, because these three lakes differ enormously in the size and form of their basins, and in the size of their catchment area. They must have general, probably climatic causes.

To recognize hydrological events in the peat of the investigated mires is rather more difficult, because peat growth was influenced to a high degree by local hydrological events and by man (cf. Rösch 1986, Rösch & Ostendorp 1988).

536

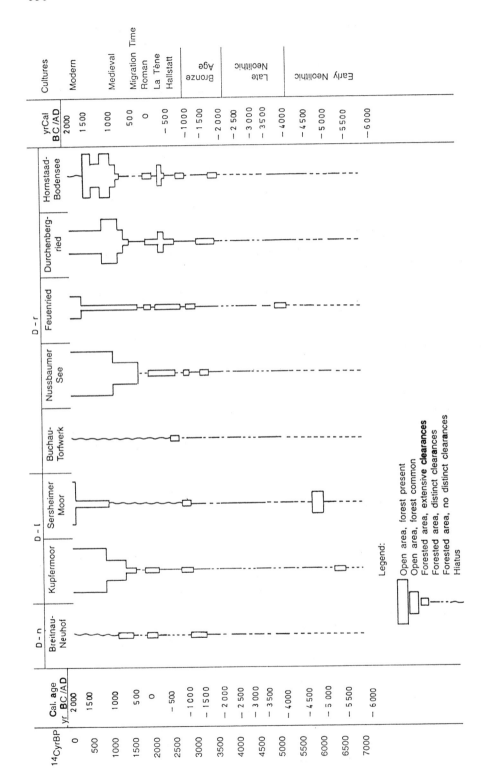

Fig. 15.16 Schematic interpretation of landscape openness based on the frequency of pollen types (including Gramineae) from dry soil habitats

Climatic events

Stable-oxygen analyses from lake marl cores of Nussbaumer Seen and of Lake Constance give hints at late Würmian and Holocene climatic fluctuations (Rösch 1983, Eicher & Rösch, unpublished, see also Fig. 15.18).

Long-term trends in climate

Ignoring short-term fluctuations, the stable-oxygen curves show rather high values between the end of the Preboreal and the Early Subboreal, reduced values between the Middle Subboreal and the Early Subatlantic, and lowest values in the Middle and Late Subatlantic. Therefore we suggest the warmest Holocene climate was from the Boreal to the Early Subboreal, and the coldest Holocene climate from the Middle Subboreal to the present.

Short-term trends in climate

(1) The climatic deterioration of the Older Dryas, which is hard to recognize in terms of vegetational changes, can be observed, as well as the climatic deterioration of the Younger Dryas. At this time the Alpine forest limit in the Black Forest decreased from above 1000 m down to 800–700 m (Lang 1971). Holocene climatic deteriorations can only be recognized by oxygen-isotope analysis, not by vegetational changes. If they correspond with climatic deteriorations known from investigations at the Alpine timberline (Zoller 1968, Bortenschlager & Patzelt 1969), these names are used (see also Fig. 15.9).

(2) In the middle of the Preboreal, therefore, there are hints of the first Holocene climatic deterioration (Schlaten-Schwankung).

(3) In the middle of the Boreal a climatic deterioration (Venediger-Schwankung) can be recognized.

(4) At the transition from the Middle to the Late Atlantic another climatic deterioration (Frosnitz-Schwankung) can be observed.

(5) In the middle of the Early Subboreal a further climatic deterioration took place (Rotmoos-Schwankung 2).

(6) Towards the end of the Early Subboreal, ^{18}O decreased again.

(7) In the middle of the Middle Subboreal, a further climatic deterioration took place (Löbben-Schwankung). This event is synchronous with a general rise of lake levels, a general expansion of *Fagus*, and a more or less general decrease in human impact.

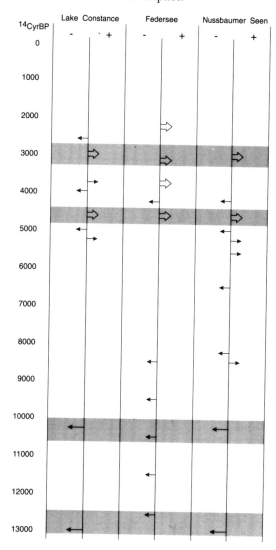

Fig. 15.17 Lake-level fluctuations of Lake Constance, Federsee and Nussbaumer Seen, according to Rösch & Ostendorp (1988), Liese-Kleiber (1988) and Rösch (1983). + and open arrows rise of lake level; – and black arrow fall of lake level

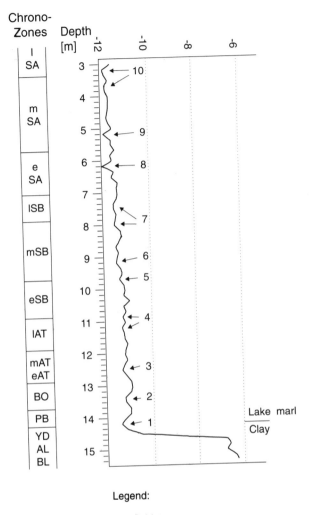

Legend:

1 Schlaten
2 Venediger
3 Frosnitz
4 Rotmoos
5 um 4000 BP ≙ 2600 BC ?
6 Löbben
7 um 3000 BP ≙ 1200 BC ?
8 Göschener Kaltphase 1
9 Göschener Kaltphase 2
10 ca. 900-1200 AD ?

Fig. 15.18 Oxygen-isotope curve of the profile Hornstaad-Bodensee. Low values with numbers: 1 corresponds to Schlaten-Schwankung (Preboreal), 2 Venediger-Schwankung (Boreal), 3 Frosnitz-Schwankung (Early Atlantic), 4 Rotmoos-Schwanhung (two parts, Late Atlantic and Early Subboreal), 5 ^{18}O-oscillation at the transition from Early to Middle Subboreal, 6 Löbben-Schwankung (Middle Subboreal), 7 Göschener Kaltphase 1 (Late Subboreal), 8 continuation of the Göschener Kaltphase 1 (Early Subatlantic), 9 Göschener Kaltphase 2 (Middle Subatlantic), 10 ^{18}O-oscillation at the transition from the Middle to Late Subatlantic

TYPE REGION D-r   LAKE CONSTANCE AND FEDERSEE REGIONS

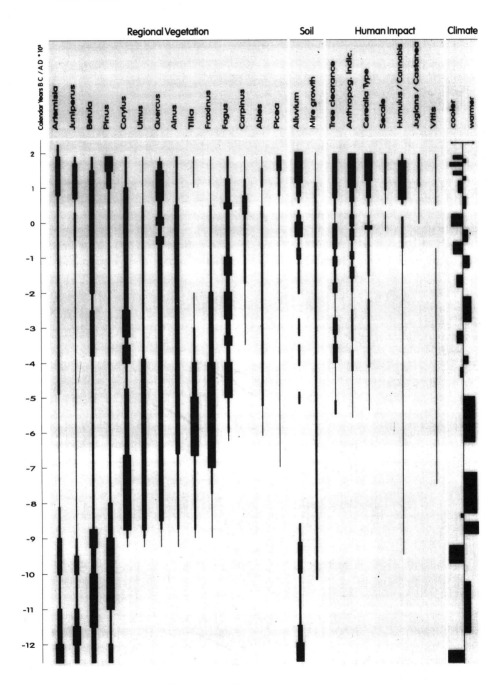

Fig. 15.19 Environmental changes in the type region D-r during the last 15 000 years

TYPE REGION D-1   NECKAR AND TAUBER REGIONS

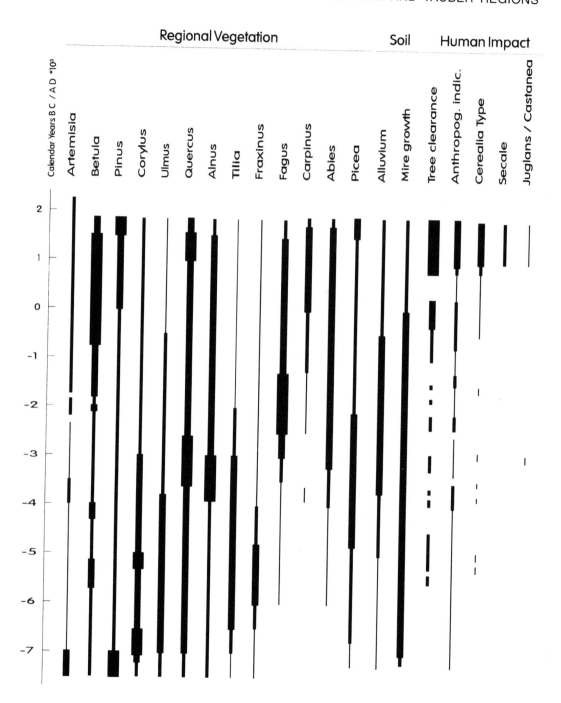

Fig. 15.20   Environmental changes in the type region D-1 during the last 9000 years

TYPE REGION D-n   BLACK FOREST

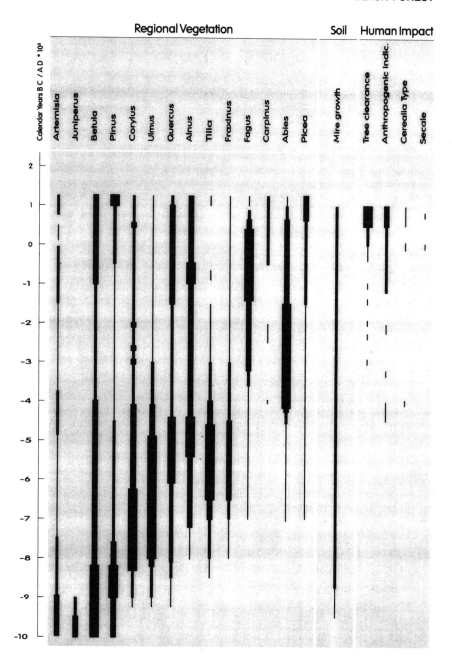

Fig. 15.21   Environmental changes in the type region D-n during the last 12 000 years

(8) At the transition from the Middle to the Late Subboreal [18]O values decreased. A general rise of lake levels and a more or less general expansion of *Fagus, Alnus* and *Betula* are synchronous. This may be the beginning of the Göschener Kaltphase 1.

(9) A subsequent δ^{18}O fluctuation in the Early Subatlantic may correspond to the continuing Göschener Kaltphase 1.

(10) [18]O values indicate a further climatic deterioration in the Middle Subatlantic (2nd to 4th century AD). At this time a marked decrease in human impact along with reforestation by *Fagus, Alnus, Betula* and *Carpinus* took place.

Soil events

Investigations have shown a close relationship between soil development, vegetation and human impact (Rösch 1983, 1986, 1987a, 1990b, Vogt 1990). Agriculture in a hilly area resulted in soil erosion and redeposition, for example in lakes and mires. This can be observed from the beginning of agriculture in the Early Neolithic period. A result was the eutrophication of lakes and mires and higher deposition rates in lakes.

CONCLUSIONS

A general overview of the region's vegetation history allows us to draw the following conclusions (see Figs 15.19, 15.20, and 15.21). The late Würmian vegetation development was more or less similar and synchronous. Only at the higher altitudes of the Black Forest was there a later reforestation and a stronger Younger Dryas deforestation. The Holocene immigration and spread of thermophilous tree species was also more or less synchronous. The early Holocene *Corylus* maximum was earlier in the Lake Constance area than in the other parts of the type region. The first occurrence of *Picea, Abies,* and *Fagus* was more or less synchronous in the Early Atlantic. The Early Atlantic replacement of *Corylus* by *Tilia* and other Mixed Oak Forest components took place at Lake Constance earlier than elsewhere. The composition of the

Atlantic Mixed Oak Forest differed from one type region and reference area to the other. The highest values of *Quercus* occur in the Neckar-Gäuplatten, the highest values of *Ulmus* at the Prealpine plateau, and the highest values of *Tilia* in the Black Forest. The spread of *Fagus* first started in the Lake Constance area about 6000 BP. This event was shortly before 5000 BP at Federsee, in the Black Forest 4500 BP, and in the Neckar-Gäuplatten 4000 BP. On the other hand the first *Tilia* decline (at about 6500 BP) and the final *Ulmus* decline (at about 5200 BP) were more or less synchronous. The reasons, however, are probably quite different. In the lowlands the *Ulmus* decline seems to have been induced by Neolithic man, while in the Black Forest the replacement of the Mixed Oak Forest by *Abies* was probably a natural succession. Later on there are differences in the vegetational development of the subregions due to different amounts of human impact. Later we can observe lake-level fluctuations and traces of climatic change but, nevertheless, the conclusion is that the human influence on vegetation here since the Neolithic period was much more important than climatic impact.

ACKNOWLEDGEMENTS

The critical reading and discussion of the manuscript by Professor Stefanie Jacomet and the revision of the English by Mr S.J. Turrell and Miss Anna Oxley is gratefully acknowledged.

TYPE REGION D-o, ALPINE FORELAND
(H. Küster)

Type region D-o lies between the Danube River to the north and the Alps to the south. It includes southern Bavaria and extends westwards into Oberschwaben, i.e. a part of the country Baden-Württemberg. The northern part is dominated by Tertiary limestone, whereas to the east the bedrock is covered by loess. Hence the eastern part is very fertile and has been intensively settled since the early Neolithic. The geomorphology of the southern part is characterized by glacial features. In the Würmian till landscape are many moraines, lakes, and bogs. There are also large sandur plains, good examples of which can be seen in the vicinity of Munich.

The geological variation is paralleled by considerable variation in topography. The region embraces plains and montane areas as well as undulating landscapes dominated by hills and valleys. Each valley appears to have its own characteristic vegetational history, making difficult the selection of a pollen profile to typify the vegetational history of the region. Profiles from Haslacher See (postglacial) and Langegger Filz (late-glacial) are suitable in the context of IGCP Project 158 B because of the many ^{14}C dates available. However, they are representative only of the vegetational development in the western part of type region D-o.

Altitude: 300 m (Danube valley) to 800 m (southern part of the Alpine Foreland), but some Tertiary nunatak regions in the south exceed 1000 m.

Climate: Mean January air temperature 2 to 4°C, July 18°C, precipitation 600 mm yr^{-1} in the north, 1800 mm yr^{-1} in the south near the Alps. More continental in the northeast, more oceanic in the south.

Geology: Tertiary limestone, loess, Würmian moraines and sandurs.

Topography: Plains and hills in the north, moraines in the south with a more or less montane, prealpine character.

Population: *ca.* 165 km^{-2}, with some large cities (e.g. Munich, Augsburg), towns, villages, and also single farms, which are very typical of the landscape.

Vegetation: The natural forests are dominated mainly by *Fagus sylvatica* in the north, *Picea abies* in the southeast, and *Abies alba* in the southwest. *Quercus* spp. forests are a feature of the loess areas in the east, and *Pinus sylvestris* woods occur on sandy areas along the Danube River. Some continental heathlands with a steppe-like vegetation were possibly never as densely wooded as other parts of the landscape.

Today most of the forests are dominated by *Picea* as a result of human management. Agriculture is based on cereal production in the northern part, whereas cattle breeding predominates in the south.

Fenland areas, such as in the vicinity of Munich, have been cultivated and support arable farming today.

Soils: Mostly brown soils, some peaty areas.

Land use: Wheat, barley, oats, rye, potatoes, maize, sugar-beet (mainly in the northeast). Pasture and hay/silage production dominate in the south.

The pollen diagrams presented (Fig. 15.22) have been made in connection with a detailed study of the Auerberg region in southern Bavaria. Langegger Filz, covering the late-glacial, is situated only 4 km south of the Haslacher See. Seventeen pollen-assemblage zones have been defined for this area, which lies west of the type region. These zones may not necessarily apply to the region as a whole.

Reference site 19. Langegger Filz
(Küster 1988)

Latitude 47°42′N, Longitude 10°46′E. Elevation 800 m. Age range 12500–0 BP (only 12500–10000 BP presented here), bog, three regional pollen-assemblage zones (paz). (Fig. 22)

1.	–12000 BP	Not forested, *Pinus* dominant within AP, *Betula* codominant. Some pollen finds of thermophilous trees probably due to contamination and/or longdistance transport.
2.	12000–11000 BP	Area already wooded. *Pinus* dominant, some *Betula*.
3.	11000–10000 BP	*Pinus* and *Betula* representation almost as high as in the previous period. No clear-cut evidence for climatic deterioration. Some pollen records of *Juniperus, Dryas, Geum*-type, Rosaceae, Cyperaceae, *Selaginella selaginoides* (not all shown in the diagram).

Reference site 20. Haslacher See
(Küster 1986, 1988, 1989)

Latitude 47°45'N, Longitude 10°47'E. Elevation
765 m. Age range 10000–0 BP, bog, 14 regional
pollen-assemblage zones (paz). (Fig. 15.22)

4. 10000–8500 BP *Pinus* dominant, *Corylus* and
    Ulmus increasing.
5. 8500–7700 BP *Pinus* decreasing, *Corylus*
    maximum, which is not as
    marked as in the western
    part of the type region and
    in the vicinity of Lake
    Constance. *Tilia* has a
    continuous pollen curve
    from the beginning. *Ulmus*
    achieves its first maximum
    at the end of the paz.
6. 7700–7200 BP *Ulmus* and *Corylus* are very
    widespread. *Ulmus* was
    very common in regions
    close to he Alps during this
    time, in contrast to the
    landscapes farther north,
    where *Quercus* was of
    greater importance. *Picea*
    is increasing. At the end of
    the paz *Abies* attains a
    continuous curve.
7. 7200–6400 BP *Ulmus* and *Picea* dominant,
    Abies increasing. At the
    end of the paz the percent-
    age representation of
    montane forest trees
    (*Picea, Abies, Fagus*)
    exceeds that of the
    Quercetum mixtum.
8. 6400–5600 BP *Ulmus* decreasing. "Elm
    declines" are normally
    found very early in this
    region. *Abies* increases as
    Ulmus declines, so it may
    be assumed that former
    Ulmus stands near the
    Alps were replaced by
    Abies. This can also be il-
    lustrated by the fact that

the middle Holocene
Ulmus woodlands and re-
cent *Abies* stands occupy
exactly the same area in
the type region.

9. 5600–5000 BP *Picea* dominant, expansion
    of both *Abies* and *Fagus*.
    At the end the *Ulmus* curve
    drops below 5%, and the
    decline of *Pinus*, traceable
    since the 9th millennium
    BP, comes to an end.
10. 5000–4500 BP *Picea* and *Abies* dominant,
    Fagus still increasing and
    attaining a maximum at
    the end, characteristic for
    the type region as a whole.
11. 4500–3900 BP *Picea, Abies,* and *Fagus*
    dominant, *Ulmus* and
    Tilia decreasing to a mini
    mum. At the end of the paz
    the *Tilia* pollen curve is in-
    terrupted for the first time.
12. 3900–2600 BP *Picea* and *Fagus* dominant,
    Abies decreasing; this
    event seems to be very
    typical for the type region.
    At the end of the paz *Abies*
    falls to a minimum.
13. 2600–1900 BP *Picea* and *Fagus* still
    dominant, *Abies* increas-
    ing again. At the end of the
    paz the beginning of
    woodland utilization by
    the Romans is recorded:
    Abies was preferentially
    selected as a source of
    good timber.
14. 1900–1600 BP Period of Roman occupa-
    tion.Deforestation, cutt-
    ing of mainly *Abies* and
    Fagus so that only *Picea*
    consolidated its role as
    dominant in the forests. At
    the end of the paz a
    marked *Betula* maximum
    shows the increase in

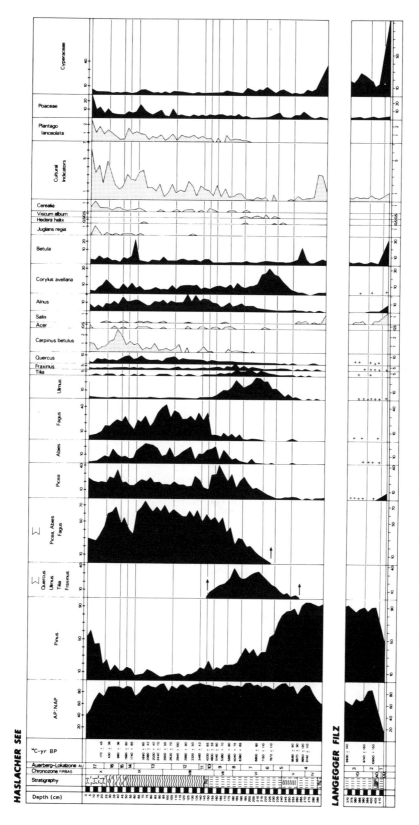

Fig. 15.22 Pollen diagrams from Langegger Filz and Haslacher See. Analyst: H. Küster

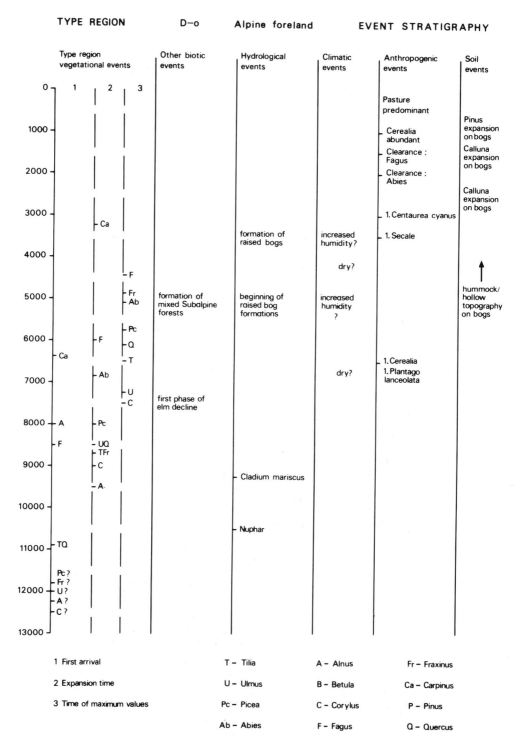

Fig. 15.23 Environmental changes in the type region D-o

fallow land as the Romans retreated. Cultural indicators show that the total landscape was not abandoned.

15. 1600–1000 BP Regeneration of the montane forests during the Dark Ages, *Fagus* and *Abies* increasing. Human impact is still visible in the curves for cultural indicators.

16. 1000–400 BP Medieval land-use. Decreases of *Abies* and *Fagus* caused by wood cutting.

17. 400–0 BP Modern land-use. *Picea* alone dominant in the forests. *Pinus* increasing, reflecting the spreading of pine onto drained bogs.

Palaeoecological patterns and events (Fig. 15.23)

(1) Late-glacial pollen grains of thermophilous trees probably result from contamination and/or long-distance transport of pollen grains.

(2) During the Late-glacial *Pinus* became the principal tree in the landscape. *Betula* was also present but never dominant. The area was wooded at an early stage.

(3) *Ulmus* forests were typical for the early and middle Holocene in the southern part of type region D-o, south of a line from Memmingen to Chiemsee. North of this "border" *Quercus* was more important.

(4) The later composition of the forest communities was mainly influenced by the different immigration pathways of *Picea*, *Abies*, and *Fagus*. *Picea*, coming from the east, is therefore dominant in the eastern part of the type region. *Picea* and *Abies*, the latter coming from the west, met in the Alpine valley of the Rhine River and Lake Constance. *Abies* therefore did not become a dominant in the forests of the type region where *Picea* arrived earlier. But *Abies* came to be of greater importance in the west than in the east.

Fagus immigrated also from the west, but its path lay farther to the north. Hence it came to dominance only in the north of the type region.

(5) Human influence can be traced to a very early date in the pollen diagrams from most parts of the type region. Some pollen diagrams show cereal and *Plantago lanceolata* pollen in the 7th millennium BP, whereas archaeologists to date have not been able to trace early Neolithic settlements in some parts of the region. Early Neolithic settlements are only known from the north and northwest of the type region.

(6) Human impact on the natural forest caused a marked decrease in mainly *Abies* and *Fagus* during the last two or three millennia. *Abies* was cut for timber and *Fagus* for producing charcoal, which was required for iron smelting and glass factories.

(7) The first *Secale* pollen grains are found at about 3000 BP (not in the Haslacher See profile). Macrofossil evidence shows that *Secale cereale* has been a weed in cereal fields since at least the early Bronze Age (beginning of the 4th millennium BP).

REFERENCES

Bakels, C.C. 1978: *Four Linearbandkeramik settlements and their environment: a palaeoecological study of Sittard, Stein, Elsloo and Hienheim. Analecta Praehistorica Leidensia 11*, 245 pp. Leiden University Press, Leiden.

Behre, K.-E. 1966: Untersuchungen zur spätglazialen und frühpostglazialen Vegetationsgeschichte Östfrieslands. *Eiszeitalter u. Gegenwart 17*, 69–84, Öhringen Württ.

Behre, K.-E. 1967: The late glacial and early postglacial history of vegetation and climate in Northwestern Germany. *Review of Palaeobotany and Palynology 4*, 149–161, Amsterdam.

Behre, K.-E. 1970: Wirkungen vorgeschichtlicher Kulturen auf die Vegetation Mitteleuropas. *n+m, Naturwiss. u. Medizin 7*(34), 15–30, Mannheim.

Behre, K.-E. 1976: Pollenanalytische Untersuchungen zur Vegetations- und Siedlungsgeschichte bei Flögeln und im Ahlenmoor (Elb–Weser–Dreieck). *Probleme der Küstenforschung im südlichen Nordseegebiet 11*, 101–118, Hildesheim.

Behre, K.-E. 1978: Die Klimaschwankungen im europäischen Präboreal. *Petermanns Geographische Mitteilungen. 2*, 97–101, Gotha/Leipzig.

Behre, K.-E. 1980: Zur mittelalterlichen Plaggenwirtschaft in Nordwestdeutschland und angrenzenden Gebieten nach botanischen Untersuchungen. *Abhandlungen Akademic der Wissenschaften. Göttingen, Phil.-Hist. Kl., 3. Folge, 116*, 30–44, Göttingen.

Behre, K.-E. & Kučan, D. 1986: Die Reflektion archäologisch bekannter Siedlungen in Pollendiagrammen verschiedener Entfernung—Beispiele aus der Siedlungskammer Flögeln, Nordwestdeutschland. *In* Behre, K.-E. (ed.) *Anthropogenic indicators in pollen diagrams*, 95–114,.Rotterdam.

Behre, K.E. & Kučan, D. 1994: Die Geschichte der Kulturlandschaft und des Ackerbaus in der Siedlungskammer Flögeln, Niedersachsen, seit der jungsteinzeit. *Probleme der Küstenforschung im südlichen Nordseegebiet 21*, 228 pp. Oldenburg.

Bertzen, G. 1987: Diatomeen in spätpleistozänen und holozänen Sedimenten des Tegeler Sees in Berlin. *Berliner Geographische Abhandlungen 45*, 1–151.

Beug, H.-J. 1976: Die spätglaziale und frühpostglaziale Vegetationsgeschichte im Gebiet des ehemaligen Rosenheimer Sees (Oberbayern). *Botanische Jahrbücher für Systematik, Pflanzengeschichte und Pflanzengeographie 95*(3), 373–400.

Beug, H.-J. 1992: Vegetationsgeschichtliche Untersuchungen über die Besiedlung im Unteren Eichsfeld, Landkreis Göttingen, vom frühen Neolithikum bis zum Mittelalter. *Neue Ausgrabungen und Forschungen in Niedersachsen 20*, 261–339.

Beug, H.-J. & Henrion, I. 1971: Eine vegetationsgeschichtliche Dokumentation in Göttingen. *Pollen et Spores 13*, 485–492.

Böcker, R. & Sukopp, H. (eds) 1985: Ökologische Karten Berlins, Beispiel Tegel und Tegeler See. *In* Hofmeister, B., Pachur, H.-J., Pape, Ch. & Reindke, G. (eds) *Berlin, Beiträge zur Geographie eines Großstadtraumes* (16 authors), Berlin.

Böcker, R. & Sukopp, H. 1987: Vegetation 1:50 000. *Umweltatlas Berlin, Teil 2*, Berlin.

Böcker, R., Brande, A. & Sukopp, H. 1986: Das Postfenn im Berliner Grunewald. *Abhandlungen aus dem Westfälischen Museum für Naturkunde 48*, 417–432, Münster.

Borcherdt, C. 1983: 1. Das Land Baden-Württemberg—ein Überblick. *In* Borcherdt, C. (ed.) Geographische Landeskunde von Baden-Württemberg. *Schriften z. polit. Landeskunde von Baden-Württemberg. 8*, 21–80, Stuttgart.

Bortenschlager, S. & Patzelt, G. 1969: Wärmezeitliche Klima- und Gletscherschwankungen im Pollenprofil eines hochgelegenen Moores (2270 m) der Venedigergruppe. *Eiszeitalter u. Gegenwart 20*, 116–122.

Böse, M. & Brande, A. 1986: Zur Entwicklungsgeschichte des Moores "Alter Hof" am Havelufer (Berliner Forst Düppel). *In* Ribbe, W. (ed.) *Berlin-Forschungen 1*, 11–42.

Brande, A. 1978/79: Die Pollenanalyse im Dienste der landschaftsgeschichtlichen Erforschung Berlins. *Berliner Naturschutzblätter 22/23*, 435–443, 469–475, Berlin.

Brande, A. 1980: Pollenanalytische Untersuchungen im Spätglazial und frühen Postglazial Berlins. *Verhandlungen des Botanischen Vereins der Provinz Brandenburg 115*, 21–72, Berlin.

Brande, A. 1985: Mittelalterlich-neuzeitliche Vegetationsentwicklung am Krummen Fenn in Berlin-Zehlendorf. *Verhandlungen des Berliner Botanischen Vereins 4*, 3–65.

Brande, A. 1986a: Mittelalterliche Siedlungsvorgänge in Berliner Pollendiagrammen. *Courier Forschungsinstitut Senckenberg 86*, 409–414, Frankfurt am Main.

Brande, A. 1986b: Stratigraphie und Genese Berliner Kleinmoore. *Telma 16*, 319–321, Hannover.

Brande, A. 1988a: Das Bollenfenn in Berlin-Tegel. *Telma 18*, 95–135, Hannover.

Brande, A. 1988b: Zum Stand der palynologischen Forschung im Berliner Quartär. *Documenta naturae 44*, 1–7, München.

Brande, A. 1990: Klimageschichte seit der Eiszeit. *In* Sukopp, H. (ed.) *Stadtökologie—das Beispiel Berlin*, 22–30. Reimer-Verlag, Berlin.

Brande, A. (in prep.): On the palynology of the Tegel lake in Berlin.

Brande, A. & Hühn, B. 1988: Zur ehemaligen Moorvegetation auf dem Teltow in Berlin (West). *Verhandlungen des Berliner Botanischen Vereins 6*, 13–39.

Brande, A., von Lührte, A. & Schumann, M. 1987: Mittelalterliche Siedlungsgeschichte und Landnutzung im Lichte der Historischen Botanik. *In* Museum für Vor- und Frühgeschichte Berlin (ed.) *Bürger Bauer Edelmann- Berlin im Mittelalter*, 56–82. Berlin.

Brande, A., Hoelzmann, Ph. & Klawitter, J. 1990: Genese und Paläoökologie eines brandenburgischen Kesselmoores. *Telma 20*, 27–54, Hannover.

Dietz, C., Grahle, H.-O. & Müller, H. 1958: Ein spätglaziales Kalkmudde-Vorkommen im Seck-Bruch bei Hannover. *Geologisches Jahrbuch 76*, 67–102, Hannover.

Dörfler, W. 1989: Pollenanalytische Untersuchungen zur Vegetations- und Siedlungsgeschichte im Süden des Landkreises Cuxhaven. *Probleme der Küstenforschung im südlichen Nordseegebiet 17*, 1–75, Hildesheim.

Firbas, F. 1949: *Spät- und nacheiszeitliche Waldgeschichte Mitteleuropas nördlich der Alpen. Bd. 1, Allgemeine Waldgeschichte*, 480 pp., Jena.

Firbas, F. 1952: *Spät- und nacheiszeitliche Waldgeschichte Mitteleuropas nördlich der Alpen. Bd. 2, Waldgeschichte der einzelnen Landschaften.* 256 pp., Jena.

Geyer, O. & Gwinner, M. 1986: *Geologie von Baden-Württemberg. 3. Aufl.* 472 pp., Stuttgart.

Geyh, M.A., Merkt, J. & Müller, H. 1971: Sediment-, Pollen- und Isotopenanalysen an den jahreszeitlich geschichteten Ablagerungen im zentralen Teil des Schleinsees. *Archiv für Hydrobiologie 69*, 366–399.

Golombek, E. B. 1980: Pollenanalytische Untersuchungen zur spät- und postglazialen Vegetationsgeschichte im Drömling (Ostniedersachsen) *Berichte naturhistorische Gesellschaft Hannover 123*, 79–157, Hannover.

Gradmann, R. 1931: *Süddeutschland. 2 vols*, 215 and 553 pp., Stuttgart.

Grenzius, R. 1985: Bodengesellschaften 1:50 000. *Umweltatlas Berlin, Teil 1*, Berlin.

Grohne, U. 1957: Zur Entwicklungsgeschichte des ostfriesischen Küstengebietes auf Grund botanischer Untersuchungen. *Probleme der Küstenforschung im südlichen Nordseegebiet 6*, 1–48, Hildesheim.

Huttenlocher, F. 1972: Naturräumliche Gliederung von Baden-Württemberg. *Historischer Atlas v. Baden-Württemberg, Erläuterungen II, 4*, Stuttgart.

Isenberg, E. 1979: Pollenanalytische Untersuchungen zur Vegetations- und Siedlungsgeschichte im Gebiet der Grafschaft Bentheim. *Abhandlungen aus dem Landesmuseum für Naturkunde zu Münster in Westfalen 41/2*, 63 pp., Münster.

Kossack, G. & Schmeidl, H. 1974/75: Vorneolithischer Getreidebau im Bayerischen Alpenvorland. *Jahresbericht der Bayerischen Bodendenkmalpflege 15/16*, 7–23.

Kramm, E. 1978: Pollenanalytische Hochmooruntersuchungen zur Floren- und Siedlungsgeschichte zwischen Ems und Hase. *Abhandlungen aus dem Landesmuseum für Naturkunde zu Münster in Westfalen 40/4*, 47 pp., Münster.

Küster, H. 1986: Werden und Wandel der Kulturlandschaft im Alpenvorland. *Germania 64(2)*, 533–559.

Küster, H. 1988: Vom Werden einer Kulturlandschaft. 214 pp. VCH, Weinheim.

Küster, H. 1989: The history of the landscape around Auerberg, Southern Bavaria: a pollen analytical study. *In* Birks, H.H., Birks, H.J.B., Kaland, P.E. & Moe, D. (eds) *The cultural landscape—past, present and future*, 301–310. Cambridge University Press, Cambridge.

Lang, G. 1955: Neue Untersuchungen über die spät- und nacheiszeitliche Vegetationsgeschichte des Schwarzwaldes. II. Das absolute Alter der Tannenzeit im Südschwarzwald. *Beiträge für naturkundlichen Forschung Südwestdeutschlands 14*, 24–31.

Lang, G. 1962: Vegetationsgeschichtliche Untersuchungen der Magdalénienstation an der Schussenquelle. *Veröffuntlichungen geobotanisches Institut ETH-Stiftung Rübel Zürich 37*, 129–154.

Lang, G. 1971: Die Vegetationsgeschichte der Wutachschlucht und ihrer Umgebung. *Die Wutach*, 323–349. Freiburg i.Br.

Lang, G. 1973: Die Vegetation des westlichen Bodenseegebiets. *Pflanzensoziologie 17*, 451 pp., Jena.

Lange, E., Illig, H., Illig, J. & Wetzel, G. 1978: Beiträge zur Vegetations- und Siedlungsgeschichte der nordwestlichen Niederlausitz. *Abhandlungen und Berichte des Naturkundemuseums Görlitz 52(3)*, 1–80, Leipzig.

Lange, E., Jeschke, L. & Knapp, H.D. 1986: Ralswiek und Rügen. Landschaftsentwicklung und Siedlungsgeschichte der Ostseeinsel. *Schriften zur Ur- und Frühgeschichte 38*, 175 pp., Akademie-Verlag, Berlin.

Lesemann, B. 1969: Pollenanalytische Untersuchungen zur Vegetationsgeschichte des Hannoverschen Wendlandes. *Flora, Abt. B 158*, 480–519, Jena.

Liese-Kleiber, H. 1988: Zur zeitlichen Verknüpfung von Verlandungsverlauf und Siedlungsgeschichte des Federsees. *Forschungen u. Berichte zur Vor- u. Frühgeschichte Baden-Württemberg 31*, 163–176.

Liese-Kleiber, H. 1991: Züge der Landschafts- und Vegetationsentwicklung im Federseegebiet. *Berichte der Römisch-Germanischen Kommission 71*, 58–83.

Litt, Th. 1992: Fresh investigations into the natural and anthropogenically influenced vegetation of the earlier Holocene in the Elbe-Saale region, Central Germany. *Vegetation History and Archaeobotany 1*, 69–86.

Mangerud, J., Andersen, S, Th., Berglund, B.E. & Donner, J.J. 1974: Quaternary stratigraphy of Norden, a proposal for teminology and classification. *Boreas 3*, 109–128.

Meynen, E., Schmithüsen, J., Gelbert, J.F., Neef, E., Müller-Miny, H. & Schultze, J.-H. (1953-)1962: *Handbuch der naturräumlichen Gliederung Deutschlands. Bundes ans falt für handeskunde, Remagen. Bundes ans falt für handeskunde, Remagen.*

Middeldorp, A.A. 1984: *Functional palaeoecology of raised bogs—an analysis by means of pollen density dating, in connection with the regional forest history —* Diss.—124 pp., Amsterdam.

Müller, H. 1970: Ökologische Veränderungen im Otterstedter See im Laufe der Nacheiszeit. *Berichte Naturhistorische Gesellschaft Hannover 114*, 33–47.

Müller, H.M. 1971: Untersuchungen zur holozänen Vegetationsentwicklung südlich von Berlin. *Petermanns Geographische Mitteilungen 115*, 37–45, Gotha.

O'Connell, M. 1986: Pollenanalytische Untersuchungen zur Vegetations- und Siedlungsgeschichte aus dem Lengener Moor, Friesland (Niedersachsen). *Probleme der Küstenforschung im südlichen Nordseegebiet 16*, 171–193, Hildesheim.

Opravil, E. 1983: *Xanthium strumarium* L. ein europäischer Archäophyt? *Flora 173*, 71–79, Jena.

Overbeck, F. 1975: *Botanisch-geologische Moorkunde.* 717 pp., Neumünster.

Pachur, H.-J. 1987: Die Sedimente in Berliner Seen als Archive der Landschaftsentwicklung. *In* Scharfe, W. (ed.) *Berlin und seine Umgebung im Kartenbild*, 73–81. Berlin.

Pachur, H.-J. & Röper, H.P. 1987: Zur Paläolimnologie Berliner Seen. *Berliner Geographische Abhandlungen 44*, 1–150, Berlin.

Pott, R. 1985: Vegetationsgeschichtliche und pflanzensoziologische. Untersuchungen zur Niederwaldwirtschaft in Westfalen. *Abhandlungen Westfälisches Museum für Naturkunde 47, H. 4*, 75 pp., Münster.

Radke, G. 1973: Landschaftsgeschichte und -ökologie des

Nordschwarzwaldes. *Hohenheimer Arbeiten 68*, 1–121, Stuttgart.

Rausch, K.A. 1975: Untersuchungen zur spät- und nacheiszeitlichen Vegetationsgeschichte im Gebiet des ehemaligen Inn-Chiemseegletschers. *Flora 164*, 235–282.

Rösch, M. 1983: Geschichte der Nussbaumer Seen (Kanton Thurgau) und ihrer Umgebung seit dem Ausgang der letzten Eiszeit aufgrund quartärbotanischer, stratigraphischer und sedimentologischer Untersuchungen. *Mitteilungen der Thurgauer Naturfoschunden Gesellschaft 45*, 110 pp.

Rösch, M. 1985a: Nussbaumer Seen—spät- und postglaziale Umweltveränderungen einer Seengruppe im östlichen Schweizer Mittelland. *In* Lang, G. (ed.) *Swiss lake and mire environments during the last 15 000 years.* *Disertationes Botanical* 87, 337–379.

Rösch, M. 1985b: Ein Pollenprofil aus dem Feuenried bei Überlingen am Ried: Stratigraphische und landschaftsgeschichtliche Bedeutung für das Holozän im Bodenseegebiet. *Materialhefte zur Vor- u. Frühgeschichte Baden-Württembergs 7*, 43–79.

Rösch, M. 1986: Zwei Moore im westlichen Bodenseegebiet als Zeugen prähistorischer Landschaftsveränderung. *Telma 16*, 83–111.

Rösch, M. 1987a: Der Mensch als landschaftsprägender Faktor des westlichen Bodenseegebiets seit dem späten Atlantikum. *Eiszeitalter u. Gegenwart 37*, 19–29.

Rösch, M. 1987b: Zur Umwelt und Wirtschaft des Jungneolithikums am Bodensee—Botanische Untersuchungen in Bodman-Blissenhalde. *Archäologische Nachrichten aus Baden 38/39*, 42–53.

Rösch, M. 1989: Pollenprofil Breitnau-Neuhof: Zum zeitlichen Verlauf der holozänen Vegetationsentwicklung im südlichen Schwarzwald. *Carolinea 47*, 15–24.

Rösch, M. 1990a: Vegetationsgeschichtliche Untersuchungen im Durchenbergried. *In Siedlungsarchäologie im Alpenvorland 2. Forsch. u. Ber. z. Vor- u. Frühgeschichte Baden.-Württemberg 37*, 9–64.

Rösch, M. 1990b: Veränderungen von Wirtschaft und Umwelt während Neolithikum und Bronzezeit am Bodensee. *Berichte der Römisch-Germanischen Kommission 71*, 161–186.

Rösch, M. 1991: Zum Stand der vegetationsgeschichtlichen Erforschung des Spätwürm und des Holozäns im Bereich Oberschwabens und der schwäbischen Alb. *In Urgeschichte in Oberschwaben und der mittleren Schwäbischen Alb. Archäologische Informationen aus Baden.-Württemberg 17*, 20–24.

Rösch, M. 1992: Human impact as registered in the pollen record—some results from western Lake Constance. *Vegetation History and Archaeobotany 1*, 101–109.

Rösch, M. 1993: Prehistoric land use as recorded in a lake-shore core at Lake Constance. *Vegetation History and Archaeobotany 2*, 213–232.

Rösch, M. & Ostendorp, W. 1988: Pollenanalytische, torf- und sedimentpetrographische Untersuchungen an einem telmatischen Profil vom Bodensee-Ufer bei Gaienhofen. *Telma 18*, 373–395.

Schmeidl, H. 1971: Ein Beitrag zur spätglazialen Vegetations- und Waldentwicklung im westlichen Salzach gletschergebiet. *Eiszeitalter und Gegenwart 22*, 110–126.

Schmeidl, H. 1972: Zur spät- und postglazialen Vegetationsgeschichte am Nordrand der bayerischen Voralpen. *Berichte der Deutschen Botanischen Gesellschaft 85* (1–4), 79–82.

Schmeidl, H. 1977: Pollenanalytische Untersuchungen im Gebiet des ehemaligen Chiemseegletschers. *Geologische Karte von Bayern 1:25 000. Erläuterungen zum Kartenblatt 8140 Prien am Chiemsee und zum Blatt 8141 Traunstein.* Geologisches Landesamt, München.

Schmeidl, H. 1980: Die Moorvorkommen des Kartenblattes 8239 Aschau i. Chiemgau. Zur spät- und postglazialen Vegetations- und Waldentwicklung in der montanen Stufe des Kartenblattes Aschau i. Chiemgau. *Geologische Karte von Bayern 1:25 000. Erläuterungen zum Kartenblatt 8239 Aschau im Chiemgau.* Geologisches Landesamt München, 111–132.

Schmeidl, H. & Kossack, G. 1967/68: Archäologische und paläobotanische Untersuchungen an der "Römerstraße" in der Rottauer Filzen, Ldkr. Traunstein. *Jahresbericht der Bayerischen Bodendenkmalpflege 8/9*, 9–36.

Schneekloth, H. 1963: Das Hohe Moor bei Scheeßel (Kr. Rotenburg/Hann.). *Beihefte Geologisches Jahrbuch 55*, 1–104, Hannover.

Smettan, H. 1985: Pollenanalytische Untersuchungen zur Vegetations- und Siedlungsgeschichte der Umgebung von Sersheim, Kreis Ludwigsburg. *Fundberichte aus Baden-Württemberg 10*, 367–421.

Smettan, H. 1988: Naturwissenschaftliche Untersuchungen im Kupfermoor bei Schwäbisch Hall—ein Beitrag zur Moorentwicklung sowie zur Vegetations- und Siedlungsgeschichte der Haller Ebene. *Forschungen u. Berichte zur Vor- u. Frühgeschichte Baden-Württemberg 31*, 81–122.

Succow, M. 1987: Zur Entstehung und Entwicklung der Moore in der DDR. *Zeitschrift für geologische Wissenschaften 15*, 373–387, Berlin.

Succow, M. & Lange, E. 1984: The mire types in the German Democratic Republic. *In* Moore, P.D. (ed.) *European mires*, 149–175. London.

Sukopp, H. & Brande, A. 1984/85: Beiträge zur Landschaftsgeschichte des Gebietes um den Tegeler See. *Sitzungsberichte der Gesellschaft Naturforschender Freunde zu Berlin (Neue Folge) 24/25*, 198–214, Berlin.

Usinger, H. 1985: Pollenstratigraphische, vegetations- und klimageschichtliche Gliederung des "Bölling–Alleröd–Komplexes" in Schleswig-Holstein und ihre Bedeutung für die Spätglazial-Stratigraphie in benachbarten Gebieten. *Flora 177*, 1–43.

Van Geel, B. 1972: Palynology of a section from the raised peat bog "Wietmarscher Moor", with special reference to fungal remains. *Acta Botanica Neerlandica. 21(3)*, 261–284, Amsterdam.

Vogt, R. 1990: Pedologische Untersuchungen im Umfeld der neolithischen Ufersiedlungen Hornstaad-Hörnle. *Berichte der Römisch-Germanischen Kommission 71*, 136–144.

Welten, M. 1982: Vegetationsgeschichtliche Untersuchungen in den westlichen Schweizer Alpen: Bern-Wallis. *Denkschriften der Schweizerischen Naturforschenden Gesellschaft 95*, 104 pp.

Wolter, K. 1992: Die postglaziale Entwicklung von Klima, terrestrischen und limnischen Prozessen am Tegeler See (Berlin). *Limnologica 22(3)*, 193–239, Jena.

Zoller, H. 1968: Postglaziale Klimaschwankungen und ihr Einfluß auf die Waldentwicklung Mitteleuropas ein-schließlich der Alpen. *Berichte der Deutschen Botanischen gesellschaft 80*, 690–696.

16

Belgium

C. Verbruggen, L. Denys and P. Kiden

INTRODUCTION

In Belgium several types of north-central European landscapes are represented on a small scale. Together with the topographic and climatic variations this results in remarkable physical diversity.

The subdivision of the country into palaeoecological type regions is based on variations in topography, geology, soils, and oceanicity of the climate. The result (Fig. 16.1) largely corresponds to the actual phytogeographic subdivision of Belgium according to De Langhe *et al.* (1983) (Fig. 16.2D).

Altitudes range from sea level in the western part of the country to almost 700 m in the Ardennes (region B-f) (Fig. 16.2A). Consequently the mean annual temperature ranges from more than 10°C in the northwestern part of the country to less than 7°C in the Ardennes. The number of frost days ranges from 45 to 120 over the same area, and mean annual precipitation from 750 to 1400 mm (Poncelet 1957).

Geologically western and central Belgium (B-a to B-d) consists largely of alternating sandy and clayey Cenozoic strata that dip gently to the north-north-east (Fig. 16.2B). These are overlain by relatively thick Quaternary sediments (up to 30 m), comprised mainly of sand in the lower part of the country (B-a, B-b, and B-c) and loess in the central low-plateau region (B-d) (Fig. 16.2C). Holocene clay and peat occur in the coastal plain (B-a) and river valleys. The plateau region south of the Rivers

Sambre and Meuse (B-e, B-f) consists mainly of folded Palaeozoic limestone, sandstone, and slate, building a varied hilly landscape with only a thin and discontinuous Quaternary cover. Ombrotrophic peat bogs were formed during the Holocene on the higher interfluvial zones in the Ardennes (B-f). The Lorraine region (B-g) in the southernmost part of the country corresponds to the northern fringe of the Paris Basin and is characterized by outcropping Mesozoic marl, clay, sandstone, and limestone.

The continental ice sheets never reached Belgium during the successive glacial stages of the Quaternary, but periglacial conditions prevailed during cold periods.

Phytogeographically that part of Belgium northwest of the Rivers Sambre and Meuse belongs to the Boreo-Atlantic sector of the Atlantic-European domain, whereas the southeastern part is in the Baltic-Rhenish sector of the Middle-European domain (Tanghe 1975) (Fig. 16.2D). According to the Council of Europe vegetation map at 1:3000000 (Ozenda *et al.* 1979), type region B-a is characterized by polder and halophytic vegetation of flat coasts and salt marshes. Acidophilous oakwoods and oak–beech woods with birch occur in type regions B-b, B-c, and B-d, whereas Atlantic beechwoods are present in B-d. Submontane neutrophilous beechwoods are found in regions B-e, B-f, and B-g. Type region B-f is characterized by submontane acidophilous beechwoods and raised bogs.

Palaeoecological Events During the Last 15000 Years: Regional Syntheses of Palaeoecological Studies of Lakes and Mires in Europe.
Edited by B.E. Berglund, H.J.B. Birks, M. Ralska-Jasiewiczowa and H.E. Wright. © 1996 John Wiley & Sons Ltd.

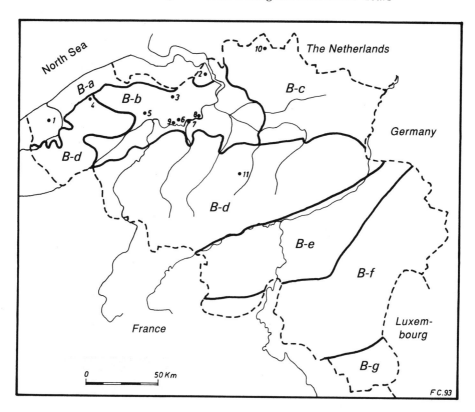

Fig. 16.1 Belgium: palaeoecological type regions and reference sites. The type regions are: B-a: Coastal plain, B-b: Sandy Flanders, B-c: Kempenland, B-d: Loess region, B-e: Mosan, B-f: Ardennes, B-g: Lorraine

The great diversity of the Belgian landscape, both in space and in Quaternary history, has attracted much research. Among the most comprehensive palaeoecological studies are those of Stockmans & Vanhoorne (1954), Allemeersch (1984) and Denys (1993) on the coastal-plain peat deposits, Coûteaux (1969) and Woillard (1975) on the Lorraine region, Munaut (1967) on the lower Scheldt valley, the Kempenland, and loess region, Bastin (1971) on loess profiles, and Dricot (1960), Woillard (1975), and Damblon (1976) on peat bogs in the Ardennes. The work of Verbruggen (1971) focused on Sandy Flanders, and more recently Beyens (1982) studied the northern Kempenland region. In addition a large number of mainly local investigations have been carried out, the most important of which are listed in the reference list.

Although the Belgian Quaternary is fairly well known, the scarcity of radiocarbon-dated sequences,

personal experience, and scientific questions compelled us to restrict our account to one type region: Sandy Flanders (B-b). However, a comparison of type region B-b with the adjacent regions B-a, B-c, and B-d is presented, and sites from these regions have been included in the reference site table for Belgium.

In Sandy Flanders conditions for organic deposition were favourable from Late-glacial to early Holocene times and during the post-Atlantic period. Continuous records with sufficient time-control are not yet available for the Atlantic period. Nevertheless, extrapolation from other sequences and comparison with adjacent regions (especially the related Kempenland) make it possible to complete our picture of the Late-glacial and Holocene vegetation evolution.

All names for stratigraphic subdivisions of the Late-glacial and Holocene periods are used here in

TOPOGRAPHY

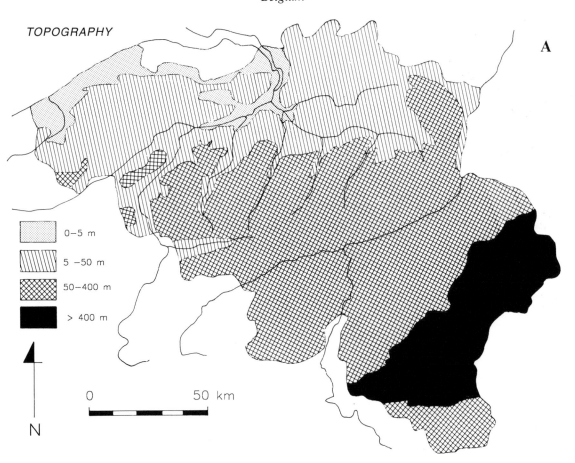

Fig. 16.2 A Belgium: Topography. Contour lines are shown at 5, 50 and 400 m above sea level

the strictly biozonal sense. The precise definition of these biozones is given in the section on Biozonation and chronozonation.

TYPE REGION B-b, SANDY FLANDERS

Altitude: 0–50 m above MSL (mean sea level).

Climate: The mean annual temperature is 9–10°C, the mean July temperature 17–18°C, and the mean January temperature 3°C. The annual precipitation amounts to 750–800 mm. The number of frost days per year is about 55 (Poncelet 1957).

Geology: Alternation of Cenozoic sandy and clayey strata dipping gently to the NNE. During the Qua-

ternary an important valley system eroded in these deposits was filled with up to 25 m of mainly sandy fluvioperiglacial sediments (Fig. 16.2B). On the interfluves, however, the Cenozoic strata are covered with only a thin layer of Quaternary sand or sand–loam.

Topography: Gently undulating hilly interfluves and low-lying sandy plains.

Population density: About 400 km^{-2}.

Vegetation: The last patches of natural vegetation were removed more than 1000 years ago. The actual dry woodlands can be classified (Vanden Berghen 1952) in the Querco roboris–Betuletum growing on nutrient-poor soils and the Quercetum atlanticum on more neutral soils. On wet, nutrient-rich

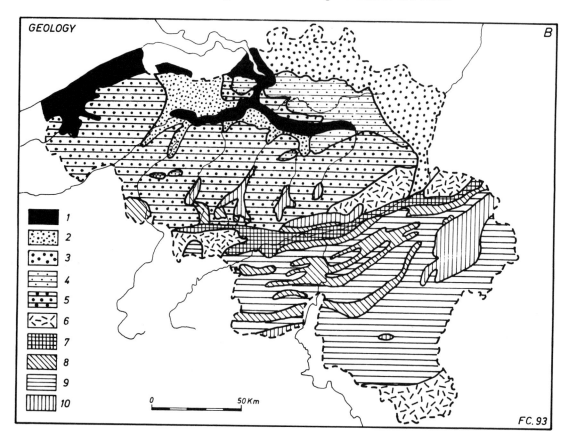

Fig. 16.2 B Belgium: Geology. 1. Holocene, 2. Upper Pleistocene, 3. Middle and Lower Pleistocene, 4. Neogene, 5. Paleogene, 6. Mesozoic, 7. Carboniferous (coal), 8. Carboniferous and Devonian limestone, 9. Devonian slates and sandstones, 10. Silurian and Cambrian

terrain, vegetation belongs to the Carici elongatae-Alnetum (=Alnetum glutinosae). *Populus x canadensis* Moench, *Pinus sylvestris* L., and *Quercus rubra* L. are the major trees in plantations.

Soils: Soils with spodic and hydromorphic characteristics.

Land use: Cropland 53%, grassland 20%, urban and industrial zones (including roads, rivers and canals) about 22%, forested area about 5%.

Reference site 3. Moerbeke (Fig. 16.3)

Latitude 51°10'17"N, Longitude 3°56'53"E. Elevation +4 m above mean sea level (MSL), large shallow depression. Three pollen-assemblage zones (paz).

1.	12000–11850 BP	Cyperaceae (*Selaginella selaginoides* present)
2.	11850–11200 BP	*Betula–Artemisia*
2a.	11850–11550 BP	*Juniperus communis–Hippophaë rhamnoides* subzone
2b.	11550–11300 BP	*Menyanthes trifoliata–Myriophyllum–Nymphaea* subzone
2c.	11300–11200 BP	*Filipendula* subzone (lower values of *Artemisia*)
3.	11200–? BP	*Pinus–Betula*

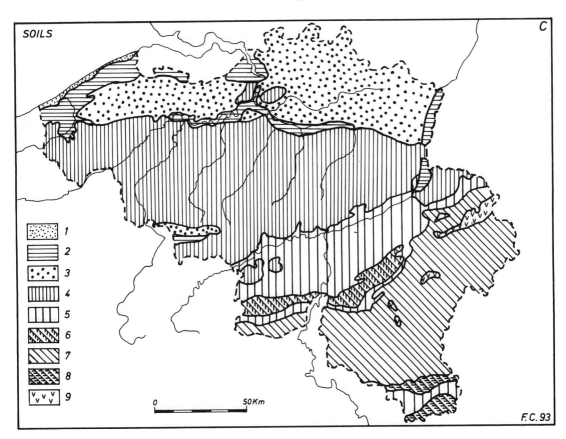

Fig. 16.2 C Belgium: Soils. 1. Dunes, 2. Alluvial soils, 3. Sandy soils, 4. Silty soils, 5. Gravelly loam soils, 6. Loamy clay soils, 7. Loamy clay soils with slates and sandstones, 8. Marly clay soils, 9. Bogs

Reference site 4. Snellegem (Fig. 16.4)

Latitude 51°11′12″N, Longitude 3°16′4″E. Elevation +12 m above MSL, shallow depression. Three pollen-assemblage zones (paz).

1. 12150–11750 BP   Cyperaceae–*Plantago*–
     Helianthemum–
     Potamogeton–*Selagin-*
   ella selaginoides
2. 11750–11200 BP   *Betula–Artemisia*
2a. 11750–11550 BP   *Juniperus communis–*
     Salix subzone
     (highest *Hippophaë*
     rhamnoides values)
2b. 11550–11200 BP   *Filipendula* subzone
3. 11200–10250 BP   *Pinus–Betula–Artemisia–*
     Filipendula

3a. 11200–11000 BP   *Filipendula* subzone
3b. 11000–10250 BP   *Artemisia–Empetrum–*
     Ranunculus subgen.
     Batrachium–Selaginella
     selaginoides subzone

Reference site 5. Vinderhoute (Fig. 16.5)

Latitude 51°04′45″N, Longitude 3°37′22″E. Elevation +4 m above MSL, abandoned river channel. Five pollen-assemblage zones (paz).

1. 12750–11300 BP   *Betula–Artemisia* (high
     values of *Nymphaea*)
2. 11300–10300 BP   *Betula–Pinus–*
     Gramineae–Cyperaceae

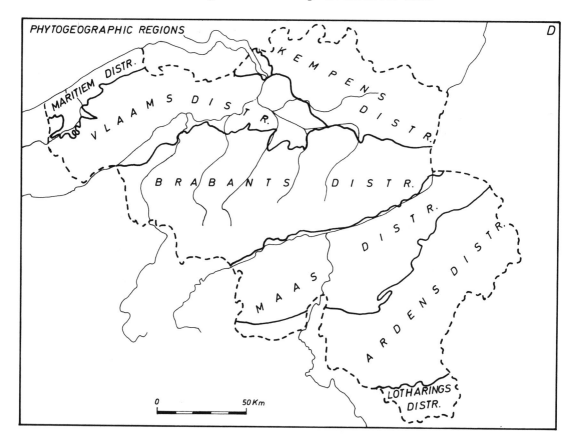

Fig. 16.2 D Belgium: Phytogeographic regions (after De Langhe *et al.* 1983)

2a.	11300–11000 BP	*Filipendula* subzone (appearance of *Viburnum*)
2b.	11000–10300 BP	*Juniperus communis– Artemisia* subzone
3.	10300–9500 BP	*Betula–Pinus*
3a.	10300–10000 BP	*Filipendula* subzone
3b.	10000–9850 BP	Gramineae subzone (marked increase of *Betula* at beginning and end of subzone)
3c.	9850–9500 BP	*Pinus* subzone
4.	9500–9100 BP	*Pinus–Betula–Corylus*
5.	9100–7500 BP	*Corylus–Pinus– Quercus–Ulmus*

Reference site 6. Berlare (Fig. 16.6)

Latitude 51°02′17″N, Longitude 3°58′54″E. Elevation +1.5 m above MSL, abandoned river channel. Two pollen-assemblage zones (paz).

1.	9500–8600 BP	*Corylus–Pinus– Quercus* (*Corylus* more important than *Quercus*)
2.	8600–7550 BP	*Pinus–Corylus–Quercus* (*Pinus* more important than *Corylus*)

General comment: *Ulmus, Viburnum, Artemisia,* and *Filipendula* are present throughout with low values.

559

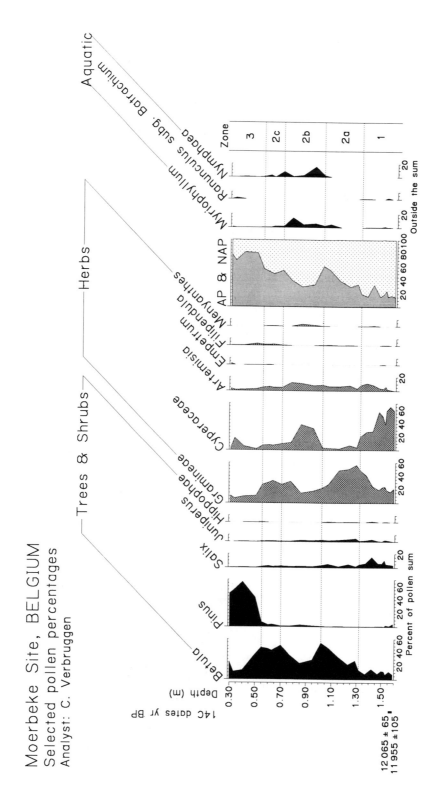

Moerbeke Site, BELGIUM
Selected pollen percentages
Analyst: C. Verbruggen

Fig. 16.3 Reference site 3 (Moerbeke): pollen diagram with selected taxa (based on Verbruggen 1971, 1979). The pollen sum does not include aquatic taxa and spores.

560

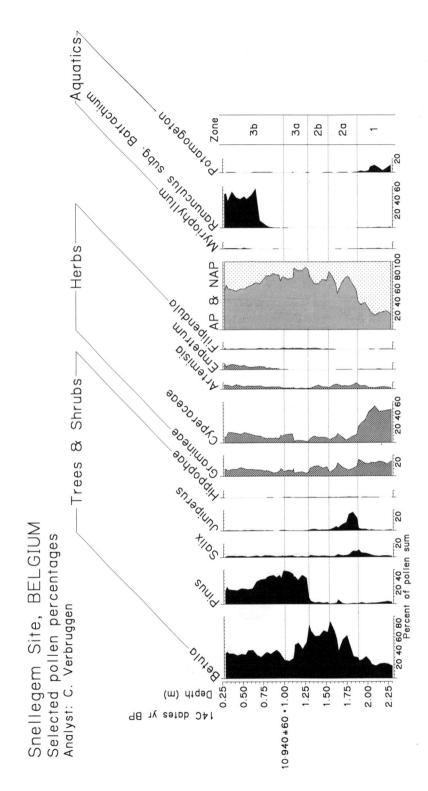

Fig. 16.4 Reference site 4 (Snellegem): pollen diagram with selected taxa (based on Verbruggen 1971, 1979). The pollen sum does not include aquatic taxa and spores

561

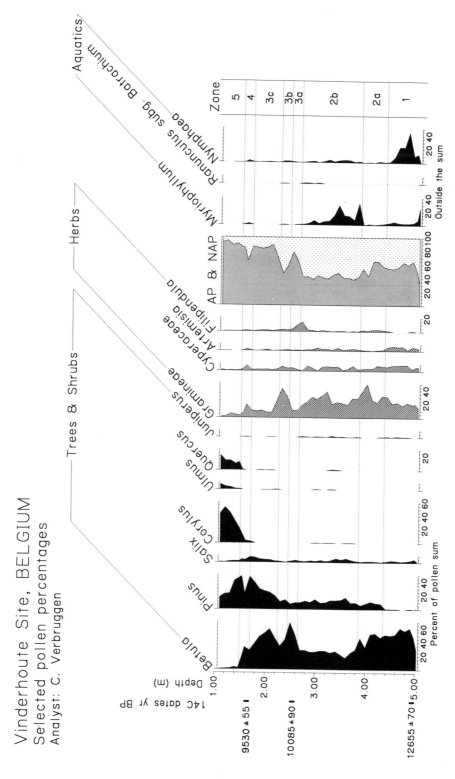

Fig. 16.5 Reference site 5 (Vinderhoute): pollen diagram with selected taxa (based on Verbruggen 1971, 1979). The pollen sum does not include aquatic taxa and spores

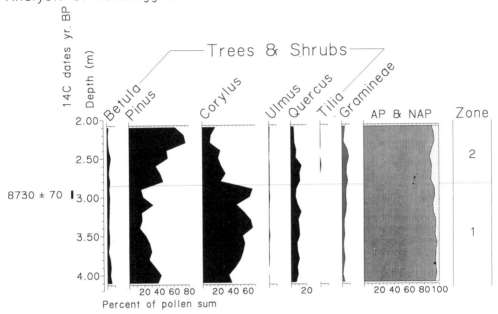

Fig. 16.6   Reference site 6 (Berlare): pollen diagram with selected taxa (based on Verbruggen 1971). The pollen sum does not include aquatic taxa and spores

Reference site 7. Dendermonde (Fig. 16.7)

Latitude 51°02′38″N, Longitude 4°05′23″E. Elevation +0.2 m above MSL, alluvial plain. Two pollen-assemblage zones (paz).

1. 5000–2250 BP   *Alnus–Quercus–Corylus–Salix*
1a. 5000–4050 BP   *Ulmus–Tilia* subzone (*Viscum alba* and *Hedera helix* present)
1b. 4050–2250 BP   Consecutively *Plantago, Fagus sylvatica, Rumex,* and cereals appear; maximum of *Fagus sylvatica* near the end of the subzone
2. 2250–1050 BP   *Quercus*–Gramineae–*Salix–Alnus* (alternating dominance of Gramineae, *Salix,* and *Alnus*;

slightly higher values for cereals and Ericaceae; *Carpinus betulus* present at the top)

Reference site 8. Moerzeke (Fig. 16.8)

Latitude 51°02′54″N, Longitude 4°10′35″E. Elevation +0.3 m above MSL, alluvial plain. Three pollen-assemblage zones (paz).

1. 5000–4000 BP   *Quercus–Alnus–Corylus–Ulmus–Tilia* (*Viscum album* and *Hedera helix* present)
2. 4000–2050 BP   *Salix–Alnus–Quercus–*Cyperaceae (continuous curves of *Plantago* and Ericaceae; *Fagus sylvatica* appears and shows highest values near the top)

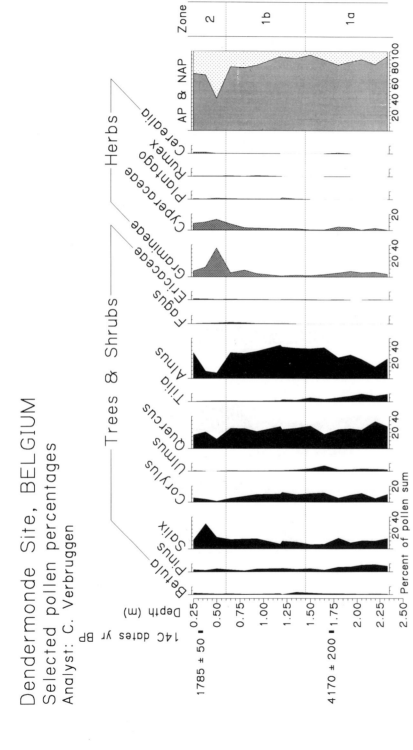

Fig. 16.7 Reference site 7 (Dendermonde): pollen diagram with selected taxa (based on Verbruggen 1971). The pollen sum does not include aquatic taxa and spores

564

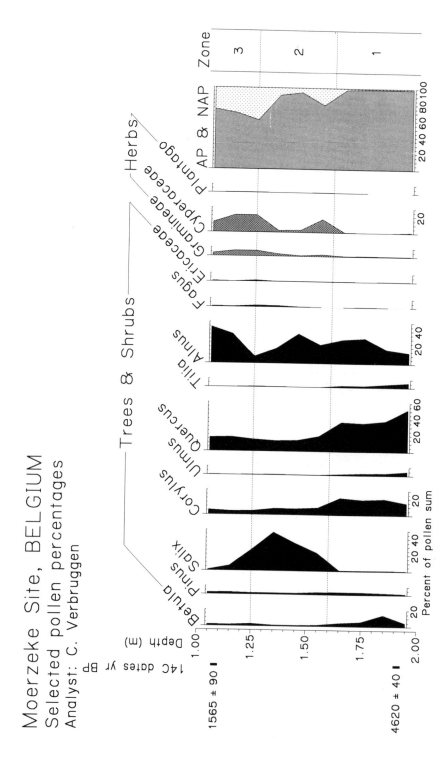

Fig. 16.8 Reference site 8 (Moerzeke): pollen diagram with selected taxa (based on Verbruggen 1971). The pollen sum does not include aquatic taxa and spores

3. 2050–1600 BP   *Alnus–Quercus*–Cyperaceae
                   (*Carpinus betulus* present)

Reference site 9. Uitbergen-Heisbroek (Fig. 16.9)

Latitude 51°01′04″N, Longitude 3°56′17″E. Elevation +2 m above MSL, shallow depression. Two pollen-assemblage zones (paz).

1. 5250–2400 BP   *Alnus–Quercus–*
                   Corylus–Tilia (*Ulmus* is
                   important in the lower-
                   most part and has low
                   values in the upper part;
                   appearance of *Fagus*
                   sylvatica in the uppermost
                   part; gradual increase of
                   Ericaceae; appearance of
                   Plantago at the top)
2. 2400–950 BP   *Alnus–Quercus* (maximum of
                   Fagus sylvatica; continuous
                   presence of *Carpinus*
                   betulus; cereals relatively
                   abundant in the upper half;
                   Ericaceae slightly
                   increasing)

Palaeoecological events

Regional vegetation development
(Figs 16.10, 16.11 and 16.13)

In the framework of the regional pollen zonation the development of the vegetation can be summarized as follows:

12100–11800 BP: Cyperaceae paz
– *Selaginella selaginoides* present.

11800–11200 BP: *Betula–Artemisia* paz
– Evolution from open to rather dense *Betula* forest.
– Initially *Juniperus communis* is important.
– Steppe elements occur (*Hippophaë rhamnoides*).
– *Nymphaea* and *Nuphar* among the aquatics.
– Increase of *Filipendula* in the final part.

11200–9500 BP: *Pinus–Betula* paz
    11200–11000 BP: *Filipendula* subzone

– Rather dense *Pinus–Betula* forest.
11000–10300 BP: *Artemisia–Juniperus com-munis–Empetrum* subzone
– Open *Pinus–Betula* forest with rich herbaceous undergrowth.
– *Ranunculus* subgen. *Batrachium*, *Myriophyllum*, and *Sparganium* prominent in the aquatic vegetation.

10300–10000 BP: *Filipendula* subzone
– *Nymphaea* and *Nuphar* reappear.

10000–9850 BP: *Betula* subzone
– Initially an almost closed *Betula* forest developing to a closed *Pinus–Betula* cover.
9850–9500 BP: *Pinus* subzone
– Appearance of *Corylus, Quercus*, and *Ulmus*.
9500–9100 BP: *Corylus–Pinus–Betula* paz
– Rapid expansion of *Corylus*.
9100–7500 BP: *Corylus–Pinus–Quercus–Ulmus* paz
– *Quercus* and *Ulmus* increase gradually.
– *Tilia* appears.
– No expansion of *Alnus*.

7500–4000 BP: *Quercus–Alnus–Corylus–Tilia–Ulmus* paz
– The vegetation is entirely controlled by local conditions: "Quercetum mixtum" woods cover the drier grounds, with Alnetum vegetation in wetter parts.
– The abundance of *Tilia* varies; whether it was a dominant tree remains questionable (Verbruggen 1984).
– Sea-level rise induced extensive fen-wood growth in the downstream alluvial plains.
– During the last millennium of this period slight human influence is observable on the driest sites.
– *Fraxinus excelsior, Hedera helix*, and *Viscum album* occur regularly.
– Sporadic occurrence of *Taxus baccata*.

4000–2250   BP: *Alnus–Quercus–Corylus–Salix*–Cyperaceae paz
– Decrease of *Ulmus* and *Tilia* at the transition from the previous period.
– *Fagus sylvatica* present and gradually increasing.
– *Carpinus betulus* present in the final part.
– Human impact on the vegetation gradually increases; ruderal herbs appear.
– Local heath development in the second half of the period.

566

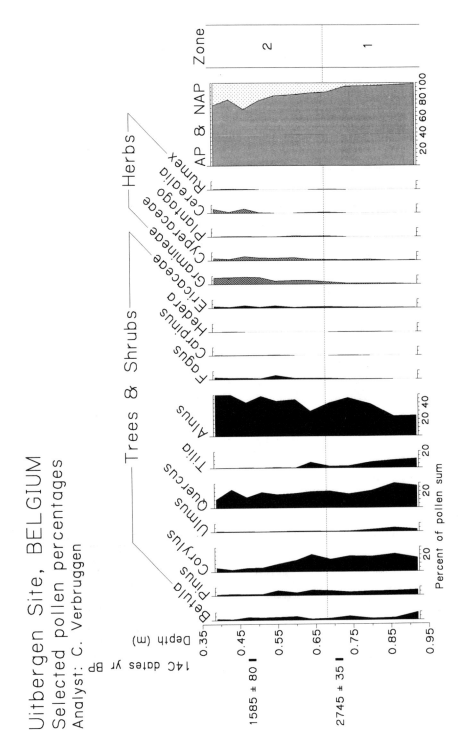

Fig. 16.9 Reference site 9 (Uitbergen–Heisbroek): pollen diagram with selected taxa (based on Verbruggen 1971). The pollen sum does not include aquatic taxa and spores

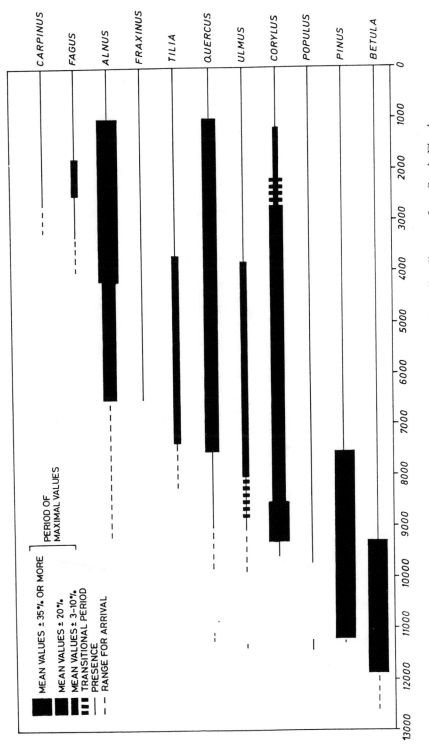

Fig. 16.10 Arrival and average abundance of major tree taxa in pollen diagrams from Sandy Flanders

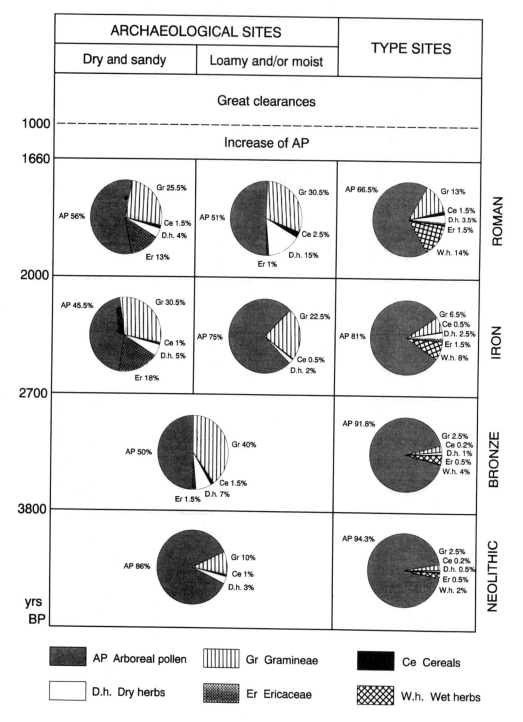

Fig. 16.11   Mean pollen percentages per archaeological period for main groups of taxa. On the left: data from 12 archaeological sites; on the right side: data from the post-Atlantic reference sites (7, 8 and 9)

2250–1000 BP: *Alnus–Quercus–Salix* Cyperaceae–Gramineae paz
- End of peat growth in alluvial plains due to increased sedimentation.
- Further increase of human influence, interrupted by Dark Ages (see below).

1000–0 BP:
- Cultural landscape: hardly any natural vegetation evolution.

Human impact (Figs 16.11 and 16.12)

- The most widespread phenomenon from Neolithic to Roman times is the gradual decrease of AP percentages.
- After the Roman period, between 1660 and 1450 BP (Dark Ages), the forest recovered to some degree; this was followed by the final clearances.
- The archaeological-site diagrams reflect local conditions. A distinction is made between "dry/sandy" sites (mainly situated in the northern part of Sandy Flanders) and "loamy/moist" sites, in order to show that heath only developed in the northern sandy part of the region. In the southern part an open grass and herb vegetation appeared instead.
- Cereals have been grown since the Neolithic Beaker-cultures (4200–3700 BP). Their cultivation was limited during the Bronze Age and then gradually increased to a maximum in the Roman period.
- *Secale cereale* has been grown at least since Roman times.
- *Fagopyrum esculentum* apparently was not cultivated before the 13th century.

Climate (Fig. 16.12)

- Whereas for the Late-glacial period mire sediments offer a detailed environmental record, the Holocene climatic evolution has to be inferred from less suited fen peats, which are mainly situated in alluvial plains and abandoned river beds. Ombrotrophic peat bogs are lacking.
- The most humid phase in the Late-glacial appears to have been the Bølling.
- The Older Dryas is characterized by colder and

drier conditions.
- The Late-glacial climatic optimum occurred in the first half of the Allerød; in the second half the climatic deterioration had already set in, culminating in the Younger Dryas.
- At the start of the Holocene, climatic amelioration was abrupt; the subsequent evolution shows some minor oscillations.
- There is a clear vegetational signal of maximum Holocene temperatures in the Atlantic.
- The increase of humidity in the Subatlantic was inferred primarily from the widespread development of ombrotrophic peat in the adjacent coastal plain region (B-a) during this period.

Groundwater (Fig. 16.12)

- The groundwater level primarily depends on precipitation and evapotranspiration.
- The high Bølling water level was caused by relatively low temperatures and increased precipitation.
- The lowering at the start of the Allerød resulted from higher temperatures and the denser vegetation cover.
- As temperatures decreased near the end of the Allerød the water level rose again and remained high during the Younger Dryas.
- The marked drop at the beginning of the Holocene was caused by increased evapotranspiration.
- Since the Atlantic a gradual recovery occurred; initially this was climatically controlled, but since the end of the Atlantic human influence became predominant, for clearances resulted in less evapotranspiration and more run-off.
- Since the middle of the Atlantic a positive trend may be observed in the Lower Schelde area due to the general sea-level rise.
- The most recent groundwater lowering was caused by artificial drainage.

Geomorphology (Fig. 16.12)

The physical landscape

- Due to the Late-glacial groundwater rise, a mire-strewn landscape originated in the northern, most sandy part of the region. In the immediate vicinity of the rivers important sand dunes were formed. Many of the mires were buried under drifting sand quite rapidly, but in more sheltered areas they persisted up to the end of the Late-glacial.

- At the start of the Holocene the mires dried up rather abruptly, and the soil surface was stabilized by a dense forest vegetation.
- Towards the last millennium of the Atlantic the alluvial plains were gradually covered with fen-woods due to the influence of man and of sea-level rise upon the groundwater level. Fens expanded and also developed in smaller river valleys and on waterlogged soils outside the alluvia. Deeper depressions filled with water. While human influence (e.g. peat digging) left its mark on the development of the wetlands, the forest disappeared step by step, first by cutting of glades, later on by large-scale clearances.

Fluvial events

- The transition from a braided to a meandering river channel pattern and the accompanying incision took place before 12000 BP; the precise date of this event remains to be determined more accurately.
- At the start of the Holocene, fluvial activity declined markedly (cf. lowering of the groundwater level).
- Anthropogenic and climatic factors induced increased floodplain sedimentation since the Atlantic. Because of the geographical distribution of Neolithic agriculture this phenomenon took place earlier in the upper reaches in the loess area than in the lower reaches. From the Bronze Age and later a phase of channel widening and deepening occurred.

Aeolian activity

- The aeolian sand transport was controlled by the availability of sand and by the vegetation.
- Both conditions were favourable in the Oldest Dryas, in the climatically unstable Older Dryas–early Allerød, and in the Younger Dryas, when exposed coversands and alluvial deposits func-

tioned as locally important sources; nevertheless, sand transport remained limited compared to more northern regions. The period of most extensive dune formation was probably associated with the transition from a braided river pattern to a meandering one. According to Heyse (1979), the predominant wind direction was north to northwest during the Late Weichselian and southwest during the Holocene (as at present).

- Holocene human activity led to locally important but short-lived events of sand movement in the Iron Age–Roman period; soil profiles show that these mainly comprised dune migration instead of neo-formation.

Biozonation and chronozonation (Fig. 16.13)

The proposed biostratigraphic subdivision of the Late Weichselian and Holocene in northern and middle Belgium is mainly based on Verbruggen (1976, 1979, 1984). For the upper boundary of the Atlantic biozone, Munaut (1967) was followed. For comparison the chronostratigraphic subdivision of the Late Weichselian and Holocene according to Mangerud *et al.* (1974) is also given.

COMPARISON WITH ADJACENT TYPE REGIONS

For a large part, the comparisons for the loess region and the Kempenland were based on fruitful discussions respectively with A.V. Munaut and with L. Beyens and A.V. Munaut.

Coastal plain (B-a)

From the geomorphological setting and the limited palynological evidence at hand, a palaeoecological evolution identical to type region B-b can be presumed prior to 7500–6000 BP. After this date, the sea invaded the region, resulting in the deposition of marine–littoral and perimarine sediments.

Between about 6000 and 2000 BP peat bogs covered the larger part of the area. Initially a eutrophic fen-peat developed, presenting a distinct *Dryopteris*

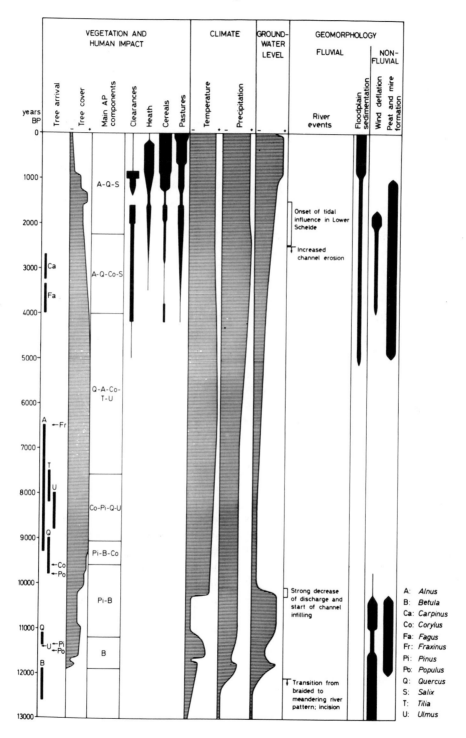

Fig. 16.12 Event stratigraphy for type region Sandy Flanders, based on Verbruggen (1971, 1976, 1979), De Laet (1974), Vanhoorne & Verbruggen (1975), Cleveringa *et al.* (1988), Kiden (1991), and present authors

phase. *Betula* and *Pinus* are the dominant trees of the mesotrophic middle part of this peat. Humified horizons show high values of *Myrica gale* pollen and probably indicate drier episodes. Macroremains of *Eriophorum* are the most typical feature of the oligotrophic upper part of the peat. The "dry" vegetation components in the pollen spectra of this part (*Fagus sylvatica*, Quercetum mixtum, etc.) were largely derived from the adjacent region B-b and show a strikingly similar evolution.

Between 2500 and 1500 BP the peat was covered by the deposits of the so-called Dunkerque transgression phases. After 1000 AD the salt marshes were reclaimed, and the peat was partly excavated in places.

Kempenland (B-c)

During the Late-glacial, the vegetation evolution was identical to that in region B-b. However, more intense dune formation took place in the Kempenland during the Younger Dryas.

During the Holocene, *Alnus* reached considerable percentages after the Preboreal (Beyens 1982). Although *Tilia* was the dominant tree during the Atlantic, its subsequent decline must be attributed to soil degradation (Munaut 1967). Beyens (1984) notes early landnams, even prior to 6000 BP.

After the Atlantic, the vegetation evolution was broadly similar to that of type region B-b. An exception to this rule is the more pronounced heathland expansion in the Kempenland, which is due to the poor sandy, acid soils. Between the 12th and 17th centuries AD heathland was the main vegetation type and plaggen-culture was widely practised. After this period large-scale reforestation with *Pinus* took place.

Loess region (B-d)

A lack of radiocarbon dates hampers the unequivocal interpretation of most presumed Late-glacial pollen spectra: these should actually be assigned to the Preboreal.

The vegetation evolution during the Preboreal and Boreal is comparable to that of region B-b. During the Atlantic, *Tilia* was the dominant element of the Quercetum mixtum.

After about 3500 BP *Fagus sylvatica* increased and attained values of up to 10% AP after 2500 BP. *Carpinus betulus* shows a more pronounced expansion since the Subatlantic than in regions B-b and B-c but never reached more than 5%. In view of this clear but limited presence, it is possible that the discontinuous record of *Carpinus betulus* pollen in regions B-b and B-c must be attributed to long-distance transport from region B-d.

The expansion of both *Fagus sylvatica* and *Carpinus betulus* in regions B-b, B-c, and B-d was favoured to some extent by human interference (clearances). For *Fagus*, the better development of this tree in region B-d compared to regions B-c and B-b may be explained by its dominance in the adjacent regions B-e and B-f, while for *Carpinus* the more favourable soil conditions (loess) may be responsible.

As in region B-b, a short period of forest recovery occurred after the Roman period.

CONCLUSIONS

The sedimentary basins that developed in Sandy Flanders after the Weichselian Pleniglacial (shallow mires, abandoned river channels, alluvial fens) provide a record of local palaeoecological conditions unobscured by extra-regional influences.

The vegetational history of the region reflects the geographical position at the transition between southern and northern regions. For the Late Weichselian there are strong similarities with the North-Central European record (southern Scandinavia to Switzerland); the dating of these events, however, shows an important discordance. During the Holocene the region was situated at the southern limit of the area of raised-bog formation. Although Sandy Flanders itself was too eutrophic to allow the growth of raised bogs, the bogs in the adjacent coastal plain (B-a) and Kempenland region (B-c) show that the climatic conditions there were also sub-optimal.

During the later part of the Atlantic the most striking development, apart from human interference, was the lateral expansion of the alluvial plains. This was due to the gradually rising groundwater level and the associated growth of alluvial fens.

In Sandy Flanders human impact remained low

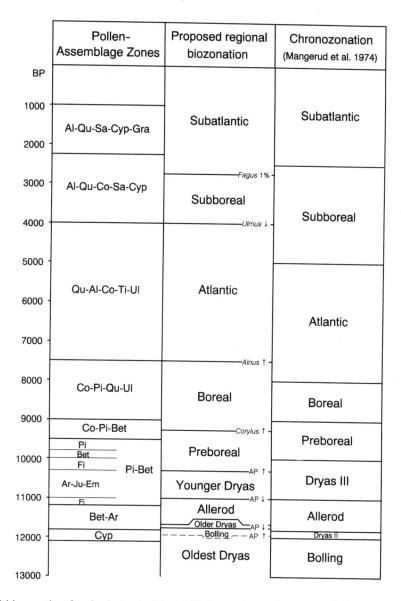

Fig. 16.13 Proposed biozonation for the Late-glacial and Holocene in northern and middle Belgium. The regional pollen zonation is more fully discussed in the text under "Regional vegetation development" and the pollen-assemblage zones correspond to the ones mentioned there. The main criterion for the delimitation of regional biozones is given on each boundary; an up-arrow indicates a rise, while a down-arrow denotes a decline in the pollen frequency. Explanation: AP: Arboreal pollen, Al: *Alnus*, Ar: *Artemisia*, Bet: *Betula*, Co: *Corylus*, Cyp: Cyperaceae, Em: *Empetrum*, Fi: *Filipendula*, Gra: Gramineae, Ju: *Juniperus*, Pi: *Pinus*, Qu: *Quercus*, Sa: *Salix*, Ti: *Tilia*, Ul: *Ulmus*

for almost a millennium longer than in the loess region, because it was not part of the expansion area of the Band-Ceramic culture. Since the Neolithic, however, man exerted a drastic and continuous pressure upon the landscape. From medieval times, the present scenery rapidly took shape.

REFERENCES

Allemeersch, L. 1984: *Het veen in het oostelijk kustgebied.* Thesis Katholieke Universiteit Leuven, 297 pp.

Baeteman, C. & Verbruggen, C. 1979: A new approach to the so-called surface peat in the Western coastal plain of Belgium. *Professional Paper Belgian Geological Survey 167*, 21 pp.

Bastin, B. 1971: Recherches sur l'évolution du peuplement végétal en Belgique durant la glaciation de Würm. *Acta Geographica Lovaniensia 9*, 136 pp.

Beyens, L. 1982: *Bijdrage tot de Holocene paleo-ecologie van het stroomgebied van de Mark in België, gebaseerd op de studie van diatomeeën, pollen en thecamoeba's.* Thesis Universitaire Instelling Antwerpen, 399 pp.

Beyens, L. 1984: Palynological and radiometric evidence for an early start of the Neolithicum in the Belgian Campine. *Notae Praehistoricae 4*, 89–95, Leuven.

Cleveringa, P., De Gans, W., Huybrechts, W. & Verbruggen, C. 1988: Outline of river adjustments in small river basins in Belgium and the Netherlands since the Upper Pleniglacial. *In* Lang, G. & Schlüchter, Chr. (eds) *Lake, mire and river environments during the last 15000 years*, 123–132. Balkema, Rotterdam.

Coûteaux, M. 1969: Recherches palynologiques en Gaume, au Pays d'Arlon, en Ardenne méridionale (Luxembourg belge) et au Gutland (Grand-Duché de Luxembourg). *Acta Geographica Lovaniensia 8*, 193 pp.

Damblon, F. 1976: *Etude paléo-écologique de tourbière en haute Ardenne.* Thèse Université de Louvain, 630 pp.

De Laet, S. 1974: *Prehistorische kulturen in het zuiden der Lage Landen.* 561 pp. Universa, Wetteren.

De Langhe, J.E., Delvosalle, L., Duvigneaud, J., Lambinon, J. & Vanden Berghen, C. 1983: *Flora van België, het Groothertogdom Luxemburg, Noord-Frankrijk en de aangrenzende gebieden (Pteridofyten en Spermatofyten).* 970 pp. Patrimonium van de Nationale Plantentuin van België, Meise.

Denys, L. 1993: *Paleoecologisch diatomeeënonderzoek van de holocene afzettingen in de westelijke Belgische kustvlakte.* Thesis Universitaire Instelling Antwerpen, 479 pp.

Dricot, E.M. 1960: Recherches palynologiques sur le plateau des Hautes-Fagnes. *Bulletin de la Société Royale Botanique de Belgique 92*, 157–196.

Heyse, I. 1979: Bijdrage tot de geomorfologische kennis van het noordwesten van Oost-Vlaanderen (België). *Verhandelingen van de Koninklijke Academie voor Wetenschappen, Letteren en Schone Kunsten van België-Klasse der Wetenschappen 41(155)*, 257 pp.

Kiden, P. 1991: The Lateglacial and Holocene evolution of the Middle and Lower River Scheldt, Belgium. *In* Starkel, L., Gregory, K.J. & Thornes, J.B. (eds) *Temperate Palaeohydrology*, 283–299. Wiley, Chichester.

Mangerud, J., Andersen, S.T., Berglund, B.E. & Donner, J. 1974: Quaternary stratigraphy of Norden, a proposal for terminology and classification. *Boreas 3*, 109–128.

Minnaert, G. & Verbruggen, C. 1986: Palynologisch onderzoek van een veenprofiel uit het Doeldok te Doel. *Bijdragen Archeologische Dienst Waasland 1*, 201–208.

Munaut, A.V. 1967: Recherches paléo-écologiques en Basse et Moyenne Belgique. *Acta Geographica Lovaniensia 6*, 191 pp.

Ntaganda, C. & Munaut, A.V. 1987: Etude palynologique et datations ^{14}C d'une couche de tourbe post-glaciaire située dans la vallée de la Lasne à Rosières (Brabant Belgique). *Bulletin de la Société Royale Botanique de Belgique 120*, 45–52.

Ozenda, P., Noirfalise, A., Tomaselli, R. & Trautmann, W. 1979: *Carte de la végétation (échelle 1:3.000.000) des états membres du Conseil de l'Europe.* Council of Europe, Strasbourg.

Poncelet, L. 1957: *Atlas van België, Verklarende tekst bij de platen 12, 13, 14: Klimaat van België.* 43 pp. Nationaal Comité voor Geografie, Brussel.

Stockmans, F. & Vanhoorne, R. 1954: Etude botanique du gisement de tourbe de la région de Pervijze (Plaine maritime belge). *Mémoires de l' Institut Royal des Sciences Naturelles de Belgique 130*, 144 pp.

Tanghe, M. 1975: *Atlas van België, Verklarende tekst bij de platen 19A en 19B: Fytogeografie.* 76 pp. Nationaal Comité voor Geografie, Brussel.

Vanden Berghen, C. 1952: *Vegetatiekaart van België, verklarende tekst bij het kaartblad Gent 55W.* 70 pp. Comité voor het opnemen van de Bodemkaart en de Vegetatiekaart van België, Brussel.

Vanhoorne, R. & Verbruggen, C. 1975: Problèmes de subdivisions du Tardiglaciaire dans la région sablonneuse du Nord de la Flandre en Belgique. *Pollen et Spores 7(4)*, 525–543.

Verbruggen, C. 1971: *Postglaciale landschapsgeschiedenis van Zandig Vlaanderen.* Thesis Rijksuniversiteit Gent, 440 pp.

Verbruggen, C. 1976: De geokronologie van het Post-Pleniglaciaal in Zandig-Vlaanderen op basis van pollenanalyses en ^{14}C-onderzoek. *Natuurwetenschappelijk Tijdschrift 58*, 233–256.

Verbruggen, C. 1979: Vegetational and palaeoecological history of the Lateglacial period in Sandy Flanders (Belgium). *Acta Universitatis Oulensis A 82 Geol. 3*, 133–142.

Verbruggen, C. 1984: Aspects des compositions et changements caractéristiques de l'évolution botanique holocène en Flandre. *Revue de Paléobiologie vol. spécial*, 231–234.

Woillard, G. 1975: Recherches palynologiques sur le Pleistocène dans l'Est de la Belgique et dans les Vosges lorraines. *Acta Geographica Lovaniensia 14*, 118 pp.

17

France

L. VISSET, S. AUBERT, J.M. BELET, F. DAVID, M. FONTUGNE,
D. GALOP, G. JALUT, C.R. JANSSEN, D. VOELTZEL and M.F.
HUAULT

INTRODUCTION (L. Visset)

The Palaeozoic Era was marked by violent Hercynian folding and the creation of high mountains composed of folded chains and crystalline massifs: the northern and central parts of the Massif Central, the Massif Armorican, the Vosges, and the Ardennes. At the end of the Palaeozoic, the Hercynian chains were worn down by erosion and reduced to peneplains.

During the Mesozoic, the sea invaded wide areas of these peneplains, depositing thick layers of sediment. Some parts of the platform gradually subsided, creating the Parisian and Aquitaine basins. A gigantic depression gradually formed on the site of the Alps.

The Tertiary was marked by thrusting first in the Pyrenees and then in the Alps and the Jura. The old massifs nearby were affected by the rise of the Alps. The Vosges and the Massif Central became higher, but some areas underwent faulting, with collapse in the Rhine and Rhône valleys and the Limagne. These movements were accompanied by extensive volcanic activity, particularly in the Massif Central.

During the Quaternary succession of great glacial ages, the high mountains were covered with glaciers that eroded cirques and valleys and built moraines in piedmont regions. Further volcanic eruptions occurred.

As a result of these geological events, the topography of France is quite varied, with vast sedimentary basins (the Parisian and Aquitaine basins), old mountains with rounded summits (the Massif Armorican, Vosges, and Massif Central), and recent steeply sloped mountains (the Jura, Pyrenees, and Alps) (Fig. 17.1).

France has more than 6000 miles of coastline on the Atlantic Ocean, the English Channel, and the Mediterranean Sea. Midway between the Equator and the North Pole, it benefits from a temperate climate and a favourable position at the western edge of the European continent. The country has three climatic zones of unequal area (oceanic, semi-continental, and Mediterranean) plus a mountain climate (Fig. 17.2). These zones correspond to four great floristic domains (Atlantic, Middle European, Mediterranean, and high-mountain) (Fig. 17.3).

In France studies carried out in five major regions are described here: the lower Seine valley (F-b) (M.F. Huault), the Vosges Mountains (F-h and F-i) (C.R. Janssen), the marshlands of the Loire Estuary (F-p) (L. Visset and D. Voeltzel), the French Pyrenees (F-r and F-zg) (G. Jalut, S. Aubert, D. Galop, M. Fontugne, and J.M. Belet) and Chaine des Hurtieres in the Internal French Alps (F-zd) (F. David) (Fig. 17.4).

The basic interest of every pollen analyst and the most immediate result from pollen analysis is the

Palaeoecological Events During the Last 15000 Years: Regional Syntheses of Palaeoecological Studies of Lakes and Mires in Europe.
Edited by B.E. Berglund, H.J.B. Birks, M. Ralska-Jasiewiczowa and H.E. Wright. © 1996 John Wiley & Sons Ltd.

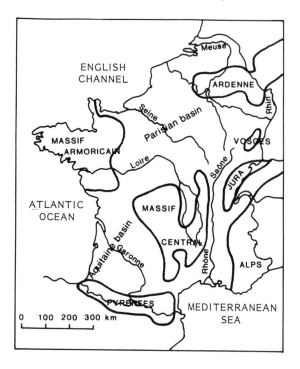

Fig. 17.1 Geomorphological regions of France

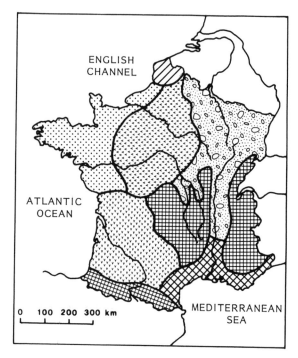

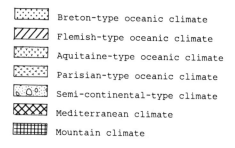

Breton-type oceanic climate
Flemish-type oceanic climate
Aquitaine-type oceanic climate
Parisian-type oceanic climate
Semi-continental-type climate
Mediterranean climate
Mountain climate

Fig. 17.2 French climatic zones

new information provided about the vegetation history of each region studied. The establishment of local pollen zonations enables the recorded events to be placed within the chronology of the Quaternary, a procedure that is the indispensable preliminary to all palaeoclimatic syntheses.

The French contribution has achieved this objective. Among the many results that could be cited, pollen analyses in the Pyrenees, for example, show that *Quercus* has been present since the Allerød and indicate the dates for the development of *Ulmus* and *Tilia*, the spread of *Alnus* and *Fraxinus* and the presence of *Castanea* between 4820 and 4130 BP. *Fagus*, which appeared around 4000 BP, owed its development not only to climatic conditions but also to the system of lansdcape exploitation. In the Vosges, the very first signs of anthropogenic activities were apparent around 5500 BP but only at low altitudes. From 2400 BP, the clearing of the valleys was evidenced by the decline of *Alnus*, *Corylus*, and *Betula*. Extensive depopulation occurred in the region during the Early Middle Ages. The study suggests how vegetation stages developed in the Internal Alps

since the Late-glacial Period, showing, in particular, that a regional synthesis cannot be based on data from scattered sites in different massifs. In the lower valleys of the Loire and the Seine, the sequences of plant communities can be explained only as responses to fluctuations in sea level. Although still far from complete, the array of results from each region constitutes a major source of data for quantitative palaeoclimatology, allowing climatic reconstructions on the basis of the fossil spectra.

The mission of IGCP 158 B was theoretically limited to the last 15 000 years of vegetation history,

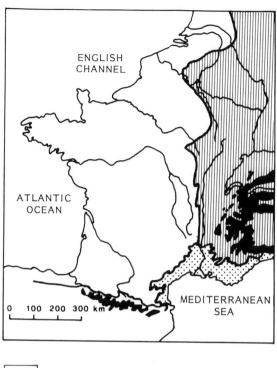

ENGLISH
CHANNEL

ATLANTIC
OCEAN

MEDITERRANEAN
SEA

0   100 200 300 km

Atlantic domain

Mediterranean domain

Middle-European domain

High mountain domain

Fig. 17.3 French floristic domains

but several pollen sites yielded detailed information about the dynamics of pioneer herbaceous communities at the end of the Würm glacial period. At the same time, these sites clarified the chronology and palaeoenvironment of the late-glacial.

The results from the sites studied for the IGCP, combined with many other findings, allow a first tentative outline of the postglacial history of some arboreal taxa. The low-altitude refuges that permitted the survival of relatively mesothermic elements during the last Pleniglacial are linked to the edaphic humidity of the valleys.

Finally, all the results show that during the first part of the Holocene human activity made no impact on the development of the vegetation, but that this was the dominant factor from the Neolithic onwards. Human impact is marked first of all by the

"landnams" made for the cultivation of cereals and the rearing of livestock, and then by the relatively widespread deforestation during the Bronze Age and more particulary in the Iron Age, associated with the formation of the first degraded heathlands. From Gallo-Roman times, human impact is everywhere very strong and is marked especially by the arrival of *Juglans*. The subsequent periods are totally governed by the ebb and flow of human dominance, with its considerable variation from region to region.

TYPE REGION F-b, LOWER SEINE VALLEY
(M.F. Huault)

Altitude: 0–150 m.

Climate: Mean January 6°C, July 19°C. Precipitation 800 mm yr^{-1}. West and especially southwest winds dominant. Atlantic climatic province.

Geology: Cretaceous chalk and upper Jurassic limestones and clays, representing the western part of the Paris basin.

Topography: The Pays de Caux to the north ranges between 100 and 150 m. In the southern part the Roumois has the same altitude as the Pays de Caux. In the Seine valley, between Rouen and the estuary, the floodplain ranges between +2 m and +10 m above sea level (Fig. 17.5).

Population: Density of population is 45–65 people km^{-2} in the Pays de Caux, 25 people km^{-2} in the Roumois and 150 people km^{-2} in the valley of the Seine, especially in the vicinity of Rouen and Le Havre.

Vegetation: Trees (mainly oak and beech) along the valley of the Seine. Salt marsh in the estuary between Le Havre and Tancarville. Peat bogs in the cut-off meanders of the Seine, such as Marais Vernier and Marais d'Heurteauville (Fig. 17.5).

Soils: Leached brown forest soils, podzolic soils, rendzinas, and peats.

Land use: Cattle grazing and open-fields (wheat, flax, oats, and beet-root) in the Pays de Caux and the Roumois. The valley of the Seine has orchards, several bogs, and forests.

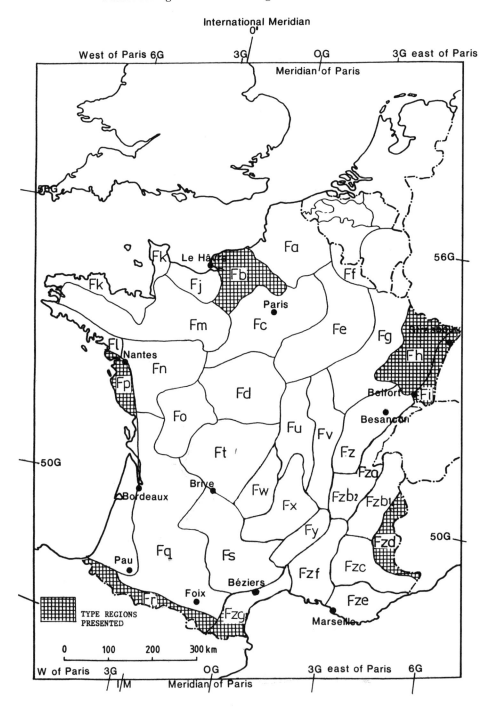

Fig. 17.4 French type regions, with those for which syntheses are presented hatched. F-b = lower Seine valley, F-h, F-i = Vosges, F-p = lower Loire valley, F-r, F-zg = Pyrenees, F-zd = internal French Alps

579

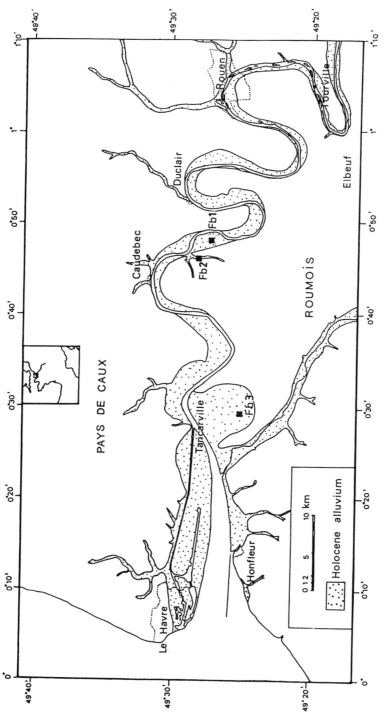

Fig. 17.5 Location map of the sites Fb1: Marais d'Heurteauville, Fb2: La Mailleraye, Fb3: Marais Vernier

Reference site 1. Marais d'Heurteauville
(Huault 1986, 1988)

Latitude 49°27′N, Longitude 0°48′E. Elevation 3 m.
Age range 6570–0 BP. Valley mire. Five local pollen-assemblage zones (paz). (Fig. 17.6)

	Age	Local paz
HE-1	?–6570 BP	*Quercus–Ulmus*
HE-2	?	*Ulmus–Tilia*
HE-3	?–4390 BP	*Quercus–Tilia*
HE-4	4390–3130 BP	*Quercus–Filicales.* Cerealia is present
HE-5	3130–0 BP	*Fagus.* The representation of Cerealia becomes continuous. Appearance of *Juglans* dated 1145 BP

Reference site 2. La Mailleraye
(Huault and Lefebvre, 1983a)

Latitude 49°28′N, Longitude 0°46′E. Elevation 3 m.
Age range 7220–0 BP. Valley mire. Five local pollen-assemblage zones (paz). (Fig. 17.7)

	Age	Local paz
LM-1	7220–? BP	*Quercus–Corylus.* Maximum frequencies of *Corylus.* The lower part of the pollen diagram is dated 7220 BP. *Quercus, Ulmus,* and *Tilia* occur
LM-2	?	*Quercus–Ulmus–Alnus.* Lower boundary placed at the decrease of *Corylus* and increase of *Alnus.* First pollen of *Fagus.* The upper boundary is placed at the decrease of *Ulmus*
LM-3	?–3675 BP	*Quercus–Tilia.* Maximum values of *Tilia. Alnus* shows a slight decrease. The end of the phase is dated 3675 BP
LM-4	?	Cyperaceae–Filicales.

	Age	Local paz
LM-5	?	Non-arboreal pollen, especially Cyperaceae and Filicales, show an expansion. This phase is characterized by the decline of *Corylus, Tilia,* and *Alnus Fagus.* The last phase is characterized by the expansion of *Fagus,* the decline of arboreal pollen such as *Ulmus, Tilia, Corylus,* and *Quercus.* Appearance of Cerealia and anthropogenic indicators (*Plantago lanceolata, Rumex,* Cruciferae, Compositae, Liguliflorae). Appearance of *Juglans*

Reference site 3. Marais Vernier
(Huault & Lefebvre 1983b, Huault 1985)

Latitude 49°24′N, Longitude 0°29′E. Elevation 2 m.
Age range 7630–0 BP. Valley mire. Five local pollen-assemblage zones (paz). (Fig. 17.8)

	Age	Local paz
MV-1	?–7630 BP	*Quercus–Corylus.* High values of Cyperaceae
MV-2	7630–6100 BP	*Quercus–Ulmus. Quercus ilex*–type very sporadic. Appearance of Chenopodiaceae and simultaneous decrease of Cyperaceae
MV-3	6100–5500 BP	*Tilia–Ulmus.* Disappearance of *Quercus ilex* type
MV-4	5500–3900 BP	*Tilia–Alnus.* Expansion of *Fraxinus*
MV-5	3900–? BP	*Fagus*–Ericaceae. Disappearance of Chenopodiaceae

581

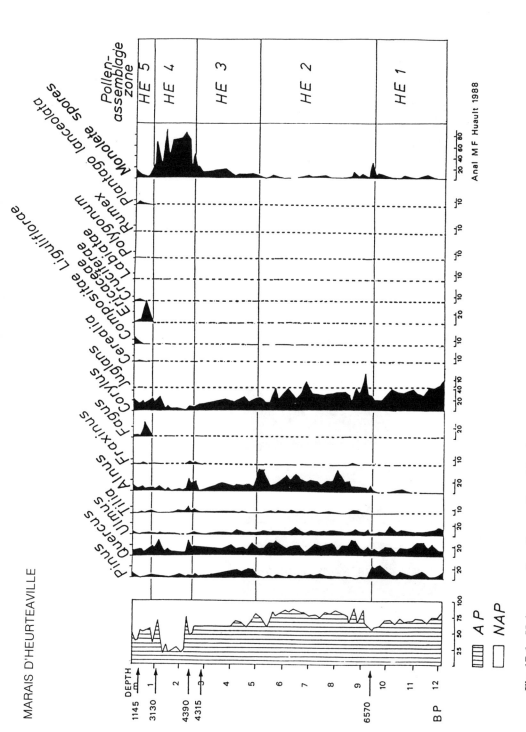

Fig. 17.6 Holocene pollen diagram from Marais d'Heurteauville. Major pollen and spore types only are shown (from Huault, 1986)

582

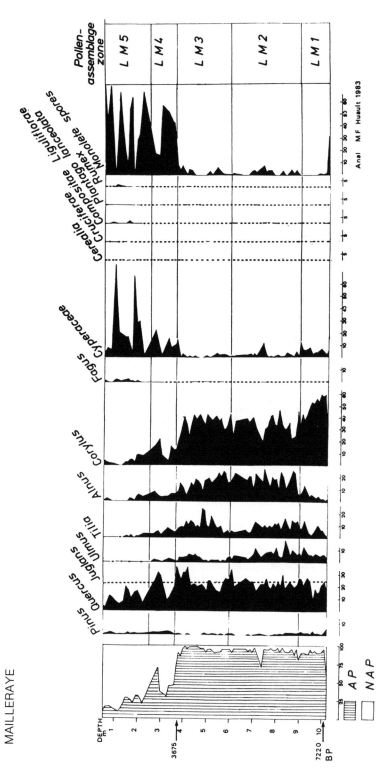

Fig. 17.7 Holocene pollen diagram from Mailleraye. Major pollen and spore types only are shown (from Huault & Lefebvre, 1983a)

583

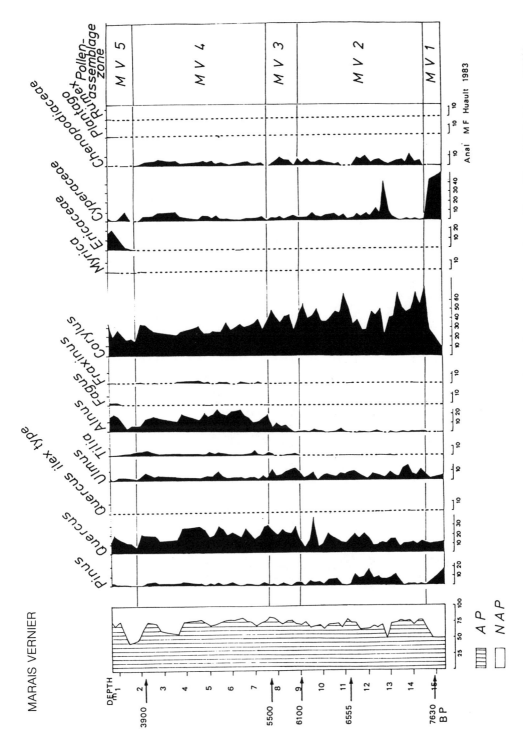

Fig. 17.8 Holocene pollen diagram from Marais Vernier. Major pollen types only are shown (from Huault & Lefebvre, 1983b)

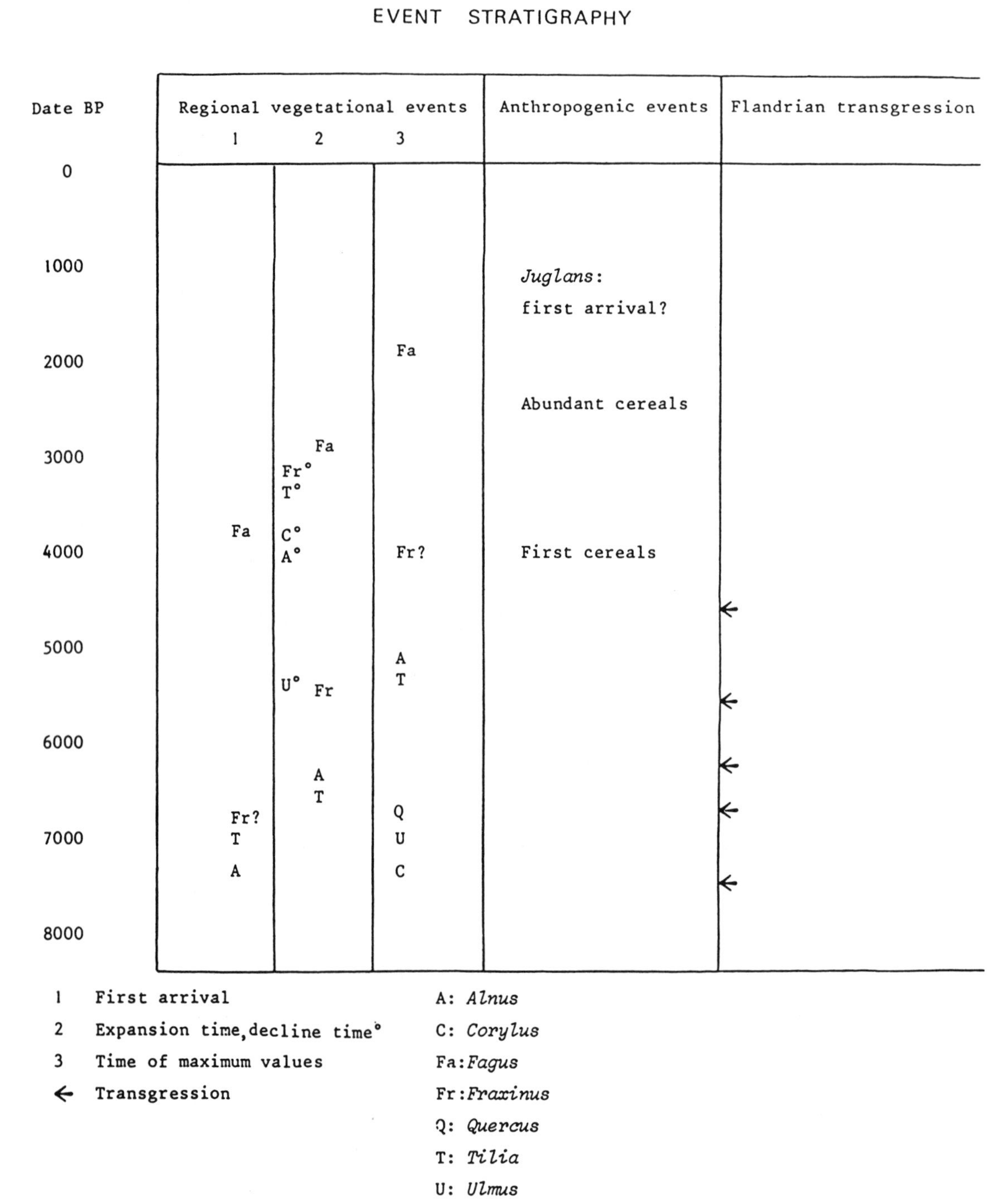

EVENT STRATIGRAPHY

Date BP	Regional vegetational events			Anthropogenic events	Flandrian transgression
	1	2	3		
0					
1000				*Juglans*: first arrival?	
2000			Fa	Abundant cereals	
3000		Fa Fr° T°			
4000	Fa	C° A°	Fr?	First cereals	←
5000			A T		←
6000		U° Fr			←
		A T	Q		←
7000	Fr? T		U		
	A		C		←
8000					

1	First arrival	A:	*Alnus*
2	Expansion time, decline time°	C:	*Corylus*
3	Time of maximum values	Fa:	*Fagus*
←	Transgression	Fr:	*Fraxinus*
		Q:	*Quercus*
		T:	*Tilia*
		U:	*Ulmus*

Fig. 17.9 Event stratigraphy in the lower Seine Valley

Palaeohydrology and Flandrian transgression
(Fig. 17.9)

Diatom investigations have been carried out on the reference sites. Since the end of the Boreal phase at Marais Vernier and Marais d'Heurteauville, five transgressive phases occurred, with high values of marine and brackish diatoms such as *Paralia sulcata, Cymatosira belgica, Cocconeis scutellum* and *Melosira westii*: two between 7600 and 6500 BP, another just before 6100 BP, and the last two during the Subboreal just after 5500 and about 4500 BP. Between these transgressive movements the diatom flora shows numerous freshwater types, such as *Cymbella, Fragilaria*, and *Epithemia*. Their great abundance indicates a stop in the rise of sea level rather than a regression, because we have not seen evidence for a general erosion in the numerous cores analysed in the valley of the Seine. The erosional data observed are probably due to a change in the channel network.

TYPE REGIONS F-h AND F-i, THE VOSGES MOUNTAINS (C.R. Janssen)

Because of intensive agriculture, no proper sites are available from Lorraine and the Alsace. However, data from the Vosges Mountains, located between areas F-h and F-i, also partly reflect the vegetation developments in these two type regions. In this synthesis conditions in a transect 60 km long in the central Vosges are reviewed, along with published and unpublished data from the northern Vosges (Firbas *et al.* 1948) and the southern Vosges (Schloss 1979). General characteristics of the Central Vosges are as follows.

Altitude: 400 m (in the west, Bruyères) to 1386 m (crest area, Hohneck) to 189 m (Colmar).

Climate: Oceanic, precipitation: west slope 1000–2000 mm yr^{-1}, depending on altitudes and local site factors; east slope 500 mm yr^{-1}. Annual temperature: west: 9.0°C, crest 3.1°C, east 10.8°C (Fig. 17.10).

Geology: Precambrian granite, flanked towards the west and east successively by a Permian sandstone (Vosges gréseuses) and Triassic limestone.

Topography: Relief asymmetric in cross-section, with altitudes slowly ascending on the west, steeply falling to the east.

Vegetation and land use: Calcareous region west: highly cultivated, with remnants of deciduous woods of Querceto–Carpinetum and Alno–Padion; sandstone area west: almost completely forested by Luzulo–Fagion woods. Area covered in last century with extensive heathlands; granitic massif west: Generally valleys at lower altitudes cultivated, upper slopes forested with montane *Abies alba–Fagus sylvatica* forest above 750 m and (planted) *Picea abies*. Above 1100 m subalpine *Fagus* forest (Aceri–Fagetum) with *Sorbus aucuparia* and *Acer pseudoplatanus*. Granitic massif, crest area: Towards the top of the mountains krummholz of *Fagus* giving way to subalpine meadows. Granitic massif east: valleys cultivated, slopes covered with montane *Fagus–Abies* forest above 500–600 m altitude. Below that level thermophilous elements like *Carpinus betulus, Juglans regia*, and *Castanea sativa*, as well as *Quercus (robur)*. Limestone/sandstone area east: vineyards along the foothills of the Vosges along the Upper Rhine Plain. Remnants of woods on limestone belong to Quercion pubescentis p.p. On sandstone many plantations of *Pinus sylvestris* and exotic tree species (e.g. *Robinia pseudoacacia*); Upper Rhine Plain: cultivated for cereals except patches of Fraxino–Ulmetum along tributaries of the Rhine River.

Soils: Subalpine meadows: rankers cryptopodzoliques. *Fagus* forest and secondary meadows: brown forest soils. *Abies/Pinus* forest on sandstone/granite: podzols.

The present-day pollen deposition

The present-day pollen deposition has been studied by means of surface sampling on both a local scale (Tamboer v.d. Heuvel & Janssen 1976, De Valk 1981, Edelman 1985) and on a regional scale (Janssen 1981). It appears that landscape regions (regions 2, 3a, 3b, 3c, 3d, 4, and 5 in Fig. 17.11) within the area of the transect can be characterized mostly by the proportions of the main tree pollen types, to a much lesser extent by herb pollen types. In

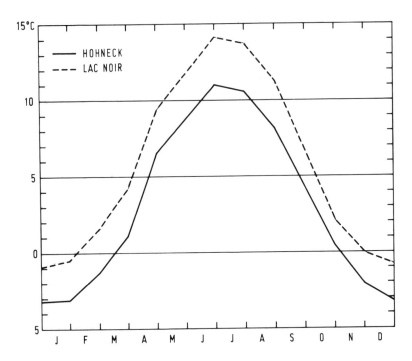

Fig. 17.10   Average monthly temperatures for the Hohneck and Lac Noir. After Rothé & Herrenschneider (1963)

this the landscape regions resemble the landform–vegetation units of Lichti-Federovich & Ritchie (1968), but on a much finer scale. In fact, scale is the biggest problem if we are to include our results in the scheme of reference areas established in the framework of the IGCP 158 B project across Europe. Berglund's (1979) concept of a reference area is much like that of landform–vegetation units. However, reference area F-h is today not characterized by a uniform regional pollen deposition as it should if we were dealing with landform–vegetation units. Rather, within the transect of only part of reference area F-h, we have already a number of sub-reference areas. This problem of scale is probably common to all mountainous regions with much diversity in all the elements that together are characteristic over rather short distances. Under these conditions it pays more to study small rather than large regions, including both low and high-altitudinal areas. Certainly it would not be useful to amalgamate the large diversity present within the transect studied in the Vosges.

Complexity of pollen diagrams

The location of 30 sites studied pollen-analytically, including 21 in the context of the Vosges project (Janssen *et al.* 1974), are shown in Figure 17.11. Radiocarbon-dated pollen diagrams of two sites (Moselotte and Goute Loiselot) are presented in a simplified form in order to facilitate comparison of the main differences. We know that differences in the pollen assemblages from site to site are either differences in time or differences in space. Time control is therefore essential in the detection of the development of the spatial arrangement of vegetation types. Such control is provided by radiocarbon dates, by which trends in the fluctuations of pollen curves can be detected for a number of sites. The trends reflect the effect of anthropogenic activities on the vegetation at lower altitudes in the mountains for the last 4000–5000 years. By a careful comparison of a number of pollen diagrams, and by taking into account accumulation rates of the deposits based on available radiocarbon dates, Kalis (1985) has estab-

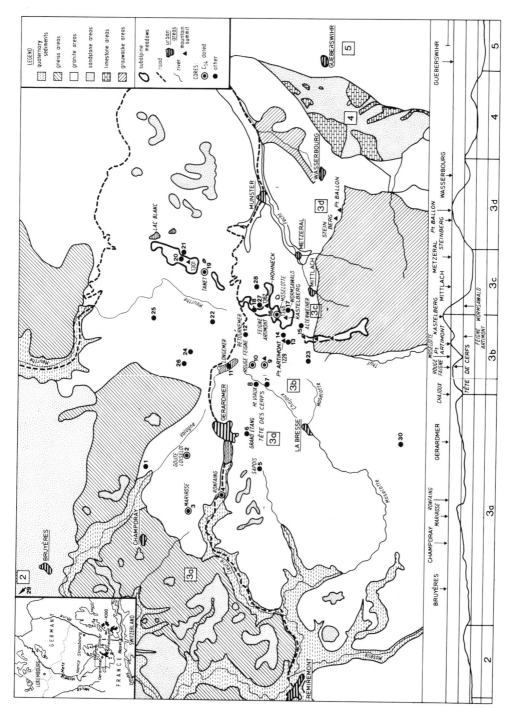

Fig. 17.11 Palynological sites, landscape regions 2, 3a, 3b, 3c, 3d, 4, and 5, and bedrock in the Central Vosges

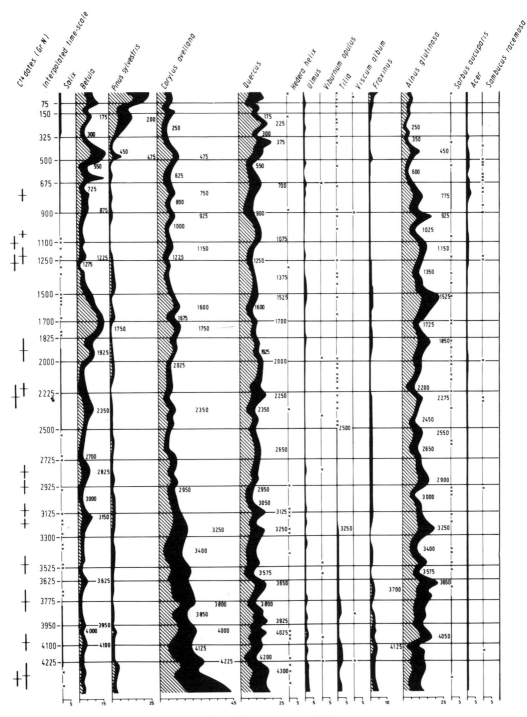

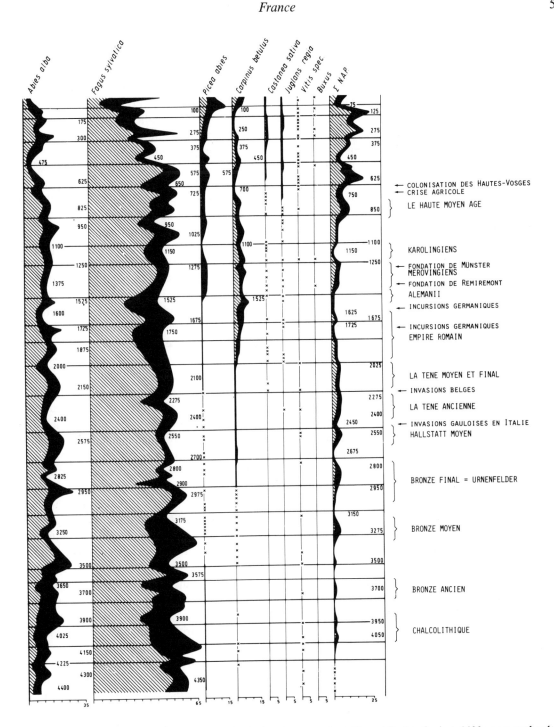

Fig. 17.12 Trends in the pollen curves from cores in landscape regions 3b and 3c for the last 4400 years and cultural phases in northeastern France. Black curves indicate maximum pollen values, hatched curves minimum pollen values.
After Kalis (1984b)

lished for the last 4000 years a detailed series of trends in the pollen curves from cores in region 3b and 3c of Figure 17.11. These trends are reproduced in Figure 17.12, along with an indication of the cultural phases that lie at their basis. Once trend curves have been established, fluctuations in the pollen curves of non-dated diagrams can be correlated by the detection of these trends. By interpolation between derived dates from an age/depth curve any level in a pollen diagram can be provided with an estimated date, allowing the synchronous comparison of pollen values for a number of sites in so-called correlation pollen diagrams. This has been done for a number of pollen types in three cores illustrating the contrast between a small peat bog and larger raised bogs (Janssen & Braber 1987).

Synchronous trends from lower altitudes in the pollen curves are more difficult to detect for the time before man began to have an impact on the vegetation large enough to be picked up in the regional and extraregional pollen deposition, i.e. the Late Glacial and early and middle Holocene. For this time span the location of the source area of the various pollen types is less clear than for the time after 4000–5000 years ago, and the establishment of synchronous events for different sites depends almost entirely on the reliability of the radiocarbon dates. Essentially the procedure applied for the late Holocene can also be followed for earlier periods, viz. construction of age/depth curves of cores on the basis of available (calibrated!) radiocarbon dates, determination of the age of pollen trends in the core by interpolation between radiocarbon dates, identification of synchronous trends common for a number of pollen diagrams, establishment of derived dates of levels of non-radiocarbon-dated cores, and finally comparison of synchronous fluctuations of pollen curves by means of correlation diagrams. This enormous task, considering the many pollen diagrams that must be recalculated and compared as described above, will take some time. Therefore, in order to arrive at a quick overview of the differences in the various pollen assemblages in time and space, the approach of Cushing (1967) will be adopted in this paper in an informal way, viz. establishment and definition of the main assemblages on the basis of the principal pollen dominants and dating of the ensuing zone

boundaries, followed by comparison of the assemblages in time and space. In Figure 17.13 the data of nine radiocarbon-dated pollen diagrams arranged from west to east in the transect are compared. The assemblage zones are informally defined. For each core the dominant pollen types are named in descending order of importance. We have refrained from defining assemblage zones common to more than one site, mainly because in this study, where the sites are located close together, there is no simple relation between pollen deposition and the vegetation in a defined area. Even for sites far apart, the source of some pollen is outside the area under consideration. The possibilities to interpret pollen assemblages in terms of vegetation in areas of defined extent diminish considerably when the sites are close together.

Under these conditions a pollen assemblage may easily reflect several vegetation types, or conversely different pollen assemblages may reflect similar vegetation types. In fact, this emphasizes that the Cushing approach is less useful for studies done spatially on a fine scale, such as is usually the case in mountain areas. Still, the effort presented in Figure 17.13 in the framework of the IGCP project 158 B is useful in that it shows at a glance the compositional relationships of the main pollen assemblages before 4000–5000 BP, allowing comparison with other reference areas within Europe.

However, a drawback is that zone boundaries as shown in Figure 17.13 do not always represent vegetation boundaries due to the effect of pollen sources in adjacent vegetation. In most cases this effect does not influence the main conclusions about vegetation development. But where it might cause confusion is when pollen of such external source is not taken into account. There a dotted line is drawn to show the time of vegetation change, i.e. at Rouge Feigne where the invasion of *Fagus* and *Abies* is blurred by pollen dispersed from the Grande Basse.

In Figure 17.13 each core has three columns: (1) GrN number of the Isotope Physics Laboratory of the University of Groningen, (2) characterization of the pollen assemblage by its dominant pollen types, (3) designation by which the site ("local") zones were named in the original publications. On the right side of the three columns the rational boundaries of the pollen curves of *Fagus* (F), *Abies* (Ab), *Ulmus* (U), *Quercus* (Q), *Tilia* (T), *Fraxinus* (Fr), *Alnus* (A), *Corylus* (C), and *Pinus* (P) are indicated.

Palaeobotanical studies on peat bogs and lakes in the Vosges

Bold numbers refer to the list of selected French reference sites (see Appendix). Other numbers refer to sites in Fig. 17.11.

Central Vosges

	1. Étang d'Orong	13000–0 BP	unpubl.
4	2. Goutte Loiselôt	11000–0 BP	Edelman 1985 (10 diagrams)
	3. Marirose	13000–4 BP	unpubl.
	4. Ronfaing	11000–4 BP	Janssen & Salomé 1974
	5. Sapois		unpubl.
	6. Grand Étang	15000?–1 BP	Teunissen & Schoonen 1973
	7. Chajoux	6000–0 BP	Janssen & Braber 1987
5	8. Ht. Viaux	3000–0 BP	Janssen & Braber 1987
6	9. Rouge Feigne	9000–0 BP	Kalis 1984a, b
7	10. Grande Basse	10000–0 BP	Kalis 1984a, b
	11. Longemer		unpubl.
	12. Retournemer		unpubl.
	13. Feigne Pt. Artimont	4000–0 BP	Kalis 1984a, b
8	14. Feigne d'Artimont	10000–0 BP	Janssen *et al.* 1975
9	15. Altenweiher	10000–0 BP	De Valk 1981
10	16. Moselotte	7000–0 BP	De Valk 1981
	17. Wormsawald	6000–0 BP	De Valk 1981
11	18. Trois Four	9000–0 BP	De Valk 1981
12	19. Tanet	3000–0 BP	Janssen & Kettlitz 1972
	20. Gazon du Faing		unpubl.

21. Lac Forlet		unpubl.
22. Belbriette	3000–1 BP	Darmois-Theobald & Denèfle 1981
23. Machey	3000–1 BP	Darmois-Theobald & Denèfle 1981
	11000–8 BP	Woillard 1974
24. Col de Surceneux	6000–0 BP	Darmois-Theobald *et al.* 1976, 1981
25. Cirque Squainfaing	7000–1 BP	Darmois-Theobald *et al.* 1976, 1981
26. Tourbiere l'Étang	10000–4 BP	Darmois-Theobald *et al.* 1976, 1981
27. Rotried	7000–0 BP	Firbas *et al.* 1948
28. Frankenthal	10000–0 BP	Firbas *et al.* 1948
29. St. Hélène	4000–1 BP	Guillet *et al.* 1972
30. Frère Joseph	12000–0 BP	Woillard 1974

Southern Vosges

Le Rahin	11000–1 BP	Dresch *et al.* 1966
Grande Goutte	6000–1 BP	Dresch *et al.* 1966
Fagnes Savou-reuse	9000–2 BP	Dresch *et al.* 1966
Rochalins		unpubl.
Grande Pile	15000–9 BP	Woillard 1978
Lac Sewen, nine diagrams	13000–1 BP	Schloss 1979

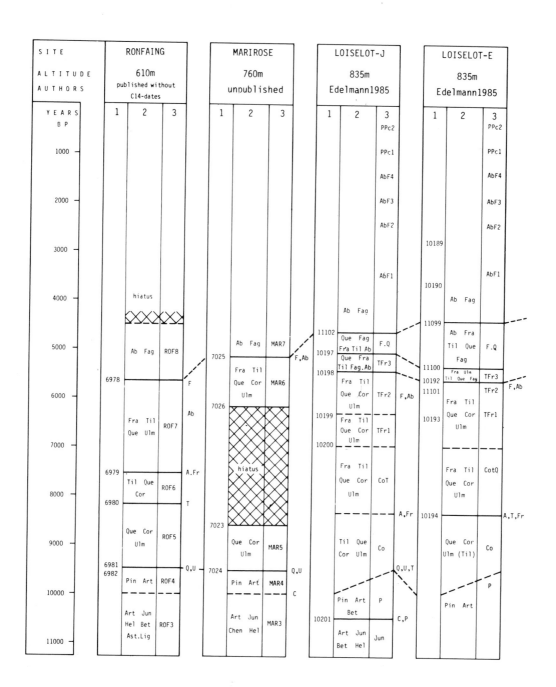

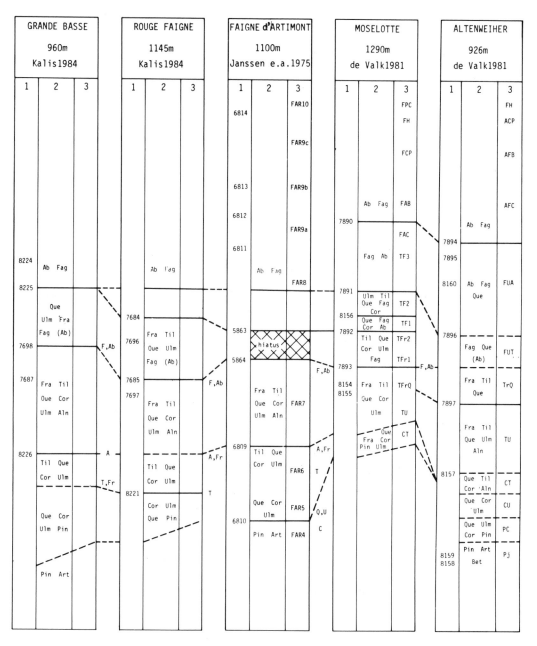

Fig. 17.13 Correlation chart of pollen assemblages since *ca.* 11000 BP for a west–east transect of nine cores in the Central Vosges showing: (1) GrN number, Isotope Physics Laboratory, Groningen, (2) pollen assemblages (paz) and (3) designation of pazs in original publication. Pollen diagrams for Moselotte and Loiselot-J are shown as Figs 17.16 and 17.17

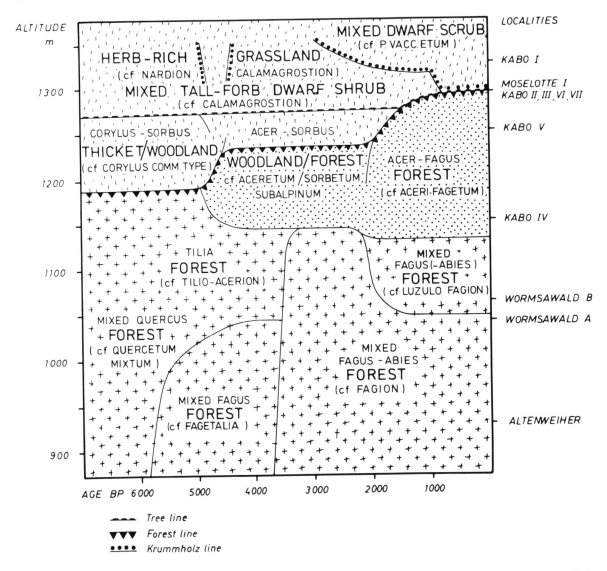

Fig. 17.14 Summary of altitudinal differentiation of vegetation belts and timber lines in the east slope area of the Kastelberg, 7000 BP to the present. For explanation see text. After DeValk (1981)

Grand Chemin	11000–9 BP	Woillard 1974	**14** Champs de Feu	8000–0 BP	Firbas *et al.* 1948 and unpubl.
Grands Pres	13000–11 BP	Woillard 1974			
			Grafenweiher	9000–0 BP	unpubl.
Northern Vosges			Eichelkopf	9000–8000, 5000–0 BP	unpubl.
13 Marais de la Max	13000–0 BP	Firbas *et al.* 1948 and unpubl.	Blanches Roches	10000–8000, 2000–0 BP	unpubl.

MOSELOTTE Ⅱ

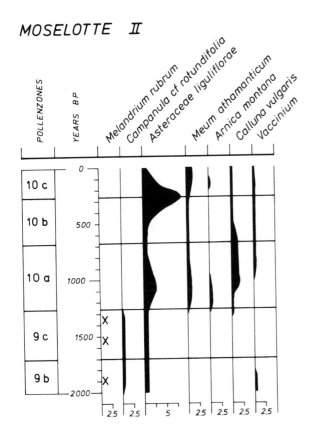

Fig. 17.15 Simplified pollen diagram from Moselotte-2, a site in the present-day subalpine vegetation belt, showing the transformation of Nardetalia–Adenostyletalia into Pulsatillo–Vaccinietum in the Middle Ages. After DeValk (1979)

| Refuge de Prayé | 3000–0 BP | unpubl. |
| Rond Perthuis | 3000–0 BP | unpubl. |

Main vegetational events

Late Glacial

Dates follow conventions of Mangerud *et al.* (1974). Deposits are known only from altitudes below 800 m.

Pre-Allerød (–11800 BP)
NAP 70–90% (*Artemisia*, Gramineae, Chenopodiaceae, *Helianthemum*, Compositae Liguliflorae), both in southern (Lac Sewen) and northern (Marais de la Max) Vosges. *Juniperus* peak in Bølling.

Allerød (11800–11000 BP)

Characterized by (open) forest. In northern, central, and southern Vosges, decline of NAP. *Betula* maximum followed by *Pinus* maximum. *Juniperus* declining to low values in *Pinus* time.

Younger Dryas (11000–10000 BP)
Characterized by open landscapes. Central Vosges: high values of *Artemisia*, Gramineae, *Juniperus*, *Anthemis* type, Asteraceae Liguliflorae, Chenopodiaceae, *Helianthemum*, *Heracleum*, *Meum*. *Pinus* values down to 10%. Northern Vosges: *Artemisia–Pinus* assemblage, low *Juniperus* values. Southern Vosges: *Pinus–Artemisia* assemblage, low *Juniperus* values.

Holocene

Pinus–Betula–Artemisia paz
From the end of the Late Glacial to *ca.* 9500 BP in the west at low altitudes, and to 9000 BP at higher altitudes.

Corylus–Quercus–Ulmus paz
Everywhere following previous paz until 8200–8500 BP, or until 7500–8000 BP in crest area.

Complexes of *Corylus, Quercus, Tilia, Alnus, Fraxinus* pazs
Recognizable everywhere, with generally diminishing *Corylus* values and increasing values of *Quercus, Ulmus, Tilia,* and *Fraxinus*, less so above 1300 m altitude. Pazs in crest area younger by 1500 years and shorter (earlier pazs) or longer-lasting (later pazs) than in deposits in west or east slope area.

- At lower altitudes *Quercus* forest and other elements of the QM. Climate warmer than today.
- *Quercus* present at 835 m, thus upper altitudinal limit higher than today.
- *Trapa natans* present at Lac Sewen (also still in Subboreal).
- *Corylus* present at high altitudes: summer temperature 10.5°C?
- At higher altitudes presence of *Corylus–Sorbus* thicket around forest limit at 1200 m altitude above a *Tilia* forest around 1100 m in the east slope area (Figure 17.14).
- Above 1250 m Calamagrostion and Nardion type of grassland present (Figure 17.14).

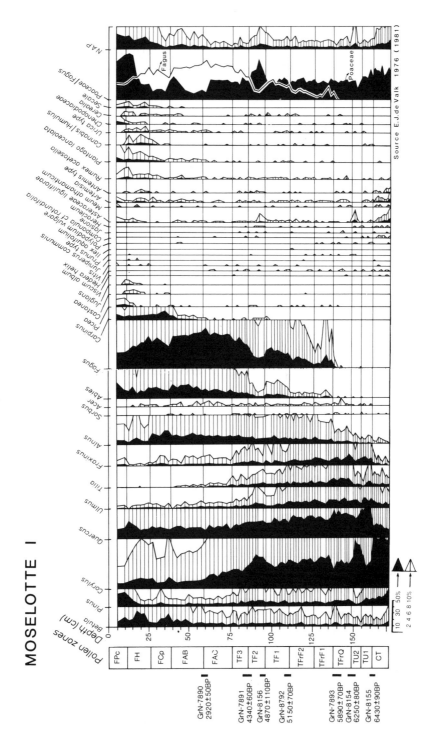

Fig. 17.16 Simplified pollen diagram for Moselotte-1. After DeValk (1981)

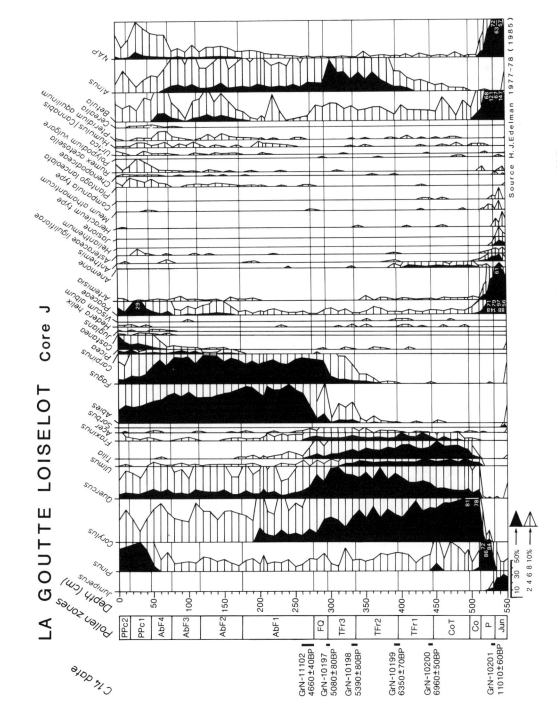

Fig. 17.17 Simplified pollen diagram for Goutte Loiselôt, Core J. After Edelman (1985)

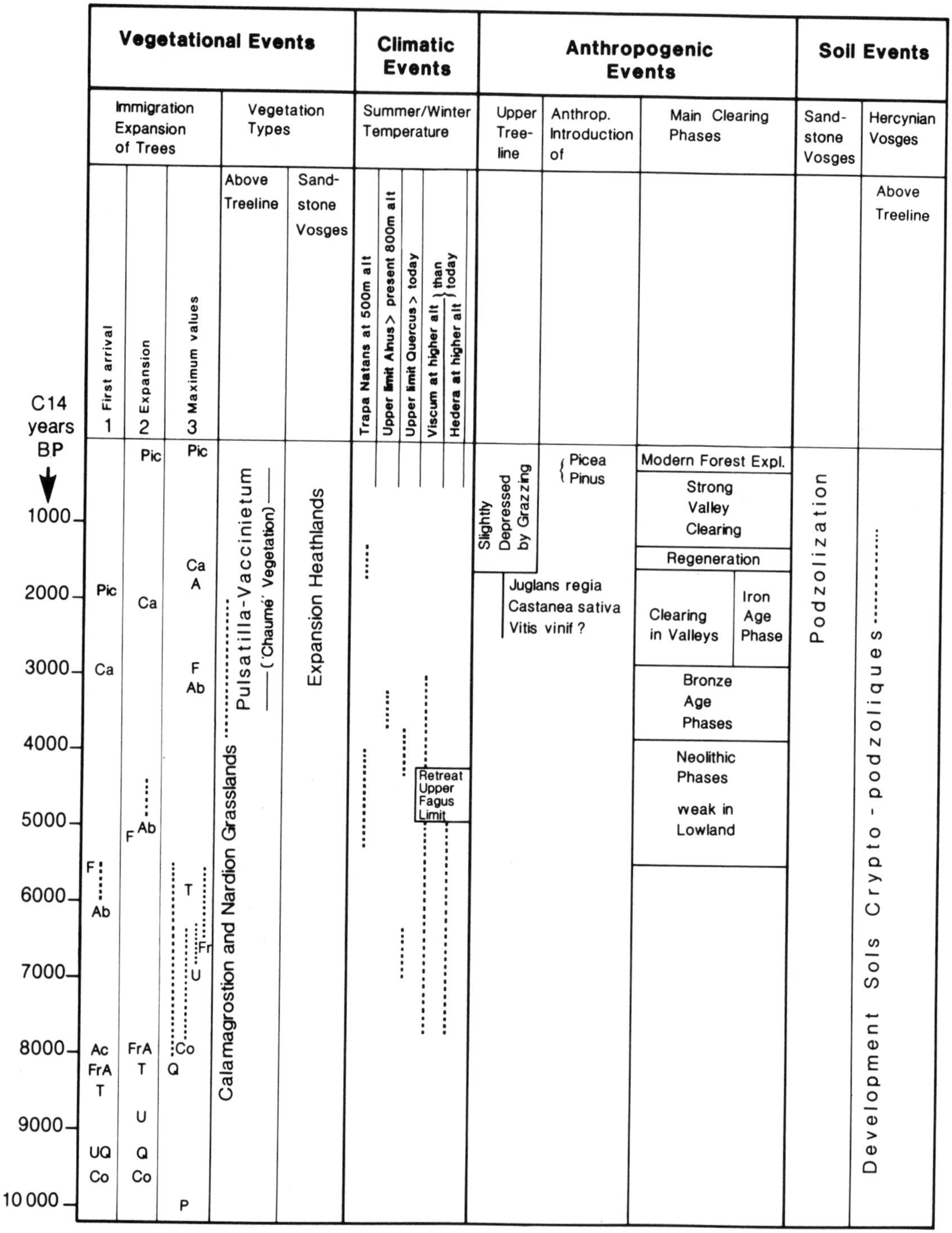

Vegetational Events				Climatic Events	Anthropogenic Events			Soil Events	
Immigration Expansion of Trees			Vegetation Types	Summer/Winter Temperature	Upper Tree-line	Anthrop. Introduction of	Main Clearing Phases	Sand-stone Vosges	Hercynian Vosges
First arrival 1	Expansion 2	Maximum values 3	Above Treeline / Sand-stone Vosges	Trapa Natans at 500m alt / Upper limit Alnus > present 800m alt / Upper limit Quercus > today / Viscum at higher alt } than / Hedera at higher alt } today					Above Treeline

C14 years BP ↓

	Pic	Pic	Pulsatilla-Vaccinietum ('Chaumé' Vegetation) / Expansion Heathlands		Slightly Depressed by Grazzing	{ Picea { Pinus	Modern Forest Expl.	Podzolization	Crypto-podzoliques
1000		Ca A					Strong Valley Clearing		
2000	Pic	Ca				Juglans regia Castanea sativa Vitis vinif?	Regeneration / Clearing in Valleys | Iron Age Phase		
3000	Ca	F Ab	Calamagrostion and Nardion Grasslands				Bronze Age Phases		
4000							Neolithic Phases		
5000	F Ab			Retreat Upper Fagus Limit			weak in Lowland		Development Sols
6000	F: Ab		T						
7000			:Fr U						
8000	Ac FrA T	FrA T	:Co Q						
9000		U							
	UQ Co	Q Co							
10 000		P							

Fig. 17.18 Event stratigraphy for type regions F-h and F-i. A: *Alnus*, Ab: *Abies*, Ac: *Acer*, Ca: *Carpinus*, Co: *Corylus*, F: *Fagus*, Fr: *Fraxinus*, Pic: *Picea*, Q: *Quercus*, T: *Tilia*, U: *Ulmus*

Complex of mixed QM–*Abies*–*Fagus* pazs
Characterized by immigration of *Fagus* and *Abies* in montane forest belt. Pazs at higher altitudes towards climatically controlled tree limit in crest area increasingly more complex and longer lasting, from 6000–4400 BP. Invasion of *Fagus* predates that of *Abies*. Pazs at lower altitudes (600–700 m) absent. Abrupt contemporaneous invasion of both *Fagus* and *Abies* around 5600 BP.

- Final establishment of subalpine *Fagus* forest or montane *Fagus*–*Abies* forest, increasingly later towards higher altitudes.
- Invasion of *Fagus* near present-day tree limit extremely slow and interrupted by temporary retreat of *Fagus* limit 4900–4300 BP. However, invasion of *Fagus* in east and west slope area was fast, within a few hundred years.
- Development of *Fagus* krummholz in lower belt of *Corylus* thicket, in upper belt of subalpine *Acer*–*Fagus* forest and in montane deciduous forest (Figure 17.14).

Complex of *Abies*–*Fagus* pazs
Characterized by full development of present-day montane *Abies*-*Fagus* forest and *Acer*-*Fagus* forest.

- Diminishing pollen values of *Abies* towards higher altitudes.
- Immigration of *Carpinus betulus* around 3000 BP below 700 m altitude.
- Around 4000 BP and 6000–7000 BP *Quercus* present at 850 m altitude along western slopes, some 200 m above present-day upper limit.
- Upper limit of *Abies* fluctuates between 1000 and 1200 m:
     below 1000 m—before 4500 BP
     around 1100 m— 4500–3500 BP
     around 1200 m—3500–2500 BP
     around 1100 m—2500–2000 BP
     around 1000 m—2000–1300 BP
- Slow transformation of herb-rich grasslands into present-day Pulsatillo–Vaccinietum (Hautes-Chaumes) from 3000 BP on, especially after 2000 BP (Figure 17.15).

- Upper tree line around 2000 BP slightly higher than today.
- Limited immigration of *Picea abies* on slopes with northeast exposure (Kaltluftschlenken) in regions 3a,b.
- In the sandstone Vosges, podzolisation of soils due to establishment of heathlands from *ca.* 2000 BP. However, podzols are rare on granite and are present only on very acid parent material, formed below forest cover.

Pinus–*Abies*–*Fagus*–*Picea* paz
Mass plantation of *Picea* and *Pinus* in montane and colline vegetation belts from *ca.* 1600 AD on. *Abies* favoured by forestry measures.

Anthropogenic events (Fig. 17.18)

- First weak signs of anthropogenic activities at 5500 BP, reflection from activities at lower altitudes. Upper reaches of mountain area essentially untouched.
- Cultural phases from Bronze Age are clearly reflected in maxima in the NAP curve (Fig. 17.12), viz. Bronze Ancien, Bronze moyen, Urnenfelder, Halstatt Moyen, La Tène Ancien, Halstatt Final–La Tène Moyen, and Final, Empire Romain, Merovingien, Haut Moyen-Age.
- In early phases only *Quercus* and *Betula* in colline vegetation belt affected.
- First recorded clearing of a valley in montane vegetation belt: 2400 BP.
- From 2400 BP on, clearing of valley floor reflected by minima in *Alnus*, *Corylus*, and *Betula* values in pollen diagrams.
- Strong depopulation of area in Early Middle Ages (Dark Ages: 300–500–600 BP).
- Large-scale exploitation of the Vosges from 1100 BP on, leading to diminished presence of *Abies* from 900–350 BP. Pollen curves of (cultivated) *Juglans*, *Castanea*, and *Vitis* (*vinifera*) continuous from 600 BP on, reflecting their expansion in eastern foothills of the Vosges.
- Introduction of exotics like *Carya amara*, *Carya alba*, and *Juglans nigra* in last 100 years.

TYPE REGION F-p, MARSHLANDS OF THE LOIRE ESTUARY (L. Visset and D. Voeltzel)

The synthesis presented here is based on the analysis of pollen, lithology, sedimentology, and radiometric dates from 22 cores from the alluvial plain of the Loire estuary and the adjacent marshlands, located downstream from Nantes. They form a triangular lowland area flanked by hills to north and south. On both sides of the river are vast wetlands, covering almost 150 km². On the north these are linked to the Brière marshlands, and to the south to the lake of Grand-Lieu via the Acheneau valley (Fig. 17.19).

The Loire estuary cuts through the metamorphic terrane of the Cornish anticline. Although the area has been subject to tectonic activity in recent times, the basic features of the relief were established at the beginning of the Tertiary. The sediments that accompanied the Flandrian marine transgression buried a very irregular landscape. The sedimentary deposits of the Loire in fact comprise two separate formations, overlying one another and differing in age and origin:

- a layer of coarse alluvium consisting of sand, gravel, and fragments of crystalline rocks, dating back to the beginning of the Weichsel period;
- a typical fluvio-estuarian complex with alternating layers of sand and muds.

Deposits of mud and later of peat have been formed since about 8000 BP in the marshlands of the alluvial plain and adjacent areas.

Pollen zones and archaeological phases at key sites in the Armorican Massif

This chronology has been discussed in previous reviews (Morzadec-Kerfourn 1974, Visset 1979, Giot *et al.* 1979a and b) and will therefore be summarized only briefly here (Fig. 17.20).

Summary of stratigraphic and radiometric evidence

Figure 17.21 brings together several of the geological sequences obtained from corings in the estuarine marshes. This diagram gives an overview of the processes of sedimentation:

- The earliest Holocene sediments are those of

Canal des Fougères III, dated to 7800±120 BP (Fig. 17.22).
- However, at most sites deposition begins only in the Atlantic period. At Canal des Fougères, 16 m of sediment were deposited during this period, and at least 20 m at Saint-Malo-de Guersac. At all sites except the lake of Grand-Lieu and Naye I, these deposits consist of muds of fluvio-marine origin, at least in part.
- During the Atlantic phase, marine influences penetrated inland along watercourses as far as the site of Petit Mars, about 70 km from the coast.
- The radiocarbon dates obtained for the earliest deposits of true peat fall into four groups:

(1) Around 6300 BP : La Basse Ville and Port-Saint-Père III in the Acheneau valley at an elevation of approximately –5 m NGF (NGF = "Nivellement Général de la France" and is roughly equivalent to present mean sea level at Marseille-France). The date from Gesvres, obtained from a sample of ash wood (5770±120 BP–Gif 5748), allows the earliest peat deposits here to be given a date close to 6000 BP, at an elevation of roughly –4 m NGF.

(2) Around 5700 BP : Lavau and La Caudelais in the marshlands on the northern shore of the estuary, at approximately –3 m NGF.

(3) Around 4300 BP : Canal des Fougères II and Ile d'Errand II (Fig. 17.23) in the Brière marshland at approximately –1 m NGF.

(4) Around 3600 BP : Port-Saint-Père II (–5.35 m NGF), Pimpenelle (–1 m NGF), and La Crôle (–2 m NGF) (Fig. 17.24).

- Only two of these sites were flooded a second time by fluvio-marine water after the formation of peat:

(1) Canal des Fougères II : the evidence here consists of muds of clearly fluvio-marine character above the level +0.40 m NGF, beginning around 2000 BP.

(2) La Caudelais : here are muddy sections a few centimetres thick and clearly fluvio-marine in character at about –0.40 m NGF (3000 BP ?), then layers of mud of less marked fluvio-marine character from +0.50 m NGF, beginning around 2000 BP.

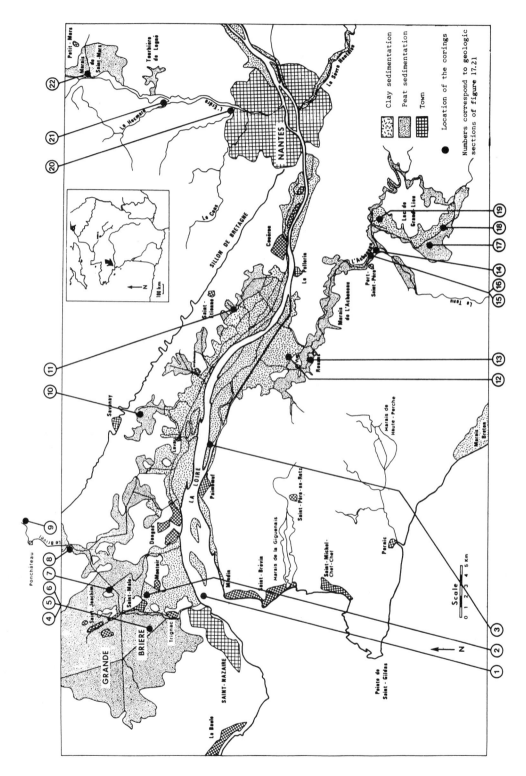

Fig. 17.19 Geographic location of the corings in the Loire estuary

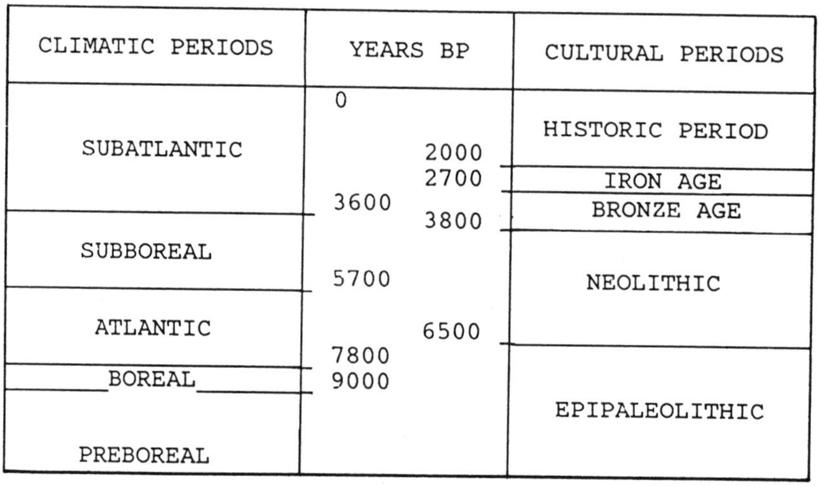

CLIMATIC PERIODS	YEARS BP	CULTURAL PERIODS
	0	
SUBATLANTIC	2000	HISTORIC PERIOD
	2700	
	3600　　3800	IRON AGE
		BRONZE AGE
SUBBOREAL	5700	NEOLITHIC
ATLANTIC	6500	
	7800	
BOREAL	9000	
		EPIPALEOLITHIC
PREBOREAL		

Fig. 17.20　Climatic and cultural periods from key sites in the Armorican massif

Chronology of marine transgressions and regressions

This chronology is based in part on radiocarbon dates for the earliest peat horizons at several sites. Only where the sedimentary hiatus between peat and underlying mud is slight do these dates give an accurate indication of the age of the mud.

The history of Holocene sedimentation in the estuarine region begins in some places at the end of the Boreal (around 8000 BP), but at most sites during the Atlantic period (7800–5700 BP). The fluvio-marine sedimentation of the Atlantic period was very active, laying down about 20 m of deposit; it corresponds to a lengthy and uninterrupted transgressive phase.

The first signs of a slackening of that transgression appear around 6300 BP in the Acheneau valley with the formation of peat at Port-Saint-Père III and La Basse-Ville. This date marks the beginning of a long episode first of a stillstand then of a fall in sea level. The episode is marked in most of the lowland sites of the region by the cessation of marine sedimentation and the beginning of peat formation. The transition

from mud to peat occurred at different times, largely depending on the distance of the site from the shoreline. By 6000 BP the fluvio-marine waters reached no farther than the site of Gesvres, and around 5700 BP no farther than the northern shore of the estuary. In Brière the deposition of muds continued, but the expansion of mudflats in the depression at the beginning of the Subboreal period is a clear sign of a stable sea level.

Around 4500 BP the sea no longer entered Brière, and a true regression phase began. Neolithic populations erected megaliths on the drying marine mud, and groves of oak trees developed around 4500–4200 BP. The establishment at the same period (4500–4000 BP) of communities of relatively non-hygrophytic plants on the estuarine marshes demonstrates the influence of sea level on the fenland water levels. Along watercourses little or no sedimentation occurred, and in some places sediments deposited earlier were eroded.

– At Carnet, a marked decline in the pollen of Chenopodiaceae during the Subboreal period shows that marine influences in the Loire sediments became insignificant. At the same time

Fig. 17.21　Lithologic description of the cores. Height NGF (top of the core) is roughly equivalent to present mean sea level at Marseille (France)

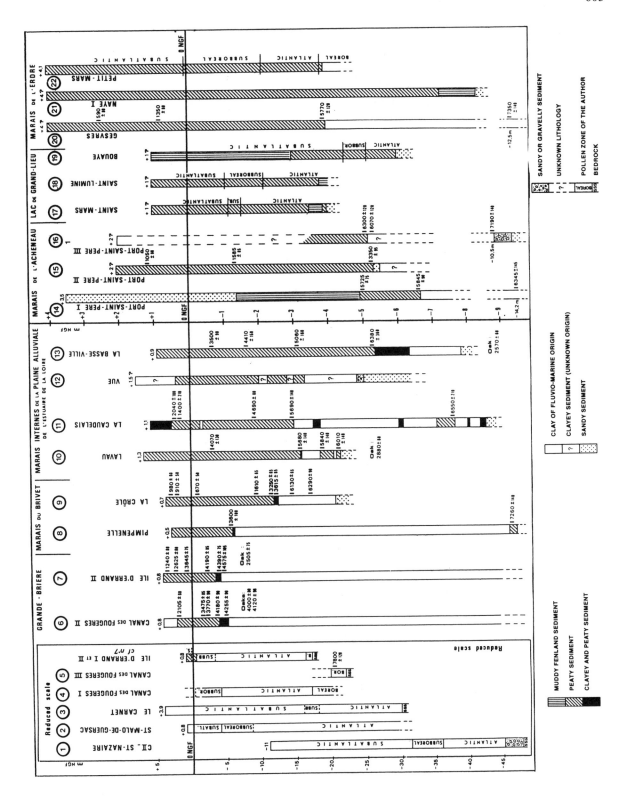

604

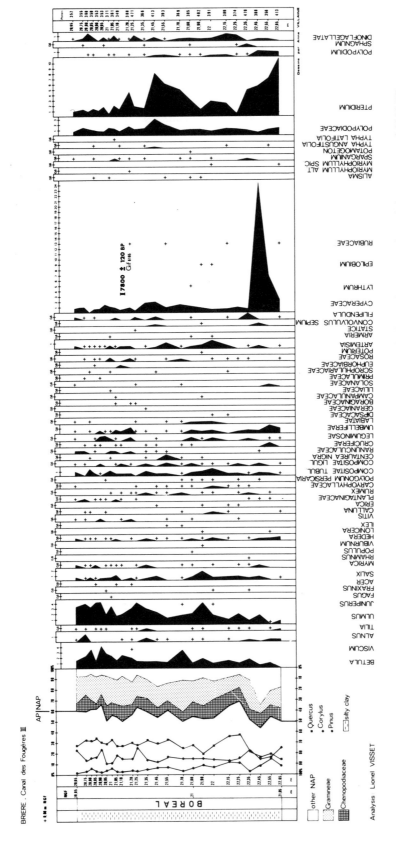

Fig. 17.22 Pollen diagram: Canal des Fougères III. Halophilous plants clearly indicate the penetration of the sea around 8000 BP

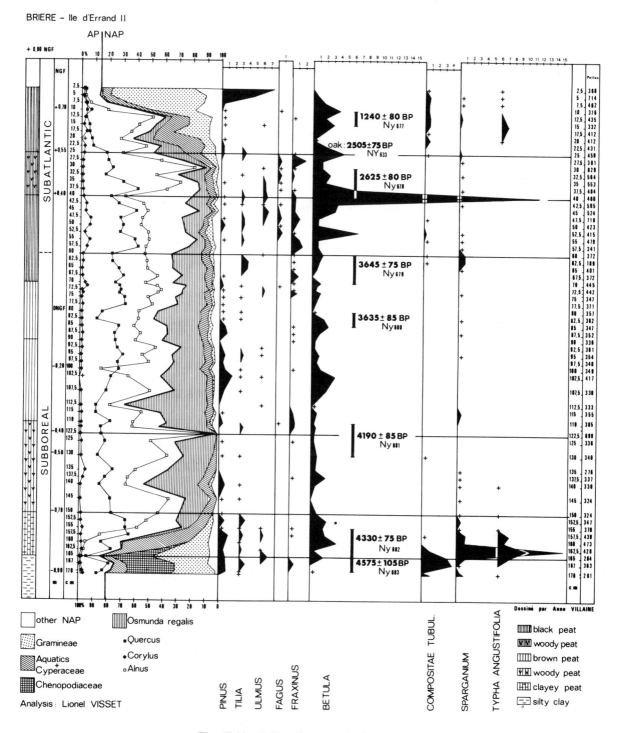

Fig. 17.23 Pollen diagram: Ile d'Errand II

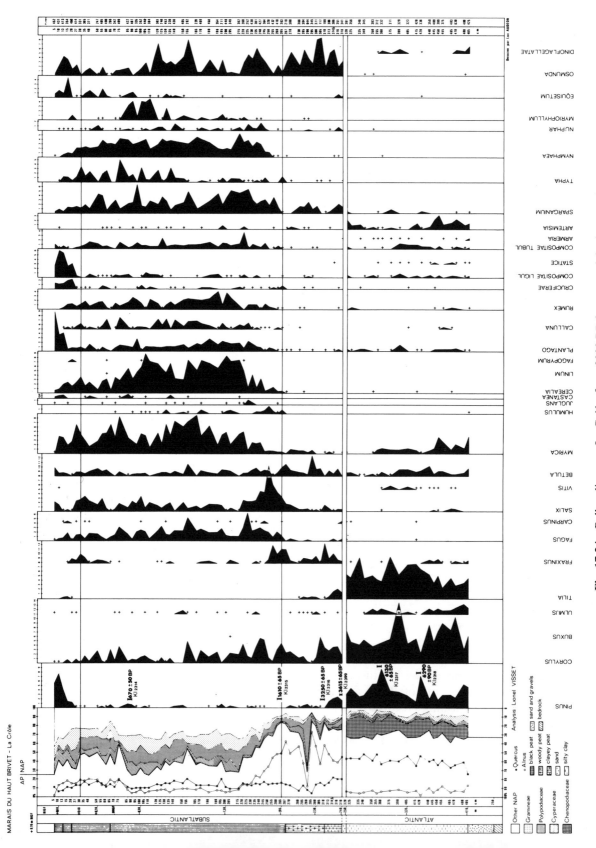

Fig. 17.24 Pollen diagram: La Crôle: from 2000 BP, human activity increased

the quantity of sediment deposited was very small. The dramatic fall in the percentages of *Tilia* pollen and the simultaneous rise of *Fagus* seem to correspond to a break in sedimentation, or even a phase of erosion of earlier sediments. These breaks are generally associated with old water channels where peat formation began later than in the rest of the marshland, and sometimes at a lower level.

- At Port-Saint-Père III, peat formation began around 6300 BP at about −5 m NGF, while less than 100 m away at Port-Saint-Père II it was only after 3400 BP that peat began to form, but at the same level and probably simply as a result of a natural shift in the stream-bed in this valley.
- Sediment deposition at the sites of Pimpenelle and La Crôle (Fig. 17.24) seems to have followed a similar pattern.

Peat formation began at the same date at three of these valley sites: 3350±95 BP—NY 790 (Port-Saint-Père II), 3616±65 BP—KI 2399 (La Crôle), and 3600±100 BP—Gif 5752 (Pimpenelle). This resumption of sediment formation can only be explained by a decrease in stream velocity caused by a rise in sea level. In Grande-Brière, the date of 3600 BP corresponds to a rise in the fresh-water level throughout the whole basin; hydrophytic and helophytic species expand, and brown peat changes to black peat. This positive fluctuation in the water table is less clearly marked in the marshes of the estuary itself, although from 4000 BP more hygrophilous species occur.

A new transgressive phase seems therefore to have commenced around 4000–3500 BP. Around 3000 BP, fluvio-marine water reached La Caudelais, where it left several layers rich in dinoflagellate cysts at about −0.40 m NGF. These deposits are the sole evidence at present of a transgression at the beginning of the Subatlantic period.

The transgression seems to have slowed between 3000 and 2600 BP; oak trees established themselves on the peat in the marshes both of Brière and the estuary, their presence indicating a relatively marked dry phase. The flooding of the southern part of Brière by fluvio-marine water around 2000 BP is the last evidence of a marine transgression in the estuary area.

The evolution of the marshland vegetation
(Fig. 17.25)

Boreal

At the close of this period the lowlands do not yet seem to have been marshy, and the deep valley bottoms were flooded by the sea. It is true that in certain restricted locations there was a fenland flora, followed by the establishment of a halophilous vegetation. But the local vegetation was still completely the same as the regional vegetation. Within the basins, as around their edges, *Corylus* was dominant with *Quercus*, and *Ulmus* was already well established, while *Tilia* made its first appearance. *Pinus* was also present. Along the rivers and watercourses were species of secondary importance, e.g. *Acer*, *Carpinus*, and *Salix*.

Atlantic

During the Atlantic period most of the marshlands of the estuary region supported a halophilous vegetation. Salt marshes or a zone of halophytic vegetation covered by the sea only at spring tides were restricted essentially to the banks of the Loire, the margins of the flooded areas of Grande-Brière, and upstream along the watercourses in the centre of the estuarine marshes. Their extent varied in response to the rhythm of changing sea level. It is probable, however, that during the Atlantic period the rapid rise in sea level never allowed any expansion in the surface area of the salt marshes.

The vegetation that became established behind these communities of Chenopodiaceae consisted in part of halophilous species (such as *Scirpus maritimus*), their abundance depending on the degree of salinity of the floodwaters. These halophilous species may have been mixed with fenland plants (e.g. *Phragmites* and *Typha*), which grew in areas fed by streams bringing freshwater from the hills. Of the marshlands that have been studied, only the lake of Grand-Lieu supported a fenland vegetation during the Atlantic period; this fenland vegetation was dominated by alder, followed by various hydrophytic and helophytic species.

608

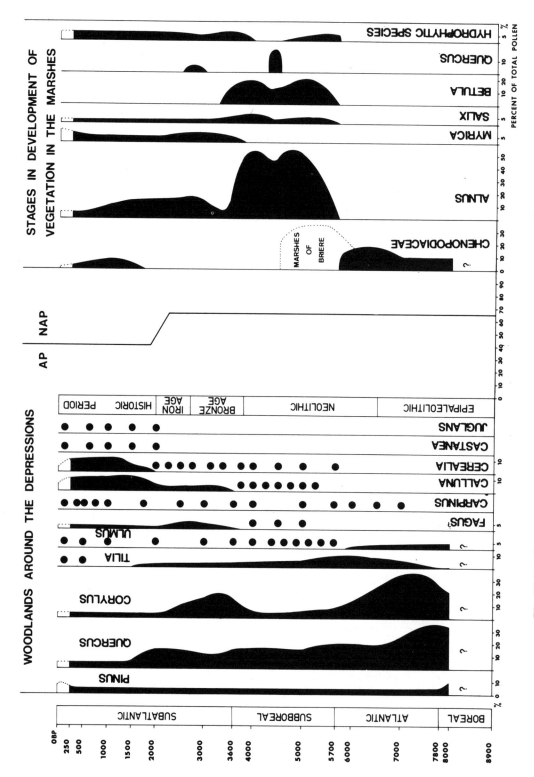

Fig. 17.25 Summary pollen diagram representing the marshlands of the Loire estuary

The slackening of the transgression at the end of the Atlantic period, from approximately 6300 BP, had a considerable effect on the marshland vegetation. The halophilous species began to decline in those areas, such as the Acheneau valley, which fluviomarine water could now only reach with difficulty. On the other hand, halophilous vegetation flourished in locations where brackish water was still to be found: in Grande-Brière, on the Loire banks, and in the marshlands on the northern side of the estuary.

Subboreal

The development that had begun at the end of the Atlantic continued in this period: fenland vegetation, which since the beginning of the Atlantic had been found only around the base of the hills or in pockets of rather limited extent, expanded to cover greater and greater areas. The fenland advanced roughly in parallel with the edges of the flooded areas.

At the beginning of the Subboreal the salt marshes reached their maximum extent in Brière, as is shown by the high percentages of Chenopodiaceae pollen at certain sites (Canal des Fougères II). In contrast, along the Loire itself regression probably occurred at this time.

The vegetation of the peat bogs was varied. The development of damp coppices in the estuarine marshes on the northern side of the river alternated with the hydrophytic vegetation of phases of flooding. The site of La Basse-Ville on the southern side had a vegetation significantly more hygrophilous in character, with *Nymphaea* especially common. The same is found in the Acheneau valley, which constitutes a continuation of the marshes of the southern side towards the lake of Grand-Lieu.

Around 4500 BP the sea ceased to penetrate into the Brière basin, and the halophilous vegetation disappeared from that region. The marine regression, which was at its maximum during the second half of the Subboreal, seems to have caused a fall in the water table in the peat bogs. All the pollen diagrams from the estuarine region show the development of a more hygrophilous vegetation, which corresponds to a relative drying of these peat bogs. This event is well dated in the estuarine marshlands: an association of heather and silver birch reaches its maximum development around 4000 BP at Lavau; heathland with heather and *Molinia* became established at La Caudelais at about 4500 BP, whereas alder woodland dominated the landscape at La Basse-Ville around 4400 BP.

In Grande-Brière, Visset (1979–1988) distinguishes three phases in the development of the vegetation: the establishment of clumps of oaks between 4500 and 4200 BP, a phase of heliophytes and the disappearance of the trees around 4200 BP, and a phase of damp coppices that begins at different dates in different places, either before peat formation or at the point when the black peat forms, the latter event dated to 4180±90 BP—Ny 494: Canal des Fougères. This third phase ends around 3600 BP.

To summarize, the vegetation of the marshes of the estuarine region was dominated during the second half of the Subboreal by damp coppices, or occasionally by clumps of *Quercus*, *Alnus*, or *Betula*. Only limited areas appear to have been suitable for the growth of hydrophytes and heliophytes.

Subatlantic

At the beginning of the Subatlantic, around 3600–3000 BP, *Alnus* and *Betula* disappeared from almost all the sites where they had been present during the Subboreal: Grande-Brière, the estuarine marshes, the lake of Grand-Lieu, and the marshland of Petit-Mars.

In Brière, the disappearance of the damp coppices coincided with the spread of heliophytes. In other marshlands *Salix* associated with *Frangula alnus* dominated the vegetation (La Caudelais, Saint-Lumine, Petit-Mars). It was at the beginning of the Subatlantic that *Myrica gale* made its appearance at most of the peat bogs of the estuary region.

These early plant communities were followed from 3000 BP until roughly 2000 BP by a vegetation of varied aspect. At some sites *Alnus* and *Betula* re-established themselves, while at others, often not far away, heliophytes and copses of *Salix* continued to flourish. *Myrica gale* was quite common in some places (Lavau). At this period the site of Petit-Mars supported a markedly hygrophilous flora (*Sparganium*, *Potamogeton*, and Cyperaceae).

The thin lenses of mud that overlie the peat at La Caudelais around 3000 BP are rich in dinoflagellate cysts but very poor in Chenopodiaceae pollen. This indicates that the salt marshes had disappeared from the banks of the Loire at this period, at least upstream of Paimboeuf.

Around 3000–2500 BP localized clumps of *Quercus* developed on the peat both in Brière (Ile d'Errand) (Fig. 17.23) and in the estuarine marshes (Lavau, La Basse-Ville). It is very difficult to determine the density and extent of these plant populations.

From about 2000 BP the hygrophilous character of the vegetation became more marked than at the beginning of the Subatlantic. Hydrophytes and heliophytes were present at most sites once again. They could even be the dominant element in the vegetation, as in the north of the Brière marshland (reed-bed) or at La Caudelais (Myriophylletum). The expansion of *Myrica gale* accelerated. This change in the vegetation was related to a rise in the water table, indirectly caused by a rise in sea level. The southern part of Brière was in fact flooded at this period by fluvio-marine water. The amounts of Chenopodiaceae pollen in the sediments laid down on top of the peat by this water, however, are fairly low (about 5–10%) and lead us to suppose that salt marshes covered only relatively limited areas along the banks of the Loire, probably downstream of Paimboeuf.

The marshlands of the estuarine region retained this kind of vegetation to the fairly recent past. The transition from this landscape to that of the present day does not figure in the diagrams, however, as the uppermost layers of sediment are often disturbed by trampling of livestock or by agricultural operations. These upper layers are oxidized and are different in nature from the underlying deposits, a fact that introduces distortions into the pollen composition.

Development of the surrounding vegetation
(Fig. 17.26)

Quercus, the dominant element in the vegetation of the region, was joined around 8000–7000 BP by *Corylus*, *Tilia* and *Ulmus*. *Carpinus* was also sometimes present in the forested landscape, if rather

discretely, from 7000 BP. *Pinus* was barely represented.

From 6000–5500 BP the forest entered a period of decline. *Fagus* appeared about 5000–4500 BP at the earliest and became progressively more important until about 2000 BP, in spite of deforestation by human communities. The decline of *Fagus* is linked to the expansion of agriculture in the Roman period.

The first evidence of human activity appears in the diagrams around 4500–4000 BP. The spread of *Corylus* and *Calluna* about 3500 BP marked the development of regressive heathland. The first significant increase in cultivated taxa and cereals coincides with the appearance of *Juglans* and thus dates to the very beginning of the present era.

Conclusions

This account shows that the history of these marshlands can be divided into two phases: a sedimentation phase during which fluvio-marine mud was deposited, followed by a phase of peat formation beginning at the earliest around 6300 BP and at the latest around 3600 BP. In general, the more a site is exposed to marine influences, the later the onset of peat formation. The only exceptions to this generalization are valley sites where peat formation depended on shifts in stream channel and velocity and only began around 3600 BP, as an indirect result of the Subatlantic marine transgression.

The establishment of relatively non-hygrophilous plant communities is dated by radiocarbon to around 4500–4000 BP in the estuarine marshlands, and in the marshes of the lake of Grand-Lieu by the more approximate method of pollen zonation to the Subboreal (5700–3600 BP) (Planchais 1971). This development suggests that the water table in these areas was strongly influenced by changes in sea level.

Each profile shows a succession of plant communities unique to itself. For this reason, the detailed reconstruction of the overall vegetation pattern of the marshy areas of the estuary at a given moment inevitably comes up against a mosaic of plant communities, even in areas where the profiles are quite closely spaced. It is to be noted, however, that certain taxa may occupy the same position in a number

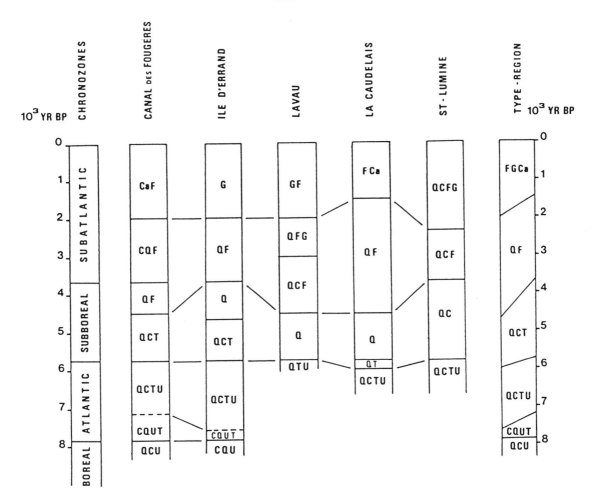

Fig. 17.26 Correlations between local pollen zones in the Loire Valley. C = *Corylus*, Ca = *Calluna*, F = *Fagus*, G = Gramineae, Q = *Quercus* (deciduous), T = *Tilia*, U = *Ulmus*

of sequences, and these sequences characterize certain phases of peatland development. An example is the sequence Gramineae–*Typha*–Polypodiaceae–*Alnus*, which occurs in several diagrams in the estuary area and characterizes the final fluvio-marine sediments and the first peat horizons.

It should be noted that we have no data on the history of these marshlands before 8000 BP. The reason for this is that the currents in the valleys and watercourses that crossed the marshes were too powerful before 8000 BP to allow sediment deposition. At the same time they completely washed away

the deposits that had been laid down previously, in particular during the interglacials. It was only at the end of the Boreal that the sea reached a sufficiently high level for the deposition of fluvio-marine muds to be possible once again.

In the present state of knowledge, within the Armorican Massif only the lake of Grand-Lieu can show a nearly continuous sequence spanning the Pleistocene and Holocene. Indeed, recent corings have demonstrated the existence of deep pockets in which sediments had begun to be deposited at the end of the Pliocene. Future studies will allow us to determine more precisely the age of these sediments of Grand-Lieu, and the conditions under which they were deposited.

ACKNOWLEDGEMENTS

We are grateful to Dr Chris Scarre (University of Cambridge) for translating this paper into English.

TYPE REGIONS F-zg AND F-r, THE NORTHERN SLOPE OF THE PYRENEES (G. Jalut, S. Aubert, D. Galop, M. Fontugne and J.M. Belet)

Running from the Atlantic Ocean to the Mediterranean Sea, the Pyrenees form the westernmost mountain chain of general E–W orientation in the south of western Europe. Such an intermediate geographical position between the Atlantic and Mediterranean climatic zones gives rise to large phytogeographical differences between southern and northern slopes, as well as a climatic gradient along the French slope. For this reason, two main regions can be distinguished: an Atlantic zone (F-r), including the western part and the Ariège region, which is climatically and botanically an intermediate zone; and the eastern part, with a Mediterranean climate. The latter belongs to a heterogenous type region F-zg, and only the Pyrenean zone is considered here.

Because of its geologic and geomorphologic structure, the Pyrenean chain shows in each of these zones a great diversity of local climatic characteristics, especially in the eastern part, where the axis is divided into three sections with different glacial histories during the last ice age. In the Atlantic area glaciers 50–70 km long spread over the piedmont, whereas in the Mediterranean Pyrenees the glacial tongues did not exceed 20 km (Hérail *et al.* 1987), being confined to the mountains. At middle and low elevations, the complex topography and the proximity of the Mediterranean provided numerous plant refuges (Jalut 1974, Jalut *et al.* 1975, Vernet 1980).

Along the northern slope of the Pyrenees five sites have been selected. Three are situated in the Atlantic area (F-r) and two in the eastern Mediterranean part of the chain (F-zg) (Fig. 17.27).

TYPE REGION F-zg, EASTERN PYRENEES, MEDITERRANEAN PYRENEES

Altitude: 0–2921 m.

Climate: Great diversity: Mediterranean type at low elevation. To the north, at middle elevation (1300–1500 m), attenuated oceanic type with high precipitation in spring, autumn, and winter. In the high valleys of the Aude and Têt (1600–1700 m), semi-continental climate (Fig. 17.27).

Geology: In the axial zone granite, crystalline schist, and gneiss, and also Devonian limestone and Silurian schist. To the north, Urgo-Aptian limestone of the Mesozoic sedimentary mantle.

Topography: Very complex. The area is divided by four main valleys: Aude, Agly, Têt, and Tech. To the north the Aude River and its tributaries cut the Urgo-Aptian limestone and are deeply enclosed. The other valleys open wide to the Mediterranean and are generally oriented E–W or NE–SW. In this region at the limit between the Atlantic and Mediterranean zones the short distance between the high summits and the Mediterranean Sea, as well as the complex topography, are the most important characteristics determining the regional plant distribution.

Population: The population is concentrated in the villages and small towns of the cultivated zones at low altitude, whereas the mountain villages have a low population.

Vegetation: Because of the complex topography, climatic diversity, and human influence, various types can be observed from Mediterranean to alpine zones. At low and middle elevations the deciduous tree vegetation consists mainly of *Quercus* and *Fagus sylvatica*, whereas *Betula* and *Corylus* are more restricted. *Abies alba* and *Pinus* are widely developed in the mountain and subalpine zones. In the Mediterranean area evergreen oaks prevail: *Quercus ilex*, *Q. coccifera*, *Q. suber*.

Soils: From alpine rankers to rendzinas and Mediterranean red soils.

Land use: In the Mediterranean area, the plains

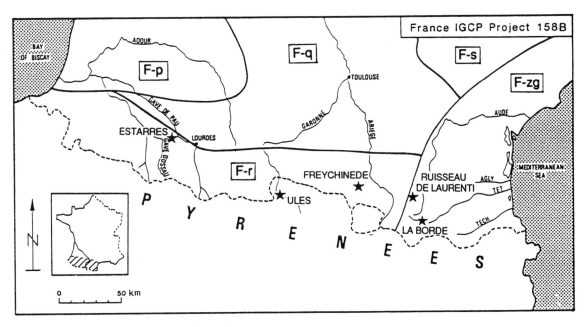

Fig. 17.27 Type regions (F-r and F-zg) and mentioned reference sites in the northern Pyrenees

zones are devoted to vineyards, orchards, and market gardens. In the mountains, forest or pasture.

Reference site 24. Peat bog of La Borde
(Jalut 1974, Reille 1990a and b; Jalut *et al.* 1992; Reille & Lowe 1993)

The site is situated in the high valley of Têt open wide to the Mediterranean. Latitude 42°32′N, Longitude 2°05′E. Elevation: 1660 m. Age range: 14000–5000 BP. 14 local pollen-assemblage zones (paz). (Fig. 17.28)

	Age	Local paz
LB 1	14000–13500 BP	*Artemisia*–Gramineae–*Pinus*–*Juniperus*
LB 2	13500–13000 BP	*Juniperus*–Gramineae–*Betula*–*Artemisia*
LB 3	13000–12000 BP	Gramineae–*Pinus*–*Juniperus*–*Betula*–*Artemisia*
LB 4	12000–11500 BP	*Pinus*–Gramineae–*Betula*–*Juniperus*–*Artemisia*
LB 5	11500–11000 BP	*Pinus*–Gramineae–

	Age	Local paz
LB 6	11000–10650 BP	*Betula*–*Artemisia* *Pinus*–Gramineae–*Artemisia*
LB 7	10650–10000 BP	*Pinus*–Gramineae–*Artemisia*–*Juniperus*
LB 8	10000–9650 BP	*Pinus*–Gramineae–*Juniperus*–*Artemisia*
LB 9	9650–9400 BP	*Pinus*–*Betula*–Gramineae
LB 10	9400–9000 BP	*Pinus*–*Betula*–Gramineae–*Quercus*–*Corylus*–*Ulmus*
LB 11	9000–8650 BP	*Pinus*–*Betula*–Gramineae–*Quercus*–*Corylus*–*Ulmus*–*Populus*
LB 12	8650–7800 BP	*Pinus*–*Betula*–*Quercus*–*Corylus*–*Abies*
LB 13	7800–6000 BP	*Pinus*–*Betula*–*Quercus*–*Abies*–*Corylus*
LB 14	6000–5000 BP	*Pinus*–*Abies*–*Betula*–*Quercus*–*Tilia*–*Fagus*

614

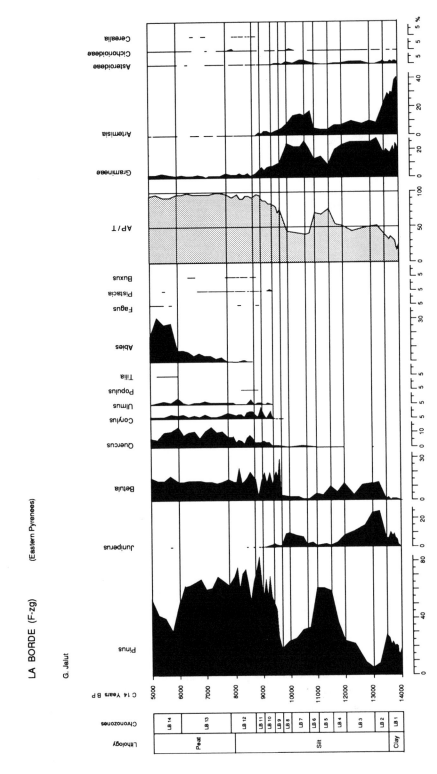

Fig. 17.28 Late-glacial and Holocene pollen diagram from the peat bog of La Borde. Only selected pollen types are shown

General remarks on the pollen diagram

- Strong increase of *Juniperus* near 14000 BP (level 699.5–700: 13630 ± 230 Gif TAN, ^{14}C AMS date on terrestrial macroremains (*Pohlia annotina* (Hedw.) Lindb.)
- Synchronous increase in percentages of *Betula* and *Juniperus* near 13500 BP and importance of *Juniperus* between 13500 and 12000 BP.
- From 12000 BP regular presence of *Quercus*.
- Low values of *Quercus* up to about 9400 BP, then increase in *Quercus* and *Corylus* values.
- *Alnus* and *Fraxinus* only represented by isolated grains.
- First presence of *Populus* dated 8700 ± 140.
- From 8700 to 8600 BP regular presence of *Pistacia*, Cerealia, *Quercus* type *ilex*, then *Buxus*, transported from the Mediterranean area.
- From 8650 BP presence of *Abies*. Then continuous curve from 8270 BP.
- *Tilia* poorly represented.
- *Fagus* present between 6010 and 5240 BP.

Reference site 23. Peat bog of Ruisseau de Laurenti

(Jalut 1974, Jalut & Vernet 1989, Reille 1990b, Jalut *et al*. 1992, Reille & Lowe 1993)

Site situated in the high valley of Aude, in northern exposure, at the limit between the Atlantic and Mediterranean climatic zones. Latitude 42°40'30" N, Longitude 2°1'46"E. Elevation: 1860 m. Age range: 14000–0 BP. 14 local pollen assemblage zones (paz). (Fig. 17.29)

	Age	Local paz
RL 1	14000–13000 BP	*Artemisia–Pinus–Gramineae*
RL 2	13000–12000 BP	*Pinus–Artemisia–Gramineae–Betula*
RL 3	12000–11000 BP	*Pinus–Betula*
RL 4	11000–10000 BP	*Pinus–Artemisia–Gramineae–Betula–Quercus*
RL 5	10000–9650 BP	*Pinus–Betula–Artemisia–Gramineae*
RL 6	9650–9400 BP	*Pinus–Betula–Quercus*

	Age	Local paz
RL 7	9400–9200 BP	*Pinus–Betula–Quercus–Corylus*
RL 8	9200–8500 BP	*Pinus–Corylus–Betula–Quercus–Ulmus*
RL 9	8500–6200 BP	*Pinus–Corylus–Quercus–Betula–Ulmus–Abies*
RL 10	6200–5000 BP	*Pinus–Betula–Corylus–Abies–Quercus–Ulmus–Alnus*
RL 11	5000–4200 BP	*Pinus–Abies–Betula– Corylus–Quercus–Fagus*
RL 12	4200–3000 BP	*Pinus–Abies–Gramineae–Alnus–Fagus–Cerealia*
RL 13	3000–2000 BP	*Pinus–Gramineae–Abies–Fagus*
RL 14	2000–0 BP	*Pinus–Gramineae–Abies– Fagus–Cerealia*

General remarks on the pollen diagram

- *Pinus* is always well represented.
- During the Late-glacial *Betula* had low values and increased only at the beginning of the Post-glacial.
- Local lack of dates for the Late-glacial phases. Limits are proposed by comparison with other sites.
- *Quercus* then *Corylus* spread early, at the beginning of the Post-glacial, but their maxima occurred near 8500–8000 BP.
- The continuous curve of *Abies* begins about 8300–8200 BP. This agrees with the results obtained in the high valley of Têt (La Borde: about 8580±150, Reille 1990a; 8270 ± 130, Jalut *et al*. 1992). But it is late in comparison with the valley of Nohède (basin of River Têt) in the same type region, where *Abies* is regularly present from about 9500 BP (Jalut 1974) and 9150 BP (Reille 1990b).
- The continuous curve of *Fagus* begins near 5000 BP, but the maximum extent of this species

616

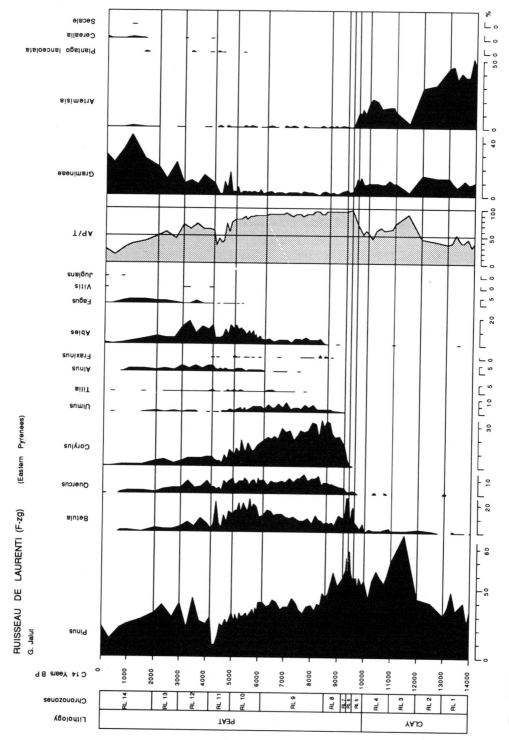

Fig. 17.29 Late-glacial and Holocene pollen diagram from the peat bog of Ruisseau de Laurenti. Only selected pollen types are shown

occurred near 4000 BP and was contemporaneous with the development of human influence.

- Locally the first anthropogenic indicators are found approximately 5000 BP. About 5 km to the northwest of the site, other indicators are observed between 6100 and 4800 BP in peat levels (peat bog of Ruisseau du Fournas: Jalut 1974, Jalut & Vernet 1989) while at the same time sheep and goats were being bred 10 km to the north (Geddes 1980). Northward (peat bog of Pinet) contemporaneous clearances are also recorded (Jalut & Vernet 1989).

- Not dated here, the presence of *Juglans* is recorded at 1100±70 in the same area (Pailhère, Galop unpublished), while in Cerdagne, not far from La Borde, its first presence is dated 760±50 (Pla de l'Orri, Galop unpublished).

Comparison of the two pollen diagrams of type region F-zg

This shows a good representation of the Late-glacial period during which *Quercus* is regularly observed, particularly during the Younger Dryas.

- The *Juniperus* phase is not recorded in the sequences of Ruisseau de Laurenti analysed by Jalut (1974) and Reille (1990b), but it is found very near, in the same high basin, at La Restanque (Reille & Andrieu 1993). Its beginning is contemporaneous with a strong decline of *Artemisia*.

- The continuous curves of *Abies* begin at the same time (La Borde, slightly before 8300±190, Jalut 1971, 1974, 8270±130, Jalut *et al*. 1992; Ruisseau de Laurenti, 8230±180, Jalut 1974; 8250±190, Reille 1990b). Similarly, the first pollen of *Tilia* is found between 7300 and 7000 BP (Jalut 1974, Jalut *et al*. 1992, Reille 1990b).

- *Fagus* is found between 6000 and 5000 BP (Jalut, 1974, Jalut & Vernet 1989, Jalut *et al*. 1992).

TYPE REGION F-r, ARIEGE

Altitude: 250–3115 m.

Climate: Northwest Atlantic disturbances with maximum rainfall in winter and spring. The complex topography induces highly contrasted local climatic conditions.

Geology: In the mountain zone, two main formations: the crystalline block of the axial chain with granite, gneiss, and micaschist, and the Urgo-Aptian limestone of the Mesozoic sedimentary mantle.

Topography: Very complex geological structure. Many deep valleys of various orientations.

Population: Very low in the mountains, where many small villages or hamlets are partially abandoned. Remaining population concentrated in small towns.

Vegetation: Great diversity, depending on geology, complex topography, altitude, and climatic situation. Up to 1400–1500 m, *Quercus* and *Ulmus* are well developed, with local *Tilia* stands and abundant *Fraxinus*. *Fagus* is extensive due to human influence. Stands of *Abies* are only present in certain valleys. On many slopes *Fagus* has been replaced by heathland or pasture. In several parts of the subalpine zone, the *Pinus* forest, especially with *Pinus uncinata*, has been completely destroyed. The calcareous rocks are commonly covered with a vegetation of Mediterranean and sub-Mediterranean type in which is observed the large stand of *Juniperus thurifera* of Quié de Lujat (Guerby 1993).

Soils: Alpine soils, forest brown soils, some rendzinas.

Land use: Large uncultivated areas, forest and pasture.

Reference site 21. Peat bog of Freychinède
(Jalut *et al*. 1982, Reille 1993a and b)

Latitude 42°48′N, Longitude 1°26′11″E. Elevation 1350 m. Age range 15000–0 BP. 14 local pollen assemblage zones (paz). (Fig 17.30)

	Age	Local paz
Fr 1	14500–14000 BP	*Artemisia*–Gramineae–*Pinus*
Fr 2	14000–13200 BP	Gramineae–*Artemisia*–*Pinus*–*Juniperus*
Fr 3	13200–12000 BP	Gramineae–*Betula*–*Artemisia*–*Pinus*–*Juniperus*
Fr 4	12000–11600 BP	Gramineae–*Pinus*–*Artemisia*–*Betula*

618

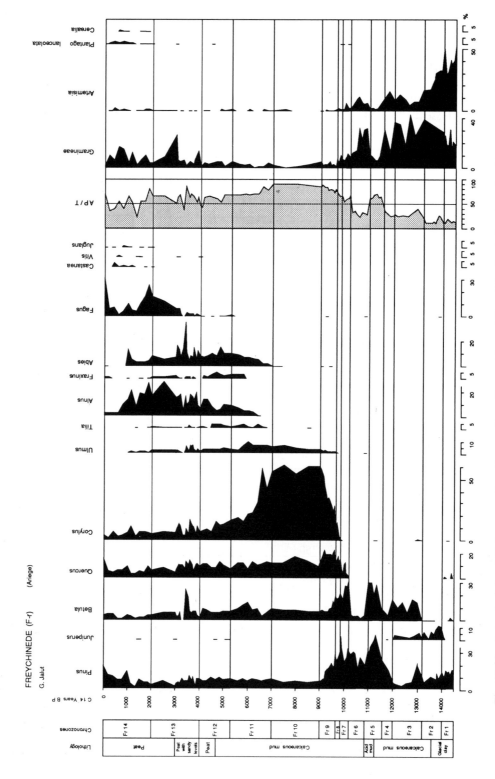

Fig. 17.30 Late-glacial and Holocene pollen diagram from the peat bog of Freychinede. Only selected pollen types are shown

	Age	Local paz
Fr 5	11600–11000 BP	*Pinus–Betula–Gramineae–Artemisia*
Fr 6	11000–10200 BP	*Pinus–Gramineae–Artemisia–Betula*
Fr 7	10200–9850 BP	*Pinus–Betula–Quercus–Gramineae–Artemisia*
Fr 8	9850–9700 BP	*Pinus–Betula–Quercus–Corylus–Gramineae*
Fr 9	9700–9000 BP	*Corylus–Quercus–Pinus–Betula–Ulmus*
Fr 10	9000–7000 BP	*Corylus–Quercus–Betula–Pinus–Ulmus*
Fr 11	7000–5300 BP	*Corylus–Quercus–Abies–Alnus–Tilia*
Fr 12	5300–4000 BP	*Abies–Alnus–Fraxinus–Fagus*
Fr 13	4000–2000 BP	*Alnus–Fagus–Abies*
Fr 14	2000–0 BP	*Alnus–Fagus–Abies–Castanea–Juglans–Cerealia*

General remarks on the pollen diagram

– From 14500 to 12000 BP the *Pinus* values are low.
– From 14500 to 14000 BP *Artemisia* pollen is more abundant than Gramineae, then Gramineae dominates.
– Near 14000 BP a *Juniperus* phase begins while the *Artemisia* values decrease. *Juniperus* is well represented up to about 12000 BP.
– From 13200 to 12000 BP *Betula* is more abundant than *Pinus*, but the landscape stays poorly forested. Between 12000 and 11000 BP the presence of stomata of *Pinus* demonstrates the local presence of this tree at the altitude of the lake (Reille & Andrieu 1993).
– The continuous curve of *Quercus* begins before 10000 BP. *Quercus* followed by *Corylus* develop and reach their maximum between 9500 and 9000 BP and 9000 and 7000 BP, respectively.
– The date of the *Abies* expansion differs according to the sedimentary sequences studied (7100–7000 BP, Jalut *et al.* 1982; near 8500 BP, Reille 1993a). In the site of l'Estagnon near Freychinède the

beginning of the continuous curve of *Abies* is dated 6690±70 (Reille 1993 a).
– *Fagus* is present from 5300 to 5200 BP, but its strong development occurred around 4000 BP. At the same time the first anthropogenic indicators are noticed.
– From 2000 BP onwards the local human impact is very strong. One of its most important consequences is the momentary local extinction of *Abies* and its substitution for *Fagus*. The beginning of the *Juglans* curve is dated at 2000 BP, and at Argentières near Freychinède at 1850±50 (Galop unpublished).

TYPE REGION F-r, CENTRAL PYRENEES

Altitude: 400–3404 m.

Climate: Northwest Atlantic impact with maximum precipitation in spring and winter. In the high valley of Garonne (Val of Aran, basin of Oo), slight continental tendency due to the sheltered situation of the valleys. In these cases, precipitation is less important in winter but increases in summer with more contrasted temperature.

Geology:
Three zones:
– The North Pyrenean zone is strongly folded, with sheets of Palaeozoic basement, with a Mesozoic cover (Triassic up to Cretaceous).
– A complex fault zone (north Pyrenean vertical fault). Zone of Mesozoic marble and schist.
– The high primary range with Hercynian basement (sedimentary and metamorphic formations) injected by intrusive granite (Aneto, 3404 m). All the geological formations concerned correspond to the Pyrenean mountain complex, at the boundary between the European plate (overlapping) and Iberian plate (overlapped).

Vegetation: Great diversity depending on geology, topography, altitude, and climatic situation. At low elevation on limestone, a *Quercus pubescens* forest is well developed with some stands of *Quercus ilex* and a stand of *Juniperus thurifera* (Marignac). Above, *Fagus* is extensive but *Abies* expands progressively.

620

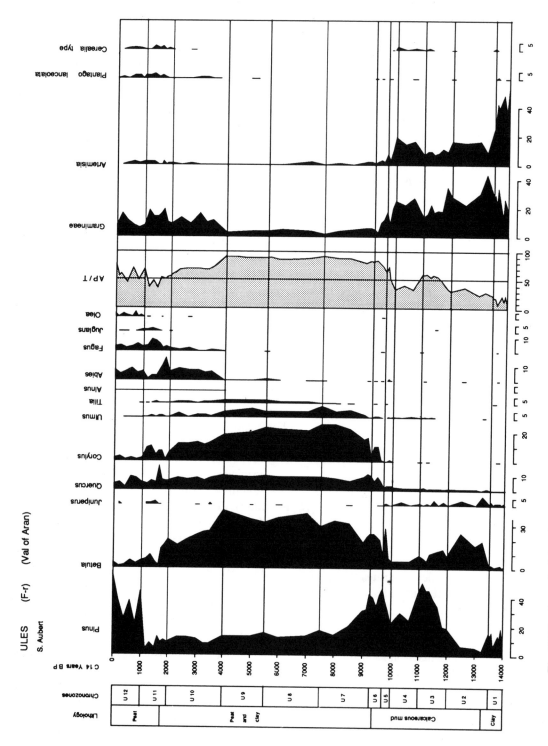

Fig. 17.31 Late-glacial and Holocene pollen diagram from the peat bog of Ules. Only selected pollen types are shown

On many slopes, meadows and coppices with hazel and birch replace the beech–fir forest. In other situations, beech–fir forest is well developed and devoted to forestry. On the south-facing slopes of the sheltered valleys a *Pinus sylvestris* forest takes the place of the beech–fir forest. Subalpine *Pinus uncinata* forest is frequently destroyed.

Soils: Alpine soils, forest brown soils, some leached soils.

Land use: Large uncultivated areas, some forestry, forests, pasture.

Reference site 22. Peat bog of La Bassa d'Ules
(Val of Aran, Spain) (Aubert 1993)

Latitude 42°44′N, Longitude 4°26′E. Elevation 1650 m. Age range 14000–0 BP. 12 local pollen assemblage zones (paz). (Fig. 17.31)

	Age	Local paz
U 1	14000–13500 BP	*Artemisia*–Gramineae–*Pinus*
U 2	13500–12000 BP	Gramineae–*Artemisia*–*Betula*
U 3	12000–11000 BP	*Pinus*–Gramineae–*Betula*–*Artemisia*
U 4	11000–10000 BP	*Pinus*–Gramineae–*Artemisia*
U 5	10000–9700 BP	*Pinus*–*Betula*–Gramineae–*Quercus*–*Artemisia*
U 6	9700–9300 BP	*Pinus*–*Betula*–Gramineae–*Corylus*–*Quercus*
U 7	9300–7500 BP	*Betula*–*Corylus*–*Pinus*–*Quercus*–*Ulmus*
U 8	7500–5500 BP	*Betula*–*Corylus*–*Pinus*–*Quercus*–*Ulmus*–*Tilia*
U 9	5500–4000 BP	*Betula*–*Corylus*–*Pinus*–*Quercus*–*Ulmus*–*Tilia*–*Abies*
U 10	4000–2000 BP	*Betula*–*Corylus*–Gramineae–*Abies*–*Pinus*–*Fagus*
U 11	2000–1000 BP	Gramineae–*Corylus*–*Pinus*–*Quercus*–
		Age Local paz

	Age	Local paz
U 12	1000–0 BP	*Abies*–*Fagus*–Cerealia–*Juglans* *Pinus*–Gramineae–*Quercus*–*Abies*–*Fagus*–*Olea*

General remarks on the pollen diagram

– Around 14000 BP absence of *Juniperus* phase but clear decrease in the *Artemisia* values.
– Continuous presence of *Quercus* and occurrence of *Ulmus* during the Younger Dryas.
– Presence of Cerealia pollen type which could be due to the local presence of Gramineae such as *Glyceria fluitans*.
– Local low representation of *Alnus* and *Fraxinus*.
– Constant local abundance of *Betula* and *Corylus* from the beginning of the Post-glacial to the beech–fir forest expansion.
– With respect to the data obtained in the more eastern sites, late expansion of *Abies* near 3600 BP synchronous with that of *Fagus*. Synchronous increase in Gramineae values and anthropogenic indicators more abundant.
– Continuous curve of *Olea* near 1000 BP (Galop, unpublished) reflects the spread of olive cultivation in Spain (Riera-Mora & Esteban-Amat 1994).

TYPE REGION F-r, WESTERN PYRENEES

Altitude: 400–3355 m.

Climate: Oceanic. High precipitation with maximum in spring, autumn, and winter.

Geology: In the central part of the chain, igneous rocks (dacite, andesite, and granite) along with Palaeozoic limestone and schist. To the north at middle elevation, Urgo-Aptian limestone of the Mesozoic sedimentary mantle.

Topography: Complex. Great diversity of situations, from high summits with some restricted glaciers to hills of the piedmont.

Population: Concentrated in villages and small towns in the valleys and in some larger towns on the piedmont.

Vegetation: A complete succession, from the low-altitude zones to the alpine zone. On the hills of the piedmont up to 1400–1500 m: deciduous oak forests. At higher elevations, *Fagus* is dominant, along with *Abies* near its western limit. In the subalpine zone, from 1700 to 2500 m, some forests and stands of pine (*Pinus uncinata*), which is also close to its western limit. Great extent of pasture.

Soils: In the mountain part, alpine and pseudo-alpine rankers as well as acid and eutrophic brown soils. Rendzinas on the Mesozoic limestone. To the north, leached soils and podzols.

Land use: Some forestry in the mountain zone but essentially dominance of pastures and uncultivated areas.

Reference site 20. Peat bog of Estarrès

(Andrieu 1987, Jalut *et al.* 1988, Jalut *et al.* 1992, Andrieu *et al.* 1993)

Latitude 43°5′43″N, Longitude 0°24′46″W. Elevation 376 m. Age range 13000–0 BP. 13 local pollen assemblage zones (paz). (Fig. 17.32)

	Age	Local paz
E 1	13000–12800 BP	Gramineae–*Artemisia*–*Betula*–*Pinus*
E 2	12800–12000 BP	Gramineae–*Betula*–*Artemisia*–*Pinus*
E 3	12000–11000 BP	Gramineae–*Pinus*–*Betula*–*Artemisia*–*Quercus*
E 4	11000–10000 BP	*Pinus*–*Betula*–Gramineae–*Artemisia*–*Quercus*
E 5	10000–9500 BP	*Pinus*–Gramineae–*Betula*–*Quercus*–*Artemisia*
E 6	9500–9000 BP	*Pinus*–Gramineae–*Quercus*–*Corylus*
E 7	9000–8200 BP	*Quercus*–*Corylus*–*Pinus*–Gramineae
E 8	8200–7000 BP	*Quercus*–*Corylus*–*Pinus*–Gramineae–*Ulmus*
E 9	7000–6000 BP	*Corylus*–*Quercus* –Gramineae–*Pinus* –*Ulmus* –*Alnus* –*Tilia*

	Age	Local paz
E 10	6000–5000 BP	*Corylus*–*Quercus*–*Alnus*–Gramineae–*Fraxinus*
E 11	5000–4000 BP	*Corylus*–*Alnus*–*Quercus*–*Abies*–*Fagus*–*Castanea*
E 12	4000–2000 BP	*Alnus*–*Corylus*–*Quercus*–*Fagus*–*Abies*–*Castanea*–*Vitis*–Cerealia
E 13	2000–0 BP	*Alnus*–Gramineae–*Corylus*–*Quercus*–*Fagus*–*Abies Castanea*–*Vitis*–*Juglans*–Cerealia–*Secale*

General remarks on the pollen diagram

– Absence of *Juniperus* phase in the pollen diagram prior to 13000 BP but presence in a new study of the site (Andrieu *et al.* 1993).

– Until 11500–11000 BP the vegetation was open. Then forest expanded, as indicated regionally by the changes in the fauna (Bahn 1982).

– The continuous curve of *Quercus* begins during E 3 paz and can be correlated with Allerød.

– Younger Dryas correlated with E 4 paz is not strongly marked. Its beginning is dated 11130 ± 200 Gif 8784. The values of *Quercus* do not decrease during this phase.

– As at the other sites, *Quercus* expands some centuries before *Corylus*.

– The beginning of the continuous curve of *Ulmus* is dated 8170 ± 80 Gif 8177. The first pollen of *Tilia* is present between 9500 and 8200 BP. Its curve is continuous near 6000 BP.

– The expansion of *Alnus* and *Fraxinus* occurs after 6000 BP.

– The continuous curves of *Abies* and *Fagus* begin between 5000 and 4000 BP, but some pollen is present near 6000 BP. Significant values of *Fagus* are recorded in the nearby site of Castet (altitude 850 m) near 7000 BP, with some pollen of *Fagus* (Jalut *et al.* 1988).

– Locally the first anthropogenic indicators appear between 5000 and 4000 BP (*Plantago lanceolata*: 4860 BP; *Castanea* : 4300 BP; Cerealia and *Vitis*:

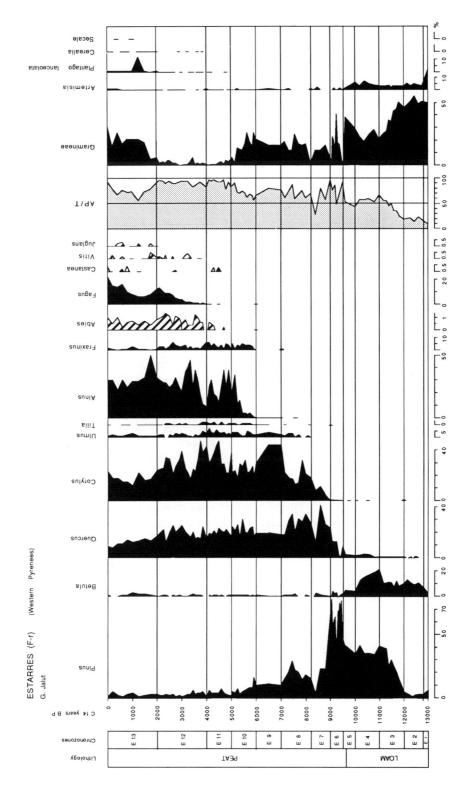

Fig. 17.32 Late-glacial and Holocene pollen diagram from the peat bog of Estarres. Only selected pollen types are shown

4000 BP). The beginning of the *Juglans* curve is dated 2060 BP.

PALAEOECOLOGICAL PATTERNS AND EVENTS WITHIN THE TYPE REGIONS F-zg AND F-r

History of the trees

Fig. 17.33 shows the relations among the five sites. However, it cannot reflect all the features of each type region in which several sites were studied by various authors and from which a more general chronological approach can be made. Taking into account all the available chronological data, Fig. 17.34 reviews the main published results.

Common patterns

– In each type region, a transition phase dominated by *Juniperus* is recorded Its beginning is well dated at La Borde (13630 ± 230 Gif TAN,) thanks to one ¹⁴C AMS date on macroremains of the moss *Pohlia annotina* (Hedw.) Lindb.). At Freychinède (Jalut *et al.* 1992) the same event is dated between 14700±800 and 14500±570 and on the piedmont of Lourdes (Biscaye, Mardonès & Jalut 1983) between 14820±240 and 13250±120. It is contemporaneous with the decrease in *Artemisia* values dated 13490±220 at Barbazan in the Garonne valley (Andrieu 1991). For the same site this *Juniperus* phase can be recorded in one sedimentary sequence and not in another (e.g. La Borde, Reille 1990a, Jalut *et al.* 1993). The *Juniperus* maximum can be prior to or contemporaneous with the expansion of *Betula* at the beginning of the Late-glacial period around 13500 to 13000 BP.
– Younger Dryas is recorded in all type regions.
– *Quercus* was present early during the Lateglacial, especially from Allerød (Jalut *et al.* 1992). Scarce pollen present during the Bølling.
– At the beginning of the Post-glacial *Quercus* expanded about two or three centuries before *Corylus*.
– The presence of *Fagus* is commonly recorded

between 5300–5200 BP and 4500 BP, but its expansion occurred around 4000 BP. If it depends chiefly on climatic conditions, local asynchronisms are due to differences in periods and systems of exploitation of the landscape (Galop & Jalut, unpubl.).

Unique patterns

– *Populus* is noticed in the eastern Pyrenees (Reille 1990b, Jalut *et al.* 1992). Its first occurrence is dated about 8700 ±140 BP (Jalut *et al.* 1992).
– The history of *Abies* is very different, depending on the areas considered.

• In some sites of the eastern Pyrenees it is found from 10000 to 9500 BP (Nohèdes, Jalut 1974) or near 9150±250 (Gourg Nègre, Reille & Lowe 1993). But its regular presence can be documented near 8650–8270 BP (Jalut *et al.* 1992). Then its expansion occurred between 8300 and 7500 BP (Figs 17.34 and 17.35).
• To the west at Freychinède and l'Estagnon, the beginning of its expansion is differently dated according to the sequences studied (Freychinede: 7030±140, Jalut *et al.* 1982; 8560±200, Reille 1993b; l'Estagnon: 6690±70, Reille 1993b) (Fig 17.36).
• In the Central Pyrenees, the dates obtained, at low and medium elevation, for the *Abies* expansion are similar: Barbazan 450 m, 3640±60 (Andrieu 1991); Ules 1650 m, 3600±65 (Aubert 1993). *Fagus* spreads at the same time (Fig. 17.37).
• In the Western Pyrenees, near its present western limit, the development of *Abies* is dated near 4900–4800 BP with some isolated pollen near 7000 BP. (Castet, Jalut *et al.* 1988). *Fagus* appears between 4800 and 4300 BP (Fig. 17.38). These data show the strong gradient of development of *Abies* between the Atlantic and the Mediterranean. On the basis of the available data, the central part of the northern Pyrenean slope seems to be a transition zone between two types of Holocene history for *Abies*.

– A first significant occurrence of *Fagus* is noticed in

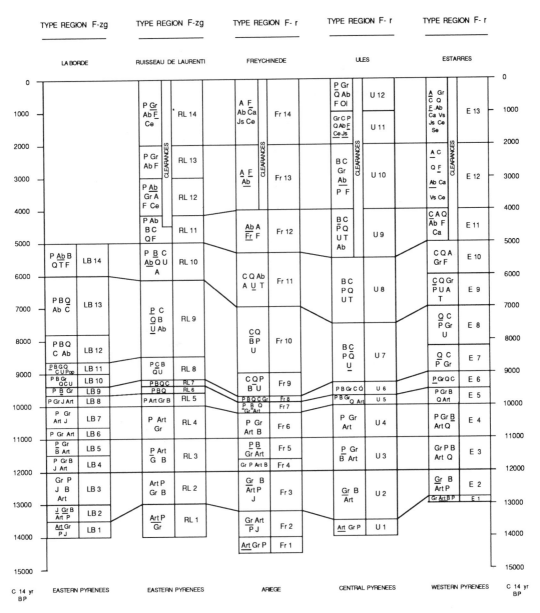

Fig. 17.33 Time–space correlation of local pollen-assemblage zones A= *Alnus*, Ab = *Abies*, Art = *Artemisia*, B = *Betula*, C = *Corylus*, Ca = *Castanea*, Ce = *Cerealia*, F = *Fagus*, Fr = *Fraxinus*, Gr = Gramineae, J = *Juniperus*, Js = *Juglans*, P = *Pinus*, Q = *Quercus*, Se = *Secale*, T = *Tilia*, U = *Ulmus*, Vs = *Vitis*. (Underlined: maximum phase)

some sites of the western Pyrenees around 7000 BP, but this species did not develop and only extended from 4500 to 4200 BP at low elevation in the oak forest and at medium elevation at the expenses of *Abies*.

Anthropogenic events

Unique patterns

The earlier anthropogenic indicators and clearances were observed on the northern slope of the eastern

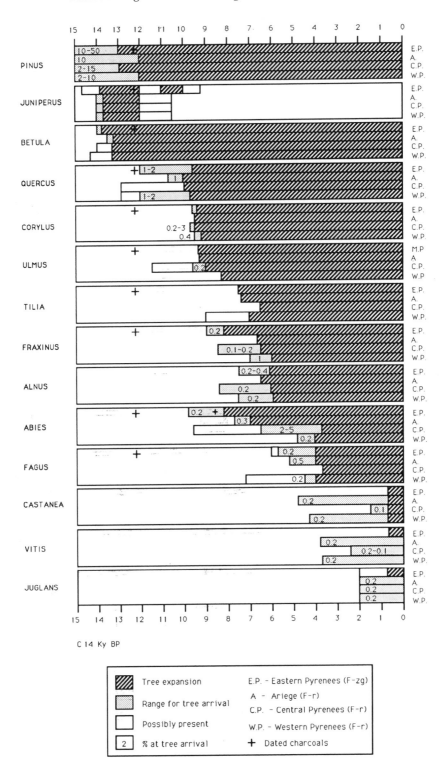

Pyrenees around 6100 BP, e.g. peat bogs of Ruisseau de Fournas 1510 m and Pinet 880 m (Jalut 1974, Jalut & Vernet 1989, Reille 1990b) and subsequently confirmed by the discovery of domestic fauna in the Cave of Dourgne in the same valley 10 km to the north (Geddes 1980) (Fig. 17.33).

In the high valley of Têt (La Borde) open wide to the Mediterranean, very regular occurrences of Cerealia pollen type are found from 8700 BP, associated with Chenopodiaceae, Cichorioideae, *Pistacia*, *Buxus*, and *Quercus ilex* type. Except for Cerealia, this pollen assemblage was also found in the first study of La Borde (Jalut 1971), with in addition an increase in *Artemisia* and *Rumex* values between 8300 and 7500 BP. It is clearly older than the first sporadic human clearances recorded between 7000 and 3000 BP on the Catalan coast (northeastern Iberian Peninsula) at the southeastern extremity of the Pyrenees (Riera-Mora & Esteban-Amat 1994). In the site of Ules, when the human impact becomes obvious in the upper zones (U 10 to U 12) with the classical presence of apophytes and anthropochores, we do not observe, as in other sites of the northern slope, the classical strong increase in *Fagus* values. On the contrary, it is at that time that *Pinus* expands. If in the northern Pyrenean slope the human activity was responsible for the *Fagus* extension (Jalut 1984), it seems in the light of this example that on the southern slope or on south-facing slopes of sheltered valleys *Pinus* was also favoured by the same phenomenon and that its present importance in such exposures must be partly related to human impact from the Bronze Age onwards.

Common patterns

In the two type regions, human influence became important between 5000 and 4000 BP. In the mountains, its general consequence was a strong decrease of *Abies* and the expansion of *Fagus* (Jalut 1984, Jalut *et al.* 1984), which continued during historical time (Fruhauf 1980).

In the Ariege and western Pyrenees, where they are well dated, the first pollen of *Castanea* is found between 4820 and 4310 BP, *Vitis* between 3800 and 3680 BP, and *Juglans* between 2060 and 2020 BP (Figs 17.35 and 17.36). Great differences exist for the beginning of the development of the same species, according to the area and the altitude, e.g. the regular presence of *Juglans* does not necessarily correspond to the Roman age (Pla de l'Orri, Cerdagne, near La Borde, 760 ± 50; Canet, Mediterranean coast, 700 ± 70 (Planchais 1985); Pailhères near Ruisseau de Laurenti, 1100 ± 70; Argentière, near Freychinède, 1850 ± 50, Galop, unpublished).

Climatic events

The end of the glacial period

The most important climatic change is reflected in the two type regions as well as in the southwestern Spanish part (Montserrat Marti 1991) by the well-dated *Artemisia* decrease and *Juniperus* expansion (La Borde: 13630±230 Gif TAN, Jalut *et al.* 1992). Due to the present distribution of *Juniperus thurifera* in northern Spain (valley of Ebro and Sierra de Alcubiere, Puerto de Somosierra to the north of Sierra de Guadarrama, northeast of Leon in the southern slope of the Cantabric Mountains), as well as in the French Pyrenees (valleys of Garonne and Ariege), it can be assumed that this species was a component of the vegetation cover during the *Juniperus* phase from about 14000 BP to about 13500–13000 and probably up to 10000 BP, with a significant presence during the Younger Dryas and the beginning of the Post-glacial in the Mediterranean Pyrenees. The Ephedro–Juniperetalia, pre-steppe formation described in the High Atlas Mountains (Morocco) (Quezel & Barbero, 1981, 1986), has its optimum development under Mediterranean climatic conditions, where annual precipitation is 250–500 mm. With respect to the phase 15000–14000 BP not described here, with low *Juniperus* values and strong representation of Chenopodiaceae, *Artemisia*, and Gramineae, the *Juniperus* phase can be considered as a period of increase in temperatures (summer and winter) and precipitation (Jalut *et al.* 1992). This climatic trend continues during the Bølling.

Fig. 17.34 Tree arrival and tree expansion in the French Pyrenees

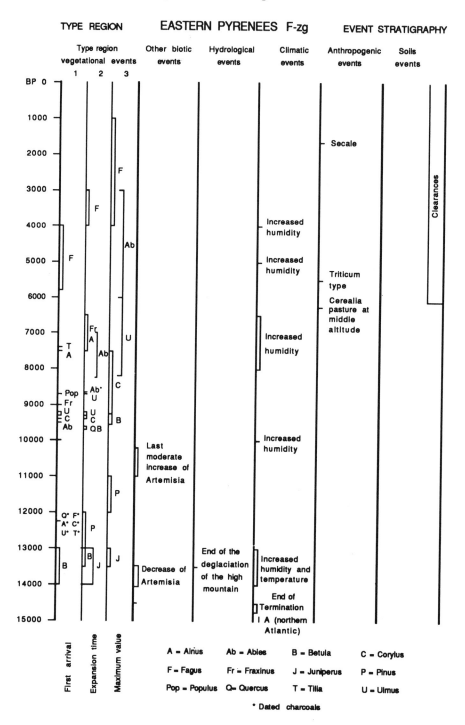

Fig. 17.35 Event stratigraphy in type region F-zg (Eastern Pyrenees)

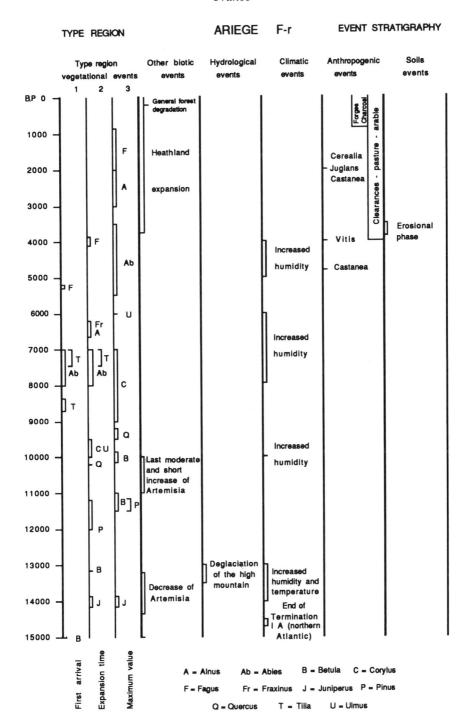

Fig. 17.36 Event stratigraphy in type region F-r (Ariege)

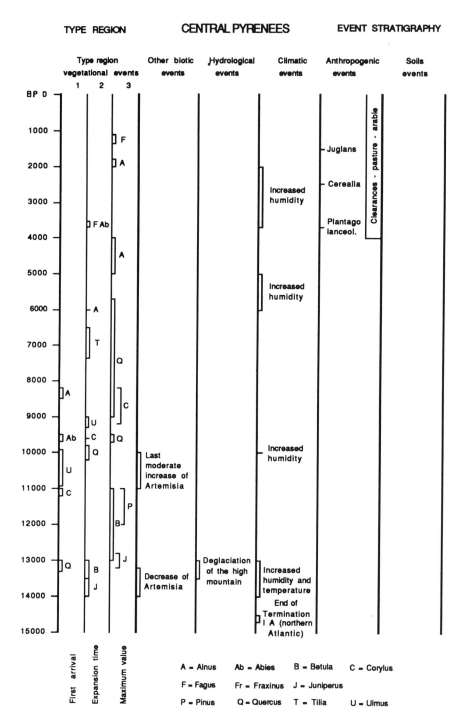

Fig. 17.37   Event stratigraphy in type region F-r (Central Pyrenees)

TYPE REGION   WESTERN PYRENEES F-r   EVENT STRATIGRAPHY

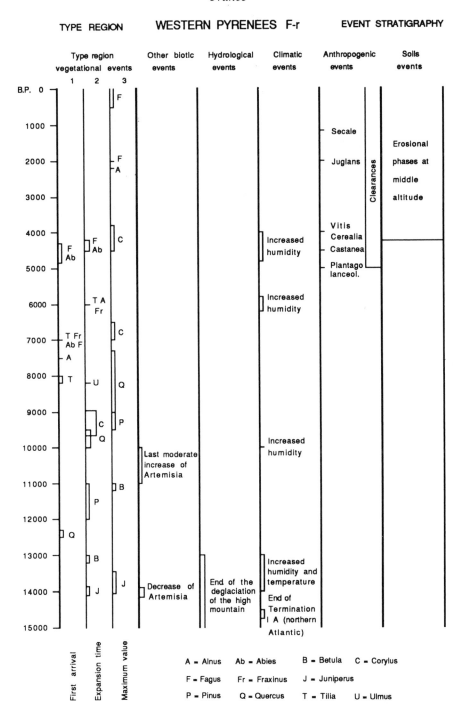

Fig. 17.38 Event stratigraphy in type region F-r (Western Pyrenees)

The Late-glacial period

Its beginning around 13500–13000 BP corresponds to the *Betula* expansion (beginning of the Bølling–Allerød period). It is followed by the spread of *Pinus* between 12200 and 11000 BP. The Younger Dryas is clearly characterized by a new development of the heliophilous taxa. Its beginning is dated 11130±200 at Estarrès. Its impact on the representation of *Quercus*, still regularly present from the Allerød period, is very limited or not visible.

The Post-glacial period

Its beginning could be dated near 10200–10100 BP (Jalut *et al.* 1982, Mardonès & Jalut 1983). *Pinus* and *Betula* are the dominant trees with steppe or heliophilous elements. *Quercus* then *Corylus* expand. The interval between the expansion of these trees (some centuries) could be related to the persistence of dry climatic conditions at the beginning of the Post-glacial, limiting the spread of other broad-leaved trees (*Corylus* and *Ulmus*) (Jalut *et al.* 1992).

An important change, interpreted as an increase in precipitation, occurred between about 8300 and 6500–6000 BP, when *Alnus*, *Fraxinus*, and *Abies* appeared or extended according to the geographical locations. In some sites of the western and Spanish Basque Country, the climatic change is marked by a short appearance of *Fagus* near 7000–6800 BP (Alimen *et al.* 1965, Penalba 1987, Jalut *et al.* 1988). No clear climatic modification is registered until 5200–4500 BP. From this period onwards *Fagus* began to expand, and *Abies* spread in the central and western Pyrenees. This reflects increasing precipitation and/or atmospheric humidity. Then, because of human influence, it is difficult to determine what in the *Fagus* development is due to climate and what results from human impact.

Soil events

In Ariège and western Pyrenees important erosional phases were recorded at middle altitudes from 4000 BP onwards (Jalut *et al.* 1982, 1988). They are always correlated with deforestation.

TYPE REGION F-zd, CHAÎNE DES HURTIÈRES (F. David)

A recent paper has synthesized the palaeoecological investigations of the last 20 years in the French Alps and the Jura Mountains (Beaulieu *et al.* 1994). It shows that the vegetation history of large areas is unknown, preventing a detailed synthesis of the vegetation history for the French Alps.

If we consider the northern French Alps only, nine important massifs extend over 140 km from north to south and 80 km from west to east (Fig. 17.39). The elevation varies between 200 m and 4800 m.

The major present ecological subdivisions in the area are linked primarily to increasing continentality (three zones: outer, intermediate and inner Alps are distinguished from west to east) and secondarily to increasing altitude. Hill, mountain, subalpine, and alpine vegetation belts have increasing altitudinal limits with increasing continentality. A complete palaeoecological reconstruction should take into account these different ecological parameters without ignoring local parameters such as exposure.

This contribution attempts to illustrate the methodological approach we need to reach such a palaeoecological synthesis, considering a whole limited mountain as a "key-massif". We present here five of the six studied sites in different vegetation belts of the Chaîne des Hurtières. This mountain extends over 15 km from southwest to northeast and 7 km from west to east. The highest point is around 2000 m, the lowest 150 m. Included in zone F-zd, the chain is at the very limit with zone F-zb1 and must be linked to the intermediate Alps (Fig. 17.40).

Climate: The mean annual temperature varies between 11°C in the valley and 4°C in the subalpine belt. Precipitation increases with altitude, around 1200 mm in the mountain belt and 1600 mm in the subalpine belt. The snow cover in the subalpine belt lasts for 3 to 5 months.

Geology: The Chaîne des Hurtières extends to the north into the crystalline Belledonne massif. The western side of the Chaîne des Hurtières presents a tectonic contact between the mica-schist axis and the Jurassic schist and the calcareous sedimentary cover (Bajocian). The relief is smooth.

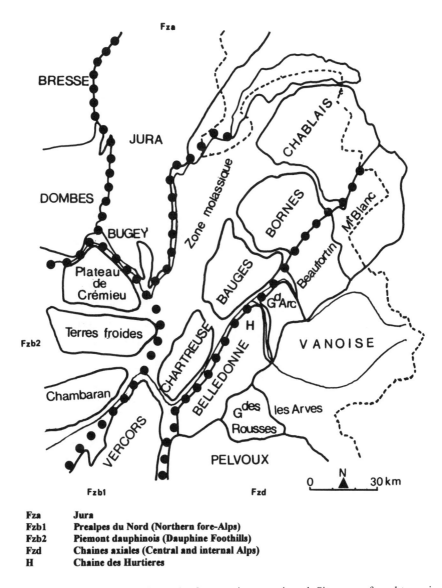

Fza	**Jura**
Fzb1	**Prealpes du Nord (Northern fore-Alps)**
Fzb2	**Piemont dauphinois (Dauphine Foothills)**
Fzd	**Chaines axiales (Central and internal Alps)**
H	**Chaine des Hurtieres**

Fig. 17.39 Type regions in the French Alps and reference sites mentioned. Sites are referred to region F-zd

Land use: Despite strong and old agricultural practices (sheeps and bovines), forest covers nearly half of the mountain and the subalpine belt and spreads over neglected patches. In the area cattle are always moved to summer pastures and as a result the forest limit (around 1700 m) is low.

Reference site 25. Le Vivier (David, 1993b)

Latitude 45°38′N, Longitude 6°18′E. Elevation 345 m. Age range *ca.* 6000–15000 BP. 5 ^{14}C dates. 998 cm sediment. 12 local pollen-assemblage zones (paz) represented (Fig. 17.41).

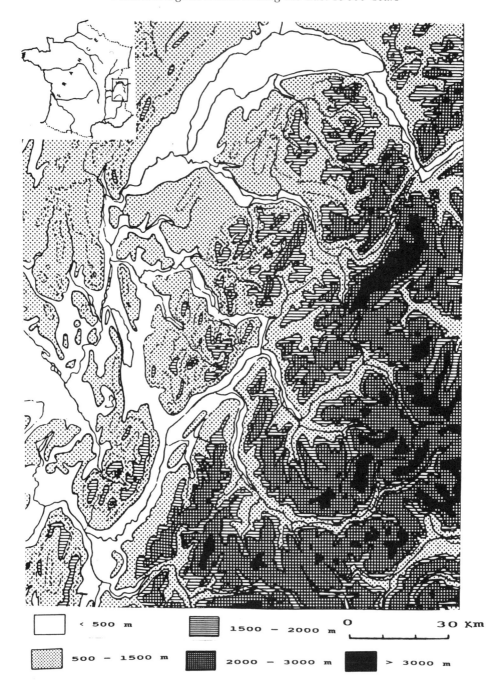

Fig. 17.40 Hypsometric map of the northern French Alps

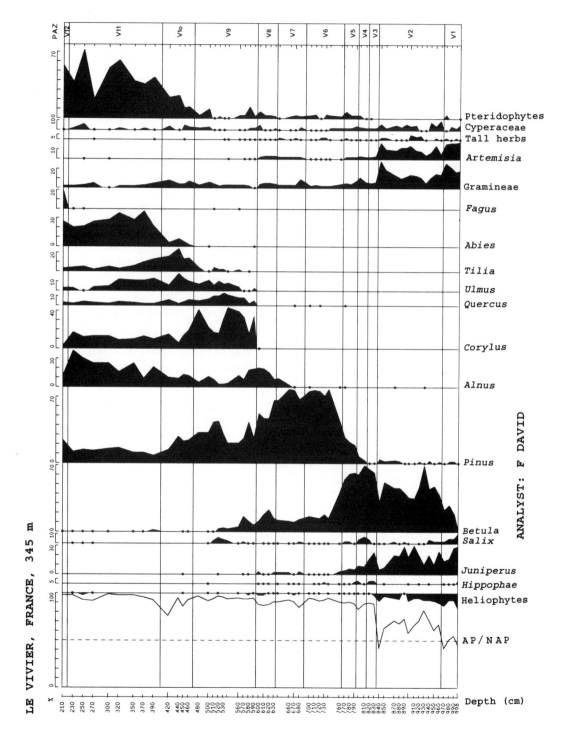

LE VIVIER, FRANCE, 345 m

ANALYST: F DAVID

Fig. 17.41 Late-glacial and Holocene pollen diagram from Le Vivier. Only selected pollen types are shown

Age		Local paz
V 1	*ca.* 15000 BP	NAP–*Juniperus*
V 2	15000–13000 BP	*Betula–Juniperus–* NAP
V 3	13000–12700 BP	*Betula–Juniperus– Hippophaë*
V 4	12700–12500 BP	*Betula–Salix*
V 5	12500–12000 BP	*Betula–Pinus*
V 6	12000–11000 BP	*Pinus–(Betula)*
V 7	11000–10000 BP	*Pinus–Alnus–Juniperus*
V 8	10000–9500 BP	*Pinus–Alnus–Betula*
V 9	9500–8400 BP	*Corylus*–Quercetum mixtum
V 10	8400–7500 BP	*Tilia–Ulmus–Abies*
V 11	7500–6400 BP	*Abies–Alnus*
V 12	< 6400 BP	*Abies–Fagus*

Reference site 26, Les Etelles (David 1993b)

Latitude 45°28′N, Longitude 6°09′E. Elevation 700 m. Age range < 15000–0 BP. Five ¹⁴C dates. 810 cm sediment. 15 local pollen-assemblage zones represented (Fig. 17.42).

Age		Local paz
ET 1	*ca.* 15000 BP	NAP–*Pinus*
ET 2	15000–13500 BP	NAP–*Juniperus*
ET 3	13500–13000 BP	NAP–*Juniperus– Betula*
ET 4	13000–12500 BP	*Juniperus– Hippophaë*
ET 5	12500–12000 BP	*Betula*
ET 6	12000–11000 BP	*Pinus–(Betula)*
ET 7	11000–10000 BP	*Pinus*–NAP
ET 8	10000–9500 BP	*Betula–Pinus*
ET 9	9500–9000 BP	*Pinus–Betula–Alnus*
ET 10	9000–8500 BP	*Corylus*
ET 11	8500–8000 BP	*Ulmus–Corylus– Quercus*
ET 12	8000–7500 BP	*Corylus–Pinus–* Quercetum mixtum
ET 13	7500–6000 BP	*Abies–Alnus*
ET 14	6000–500 BP	NAP–*Fagus– Quercus*
ET 15	500–0 BP	*Picea–Pinus*

Reference site 27. La Coche (David 1993b)

Latitude 45°29′N, Longitude 6°15′E. Elevation 985 m. Age range *ca.* 12500–0 BP. Four ¹⁴C dates. 390 cm sediment. 10 local pollen-assemblage zones represented (Fig. 17.43).

Age		Local paz
LC 1	*ca.* 12500–12000 BP	*Betula*
LC 2	12000–11000 BP	*Pinus*
LC 3	11000–10000 BP	*Pinus*–NAP
LC 4	10000– 9500 BP	*Betula–Pinus*
LC 5	9500–9000 BP	*Pinus–Alnus*
LC 6	9000–8000 BP	*Corylus*
LC 7	8000–7500 BP	Quercetum mixtum–*Abies*
LC 8	7500–2000 BP	*Abies*
LC 9	2000–500 BP	*Fagus–Alnus– Picea*
LC 10	500–0 BP	*Betula–Alnus–Picea*

Reference site 28. Montendry (David 1993b)

Latitude 45°30′N. Longitude 6°15′E. Elevation 1335 m. Age range *ca.* 15000–0 BP. No ¹⁴C dates. 820 cm sediment. 12 local pollen-assemblage zones represented (Fig. 17.44).

Age		Local paz
MT 1	*ca.* 14000 BP ?	NAP–*Pinus*
MT 2	14000–13000 BP	NAP–*Juniperus*
MT 3	13000–12000 BP	*Betula–Juniperus*
MT 4	12000–11000 BP	*Pinus–(Betula)*
MT 5	11000–10000 BP	*Pinus*–NAP
MT 6	10000–9500 BP	*Pinus–Betula*
MT 7	9500–9000 BP	*Betula–Pinus–Corylus*
MT 8	9000–8500 BP	*Corylus*–Quercetum mixtum
MT 9	8500–7500 BP	Quercetum mixtum– *Abies*
MT 10	7500–2000 BP	*Abies*
MT 11	2000–200 BP	*Abies–Picea–Fagus*
MT 12	200–0 BP	NAP–*Pinus–Picea*

Reference site 29. Le Grand Leyat

Latitude 45°28′N. Longitude 6°13′E. Elevation 1660 m. Age range *ca.* 13000–100 BP. Three ¹⁴C dates. 410 cm sediment. 10 local pollen-assemblage zones represented (Fig. 17.45).

637

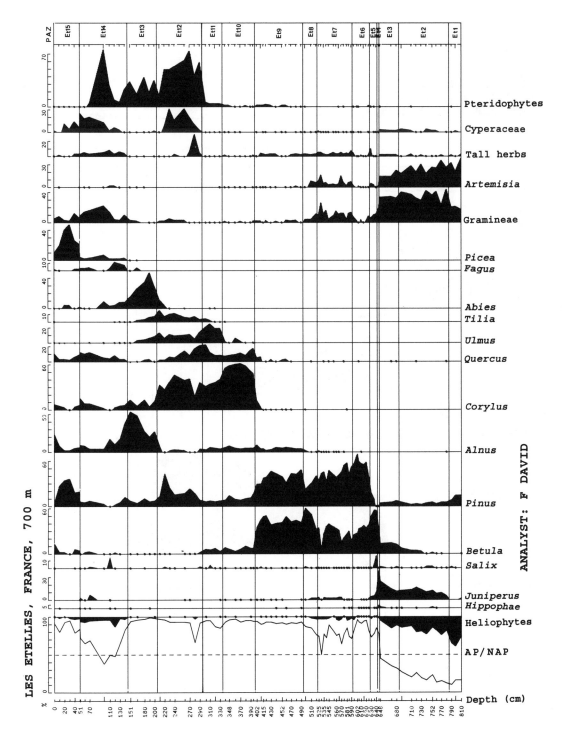

LES ETELLES, FRANCE, 700 m

ANALYST: F DAVID

Fig. 17.42 Late-glacial and Holocene pollen diagram from Les Etelles

638

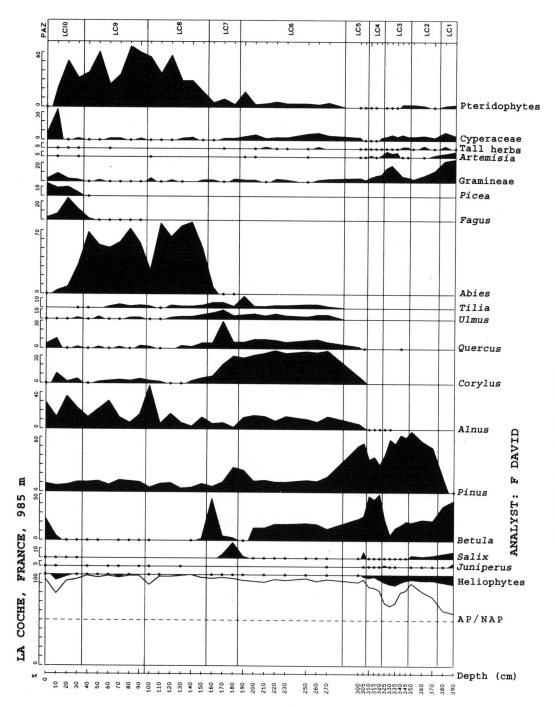

Fig. 17.43 Late-glacial and Holocene pollen diagram from La Coche

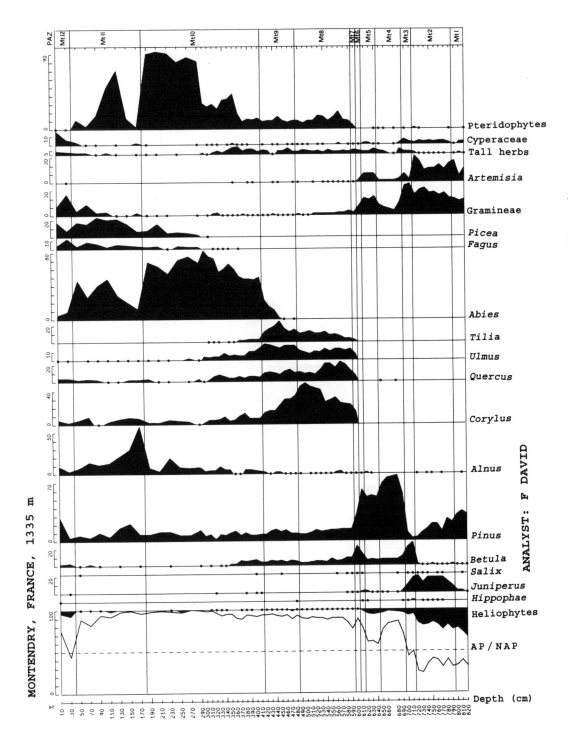

MONTENDRY, FRANCE, 1335 m

ANALYST: F DAVID

Fig. 17.44 Late-glacial and Holocene pollen diagram from Montendry

640

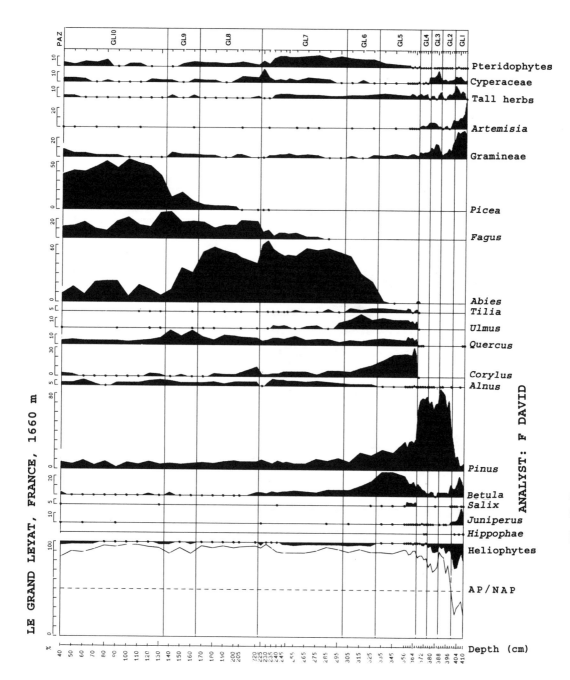

Fig. 17.45 Late-glacial and Holocene pollen diagram from Le Grand Leyat

	Age	Local paz
GL 1	13000–12000 BP	NAP– *Juniperus*– *Betula*
GL 2	12000–11000 BP	*Pinus*–(*Betula*)
GL 3	11000–10000 BP	*Pinus*–NAP
GL 4	10000–9000 BP	*Pinus*–*Betula*
GL 5	9000–8000 BP	*Betula*–*Corylus*
GL 6	8000–7000 BP	*Abies*–Quercetum mixtum
GL 7	7000–4500 BP	*Abies*–*Fagus*
GL 8	4500–2000 BP	*Abies*–*Fagus*–*Picea*
GL 9	2000–1500 BP	*Fagus*–*Abies*– *Picea*
GL 10	1500–100 BP	*Picea*–*Fagus*–*Abies*

Vegetation history

Despite the uncertainty of the absolute chronology caused by bulk dates, pollen percentages of different taxa undeniably vary according to elevation and indicate an early altitudinal effect on vegetation history. 700 m appears to be the best altitudinal level for both comparisons in the different belts of our area and comparisons with adjacent areas.

Oldest Dryas, ca. 15000–13000 BP

Herbaceous communities followed one another, first dominated by *Artemisia* and Chenopodiaceae next by Gramineae and *Helianthemum*. A similar succession has been described in Moyen Pays Romand, Switzerland (Gaillard 1984). Earlier spreading of *Juniperus* and *Betula* is recorded up to 700 m in Chaîne des Hurtières as in the southern Alps (Beaulieu 1977). A ^{14}C date of 20000 BP for the first *Juniperus* expansion appears too old. As we wait for new AMS dates, 15000 BP is assumed for the beginning of the Oldest Dryas period (Beaulieu & Reille 1983).

Bølling, 13000–12000 BP

A second *Juniperus* maximum related to *Hippophaë* has been recorded at each elevation. It corresponds also with a new *Betula* increase at the lowest elevation. The date 13000 BP agrees with bulk dates of adjacent areas, but the same episode is AMS dated

12600 BP in Switzerland (Ammann & Lotter 1989). New AMS dates could confirm the younger age in our area. The next stage is dominated by *Betula* with *Salix* and later *Pinus*. During the *Betula* and *Pinus* phase, one or two short events occurred, underlined by more clastic sediments and *Artemisia* and Gramineae. Linking one of those events to the Older Dryas is tempting. The last one always occurred at the transition with the next phase.

Allerød, 12000–11000 BP

The period is dominated at each elevation by *Pinus*. Absolute concentration data clearly demonstrate that the low *Betula* percentage is a statistical effect due to a more abundant *Pinus* pollen production. *Betula* stands did not regress with a *Pinus* expansion. High *Pinus* percentages at both lowest and highest elevation, regularly decreasing *Betula* percentages with increasing altitude, and the lowest *Pinus* percentages at intermediate belt around 1300 m clearly demonstrate the overrepresentation of *Pinus* pollen above 1300 m.

Younger Dryas, 11000–10000 BP

The period is clearly recorded at each elevation, but absolute concentrations only permit its recognition at low elevation. *Pinus* stands did not regress below 1000 m. *Betula* stands seemed more affected at least twice, at the beginning and the end of the period.

Preboreal, 10000–9000 BP

Holocene climatic improvement is clearly recorded first by *Betula* prior to *Pinus* up to 1300 m and by *Pinus* above 1300 m. Earlier *Pinus* pollen increase at higher elevation reflects in this deforested belt the growth of the lowest *Pinus* stands. Up to 700 m early occurrence of *Alnus glutinosa* and *A. incana* are typical features in the area (David 1993 a). *Corylus* also appeared in the lowest sites with *Populus*.

Boreal, 9000–8000 BP

Corylus and Quercetum mixtum dominated up to 1300 m. *Ulmus* was particularly abundant in the area. Above 1300 m, a *Betula* belt developed up to 1800–2000 m at the upper forest limit.

Atlantic, 8000–4000 BP

First appearance of *Abies* corresponds to highest *Tilia* percentages in oak forest below 1300 m. *Betula* reached its maximum expansion above 1300 m. *Pinus cembra* settled with *Betula*. Around 7000 BP *Abies* forest constituted a unique vegetation belt from 1000 m up to the top at the expense of *Betula* stands. Chaine des Hurtières was not high enough for a large *Pinus cembra* belt above the *Abies* belt. Nevertheless, few stands spread out when *Fagus* appeared in the *Abies* forest around 6000 BP. The records of *Fagus* occurrence are not synchronous. At the highest site, above 1800 m, the first western long-distance *Fagus* grains are recorded with the first *Abies*. On the eastern side of the mountain *Fagus* appeared later, around 5000 BP.

Subboreal, 4000–2500 BP

First *Picea* pollen is recorded with a new *Fagus* and ultimately a *Pinus cembra* expansion. *Abies* remained the dominant tree up to 1800 m. Below 1000 m the chronology of changes is not established in the area, but we can see a large expansion of *Alnus glutinosa* and *A. incana*.

Subatlantic, 2500–0 BP

Picea and *Fagus* are the dominant trees. *Abies* forests disappeared except at 1300 m where remnants persist today. Vegetation history during this period mainly depended on human history. *Picea* was largely favoured by man until today. The continuous *Juglans* curve is a typical feature in the area at least since the Roman time.

Conclusion

Such type of investigations in a restricted area demonstrate the overrepresentation of easily wind-transported pollen at the highest sites or in deforested belts. Comparison between mountainous areas involves comparison between the chosen sites at the

same altitude. Therefore, comparison between sites at different levels in different regions is ecologically and palaeoecologically difficult and of little value.

The chronology used here is based entirely on conventional bulk-sample ^{14}C assays. A new chronology based on AMS dates is in progress.

REFERENCES

Alimen, H., Florschütz, F. & Menendez Amor, J. 1965: Etude géologique et palynologique sur le Quaternaire des environs de Lourdes. *In Actes 4e Congrès International d'Etudes Pyrénéennes, Pau-Lourdes*, 1962, 1, 7–26.

Ammann, B. & Lotter, A. 1989: Late-Glacial radiocarbon- and palynostratigraphy on the Swiss Plateau. *Boreas 18*, 109–126.

Andrieu, V. 1987: Le paléo-environnement du piémont nord-pyrénéen occidental de 27 000 BP au Postglaciaire: la séquence de l'Estarrès (Pyrénées-Atlantiques, France) dans le bassin glaciaire d'Arudy. *Comptes Rendus de l'Académie des Sciences 304*, 103–108.

Andrieu, V. 1991: *Dynamique du paléoenvironnement de la vallée montagnarde de la Garonne (Pyrénées centrales, France) de la fin des temps glaciaires à l'actuel*. Thèse, Université Toulouse 2, 330 pp.

Andrieu, V., Eicher, U. & Reille, M. 1993: La fin du dernier Pléniglaciaire dans les Pyrénées (France): données polliniques, isotopiques et radiométriques. *Compte Rendus de l'Académie des Sciences 316*, 245–250.

Aubert, S. 1993: *Etude palynologique de la tourbière de la Bassa d'Ules (Val d'Aran, Espagne)*. DEA Géographie et Aménagement, Université de Toulouse le Mirail, 34 pp.

Bahn, P. 1982: L'économie paléolithique et mésolithique du Béarn. *Revue de Pau et du Béarn*, 127–140.

Beaulieu, de, J.L. & Reille M. 1983: Paléoenvironnement tardiglaciaire et holocène des lacs de Pelleautier et Siguret (Hautes Alpes, France). *Ecologia Mediterranea IX*, 19–36.

Beaulieu, de, J.L., Clerc J., Richard H. & Ruffaldi P. 1994: History of vegetation in the french Alps and the Jura over the last 15000 years. *Dissertationes Botanicae 234*, 1–23.

Berglund, B.E. 1979: Presentation of the IGCP project 158b. Palaeohydrological changes in the temperate zone in the last 15000 years—Lake and mire environments. *Acta Univ. Oulensis Series A/Scient. Rerum*, Nat Nr 82. Geol Nr 3.

Bick, H., Carbiener, R., Edelman, H.J., Janssen, C.R., Kalis, A.J., Salome, A.I., Tamboer- van de Heuvel, G. & de Valk, E.J. 1974: Vogezen Symposium. *Laboratory of Palaeobotany and Palynology Report*, 74 pp.

Cushing, E.J. 1967: Late-Wisconsin pollen stratigraphy and the glacial sequence in Minnesota. *In* Cushing, E.J. & Wright, H.E. Jr (eds) *Quaternary paleoecology,* 59–88. Yale University Press, New Haven, Connecticut.

Darmois-Theobald, M. & Denèfle 1981: Observations sur la végétation holocène des Vosges Méridionales et de leur piémont Haut-Saonois (France). *Annales Scientifiques Univ. Franche-Comté-Besancon.* Geologie fasc. 3. 4e série, 3–12.

Darmois-Theobald, M., Denèfle, M. & Menillet, F. 1976: Tourbières de moyenne altitude de la fôret de Haute-Meurthe (Vosges, France). *Bulletin de l'association francaise Etude du Quaternaire,* 1976–2, 99–107.

David, F. 1993a: Développement des aulnes dans les Alpes françaises du Nord. *Comptes Rendus de Académie des Sciences, Paris, 316, Série II,* 1815,1822.

David, F. 1993b: Altitudinal variation in the response of the vegetation to Late-glacial climatic events in the northern French Alps. *New Phytologist 125,* 203–220.

DeValk, E.J. 1979: Pollenanalytical contributions towards Late-Holocene history of the subalpine plant communities of the Kastelberg (Vosges, France). *In* Werden und Vergehen von Pflanzengesellschaften. *Berichte Internationale Symposium Internationale Vereinigung für Vegetationskunde,* pp. 269–284, Cramer, Vaduz.

DeValk, E.J. 1981: *Late Holocene and Present vegetation of the Kastelberg (Vosges, France).* PhD thesis, University Utrecht, 294 pp.

Dresch, J., Elhai, H. & Denèfle-Labiole, M. 1966: Analyse pollinique de quatre tourbières du Ballon d'Alsace (Vosges, France). *Compte Rendu Soc. Biogeogr.,* 376, 78–89.

Edelman, H.J. 1985: *Late Glacial and Holocene vegetation development of la Goutte Loiselot (Vosges, France).* PhD thesis, University Utrecht, 197 pp.

Firbas, F., Grünig, G., Weischedel I. & Worzel, G. 1948: Beitrage zur spät- und nacheiszeitlichen Vegetationsgeschichte der Vogesen. *Bibliotheca Botanica 121.*

Fruhauf, Ch. 1980: *Forêt et Société,* 302 pp. CNRS, Toulouse.

Gaillard, M.J. 1984: Etude palynologique de l'évolution tardi- et postglaciaire de la végétation du Moyen-Pays Romand (Suisse). *Dissertationes Botanicae 77,* 1–322.

Galop, D. & Jalut, G. (in press): Differentiated human impact and vegetation history in adjacent Pyrenean valleys from 3000 BP to Present (Ariege basin, Southern France). *Vegetation History and Archaeobotany.*

Geddes, D. 1980: De la chasse au troupeau en Méditerranée occidentale. *Archives d'Ecologie Préhistorique 5,* 145 pp.

Giot, P.R., Briard, J. & Pape, L. 1979a: *Protohistoire de la Bretagne. Ouest-France 437* pp. Rennes.

Giot, P.R., L'Helgouach, J. & Monnier, J.L. 1979b: *Préhistoire de la Bretagne. Ouest-France,* 443 pp. Rennes.

Guerby, L. 1993: Une nouvelle station de Genévrier thurifère dans les Pyrénées: le Quié de Lujat (Ussat les Bains, Ariège). *Le Monde des Plantes 447,* 26–27.

Guillet, B. 1971a: Étude palynologique des podzols. II. La podzolisation sur les versants secs gréseux des basses-Vosges. *Pollen et Spores 13(2),* 233–254.

Guillet, B. 1971b: Étude palynologiqe des podzols: III. La podzolisation sur granite dans les Vosges hercyniennes de l'étage montagnard. Comparaison avec la podzolisation dans les basses Vosges gréseuses et sur le plateau Lorrain. *Pollen et Spores, 13(3),* 421–446.

Guillet, B., Hassko, B. & Jaegy, R. 1972: Approche palynologique de la limite spontanée du sapin sur la bordure orientale du Plateau Lorrain. *Compte Rendu Acad. Sc. Paris,* série D. 274, 2966–2968.

Guillet, B., Janssen, C.R., Kalis, A.J. & de Valk, E.J. 1976: La végétation pendant le Post-Glaciaire dans l'est de la France. *In* Guilaine, J. (ed.) *La Préhistoire française, 2: Civilisations néolithiques et protohistoriques,* 82–87.

Hérail, G., Hubschman, J. & Jalut, G. 1987: Quaternary glaciation in the French Pyrenees. *In* Sibrava, V., Bowen, D.Q. & Richmond, G.M. (eds) Quaternary glaciations in the Northern Hemisphere, IGCP Project 24, Final Report. *Quaternary Science Reviews 5,* 397–402.

Jalut, G. 1971: Analyse pollinique de sédiments des Pyrénées orientales. Tourbière de La Borde: Haute vallée de la Têt, alt. 1660m; gisement des Estables I: Haut Vallespir, alt. 1750m. *Bulletin de l'Association Française pour l'Etude du Quaternaire 27,* 91–110.

Jalut, G. 1974: *Evolution de la végétation et variations climatiques durant les quinze derniers millénaires dans l'extrémité orientale des Pyrénées.* Thèses, Université de Toulouse III, 181 pp.

Jalut, G. 1984: L'action de l'homme sur la forêt montagnarde dans les Pyrénées ariègeoises et orientales depuis 4 000 BP d'après l'analyse pollinique. *In Actes 106e Congrès National des Sociétés Savantes, Perpignan, 1981, Section. Géographie,* 163–172.

Jalut, G. & Vernet, J.L. 1989: Végétation et climat du Pays de Sault depuis 15000 ans. Réinterprétation des données palynologiques et anthracologiques. *In Pays de Sault: Espaces, Peuplement, Population,* 23–35. CNRS, Paris.

Jalut, G., Sacchi, D. & Vernet, J.L. 1975: Mise en évidence d'un refuge tardiglaciaire à moyenne altitude sur le versant nord-oriental des Pyrénées (Belvis, alt. 960m, Aude). *Comptes Rendus de l'Académie des Sciences 280,* 1781–1784.

Jalut, G., Delibrias, G., Dagnac, J., Mardonès, M. & Bouhours, M. 1982: A palaeoecological approach to the last 21 000 years in the Pyrenees: the peat bog of Freychinede (alt. 1350m, Ariege, South France). *Palaeogeography, Palaeoclimatology, Palaeoecology 40,* 321–359.

Jalut, G., Esteban-Amat, A., Pagès, Ph. & Mardonès, M.

1984: Quelques aspects de l'action de l'homme sur le milieu montagnard pyrénéen: conséquences phyto-géographiques. *In* Ecologie des milieux montagnards et de haute altitude. *Documents d'Ecologie Pyrénéenne 3–4*, 503–509.

Jalut, G., Andrieu, V., Delibrias, G., Fontugne, M. & Pagès, Ph. 1988: Paleoenvironment of the valley of Ossau (French western Pyrenees) during the last 27 000 years. *Pollen et Spores 30*, 357–394.

Jalut, G., Monserrat Marti, J., Fontugne, M., Delibrias, G., Vilaplana, J. & Julia, R. 1992: Glacial to interglacial vegetation changes in northern and southern Pyrenees: deglaciation, vegetation cover and chronology. *Quaternary Science Reviews 11*, 449–480.

Janssen, C.R. 1979: Pollenassoziation als Ausdruck der Vegetation. *In* Werden und Vergehen von Pflanzengesellschaften. *Berichte Internationale Symposium Internationale Vereinig für Vegetationskunde*, 253–261, Cramer, Vaduz.

Janssen, C.R. 1981: Contemporary pollen assemblages from the Vosges (France). *Review of Palaeobotany and Palynology, 33*, 183–313.

Janssen, C.R. & Braber, F.I. 1987: The present and past grassland vegetation of the Chajoux and Moselotte valleys (Vosges, France). 2. Dynamic aspects and origin of grassland vegetation in the Chajoux valley, interpreted from the contrast between regional and local pollen deposition of dominant pollen types and the distribution of pollen indicators in pollen diagrams. *Koninklijke Akademie Wetenschappen Series C 90(2)*, 115–138.

Janssen, C.R. & Kettlitz, L.M. 1972: A post-atlantic pollen sequence from the tourbière du Tanet (Vosges, France). *Pollen et Spores 14*, 65–76.

Janssen, C.R. & Salomé, A.I. 1974: Voorlopige resultaten van een onderzoek in Le Ronfaing, Cleuridal (Vogezen). *Berichten Fysisch Geografische Afdeling*, Nr. 8, 21–24.

Janssen, C.R., Kalis, A.J., Tamboer-v.d. Heuvel, G. & de Valk, E.J. 1974: Palynological and palaeo-ecological investigations in the Vosges (France): a research project. *Geologie en Mijnbouw 53*, 406–414.

Janssen, C.R., Cup-Uiterwijk, M.J.J., Edelman, H.J., Mekel-te Riele, J. & Pals, J.P. 1975: Ecologic and paleoecologic studies in the Feigne d'Artimont (Vosges, France). *Vegetatio 30*, 165–178.

Janssen, C.R., Braber, F.I., Bunnik, F.P.M., Delibrias, G., Kalis, A.J. & Mook, W.G. 1985: The significance of chronology in the ecological interpretation of pollen assemblages of contrasting sites in the Vosges. *Ecologia Mediterranea 11 (1)*, 39–43.

Kalis, A.J. 1979: Ergebisse pollenanalytischer und Vegetationskundlicher Untersuchungen zur holozänen Waldgeschichte der westlichen Vogesen (Frankreich). *In* Werden und Vergehen von Pflanzengesellschaften. *Berichte Internationale Symposium Internationale*

Vereinigung für Vegetationskunde, 263–268. Cramer, Vaduz.

Kalis, A.J. 1984a: L'indigénat de l'épicea dans les Hautes Vosges. *Revue de Paléobiologie* Special Volume, 103–115.

Kalis, A.J. 1984b: *Foret de la Bresse (Vosges). Phytosociological and palynological investigations on the forest-history of a central-European mountain range.* PhD. thesis, University Utrecht, 349 pp. (in Dutch).

Kalis, A.J. 1985: Un miroir éloigné: les défrichements anthropogènes dans les zones de basse altitude, réfléchis dans des diagrammes polliniques montagnards. *In* Renault-Miskovski, J., Bui-Thi-Mai and Girard, M. (eds) Palynologie archéologique, actes des journées du 25–26–27 Janvier 1984. *CNRS Centre de Recherches Archéologiques. Notes et Monographies techniques*, Nr. 17, 195–211.

Lemee, G. 1963: L'évolution de la végétation et du climat des Hautes Vosges centrales depuis la dernière glaciation. *In* Le Hohneck (ed.) Aspects physiques, biologiques et humains. *L'association Philomatique d'Alsace et de Lorraine*, 185–192.

Mangerud, J., Andersen, Sv.T., Berglund, B.E. & Donner, J.J. 1974: Quaternary stratigraphy of Norden, a proposal for terminology and classification. *Boreas 3(3)*, 109–128.

Mardonès, M. & Jalut, G. 1983: La tourbière de Biscaye (alt. 409m, Hautes Pyrénées): Approche paléoécologique des 45000 dernières années. *Pollen et Spores 25*, 163–212.

Montserrat Marti, J. 1991: *Evolucion glaciar y postglaciar del clima y la vegetacion en la verdiente sur del Pirineo: estudio palinologico.* Thesis, Barcelona, 115 pp.

Morzadec-Kerfourn, M.T. 1974: Variations de la ligne de rivage armoricaine au Quaternaire. Analyses polliniques de dépôts organiques littoraux. *Mémoires Société géologique minéralogique de Bretagne 17*, 208 pp. Rennes.

Penalba, M.C. 1987: Analyse pollinique de quatre tourbières du Pays Basque Espagnol. *Institut Français de Pondichéry, Travaux Section Sciences et Techniques 25*, 65–71.

Planchais, N. 1967: Palynologie du Lac de Grand-Lieu. *Compte Rendu Société de Biogéographie 387*, 81–90.

Planchais, N. 1971: *Histoire de la végétation post-würmienne des plaines du bassin de la Loire, d'après l'analyse pollinique.* Thèse d'Etat, Montpellier, 115 pp.

Planchais, N. 1985: Analyses polliniques du remplissage holocène de la lagune de Canet (plaine du Roussillon, Pyrénées orientales). *Ecologia Mediterranea 11*, 117–127.

Quezel, P. & Barbero, M. 1981: Contribution à l'étude des formations pré-steppiques à genévriers du Maroc. *Bulletin de la Société Broteriana 53*, 1137–1160.

Quezel, P. & Barbero, M. 1986: Aperçu syntaxonomique de la connaissance actuelle de la classe des Quercetalia ilicis au Maroc. *Ecologia Mediterranea 13*, 105–111.

Reille, M. 1990a: La tourbière de La Borde (Pyrénées orientales, France): un site clé pour l'étude du Tardiglaciaire sud européen. *Comptes Rendus de l' Académie des Sciences 310*, 823–829.

Reille, M. 1990b: Recherches pollenanalytiques dans l'extrémité orientale des Pyrénées: données nouvelles, de la fin du Glaciaire à l'Actuel. *Ecologia Mediterranea 16*, 317–357.

Reille, M. 1993a: New pollen-analytical researches at Freychinède, Ariège, Pyrénées, France. *Dissertationes Botanicae 196*, 377–386.

Reille, M. 1993b: Nouvelles recherches pollenanalytiques dans la région de Freychinède, Pyrénées ariègeoises, France. *Palynosciences 2*, 109–131.

Reille, M. & Andrieu, V. 1993: Variations de la limite supérieure des forêts dans les Pyrénées (France) pendant le Tardiglaciaire. *Compte Rendus de l'Académie des Sciences 316*, 547–551.

Reille, M. & Lowe, J.J. 1993: A re-evaluation of the vegetation history of the eastern Pyrenees (France) from the end of the last Glacial to the present. *Quaternary Science Reviews 12*, 47–77.

Riera-Mora, S. & Esteban-Amat, A. 1994: Vegetation history and human activity during the last 6000 years on the central Catalan coast (northeastern Iberian Peninsula). *Vegetation history and Archaeobotany 3*, 7–23.

Rothé, J.P. & Herrenschneider, A. 1963: Climatologie du Massif du Hohneck. *In* Le Hohneck: Aspects physiques, biologiques et humains. *L'Association Phylomatique d'Alsace et de Lorraine*, 63–93. Strasbourg.

Schloss, S. 1979: Pollenanalytische und stratigraphische Untersuchungen im Sewensee-Ein Beitrag zur spät- und postglazialen Vegetationsgeschichte der Sudvogesen. *Dissertationae Botanicae 52*, 120 pp.

Tamboer-v.d. Heuvel & Janssen, C.R. 1976: Recent pollen-assemblages from the crest region of the Vosges (France). *Review of Palaeobotany and Palynology 21*, 219–240.

Teunissen, D. & Schoonen, J.M.C.P. 1973: Vegetationssedimentationsgeschichtliche Untersuchungen am Grand Étang bei Gerardmer. *Eiszeitalter und Gegenwart 23*, 63–75.

Toussaint, E. & Toussaint, J. 1969: Juglans et Carya's en Alsace. *Revue Forestière francaise 21*, 537–546.

Vernet, J.L. 1980: La végétation du bassin de l'Aude, entre Pyrénées et Massif Central, au Tardiglaciaire et au Postglaciaire d'après l'analyse anthracologique. *Review of Palaeobotany and Palynology 30*, 33–55.

Visset, L. 1979: Recherches palynologiques sur la végétation pléistocène et holocène de quelques sites du district phytogéographique de Basse-Loire. *Société Sciences naturelles Ouest France* 282 pp. Nantes.

Visset, L. 1982a: Nouvelles recherches palynologiques (Boréal-Actuel) dans les marais de Brière: Ile d'Errand en Saint-Malo-de-Guersac (Loire-Atlantique-France). *Bulletin Association française Etudes du Quaternaire 1*, 29–38.

Visset, L. 1982b: *Carte géologique de la France au 1/ 50000ème. Notice explicative de la feuille de Saint-Philbert-de-Grand-Lieu XII-24, No 508, Lac de Grand-Lieu* (en collaboration, sous la direction de M. TERS), Bureau de Recherches géologiques et Minières, 49–53.

Visset, L. 1985: Dernières données pollenanalytiques et radiométriques du golfe briéron (Massif armoricain France). *Ecologia Mediterranea 11(1)*, 107–116.

Visset, L. 1988: The Brière marshlands: a palynological survey. *New Phytologist 110*, 409–424.

Voeltzel, D. 1987: *Recherches pollenanalytiques sur la végétation holocène de la plaine alluviale de l'estuaire de la Loire et des coteaux environnants*. Thèse, 175 pp. Nantes-Marseille.

Woillard, G. 1974: *Exposé des recherches palynologiques sur le Pleistocène dans l'est de Belgique et dans les Vosges Lorraines*. Travaux du Laboratoire de Palynologie et Phytosociologie. Université Catholique Louvain, 19 pp.

Woillard, G. 1978: Grande Pile Peat Bog: A continuous pollen record for the last 140,000 years. *Quaternary Research 9*, 1–21.

18

Switzerland

B. AMMANN, M.-J. GAILLARD and A.F. LOTTER

INTRODUCTION

Switzerland is a mountainous country in which 60% of the surface is in the Alps (CH-c, including the Prealps), 10% in the Jura Mountains (CH-a), and 30% in the Swiss Plateau between the two (CH-b) (Fig. 18.4). About a fifth of the entire Alps belongs to Switzerland. The Gotthard Massif (M3 in Fig. 18.1) in its central part forms a continental divide: the Rhône River drains to the western Mediterranean, the Ticino River to the Adriatic, the Rhine River to the North Sea, and the Inn River to the Danube and ultimately to the Black Sea.

The Alps have a very complex geology and structure of massifs and sedimentary overthrust nappes that were folded and lifted between the European and the African plates (Fig. 18.1). The Jura Mountains were mostly folded but in the north also faulted during the Alpine orogeny by the push from the southeast against the old massifs of the Massif Central, the Vosges, and the Black Forest. The Plateau between Lake Geneva and Lake Constance was a basin during the Tertiary in which sand and gravel from the young Alps were deposited; these deposits of conglomerate, sandstone, and marl (Molasse) today form the Plateau.

During the Last Glaciation most of Switzerland was covered by Alpine ice (see Fig. 18.2), and moraines and outwash plains are therefore common.

The pattern of the natural potential vegetation is controlled not only by regional climate and local soils but mainly by altitude. Instead of a map (see Schmid 1949, Hess *et al.* 1967, Hegg *et al.* 1992) we give a simplified transect from NW to SE (Fig. 18.3). The vegetation of the Jura Mountains resembles that in other European medium–high mountains (Mittelgebirge) like the Vosges or the Massif Central. The Plateau is heavily settled and intensively cultivated today. It is situated at the limit between the colline oak–hornbeam and the montane beech belt. The timberline rises from about 1800 m asl in the northern Alps to 2000–2500 m asl in the central Alps; the snowline rises in a similar way from 2400–2700 m asl to 2700–3200 m asl (i.e. the effect of mass elevation). The highly continental central alpine valleys like Valais and Engadin have no beech and very little fir, and the timberline is formed by larch and Swiss stone pine. In the northern Alps the timberline is formed mainly by spruce, whereas in the southern Alps it is spruce or larch and farther south even beech.

Switzerland was subdivided into seven IGCP type regions (Fig. 18.4). The selected reference sites for each type region are listed in the Appendix. In the following contribution we concentrate on a synthesis for the Swiss Plateau (Schweizer Mittelland, Plateau Suisse), type region CH-b. The lowland area of the Swiss Plateau permits a comparison in a "horizontal" European synthesis (Ammann *et al.* 1994,

Palaeoecological Events During the Last 15000 Years: Regional Syntheses of Palaeoecological Studies of Lakes and Mires in Europe.
Edited by B.E. Berglund, H.J.B. Birks, M. Ralska- Jasiewiczowa and H.E. Wright.   © 1996 John Wiley & Sons Ltd.

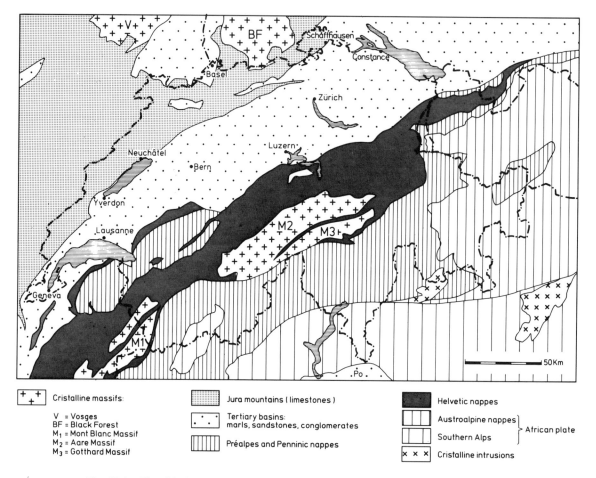

Fig. 18.1   Simplified geological map of Switzerland (modified after Heitzmann 1991)

NASP members 1994) whereas a synthesis along a cross-section (Lang 1985, Burga 1988, Ammann *et al.* 1993) covers the environmental history along an altitudinal gradient crossing the Alps. Moreover, a synthesis of the Swiss pollen-analytical data is in preparation (Burga, in prep.).

TYPE REGION CH-b, THE SWISS PLATEAU

The type region CH-b comprises the 300-km-long Tertiary trough between Lake Geneva and Lake Constance, delimited by the Jura Mountains in the north and the Alps in the south (Fig. 18.4). Its width increases from 10 km in the southwestern part to 70 km in the northeast. In the southwest the drainage is

towards Lake Geneva and the Rhône River, whereas the rest of the Swiss Plateau drains into the Rhine River. During the last glaciation the major part of the area was covered by ice of different alpine glaciers.

The type region CH-b can be subdivided into a narrow hilly southwestern subregion (CH-bs) and a broad, less hilly northeastern subregion (CH-bn, Fig. 18.4). This subdivision is further justified by slightly different vegetation histories.

Altitude: CH-bn 300–600 m above sea level, average 400–500 m (hills up to 1400 m); CH-bs 350–850 m above sea level, average 600 m (hills up to 1500 m).

Climate: Annual mean temperature 7–12°C, mean January −2 to 0°C, mean July 17–18°C. Precipit-

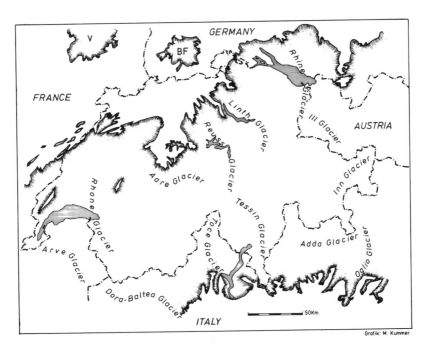

Fig. 18.2 Map of the ice extension during the last two glaciations (modified after Jäckli *et al.* 1970). For simplification the large number of nunataks in the Alps is not shown. The piedmont lakes were all covered with ice but are given here for orientation

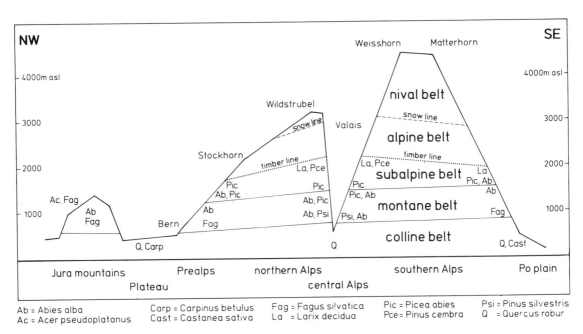

Fig. 18.3 Vegetational belts on a cross-section from the Jura Mountains (NW) to the Po Plain (SE), modified after Hess *et al.* 1967

ation 900–1200 mm yr^{-1} (approx. 10% as snow). The driest climate is found in two areas of the central part of the Plateau, between Lausanne and Yverdon, and between Schaffhausen and Constance (Fig. 18.1). Higher precipitation is found along the Prealps. Humid and temperate lowland climate, transitional between oceanic and continental.

Geology: Tertiary Molasse (sandstone, marl, conglomerate), mostly covered by Quaternary till and moraines rich in carbonates.

Topography: Northern part flat, southern part and northern alpine foreland hilly.

Population: 100 to 500 inhabitants km^{-2} in the countryside, 500 to 5000 inhabitants km^{-2} in urban areas.

Vegetation: Potential natural vegetation mainly *Fagus* and *Quercus–Carpinus* forests, at higher elevations *Fagus–Abies* forests. On wet and nutrient-rich soils *Acer–Fraxinus* forests. This vegetation has been replaced by arable land and meadows, and the modern forests consist mainly of *Picea* plantations. Mires and riparian forests, with a few exceptions, have been destroyed. The lakes are carbonate-rich and eutrophic.

Soils: Mainly fertile brown earths. In hilly areas

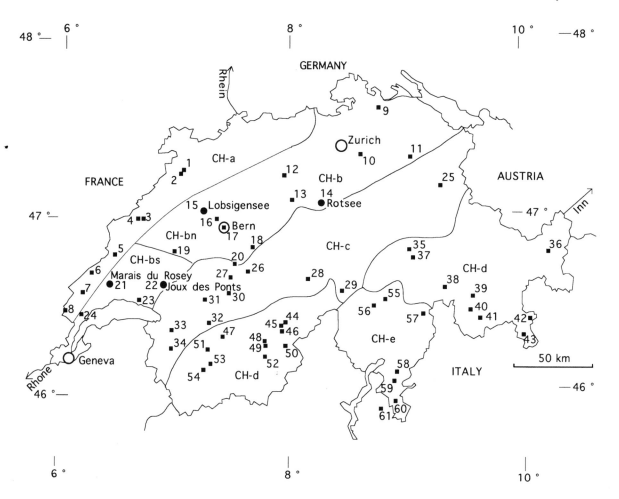

Fig. 18.4 Map of Switzerland showing the three main type regions: the Jura (CH-a), the Plateau (CH-b), and the Alps (northern, central and southern Alps referred to as CH-c, CH-d, and CH-e, see Lang 1985b). The southern tip of the Ticino touches the Po Plain (CH-f). Reference sites indicated with site numbers according to Appendix list

with higher precipitation tendency to podzolization.

Land use: Intensive agriculture, industrial areas, roads, and towns cover most of the area. All forests are managed.

Four reference sites were selected for this presentation (see Appendix for coordinates and elevations):

(1) Lobsigensee (primary reference site, Ammann & Tobolski 1983, Ammann *et al.* 1985, Ammann 1989), kettle-hole lake formed by the Rhône Glacier. Ten Late-Glacial and 20 Holocene local pollen-assemblage zones (Tables 18.1 and 18.2).
(2) Rotsee (Lotter 1988), lake formed by the Reuss Glacier. Ten Late-Glacial and 16 Holocene local pollen-assemblage zones (Tables 18.1 and 18.2).
(3) Joux des Ponts (Gaillard 1984, 1985), pasture land, formerly peat bog (8500–0 BP) and lake (15000–8500 BP). The site was formed by the Rhône Glacier. Seven Late-Glacial and three Holocene local pollen-assemblage zones (Tables 18.1 and 18.2).
(4) Marais du Rosey (Gaillard 1984, 1985). Pasture land, formerly peat bog (7000–0 BP) and lake (15000–7000 BP). The site was formed by the Rhône Glacier. Seven Late-Glacial and four Holocene local pollen-assemblage zones (Tables 18.1 and 18.2).

Pollen diagrams

Pollen diagrams from four sites are presented for type region CH-b. The pollen diagrams of Lobsigensee (primary reference site, Figs 18.5 and 18.8) represent the typical Late-Glacial (LQ-150 littoral) and Holocene (LQ-120 profundal) pollen succession for the Swiss Plateau. In addition to the local paz (for details see Ammann 1989) and the Firbas zones (1949, 1954), the regional pollen-assemblage zones are given.

The Late-Glacial

The chronology of the Late-Glacial vegetation development of the Swiss Plateau is based primarily upon the well-dated Late-Glacial sequences of Lobsigensee (Fig. 18.5) and Rotsee RL-300 (Fig.

18.6) consisting of over 90 AMS ^{14}C dates on terrestrial plant remains. (Lotter 1988, Ammann 1989, Ammann & Lotter 1989, Lotter 1991). The pollen diagram from Joux des Ponts (Fig. 18.7) illustrates the typical Late-Glacial pollen succession for sites at higher altitude and close to the Swiss Jura or the Northern Alps. The Late-Glacial part of the pollen diagram from Marais du Rosey (Table 18.1, Gaillard 1985) is not presented here as it is very similar to the Lobsigensee sequence.

Correlation of the Late-Glacial local pollen-assemblage zones for the four reference sites and the proposed regional pollen-assemblage zones (paz) for type region CH-b is presented in Table 18.1.

Regional paz	Approximate age BP	Vegetation phases
CHb-1	?–12600	*Artemisia* paz
CHb-1a	?	*Pinus* sub paz, lower and upper boundaries undated, not reached in all cores, rich in reworked and long-distance transported grains; sandy clay over till. Tree-less, sparse pioneer herb vegetation
CHb-1b	–13300	*Helianthemum* sub paz, tree-less species-rich "steppe tundra"
CHb-1c	13300–12600	*Betula nana* sub paz, shrub tundra
CHb-2	12600–12500	*Juniperus–Hippophaë* paz, juniper–scrub phase of the reforestation. At some sites tree-birch expands
CHb-3	12500–12000	*Betula* paz, tree-birch woodland
CHb-3a	12500–12100	*Salix* sub paz, tree-birch wood land with willow shrubs

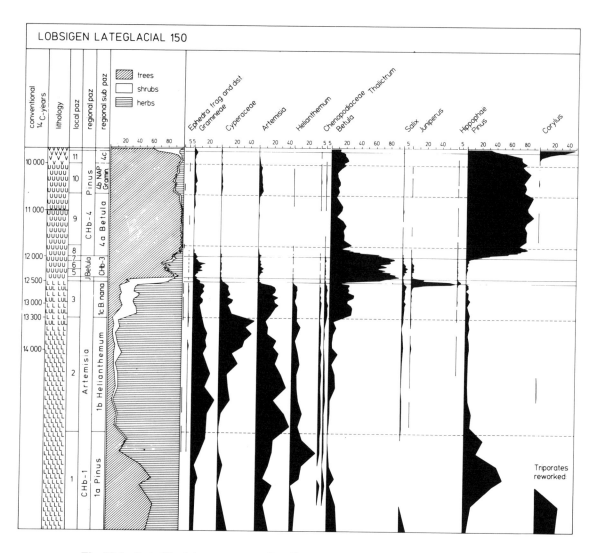

Fig. 18.5　Late-Glacial percentage pollen diagram from Lobsigensee (Ammann 1989)

Regional paz	Approximate age BP	Vegetation phases
CHb-3b	12100–12000	Gramineae–*Artemisia* sub paz, tree-birch woodland (opening? see discussion in Gaillard 1984, 1985, Ammann 1989, Lotter *et al.* 1992, Ammann *et al.* 1994)

Regional paz	Approximate age BP	Vegetation phases
CHb-4	12000–9500	*Pinus* paz
CHb-4a	12000–10700	*Betula* sub paz, forests of birch and pine
CHb-4b	10700–10000	Gramineae–NAP sub paz, pine forests with increased number of heliophilous taxa

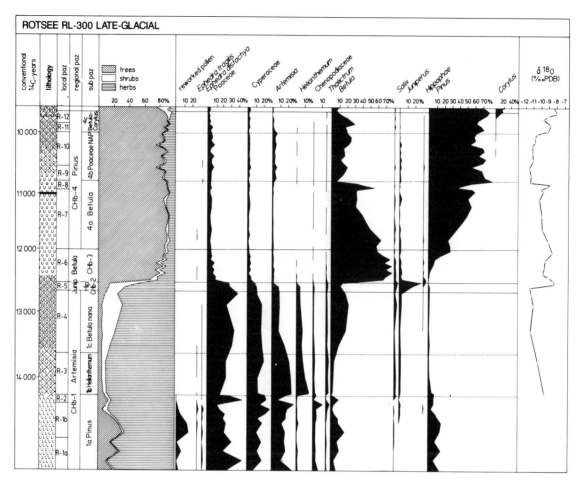

Fig. 18.6 Late-Glacial percentage pollen diagram from Rotsee (Lotter 1988)

General patterns

(1) The glacial clay sediments commonly contain many reworked pollen types (e.g. *Abies*, *Picea*, *Quercus*, *Ulmus*, *Tilia*). The high percentages of *Betula* and *Pinus* pollen during paz CHb-1a are due to long-distance transport and/or reworking.

(2) Heliophilous herbs (*Artemisia*, *Helianthemum*, Chenopodiaceae) prevail before 12600 BP (CHb-1).

(3) At about 13500 BP a shrub-phase starts with *Betula nana*, *Salix*, and *Juniperus* and lasts until

ca. 12600 BP (CHb-1c), when a sharp rise in *Juniperus* and *Hippophaë* pollen indicates the onset of reforestation (CHb-2).

(4) With the expansion of *Betula* (tree-birch) most of the heliophilous herbs decrease rapidly (CHb-2 and CHb-3). *Pinus* migrates into the Swiss Plateau at 12000 BP (CHb-4a), in the northeastern parts even earlier.

(5) The tephra layer from the last eruption (11000 BP) of the Laacher See volcano is found in most lacustrine deposits.

(6) Increased values of heliophilous herbs, mainly of *Artemisia* and Gramineae, mark the onset of

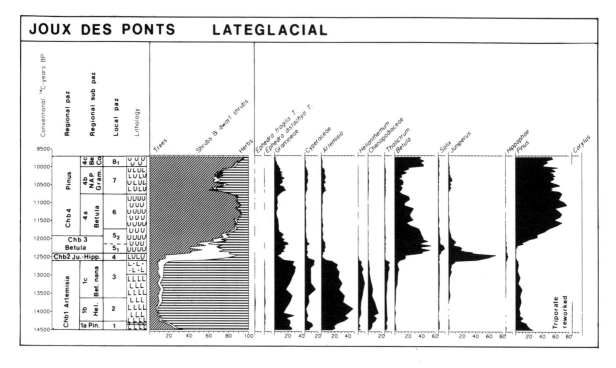

Fig. 18. 7 Late-Glacial percentage pollen diagram from Joux des Ponts (Gaillard 1984, 1985)

a regressive phase in the vegetation development, lasting from *ca.* 10700 BP to *ca.* 10000 BP (CHb-4b).

(7) Three characteristic shifts in the oxygen-isotope ratio ($\delta^{18}O$), reflecting major temperature changes, can be detected in most of the carbonate-rich lake sediments (see Eicher & Siegenthaler 1976, Eicher 1987, Kaiser & Eicher 1987, Lotter *et al.* 1992, and Fig. 18.6).

The Holocene

Holocene pollen diagrams from the Swiss Plateau are fairly similar to each other and to Central European sequences before about 5000 BP, but they may differ after this date due to a variable degree of human impact on the flora and vegetation and/or due to different migration routes of individual species.

Radiocarbon dating is hampered at many sites by hard-water errors (see e.g. Lotter 1988, table 7, and Ammann 1989, table 19). The chronology of the three selected pollen diagrams is based upon the available ^{14}C dates from bulk samples at each site. At Lobsigensee nine ^{14}C dates show a linear age–depth relationship, but it may be affected by a hard-water error (Ammann 1989, Fig. 23). At Rotsee, 14 ^{14}C dates show a linear age–depth relationship for the period of 9000 to *ca.* 2000 BP (Lotter 1988, fig. 21). The younger dates seem to be too old as a result of inwash of old organic carbon by soil erosion. The chronology of Marais du Rosey relies on seven ^{14}C dates from slightly calcareous gyttja (9800, 9400, 8300, and 7800 BP) and peat (5300, 4700, and 4300 BP). The four oldest dates may be affected by hard-water errors. By interpolation of these dates, the Firbas zone boundary VI/VII is dated to *ca.* 6000

Table 18.1 Late-Glacial local and regional pollen zones for the Swiss Plateau

Local pollen-assemblage zones				Conventioanl ^{14}C age BP	Regional pollen zones			
La Joux des Pont	Le Marais du Rosey	Lobsigensee	Rotsee		Zones	Subzones	Zones	Subzones
Jx P1	Gr1	L1	R1a+b	?15000	*Artemisia*	*Pinus*	CHb-1	1a
Jx P2	Gr2	L2a+b	R2, R3	-13300		*Helianthemum*		1b
Jx P3	Gr3	L3a+b	R4	13300–12600		*Betula nana*		1c
Jx P4	Gr4	L4	R5	12600–12500	*Juniperus–Hippophaë*		CHb-2	
Jx P5$_1$+5$_2$	Gr5$_1$+5$_2$	L5–L7	R6	12500–12000	*Betula–Gramineae*	*Salix*	CHb-3	3a
						Gramineae–Artemisia		3b
Jx P6	Gr6	L8, L9	R7, R8	12000–10700	*Pinus*	*Betula*	CHb-4	4a
Jz P7	Gr7	L10	R9, R10	10700–10000		*Gramineae–NAP*		4b
Jx P8$_{1+8_2}$	Gr8	L11	R11,R12	10000–9500		*Betula–Corylus*		4c

BP, which is in good agreement with the chronology of Lobsigensee.

In Switzerland the Firbas zone system (Firbas 1949, 1954) has been used for a long time and is still used as a chronozone system. However, the Holocene zone boundaries are still badly dated, and large discrepancies in the ages obtained are observed, especially for the boundaries V/VI (CHb-5/CHb-6), VI/VII (CHb-6/CHb-7), and VII/VIII (see discussion in , for example, Gaillard 1984, Ammann 1989, Richoz *et al.* 1994). These age differences may suggest that the Firbas zones are not synchronous on the Swiss Plateau, or that the zone boundaries are difficult to define in a consistent way. Moreover, as the Firbas zones are basically pollen zones, they cannot be expected to be necessarily synchronous on a regional scale. Further efforts in dating the Holocene pollen zones by AMS dates of terrestrial plant macroremains will be essential for the establishment of reliable Holocene chronologies for the Swiss Plateau (Hajdas *et al.* 1993).

The major difference in the pollen stratigraphy from Rotsee (Fig.18.9) in comparison with other diagrams of the Plateau is the dominance of *Abies* over *Fagus* during the middle Holocene (local paz R-17 to R-19). This is a characteristic feature found in many pollen records from sites on the northeastern Swiss Plateau situated close to the Prealps. The pollen profile from Marais du Rosey (Fig. 18.10) illustrates the expansion and dominance of *Abies* during CHb-7 (Gr 10$_2$), which is typical for the southwestern part of the Swiss Plateau as well as the relative importance of *Quercus* and *Hedera* during CHb-5 (Gr 9), a characteristic feature of the central, driest part of the Swiss Plateau.

In the example given from Lobsigensee (Fig.18.8) the local pollen-assemblage zone L19 corresponds to the settlement of the Neolithic Cortaillod culture at the lake shore, and L23 to the Roman period. During the High Middle Ages (L26) the lake was used for *Cannabis* retting.

For the Bronze and Iron Ages, as well as for the post-Roman periods, the subdivision of the pollen diagrams presented is often rather difficult. Therefore, further research in collaboration with archaeologists and historians is needed. Human impact

phases from Early Neolithic until today are described in terms of pollen-stratigraphical data from other sites of the Swiss Plateau (Rösch 1985, 1991, Richoz & Gaillard 1989, Hadorn 1992, Richoz *et al.* 1994).

Correlation of the Holocene local pollen assemblage zones for the three reference sites, and the proposed regional pollen assemblage zones for type region CH-b are presented in Table 18.2.

Regional paz	Approximate age BP	Vegetation phases
CHb-4c	10000–9500	*Betula–Corylus* sub paz, pine and birch forests; *Corylus, Ulmus,* and *Quercus* immigrate during the second part of the subzone
CHb-5	9500–8000	*Corylus*–Quercetum mixtum paz
CHb-5a	9500–9000	*Ulmus–Quercus* sub paz, *Corylus* scrub with *Quercus* and *Ulmus*
CHb-5b	9000–8000	*Fraxinus–Hedera* sub paz, *Hedera* and *Fraxinus* expansion
CHb-6	8000–6500	Quercetum mixtum–*Corylus* paz
CHb-6a	8000–7500	*Corylus* sub paz, expansion of mixed *Quercus* forests
CHb-6b	7500–6500	*Fagus–Abies* sub paz, mixed *Quercus* forests, arrival of *Fagus* and *Abies*
CHb-7	6500–1000	*Fagus–Abies–Alnus* paz

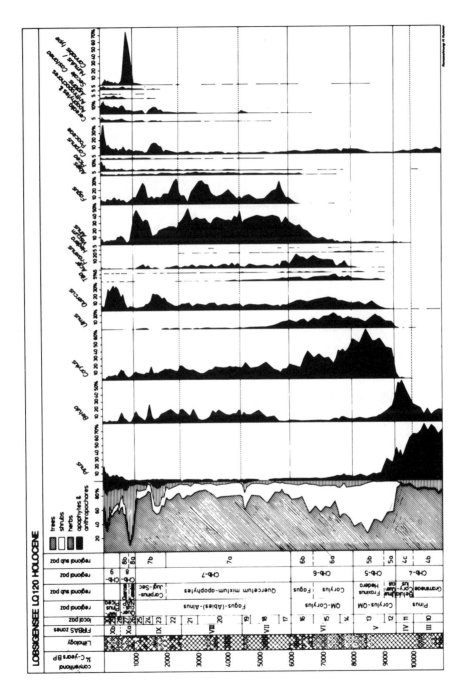

Fig. 18.8 Holocene percentage pollen diagram from Lobsigensee (Ammann 1989). Note that *Humulus/Cannabis* is included in the calculation sum (terriphytic spermatophytes AP+NAP)

658

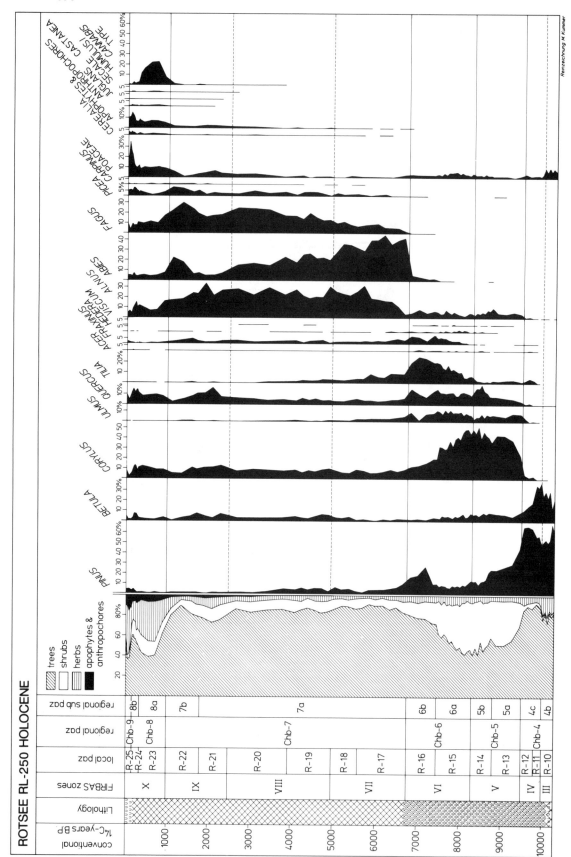

Fig. 18.9   Holocene percentage pollen diagram from Rotsee (Lotter 1988). Note that *Humulus/Cannabis* is included in the calculation sum (terriphytic spermatophytes AP+NAP)

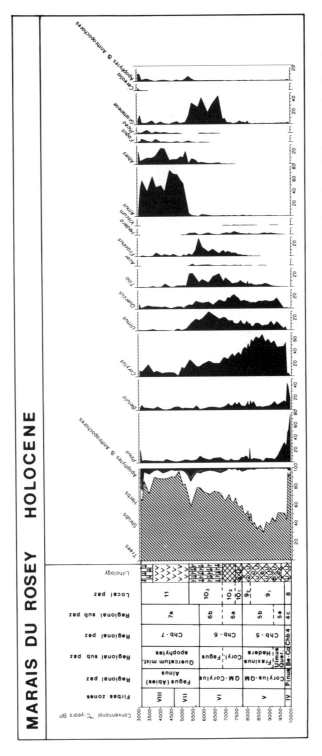

Fig. 18.10 Holocene percentage pollen diagram from Marais du Rosey (Gaillard 1984, 1985). To avoid distortion of all curves, the Gramineae were excluded from the calculation sum, as their high values during biozone VI are related to the overgrowing of the small lake by *Phragmites*

660

Table 18.2 Holocene local and regional pollen zones for the Swiss Plateau

Local pollen-assemblage zones			Conventional ^{14}C ages BP	Regional pollen zones			
Marais du Rosey	Lobsigensee	Rotsee		Zones	Subzones	Zones	Subzones
Gr8	L11	R11, R12	10000–ca.9500	Pinus	Betula–Corylus	CHb-4	4c
Gr9₁ + 9₂	L12	R13, R14	ca. 9500–ca. 8000 (8300)	Corylus–QM	Ulmus–Quercus Fraxinus–Hedera	CHb-5	5a 5b
Gr10₁ + 10₂ Gr10₃	L14(L15pp) L16 (L15pp)	R15 R16	ca.8000 (8300)– ca 6000 (6800)	QM–Corylus	Corylus Fagus–Abies	CHb-6	6a 6b
Gr11	L17–L21 L22–L25	R17–R20 R21,R22	ca. 6000 (6800)– ca. 1000	Fagus (Abies)– Alnus	QM–apophytes Carpinus– Juglans–Secale	CHb-7	7a 7b
	L26	R23	ca.1000–ca.600	Quercus–NAP	Humulus– Cannabis Quercus	CHb-8	8a
	L27	R24					8b
	L28–L30	R25,R26	ca. 600–0	NAP–Pinus– (Picea)		CHb-9	

Regional paz	Approximate age BP	Vegetation phases
CHb-7a	6500–2000	QM–apophyte sub paz, long sequence of *Fagus*–dominated forests (near the Prealps more or less rich in *Abies*), with many local variations, partly controlled by human impact (local deforestations since 6000–5000BP)
CHb-7b	2000–1000	*Carpinus–Juglans–Secale* sub paz, records of *Juglans, Castanea*, and *Secale* and higher frequencies of *Carpinus*
CHb-8	1000–600	*Quercus*–NAP paz
CHb-8a	1000– ?	*Humulus–Cannabis* sub paz, local hemp retting
CHb-8b	?–600	*Quercus* sub paz, *Quercus*–forests used as wood pasture
CHb-9	600–0	NAP–*Pinus*–(*Picea*) paz, deforestation and plantation of *Pinus* and *Picea* are widespread but vary in their local importance

General patterns

(1) The transition from the Younger Dryas (CHb-4b = Firbas zone III) to the Preboreal (CHb-4c = Firbas zone IV) is characterized by an expansion of *Betula* and the decrease of *Pinus* and NAP (especially *Artemisia* ≤2%). Note that both *Pinus*-pollen concentrations and influx increase (see e.g. Gaillard 1985, Ammann 1989). Moreover, at some sites, the transition CHb-4b/CHb-4c is characterized by an increase in *Pinus* percentages, while *Betula* is not, or only slightly increasing (e.g. Gaillard 1984, 1985, Ammann 1989).

(2) During the Preboreal (CHb-4c), a *Betula* peak (first or second half of this biozone) occurs, as well as the empirical limits of *Corylus*, *Quercus*, *Ulmus*, and *Alnus*.

(3) The rise in *Corylus*, generally used as the limit between the Preboreal (CHb-4c) and the Boreal (CHb-5a = Firbas zone V) biozones, is extremely rapid, and its radiocarbon age varies considerably (9300–8600 BP, see Gaillard 1984). The date 9000 BP, traditionally given to the transition between the chronozones Preboreal and Boreal, is found at many sites within the Boreal biozone (= dominance of *Corylus*), where it often coincides with the empirical limits of *Hedera* and/or *Tilia*, *Fraxinus*, and *Viscum* (CHb-5b).

(4) Around 8000 BP the *Corylus* values decrease and the elements of the QM (= Quercetum mixtum) increase, i.e. *Ulmus*, *Quercus*, *Tilia*, *Acer*, and *Fraxinus* (CHb-6a).

(5) During the biozones of the Atlantic (CHb-6 and CHb-7a pp.= Firbas zones VI and VII), dominated by mixed oak forests, the values of *Ulmus*, *Quercus*, and *Tilia* decrease in several small steps, whereas *Alnus* increases. During the Younger Atlantic (VII) the main elm decline is recorded, formerly used as the onset of the Subboreal biozone (VIII in Firbas 1949). However, the elm decline of the Swiss Plateau cannot be correlated with the Northwest European elm decline, as it is a gradual and stepwise decline between *ca.* 6000 and 4500 BP (Ammann 1989). The age discrepancies between Rotsee and the two other sites are discussed above. The possible causes behind this stepwise decline are discussed by, for example, Heitz-Weniger (1976), Ammann (1989), Richoz *et al.* 1994.

(6) *Fagus* expanded at many sites at or before 6000 BP. Later it is affected by phases of deforestation (see Müller 1962, Ammann 1988, Richoz & Gaillard 1989, Clark *et al.* 1989, Richoz *et al.* 1994).

Arrival and expansion of main taxa on the Central Swiss Plateau

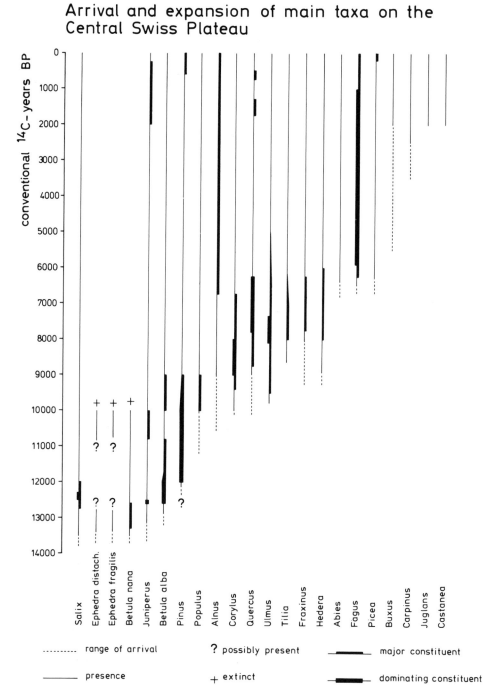

Fig. 18.11 Arrival and expansion of main taxa on the Swiss Plateau. Note that the trends shown for *Fagus* and *Abies* are valid for the central Swiss Plateau. *Abies* is a major or dominant constituent from *ca.* 7000 BP to 4500 BP at sites closer to the Prealps in the northeast, and from *ca.* 5500 BP to 3000 BP in the southwestern part of the Swiss Plateau up to the lake of Neuchâtel

EVENT STRATIGRAPY: TYPE REGION CH-B - SWISS PLATEAU

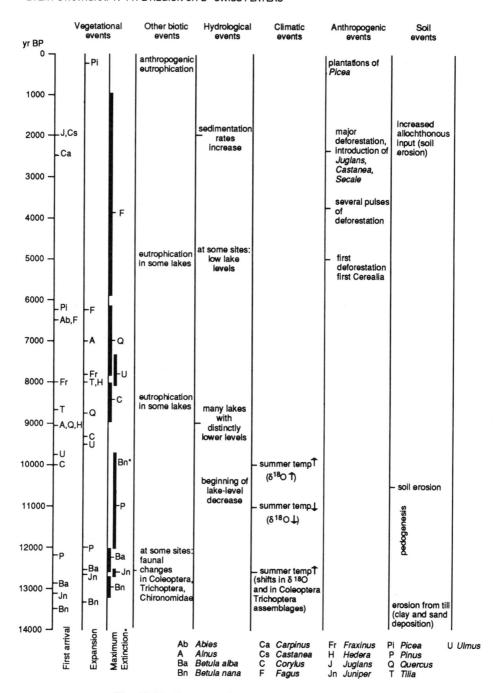

Fig. 18.12 Event stratigraphy for the Swiss Plateau

(7) At sites closer to the Prealps (Rotsee, Fig. 18.9) as well as in the southwestern part of the Swiss Plateau (Marais du Rosey, Fig. 18.10) up to the Lake of Neuchâtel, pollen of *Abies* plays a major role and reflects the expansion of fir between *ca.* 7000 and 6000 BP (Rotsee), or between *ca.* 5500 and 4500 BP (Marais du Rosey).

(8) Neolithic traces such as grains of Cerealia, *Plantago lanceolata*, *Urtica*, Chenopodiaceae and other NAP are recorded in many diagrams. The Cortaillod culture (*ca.* 5060–4680 BP) is particularly distinct in the pollen diagram from Lobsigensee (Fig. 18.8). Late Neolithic and Bronze Age phases of human activity are well recorded in other pollen diagrams of the Swiss Plateau (e.g. Rösch 1985, 1991, Richoz & Gaillard 1989, Hadorn 1992, Richoz *et al.* 1994).

(9) The beginning of Roman colonization (*ca.* 2000 BP) is indicated by the absolute limits of *Juglans*, *Castanea*, and *Secale* as well as by a pronounced increase in NAP.

Palaeoecological events

Tree arrival

Fig. 18.11 summarizes the arrival times of major taxa during the Late-Glacial and Holocene for the Swiss Plateau. The time of arrival depends mainly on the geographical location of the site. Location of the refugia, migration routes, and rate of spread account for most of the differences in the spread of some trees. *Pinus*, for example, increases earlier in the northeastern parts of the Swiss Plateau, and *Abies* seems to migrate and spread faster along the Prealps, along the southwestern part of the Jura, and in the southernmost area of the Swiss Plateau. During the last 2000 years human impact accounts for much of the spread of, for example, *Juglans* and *Castanea*.

Event stratigraphy

The major events of environmental history on the Swiss Plateau are summarized in Fig. 18.12. The first

three columns show vegetational changes, i.e. the arrival, expansion, and maxima of major taxa (for *Betula nana* also the extinction). Biotic events such as faunal changes or eutrophication are studied at few sites only. Among the hydrological events three periods of decreasing lake-level are often found: 11000–10000 BP, 10000–9000 BP, and 6000–5000 BP (e.g. Ammann 1989, Gaillard & Moulin 1989).

Climatic events are of a large amplitude in the Late-Glacial and at the transition to the Holocene but small during the Holocene. Anthropogenic events consist of deforestation, introduction of crops (Cerealia by the Neolithic people, *Juglans, Castanea,* and *Secale* by the Romans), increase of apophytes and modern spruce plantations. Soil events are mainly related to hydrologic, climatic, and anthropogenic changes.

REFERENCES

Ammann, B. 1988: Palynological evidence of prehistoric anthropogenic forest changes on the Swiss Plateau. *In* Birks, H.H., Birks, H.J.B., Kaland, P.E. and Moe, D. (eds) *The cultural landscape*, 289–299. Cambridge University Press, Cambridge.

Ammann, B. 1989: Late-Quaternary palynology at Lobsigensee. Regional vegetation history and local lake development. *Dissertationes Botanicae 137*, 157 pp.

Ammann, B. and Lotter, A.F 1989: Late-Glacial radiocarbon- and palynostratigraphy on the Swiss Plateau. *Boreas 18*, 109–126.

Ammann, B. & Tobolski, K. 1983: Vegetational development during the Late-Würm at Lobsigensee (Swiss Plateau). Studies in the Late Quaternary of Lobsigensee 1. *Revue de Paléobiologie 2*, 263–180.

Ammann, B., Andrée, M., Chaix, L., Eicher, U., Elias, S.A., Hofmann, W., Oeschger, H., Siegenthaler, U., Tobolski, K., Wilkinson, B. and Züllig, H. 1985: An attempt at a paleoecological synthesis. *Dissertationes Botanicae 87*, 165–170.

Ammann, B., Birks, H.J.B., Drescher-Schneider, R., Juggins, S., Lang, G. & Lotter, A.F. 1993: Patterns of variation in late-glacial pollen stratigraphy along a North-West–South-East transect through Switzerland—a numerical analysis. *Quaternary Science Reviews 12*, 277–286.

Ammann, B., Eicher, U., Gaillard, M.-J., Haeberli, W., Lister, G., Lotter, A.F., Maisch, M., Niessen, F., Schlüchter, Ch. & Wohlfarth, B. 1994. The Würmian Late-glacial in Lowland Switzerland. *Journal of Quaternary Science 9*, 119–125.

Burga, C.A. 1980: Pollenanalytische Untersuchungen zur Vegetationsgeschichte des Schams und des San Bernardino-Passgebietes (Graubünden, Schweiz). *Dissertationes Botanicae 26*, 165 pp.

Burga, C. A. 1985: Lai da Vons—Spät- und postglaziale Umweltsveränderungen eines Sees an der Baumgrenze in den östlichen Schweizer Alpen—mit einem Beitrag von U. Eicher. *Dissertationes Botanicae 87*, 381–428.

Burga, C.A. 1987: Gletscher- und Vegetationsgeschichte der Südrätischen Alpen seit der Späteiszeit (Puschlav, Livorgno, Bormiese). *Denkschriften der Schweizerischen Naturforschenden Gesellschaft 101*, 162 pp.

Burga, C. A. 1988: Swiss vegetation history during the last 18000 years. *New Phytologist 110*, 581–602.

Burga, C.A. (in prep.) *Quartäre Paläoökologie der Schweiz. Atlas zur Entwicklung von Flora, Vegetation und Klima der letzten 300.000 Jahre.* Ott-Verlag, Thun.

Clark, J.S., Merkt, J. & Müller, H. 1989: Post-glacial fire, vegetation, and human history on the northern alpine forelands, south-western Germany. *Journal of Ecology 77*, 897–925.

Eicher, U. 1987: Die spätglazialen sowie die frühpostglazialen Klimaverhältnisse im Bereiche der Alpen: Sauerstoffisotopenkurven kalkhaltiger Sedimente. *Geographica Helvetica 42*, 99–104.

Eicher, U. & Siegenthaler, U. 1976: Palynological and oxygen isotope investigations on Late-Glacial sediment cores from Swiss lakes. *Boreas 5*, 109–117.

Firbas, F. 1949: *Spät- und nacheiszeitliche Waldgeschichte Mitteleuropas nördlich der Alpen.* Vol. 1, 488 pp. G. Fischer, Jena.

Firbas, F. 1954: Die Synchronisierung der mitteleuropäischen Pollendiagramme. *Danmarks Geologiske Undersøgelse II 80*, 12–21.

Gaillard, M.-J. 1984: Etude palynologique de l'Evolution Tardi- et Postglaciaire de la Végétation du Moyen-Pays Romand (Suisse). *Dissertationes Botanicae 77*, 322 pp.

Gaillard, M.-J. 1985: Late-glacial and Holocene environments of some ancient lakes in the Western Swiss Plateau. *Dissertationes Botanicae 87*, 273–336.

Gaillard, M.-J. & Moulin, B. 1989: New results on the Late-Glacial history and environment of the Lake of Neuchâtel (Switzerland). Sedimentological and palynological investigations at the Palaeolithic site of Hauterive-Champréveyres. *Eclogae geologiae Helveticae 82*, 203–218.

Hadorn, P. 1992: *Vegetationsgeschichtliche Studie am Nordufer des Lac de Neuchâtel.* Thesis, University of Bern, 112 pp.

Hajdas, I., Ivy, S.D., Beer, J., Bonani, G., Imboden, D., Lotter, A.F., Sturm, M. & Suter, M. 1993: AMS radiocarbon dating and varve chronology of Lake Soppensee: 6000 to 12000 ^{14}C years BP. *Climate Dynamics 9*, 107–116.

Heeb, K. & Welten, M. 1972: Moore und Vegetationsgeschichte der Schwarzenegg und des Molassevorlandes zwischen dem Aaretal unterhalb Thun und dem oberen Emmental. *Mitteilungen der Naturforschenden Gesellschaft in Bern NF 29*, 2–54.

Hegg, O., Béguin, C. & Zoller, H. 1992: *Atlas schutzwürdiger Vegetationstypen der Schweiz.* Bundesamt für Wald und Landschaft, Bern.

Heitz, Ch. 1975: Vegetationsentwicklung und Waldgrenzschwankungen im Spät- und Postglazials im Oberhalbstein (GR) mit besonderer Berücksichtigung der Fichteneinwanderung. *Beiträge Geobotanische Landesaufnahme Schweiz 55*, 63 pp.

Heitzmann, P. 1991: Europäische und Afrikanische Platte in den Schweizer Alpen verzahnt. *Spektrum der Wissenschaft, Sept. 1991*, 23–26.

Heitz-Weniger, A. 1976: Zum Problem des mittelholozänen Ulmenabfalls im Gebiet des Zürichsees (Schweiz). *Bauhinia 5(4)*, 215–229.

Hess, H.E., Landolt, E. & Hirzel, R. 1967: *Flora der Schweiz und angrenzender Gebiete.* 3 vols. Birkhäuser, Basel.

Hubschmid, F. & Lang, G. 1985: Les Embreux—Holocene environments of a mire in the Swiss Jura mountains. *Dissertationes Botanicae 87*, 115–126.

Jäckli, H., Hantke, R., Imhof, E. & Lenzinger, H. 1970: *In* Atlas der Schweiz, map no. 6.

Kaiser, K.F. & Eicher, U. 1987: Fossil pollen, molluscs, and stable isotopes in the Dättnau valley, Switzerland. *Boreas 16*, 293–303.

Kleiber, H. 1974: Pollenanalytische Untersuchungen zum Eisrückzug und zur Vegetationsgeschichte im Oberengadin I. *Botanische Jahrbücher Systematik 94*, 1–53.

Küttel, M. 1977: Pollenanalytische und geochronologische Untersuchungen zur Piottino-Schwankung (Jüngere Dryas). *Boreas 6*, 259–274.

Küttel, M. 1979: Pollenanalytische Untersuchungen zur Vegetationsgeschichte und zum Gletscherrückzug in den westlichen Schweizeralpen. *Berichte Schweizerische Botanische Gesellschaft 89*, 9–62.

Küttel, M. 1989: Züge der jungpleistozänen Vegetations- und Landschaftsgeschichte der Zentralschweiz. *Revue de Paléobiologie 8*, 525–614.

Küttel, M. 1990: Zur Vegetationsgeschichte des Gotthardgebietes. *Mitteilungen der Naturforschenden Gesellschaft Luzern 31*, 99–111.

Lang, G. 1985: Palynological research in Switzerland 1925–1985. *Dissertationes Botanicae 87*, 11–82.

Lang, G. & Tobolski, K. 1985: Hobschensee—Lateglacial and Holocene environments of a lake at the timberline in the Central Swiss Alps. *Dissertationes Botanicae 87*, 209–228.

Lotter, A. 1985: Amsoldingersee—Late Glacial and Holocene environments of a lake at the southern edge of the Swiss Plateau. *Dissertationes Botanicae 87*, 185–208.

Lotter, A. 1988: Paläoökologische und paläolimnologische Studie des Rotsees bei Luzern. Pollen-, grossrest-, diatomeen- und sedimentanalytische Untersuchungen. *Dissertationes Botanicae 124*, 187 pp.

Lotter, A. F. 1989: Evidence of annual layering in Holocene sediments of Soppensee, Switzerland. *Aquatic Sciences 51*, 19–30.

Lotter, A.F. 1990: Die Entwicklung terrestrischer und aquatischer Ökosysteme am Rotsee (Zentralschweiz) im Verlauf der letzten 15000 Jahre. *Mitteilungen der Naturforschenden Gesellschaft Luzern 31*, 81–97.

Lotter, A.F. 1991: Absolute dating of the Late-Glacial period in Switzerland using annually laminated sediments. *Quaternary Research 35*, 321–330.

Lotter, A.F., Eicher, U., Siegenthaler, U. & Birks, H.J.B. 1992: Late-glacial climatic oscillations as recorded in Swiss lake sediments. *Journal of Quaternary Science 7*, 187–204.

Markgraf, V. 1969: Moorkundliche und vegetationsgeschichtliche Untersuchungen an einem Moorsee an der Waldgrenze im Wallis. *Botanische Jahrbücher 89*, 1–63.

Matthey, F. 1971: Contribution à l'étude de l'évolution tardi- et postglaciaire de la végétation dans le Jura central. *Beiträge Geobotanische Landesaufnahme Schweiz 53*, 86 pp.

Müller, H. 1962: Pollenanalytische Untersuchung eines Quartärprofils durch die spät- und nacheiszeitlichen Ablagerungen des Schleinsees (Süd-Westdeutschland). *Geologisches Jahrbuch 79*, 493–526.

Müller, H.J. 1972: Pollenanalytische Untersuchungen zum Eisrückzug und zur Vegetationsgeschichte im Voderrein - und Lukmaniergebiet. *Flora 161*, 333–382.

NASP members: Executive Group (J.J. Lowe, B. Ammann, H.H. Birks, S. Björck, G.R. Coope, L. Cwynar, J.L. de Beaulieu, R.J. Mott, D.M. Peteet & M.J.C. Walker): 1994 Climatic changes in areas adjacent to the North Atlantic during the last glacial–interglacial transition (14–9 ka BP): a contribution to IGCP-253. *Journal of Quaternary Science 9*, 185–198.

Reille, M. 1991: L'origine de la station de pin à crochets de la tourbière de Pinet (Aude) et de quelques stations isolées de cet arbre dans les Vosges et le Jura. *Bulletin Société botanique de France 138*, 123–148.

Richoz, I. & Gaillard, M.-J. 1989: Histoire de la végétation de la région neuchâteloise de l'époque néolitique à nos jours. Analyse pollinique d'une colonne sédimentaire prélevée dans le lac de Neuchâtel (Suisse). *Bulletin de la Société Vaudoise des Sciences Naturelles 79*, 355–377.

Richoz, I., Gaillard, M.-J. & Magny, M. (1994) The influence of human activities and climate on the development of vegetation at Seedorf, southern Swiss Plateau during the Holocene: a case study. *In* Lotter, A.F. and Ammann, B. (eds) *Festschrift Gerhard Lang*. *Dissertationes Botanicae 234*, 423–445.

Rösch, M. 1985: Nussbaumer Seen–Spät- und postglaziale Umweltveränderungen einer Seengruppe im östlichen Schweizer Mittelland. *Dissertationes Botanicae 87*, 337–379.

Rösch, M. 1991: Veränderung der Wirtschaft und Umwelt während Neolithikum und Bronzezeit am Bodensee. *Bericht der Römisch-Germanischen Kommission 71*, 161–186.

Schmid, E. 1949: *Vegetationskarte der Schweiz*. Pflanzengeographische Kommission der Schweizerische Naturforschenden Gesellschaft. Hans Huber, Bern.

Schneebeli, M., Küttel, M. & Fäh, J. 1989: Die dreidimensionale Entwicklung eines Hanghochmoores im Toggenburg, Schweiz. *Vierteljahrsschrift der Naturforschenden Gesellschaft in Zürich 134(1)*, 1–32

Schneider, R. & Tobolski, K. 1983: Palynologische und stratigraphische Untersuchungen im Lago di Ganna (Varese, Italien). *Botanica Helvetica 93*, 115–122.

Wegmüller, H.P. 1976: Vegetationsgeschichtliche Untersuchungen in den Thuralpen und im Faningebiet (Kantone Appenzell, St-Gallen, Graubünden, Schweiz). *Botanische Jahrbücher Systematik 97*, 226–307.

Wegmüller, S. 1966: Ueber die spät- und postglaziale Vegetations-geschichte des südwestlichen Juras. *Beiträge Geobotanische Landesaufnahme Schweiz 48*, 143 pp.

Welten, M. 1944: Pollenanalytische, stratigraphische und geochronologische Untersuchungen aus dem Faulenseemoos bei Spiez. *Veröffentlichungen Geobotanisches Institut Rübel Zürich 21*, 201 pp.

Welten, M. 1982: Vegetationsgeschichtliche Untersuchungen in den westlichen Schweizer Alpen: Bern-Wallis. *Denkschriften Schweizerische Naturforschende Gesellschaft 95*, 1–104.

Wick, L. 1988. *Palynologische Untersuchungen zur spät- und postglazialen Vegetationsgeschichte am Greifensee bei Zürich (Mittelland)*. Diplomarbeit, Botanical Institute, University of Bern.

Wick, L. 1989: Pollenanalytische Untersuchungen zur spät- und postglazialen Vegetationsgeschichte am Luganersee (Südtessin, Schweiz). *Eclogae geologicae Helveticae 82*, 265–276.

Zoller, H. 1960: Pollenanalytische Untersuchungen zur Vegetationsgeschichte der insubrischen Schweiz. *Denkschriften der Schweizerischen Naturforschenden Gesellschaft 83*, 180 pp.

Zoller, H. & Kleiber, H. 1971: Vegetationsgeschichtliche Untersuchungen in der montanen und subalpinen Stufe der Tessintäler. *Naturforschenden Gesellschaft Basel 81*, 1–154.

Zwahlen, R. 1985: Lörmoos—Late-Glacial and Holocene environments of an ancient lake on the Central Swiss Plateau. *Dissertationes Botanicae 87*, 171–184.

19

Austria

S. Bortenschlager, K. Oeggl and N. Wahlmüller

INTRODUCTION

The mountain landscape of Austria is dominated by folded Tertiary strata. From the geological point of view, the Alps are a very complicated mountain mass. Because they consist of rocks of different kinds and ages, the local relief changes rapidly over short distances. The basic geological construction of the Alps is roughly symmetrical: limestones in the north, crystalline rocks in the central parts, and further limestones in the south. The mountains are bordered by different basins: the so-called Alpine Foreland to northeast and the Pannnian Lowland to the east. The northernmost parts of Austria border the Bohemian and Moravian Uplands, which are a secondary chain of lower mountains consisting of Palaeozoic granite and gneiss.

The present-day vegetation is influenced by three factors: the bedrock, the relief, and the climate. Eight type regions can thereby be distinguished (Fig. 19.1).

Because the landscape differs locally in the available data it is only possible to give a synthesis of the type regions A-e, A-g, and A-h, represented, respectively, by the reference sites Halleswiessee (A4), Schwemm (A10), and Grosses Überling Schattseit Moor (A11) and Dortmunder Hütte (A14), as shown on Fig. 19.1.

The type regions A-e and A-g, which are the valleys and montane woodlands of the Eastern and Western Alpine regions, respectively, are situated north of the central Alpine chain. They both form parts of the Northern Limestone Alps (Fig. 19.2). The climate is Central European in character, with a rainfall maximum during the summer. The blocking effect of the Alps to the passage of air masses is clearly evident from the pattern of precipitation, which decreases from west (>1750 mm yr^{-1}) to east (<1750 mm yr^{-1}). The vegetation cover of the deep valley bottoms at the present-day is predominantly cultivated land. The forests are generally composed of *Picea, Abies*, and *Fagus* (Abieti–Fagetum), except in the inner-alpine regions with the more continental climate (precipitation about 1000 mm yr^{-1}), where *Picea–Abies* forest (Abietetum) is predominant.

The type region A-h comprises the subalpine and alpine zones of the central chain of the Alps. The climate is more continental in character. The forest limit is formed by *Pinus mugo* in areas of basic rocks and by *Picea abies, Pinus cembra*, and *Larix decidua* in areas of acidic rocks. As a result of anthropogenic influence the forest limit lies at altitudes of 1800 to 2200 m above sea level (asl). At higher altitudes, a dwarf-shrub heath vegetation is followed by alpine heathy grassland with patches of snow-bed vegetation.

Palaeoecological Events During the Last 15000 Years: Regional Syntheses of Palaeoecological Studies of Lakes and Mires in Europe.
Edited by B.E. Berglund, H.J.B. Birks, M. Ralska-Jasiewiczowa and H.E. Wright. © 1996 John Wiley & Sons Ltd.

668

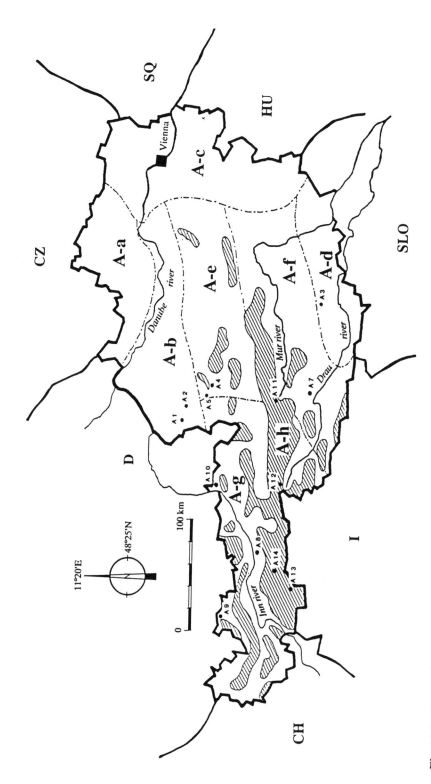

Fig. 19.1 Type regions of Austria: A-a The Bohemian and Moravian Uplands, A-b The Alpine Foreland, A-c The region of Pannonian Influence, A-d The region of Illyrian–Dinaric Influence, A-e The valleys and montane woodlands of the Eastern Alpine region, A-f The valleys and montane woodlands of the southeastern Alpine region, A-g The valleys and montane woodlands of the Western Alpine region, A-h The region above the forest limit. The hatched areas lie above the forest limit. The reference sites mentioned in the text are A4 Halleswiessee, A10 Schwemm, A11 Grosses Überling Schattseit Moor, A14 Dortmunder Hütte

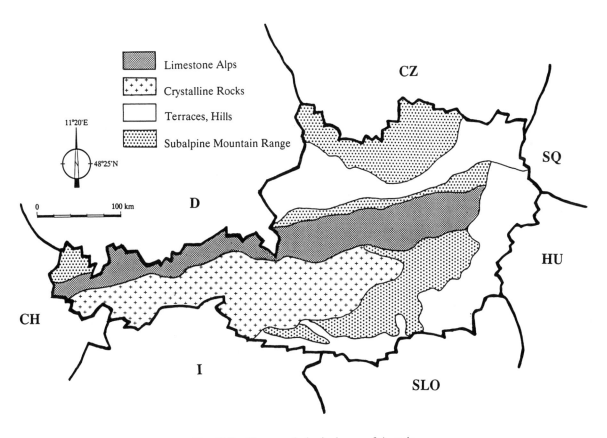

Fig. 19.2 Geomorphological map of Austria

TYPE REGION A-e, THE VALLEYS AND MONTANE WOODLANDS OF THE EASTERN ALPINE REGION
(M. Handl, Limnologisches Institut der Österreichischen Akademie der Wissenschaften Mondsee)

Altitude: 400–1400 m.

Climate: Mean temperatures: January −1.3°C, July 16°C; precipitation 1700 mm yr^{-1}.

Geology: Triassic limestones.

Topography: Karst landscape on the margins of the Northern Limestone Alps, which form mountain ranges of relatively minor elevation (Breitenberg 1412 m asl).

Population: 75 people km^{-2}, mainly in small villages.

Vegetation: Predominantly Abieti–Fagetum and Piceetum montanum woodlands.

Soils: Brown earths (Fig. 19.3).

Land use: Forestry and alpine pasturage.

Reference site 4. Halleswiessee

Latitude 47°45′N, Longitude 13°32′E. Elevation 781 m. Age range *ca.* 11000–0 BP. Mire. One pollen diagram (Fig. 19.4) presented. No local pollen-assemblage zones are defined.

Event stratigraphy

The major palaeoevents are shown in Fig. 19.6.

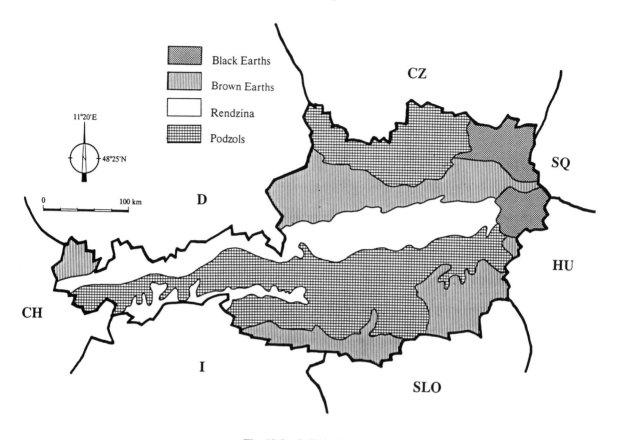

Fig. 19.3 Soils in Austria

General patterns

- The high NAP values from 608 to 595 cm belong to the Younger Dryas.
- The *Corylus* phase lasts from 8600 to 7300 BP.
- The expansion of *Abies* and *Fagus* starts between 6500 and 6000 BP.
- *Carpinus* arrives at 4000 BP.
- First traces of anthropogenic indicators are found around 4000 BP (Fig. 19.5).

TYPE REGION A-g, THE VALLEYS AND MONTANE WOODLANDS OF THE WESTERN ALPINE REGION

Altitude: 475–1400 m.

Climate: Mean temperature: January −3°C, July 18°C; 111 frost days yr⁻¹; precipitation 800–1800 mm yr⁻¹, 119 days yr⁻¹ with precipitation; snow cover 76 days yr⁻¹.

Geology: The valley bottoms are filled with alluvial and diluvial gravels. The valley slopes are formed of Jurassic and Triassic limestones in the north, gneiss and crystalline slate in the south.

Topography: Very varied. In the Northern Limestone Alps the valleys are characterized by steep slopes of the main mountain chains; in the Palaeozoic areas they form the lower slopes of secondary mountain chains.

Population: 25–200 people km⁻², mainly in towns.

Vegetation: Because the cultivable area is limited to the valleys the original vegetation there has been almost wholly destroyed. The valley bottoms are

671

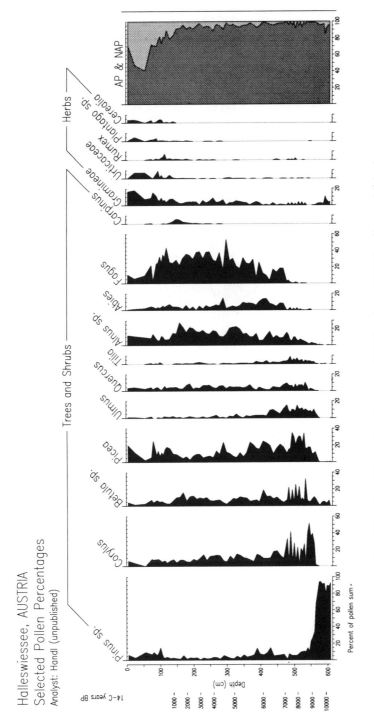

Fig. 19.4 Simplified pollen diagram of Halleswiessee (M. Handl. unpublished)

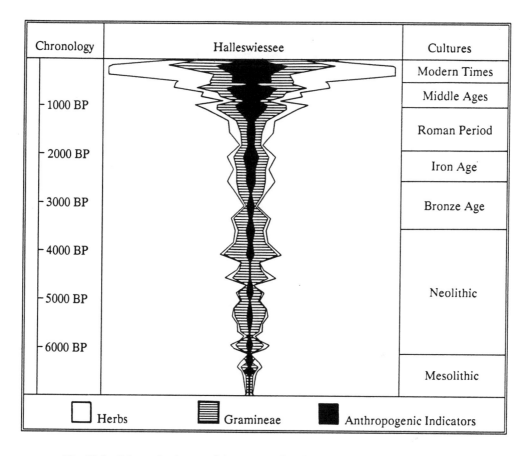

Fig. 19.5   Schematic picture of the course of anthropogenic events at Halleswiessee

intensively used for agriculture, so the vegetation is characterized by arable land and pasture. On riverbanks, remnants of Alnetum glutinosae and Salicetum albae–Populetum nigrae woodlands can be found. The valley slopes, terraces, and secondary mountain chains are occupied by various types of Abietetum and Abieti–Fagetum, with Erico–Pinetum sylvestris on sandy soils.

Soils: Brown earths, semi-podzols (Fig. 19.3).

Land use: Tillage and stock-breeding.

Reference site 10. Schwemm (Oeggl 1988a)

Latitude 47°39′N, Longitude 12°18′E. Elevation 664 m. Age range *ca.* 14000–0 BP. Mire. One general diagram is presented (Fig. 19.7). Local pollen-assemblage zones (paz) are given, whereas regional

pollen zones have not yet been defined. 12 local pollen-assemblage zones.

Local paz	Age	Pollen-assemblage zone
1.	–13300 BP	*Artemisia*–Cyperaceae–Gramineae paz
2.	13300–13000 BP	*Juniperus–Salix* paz
3.	13000–9500 BP	*Pinus–Betula* paz
4.	9500–8200 BP	*Corylus*–Quercetum mixtum paz
5.	8200–7100 BP	Quercetum mixtum paz rich in *Picea*
6.	7100–5200 BP	*Picea* paz rich in Quercetum mixtum
7.	5200–4200 BP	*Picea–Fagus–Abies* paz
8.	4200–3800 BP	*Abies–Fagus–Picea* paz

Age kyr	Type regional vegetational events			Hydrological events	Climatic events	Pedology events	Anthropogenic events
	1	**2**	**3**				
0							
1				wetter, increased humidity	Medieval times?	erosion	
2			-Car			erosion	Cerealia abundant
3				warm,dry			
4	-Car		-F	wetter, increased humidity			1. Plantago
5				dry, warm	Rotmoos?	erosion	
6		-A,F	-A	wetter, increased humidity	Frosnitz	erosion	
7	-F						
8			-Pi			erosion	
9		-Al Q,T U	-C,Q, T,U	wetter, increased humidity		erosion	
10	-Pi C,Q,T U,Al	-C					
11							

(Table title: Event stratigraphy of type region A-e: The valleys and montane woodlands of the Eastern Alpine region)

1	First arrival	A	- *Abies*	P	- *Pinus*
		Al	- *Alnus*	Pi	- *Picea*
2	Expansion time	B	- *Betula*	Q	- *Quercus*
		C	- *Corylus*	T	- *Tilia*
		Car	- *Carpinus*	U	- *Ulmus*
3	Time of maximum values	F	- *Fagus*		

Fig. 19.6 Event stratigraphy of type region A-e

674

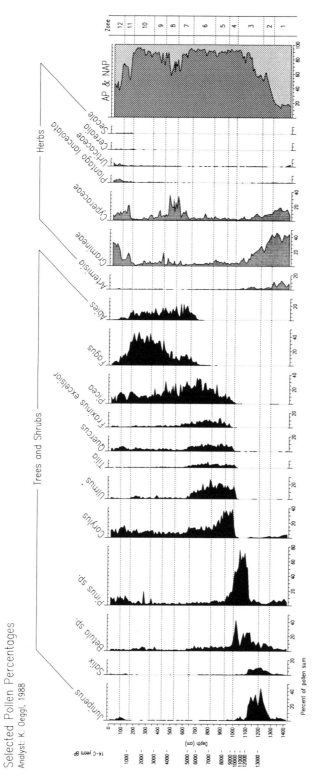

Schwemm, AUSTRIA
Selected Pollen Percentages
Analyst: K. Oeggl, 1988

Fig. 19.7 Simplified pollen diagram for the reference site of Schwemm (Oeggl 1988a)

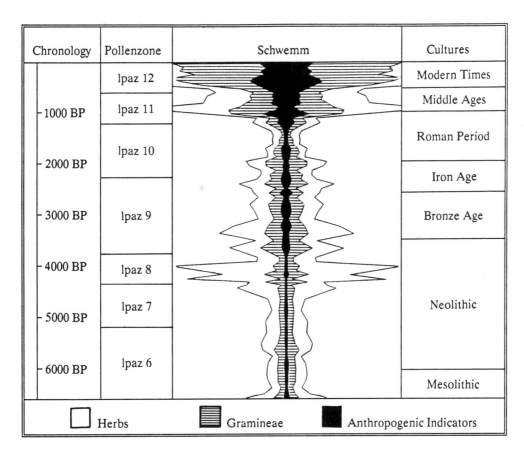

Chronology	Pollenzone	Schwemm	Cultures
	lpaz 12		Modern Times
1000 BP	lpaz 11		Middle Ages
	lpaz 10		Roman Period
2000 BP			
3000 BP	lpaz 9		Iron Age
			Bronze Age
4000 BP	lpaz 8		
5000 BP	lpaz 7		Neolithic
6000 BP	lpaz 6		
			Mesolithic

☐ Herbs ▤ Gramineae ■ Anthropogenic Indicators

Fig. 19.8 Schematic picture of the course of anthropogenic events at Schwemm

Local paz	Age	Pollen-assemblage zone
9.	3800–2500 BP	*Fagus* paz
10.	2500–1200 BP	*Fagus* paz rich in *Betula, Carpinus,* and *Ulmus*
11.	1200–500 BP	NAP paz rich in *Fagus*
12.	500–0 BP	NAP paz rich in *Picea*

Hydrological events

Following a climatic deterioration (Frosnitz fluctuation) a rise in lake-level is detectable at 6500 BP.

Climatic events

(1) The climatic deterioration of the Younger Dryas is evidenced both in the pollen diagram as an NAP peak in the *Pinus–Betula* paz and in the high negative values of the oxygen-isotope curve (Oeggl & Eicher 1989).

(2) Around 6500 BP a climatic deterioration—the Frosnitz fluctuation (Patzelt & Bortenschlager 1973)—caused vegetational changes on a regional scale.

Anthropogenic events

First traces of anthropogenic indicators are found in the Atlantic (Wahlmüller 1985a). A general picture of the course of anthropogenic events is given in Fig. 19.8. The initial occurrences of anthropogenic indicators in the pollen diagram appear during the Neolithic period. The montane woodlands show signs of being thinned and grazed from 4000 BP

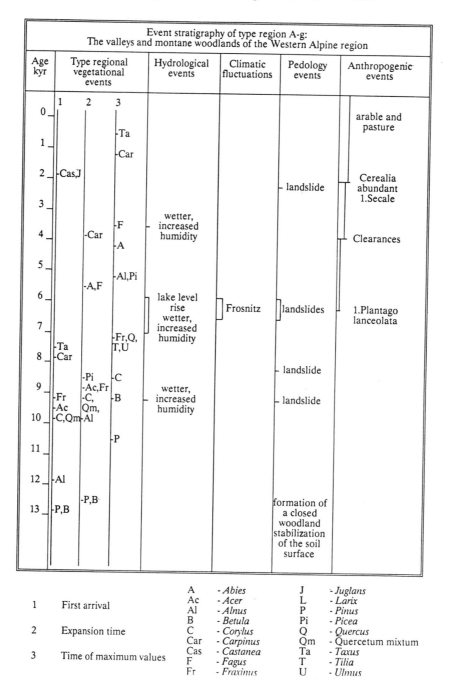

Age kyr	Type regional vegetational events			Hydrological events	Climatic fluctuations	Pedology events	Anthropogenic events
	1	2	3				
0	Cas,J		Ta				arable and pasture
1			Car				
2						landslide	Cerealia abundant 1.Secale
3				wetter, increased humidity			
4		Car	F				Clearances
			A				
5		A,F	Al,Pi				
6				lake level rise	Frosnitz	landslides	1.Plantago lanceolata
7	Ta		Fr,Q, T,U	wetter, increased humidity			
8	Car					landslide	
9	Fr	Pi Ac,Fr C,	C B	wetter, increased humidity		landslide	
10	Ac C,Qm	Qm Al				landslide	
11			P				
12	Al						
13	P,B	P,B				formation of a closed woodland stabilization of the soil surface	

1	First arrival	A	- *Abies*	J	- *Juglans*
		Ac	- *Acer*	L	- *Larix*
		Al	- *Alnus*	P	- *Pinus*
		B	- *Betula*	Pi	- *Picea*
2	Expansion time	C	- *Corylus*	Q	- *Quercus*
		Car	- *Carpinus*	Qm	- Quercetum mixtum
3	Time of maximum values	Cas	- *Castanea*	Ta	- *Taxus*
		F	- *Fagus*	T	- *Tilia*
		Fr	- *Fraxinus*	U	- *Ulmus*

Fig. 19.9   Event stratigraphy of type region A-g

onwards. The side valleys, too, were colonized at the start of the Iron Age. The major clearance of the forests and their transformation into a cultural landscape took place during the Early Mediaeval period (Oeggl 1988b).

Event stratigraphy

A short summary of the important palaeohistoric events in the A-g type region is given in Fig. 19.9.

General patterns

- formation of *Pinus* and *Betula* "*alba*" (tree birches) woodland in the Bölling (Bortenschlager 1984 a,b).
- Immigration of *Corylus*, *Quercus*, *Tilia*, *Ulmus*, and *Picea* in the Preboreal.
- From 3800 BP onwards *Fagus* predominates in the montane forests.

Unique patterns

- Occasional finds of *Cerealia* pollen, together with *Plantago lanceolata*, in the early Atlantic (Wahlmüller 1985a,b).
- Evidence of the climatic deterioration of the Younger Dryas from oxygen-isotope analyses (Oeggl & Eicher 1989).

TYPE REGION A-h, THE REGION ABOVE THE FOREST LIMIT

Altitude: 1800–3500 m.

Climate: Mean temperatures: January −6.4°C, July 9.8°C, 204 frost days yr⁻¹; precipitation 1500–2000 mm yr⁻¹, 126 days yr⁻¹ with precipitation, snow cover 163 days yr⁻¹, local snow-line (mean annual values for the equilibrium line altitude from 1962 to 1975) 2720–3180 m asl.

Geology: Limestone in the north and south; granite, gneiss and crystalline slate in the Central Alps.

Topography: The Northern Limestone Alps are characterized by steep slopes, rocky mountain tops, and cirques. The Central Alps are more gently

sloping. The highest mountain tops are covered by glaciers.

Population: Fewer than 25 people km⁻¹. This region of the Alps is not permanently inhabited.

Vegetation: Dwarf-shrub communities, such as Rhododendretum hirsuti, Ericetum carneae, Dryadetum octopetalae, and Salicetum waldsteinianae, glabrae, and retusae occur on the basic soils, followed at higher altitudes by alpine grassland communities of Caricetum ferrugineae and firmae, Seslerio–Caricetum sempervirentis, Elynetum myosuroides, and Salicetum reticulatae.

The dwarf-shrub communities Rhododenretum ferruginei, and Vaccinietum myrtilli and uliginosi are found on acid soils. The alpine grassland is here represented by Caricetum curvulae, Nardetum, and Salicetum herbaceae.

Soils: Podzols (Fig. 19.3).

Land use: Alpine pasturage, tourism.

Reference site 11. Grosses Überling Schattseit Moor (R. Krisai, Institut für Botanik der Universität Salzburg)

Latitude 47°10′N, Longitude 13°54′E. Elevation 1730 m. Age range 14000–0 BP. Mire. Simplified diagram shown in Fig. 19.10. Four local pollen-assemblage zones with five subzones are defined.

Local Age paz		Pollen-assemblage zone
1.	−13000 BP	*Artemisia–* Chenopodiaceae– NAP paz
2.	13000–8000 BP	*Pinus* paz
2a.	13000–11000 BP	*Pinus–Betula* subzone
2b.	11000–10000 BP	*Pinus–*NAP subzone
2c.	10000–8000 BP	*Pinus–Picea* subzone
3.	8000–1200 BP	*Picea* paz
3a.	8000–6500 BP	*Picea* subzone
3b.	6500–1200 BP	*Picea–Abies–Fagus* subzone
4.	1200–0 BP	*Pinus–*cultural indicators paz

678

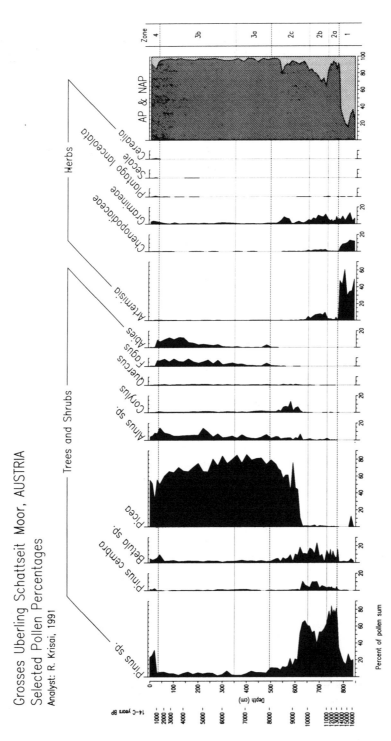

Fig. 19.10 Simplified pollen diagram for the reference site of Grosses Uberling Schattseit Moor (Krisai *et al.* 1991)

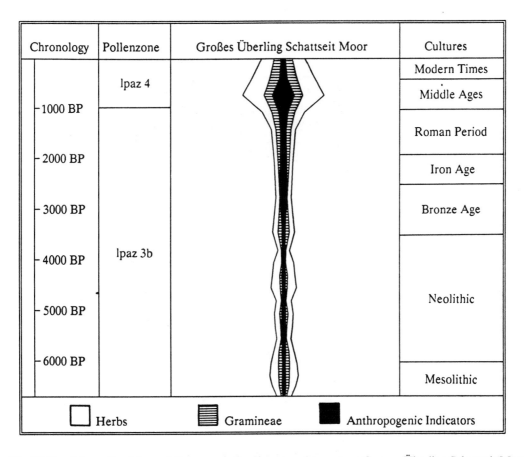

Chronology	Pollenzone	Großes Überling Schattseit Moor	Cultures

Fig. 19.11 Schematic picture of the course of anthropogenic events at Grosses Überling Schattseit Moor

Climatic events

Distinct climatic deterioration in the Younger Dryas.

Anthropogenic events

(1) First traces of human activities about 4500 BP: single cereal pollen grains, local forest clearances for grazing (Fig. 19.11).
(2) After 1200 BP, distinct changes in the bog vegetation caused by grazing. Massive clearances and changes in the forest composition due to human activities (Krisai *et al*. 1991).

Reference site 14. Dortmunder Hütte
(Hüttemann & Bortenschlager 1987)

Latitude 47°12′N, Longitude 11°1′E. Elevation 1880 m. Age range 9000–0 BP. Mire. Three local pollen-assemblage zones and seven subzones are defined (Fig. 19.12).

Local paz	Age	Pollen-assemblage zone
1.	9000–6800 BP	*Pinus* paz
1a.	9000–7800 BP	*Pinus*–NAP subzone
1b.	7800–6800 BP	*Pinus*–*Picea*–Ericaceae subzone
2.	6800–1700 BP	*Picea* paz
2a.	6800–6200 BP	*Picea*–NAP–Cyperaceae subzone
2b.	6200–4400 BP	*Picea* subzone
2c.	4400–3800 BP	*Picea*–Gramineae subzone
2d.	3800–1700 BP	*Picea*–*Abies*–*Fagus* subzone
3.	1700–0 BP	*Pinus*–anthropogenic indicators paz

680

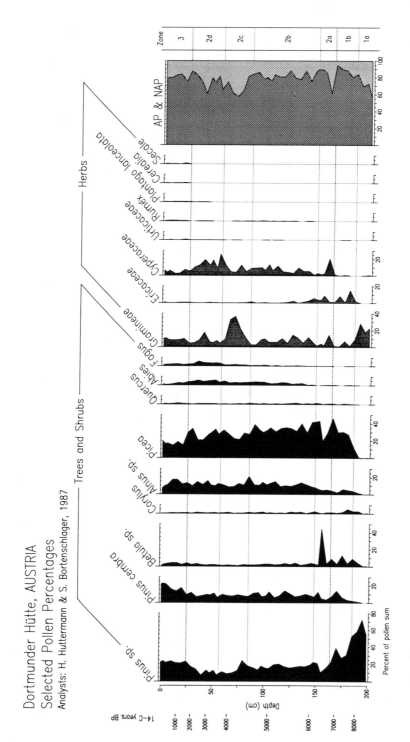

Fig. 19.12 Simplified pollen diagram for the reference site of Dortmunder Hütte (Hüttemann & Bortenschlager 1987)

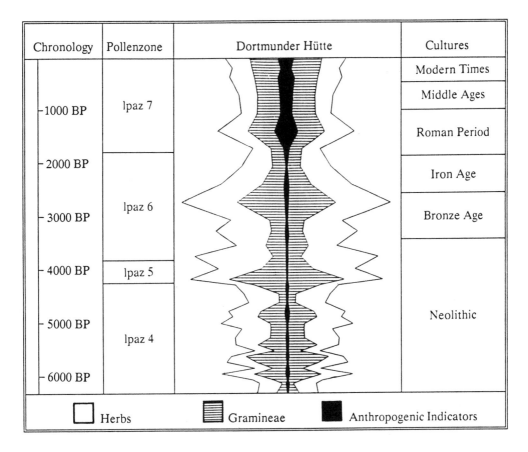

Fig. 19.13   Schematic picture of the course of anthropogenic events at Dortmunder Hütte

Climatic events

(1) Evidence for four separate deteriorations in climate (Venediger, Frosnitz, Rotmoos, Löbben).

(2) One local climatic fluctuation found that it cannot be correlated with any of the other ones known so far in the Eastern Alps.

Anthropogenic events

A general picture of the course of anthropogenic events is given in Fig. 19.13.

(1) *Pinus* increases because of the human impact in the Roman period.

(2) Anthropogenic depression of the forest limit in the Middle Ages.

Event stratigraphy

The major palaeoecological events of the type region A-h are shown in Fig. 19.14.

General patterns

– Four regional pollen-assemblage zones can be defined:

2000–0 BP	*Pinus*–cultural indicators paz
7500–2000 BP	*Picea* paz
13000–7500 BP	*Pinus* paz
–13000 BP	NAP paz

– A total of eight post-glacial climatic fluctuations have been recognized (Patzelt & Bortenschlager

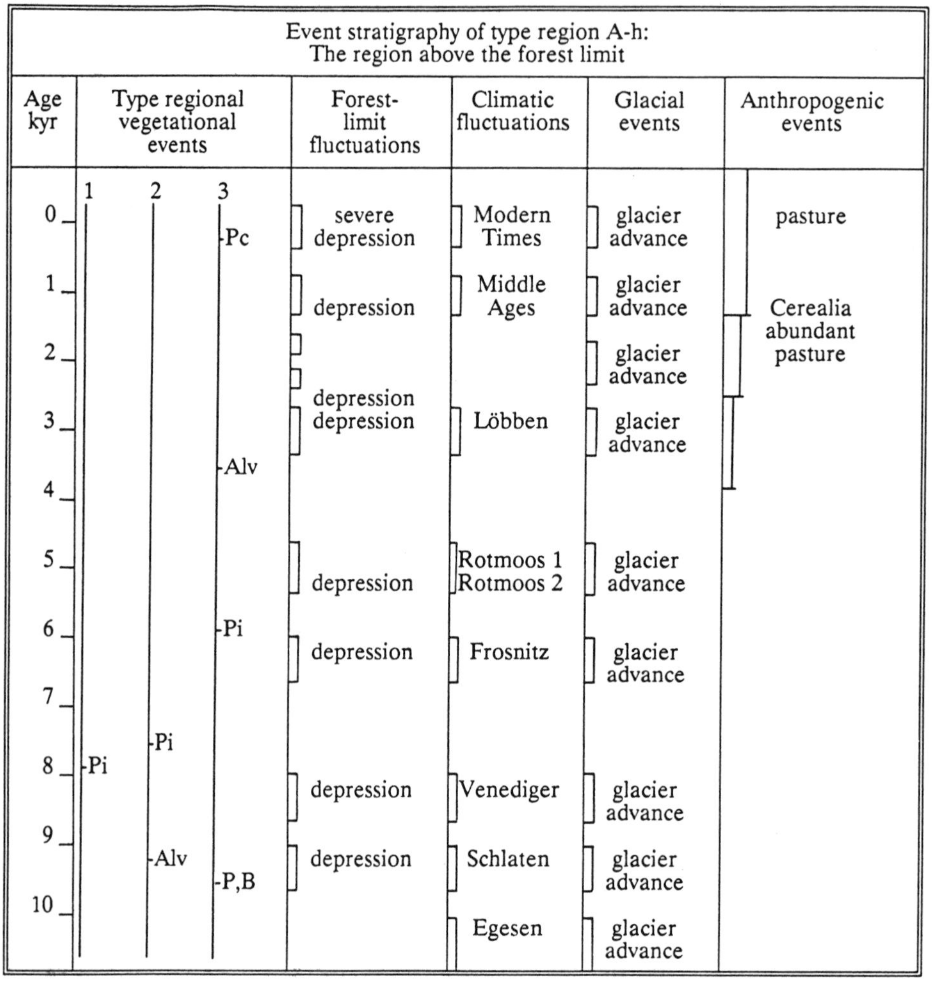

Event stratigraphy of type region A-h: The region above the forest limit					
Age kyr	Type regional vegetational events	Forest-limit fluctuations	Climatic fluctuations	Glacial events	Anthropogenic events
0	1 2 3	severe depression	Modern Times	glacier advance	pasture
	Pc				
1		depression	Middle Ages	glacier advance	Cerealia abundant pasture
2				glacier advance	
3		depression depression	Löbben	glacier advance	
4	Alv				
5		depression	Rotmoos 1 Rotmoos 2	glacier advance	
6	Pi	depression	Frosnitz	glacier advance	
7					
8	Pi Pi	depression	Venediger	glacier advance	
9	Alv	depression	Schlaten	glacier advance	
10	P,B		Egesen	glacier advance	

1   First arrival

2   Expansion time

3   Time of maximum values

Alv   - *Alnus viridis*
B   - *Betula*
P   - *Pinus*
Pc   - *Pinus cembra*
Pi   - *Picea*

Fig. 19.14 Event stratigraphy of type region A-h

1973, Bortenschlager 1982, Weirich & Bortenschlager 1980).

– Forest limit rises up to 1800 m asl in the Alleröd.
– Human impact can be demonstrated above the forest limit throughout the Eastern Alps after 4000 BP.

CONCLUSIONS

The basic vegetational succession described for the late-glacial and Holocene in Central Europe has been confirmed in the Eastern Alps. The course of events has differed locally in relation to climate, soil

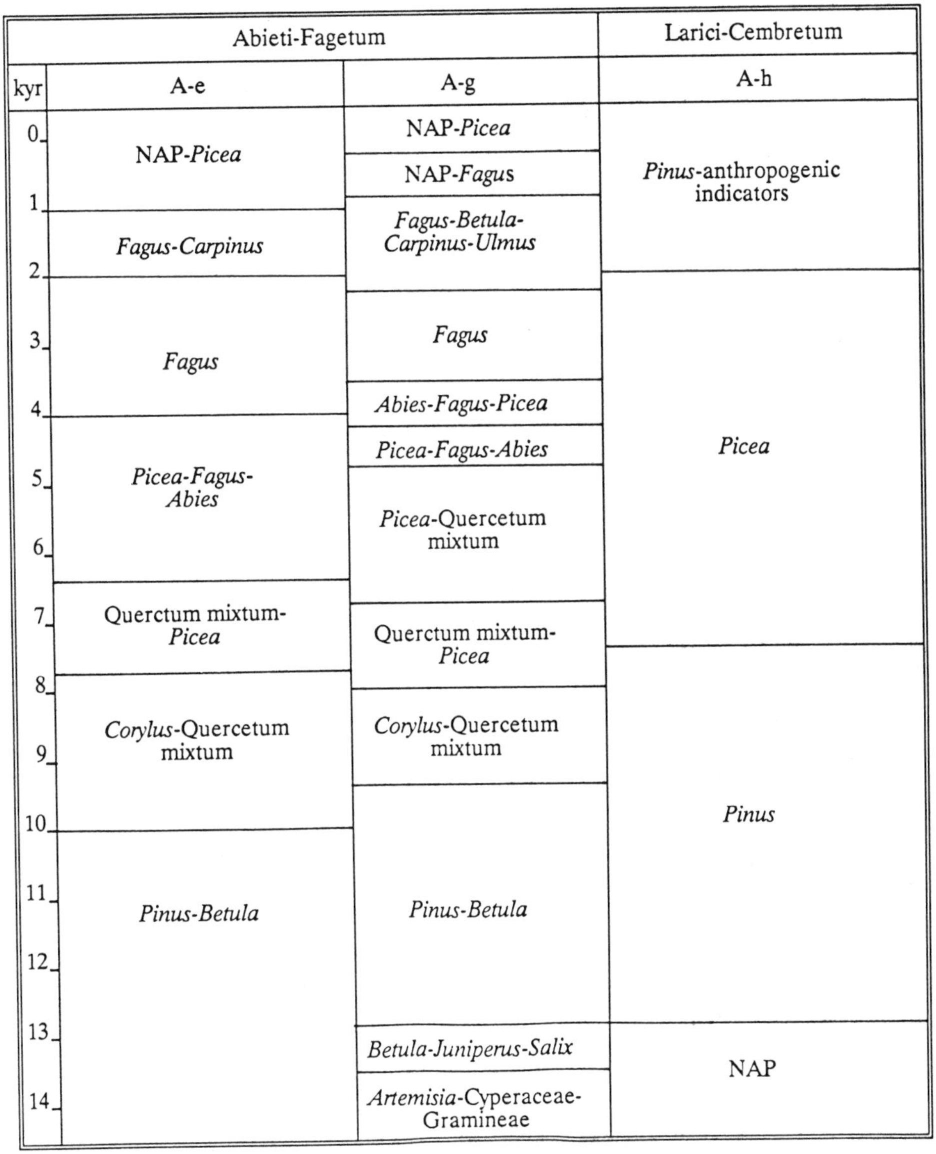

	Abieti-Fagetum		Larici-Cembretum
kyr	A-e	A-g	A-h
0	NAP-*Picea*	NAP-*Picea*	*Pinus*-anthropogenic indicators
1		NAP-*Fagus*	
	Fagus-Carpinus	*Fagus-Betula-Carpinus-Ulmus*	
2			
3	*Fagus*	*Fagus*	*Picea*
4		*Abies-Fagus-Picea*	
5	*Picea-Fagus-Abies*	*Picea-Fagus-Abies*	
		Picea-Quercetum mixtum	
6			
7	Querctum mixtum-*Picea*	Querctum mixtum-*Picea*	
8			
	Corylus-Quercetum mixtum	*Corylus*-Quercetum mixtum	
9			
10			*Pinus*
11	*Pinus-Betula*	*Pinus-Betula*	
12			
13		*Betula-Juniperus-Salix*	NAP
14		*Artemisia*-Cyperaceae-Gramineae	

Fig. 19.15 Time–space correlation of the pollen-assemblage zones in relation to recent woodland vegetation from the montane to the alpine region in the Eastern Alps

conditions, and topography. The oceanic influence is more marked in the valleys and montane woodlands of the Eastern and Western Alpine regions, A-e and A-g, whereas in the inner alpine region, A-h, the effect of greater continentality is clearly shown. A schematic picture of the development of the vegeta-tion in the different zones (montane to alpine) is shown in Fig. 19.15.

A herb-rich pioneer vegetation, characterized by *Artemisia*, Cyperaceae, and Gramineae, occupied the ice-free parts of the montane region up to 13300 BP, followed by an initial shrub phase with *Betula nana, Hippophae, Juniperus*, and *Salix*. This was not present in the subalpine and alpine zones of the Central Alps (see reference site A 11). The re-establishment of a forest cover during the Bölling was achieved by *Betula "alba"* and *Pinus* (Bortenschlager 1984a,b, Wahlmüller 1985a,b). According to oxygen-isotope analyses the major late-glacial vegetational changes (initial shrub-phase, reforestation, etc.) occurred synchronously within the Western and Eastern Alps (Oeggl 1992). The *Pinus–Betula* phase in the montane zone lasted up to 9500 BP and in the subalpine zone in places up to 7000 BP (see reference site A 14).

Regional variations have been found in the vegetational succession during the Holocene. In the valleys and montane woodlands of the Eastern and Western Alpine regions (A-e, A-g), which are more oceanic in character, the pollen values for the deciduous trees attained higher values than in the Inner Alpine region (A-h), where coniferous species predominated. *Corylus, Quercus, Tilia, Ulmus,* and *Picea* immigrated into the montane woodlands of the Eastern and Western Alpine regions (A-e, A-g) about 9500 BP. These tree species predominated in the forests up to the middle of the Atlantic period, when *Abies* and *Fagus* immigrated. The spread of these two tree species occurred at different times in different places. The dominance of *Fagus* in the montane forests seems to be related to anthropogenic influences. During these times *Picea* spread from the east to the west in the inner alpine regions and became the dominant tree in the montane and subalpine forests.

Eight climatic fluctuations have been documented at the forest limit during the Holocene. The Frosnitz fluctuation was so marked that it has been recognized in all the regions investigated.

The earliest finds of *Cerealia*-type pollen grains, together with *Plantago lanceolata* grains, have been made in the middle Atlantic period (Wahlmüller 1985a). Anthropogenic influences in the montane zone have been noted in many pollen diagrams from the Neolithic period onwards (e.g. I. & S. Bortenschlager 1981). Human activity above the forest limit has been documented by finds of cultural-indicator pollen more or less generally already at the end of the Neolithic period (Hüttemann & Bortenschlager

1987, Oeggl 1991). Intensive colonization and forest clearance began during the Early Medieval period, leading to drastic changes in the natural vegetation. The coniferous species were preferentially aided by large-scale forest clearance, whereby *Picea* has become the dominant species in the former Abieti–Fagetum woodlands of the montane zones. At the forest limit, however, *Picea* became replaced by *Pinus cembra* and *Larix*.

REFERENCES

Bortenschlager, I. & Bortenschlager, S. 1981: Pollenanalytischer Nachweis früher menschlicher Tätigkeit in Tirol. *Veröffentlichungen des Museum Ferdinandeum Innsbruck 61*, 5–12.

Bortenschlager, S. 1982: Chronostratigraphic subdivisions of the Holocene in the Alps. *Striae 16*, 75–79.

Bortenschlager, S. 1984a: Beiträge zur Vegetationsgeschichte Tirols I: Ötztal—Unteres Inntal. *Berichte des naturwissenschaftlich-medizinischen Vereins Innsbruck 71*, 19–56.

Bortenschlager, S. 1984 b: Die Vegetationsentwicklung im Spätglazial: das Moor Lanser See III, ein Typprofil für die Ostalpen. *Dissertationes Botanicae 72*, 71–79.

Bortenschlager, S. in press: Beiträge zur Vegetationsgeschichte Tirols VIII: Das Salobermoor. *Berichte des naturwissenschaftlich-medizinischen Vereins Innsbruck*.

Hüttemann, H. & Bortenschlager, S. 1987: Beiträge zur Vegetationsgeschichte Tirols VI: Riesengebirge, Hohe Tatra-Zillertal, Kühtai. *Berichte des naturwissenschaftlich-medizinischen Vereins Innsbruck 74*, 81–112.

Krisai, R., Burgstaller, B., Ehmer-Künkele, U., Schiffer, R. & Wurm, E. 1991: *Moore des Ostlungaus Sauteria 5*, 240 pp.

Oeggl, K. 1988a: Beiträge zur Vegetationsgeschichte Tirols VII: Das Hochmoor Schwemm bei Walchsee. *Berichte des naturwissenschaftlich-medizinischen Vereins Innsbruck 75*, 37–60.

Oeggl, K. 1988b: Pollenanalytische Untersuchungen zum Nachweis anthropogener Vegetationsveränderungen in einem großen Sedimentationsbecken. *Berichte des naturwissenschaftlich-medizinischen Vereins Innsbruck, Supplementum 2*, 59–72.

Oeggl, K. 1991: Botanische Untersuchungen zur menschlichen Besiedlung im mittleren Alpenraum während der Bronze- und Eisenzeit. *In Die Räter. Schriften der Arbeitsgemeinschaft der Alpenländer*, Athesia Verlag, Bozen .

Oeggl, K. 1992: Sediment- und Makrofossilanalysen aus dem Lanser See in Tirol (Austria): Ein Beitrag zur spätglazialen Bio- und Chronostratigraphie der Ostalpen. *Flora 186*, 317–339.

Oeggl, K. & Eicher, U. 1989: Pollen- and oxygen-isotope analyses of late- and postglacial sediments from the Schwemm raised bog near Walchsee in Tirol, Austria. *Boreas 18*, 245–253.

Patzelt, G. & Bortenschlager, S. 1973: Die postglazialen Gletscher-und Klimaschwankungen in der Venediger Gruppe (Hohe Tauern, Ostalpen). *Zeitschrift für Geomorphologie, Supplementum 16*, 185–197.

Wahlmüller, N. 1985 a: Der vorgeschichtliche Mensch in Tirol. Neue Aspekte aufgrund der Pollenanalyse. *Veröffentlichungen des Museum Ferdinandeum 65*, 105–120.

Wahlmüller, N. 1985b: Beiträge zur Vegetationsgeschichte Tirols V: Nordtiroler Kalkalpen. *Berichte des naturwissenschaftlich-medizinischen Vereins Innsbruck 72*, 101–144.

Weirich, J. & Bortenschlager, S. 1980: Beiträge zur Vegetationsgeschichte Tirols III: Stubaier Alpen –Zillertaler Alpen. *Berichte des naturwissenschaftlichen Vereins Innsbruck 67*, 7–30.

20

Slovenia

M. Culiberg and A. Šercelj

INTRODUCTION

Slovenia covers 20250 km^2. On the north it is bounded by the ridge of the southeastern Alps, on the south by the Adriatic Sea. On the west it opens towards Italy, on the east towards the Pannonian Plain. Therefore it is a typical transitional territory, the southern gate between the East and the West.

Slovenia is divided into six phytogeographic type regions, according to M. Wraber (1969). They were established on the basis of geographical situation and forest vegetation types. The IGCP 158 type regions follows this subdivision (Fig. 20.1).

SLO-a   the Julian Alps and Karavanke Mountains
SLO-b   the Subalpine Promontory
SLO-c   the Subpannonian Headland
SLO-d   the Praedinaric
SLO-e   the Dinaric
SLO-f   the Submediterranean

Geologically Slovenia belongs to the Alpine orogeny. The east of the country descends to the Panonnian Plain. On the southeast, however, it continues into the Dinaric Mountains. This deviation from the Alpine tectonics to the Dinaric one is well marked by the dislocation lines on the geological map (Fig. 20.2).

The constituents of the Alpine orogeny are mostly sedimentary rocks, much less plutonic ones, e.g. Pohorje Mts. The sediments are partly of Palaeozoic age, but most of them are Mesozoic. Limestone and dolomite prevail. In these areas various surface and underground karst features are developed, with *ca.* 7000 caves registered. The Submediterranean area is exclusively sedimentary, consisting of Cretaceous and Tertiary limestone and sandstone (flysch). The Pannonian Plain is covered by Tertiary sediments in the bordering areas, whereas in the centre Quaternary sediments prevail, including loess (Fig. 20.2).

The relief, due to the younger and still active tectonics (frequent earthquakes) is rough, with gorges, cliffs, and canyons and with underground streams and caves. Hypsometrically, about 9% of the territory is lowland, including hills and benches less than 200 m in elevation. The elevations between 200 and 500 m occupy 44% of the area, between 500 and 1500 m elevation is 45%, while only 2% of the surface exceeds 1500 m in elevation.

Most of the rivers belong to the Danube or to the Black Sea drainage area. The watershed between the Black Sea and the Adriatic drainage systems lies less than 50 km from the Adriatic, so most rivers flow north to the Sava River, which is a tributary of the Danube. A few rivers, Soča (Isonzo), Dragonja, flow into the Adriatic, and the river Reka discharges through underground streams as Timavo into the Adriatic near Trieste.

The soils range from lithosols on bare limestone or sandstone and regosols on colluvial substrata to rendzinas and on loess in the Pannonian region. In the central parts cambisols and brown soils prevail,

Palaeoecological Events During the Last 15000 Years: Regional Syntheses of Palaeoecological Studies of Lakes and Mires in Europe.
Edited by B.E. Berglund, H.J.B. Birks, M. Ralska-Jasiewiczowa and H.E. Wright. © 1996 John Wiley & Sons Ltd.

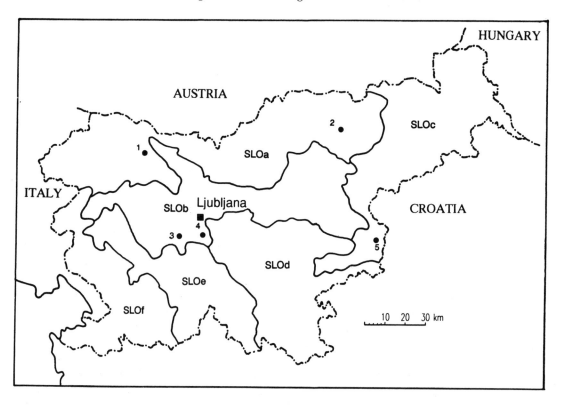

Fig. 20.1   IGCP 158 type regions of Slovenia

mostly covered by forests, while remnants of terra rossa are frequent on limestone. In the forested regions and on wet lowlands gleys and pseudogleys are frequent.

Corresponding to the geographical situation, two climatic zones can be distinguished, although not precisely delineated. The Submediterranean zone occurs along the Adriatic and into the interior along the river Soča, with dry summers and humid autumns. Winds from the east prevail during the winter, and winds from the west and southwest during warm periods. The Alpine–Dinaric belt covers the central part of the country. Its climate is moderately warm and wet, with 2000–3000 mm (4000 max.) precipitation, distributed almost uniformly through the year. Because of the rough relief the microclimatic conditions are subject to radical variations. The Subpannonian part of the country has the most continental character and consequently greater temperature gradients, with hot dry summers and in-

creasingly cold winters towards the east, with 700–1000 mm precipitation per year.

The vegetation is very heterogeneous, because the variable relief, geological structure, climate, and precipitation create an almost infinite variety of habitats (Fig. 20.3). The main part (53%) of the land is forested, except for the eastern part of the Pannonian region, which has a rather steppic character. The *Fagus–Abies* forests in numerous variants are characteristic of mountainous regions, which are less influenced by man. Lowlands are mostly cultivated and to a lesser extent covered by forests of Querco–Carpineta. Because of varied biotopes and numerous ecological niches some plant taxa, e.g. *Pinus*, *Picea*, *Fagus*, and *Fraxinus*, found favourable refuges during the Pleistocene glaciations. Here they avoided extinction and later some developed new forms confined to small areas as endemics or relics (Horvat *et al.* 1974). On this territory the elements of the alpine, mediterranean, and southeasteuro-

Fig. 20.2 Geological map of Slovenia, simplified (Geological Survey of Slovenia)

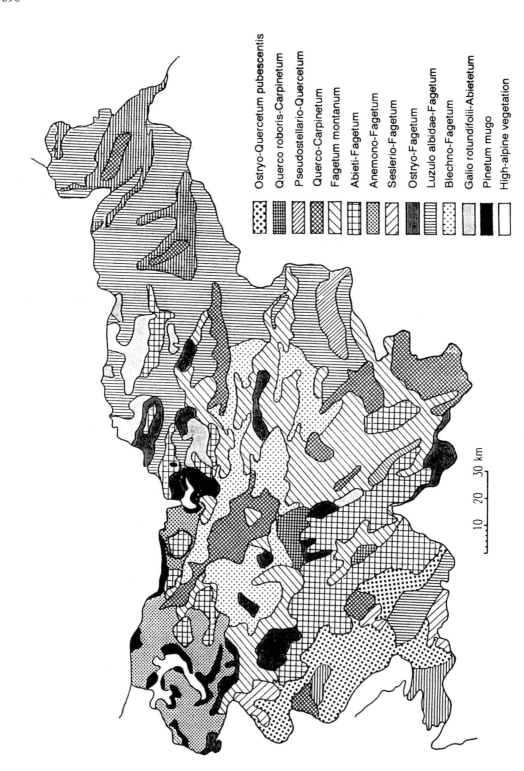

Ostryo-Quercetum pubescentis
Querco roboris-Carpinetum
Pseudostellario-Quercetum
Querco-Carpinetum
Fagetum montanum
Abieti-Fagetum
Anemono-Fagetum
Seslerio-Fagetum
Ostryo-Fagetum
Luzulo albidae-Fagetum
Blechno-Fagetum
Galio rotundifolii-Abietetum
Pinetum mugo
High-alpine vegetation

10 20 30 km

Fig. 20.3   Natural potential vegetation map of Slovenia (Zupančič *et al.* 1989)

pean-illyrian floras meet and interconnect.

Type-region syntheses are presented for the following regions: SLO-a, SLO-b and SLO-c:

TYPE REGION SLO-a, JULIAN ALPS AND KARAVANKE MOUNTAINS

Altitude: 500–2800 m.

Climate: Mean January −3 to −6°C, July 15–18°C. Precipitation 2000–3000 mm yr^{-1}, 4000 mm yr^{-1} max.

Geology: Prevailing Mesozoic limestone, dolomite, and sandstone, glacial till on the plateaux and in the valleys.

Topography: River valleys narrow, high plateaux passing over into mountain ridges of over 2000 m elevation.

Population: Mostly small towns, scattered farms in the mountains.

Vegetation: *Picea*, Abieti–Fagetum, *Pinus mugo*, alpine peat bogs, sparse alpine pastures above the tree-line.

Soils: Brown earths, rendzinas, podzols.

Land use: Forests, pastures, agriculture in the valleys, industry.

Reference site 1. Ledine, Jelovica
(Culiberg *et al.* 1981)

Latitude 46°16′N, Longitude 14°07′E. Elevation 1100 m. Age range 10000–0 BP. Peat bog. 10 local pollen-assemblage zones (paz). (Fig. 20.4)

1.	10000 BP	*Pinus–Picea–Betula–Artemisia*
2.	10000–8000 BP	*QM–Corylus–Betula–Picea–Pinus*
3.	8000–7000 BP	*Corylus–QM–Fagus–Picea*
4.	7000–4000 BP	*Picea–Fagus–Abies–Corylus–QM*
5.	4000–3600 BP	*Picea–Fagus–Abies–QM*
6.	3600–3000 BP	*Picea–Abies–Fagus–QM*
7.	3000–2000 BP	*Abies–Fagus–Picea*
8.	2000–1000 BP	*Picea–Abies–Fagus*
9.	1000–100 BP	*Pinus–Picea–Abies–Fagus*
10.	100–0 BP	*Fagus–Picea–Abies–Pinus*

General patterns (within each paz)

(1) The late-glacial pollen assemblage *Pinus–Picea–Betula–Artemisia* declines. Sporadic appearance of pollen of mesophytic broad-leaved taxa such as *Ulmus, Tilia, Quercus*, and *Fagus* is characteristic.

(2) The forest phase of Quercetum mixtum (*Quercus, Tilia, Ulmus*) replaces *Pinus* and *Betula*.

(3) There follows an intermediate *Corylus* phase with pollen values up to 40%. *Pinus* and *Picea* are subordinate. Note the inversion of QM–*Corylus* phase.

(4) At about 7000 years BP *Abies* joins the *Fagus* forests and together they form the climax forest Abieti–Fagetum, which remains the dominant formation. However, it is surpassed for a certain time by *Picea*, but as a formation it remains dominant until the Middle Ages.

(5–9) After 900 BP a radical clearing of the forest is evident from the pollen diagram (colonization!). *Abies* and *Fagus* were most affected. *Fagus* was used for charcoal for the iron works and *Abies* for timber. *Pinus* assumed again the dominace as a pioneer in forest regeneration.

(10) Abieti–Fagetum expands recently, *Picea* persists.

Note: A nearly identical pollen diagram derives from the nearby high plateau Pokljuka (elevation 1100 m) with first arrival of *Abies* at 6890±45 yr BP (Šercelj 1972).

Reference site 2. Lovrenško barje, Pohorje Mts
(Culiberg 1986)

Latitude 46°30′N, Longitude 15°19′E. Elevation 1500 m. Age range 4500–0 BP. Slope peat bog. Five local pollen-assemblage zones (paz). (Fig. 20.5)

1.	4500–4000 BP	*Corylus–Picea–Abies–Tilia–Fagus*
2.	4000–3700 BP	*Fagus–Abies–Picea–Betula–Carpinus–QM–Gramineae*
3.	3700–2400 BP	*Fagus–Abies–Picea–Carpinus–Betula–Pinus*

Ledine – Jelovica, SLOVENIA
Selected pollen percentages
Analyst: M. Culiberg

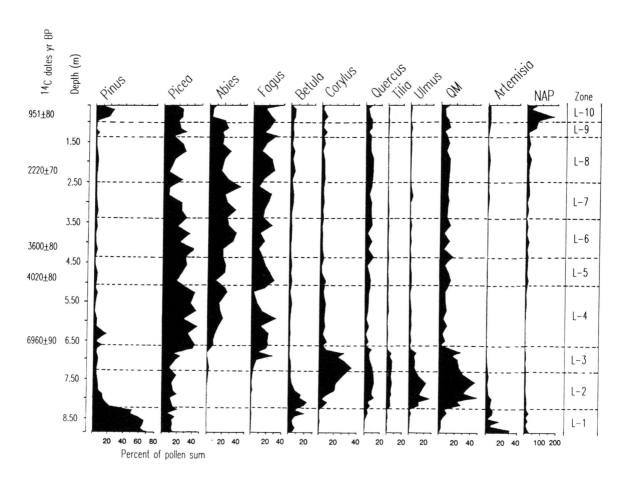

Fig. 20.4 Summary pollen diagram from Ledine, Jelovica

4. 2400–1670 BP *Fagus–Abies–Picea–*
   QM–Gramineae
5. 1670–0 BP *Pinus–Picea–Betula–*
   Fagus–Corylus–Carpinus–
   Gramineae–Cyperaceae

General patterns (within each paz)

(1) Secondary *Corylus* phase dominated before 4000 BP, while *Fagus* and *Abies* were previously much reduced—pasture?

(2) *Fagus* curve increases abruptly, as do also those of *Abies* and *Picea*. *Corylus* frequency is in decline. *Carpinus* and *Quercus* are moderately present. *Tilia* disappears from the diagram. *Pinus* curve increases slightly at 2500 BP.

(3–4) *Fagus*, *Abies*, and *Picea* dominate until about 1600 BP. The curves of *Carpinus, Corylus,* and QM slightly increase.

(5) *Pinus (mugo?)* expands on the peat bog. The curves of *Abies* and *Fagus* decline, mean-

Lovrenško barje – Pohorje Mts., SLOVENIA
Selected pollen percentages
Analyst: M. Culiberg

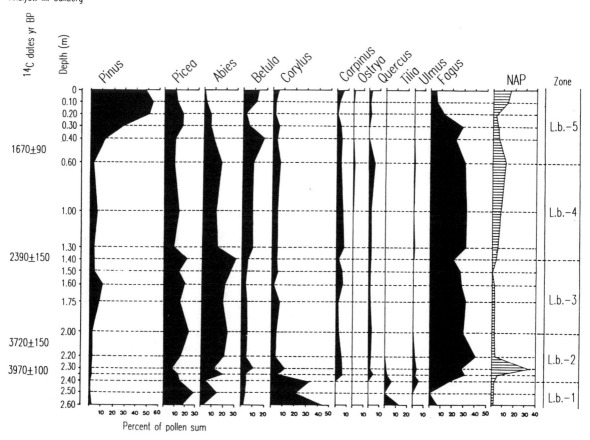

Fig. 20.5　Summary pollen diagram from Lovrenško barje, Pohorje Mts

while *Picea* and *Betula* slightly expand, as do Gramineae and Cyperaceae.

TYPE REGION SLO-b, SUBALPINE PROMONTORY

Altitude: 300–1000 m.

Climate: Mean January −1 to −3°C, July 18–20°C. Precipitation 1200–2000 mm yr^{-1}.

Geology: Palaeozoic and Mesozoic limestone, dolomite, and shale, fluvioglacial gravels, Quaternary lake sediments.

Topography: Undulating hills and mountains.

Population: The most densely populated area of the region, mostly in towns, more than 100 inhabitants km^{-2} in places. Industry.

Vegetation: Forests cover about 40% of the area. On mountainous areas not excessively influenced by man, Abieti-Fagetum prevails as the "zonal" vegetation. In lower positions and valleys on more densely settled and cultivated surfaces Querco–Carpineta prevail. Pioneer formations of *Pinus–Betula* occupy extremely degraded soils as regeneration stages.

Soils: Brown earths, rendzinas, terra rossa in karst areas.

Land use: Agriculture in valleys along the rivers and in lowlands, vineyards and fruit growing on hilly areas, cattle breeding and forestry in mountainous parts.

Reference site 3. Zamedvejca, Ljubljana Moor
(Culiberg 1991)

Latitude 45°59′N, Longitude 14°25′E. Elevation 300 m. Age range 14000–9000 BP. Mire. Six local pollen-assemblage zones (paz). (Fig. 20.6)

1.	14000 BP	*Pinus–Picea–Artemisia–Ephedra–Selaginella selaginoides*
2.	14000–13000 BP	*Pinus–Betula–Picea–Ulmus*
3.	13000–12000 BP	*Pinus–Betula–Picea–Tilia–Ulmus*
4.	12000–11000 BP	*Pinus–Picea–Betula–Tilia–Ulmus*
5.	11000–10000 BP	*Pinus–Betula–Tilia–Picea*
6.	10000–9000 BP	*Betula–Tilia–Picea–Pinus*

General patterns (within each paz)

(1) *Pinus*, *Picea*, and *Artemisia* dominate. High NAP values indicate scarce forest vegetation. *Selaginella selaginoides* is present.
(2) *Pinus* and *Betula* are present in nearly equal proportions. *Tilia* and *Ulmus* appear sporadically.
(3) *Pinus* and *Betula* and more *Picea*. Quercetum mixtum joins the vegetation.
(4) *Pinus* curve remains at nearly the same level, while the curve of *Betula* declines. The curves of *Tilia* and *Ulmus* rise.
(5) At about 11000 radiocarbon years *Pinus–Betula* and QM dominate. *Tilia* especially expands.
(6) *Betula–Pinus* curves cross one another. *Betula* becomes dominant. *Tilia* persists.

Reference site 4. Lake dwellers' site Parti,
Ljubljansko barje (Culiberg & Šercelj 1978)

Latitude 45°59′N, Longitude 14°32′E. Elevation 300 m. Age range 7000–800 BP. Mire. Six local pollen-assemblage zones (paz). (Fig. 20.7)

1.	–7000 BP	*Fagus–Corylus–Picea–Abies* (first arrival of *Abies*)
2.	7000–5000 BP	*Fagus–Abies–Corylus–QM–Carpinus*
3.	5000–4000 BP	*Fagus–Corylus–Abies–QM–Cerealia* (Eneolithic stratum)
4.	4000–3000 BP	*Fagus–QM–Abies–Carpinus–Corylus–Cerealia*
5.	3000–2100 BP	*Abies–Fagus–QM–Carpinus–Corylus*
6.	2100–800 BP	*Abies –Fagus–Picea –QM–Pinus*

General patterns (within each paz)

(1) The entire period until the Neolithic lake dwellers' culture is dominated by Abieti–Fagetum, yet the *Corylus* curve rises a little, as a possible indication of cattle breeding by Neolithic settlers. In the Eneolithic cultural stratum, although barren of tree pollen, a lot of cereal pollen and even some charred grains of wheat have been found. Results of two radiocarbon measurements of this stratum are 3910 ± 100 yr BP and 4000 ± 100 yr BP. The state of the neighbouring forests, as affected by the lake dwellers, is reflected in the temporary rise of the QM curve. The general composition is Abieti–Fagetum, however. But the results of wood analysis of the poles from the lake villages show that the forests around the lake were degraded Quercetum mixtum. Of the more than 2000 poles analysed, 897 were identified as belonging to *Quercus* (three species), 665 to *Fraxinus*, 231 to *Sorbus*, 60 to *Corylus*, 64 to *Carpinus*, 6 to *Ostrya*, and the remaining 110 pieces to different trees and shrubs, among them only 12 to *Fagus* (Culiberg and Šercelj 1991).
(2) After the decline of the lake dwellers' culture, the distant forests of Abieti–Fagetum were still influenced by man. The *Fagus* curve, it is true, remains at the same level, but the curve of *Abies* rises much more than would be expected. It is reasonable to suppose that the changes in the Abieti–Fagetum were caused by the people who

Zamedvejca – Ljubljana Moor, SLOVENIA
Selected pollen percentages
Analyst: M. Culiberg

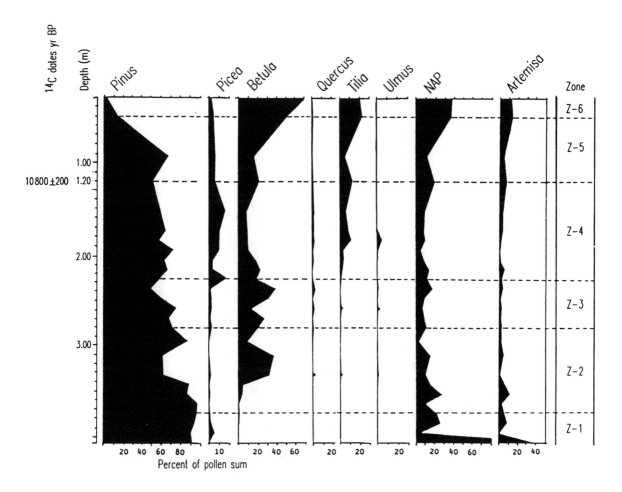

Fig. 20.6   Summary pollen diagram from Zamedvejca, Ljubljana

left the lake and settled on dry land, also some distance from the lake. In the regeneration process *Abies* was faster to occupy open sites. It does not seem reasonable to interpret this phenomenon as a consequence of climatic change to more oceanic conditions in the "dry" Subboreal.

(3) After the regeneration of Abieti–Fagetum, the *Pinus* curve, which hitherto remained below 10% in the upper horizon, rose to over 20%. This can be interpreted as a sign of intensive

degradation of the soil on the surrounding hills, probably also as a consequence of human activity, i.e. forest clearances.

TYPE REGION SLO-c, SUBPANNONIAN HEADLAND

Altitude: 200–600 m.

Climate: Continental, mean January temperatures

Parti – Ljubljana Moor, SLOVENIA
Selected pollen percentages
Analyst: M. Culiberg

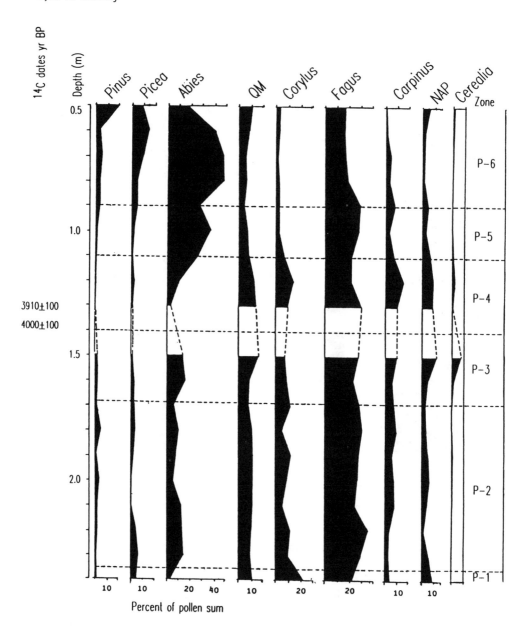

Fig. 20.7 Summary pollen diagram from lake dwellers' site Parti, Ljubljana

−2 to −4°C; mean July temperatures 18–23°C. Precipitation 700–1200 mm yr⁻¹.

Geology: Prevailing Tertiary limestone and sandstone, Quaternary alluvial and eolian sediments.

Topography: Tertiary hills, passing over to the east in lowland of the Pannonian Plain.

Population: Densely populated, mainly farming area, 80–100 people km⁻².

Vegetation: Querco–Carpineta, pioneer pine forests, meadows.

Soils: Gley, pseudogley, brown earth.

Land use: Agriculture, viniculture on hilly slopes, cattle breeding.

Reference site 5. Kaznarice (Šercelj 1984)

Latitude 46°05′N, 15°39′E. Elevation 500 m. Age range 15000–6000 BP. Seven local pollen-assemblage zones (paz). (Fig. 20.8)

1. 15000–13000 BP *Pinus–Selaginella selaginoides*
2. 13000–12000 BP *Pinus–Betula–Artemisia– Gramineae– Chenopodiaceae*
3. 12000–11000 BP *Pinus–Betula–Tilia– Quercus*
4. 11000–10000 BP *Pinus–Betula–Artemisia– Gramineae–Cyperaceae*
5. 10000–9000 BP *Pinus–Betula–Picea– Tilia–Quercus*
6. 9000–7700 BP *Tilia –Pinus –Picea – Fagus –Corylus–Ulmus*
7. 7700–6000 BP *Fagus –Tilia–Corylus– Quercus–Picea*

General patterns (within each paz)

(1–2) The early late-glacial vegetation in the region was like taiga, with only *Pinus* stands at the beginning, later on joined by *Betula* and *Juniperus*. Characteristic are high values of *Selaginella selaginoides* and later on the herbs *Artemisia*, Chenopodiaceae, Gramineae, Cyperaceae.

(3) Domination of *Pinus–Betula* continues. Pollen of *Tilia* and *Quercus* appear for the first time. Pollen frequency of herbs decreases as a consequence of increasing forest density. This stage could be correlated to the Alleröd interstadial.

(4) *Pinus–Betula* phase continues to dominate and also the pollen frequency of herbs increases for a time.

(5) *Picea* and *Tilia* join the forest, followed by *Ulmus* and *Quercus*. The *Pinus* curve, in contrast, is steeply decreasing. Onset of the Holocene.

(6) *Corylus* and *Fagus* join the existing forest community. Noteworthy is a temporary rise of *Pinus* curve at about 7700 BP, at the expense of *Fagus*. Possibly it is due to a disturbance by man.

(7) *Fagus* prevails again with absolute domination. *Abies* failed to join it to form Abieti–Fagetum, usual in the Alpine and Dinaric regions.

PALAEOECOLOGICAL PATTERNS AND EVENTS

Quaternary sediments are very scarce, and few of them are polleniferous. In glaciated or periglacial areas clays and marls are more frequent. The cause for this lack of appropriate sediments lies in the local lithological (karst) composition and tectonics. Under such conditions it is understandable that very few sediment profiles can be found that would represent an uninterrupted succession for a long period, e.g. the whole late-glacial and Holocene. In general we have to deal with incomplete sections, which must be synthesized to provide a more complete series of vegetation phases. The minerogenic provenience of the greater part of polleniferous sediments is also disadvantageous. So we have little opportunity to check the pollen-stratigraphic classification by radiocarbon dating, unless pieces of wood or similar plant material are incorporated in the sediment.

The previous investigations permit an attempt to present the succession series and their chronological sequence only for the Alpine and Prealpine regions (Fig. 20.9), and partly for the Subpannonian region.

Kaznarice, SLOVENIA
Selected pollen percentages
Analyst: A. Šercelj

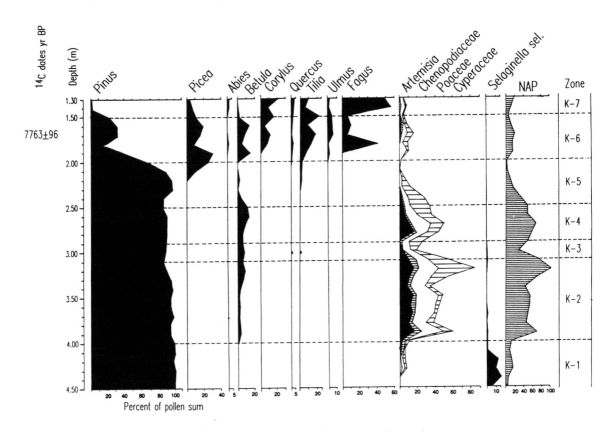

Fig. 20.8 Summary pollen diagram from Kaznarice

Vegetation events

(1) The late-glacial vegetation consisted of open *Pinus–Betula* stands during stadial periods, partly even from taiga or tundra. However, during the interstadial periods some mesophilic broad-leaved trees joined the pioneer forest, for instance *Tilia, Quercus, Ulmus, Fraxinus,* and, in some places, *Fagus,* e.g. around Ljubljansko barje, then a lake of about 160 km² surface. This fact provides support for the idea that there were some refugia for these taxa in this region, e.g. *Fagus* pollen in the sediments of Sečovlje (Šercelj 1981) and charcoal of *Picea, Fraxinus,* and *Fagus* (12580 ± 250 ^{14}C yr BP) in

Palaeolithic site Lukenjska jama near Novo mesto (Culiberg 1991).

(2) The representatives of the Quercetum mixtum (*Tilia–Quercus–Ulmus–Fraxinus*) spread in some areas by the end of the late-glacial and (or) at the very beginning of the Holocene. During the Preboreal–Boreal period the QM phase dominated.

(3) During the Boreal the *Corylus* phase also developed but was generally of short duration and with low pollen frequency. Nowhere did *Corylus* exceed values of 40%.

(4) Very soon, still during the Boreal, the aggressive *Fagus* supplanted the preceding phases to establish an absolute *Fagus* dominance.

TYPE REGION SLO-b PRAEALPINE /LJUBLJANA MOOR/ EVENT STRATIGRAPHY

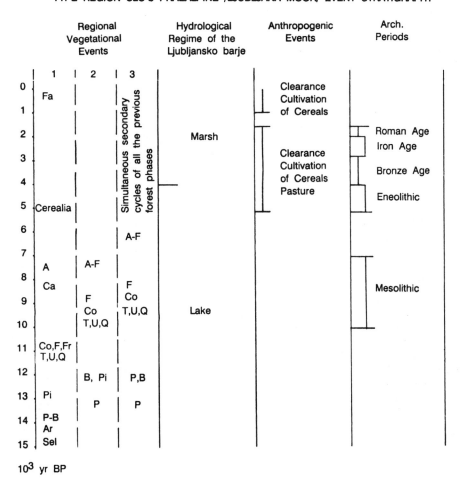

1 First arrival
2 Expansion time
3 Maximum values

Fa = *Fagopyrum*	F = *Fagus*	B = *Betula*
Se = *Secale*	Fr = *Fraxinus*	Pi = *Picea*
Ce = *Cerealia*	T = *Tilia*	Ar = *Artemisia*
A = *Abies*	U = *Ulmus*	Se = *Selaginella*
Ca = *Carpinus*	Q = *Quercus*	A-F = Abieti-Fagetum
Co = *Corylus*	P = *Pinus*	

Fig. 20.9 Event stratigraphy for type region SLO-b, the Prealpine area

(5) At about 7000 years BP, *Abies* joined *Fagus* and together they formed the climax Abieti–Fagetum, which has been dominant until now in the mountainous areas.

(6) After that time all the forest phases existed simultaneously as a consequence of regeneration of individual phases under natural trends (endodynamics) and (or) under natural and human-anthropozoogenic influence (exodynamics) as well.

Differences from the central European vegetation development

(1) The inversion of *Corylus*–Quercetum mixtum phase and very low values of the *Corylus* phase compared to those in Middle Europe (Šercelj 1972).

(2) The much earlier appearance and development of individual forest phases.

(3) The NAP usually remained at low values except in rare cases and for very short times. This means that no extensive clearances of the forests occurred and that the population density was nowhere especially high. Other anthropogenic events are also more or less locally restricted, as evident from the palaeoecological investigations of the archaeological sites.

Historical considerations

The geomorphic and geographic situation of Slovenia provided a great number of convenient places for different plants, hence the much earlier spread and first settlement of some broad-leaved trees in the late-glacial. The rugged configuration, dissected mountain systems, cliffs, and gorges open to the south created ecological niches where individual plant taxa found their refugia during cold periods. From there, with the first onsets of warm phases they started to migrate north. Hence, according to Trinajstič (1986), the phytogeographical classification of the potential natural vegetation into generally adopted regions cannot be exactly delineated, as all the regions interconnect.

REFERENCES

Culiberg, M. 1986: Palinološka raziskovanja na Lovrenškem in Ribniškem barju na Pohorju. (Palynological investigations of two peat-bogs, Lovrenško and Ribniško barje from the Pohorje Mts). *Biološki vestnik 34*, 1–14.

Culiberg, M. 1991: Late-Glacial vegetation in Slovenia. *Dela SAZU, razred 4, 29*,1–52.

Culiberg, M. & Šercelj, A. 1978: Ksilotomske in palinološke analize rastlinskih ostankov s kolišča na Partih pri Igu—izkopavanja leta 1977. (Xylotomische und palynologische Analysen der Pflanzenüberreste aus dem Pfahlbau Parti bei Ig, Ausgrabung 1977). *Poročilo o raziskovanju paleolita, neolita in eneolita v Sloveniji 6*, 95–98.

Culiberg, M. & Šercelj, A. 1991: Razlike v rezultatih raziskav makroskopskih rastlinskih ostankov s kolišč na Ljubljanskem barju in pelodnih analiz—dokaz človekovega vpliva na gozd. (Die Unterschiede zwischen den Resultaten der Untersuchungen der makrosko-pischen Reste aus den Pfahlbauten und den Pollenanalysen—ein Beweis für den Einfluß des Menschen auf die Wälder). *Poročilo o raziskovanju paleolita, neolita in eneolita v Sloveniji 19*, 249–256.

Culiberg, M., Šercelj, A. & Zupančič, M. 1981: Palynologische und phytozönologische Untersuchungen auf den Ledine am Hochplateau Jelovica. *Razprave SAZU, razred 4, 23/6*, 175–193.

Horvat, I., Glavač, V. & Ellenberg, H. 1974: Vegetation Südosteuropas. *Geobotanica selecta 4*.

Šercelj, A. 1971: Postglacialni razvoj gorskih gozdov v severozahodni Jugoslaviji. (Die postglaziale Entwicklung der Gebirgswälder im nordwestlichen Jugoslawien.) *Razprave SAZU 14/9*, 265–294.

Šercelj, A. 1972: Verschiebung und Inversion der postglazialen Waldphasen am südöstlichen Rand der Alpen. *Berichte der Deutchen Botanischen Gesellschaft 85*, 123–128.

Šercelj, A. 1981: Pelod v vzorcih jedra iz vrtine V-6/79. *In* Ogorelec, B. *et al.*: Sediment sečoveljske soline. (Sediment of the salt marsh of Sečovlje). *Geologija- razprave in poročila 24/2*, 179–216

Šercelj, A. 1984: Razvoj kasnoglacialne vegetacije v obrobju Panonske nižine. *Radovi Akademije Nauka Bosne i Hercegovine 76*, 165–170.

Trinajstič, I. 1986: Basic phytogeographical division of the potential natural vegetation of Yugoslavia. *Natural potential vegetation of Yugoslavia*, 76–77.

Wraber, M. 1969: Pflanzengeographische Stellung und Gliederung Sloweniens. *Vegetatio 17*, 176–199.

Zupančič, M., Puncer, I. *et al.* 1989: Map: Potencialna naravna vegetacija Slovenije. In: *Enciklopedija Slovenije 3*, 118–119.

21

Bulgaria

E. Bozilova, M. Filipova, L. Filipovich and S. Tonkov

INTRODUCTION (E. Bozilova)

Bulgaria is located in the centre of the Balkan peninsula (southeastern Europe), where the geologic, climatic, and vegetational characteristics are rather diverse. The territory has been land since the beginning of the Pleistocene, subjected to a general uplift, with the exception of the Black Sea coast. The highest mountain ridges were covered by ice during the Riss and Würm glaciations.

Taking into consideration the presence of suitable sites (lakes and mires) for detailed palaeoecological research and the richness and diversity of the recent flora and vegetation, the country was divided into nine type regions with 16 reference sites (Fig. 21.1):

BG-a   Danube plain and Dobrudza
BG-b   Black Sea coast
BG-c   Stara Planina foothills
BG-d   Stara Planina and Sredna Gora Mountains
BG-e   Vitosha, Rila, Pirin, and Rhodopes
       Mountains
BG-f   Znepole
BG-g   Thracian plain and Tundza hilly region
BG-h   Western frontier Mountains and Struma
       Valley
BG-i   Strandza Mountains

Type-regions syntheses are presented for the following four regions:

BG-b   Black Sea coast
BG-d   Stara Planina and Sredna Gora Mountains
BG-e   Vitosha, Rila, Pirin, and Rhodopes
       Mountains
BG-f   Znepole

No syntheses are made for the rest of the type regions where the palaeoecological data either are not sufficient at the moment or are still completely lacking.

In the region BG-a (Danube plain and Dobrudza) sediments of Lake Srebarna were studied by means of pollen analysis. This UNESCO Biosphere Reserve close to the Danube river is famous for its animal and plant life.

The most extensive palaeoecological information was collected from the northern part of type region BG-b (Black Sea coast). The results from pollen analysis of sediments of Lake Durankulak, Lake Shabla–Ezeretz, and Lake Varna are supplemented with data from molluscs, diatoms, plant macrofossils, fungal remains, and radiocarbon dating. The region has been populated since the Neolithic (8000 BP).

In the mountainous areas of type region BG-d (Stara Planina and Sredna Gora Mountains) several peat bogs were investigated, and the main forest phases for the last 8000 years are described.

The recent vegetation in type region BG-e (Vitosha, Rila, Pirin, and Rhodopes Mountains) is dominated by coniferous forests. The continuous Late glacial and Holocene vegetation history was recorded for

---

Palaeoecological Events During the Last 15 000 Years: Regional Syntheses of Palaeoecological Studies of Lakes and Mires in Europe.
Edited by B.E. Berglund, H.J.B. Birks, M. Ralska-Jasiewiczowa and H.E. Wright.   © 1996 John Wiley & Sons Ltd.

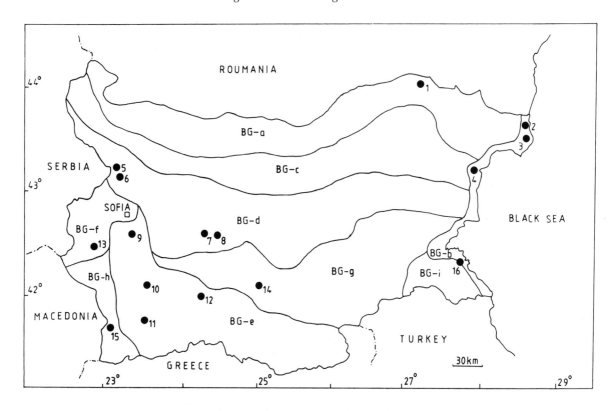

Fig. 21.1 Map of type regions. Reference sites are listed in the Appendix

two lakes for Rila and Rhodopes Mountains. Additional information for tree-limit fluctuations and tree arrival was gathered from the analysis of plant macrofossils.

The vegetational successions in the western and southern central lowlands were traced for two mires in type region BG-f (Znepole) and type region BG-g (Thracian plain).

In type-region BG-i (Strandza Mountains), which coincides with a separate floristic region in the southeasternmost corner of the country, pollen and plant macrofossils were analyzed from the Holocene sediments of Lake Arkutino.

TYPE REGION BG-b, BLACK SEA COAST
(E. Bozilova, M. Filipova)

Altitude: 0–250 (400 m).

Climate: Black Sea climatic region; mean tempera-

tures January 1°C, July 22°C; precipitation 550 mm yr^{-1}.

Geology: Sarmatian limestone, Pleistocene loess, Holocene deposits.

Topography: Coastal plain, plateaux, hills, cliffs.

Population: Approximately 40–100 people km^{-2} in small villages, resorts, and two big cities with over 300000 people.

Vegetation: Xerothermic herb and steppe vegetation in the northern part; broad-leaved mixed deciduous *Quercus* forests in the southern part; psammophytic and halophytic communities on sand dunes and saline soils.

Soils: Chernozems, humic carbonate, eroded carbonate, grey forest, alluvial.

Land use: Arable land in areas of former forests; crops and maize cultivation.

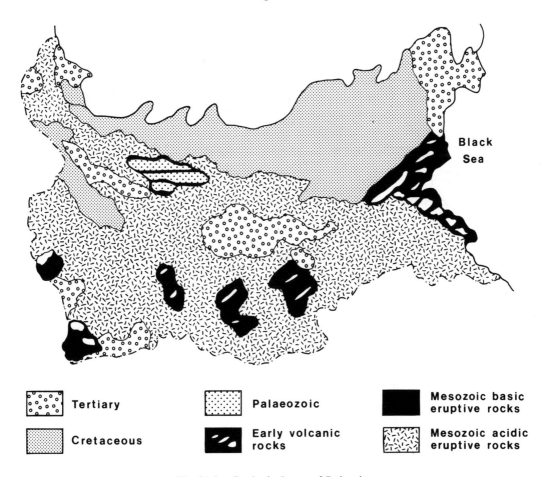

Fig. 21.2 Geological map of Bulgaria

Reference site 2. Lake Durankulak (Bozilova &
Tonkov 1985, Bozilova & Atanassova 1990)

Latitude 43°15′N, Longitude 28°23′E. Elevation
0 m. Age range 9000–0 BP. Lagoon lake. Two pollen
diagrams were produced, one of them with selected
tree and herb taxa (Fig. 21.4). The diagram is divided into four pollen-assemblage zones and four
radiocarbon dates are available. Additional information was obtained from the analysis of molluscs,
diatoms, and archaeological evidence.

D-1   9000–7000 BP   Chenopodiacea–Compo-
                     sitae–Umbelliferae
D-2   7000–4500 BP   *Quercus–Carpinus betu-
                     lus*–Chenopodiaceae
D-3   4500–2800 BP   *Quercus–Carpinus*

D-4   2800–0 BP   *betulus*–Gramineae
                     *Quercus–Carpinus
                     betulus–Fagus*–
                     Gramineae

General patterns

(1) The early-Holocene pollen assemblages were
    dominated by herbs of Chenopodiaceae, Com-
    positae, and Gramineae as components of a
    xerophilous steppe vegetation.

(2) The forest–steppe character of the vegetation
    was preserved in the vicinity of the lake between
    7000 and 5000 BP.

(3) *Fagus* is continuously present after about 6000 BP.

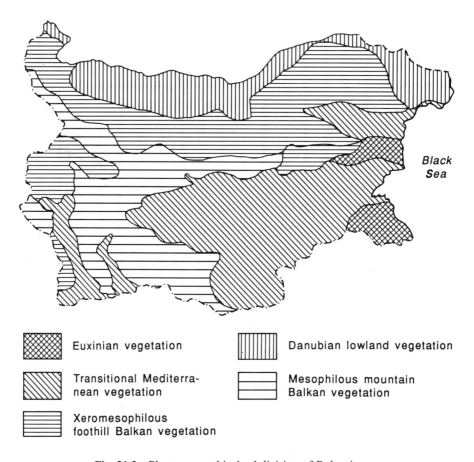

Euxinian vegetation

Transitional Mediterra-
nean vegetation

Xeromesophilous
foothill Balkan vegetation

Danubian lowland vegetation

Mesophilous mountain
Balkan vegetation

Fig. 21.3 Phytogeographical subdivision of Bulgaria

(4) The Subboreal pollen assemblages (paz D-3) are characterized by *Quercus* and *Carpinus betulus*.

(5) The first find of Cerealia pollen (*Triticum, Hordeum*) and anthropophytes (*Plantago lanceolata, Rumex*) is explained by the existence of Neolithic settlements (Fig. 21.5).

(6) The second increase of *Cerealia* pollen is in the Subatlantic (paz D-4).

Reference site 3. Lake Shabla-Ezeretz (Filipova 1985, Bozilova & Filipova 1986, Filipova & Bozilova 1990)

Latitude 43°35′N, Longitude 28°20′E. Elevation 0 m. Age range 7000–0 BP. Lagoon lake. Time scale pollen diagram with the major taxa (Fig. 21.6) is

divided into five pollen-assemblage zones. Information is also available from the analysis of molluscs, diatoms, archaeological evidence, and radiocarbon dating.

ShE-1 7000–6000 BP Gramineae–*Artemisia*–
Chenopodiaceae–
Quercus–Corylus

ShE-2 6000–5700 BP *Quercus–Carpinus*
betulus–Ulmus–
Corylus-Gramineae–
Chenopodiaceae–
Artemisia

ShE-3 5700–5000 BP *Carpinus betulus–*
Quercus–Corylus–
Gramineae–*Artemisia*–
Chenopodiaceae

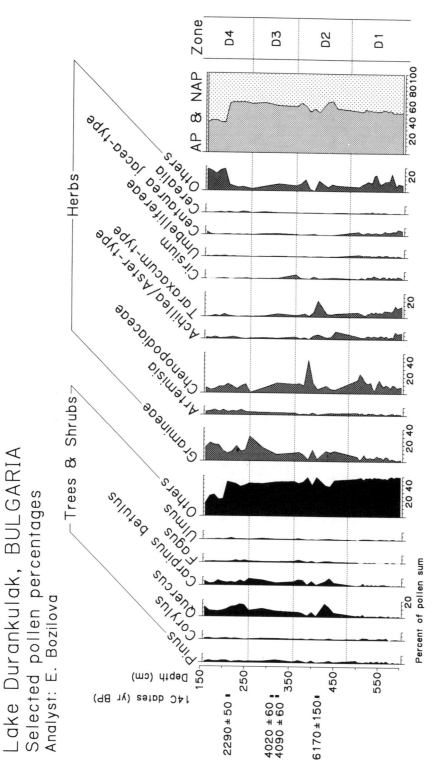

Fig 21.4 Pollen diagram with selected taxa from Lake Durankulak-2

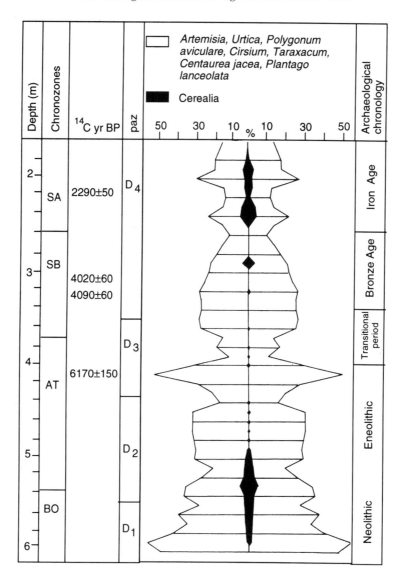

Fig. 21.5   Indicators of human influence selected from Lake Durankulak-2

ShE-4 5000–3000 BP   Gramineae–*Artemisia*–
                       Chenopodiaceae–
                       *Quercus–Carpinus
                       betulus–Corylus*

ShE-5 3000–0 BP   Gramineae–*Artemisia*–
                       Chenopodiaceae–
                       *Quercus–Corylus–
                       Carpinus betulus*

General patterns

(1) The arrival of *Quercus, Carpinus betulus,
     Ulmus*, and *Corylus* is in the early Holocene.

(2) The expansion of *Quercus, Corylus*, and *Ulmus*
     date after 6800 BP.

(3) A weak *Ulmus* decline is registered at approxi-
     mately 5600 BP (Fig. 21.6).

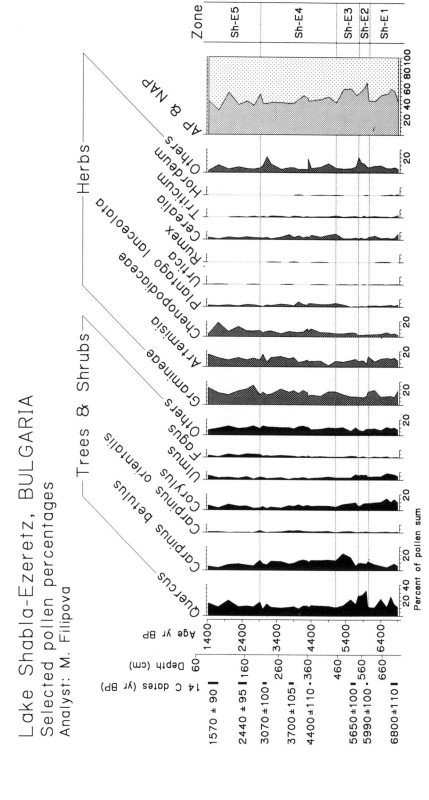

Fig 21.6 Pollen diagram with selected taxa from Lake Shabla-Ezeretz. Pollen data plotted against time

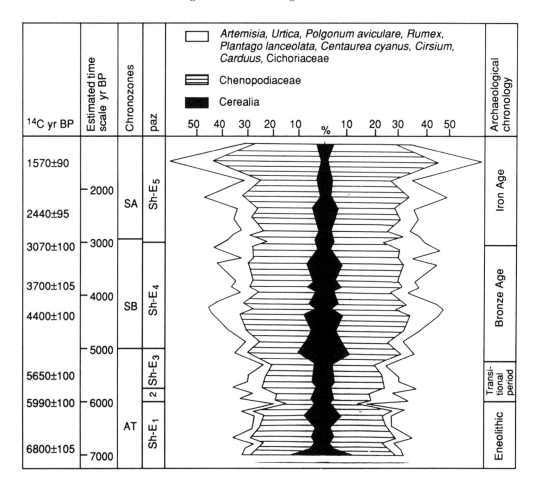

Fig. 21.7 Indicators of human influence selected from Lake Shabla-Ezeretz

(4) The increase of *Carpinus betulus* is around 5700–5500 BP.

(5) The constant presence of *Fagus* pollen begins after 6000 BP (Fig. 21.6).

(6) The decline of *Corylus* is after 5000 BP (Fig. 21.6).

(7) Cerealia pollen grains are found throughout with a three-fold increase at the beginning of the Bronze Age.

(8) Small values of *Plantago lanceolata* pollen are present in the period 6000–5600 BP (Figs 21.6 and 21.7).

Unique patterns

(1) The early-Holocene pollen assemblages are dominated by Gramineae, *Artemisia*, and Chenopodiaceae.

(2) Forest-steppe communities were already established by 7000 BP.

Reference site 4. Lake Varna (Tschakalova & Bozilova 1984, Bozilova & Ivanov 1985, Bozilova 1986, Bozilova & Filipova 1986)

Latitude 43°12′N, Longitude 27°50′E. Elevation

0 m. Age range 9000–0 BP. Estuary lake. Four pollen diagrams were produced, and one of them with selected tree and herb taxa is presented (Fig. 21.8). Three radiocarbon dates are used to date six pollen-assemblage zones. Information is also available from the analysis of molluscs, plant macrofossils, and from archaeological evidence.

B-1 and B-2	9000–8000 BP	Gramineae–Artemisia–Quercus–Corylus–
B-3	8000–6000 BP	Quercus–Corylus–Tilia–Artemisia–Chenopodiaceae
B-4	6000–5000 BP	Quercus–Carpinus betulus–Juniperus–Chenopodiaceae
B-5	5000–2800 BP	Quercus–Carpinus betulus–Fraxinus excelsior–Gramineae
B-6	2800–0 BP	Quercus–Carpinus betulus–Chenopodiaceae

General patterns

(1) The early-Holocene pollen assemblages are dominated by Gramineae and *Artemisia* with some *Quercus* and *Corylus* .
(2) *Ulmus* and *Tilia* are continuously present since 7500 BP.
(3) The maximal distribution of *Quercus* and *Carpinus betulus* was characteristic for the late Atlantic and early Subboreal (paz B-4, B-5).
(4) The increase of Chenopodiaceae in the Subatlantic (paz B-6) reflects the spreading of halophytic communities on fallow land.

Unique patterns

(1) The presence of anthropophytes since the Eneolithic (5500 BP) is explained by the human settlements around the lake (Fig. 21.9).

(2) Seeds of *Triticum monococcum, T. dicoccum, Hordeum vulgare var. nudum, Setaria*, and *Vitis*, and charcoal of *Ulmus, Fraxinus, Quercus, Carpinus*, and *Acer* were found in the Early Bronze Age (Fig. 21.10).

Palaeoecological events in the northern Black Sea subregion (Fig. 21.11)

Landscape patterns

(1) Open oak forests were distributed along the Black Sea coast in the early Holocene. In the lake area of Durankulak–Shabla–Ezeretz steppe vegetation persisted. Xerophytic herb communities played a dominant role with stands of *Quercus, Corylus, Ulmus, Tilia*, and *Betula*.
(2) During the Atlantic in the coastal area of Lake Varna, balanced mesoxerophilous oak and hornbeam forests were already established, while to the north the forest–steppe communities retained a dominant position.
(3) Expansion of the steppe communities in the late Subboreal and Subatlantic was due mainly to climatic change and anthropogenic activity.

Hydrological events

The hydrological events in the coastal lakes were studied through analysis of molluscs and diatoms and changes in the aquatic vegetation.

Common patterns

(1) The coastal lakes were formed around 9000 BP when water from the Mediterranean penetrated into the Black Sea.
(2) The constant inflow of sea water and the establishment of brackish conditions in the lakes has been characteristic since the Atlantic.

Unique patterns

(1) One short period of low water level in Lake Shabla-Ezeretz around 6800 BP is confirmed by the absence of pollen and the find of terrestrial gastropods.

710

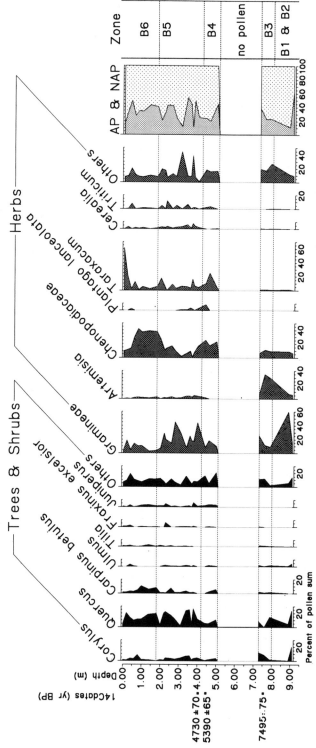

Fig 21.8 Pollen diagram with selected taxa from Lake Varna-Arsenala

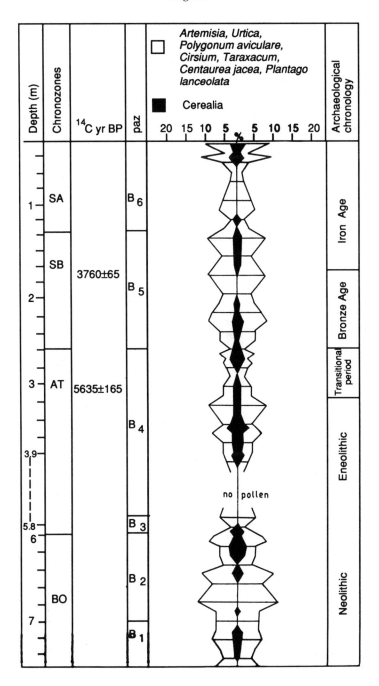

Fig. 21.9 Indicators of human influence selected from Lake Varna-Poveljanovo-2

ARCHAEOLOGICAL CHRONOLOGY	Lake Durankulak (Boeva, unpubl.)		Lake Varna (Tschakalova & Bozilova 1984)	
	Seeds	Wood	Seeds	Wood
NEOLITHIC	Triticum spelta Triticum sp. Hordeum sp. Setaria viridis Lens esculenta Panicum miliaceum Polygonum sp. Rumex acetosella Sapponaria officinalis			
ENEOLITHIC	Tr. monococcum Tr. dicoccum Hordeum vulgare Lens esculenta			
TRANSITIONAL PERIOD				
BRONZE AGE	Tr. monococcum Tr. dicoccum Tr. aestivum Tr. spelta Tr. beoticum Hordeum sp. Vicia ervilia Panicum italicum	Sambucus nigra Cornus mas Prunus avium	Tr. monococcum Tr. dicoccum Hordeum vulgare var. nudum H. vulgare Setaria sp. Lens sp. Galium sp.	Quercus Acer Alnus Fraxinus Carpinus Cornus mas Ulmus Vitis Euonymus
IRON AGE				
HISTORICAL TIME MIDDLE AGE	Vitis sylvestris Pisum sp. Tr. durum Tr. spelta			

Fig. 21.10 Plant macrofossil remains for Lake Durankulak and Lake Varna

(2) The link between Lake Shabla-Ezeretz and the Black Sea was disconnected between 3100 and 2000 BP and around 1500 BP. The evidence comes from the presence of fresh-water molluscs and the absence of diatoms.

Anthropogenic events

Early human influence on the natural vegetation started with the end of the Neolithic and increased during the Eneolitic and Bronze Age.

Climatic events

(1) The climate in the lake area of Durankulak–Shabla–Ezeretz was dry at the beginning of the Holocene due to the prevailing northerly winds. In the Lake Varna district the general improvement of the climate caused replacement of xerophytic steppe communities by forests.

(2) The distribution of forest-steppe communities in the lake area of Durankulak-Shabla–Ezeretz since 7000 BP was determined by the specific climatic conditions along the sea coast.

713

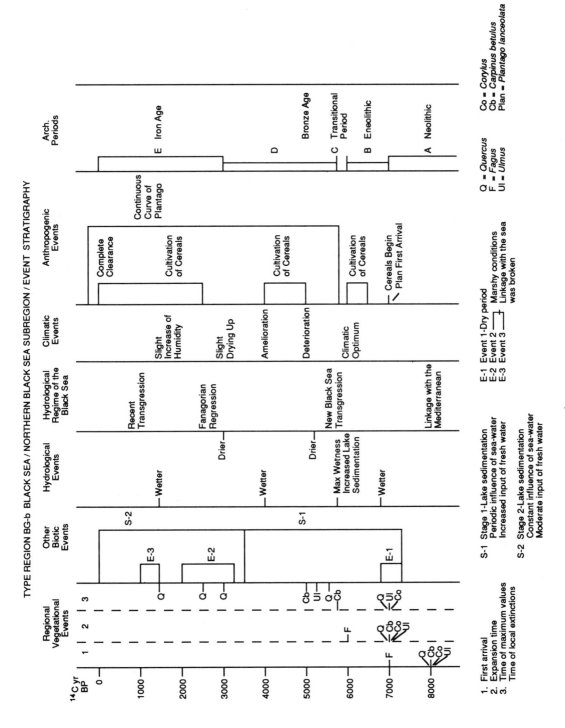

Fig. 21.11 Event stratigraphy for type region BG-b, Black Sea coast (northern subregion)

TYPE REGION BG-b

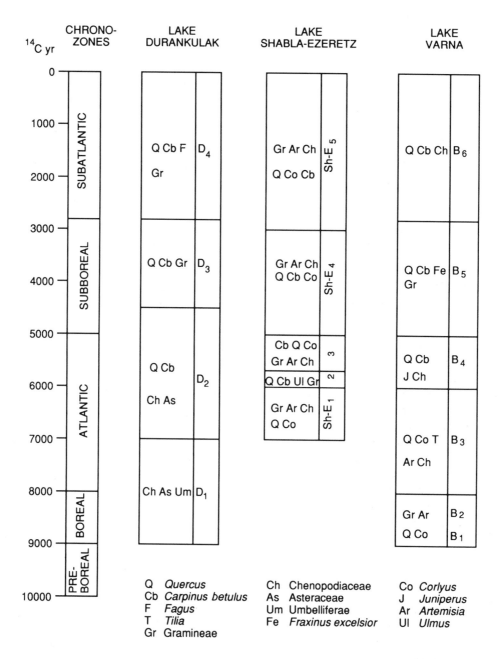

Fig. 21.12   Time-space correlation of local pollen-assemblage zones from type region BG-b, Black Sea coast (northern subregion).

(3) An increase in humidity, particularly in the Lake Varna district, occurred during the Subatlantic, influencing the formation of the characteristic flooded forests.

Soil events

Podzolization and degradation of the soils was synchronous with periods of stronger human activities.

TYPE REGION BG-d, STARA PLANINA AND SREDNA GORA MOUNTAINS
(L. Filipovich)

Altitude: 500–2200 m.

Climate: Moderate continental–mountainous, mean annual temperature 4–5°C; precipitation 800 mm yr^{-1} or more.

Geology: Palaeozoic and Mesozoic shale, granite, and syenite.

Topography: Mountainous.

Population: Approximately 43–66 people km^{-2}, in villages and towns.

Vegetation: Mixed *Quercus* forests up to 700–800 m, including on northern slopes scattered forests of *Carpinus betulus*. The belt of *Fagus sylvatica* covering the slopes above 800 m forms the upper tree-line around 1700 m. A narrow strip of conifers (50–100 m) is situated on the northern slopes under the highest peaks in western and middle Stara Planina. Herb vegetation and communities of *Juniperus sibirica*, *Vaccinium myrtillus*, *V. uliginosum*, and *Bruckenthalia spiculifolia* occupy the highest slopes. The natural vegetation has been seriously damaged by human activity.

Soils: Brown mountain-forest, and mountain-meadow soils.

Land use: Pasture land, arable land around the villages, fruit-growing, forestry.

Reference site 7. Tschumina Peat Bog
(Filipovich 1977, Petrov & Filipovich 1987)

Latitude 42°37′N, Longitude 24°22′E. Elevation 1300 m. Age range 4500–0 BP. One pollen diagram with four radiocarbon dates is presented. Three main pollen assemblage zones are defined (Fig. 21.13).

X   4500–4000 BP   *Corylus–Tilia–Ulmus–Quercus*
Y   4000–1700 BP   *Carpinus betulus–Abies*
Za   1700–400 BP   *Fagus–Abies*
Zb   400–0 BP   *Fagus*

Pollen-analytical investigations of 15 peat bogs located between 1400 and 1800 m have been carried out. The radiocarbon dates for eight samples from three profiles allow the correlation of the pollen diagrams. The larger parts of these peat bogs were formed about 2000 years ago (site 6, Stara Planina). Two of them reflect a period of about 4500 years (site 5, Stara Planina; 7, Sredna Gora) and one about 8000 years (site 8, Sredna Gora). Four regional pollen-assemblage zones are established (Figs 21.14 and 21.15).

W – *Betula, Salix,* and Gramineae dominant. Upper zone border at the increase of *Ulmus, Tilia,* and *Quercus*.
X – *Ulmus, Tilia,* and *Quercus* dominant. Upper zone border at the expansion of *Abies, Picea,* and *Carpinus betulus*.
Y – Conifers and *Carpinus betulus* dominant. Upper zone border at the beginning of the *Fagus* expansion.
Z – *Fagus* dominant, conifers and *Carpinus betulus* decreasing. The subzone border Za/Zb is at the AP decrease and the increase of *Juniperus, Carpinus orientalis,* and all anthropogenic taxa.

General patterns

(1) The first pollen grains of *Picea* and *Abies* appear about 7000 BP.
(2) The highest values of *Picea* and *Abies* as well as of *Carpinus betulus* occur between 4000 and 2000 BP.
(3) The first appearance of *Fagus* grains is about 4000 BP, and its expansion begins about 2000 BP.
(4) *Corylus* has maximal percentages at about 5000–4000 BP.
(5) The period between 8000 and 4000 BP is characterized by the highest percentage values of *Tilia, Ulmus,* and *Quercus* and relatively high

716

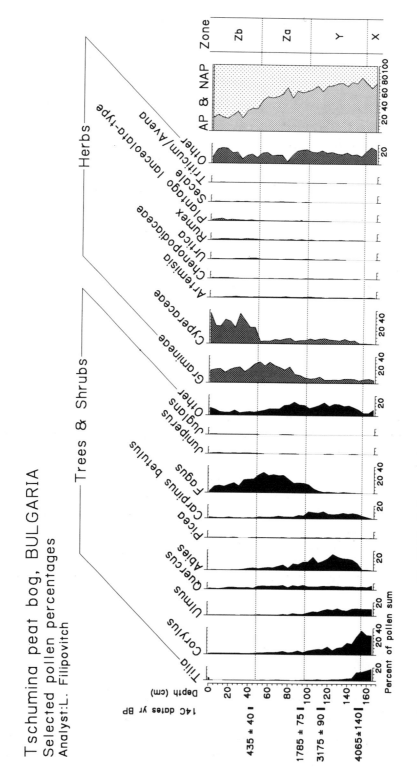

Fig. 21.13 Pollen diagram with selected taxa from Tschumina peat bog

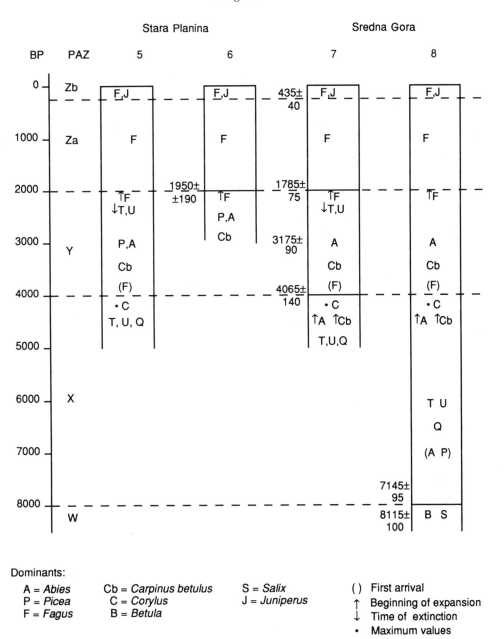

Fig. 21.14 Correlation of some pollen diagrams from type region BG-d

^{14}C date BP	PAZ	Forest phases	Main vegetational events	Supposed climatic conditions	Anthropogenous influence
0	b	Beech-coniferous forests	Expansion of *Juniperus* Lowering of upper forest boundry Deforestation		Very strong
435±40	Z				
1000					
1785±75	a		Formation of modern vegetational belts	Slight lowering of temperature	Cerealia, *Pl. lanc.*, *Pl. media*
2000					
3000	Y	Hornbeam coniferous forests	Beginning of beech expansion Formation of many peat bogs		Pasture
3175±90					
4000			Appearance of *Fagus*	Increase of humidity	*Pl. lanceolata* *Pl. major*
4065±140			Maximum of *Corylus*		
5000					
	X	Mixed deciduous forests			
6000				Higher temperature	Summer pasture ?
7000			Appearance of *Abies, Picea*		
7145±95					
8000					
8115±100	W	Open vegetation			

Fig. 21.15   Vegetational changes in Sredna Gora Mountains

amounts of *Acer* and *Fraxinus excelsior*.

(6) Destruction of the forests is marked about 430 BP, as indicated by the increase of *Juniperus*, *Carpinus orientalis*, and all herbs.

(7) The comparison of the pollen diagrams from

Stara Planina and Sredna Gora mountains shows a similar behaviour of the pollen curves. The only difference is the prevailence of *Abies* and almost no participation of *Picea* in the forests of Sredna Gora.

TYPE REGION BG-e, VITOSHA, RILA, PIRIN AND RHODOPES MOUNTAINS
(E. Bozilova)

Altitude: 300–2925 m.

Climate: Mountainous, mean temperatures January −6 to 0°C, July 8–22°C; precipitation 800 mm yr⁻¹.

Climate: Mountainous, mean temperatures January −6 to 0°C, July 8–22°C; precipitation 800 mm yr^{-1}.

Geology: Mainly Proterozoic and Palaeozoic metamorphic rocks (crystalline schist and marble) and intrusive rocks (granite), and plutonites (Vitosha).

Topography: Mountainous.

Population: Approximately 30 people km^{-2} in villages and small towns.

Vegetation: Approximately 60 % of the area is occupied by forests (50 % of them are coniferous forests of *Pinus sylvestris, P. nigra, P. peuce, P. heldreichii, P. montana, Picea abies, Abies alba*). The main components of the deciduous forests are *Quercus* spp., *Fagus sylvatica*, and *Carpinus betulus*.

Soils: Brown forest, cinnamonic forest, mountain meadow.

Land use: Part of the deforested area is cultivated, mountain meadows are used for stockbreeding.

Reference site 10. Lake Sucho Ezero
(Bozilova & Smit 1979, Bozilova *et al.* 1990)

Latitude 42°04′N, Longitude 23°35′E. Elevation 1900 m. Age range 12000–0 BP. Peat bog on former glacial lake. Two pollen diagrams and one macrofossil diagram are available. One simplified pollen diagram with two radiocarbon dates is presented and seven pollen-assemblage zones are defined (Fig. 21.16). A tree-arrival diagram is also presented (Fig. 21.17).

R-1	12000–10200 BP	*Artemisia*–Chenopodiaceae–*Pinus* Diploxylon–*Betula*
R-2	10200–9000 BP	*Betula*–*Pinus* Diploxylon–Gramineae
R-3	9000–7000 BP	*Betula*–*Quercus*–*Pinus* Diploxylon
R-4	7000–5000 BP	*Pinus*–*Abies*–*Quercus*
R-5	5000–2800 BP	*Pinus*–*Abies*
R-6	2800–1600 BP	*Pinus*–*Picea*–*Fagus*
R-7	1600–0 BP	*Picea*–*Pinus*–*Fagus*

General patterns

(1) The Late-Glacial pollen assemblages are dominated by *Artemisia*, Chenopodiaceae, and Gramineae with some *Pinus* and *Betula*.

(2) The transition to the Holocene is marked by rise of *Pinus* Diploxylon and *Betula* pollen values.

(3) The Preboreal and Boreal pollen assemblages are dominated by *Betula* and *Quercus*.

(4) The continuous pollen curves of *Pinus peuce* and *Abies* appear after 6000 BP.

(5) The Subboreal pollen assemblages are dominated by *Pinus* Diploxylon, *Pinus peuce,* and *Abies*.

(6) The expansion of *Picea* starts after 2800 BP in the Subatlantic.

Reference site 12. Kupena (Bozilova *et al.* 1989)

Latitude 41°59′N, Longitude 24°20′E. Elevation 1300 m. Age range 12000–0 BP. Mire over infilled lake. One pollen diagram with three radiocarbon dates is presented, and six pollen-assemblage zones are defined (Fig. 21.18). A tree-arrival diagram is also constructed (Fig. 21.19).

Rd-1	12000–10200 BP	*Artemisia*–Gramineae–Chenopodiaceae–*Pinus* Diploxylon–*Juniperus*
Rd-2	10200–9300 BP	*Artemisia*–Gramineae–Chenopodiaceae–*Pinus* Diploxylon
Rd-3	9300–7500 BP	*Quercus*–*Betula*–*Ulmus*
Rd-4	7500–5400 BP	*Quercus*–*Ulmus*–*Fraxinus excelsior*
Rd-5	5400–2700 BP	*Quercus*–*Corylus*–*Abies*
Rd-6	2700–0 BP	*Pinus* Diploxylon–*Fagus*–Gramineae

General patterns

(1) The Late-Glacial pollen assemblages are dominated by *Artemisia*, Chenopodiaceae, and

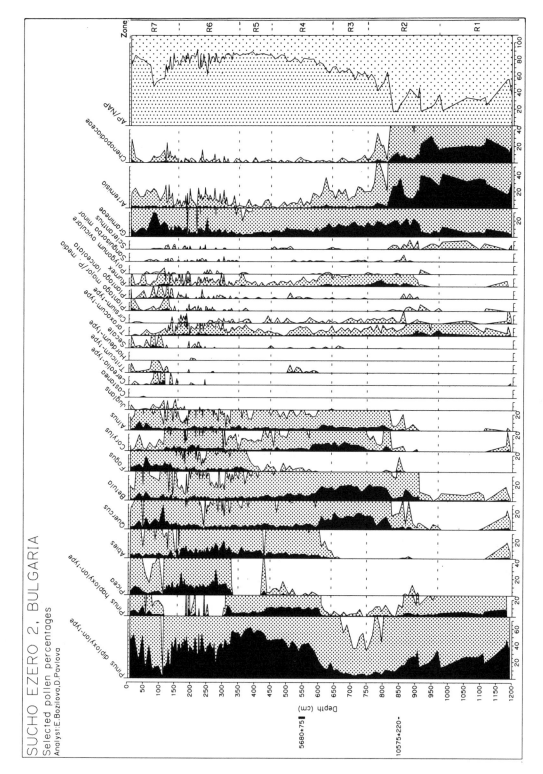

SUCHO EZERO 2, BULGARIA
Selected pollen percentages
Analyst:E.Bozilova,D.Pavlova

Fig. 21.16 Pollen diagram with selected taxa from Lake Sucho Ezero

TREE ARRIVAL AND PRESENCE IN WESTERN RHODOPES MOUNTAINS

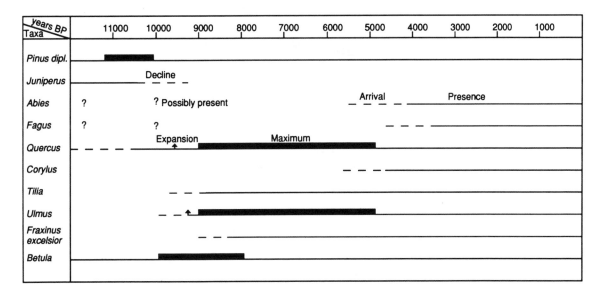

Fig. 21.17 Tree arrival and tree presence in Rila Mountains

Gramineae, with *Pinus* Diploxylon, *Juniperus,* and some *Betula.*

(2) The Boreal starts with the sharp rise of *Quercus* and *Ulmus* values.

(3) The pollen curves of *Corylus, Abies,* and *Fagus* appear, respectively, after 6500–5500, and 4500 BP.

(4) Pollen grain of *Hordeum* are found in the late Subboreal and in the Subatlantic (Figs. 21.18 and 21.20).

Common patterns in type region BG-e
(Fig. 21.21)

(1) The Late-Glacial pollen assemblages are dominated by *Artemisia* and Chenopodiaceae with *Pinus* Diploxylon, *Juniperus*, and *Betula*.

(2) Characteristic for the Preboreal is the pioneer *Betula* phase and the presence of *Pinus* Diploxylon.

(3) The spreading of Quercetum mixtum at lower elevations occurred during the Boreal and early Atlantic.

(4) The formation of the coniferous belt with *Pinus* and *Abies* started at the Atlantic after 6000 BP.

(5) In the Subboreal and Subatlantic coniferous and *Fagus* forests developed.

Unique patterns in type region BG-e

(1) In the Rhodopes Mountains *Pinus peuce* appeared in the Late Glacial together with deciduous trees, which presume the existence of local refugia.

(2) In the Rila Mountains the expansion of *Picea* started rather late after 2800 BP.

(3) In the Rhodopes Mountains the increase of *Fagus* was in the Subatlantic.

TYPE REGION BG-f, ZNEPOLE (S. Tonkov)

Altitude: 500–1700 m.

Climate: Continental, mean temperatures January –2°C, July 18°C; precipitation 700 mm yr[-1].

Geology: Triassic limestone and Jurasic conglomerate.

Topography: Valleys and lower mountains.

722

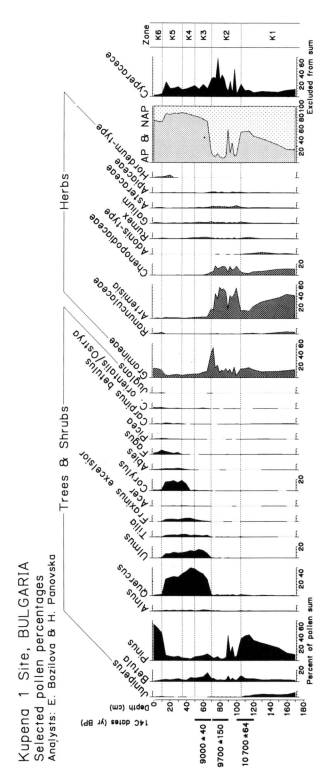

Fig. 21.18 Pollen diagram with selected taxa from Kupena-1

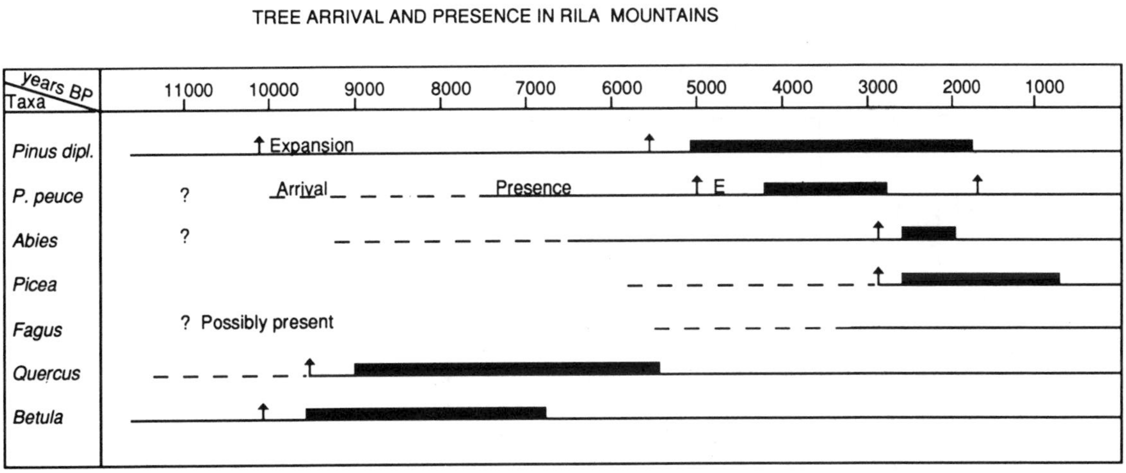

TREE ARRIVAL AND PRESENCE IN RILA MOUNTAINS

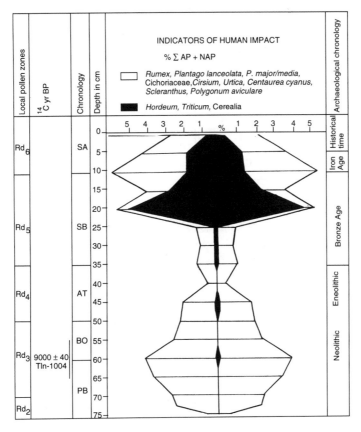

Fig. 21.19 Tree arrival and tree presence in Western Rhodopes Mountains

Fig. 21.20 Indicators of human influence selected from Kupena-1

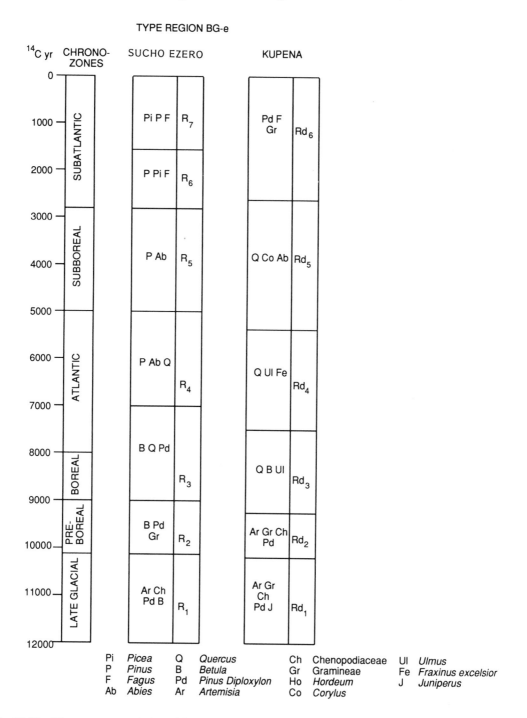

Fig. 21.21   Time–space correlation of local pollen-assemblage zones from reference sites BG-10 and BG-12

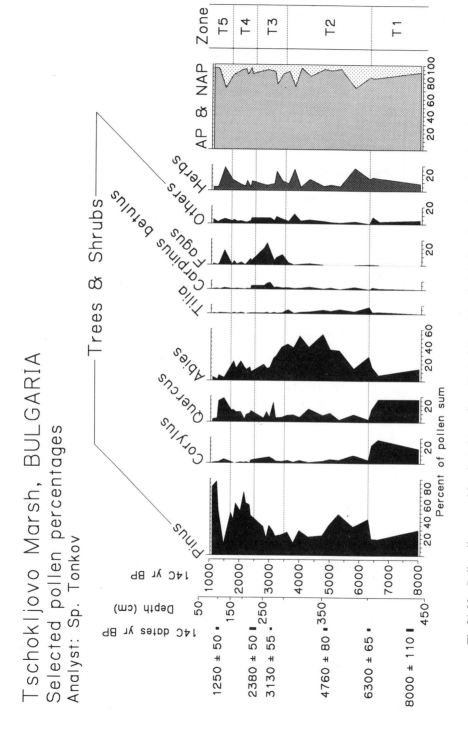

Fig 21.22 Pollen diagram with selected taxa from Tschokljovo Marsh-1. Pollen data plotted against time.

TREE ARRIVAL AND PRESENCE IN KONJAVSKA MOUNTAINS

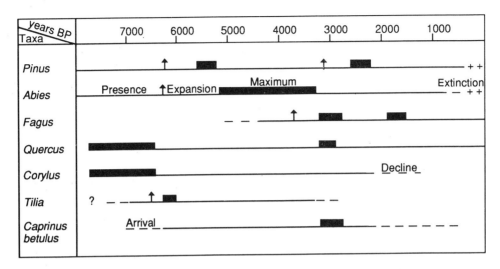

Fig 21.23 Tree arrival and tree presence in Konjavska Mountains

Population: Approximately 60–80 people km⁻² in villages and towns.

Population: Approximately 60–80 people km^{-2} in villages and towns.

Vegetation: Approximately 20% of the territory is occupied by deciduous forests of *Quercus* spp., *Carpinus betulus,* and *Fagus sylvatica.* In many places the natural vegetation is strongly influenced by anthropogenic activity.

Soils: Podzols and eroded cinnamonic forest soils, smolnitzas, and brown forest soils.

Land use: Most of the area is arable land.

Reference site 13. Tschokljovo Marsh
(Tonkov 1988, Tschakalova *et al.* 1990, Tonkov & Bozilova 1992)

Latitude 42°22′N, Longitude 22°50′E. Elevation 870 m. Age range 8000–0 BP. Marsh over infilled lake. Two pollen diagrams and one macrofossil diagram were produced. Time-scale pollen diagram with selected tree taxa is presented (Fig. 21.22). Five pollen assemblage zones are delimited and dated with the help of six radiocarbon dates. A tree-arrival diagram is also constructed (Fig. 21.23).

T-1	8000–6300 BP	*Quercus–Corylus–Pinus–Abies*
T-2	6300–3500 BP	*Abies–Pinus*
T-3	3500–2300 BP	*Pinus–Fagus–Abies*
T-4	2300–1700 BP	*Pinus–Abies–Quercus*
T-5	1700–0 BP	*Quercus–Fagus–Pinus*

General patterns

(1) The early Atlantic pollen assemblages are dominated by *Corylus* and *Quercus*, followed by *Pinus* and some *Abies*.
(2) The pollen values of *Abies* increase in the late Atlantic and are dominant in the early Subboreal about 4700–4300 BP.
(3) The pollen curve of *Fagus* starts at 4400 BP.
(4) The presence of *Carpinus betulus* and *Tilia* is recorded for the late Atlantic and the Subboreal.
(5) The decline of all coniferous and deciduous trees occurs after 1250 BP.

CONCLUSIONS

The vegetation development in Late-Glacial and Holocene was determined by the geomorphic and edaphic characteristics and by dynamic changes, as well as by the location of glacial refugia for trees and by the millennial human impact on the environment.

In the Late-Glacial steppe communities were distributed in the lowlands, composed of Chenopodiaceae, *Artemisia*, and stands of trees. At higher elevations (500–800 m) in isolated habitats with sufficient moisture, groups of deciduous trees were present. The dominant vegetation was the herbaceous xerophytic steppe communities of Chenopodiaceae, *Artemisia*, and Gramineae with *Juniperus*, *Ephedra*, and some *Betula* and *Pinus* during the interstadials.

A long dynamic period of vegetation development after the final improvement of the climate was characteristic for the Rila, Pirin, and Rhodopes Mountains (BG-e), expressed by the alternation of *Betula*, *Pinus–Betula*, *Pinus–Abies–Fagus*, and *Picea* phases at higher elevations and a vast distribution of Quercetum mixtum forests at lower levels.

In the lower mountains this dynamic period was rather shorter, with the establishment and succession of the *Betula*, *Pinus–Abies*, *Carpinus betulus*, and *Fagus* phases (BG-d and BG-f).

The vegetation development in the lowlands of eastern Bulgaria along the coastal area (BG-b) took three different directions. The first one is characteristic for the lake area Durankulak–Shabla Ezeretz, with primary steppe vegetation and occurrence of forest–steppe communities during the climatic optimum.

In the plateau area around Lake Varna, *Quercus* and later *Carpinus betulus* forests dominated the greater part of the Holocene.

The third line of vegetation development is established in the area of Lake Arkutino (BG-i) in the foot-hills of the Strandza Mountains. Forests of Quercetum mixtum and *Fagus orientalis* were present since the early Holocene, followed by the expansion of *Carpinus betulus*. The occurrence of specific flood-plain forests with *Fraxinus, Ulmus, Carpinus betulus, Alnus,* and climbers dates back to 3000 BP.

Signs of human influence are present in all parts of the country since the Neolithic (8200–7500 BP). The first Cerealia pollen dates from around 8000 BP. The first clearances and the effects of grazing are visible in all type regions, particularly in the mountain areas.

ACKNOWLEDGEMENTS

The authors are thankful to the leader of the IGCP Project 158B, Prof. Björn Berglund for his comprehensive help in organizing the research activities in Bulgaria.

Thanks are due to the Bulgarian Committee of IGCP for assistance in the elaboration of the national project programme.

REFERENCES

Bozilova, E. 1986: *Palaeoecological conditions and changes of the vegetation in eastern and southwestern Bulgaria during the last 15000 years.* DSc Thesis, Sofia University Press, Sofia 318 pp.

Bozilova, E. & Atanassova, J. 1990: Palaookologischen Bedingungen und die Pflanzengeschichte der Umgebung von Durankulak. *In* Todorova, H. (ed.) *Durankulak I*, 197–205. Bulgarian Academy of Sciences, Sofia.

Bozilova, E. & Filipova, M. 1986: Palaeoecological environment in northeastern Black Sea during Neolithic, Eneolithic and Bronze periods. *Studia Praehistorica 8*, 160–166. Sofia.

Bozilova, E. & Ivanov, I. 1985: Palaeoenvironment in the area of the Varna lake during Eneolithic and Bronze age on the basis of palynological, palaeoethnobotanical and archaeological evidence. *Bulletin Musee National Varna XXI*, 43–50.

Bozilova, E. & Smit, A. 1979: Palynology of Lake "Sucho Ezero" from South Rila Mountain (Bulgaria). *Fitologija 11*, 54–67.

Bozilova, E. & Tonkov, Sp. 1985: Vegetational development in the mountainous areas of southwestern Bulgaria. I. Palynological investigations and reconstruction of past vegetation. *Ecologia Mediterranea 11*, 33–37.

Bozilova, E., Panovska, H. & Tonkov, Sp. 1989: Pollenanalytical investigations in the Kupena National Reserve, West Rhodopes. *Geographica Rhodopica 1*, 186–190. Sofia.

Bozilova, E., Tonkov, Sp. & Pavlova, D. 1990: Pollen and plant macrofossil analyses of the Lake Sucho Ezero in the South Rila mountain. *Annuaire de'l Universite de Sofia, Faculte de Biologie 80 (2)*, 48–57.

Filipova, M. 1985: Palaeoecological investigations of Lake Shabla-Ezeretz in north-eastern Bulgaria. *Ecologia Mediterranea 11*, 147–158.

Filipova, M. & Bozilova, E. 1990: Palaeoecological investigations of Lake Shabla-Ezeretz along the Black Sea coast.
In Krastev, T. (ed.): *Geological evolution of the western part
of the Black Sea depression in Neogen-Quaternery*, 41–87.
Bulgarian Academy of Sciences, Sofia.

Filipovich, L. 1977: Postglacial forest phases on the high
slopes of the Balkan range (Bulgaria). *In Proceedings of
Working Session of Commission of Holocene*—INQUA,
173–178, Bratislava.

Petrov, S. & Filipovich, L. 1987: Postglacial changes of the
vegetation on the slopes of Sredna Gora mountain. In:
Proceedings of Fourth National Botanical Conference,
399–406. Sofia.

Tonkov, Sp. 1988: Sedimentation and local vegetation development of a reference site in southewestern Bulgaria. *In*

Lang, G. & Schluchter, Ch. (eds.) Lake, Mire and
River environment during the last 15000 years, 99–
101. Balkema, Rotterdam.

Tonkov, Sp. & Bozilova, E. 1992: Palaeoecological investigation of Tschokljovo marsh (Konjavska mountain). *Annuaire de'l Universite de Sofia, Faculte de
Biologie 83 (2)*, 5–16.

Tschakalova, E. & Bozilova, E. 1984: Subfossil material from Early Bronze. *Annuaire de'l Universite de
Sofia, Faculte de Biologie 74 (2)*, 18–27.

Tschakalova, E., Stojanova, D. & Tonkov, Sp. 1990:
Plant macrofossil remains from Tschokljovo marsh
(Konjavska mountain). *Annuaire de'l Universite de
Sofia, Faculte de Biologie 80 (2)*, 41–47.

Appendix: Biostratigraphical Reference Site Tables

AUSTRIA

Type region	Reference site No., name, lat./long., elevation	Lake	Mire	Time span 14C yrs BP	14C dates, no.	Pollen %	Pollen conc.	Macrofossils Bot.	Macrofossils Zool.	Chem. anal.	Publ. year	Main scientists
A-e	4 Halleswiessee, 47°45'N, 13°32'E, 781 m	X		11000-0	2	X	X				unpubl.	Handl
A-g	8 Lanser See, 47°14'N, 11°25'E, 840 m	X		14000-9500	4	X	X	B	Z		1984 a,b; 1992	Bortenschlager; Oeggl
A-g	9 Salober Moor, 47°35'N, 10°35'E, 1089 m		X	14000-0	4	X	X				in press	Bortenschlager
A-g	10 Schwemm, 47°39'N, 12°18'E, 664 m		X	14000-0	26	X	X	B		X	1988a	Oeggl
A-h	11 Großes Überling Schattseit Moor, 47°10'N, 13°54'E, 1730 m		X	14000-0	5	X	X				1991	Krisai et al.
A-h	14 Dortmunder Hütte, 47°12'N, 11°1'E, 1880 m		X	9000-0	6	X	X				1987	Hüttemann

BELGIUM

Type region	Reference site No., name, lat./long., elevation	Lake	Mire	¹⁴C yrs BP Time span	¹⁴C dates, no.	Pollen %	Pollen conc.	Microfossils Diatoms	Publ. year	Main scientists
B-a	1 Avekapelle, 51°04'55"N, 2°44'10"E, 0.7 m		X	5000–3300	2	X			1979	Baeteman & Verbruggen
B-a	2 Doel, 51°18'15"N, 4°15'10"E, –0.3 m		X	5500–2000	4	X			1986	Minnaert & Verbruggen
B-b	3 Moerbeke, 51°10'17"N, 3°56'5"E, 4 m		X	12000–11000	2	X	X	D	1979	Verbruggen
B-b	4 Snellegem, 51°11'12"N, 3°16'48"E, 12 m	X		12200–10300	4	X	X	D	1979	Verbruggen
B-b	5 Vinderhoute, 51°04'45"N, 3°37'22"E, 4 m	X		12800–7500	3	X			1979	Verbruggen
B-b	6 Berlare, 51°02'17"N, 3°58'54"E, 1.5 m		X	9500–7600	1	X			1971	Verbruggen
B-b	7 Dendermonde, 51°02'38"N, 4°05'23"E, 0.2 m		X	5000–1600	2	X			1971	Verbruggen
B-b	8 Moerzeke, 51°02'54"N, 4°10'35"E, 0.3 m	X		5000–1600	2	X			1971	Verbruggen
B-b	9 Uitbergen-Heisbroek, 51°01'04"N, 3°56'17"E, 2 m		X	5300–1000	2	X			1971	Verbruggen
B-c	10 Wortel, 51°23'55"N, 4°47'38"E, 14 m		X	10000–1500	5	X		D	1982	Beyens
B-d	11 Rosières, 50°44'37"N, 4°33'29"E, 43 m		X	10300–1500	4	X			1987	Ntaganda & Munaut

BULGARIA

Type region	Reference site No., name, lat./long., elevation	Lake	Mire	Time span 14C yrs BP	14C dates, no.	Pollen %	Pollen conc.	Micro-fossils Diatoms	Micro-fossils X-others	Macro-fossils Bot. Zool.	Chem. anal.	Chem. year	Main scientists
BG-a	1 Lake Srebarna, 44°05'N, 27°07'E, 20 m	X		8000–0		X					X	1989	Lazaarova & Bozilova
BG-b	2 Lake Durankulak, 43°15'N, 28°23'E, 0 m	X		9000–0	6	X		D		B,Z		1985,1990	Bozilova
BG-b	3 Lake Shabla-Ezeretz, 43°35'N, 28°20'E, 0 m	X		7000–0	10	X		D		Z		1985,1990	Filipova
BG-b	4 Lake Varna, 43°12'N, 27°50'E, 0 m	X		9000–0	5	X				B,Z		1986	Bozilova
BG-d	5 Kom, 43°11'N, 23°03'E, 1800 m		X	4500–0		X						1981	Filipovitch
BG-d	6 Petrohan, 43°08'N, 23°09'E, 1450 m		X	2000–0	2	X						1981	Filipovitch
BG-d	7 Tschumina, 42°37'N, 24°22'E, 1300 m		X	4500–0	4	X						1977,1987	Filipovitch
BG-d	8 Bogdan, 42°37'N, 24°28'E, 1500 m		X	8000–0	2	X						1987	Filipovitch
BG-e	9 Vitosha Mts, 42°35'N, 23°17'E, 1400–2200 m		X	6000–0	2	X						1988	Filipovitch
BG-e	10 Lake Sucho Ezero, 42°04'N, 23°35'E, 1900 m		X	12000–0	4	X				B		1979,1990	Bozilova
BG-e	11 Lake Bezbog, 41°45'N, 23°32'E, 2230 m	X		9000–0	4	X							Stefanova & Bozilova
BG-e	12 Kupena, 41°59'N, 24°20'E, 1300 m		X	12000–0	3	X						1989	Bozilova & Panovska
BG-f	13 Tschokljovo Marsh, 42°22'N, 22°50'E, 870 m		X	8000–0	9	X				B	X	1988,1992	Tonkov & Bozilova
BG-g	14 Sadovo, 42°10'N, 24°54'E, 150 m		X	4000–0	3	X						1990	Filipovitch
BG-h	15 Maleshevska Mts peat-bog, 41°42'N, 23°02'E, 1720 m		X	7000–0	4	X						1992	Tonkov & Bozilova
BG-i	16 Lake Arkutino, 42°22'N, 27°44'E, –4 m	X		8000–0	11	X		X		B,Z		1992	Bozilova & Beug

CZECH AND SLOVAK REPUBLICS

Type region	Reference site: No., name, lat./long., elevation	Lake	Mire	Time span ^{14}C yrs BP	^{14}C dates no.	Micro-fossils Bot.	Publ. year	Main scientists
CS-b	1 Komořanské jezero, 50°30'N, 13°32'E, 230 m	X	X	9000–0	4		1983,1984b 1988b	Jankovská
CS-c	2 Řežabinec, 49°15'N, 14°05'E, 370 m	X	X	9500–0 H	9	B	1985,1987	Rybníčková & Rybníček
CS-c	3 Zbudov, 49°05'N, 14°19'E, 380 m		X	11000–0 H	4	B	1975	Rybníčková et al.
							1982,1987	Rybníčková & Rybníček
CS-c	4 Červené blato, 48°52'N, 14°48'E, 470 m		X	12500–0		B	1980	Jankovská
CS-c	5 Mokré louky, 49°01'N, 14°46'E, 425 m		X	10000–0	4	B	1987	Jankovská
CS-c	6 Borkovice, 49°16'N, 14°36'E, 415 m		X	12000–0	4	B	1980	Jankovská
CS-c	7 Bláto, 49°03'N, 15°12'E, 465 m		X	11200–0	2	B	1968	Rybníček & Rybníčková
							1974	Rybníčková
CS-c	8 Řásná, 49°14'N, 15°23'E, 680 m		X	10100–0	1		1974	Rybníčková
CS-c	9 Loučky, 49°19'N, 15°32'E, 580 m		X	10200–0 H	1		1974	Rybníčková
							1987	Rybníček & Rybníčková
CS-c	10 Kameničky, 49°44'N, 16°03'E, 625 m		X	11300–0	8	B	1988	Rybníčková & Rybníček
CS-d	11 Pančická louka, 50°46'N, 15°32'E, 1325 m		X	7700–0	4		1987	Hüttemann & Bortenschlager
CS-d	12 Verněřovice, 50°33'N, 16°21'E, 450 m		X	11800–0	9		1979	Peichlová
CS-f	13 Svatobořice-Mystřín, 48°57'N, 17°04'E, 175 m	X	X	8000–0	5		1989	Svobodová
CS-f	14 Vracov, 48°58'N, 17°13'E, 190 m	X	X	12000–0	6	B	1972	Rybníčková & Rybníček
CS-f	15 Cerová-Lieskové, 48°14'N, 17°21'E, 120 m	X	X	12000–0			1965	Krippel
CS-g	16 Kameničná-Pusté Ulany, 48°14'N, 17°32'E, 120 m	X		10600–0			1963	Krippel
CS-i	17 Zlatnická dolina, 49°29'N, 19°16'E, 900 m		X	8000–0	5		1989a	Rybníčková & Rybníček
CS-j	18 Bobrov, 49°27'N, 19°34'E, 640 m		X	10400–0	12		1989b	Rybníčková & Rybníček
CS-j	19 Ivančiná, 48°55'N, 18°49'E, 455 m		X	7000–0			1974	Krippel
CS-j	20 Liptovský Ján, 49°03'N, 19°41'E, 660 m		X	13000–0	3		unpubl.	Dolejšova
CS-j	21 Hozelec, 49°02'N, 20°21'E, 685 m		X	11500–4500	3		1988a	Jankovská
CS-j	22 Sivár,a, 49°19'N, 20°36'E, 615 m		X	14000–700	4	B	1984a	Jankovská
CS-j	23 Trojrohé pleso, 49°14'N, 20°15'E, 1650 m		X	6000–0	5		unpubl.	Jankovská
							1987	Hüttemann & Bortenschlager
CS-m	24 Hypka,a, 48°55'N, 22°09'E, 820 m	X	X	10300–0*			1971	Krippel
CS-n	25 Rad, 48°28'N, 21°52'E, 102 m	X	X	(4500–0)			unpubl.	Krippel

H Hiatus * Correlation to data by Ralska-Jasiewiczowa (1980, 1989)

DENMARK

Type region	Reference site No., name, lat./long., elevation	Lake	Mire	Time span ¹⁴C yrs BP	¹⁴C dates, no.	Pollen %	conc.	Micro-fossils Diatoms	Macro-fossils Bot.	Zool.	Chem. anal.	Publ. year	Main scientists
DK-a	1 Solsø, 56°08'N, 8°37'E, 40 m	X		$10500-0$	41	X	X	X			X	1994	Odgaard
DK-a	2 Skånsø, 56°31'N, 8°50'E, 8 m	X		$10500-0$	37	X	X	D	B		X	1994	Odgaard
DK-a	3 Kragsø, 56°15'N, 9°06'E, 45 m	X		$9000-0$	19	X	X				X	1994	Odgaard
DK-b	4 Holmegaard Mose, 55°18'N, 11°48'E, 31 m	X	X	$10000-0$	30	X	X				X	1983	Aaby
DK-b	5 Fuglsø Mose, 56°29'N, 10°35'E, 31 m		X	$8000-400$	50	X	X				X	1985	Aaby
DK-b	6 Elsborg Mose, 56°26'N, 10°28'E, 60 m	X	X	$8000-0$	15	X						1984	Andersen
DK-b	7 Abkær Mose, 55°11'N, 9°21'E, 50 m		X	$9000-0$	30	X	X		B		X	1988	Aaby
DK-b	8 Gudme Sø, 55°09'N, 10°45'E, 49 m	X		$5500-0$	18	X	X		B			in press	Rasmussen
DK-b	9 Hassing Huse Mose, 56°49'N, 8°26'E, 5 m	X	X	$10000-0$	22	X					X	in press	Andersen

735

ESTONIA

Type region	No., name, lat./long., elevation	Lake	Mire	Time span 14C yrs BP	14C dates, no.	Pollen %	conc.	Micro-fossils Diatoms	Macro-fossils Bot.	Chem. anal.	Publ. year	Main scientists
E-a	1 Ülemiste, 59°24'N, 27°47'E, 33 m	X		9000–0	2	X		D		X	1986	Kessel et al.
E-a	2 Maardu, 59°27'N, 25°00'E, 32 m	X		9600–0	9	X		D	B	X	1992	Veski
E-b	3 Pelisoo, 58°26'N, 22°13'E, 32.5–34.3 m		X	9000–0	5	X				X	1990	Saarse et al.
E-b	4 Karujärv, 58°22'N, 22°07'E, 32 m	X		8000–0	10	X		D		X	1990	Saarse et al.
E-b	5 Ermistu, 58°22'N, 23°58'E, 17.8 m	X		9700–0	5	X		D		X	unpubl.	Veski & Heinsalu
E-c	6 Viitna, 59°27'N, 26°01'E, 74.4 m	X		10200–0	3	X				X	unpubl.	Pirrus
E-c	7 Nigula, 58°00'N, 24°40'E, 59–61 m		X	9000–0	26	X			B	X	1976	Sarv & Ilves
E-d	8 Raigastvere, 58°35'N, 26°39'E, 51.8 m	X		10000–0	11	X				X	1987a	Pirrus et al.
E-e	9 Saviku, 58°26'N, 27°15'E, 30–31 m		X	10200–0	13	X			B		1975	Sarv & Ilves
E-e	10 Kalina, 59°16'N, 27°16'E, 70 m		X	10000–0	12	X			B		1969	Ilves & Sarv
E-f	11 Päidre, 58°16'N, 25°30'E, 50.6 m	X		11000–0	22	X				X	unpubl.	Saarse et al.
E-g	12 Tuuljärv, 57°41'N, 27°08'E, 257 m	X		10500–0	9				B		1987	Ilves & Mäemets
E-g	13 Vaskna, 57°43'N, 27°04'E, 244 m	X		11000–0	17	X			B		1987	Ilves & Mäemets
E-g	14 Kirikumäe, 57°40'N, 27°15'E, 183 m	X		11000–0	9	X				X	unpubl.	Saarse et al.

FAROE ISLANDS

Type region	Reference site No., name, lat./long., elevation	Lake	Mire	Time span ^{14}C yrs BP	^{14}C dates, no.	Pollen %	Chem. anal.	Publ. year	Main scientists
FR	1 Saksunarvatn, 62°14', 7°9'W, 25m	X		10000–2900	16	X	X	1982	Jóhansen
	2 Hoydalar, 62°1'N, 6°46'W, 14 m		X	10000–3600	7	X		1975	Jóhansen
	3 Hovi, 61°30'N, 6°46'W, 12 m		X	3000–0	7	X		1982	Jóhansen
	4 Tjørnuvík, 62°18'N, 7°9'W, 5.5 m		X	2000–0	5	X		1971	Jóhansen
	5 Lambi, 62°6'N, 7°39'W, 50 m		X	1400–0	7	X		1979	Jóhansen
	6 North of Uldalid, 62°6'N, 7°38'W, 190 m		X	2000–0	6	X		1979	Jóhansen

FINLAND

Type region	Reference site No., name, lat./long., elevation	Lake	Mire	Time span ^{14}C yrs BP	^{14}C dates, no.	Pollen %	conc.	Micro-fossils Diatoms / X-others	Macro-fossils Bot.	Zool.	Chem. anal.	Publ. year	Main scientists
SF-a	1 Tjärnbergen, Saltvik, 60°20'N, 20°06'E, 29.5 m	X		4000–0	2	X		D				1978	Glückert
SF-a	2 Dalkarbyträsk, 60°08'N, 19°56'E, 15 m	X		3000–0	2	X						1961 1963	Fries
SF-a	3 Kolmilaträsk, 60°17'N, 20°09'E, 11.6 m	X		2200–0	5	X	X					1991	Sarmaja-Korjonen et al.
SF-a	4 Lalax-kärret, 60°09'N, 21°52'E, 20 m		X	3700–0	3	X	X				humus	1990a	Vuorela
SF-b	5 Lake Littoinen, 60°27'N, 22°24'E, 35.8 m	X		6000–0	5	X	X	X			X	1992	Glückert et al.
SF-b	6 Hiittenmäensuo, 60°26'N, 23°00'E, 89 m		X	9500–0	1	X		D				1976	Glückert
SF-b	7 Vitsjön, 59°58'N, 23°19'E, 16.3 m	X		4500–0	6	X	X	D			X	1988	K. & M. Tolonen
SF-b	8 Nälköönsuo, 60°18'N, 24°12'E, 79–80 m		X	10000–0	25	X		D				1958 1970 1976 1972	Sauramo Glückert Tolonen & Ruuhijärvi Donner
SF-b	9 Vakojärvi, 60°20'N, 24°36'E, 82 m	X		9000–0	7	X	X						
SF-b	10 Pilisuo, 60°17'N, 24°11'E, 78.6 m		X	10500–2000	1	X						1970	Glückert
SF-b	11 Majassuo and Majaslampi, 60°19'N, 24°36'E, 97–101 m	X		9500–0	5	X	X	D, Cladocera				1991 1992 1992 1993	Korhola & Tikkanen Korhola Korhola & Tikkanen Tikkanen & Korhola
SF-c	12 Nääverpakanneva, 63°22'N, 23°07'E, 58 m		X	3500–0	3	X						1982	Vuorela
SF-c	13 Lintunemossen, 63°07'N, 22°12'E, 17 m		X	1200–0	4	X	X					1988	Miettinen & Vuorela
SF-c	14 Marjenemossen, 63°08'N, 22°18'E, 27 m		X	2000–0	5	X	X					1988	Miettinen & Vuorela
SF-c	15 Inmossen, 62°52'N, 21°33'E, 17.5 m		X	2000–0	3	X	X	D				1986 1988	Vuorela Miettinen & Vuorela

FINLAND (continued)

Type region	Reference site No., name, lat./long., elevation	Lake	Mire	Time span ^{14}C yrs BP	^{14}C dates, no.	Pollen %	Pollen conc.	Micro-fossils Diatoms	Micro-fossils X-others	Macro-fossils Bot.	Macro-fossils Zool.	Chem. anal.	Publ. year	Main scientists
SF-c	16 Kaluneva, 62°40'N, 22°00'E, 66.8 m		X	3800–0	4	X							unpubl.	Hyvärinen, R.
SF-c	17 Pesänsuo, 60°46'N, 22°56'E, 87 m		X	8500–0	211	X		D	Rhizop.	B			1993 1993 1993	Ikonen Kankainen Stén
SF-c	18 Haukkasuo, 60°54'N, 26°57'E, 55 m		X	9700–0	12	X		D					1976	Tolonen, K. & Ruuhijärvi
SF-c	19 Konnunsuo, 61°02'N, 28°27'E, 50–53 m		X	9800–0	13	X		D					1976	Tolonen, K. & Ruuhijärvi
SF-d	20 Piilonsuo, 60°47'N, 24°39'E, 92–94 m		X	10000–0	12	X		D		B	Z		1973	Koponen & Nuorteva
SF-d	21 Kaurastensuo-Kantosuo-Laaviosuo, 61°01'N, 24°58'E, 94 m		X	9000–0	27	X			Rhizop.	B			1976 1987	Tolonen, K. & Ruuhijärvi Tolonen, K.
SF-d	22 Ahvenainen, 61°02'N, 25°07'E, 122.2 m	X		5000–0	13	X	X	D		B		X	1978 a,b,c	Tolonen, M.
SF-d K.	23 Lovojärvi, 61°05'N, 25°02'E, 108.2 m	X		9000–0	16 + varves	X	X	D				X	1972 1977	Kukkonen & Tynni Huttunen, P. & Tolonen,
SF-d	24 Työtjärvi, 60°59'N, 25°28'E, 142.8 m	X		9000–0	17	X	X	D,Cl					1977 1980 1978	Saarnisto et al. Huttunem, P. Donner et al.
SF-d	25 Varrassuo, 60°59'N, 25°29'E, 144–148 m		X	9000–0	23	X	X	D					1978	Donner et al.
SF-d	26 Linnasuo, 61°38'N, 25°12'E, 135 m		X	9000–0	9	X	X						1990	Tolonen, M.
	Linnajärvi, 61°38'N, 25°13'E, 130 m	X		7000–0	4	X	X						1990	Tolonen, M.
SF-d	27 Aholammi, 61°53'N, 25°13'E, 114.4 m	X		8000–0	3	X	X						1987	Koivula
SF-d	28 Mäyrälampi (Hankasalmi), 62°20'N, 26°14'E, 118.4 m	X		8500–0	6	X	X						1994 1987 1994	Koivula et al. Koivula Koivula et al.
SF-d	29 Syrjälänsuo, 61°10'N, 28°10'E, 83 m		X	9700–0	7	X	X		X				1993b 1995	Vuorela Vuorela

			Age	No.						Year	Reference
SF-d	30 Koivusilta, 61°30'N, 29°30'E, 110 m	X	10000–9000	9	X	X		B,Z		1994	Bondestam et al.
SF-d	31 Puohtiinsuo, 62°45'N, 31°03'E, 145–152 m	X	10000–0	2	X	X	D			1963, 1967	Tolonen, K.
SF-d	32 Laukunlampi, 62°40'N, 29°10'E, 84.0 m	X	10000–0	6	X	X	D		X	1980, 1984	Battarbee et al. / Simola et al.
SF-d	33 Iso Hanhilampi, 63°37'N, 27°04'E, 121.8 m	X	9000–0	4	X	X	D	B	X	1983	Ruohomäki
SF-e	34 Ylimysneva, 62°08'N, 22°52'E, 171–178 m	X	8100–0	4	X			B		1990	Huttunen, A.
SF-e	35 Ahmasjärvi, 64°29'N, 26°27'E, 98.5 m	X	8400–0	9	X					1980, 1974	Reynaud & Hjelmroos / Eronen
SF-e	36 Sotkasuo, 64°50'N, 26°25'E, 81 m	X	6000–2500	4	X					1980	Reynaud & Hjelmroos
SF-e	37 Järvenpäänsuo, 64°50'N, 26°40'E, 102 m	X	7500–0	5	X	X		B	X	1976	Holappa
SF-e	38 Kittilä, 65°01'N, 24°41'E, 8 m	X	900–0	1+rate of land uplift	X					1988, 1992	Hicks, Hicks
SF-e	39 Kiimisuo, 65°02'N, 24°42'E, 12 m	X	1100–0	2	X	X		B	X	1983	Rönkä
SF-e	40 Merijänjärvi, 65°17'N, 25°30'E, 30 m	X	2800–0	7	X	X				1980	Reynaud & Hjelmroos
SF-e	41 Sammakkolampi, 65°15'N, 27°05'E, 121 m	X	8000–0	5	X	X				1982	Haapalahti
SF-f	42 Pappilanlampi, 63°18'N, 30°55'E, 200 m	X	10000–0	20	X	X	D	B		1966, 1967, 1975	Hyvärinen, H. / Tolonen, K. / Vourinen & Tolonen, K.
SF-f	43 Hamunen, 63°40'N, 28°15'E 198.6 m	X	9300–0	3	X	X		B		1984	Nykänen

FINLAND (*continued*)

Type region	Reference site No., name, lat./long., elevation	Lake	Mire	Time span ^{14}C yrs BP	^{14}C dates, no.	Pollen %	conc.	Micro-fossils Diatoms X-others	Macro-fossils Bot. Zool.	Chem. anal.	Publ. year	Main scientists
SF-g	44 Vasikkasuo, 64°41'N, 27°52'E, 270 m	X		9000–0	6	X		X			1990b 1991	Vuorela Vuorela & Kankainen
SF-h	45 Iso Mustajärvi, 66°13'N, 24°48'E, 75 m	X		5400–0	2	X					1978	Hjelmroos
SF-h	46 Kivilompolon jänkä, 66°18'N, 24°17'E, 110 m		X	8000–0	2	X		D			1980 1974	Reynaud & Hjelmroos Eronen
SF-h	47 Kuprujänkä (Pisavaara), 66°15'N, 25°08'E, 91 m		X	5400–0	4	X	X		B		1982	Juola-Helle
SF-h	48 Ylempi Silmäslampi, 66°39'N, 25°58'E, 206.7 m	X		9000–0	3	X					1981	Saarnisto
SF-h	49 Valkiajärvi, 66°48'N, 24°07'E, 188.0 m	X		9260–2500	3	X					1981	Saarnisto
SF-i	50 Liippasuo, Kuusamo, 66°07'N, 29°22'E, 357 m	X		8000–0	8	X					1985	Seppälä & Koutaniemi
SF-i	51 Kolmiloukkonen, Posio, 66°14'N, 28°29'E, 344 m	X		9300–0	5	X	X		B		1962	Vasari
SF-i	52 Keski-Pohjassuo, Posio, 66°14'N, 28°31'E, 410 m		X	6900–0	5	X	X		B		1987 1987	Huttunen, A. Huttunen, A.
SF-j	53 Aapalampi, 66°48'N, 28°32'E, 204 m	X		9100–0	5	X			B		1964 1965	Sorsa Sorsa
SF-j	54 Kaakkurilampi, 67°03'N, 28°56'E, 180 m	X		9100–0	3	X					unpubl. 1965	Vasari *et al.* Sorsa
SF-j	55 Sudenvaaranaapa, 67°12'N, 27°35'E, 157 m		X	9000–3500	2	X					1970	Lappalainen
SF-k	56 Tankavaara, 68°11'N, 27°14'E, 335 m	X		8000–0	4	X					1965	Sorsa
SF-k	57 Kaunispää, 68°25'N, 27°25'E, 300 m	X		8500–0	3	X					1965	Sorsa
SF-k	58 Virtaniemi, 68°53'N, 28°24'E, 121 m	X		8900–0	6	X					1965	Sorsa
SF-k	59 Akuvaara,	X		9000–0	5	X	X				1975	Hyvärinen, H.

		Location		Range						Year	Reference
SF-l	60	Mukkavaara, 68°55'N, 21°00'E 535 m	X	9000–0	9	X	X			1982	Eronen & Hyvärinen, H.
SF-l	61	Isohattu, 68°38'N, 23°36'E, 386 m	X	8000–0	5	X	X	X		1993 unpubl.	Hyvärinen Mäkelä
SF-l	62	Suovalampi, 69°35'N, 28°50'E, 104 m	X	9000–0	5	X	X		X	1994 1975	Hyvärinen, H. & Alhonen Hyvärinen, H.
SF-m	63	Masehjavri, 69°03'N, 20°59'E, 680 m	X	10000–1500	6	X				1993	Hyvärinen, H.

FRANCE

Type region	Reference Site No., name, lat./long.	Lake	Mire	Time span ¹⁴C yrs BP	¹⁴C dates, no.	Pollen %	Pollen conc.	Micro-fossils Diatoms / X-others	Macro-fossils Bot. / Zool.	Chem. anal.	Publ. year	Main scientists
F-b	1 Heurteauville, 49°27'N, 00° 48'E		X	6500–0	4	X		D			1986, 1988	Huault
	2 La Mailleraye, 49°28'N, 00°46'E		X	7200–0	2	X		D	Z		1983a	Huault & Lefebvre
	3 Marais Vernier, 49°24'N 00°29'E		X	7600–0	4	X		D,X			1983b 1985	Huault & Lefebvre
F-h/ F-i	4 Goutte Loiselot, 48°00'40''N, 06°51'48''E		X	11000–0	6						1985	Edelman
	5 Haute-Viaux, 48°02'40''N, 06°55'52''E		X	3000–0		X					1987	Janssen & Braber
	6 Rouge Feigne, 48°02'40''N,06°56'08''E		X	9000–0	5	X					1984a,b	Kalis
	7 Grande Basse, 48°02'50''N, 06°57'10''E		X	10000–0	6	X					1984a,b	Kalis
	8 Feigne d'Artimont, 48°01'32''N,06°59'05''E	X		10000–0	8	X			B		1975	Janssen et al.
	9 Altenweiher, 48°00' 48''N,06°59'40''E		X	10000–0	9	X					1981	Devalk
	10 Moselotte, 48°01'55''N,07°00'00''E		X	7000–0	7	X					1981	Devalk
	11 Trois Four, 48°03'18''N,07°01'25''E		X	9000–0		X					1981	Devalk
	12 Tanet, 48°06'05''N, 07°06'15''E		X	3000–0	6	X					1972	Janssen & Kettlitz
	13 la Max, 48°07'48''N, 07°27'50''E		X	13000–0	14	X	X				unpubl.	Janssen

	Site								1948 / unpubl.	Firbas / Janssen
	14 Champs de Feu, 48°24'30"N, 07°17'10"E	X	X	8000–0	10	X	X			
F-p	15 Canal des Fougéres, 47°20'N, 02°12'W	X	X	8000–0	8	X			1979,1982a	Visset
	16 Ile d'Errand, 47°22'N, 02°09'W	X	X	8000–0	8	X			1982a	Visset
	17 Lavau, 47°19'N,01°55' W	X	X	6000–0	5	X			unpubl.	Voeltzel
	18 La Caudelais, 47°15'N, 01°48'W	X	X	7000–0	5	X			unpubl.	Voeltzel
	19 St Lumine de Coutais 47°04'N, 03°02'W	X	X	6500–0	5	X			unpubl.	Visset
F-r	20 Estarres, 43°05'N,00°24'W	X	X	13000–0	4	X	X	B	1987,1988 / 1988	Andrieu-Jalut / Jalut
	21 Freychinede, 42°48'N 01°26'E	X	X	15000–0	23	X	X		1982,1993	Jalut-Reille
	22 Ules (in Spain), 42°44'N, 04 26'E	X		14000–0	2	X			1993	Aubert
F-zg	23 Laurenti, 42°40'N, 02°01'E	X	X	14000–0	22	X	X		1974,1989,1990	Jalut-Reille
	24 La Borde, 42°32'N,02°05'E	X		14000–5000	15	X			1974,1990	Jalut-Reille
F-zd	25 Le Vivier, 45°38'15" N 6°18'34"E	X		15000–6000	5	X		Z,B	unpubl.	David
	26 Les Etelles, 45°28'25"N, 6°09'23"E	X		15000–0	5	X		Z,B	unpubl.	David
	27 La Coche, 45°29'06"N, 6°15'16"E	X		12500–0	4	X		Z,B	unpubl.	David
	28 Montendry, 45°30'59"N, 6°15'45"E	X		15000–0	5	X		Z,B	unpubl.	David
	29 Grand Leyat, 45°28'06"N, 6°13'38"E	X		13000–100	3	X		Z,B	unpubl.	David

GERMANY

Type region	Reference site — No. name, lat/long., elevation	Lake	Mire	Time span ^{14}C years	^{14}C dates, no.	Pollen %	Pollen conc.	Micro-fossils Diatoms	Macro-fossils Bot. Zool.	Chem. anal.	Publ. year	Main scientists
D-d	1 Westrhauderfehn I, 53°07'N, 7°33'E, 3 m	X		12500–7500	1*	X					1966	Behre
D-d	2 Seckbruch SB1, 52°24'N, 9°52'E, 57 m	X		13000–9500		X			B, Z		1958	Dietz, et al.
D-d	3 Kaiserwinkel KaBI, 52°31'N, 10°58'E, 58 m	X		13000–8500	1	X			B		1980	Golombek
D-d	4 Hahnenmoor, 52°38'N, 7°38'E, 22 m		X	5800–500	23	X	X		B	X	1978	Kramm
	5 Spolsener Moor, 53°23'N, 7°54'E, 15 m		X	5500–0	7	X					1984	Middeldorp
D-d	6a Swienskuhle, 53°40'N, 8°43'E, 6 m		X	7000–200	9	X	X		B		1986	O'Connell
D-d	6b Flögelner Holz, 53°39'N, 8°46'E, 6 m		X	5600–0	9	X			B		1986	Behre & Kucan
D-d	7 Altes Moor WA 3, 53°26'N, 8°52'E, 17 m	X	X	10500–200	7	X	X		B		1986	Behre & Kucan
D-d	8 Hohes Moor A, 53°11'N, 9°27'E, 37 m	X	X	11500–1000	5	X					1989	Dörfler
									B	X	1963	Schneekloth
D-s	9 Tegeler See, 52°35'N, 13°13'E, 31 m	X		11400–0	7*	X		D		X	1980	Brande
											1988b	Brande
D-s	10 Pechsee, 52°29'N, 13°13'E, 31 m		X	12500–11400	2*	X					1980	Brande
											1988b	Brande
D-n	11 Breitnau-Neuhof, 47°58'N, 8°12'E, 985 m		X	10000–1500	15	X	X		B		1989	Rösch
D-l	12 Sersheimer Moor, 48°58'N, 9°03'E, 234 m	X		8500–0	16	X					1985	Smettan
D-l	13 Kupfermoor, 49°07'N, 9°43'E, 275 m	X		8200–0	34	X			B,Z		1988	Smettan
D-o	14 Buchau-Torfwerk, 48°03'N, 9°38'E, 584m		X	8000–2000	22	X					1991	Liese-Kleiber
D-r	15 Durchenbergried, 47°44'N, 8°59'E, 432 m	X		13000–0	50	X	X				1986	Rösch
											1990a	Rösch
D-r	16 Hornstaad Bodensee, 47°40'N, 9°01'E, 395 m		X	13000–400	25	X	X				1992	Rösch
											1993	Rösch
D-r	17 Feuenried, 47°45'N, 8°54'E, 407 m	X		13000–0	33	X	X		B		1985b	Rösch
											1983	Rösch
D-r	18 Nussbaumer Lakes, 47°37'N, 8°50'E, 434 m	X		13000–0	15	X	X		B	X	1985a	Rösch
D-o	19 Langegger Filz, 47°42'N, 10°46'E, 800 m	X		12500–10000(-0)	3	X					1988	Küster
											1986	Küster
D-o	20 Haslacher See, 47°45'N, 10°47'E, 765 m	X		10000–0	29	X					1988	Küster
											1989	Küster

* from adjacent sites

GB—ENGLAND

Type region	Reference site — No., name, lat./long., elevation	Lake	Mire	Time span ^{14}C yrs BP	^{14}C dates, no.	Pollen %	Pollen conc.	Microfossils Bot. Zool.	Publ. year	Main scientists
GB-a	1 Blacka Brook, 50°30'N, 4°01'W, 270 m		X	9000–0	7	X	CO	B	1981	Beckett
GB-a	2 Chains (Exmoor) 51°10'N, 3°50'W, 250 m		X	5000–0	5	X			1974	Merryfield
GB-a	3 Hawks Tor 50°29'N, 5°25'W, 229 m		X	13000–0	11	X	X		1977	Brown
GB-b	4 Winchester 51°4'N, 1°19'W, 40 m		X	8700–0	1	X			1982	Waton
GB-b	5 Rimsmoor 50°44'N, 1°21'W, 30 m		X	8000–0	6	X			1982	Waton
GB-b	6 Meare Heath 51°12'N, 2°57'W, 20 m		X	4000–1400	8	X	X		1978	Beckett
GB-c	7 Gatcombe 50°40'N, 1°17'W, 30 m		X	10000–0	4	X		B	1982	Scaife
GB-d	8 Pannel Bridge 50°56'N, 00°45'E, 4 m		X	10000–0	8	X		B	1987	Waller
GB-d	9 Amberley, 50°57'N, 0°26'E, 30 m		X	3000–500	2	X			1982	Waller
GB-i	10 Stafford, 52°48'N, 2°06'W, 75 m	X		13000–1000	14	X	X	B,Z	1990	Bartley
GB-i	11 Cookley, 52°25'N, 2°16'W, 60 m		X	10000–400	2	X		B,Z	in prep.	Greig
GB-j	12 Sidlings Copse, 51°47'N, 1°12'W, 80 m		X	9500–0	5	X			1991	Day
GB-k	13 Tilbury, 51°27'N, 0°22'E, 1 m		X	8200–2000	8	X		B,Z	1979	Devoy
GB-k	14 West Heath Spa 51°35'N, 0°7'W, 120 m		X	6000–0		X			1990	Greig
GB-l	15 Hockham Mere, 52°30'N, 0°50'E, 33 m	X		12600–1600	19	X			1983	Bennett
GB-l	16 Diss Mere, 52°22'N, 1°6'E, 29 m	X		7000–0		X			1993	Peglar
GB-l	17 Quidenham Mere, 52°26'N, 1°0'E, 25 m	X		12500–0		X			1992	Peglar
GB-m	18 Haddenham, 52°20'N, 0°3'E, 0 m		X	10650–3850	4	X			1993	Peglar
GB-m	19 Holme Fen, 52°29'N, 0°15'W, –2 m		X	4500–0	4	X			1974	Vishnu-Mittre
GB-n	20 Knowsley Park, 53°28'N, 2°55'W, 70 m		X	10000–0	16	X			1994	Innes
GB-n	21 Fenton Cottage, 53°54'N, 2°55'W, 9 m		X	4900–0	19	X			1992	Huckerby
GB-o	22 White Moss 54°17'N, 3°12'W, 190 m		X	9900–500	4	X		B,Z	1990	Bartley
GB-o	23 Leash Fen, 53°16'N, 1°34'W, 320 m		X	4300–400	9	X			1971	Hicks
GB-o	24 Valley Bog, 54°42'N, 2°23'W, 541 m		X	8000–0	8	X			1978	Chambers
GB-p	25 Roos, 53°42'N, 0°7'W, 5 m		X	13000–2000	2	X		B,Z	1981	Beckett
GB-p	26 Willow Garth, 54°05'N, 0°17'W, 18 m		X	9500–0	10	X			1988	Bush
GB-q	27 Low Wray Bay, 54°24'N, 2°58'W, 33 m	X		14000–10000	15	X		B,Z	1977	Pennington
GB-q	28 Blelham Tarn, 54°24'N, 2°59'W, 50 m	X		14000–0	10	X			1979	Pennington
GB-r	29 Fen Bogs, 54°22'N, 0°42'W, 164 m		X	9000–0	6	X			1976	Atherden
GB-s	30 Neasham Fen, 54°30'N, 1°24'W, 46 m		X	10000–0	11	X			1976	Bartley
GB-t	31 Scaleby Moss, 54°57'N, 2°52'W, 30 m		X	10000–0	16	X			1966	Walker
GB-t	32 Glasson Moss 54°50'N, 3°03'W, 30 m		X	5000–0	3	X			1994	Dumayne

GB—SCOTLAND

Type region	Reference site No., name, lat./long., elevation	Lake	Mire	Time span ^{14}C yrs BP	^{14}C dates, no.	Pollen % 	Pollen conc.	Micro-fossils Diatoms	Macro-fossils Bot.	Chem. anal.	Publ. year	Main scientists
GB-v	1 Loch Dungeon, Loch Dungeon Peat, 55°8'N, 4°19'W, 305 m	X	X	10200–150	2	X			B		1972 1975	H.H. Birks H.H. Birks
GB-v	2 The Round Loch, of Glenhead, 55°6'N, 4°25'W, 330 m	X		10000–0	25	X	X	D			1989 1993 1990	Jones et al. Jones et al. Stevenson et al.
GB-w	3 Din Moss, 55°34'N, 2°16'W, 170 m		X	12200–0	18	X					1976	Hibbert & Switsur
GB-x	4 Machrie Moor, 55°28'N, 5°12'W, 70 m	X		9000–0	10	X	X		B		1988	Robinson & Dickson
GB-x	5 Loch a'Mhuilinn, 55°42'N, 5°16'W, 28 m	X		8700–0	9	X	X		B		1987	Boyd & Dickson
GB-x	6 Loch Cill an Aonghais, 55°48'N, 5°35'W, 20 m	X		12000–0	8	X	X				unpubl.	Peglar
GB-x	7 Loch Cholla, 56°2'N, 6°12'W, 30 m	X		10000–0	4	X					1987	Andrews
GB-x	8 Lochan a'Builg, 56°23'N, 5°27'W, 60 m	X		10000–0	5	X					1957	Donner
GB-x	9 Salen, 56°44'N, 5°47'W, 25 m	X		10000–0	10	X	X				unpubl.	Williams
GB-x	10 Lochan Doilead, 56°59'N, 5°48'W, 39 m	X		13000–0	12	X	X				unpubl.	Williams
GB-x	11 Loch Meodal, 57°8'N, 5°5'W, 110 m	X		10300–0	11	X	X				1977	Williams
GB-x	12 Loch Ashik, 57°15'N, 5°50'W, 40 m	X		13000–0	16	X	X				1977	Williams
GB-x	13 Dubh Lochan, 56°8'N, 4°36'W, 30 m	X		9000–0	6	X	X		B		1984	Stewart et al.
GB-y	14 Clashgour, 56°32'N, 5°52'W, 293 m		X	9500–2000	6	X					1990	Bridge et al.
GB-y	15 North Mains, 56°20'N, 3°45'W, 30 m		X	9500–0	4	X			B		1985	Hulme & Shirriffs
GB-y	16 Coire Fee, 56°51'N, 3°13'W, 450 m		X	9500–0	7	X	X				1981	Huntley
GB-y	17 Caenlochan Glen, 56°52'N, 3°30'W, 550 m		X	9100–0	5	X	X				1981	Huntley
GB-z	18 Abernethy Forest, 56°14'N, 3°43'W, 220 m	X		12200–0	7	X	X		B		1970 1978	H.H. Birks H.H. Birks & Matthewes
GB-z	19 Loch Garten, 57°14'N, 3°42'W, 220 m	X		8500–0	3	X					1974	Sullivan
GB-z	20 Loch Pityoulish, 57°12'N, 3°47'W, 210 m	X		9500–0	8	X					1976	Sullivan
GB-za	21 Braeroddach Loch, 57°5'N, 3°51'W, 195 m	X		10500–0	18	X	X			X	1980	Edwards & Rowntree
GB-za	22 Allt na Feithe, Sheilich, 57°19'N, 3°54'W, 595 m		X	9500–0	2	X	X		B		1975	H.H. Birks
GB-zb	23 Loch Lang, 57°15'N, 7°18'W, 80 m	X		10500–0	6	X	X			X	1990	Bennett et al.
GB-zb	24 Gleann Mor, 57°50'N, 8°36'W, 90 m		X	8000–1500	4	X					1984	Walker
GB-zc	25 Kinloch, 57°1'N, 6°18'W, 25 m	X	X	8000–0	5	X	X				1990	Hirons & Edwards
GB-zc	26 Loch Cleat, 57°41'N, 6°20'W, 38 m	X		10400–0	10	X	X				1977	Williams

Code	Site		Age range	n						Date	Reference
GB-zd	27 By Loch Coultrie, 57°26'N, 5°35'W, 65m		10000–0	8	X	X			X	unpubl.	H.H. Birks
GB-zd	28 Loch Clair, 57°34'N, 5°21'W, 80m	X	10000–0	7	X	X				1972	Pennington *et al.*
GB-zd	29 Loch Maree, 57°40'N, 5°30'W, 20m	X	9500–0	6	X			B		1972	H.H. Birks
GB-zf	30 Little Loch Roag, 58°8'N, 6°53'W, 27m	X	9000–0	6	X	X			X	1979	H.J.B. Birks & B.J. Madsen
GB-zg	31 Loch Sionascaig, 58°4'N, 5°12'W, 100m	X	10000–0	7	X		D		X	1972	Pennington *et al.*
GB-zg	32 Lochan Dubh, 58°4'N, 5°15'W, 92m	X	9500–0	13	X	X	X			1982	Kerslake
GB-zg	33 By Loch Assynt, 58°10'N, 5°4'W, 80m	X	10000–0	9	X	X	X			unpubl.	H.H. Birks
GB-zg	34 An Druim, Eriboll, 58°28'N, 4°42'W, 40m	X	12500–0	11	X	X	X	B		unpubl.	H.H. Birks
GB-zg										1984	H.H. Birks
GB-zi	35 Loch of Winless, 58°28'N, 3°13'W, 9m	X	13000–0	10	X			B		1979	Peglar
GB-zi	36 Glims Moss, 59°7'N, 3°13'W, 40m	X	7000–0	7	X					1979	Keatinge & Dickson
GB-zj	37 Murraster, 60°15'N, 1°20'W, 15m	X	10000–0	4	X					1975	Johansen
GB-zj	38 Dallican Water, 60°23'N, 1°6'W, 56m	X	9500–0	6	X	X			X	1992	Bennett *et al.*

GB—WALES

Type region	Reference site — No., name, lat./long., elevation	Lake	Mire	Time span ¹⁴C yrs BP	¹⁴C dates, no.	Pollen %	Pollen conc.	Micro-fossils Diatoms	Macro-fossils Bot.	Macro-fossils Zool.	Chem. anal.	Publ. year	Main scientists
GB-e	1 Tregaron (Southeast)Bog, 52°15'N, 3°55'W, 165 m		X	10200–0	18	X	X		B			1976 1964 1938	Hibbert & Switsur Turner Godwin & Mitchell
GB-f	2 Waun-Fignen-Felen (16 profiles), 51°51'N, 3°42'W, 488 m	X	X	10000–1000 (overall)	88 (between 16 profiles)	X	X					1988	Smith & Cloutman
	3 Llanilid, 51°31'N, 3°27'W, 60 m	X	X	13200–9300	12	X						1990	Walker & Harkness
GB-g	4 Cefn Gwernffrwd, 52°07'N, 3°50'W, 395 m	X	X	10000–0	9	X						1982	Chambers
	5 Carneddau, 52°35'N,3°29'W, 400 m		X	9000–0	8(between 3 profiles)	X						1993	Walker
GB-h	6 Nant Ffrancon, 53°08'N, 4°02'W, 198 m	X	X	>10000–0	20	X						1976	Hibbert & Switsur
GB-i	7 Brecon Beacons, 51°52'N, 3°23'W, 715 m		X	4295–0	9	X						1982	Chambers
	8 Rhosgoch Common, 52°07'N, 3°10'W, 230 m	X	X	12000–2500	none	X						1960	Bartley
	9 Llangorse Lake, 51°55'N, 3°15'W, 230 m	X		>10000–0	9	X		D		Z	X	1985 1993	Jones et al. Walker et al.
	Additional reference sites in England												
GB-n	No reference site in Wales—see entry for England												

IRELAND

Type region	Reference site No., name, lat./long., elevation	Lake	Mire	Time span ^{14}C yrs BP	^{14}C dates, no.	Publ. year	Main scientists
IRL-a 1	Namackanbeg, 53°16'N, 9°17'W, 90 m	X		9200–0	11	1988	O'Connell et al.
IRL-a 2	Corslieve, 54°03'N, 9°37'W, 320 m	X		10000–0	4	1987	Bradshaw & Browne
IRL-a 3	Camclaun, 52°12'N, 10°9'W, 244 m	X		9500–0	5	1990	Dodson
IRL-a 4	Sheeauns, 53°33'N, 10°3'W, 18 m	X		9400–1500	13	1991	Molloy & O'Connell
IRL-b 5	Sluggan Bog, 54°46'N, 6°18'W, 65 m		X	12350–0	23	1991	Smith & Goddard
IRL-b 6	Slieve Gallion, 54°43'N, 6°45'W, 320 m		X	9400–2000	12	1973	Pilcher
IRL-c 7	Red Bog, 53°57'N, 6°38'W, 50 m		X	6500–0	11	1985	Watts
IRL-c 8	Arts Lough, 52°58'N, 6°25'W, 490 m	X		10000–0	5	1988	Bradshaw & McGee
IRL-c 9	Belle Lake, 52°11'N, 7°02'W, 33 m	X		12000–5500	8	1978	Craig

LITHUANIA

Type region	Reference site No., name, lat./long., elevation	Lake	Mire	Time span ^{14}C yrs **BP**	^{14}C dates, no.	Pollen %	Pollen conc.	Micro-fossils Diatoms	Micro-fossils X-others	Macro-fossils Bot.	Macro-fossils Zool.	Chem. anal.	Publ. year	Main scientists
LT-a	1 Šventelė, 55°29'N, 21°17'E, 3 m		X	12000–0		X		D		B		X	1959,1973	Kabailienė
LT-a	2 Šventoji, 56°00'N, 21°05'E, 2 m		X	5700–0	12	X		D					1979	Dvareckas, Kunskas, Savukynienė
LT-b	3 Tytuvėnu tyrelis, 55°35'N, 23°18'E, 125 m		X	12000–0		X		D	X	B,Z			1979	Kabailienė
LT-c	4 Žuvintas, 54°59'N, 23°37'E, 90 m	X	X	11000–0		X	X	D	X	B,Z		X	1962 1968	Kunskas Kabailienė
LT-d	5 Šepeta, 55°47'N, 25°10'E, 110 m		X	10000–0		X				B			1971	Kabailienė
LT-d	6 Bebrukas, 54°35'N, 24°38'E, 160 m	X	X	11800–0	10	X		D	X	B		X	1965,1986,1987a	Kabailienė
LT-e	7 Čepkeliai, 54°00'N, 24°32'E, 130 m	X	X	8820–0	4	X				B			1976	Savukynienė
LT-e	8 Glūkas, 54°17'N, 24°34'E, 109 m	X	X	12000–0	3	X	X	D		B		X	1986,1987b Unpubl.	Kabailienė Kunskas

NORWAY

Type region	No., name, lat./long., elevation	Lake	Mire	Time span ^{14}C yrs BP	^{14}C dates, no.	Pollen %	conc.	Publ. year	Main scientists
N-f	1 Korsegården, 59°42'N, 10°43'E, 85 m		X	8500–0	6	X		1990	Sørensen
N-f	2 Kjeldemyr, 59°10'N, 10°15'E, 87 m		X	10000–0	5	X		1980	Henningsmoen
N-j	3 Sandvikvatn, 59°17'N, 5°30'E, 128 m	X		14000–0	13	X	X	1982	Eide
N-k	4 Hadlemyrane, 60°24'N, 07°17'E, 1000 m		X	9500–0	2	X		1978	Moe
								1980	Simonsen
N-k	5 Nordmannslågen, 60°00'N, 07°28'E, 1245 m		X	9000–0	3	X		1978	Moe
N-k	6 Våtenga, 61°30'N, 8°29'E, 1480 m		X	4000–0	2	X		1984	Caseldine
N-mo	7 Grasvatn, 63°42'N, 8°42'E, 45 m	X		11000–0	4	X		1982	Paus
N-mo	8 Krokvatn, 63°42'N, 8°45'E, 28 m	X		9200–0	2	X		1982	Paus
N-si	9 Kalvikmyr, 64°27'N, 13°52'E, 430 m		X	8700–0	4	X		1985	Selvik
N-n	10 Formofoss, 64°23'N, 12°21'E, 175 m		X	7000–0	2	X		1985	Selvik
N-mi	11 Koltjønn, 64°31'N, 11°44'E, 159 m	X		8500–3000	3	X		1969	Vorren
N-mo	12 Blåvasstjønn/Gorrtjønn, 64°54'N, 11°38'E, 92/66 m	X		8950–0	7	X		1979,1982	Ramfjord
N-u	13 Dønvoldmyra, 68°08'N, 13°35'E, 10 m	X		8600–0	6	X	X	1983	Nilssen
N-v	14 Endletvatn/Æråsvatn, 69°16'N, 16°05'E, 36 and 35 m	X		19500–11000	21	X	X	1993	Alm,
	Endletvatn, 69°16'N, 16°05'E, 36 m	X		18000–0	14	X		1978	Vorren
N-x	15 Prestvannet/Tjernet, 69°44'N, 18°57'E, 95.5/101 m	X		12300–0	11	X	X	1980	Fimreite
N-z	16 Bruvatnet, 70°11'N, 28°25'E, 119 m	X		>10280–0	5	X	X	1975	Hyvärinen
N-æ	17 Domsvatnet, 70°19'N, 31°02'E, 120 m	X		>8570–0	5	X	X	1976	Hyvärinen

POLAND

Type region	Reference site No., name, lat./long., elevation	Lake	Mire	Time span ^{14}C yrs BP	^{14}C dates, no.	Pollen %	conc.	Micro-fossils Diatoms / X-others	Macro-fossils Bot. / Zool.	Chem. anal.	Publ. year	Main scientists
P-a	1 Puścizna Rękowiańska, 49°29'N, 19°49'E, 656 m		X	10000–0	8	X	X	X	B	X	1989,1990	Obidowicz
P-c	2 Szymbark, 55°00'N, 21°06'E, 465 m		X	8600–0	5	X			B		1974 / 1989a	Szczepanek in Gil et al. / Szczepanek
P-c	28 Jasiel, 49°22'N, 21°53'E, 670–680 m		X	10350–0	6	X	X		B		1987,1989a	Szczepanek
P-d	3 Tarnowiec, 49°42'N, 21°37'E, 240 m		X	12000–ca. 1000	8	X	X	X	B,Z	X	1987 / 1989	Harmata
P-d	4 Roztoki, 49°43'N, 21°35'E, 232 m		X	12000–4000	4	X	X	X	B,Z	X	1987,1989	Harmata
P-d	36 Jasło-unpublished										unpubl.	Harmata
P-e	5 Tarnawa Wyżna, 49°07'N, 22°50'E, 670 m		X	12000–0	9	X	X		B		1980,1989a	Ralska-Jasiewiczowa
P-f	6 Zieleniec, 50°21'N, 16°24'E, 750–760 m		X	9000–6	6	X	X		B	X	1989	Madeyska
P-h	7 Wolbrom, 50°23'N, 19°46'E, 375 m		X	13000–5500 and ca. 2500–2300	19	X	X	X	B,Z		1976,1988a / 1989a	Latałowa
P-i	29 Czajków, 50°05'N, 21°19'E, 205–207 m		X	12000–0		X			B		1971	Szczepanek
P-j	8 Stopiec, 50°47'N, 20°47'E, 248 m		X	10200–0	17	X	X	X	B	X	1982,1989b	Szczepanek
P-m	9 Łukcze Lake, 51°30'N, 22°57'E, 163 m		X	12800–0	10	X	X	X	B	X	1989,1990	Bałaga
P-m	35 Krowie Bagno-preliminary publication, full data unpubl.										1980/81	Bałaga et al.
P-n	10 Maliszewskie Lake and Wizna mires, 53°10'N, 22°30'E, 104.1 m	X		11800–0	5	X	X	X	B	X	1987,1989	Balwierz & S. Żurek
P-n	11 Błędowo Lake, 52°32'N, 20°40'E, 78m	X		11800–0	7	X	X	D,X	B	X	1989 / 1991	Binka & Szeroczyńska / Binka et al.
P-n	13 Witów, 52°00'N, 19°31'E, 115 m	X		13000–3000	7	X		X	B		1964,1978	Wasylikowa

	Site			Time range	No.					Year	Author
P-p 12	Węglewice-type region not discussed in the synthesis									1966	Tobolski
P-q 31	Pomorsko-palynological investigation planned-not executed									1983	Alexandrowicz & Nowaczyk
P-r 14	Skrzynka Lake, 52°15'N, 16°47'E, 65.5 m	X		12000–0	11	X	X		X	1987 / 1966 / 1989	Okuniewska-Nowaczyk / Tobolski / Tobolski & Okuniewska-Nowaczyk
P-r 15	Skrzetuszewskie Lake, 52°33'N, 17°23'E, 109.1 m	X		9000–0		X	X	B		1989	Tobolski
P-r	Dziekanowice, 52°32'N, 17°22'E, 110 m		X	13000–8500?	4	X	D	B	X	1988	Litt
P-r 16	Gopło Lake, 52°23'N, 18°22'E, 77 m	X		13000–0		X		B		1980	Jankowska
P-r 30	Wonieść, full data unpubl.	X		12000–0	2	X				1985	Okuniewska-Nowaczyk
P-s 17	Wielkie Gacno Lake, 53°47'N, 17°30'E, 130 m	X		12000–0	23	X	D	B	X	1981	Hjelmroos-Ericsson
P-s 34	Fletnowo-not reinvestigated									1968	Kępczyński & Noryśkiewicz
P-s 37	Mały Suszek, 53°43'N, 17°46'E, 115 m	X		12000–250	10	X	D, X		X	1989, 1992	Miotk-Szpiganowicz
P-t 18	Darżlubie, 52°42'N, 18°10'E, 40 m		X	10000–?800	9	X	X	B	X	1982a,b, 1989b	Latałowa
P-t 19	Żarnowiec lake and peat bog, 54°43'N, 18°07'E, 5 m	X		11000–1500	8	X	D	B	X	1982a,b	Latałowa
P-t 32	Żurawiec, 54°25'N, 16°30'E, 33 m		X	5000–0	6	X	X	B	X	1989b	Latałowa

POLAND (*continued*)

Type region	Reference site No., name, lat./long., elevation	Lake	Mire	Time span ^{14}C yrs BP	^{14}C dates, no.	Pollen %	Pollen conc.	Micro-fossils Diatoms / X-others	Macro-fossils Bot./Zool.	Chem. anal.	Publ. year	Main scientists
P-u	20 Racze Lake, 53°55'N, 11°40'E, 8 m	X		11000–0	9	X	X	D, X	B	X	1992	Latałowa
P-u 21	Niechorze, 54°0'N, 15°03'E, 5 m	X		12000–10000 6000–3000	9	X	X	D, X	B		1987	Ralska-Jasiewiczowa & Rzętkowska
P-u 22	Kluki, 54°40'N, 17°19'E, 2.1 m		X	10000–0	20	X	X		B	X	1982, 1987	Tobolski
P-u 38	Kołczewo, 53°55'N, 14°40'E, 15 m	X	X	12000–800	15	X	X	X	B	X	1989c, 1992	Latałowa
P-u 39	Wolin II, 53°50'N, 14°40'E, *ca.* 5m		X	7300–0	8	X	X		B		1989c, 1992	Latałowa
P-v 23	Druzno Lake, 54°07'N, 19°28'E, −1.8 m	X		11300–0	4	X	X	D, X	Z	X	1982 1987	Zachowicz et al. Zachowicz & Kępińska
P-w 24	Steklin Lake, 52°56'N, 19°0'E, 73.7 m	X		11600–0	1	X	X		B	X	1982	Noryśkiewicz
P-w 25	Strażym Lake, 53°20'N, 19°27'E, 71m	X		12000–0	7	X	X	X		X	1987	Noryśkiewicz
P-w 26	Woryty, 53°45'N, 20°12'E, 105 m	X		11800–200	18	X	X	D, X	B	X	1982	Ralska-Jasiewiczowa in Pawlikowski et al.
P-w 33	Rudnickie Małe-full data unpubl.										1974	Noryśkiewicz in Drozdowski
P-x 27	Mikołajki Lake, 53°46'N, 21°35'E, 116 m	X		13000–0	5	X	X	D, X	B	X	1966, 1989	Ralska-Jasiewiczowa

RUSSIA KARELIA

Type region	No., name, lat./long., elevation	Lake	Mire	Time span ^{14}C yrs BP	^{14}C dates, no.	Pollen % conc.	Micro-fossils Diatoms X-others	Macro fossils Bot. Zool.	Chem. anal.	Publ. year	Main scientists
K-a	1 Ptichje, 66°20'N, 30°30'E, 120 m		X	8600–0	2	X		B		1981	Elina & Kuznetsov
K-a	2 Neinasuo, 66°20'N, 30°30'E, 111 m		X	8700–0	2	X	D	B		1981	Elina & Kuznetsov
K-c	3 Zapovednoe, 65°07'N, 32°10'E, 115 m	X		8990–0	2	X	D	B	X	1979	Kuznetsov & Elina
K-c	4 Shombashuo, 65°05'N, 33°03'E, 97 m		X	11500–0	2	X		B	X	1981	Elina & Kuznetsov
K-c	5 Solnechnoe, 65°46'N, 34°30'E, 10 m	X	X	11000–10000 6500–0	2	X	D	B	X	1989	Elina & Lak
K-c	6 Zarutzkoe, 63°50'N, 36°20'E, 20 m		X	8500–0	2	X		B		1981	Elina
K-c	7 Nyuhchinky moh, 63°50'N, 36°20'E, 18–20 m		X	5800–0	1	X		B		1981	Elina
K-c	8 Primorskoe, 63°48'N, 36°21'E, 10–12 m			7500–0	0	X		B		1981	Elina
K-d	9 Nosuo, 64°35'N, 30°35'E, 165m	X	X	9500–0	2	X		B		1981	Elina
K-d	10 Mini-Tumba, 63°18'N, 33°00'E 150 m	X	X	10100–0	0	X		B		1981	Elina
K-e	11 Rugozero, 65°05'N, 32°00'E 130 m	X	X	10000–0	2	X		B		1981	Elina
K-g	12 Punozerka, 62°46'N, 33°50'E, 158 m	X	X	11500–0		X		B	X	unpubl.	Filimonova
K-g	13 Sovdozerskoe, 62°45'N, 33°28'E, 142–145 m	X	X	10000–0		X		B		1985	Filimonova
K-g	14 Bezdonnoe, 62°00'N, 32°30'E, 123 m		X	9880–0	3	X		B		1981	Elina
K-g	15 Nenazvannoe,	X	X	8500–0	1	X	D	B	X	1981	Elina

RUSSIA KARELIA (*continued*)

Type region	Reference Site No., name, lat./long., elevation	Lake	Mire	Time span ^{14}C yrs **BP**	^{14}C dates, no.	Pollen %	Pollen conc.	Micro-fossils Diatoms / X-others	Macro-fossils Bot. / Zool.	Chem. anal.	Publ. year	Main scientists
K-h	17 U. Verstovoi Gorki, 62°20'N,37°05'E, 140 m	X	X	10500–0		X			B	X	1981	Elina
K-j	18 Gotnavolok, 62°10'N, 33°45'E, 88 m	X	X	11500–0	3	X			B		1981	Elina
K-j	19 Diinnoe, 62°20'N, 33°55'E, 63–66 m	X	X	11000–0	3	X			B	X	1988	Filimonova & Elovicheva
K-j	20 Koppalasuo, 62°25'N, 33°45'E, 90 m	X	X	10500–0		X			B	X	1986	Filimonova
K-k	21 Niilisuo, 61°40'N,34°05'E, 100 m		X	8500–0	1	X			B		1981	Elina
K-e	22 Uzkoe, 66°10'N,33°45'E, 80–82 m	X	X	11500–0	3	X			B	X	1992	Elina & Lebedeva
K-e	23 Kamennyi moh, 63°30'N, 36°30'E, 230–234 m		X	9100–0	3	X			B	X	1988	Elina & Juskovskaj
K-j	24 Razlomnoe, 62°25'N, 34°30'E, 52–55 m	X	X	11500–0	10	X			B	X	1988	Elina & Chomutova
K-j	25 Chechkino, 62°18'N, 34°00'E, 52–54 m		X	11500–0	3	X	X	D	B,Z	X	1988	Filimonova & Eloricheva
K-j	26 Moshkarnoe, 62°18'N, 33°55'E, 54–57 m	X	X	11500–0	9	X			B,Z	X	1994	Elina *et al.*

SLOVENIA

Type region	Reference site No., name, lat./long., elevation	Lake	Mire	Time span ¹⁴C yrs BP	¹⁴C dates, no.	Pollen %	Pollen conc.	Macro-fossils Bot. Zool.	Publ. year	Main scientists
SLO-a 1	Ledine, 46°16'N, 14°07'E, 1100 m		X	10000–0	5	X			1981	Culiberg & Šercelj
SLO-a 2	Lovrenško barje, 46°30'N, 15°19'E, 1500 m		X	4500–0	4	X			1986	Culiberg
SLO-b 3	Zamedvejca, 45°59'N, 14°25'E, 300 m		X	14000–9000	1	X		B	1991	Culiberg
SLO-b 4	Parti, 45°59'N, 14°32'E, 300 m		X	7000–800	2	X		B, Z	1978	Culiberg & Šercelj
SLO-c 5	Kaznarice, 46°05'N, 15°39'E, 500 m		X	15000–6000	1	X			1984	Šercelj

SWEDEN

Type region	Reference site — No., name, lat./long., elevation	Lake	Mire	Time span ¹⁴C yrs BP	¹⁴C dates, no.	Pollen %	Pollen conc.	Micro-fossils Diatoms	Macro-fossils Bot. Zool	Chem. anal.	Publ. year	Main scientists
S-a	1 Håkulls Mosse, 56°17′N, 12°31′E, 125 m	X	X	12300–9300	20	X	X		B,Z	X	1971 1971 1988 1988	Berglund Berglund & Malmer Lemdahl Liedberg Jönsson
S-a	2 Krageholmssjön, 55°30′N, 13°45′E, 43 m	X		10000–0	0	X	X	D	B	X	1984	Gaillard
S-a	3 Färskesjön, 56°10′N, 15°52′E, 14 m	X		10800–0	0	X					1989 1966b	Regnéll Berglund
S-a	4 Ageröds Mosse, 55°55′N, 13°26′E, 55 m	X	X	10000–0	33	X					1964	Nilsson
S-b	5 Sämbosjön, 57°10′N, 12°25′E, 38 m	X		10000–0	15	X		D		X	1982 1993	Digerfeldt Digerfeldt
S-c	6 Sandsjön, 56°45′N, 13°25′E, 155 m	X		11300–0	15	X			B		1989	Thelaus
S-d	7 Immeln, 56°17′N, 14°17′E, 81 m	X		10000–0	20	X		D	B	X	1974	Digerfeldt
S-d	8 Trummen, 56°52′N, 14°50′E, 161 m	X		10000–0	30	X		D	B	X	1975 1972	Digerfeldt Digerfeldt
S-d	9 Striern–Vån, 58°05′N, 15°47′E, 87 m	X		10000–0	26	X					1977	Göransson
S-e	10 Gladvattnet, 56°46′N, 16°37′E, 43 m	X	X	10000–0	0	X					1968	Königsson
S-e	11 Broträsk, 57°19′N, 18°26′E, 40 m	X		10000–0	0	X					1977	Påhlsson
S-f	12 Sjömyretjärn, 58°48′N, 12°05′E, 152 m	X		10000–0	11	X		D		X	unpubl.	Digerfeldt
S-g	13 Flarken, 58°35′N, 13°40′E, 109 m	X		10000–0	13	X		D		X	1977	Digerfeldt
S-g	14 Dagsmosse, 58°20′N, 14°42′E, 99 m	X	X	7000–0	14	X					1986	Göransson
S-g	15 Skyttasjön, 58°57′N, 14°45′E, 189 m	X		10000–0	0	X					1985	Påhlsson
S-g	16 Långa Getsjön, 59°14′N, 14°43′E, 120 m	X		10000–0	7	X		D			1969,1977	Florin

Code	Site, coordinates, altitude			Age range (yr B.P.)	¹⁴C dates		D	B		Year	Reference
S-h	17 Borsöknasjön, 59°20'N, 16°26'E, 24 m	X		4600–0	7	X			X	1986,1989	Hammar
S-h	18 Ådran, 59°10'N, 18°01'E, 45 m	X		10000–0	6	X	D		X	1989	Risberg & Karlsson
S-h	19 Laduviken, 59°21'N, 18°04'E, 0.5 m	X		2700–0	6	X	D		X	1980 / 1982	Digerfeldt et al. / Miller & Robertsson
S-i	20 Lilla Gloppsjön, 59°48'N, 14°37'E, 198 m	X		9600–0	11	X	D		X	1994	Almquist-Jacobson
S-i	21 Styrsjön, 60°45'N, 14°50'E, 174 m	X		6000–0	4	X				unpubl.	Påhlsson
S-j	22 Hög, 61°21'N, 17°10'E, 50 m		X	3500–0	3	X				1985	Engelmark & Wallin
S-j	23 Rudetjärn, 62°22'N, 17°00'E, 45 m	X		4600–0	4 (annual lamin.)	X	D	B	X	1978 / 1978	Engelmark / Renberg
S-k	24 Bymyren, 62°27'N, 15°22'E, 320 m		X	8000–0	3	X			X	1978	Engelmark
S-l	25 Hästlidmyren, 64°30'N, 18°45'E, 220 m		X	8500–0	0	X				unpubl.	Engelmark
S-l	26 Garaselet, 65°22'N, 19°58'E, 293 m	X	X	9000–0	21	X	D	B	X	1985	Robertsson & Miller
S-l	27 Kroktjärn, 66°14'N, 20°51'E, 58 m	X		3000–0	0 (annual lamin.)	X X				1990	Segerström
S-m	28 Prästsjön, 63°50'N, 20°10'E, 32 m	X	X	3000–0	3	X	D	B	X	1976 / 1976	Engelmark / Renberg
S-m	29 Hömyren, 64°00'N, 20°05'E, 100 m	X		6500–0	3	X				unpubl.	Engelmark
S-n	30 Ändsjön, 63°15'N, 14°32'E, 312 m	X		8000–0	4	X	D	B	X	1986	Hemmendorf & Påhlsson
S-p	31 Strömsund, 65°20'N, 16°35'E, 350 m		X	8500–0	0	X				unpubl.	Engelmark
S-r	32 Hemavan, 65°50'N, 15°00'E, 450 m		X	8500–0	3	X				unpubl.	Engelmark
S-r	33 Voulep Njakajaure, 68°20'N, 18°45'E, 408 m	X		8500–0	42 (annual lamin.)	X X	D	B		1974 / unpubl.	Sonesson / Barnekow
S-s	34 Leveäniemi, 67°38'N, 21°01'E, 360 m		X	8000–0	5	X	D			1971 / 1971	Miller / Robertsson

SWITZERLAND

Type region	Reference site — No., name, lat./long., elevation	Lake	Mire	Time span 14C yrs BP	14C dates, no.	Pollen %	Pollen conc.	Microfossils Diatoms	Macrofossils Bot. Zool.	Chem. anal. 0=δO¹⁸	Publ. year	Main scientists
CH-a	1 Les Embreux, 47°15'50"N, 7°07'06"E, 1017 m		X	10000–0	0	X			B		1985	Hubschmid
CH-a	2 Les Veaux, 47°14'32"N, 7°05'46"E, 1020m		X	7000–0	1	X					1991	Reille
CH-a	3 Sur les Bieds, 46°58'52"N, 6°44'22"E, 1007 m		X	11000–0	4	X					1971	Matthey
CH-a	4 Sous Martel Dernier, 46°58'51"N, 6°42'48"E, 1010 m		X	11000–0	3	X					1971	Matthey
CH-a	5 Marais de Rances, 46°46'24"N, 6°30'46"E, 607 m		X	15000–0	3	X	X				1984	Gaillard
CH-a	6 Les Cruilles, 46°39'48"N, 6°18'38"E, 1035 m		X	13000–5000	4	X					1966	Wegmüller
CH-a	7 Le Marais des Amburnex, 46°33'60"N, 6°14'02"E, 1300 m		X	12000–0	4	X					1966	Wegmüller
CH-a	8 La Pile, 46°26'50"N, 6°05'42"E, 1220 m		X	13000–0	3	X					1966	Wegmüller
CH-b	9 Nussbaumer Seen, 47°36'52"N, 8°49'54"E, 434 m	X		15000–0		X	X			X	1985	Rösch
CH-b	10 Greifensee, 47°21'01"N, 8°40'14"E, 435 m	X		15000–0	0	X	X			X	1988	Wick
CH-b	11 Gonten, 47°19'60"N, 9°06'22"E, 920 m		X	10000–2000	6	X					1976	Wegmüller
CH-b	12 Uffikon, 47°12'33"N, 8°01'32"E, 500 m		X	15000–9000		X					1989	Küttel
CH-b	13 Soppensee, 47°05'30"N, 8°04'54"E, 596 m	X		11500–9500		X					1989,1991 1992	Lotter
CH-b	14 Rotsee, 47°04'33"N, 8°19'33"E, 419 m	X		15000–0	27	X	X	D	B	X	1988,1989, 1990	Lotter
CH-b	15 Lobsigensee, 47°01'53"N, 7°18'05"E, 514 m	X		15000–0	19	X	X	D	B, Z	0,X	1989	Ammann
CH-b	16 Lörmoos, 46°59'02"N, 7°24'48"E, 583 m		X	15000–0	0	X			B	X	1982,1985	Welten, Zwahlen
CH-b	17 Murifeld, 46°56'19"N, 7°28'38"E, 554 m		X	15000–8000	15	X					1982	Welten
CH-b	18 Wachseldorn, 46°49'19"N, 7°44'06"E, 980 m		X	13000–0	75	X					1972,1982	Heeb & Welten, Welten
CH-b	19 Lac de Seedorf, 46°47'48"N, 7°02'33"E, 609 m	X		10000–0		X				X	1994	Richoz et al.
CH-b	20 Amsoldingersee, 46°43'22"N, 7°34'37"E, 641 m	X		15000–0		X	X			0,X	1985	Lotter

Site	Locality, coordinates, altitude		Age span (yr BP)	n			B		Year	Reference
CH-b 21	Le Marais du Rosey, 46°36′14″N, 6°28′26″E, 565 m	X	14000–4000	10	X	X			1984	Gaillard
CH-b 22	La Joux des Ponts, 46°36′21″N, 6°56′44″E, 837 m	X	15000–8000	0	X	X			1984	Gaillard
CH-b 23	Le Tronchet, 46°31′60″N, 6°43′41″E, 715 m	X	15000–8500	11	X				1984	Gaillard
CH-b 24	La Tourbière, 46°25′41″N, 6°13′54″E, 480 m	X	13000–0	4	X				1966	Wegmüller
CH-c 25	Gamperfin, 47°10′12″N, 9°21′56″E, 1320 m	X	10000–0	10	X				1989	Schneebeli et al.
CH-c 26	Faulenseemoos, 46°40′49″N, 7°41′41″E, 590 m	X	15000–0	0	X				1944, 1982	Welten
CH-c 27	Egelsee, 46°38′55″N, 7°32′37″E, 989 m	X	9000–0	16	X			X	1982, 1991	Welten, Lotter
CH-c 28	Plidutscha, 46°38′26″N, 8°40′50″E, 2128 m	X	10500–1000	12	X				1990	Küttel
CH-c 29	Höhenbiel, 46°34′15″N, 8°29′46″E, 1970 m	X	10500–1000	3	X				1990	Küttel
CH-c 30	Obergurbs, 46°33′33″N, 7°31′45″E, 1910 m	X	11000–6500		X				1979	Küttel
CH-c 31	Saanenmöser, 46°31′07″N, 7°18′42″E, 1256 m	X	14000–9900	0	X				1982	Welten
CH-c 32	Kühdungel, 46°23′17″N, 7°21′02″E, 1800 m	X	11000–6000		X				1979	Küttel
CH-c 33	Leysin, Les Léchières, 46°20′50″N, 7°01′18″E, 1255 m	X	14000–8500	0	X				1982	Welten
CH-c 34	Etang de Luissel, 46°14′14″N, 7°01′03″E, 540 m	X	5300–0	3	X				1982	Welten
CH-d 35	Brigels, 46°48′19″N, 9°05′26″E, 1530 m	X	10000–0	5	X		B		1972	Müller
CH-d 36	Lai Nair, 46°46′41″N, 10°04′43″E, 1546 m	X	12500–0	5	X				1982	Welten
CH-d 37	Affeier, 46°45′20″N, 9°06′54″E, 1300 m	X	11000–0	1	X		B		1972	Müller
CH-d 38	Lai da Vons, 46°35′12″N, 9°23′12″E, 1991 m	X	13000–0	3	X		B	X	1980, 1985	Burga
CH-d 39	Sur, 46°31′53″N, 9°37′51″E, 1780 m	X	13000–0	8	X				1975	Heitz
CH-d 40	Stallenberg, 46°27′10″N, 9°36′06″E, 2450 m	X	11000–0	6	X				1975	Heitz
CH-d 41	Maloja-Riegel I, 46°24′23″N, 9°41′27″E, 1870 m	X	12000–0	3	X		B		1974	Kleiber
CH-d 42	Plansena, 46°23′38″N, 10°00′34″E, 1892 m	X	11000–0	2	X				1987	Burga

SWITZERLAND (*continued*)

Type region	Reference site No., name, lat./long., elevation	Lake	Mire	Time span ^{14}C yrs BP	^{14}C dates, no.	Pollen %	Pollen conc.	Micro-fossils Diatoms	Macro-fossils Bot. Zool.	Chem. anal. O=δO^{18}	Publ. year	Main scientists
CH-d 43	Selva, 46°18'12"N, 10°00'36"E, 1440 m		X	13000–7500	3	X			B		1987	Burga
CH-d 44	Aletschwald, 46°23'23"N, 8°01'32"E, 2017 m		X	9000–500	14	X					1982	Welten
CH-d 45	Eggen ob Blatten, 46°22'19"N, 7°59'26"E, 1645 m		X	10000–2000	18	X					1982	Welten
CH-d 46	Bitsch-Naters, 46°20'28"N, 7°59'30"E, 1030 m		X	9500–500	8	X					1982	Welten
CH-d 47	Montana-Xirès, 46°18'33"N, 7°28'20"E, 1445 m		X	13000–4500	1	X					1982	Welten
CH-d 48	Zeneggen-Hellelen, 46°16'59"N, 7°50'39"E, 1520 m		X	13500–0	10	X					1982	Welten
CH-d 49	Böhnigsee, 46°15'18"N, 7°50'58"E, 2095 m		X	11000–0	17	X					1969	Markgraf
CH-d 50	Hobschensee, 46°15'12"N, 8°01'26"E, 2017 m	X		13000–0		X	X	D	B, Z	X	1979, 1982, 1985	Küttel, Welten; Lang & Tobolski
CH-d 51	Lac du Mont d'Orge, 46°14'04"N, 7°20'29"E, 640 m	X		14000–0	11	X					1982	Welten
CH-d 52	Grächen-See, 46°11'50"N, 7°50'46"E, 1710 m	X		8000–0	18	X					1982	Welten
CH-d 53	Mont Carré, 46°09'14"N, 7°22'07"E, 2290 m		X	10500–3500	3	X					1982	Welten
CH-d 54	Tortin, 46°06'59"N, 7°18'20"E, 2039 m		X	10500–6000		X					1979	Küttel
CH-e 55	Campra, 46°31'13"N, 8°52'24"E, 1420 m		X	13000–0	5	X			B		1972	Müller
CH-e 56	Bedrina (M. Piottino), 46°29'03"N, 8 46°26'E, 1235 m		X	13000–0	27	X					1960,1977	Zoller, Küttel
CH-e 57	Suossa, 46°26'12"N, 9°11'56"E, 1700 m		X	13000–0	7	X			B		1971	Zoller & Kleiber
CH-e 58	Gola di Lago, 46°06'19"N, 8°57'57"E, 960 m		X	13000–0	3	X			B		1971	Zoller & Kleiber
CH-e 59	Lago d'Origlio, 46°03'06"N, 8°56'42"E, 421 m		X	13000–0	4	X					1960	Zoller
CH-e 60	Lago di Lugano, 45°56'30"N, 8°57'09"E, 271 m	X		15000–0	6	X	X			X	1989	Wick
CH-e 61	Lago di Ganna, 45°53'54"N, 8°49'39"E, 452 m	X		15000–0	6	X	X		B,Z	X	1983	Schneider & Tobolski

Index of Site Names

Numbers in **bold** refer to pollen diagrams